SEAWEEDS

SEAWEEDS
Their Environment, Biogeography, and Ecophysiology

Klaus Lüning
Biologische Anstalt Helgoland
Hamburg
Federal Republic of Germany

English language edition edited by:

Charles Yarish
University of Connecticut
Stamford, Connecticut

Hugh Kirkman
CSIRO Fisheries
Marmion Laboratories
North Beach
Australia

A WILEY-INTERSCIENCE PUBLICATION
John Wiley & Sons, Inc.
NEW YORK / CHICHESTER / BRISBANE / TORONTO / SINGAPORE

Copyright © 1990 by John Wiley & Sons, Inc.

All rights reserved. Published simultaneously in Canada.

Reproduction or translation of any part of this work
beyond that permitted by Section 107 or 108 of the
1976 United States Copyright Act without the permission
of the copyright owner is unlawful. Requests for
permission or further information should be addressed to
the Permissions Department, John Wiley & Sons, Inc.

Authorized translation from the German language edition *Meeresbotanik: Verbreitung,
Ökophysiologie und Nutzung der marinen Makroalgen* by Klaus Lüning, © 1985, Georg
Thieme Verlag, Stuttgart. English language edition translated and revised by Klaus Lüning
and edited by Charles Yarish and Hugh Kirkman.

Library of Congress Cataloging in Publication Data:

Lüning, Klaus.
   [Meeresbotanik. English]
   Seaweeds. Their environment, biogeography, and ecophysiology / Klaus Lüning.
     p. cm.
   Rev. translation of: Meeresbotanik.
   "An Interscience publication."
   Includes bibliographical references.
   ISBN 0-471-62434-9
   1. Marine algae--Geographical distribution. 2. Marine algae—
Physiological ecology. I. Title.
QK570.2.L8613   1990
589.4'5--dc20   89-22633
                                                                                                              CIP

Printed in the United States of America

10 9 8 7 6 5 4 3 2 1

PREFACE

Anyone attempting to describe the seaweed vegetation of the world has to face the problem of how to deal with the existing terminologies of biogeographical regions, community structure, and vertical zonation. I have adopted the detailed system of biogeographical regions that existed in marine benthic zoology, following the work of S. Ekman, J. W. Hedgpeth, and J. C. Briggs. After a promising start by F. R. Kjellman and W. A. Setchell at the turn of the century, the biogeography of seaweeds fell into a twilight sleep from which it was revived mainly by the recent efforts of C. van den Hoek and his coworkers. Their phycogeographical analyses bridged the gap to modern marine zoogeography and largely supported the zoological system of biogeographical regions.

I have respected the original authors' opinions on how to describe community structure in order to preserve this more inherent aspect of their original work. Consequently, the reader will find a phytosociological terminology in the chapter about Mediterranean seaweed vegetation but simple descriptive terms in other chapters. For zonation terminology the reader will find himself or herself in the Procrustean bed of the classical European terminology (supra-, eu-, and sublittoral zones), but I do not think this will pose a major difficulty, especially since translations have been used.

Marine biogeography bears a fascinating experimental aspect, so a third of this book is devoted to the impact of abiotic and biotic factors on seaweed performance. I have presented a rather short and selective treatment on the environment, with an emphasis on light and temperature.

H. Kirkman and C. Yarish corrected, with patience and many suggestions, the English of my translation and of the newly written parts for the English edition. My sincere thanks are due to both of them, and to those colleagues who read and corrected parts of the final manuscript: J. I. Bolton, A. M. Breeman, A. R. O. Chapman, M. N. Clayton, I. tom Dieck, M. J. Dring, L. D. Druehl, M. D. Guiry, K. M. Khailov, J. T. O. Kirk, T. Larkum, J. McLachlan, W. Nultsch, D. Nürnberg, E. C. de Oliveira Filho, A. Peters, W. F. Prud'homme van Reine, J. A. Raven, J Rueness, G. Russell, D. Schnitker, E. H. Schulte, R. B. Searles, G. R. South, S. Srivastava, J. Thiede, E. A. Titlyanov, R. L. Vadas, J. R. Waaland, R. Westermeier, C. Wiencke, W. J. Woelkerling, and H. B. S. Womersley. They all provided valuable suggestions, and I have tried to incorporate their professional knowledge in the text as far as possible. Any inconsistencies that remain in the book are exclusively my responsibility.

Numerous facts presented in this book come from the difficult and sometimes dangerous fieldwork of many phycologists, and I am sure they would consent to the dedication of this book to four eminent diving pioneers. First to Mats Waern (Sweden), who, in the 1940s as one of the Nestors of the Art, dove with a diving helmet to study the seaweed vegetation of the Baltic Sea. Next, to the early deceased Julius Ernst (Austria), who worked on the sublittoral vegetation of the Adriatic Sea and the coast of Brittany. Then to Per Svendsen (Norway), who investigated the seaweed vegetation of Spitsbergen and lost his life in a diving accident off the coast of Norway. And finally to E. Yale Dawson (U.S.A.), who, after his fundamental work on the American Pacific algal flora, died off the coast of the Red Sea.

Hamburg, 1990   KLAUS LÜNING

EDITORS' PREFACE

In 1985 Klaus Lüning's book *Meeresbotanik* was published in German. The book concentrated on the biogeography of seaweeds and brought together information from the thousands of published papers that constituted the marine botanists' inventory of the world's seaweeds. *Meeresbotanik* was an immediate success, and one of us, Yarish, persuaded Lüning to translate his book into English. Lüning proceeded to prepare a preliminary translation and Yarish was joined by Kirkman in correcting and editing this English translation of the book. The book is intended to serve as a reference for marine biologists on the biogeography of marine macroalgae and as a text for students in the field at the advanced undergraduate and graduate levels.

A knowledge of the biogeography of seaweeds and the paleobiogeography of their ancestors is essential to all marine biologists including ecologists, developmental biologists, phylogenetic systematists, paleoecologists, and aquaculturists. Not only does this book cater to the specialists, but it is also intended to make students aware of the tremendous range and diversity of disciplines within or dependent on the biogeography of seaweeds and their environment. No one interested in marine biology can be an expert in all these disciplines, but no one can afford to be ignorant of them either.

As seaweeds are exploited for their food or chemical properties, the temptation to introduce commercially viable species to sites where they may never have grown will be difficult to resist. This volume's detailed descriptions of continental drift, oceanic currents, and historical factors leading to the distribution of algae around the world will help the reader understand the implications of chance or intentional introduction of algal species to nonnative sites.

The earth's environment has fluctuated thoughout geologic time, and the patterns of marine algal zonation have adapted to it. However, there is now ample evidence of global climatic changes leading to possible rises in sea level of up to 2 m in the next 40 years, and the destruction of the ozone layer, which will lead to increased exposure to ultraviolet radiation. These two global climatic events will result in rapid changes in the zonation of seaweeds on the shore, probably resulting in different marine communities in a very short timespan. This book may thus also serve as a timely record of seaweed dominated communities as they exist presently.

Stamford, Connecticut CHARLES YARISH
Perth, Western Australia, Australia HUGH KIRKMAN

CONTENTS

PART ONE DISTRIBUTION AND STRUCTURE OF SEAWEED VEGETATION

1 Introduction to Vertical and Geographical Distribution — 3

   1.1 Vertical zonation — 3
       1.1.1 Subdivisions of the euphotic zone — 4
       1.1.2 Comparison of different zonation systems — 8
   1.2 Geographical distribution — 10
       1.2.1 Marine biogeographical regions — 11
       1.2.2 Temperature and boundaries of species distribution — 13

2 Seaweed Vegetation of the Cold and Warm Temperate Regions of the Northern Hemisphere — 22

   2.1 Paleobiogeography and the possible history of the marine-benthic flora — 22
   2.2 Distributions of characteristic seaweed species of the North Atlantic — 40
       2.2.1 Arctic–cold temperate amphioceanic species — 40
       2.2.2 Arctic–cold temperate North Atlantic species — 53
       2.2.3 Cold temperate North Atlantic species — 55
   2.3 Europe: cold temperate region (northern Norway and Iceland to northern France) — 60
       2.3.1 Supralittoral zone — 69
       2.3.2 Eulittoral zone — 70
       2.3.3 Sublittoral zone — 73
       2.3.4 Maerl vegetation — 79
       2.3.5 Significance of water motion in the sublittoral zone — 80

		2.3.6 Brackish water areas: fjords, estuaries, and the Baltic Sea	82
	2.4	Europe–Northwest Africa: warm temperate Mediterranean–Atlantic region	85
		2.4.1 Western Ireland–northern France to Senegal (Lusitania province)	85
		2.4.2 Canary province	90
		2.4.3 Seaweed migrations and long-distance dispersal	92
		2.4.4 Introduced seaweeds	94
		2.4.5 Mediterranean province: introduction	98
		2.4.6 Mediterranean province: distribution groups and floral history	102
		2.4.7 Mediterranean province: depth distribution and vegetation types	106
		2.4.8 Black, Caspian, and Aral seas	121
	2.5	North America: cold and warm temperate regions in the North Atlantic	123
		2.5.1 American province of the cold temperate North Atlantic region (Newfoundland to Cape Hatteras)	125
		2.5.2 Warm temperate Carolina region (Cape Hatteras to Cape Kennedy)	132
	2.6	North America: cold and warm temperate regions in the North Pacific	134
		2.6.1 Cold temperate northeastern Pacific region (Alaska to central California)	139
		2.6.2 Warm temperate California region (Point Conception to Baja California)	145
	2.7	Asia: cold and warm temperate regions in the North Pacific	154
3	**Seaweeds of the Arctic**		**164**
	3.1	Limits of the Arctic region	165
	3.2	Seaweed environment in the Arctic region	166
	3.3	Distributions of Arctic-endemic species	173

		3.4	Local seaweed floras and their vegetation structure	174
			3.4.1 Greenland	174
			3.4.2 Spitsbergen, Franz Josef Land, Bear Island, and Jan Mayen	180
			3.4.3 Russian Arctic	182
			3.4.4 North American Arctic	185
4	**Seaweed Vegetation of the Tropical Regions**			**189**
	4.1	Dominant seaweeds and seaweed paleobiogeography		189
	4.2	Fossil calcified algae and the evolution of algae and seagrasses		197
	4.3	Eastern Atlantic tropical region (tropical West Africa)		203
	4.4	Tidal mangrove forests and their seaweeds		208
	4.5	Coral reefs and the role of algae		209
	4.6	Western Atlantic tropical region (tropical-Atlantic America)		216
	4.7	Sargasso Sea		222
	4.8	Indo–West Pacific tropical region		225
	4.9	Eastern Pacific tropical region		233
5	**Temperate and Polar Seaweed Vegetation of the Southern Hemisphere**			**235**
	5.1	Delimitation of the regions and paleobiogeography		235
		5.1.1 Convergences, currents, and regions		235
		5.1.2 Paleoceanography and paleobiogeography of the Southern Hemisphere		237
	5.2	Comparison of the two polar regions and bipolar distributed seaweeds		240
	5.3	Antarctic region		245
	5.4	Sub-Antarctic (cold temperate) islands region		250

5.5	South America		252
5.6	Southern Africa		258
5.7	Southern Australia and New Zealand		265

PART TWO ECOPHYSIOLOGY OF SEAWEEDS

6 Light — 277

6.1	Spectral distribution in the euphotic zone		277
6.2	Algal depth limits in relation to naturally occurring light levels		281
6.3	Light demands for photosynthesis and growth		287
	6.3.1	Photosynthetic pigments, action spectra, and vertical distribution of seaweeds	287
	6.3.2	Quantitative light demands for photosynthesis and growth	296
	6.3.3	Photoacclimation of pigment content	305
6.4	Light as an environmental signal		308
	6.4.1	Photoperiodism	308
	6.4.2	Photomorphogenetic reactions in seaweeds	316
	6.4.3	Further signal effects of light	317

7 Temperature, Salinity, and Other Abiotic Factors — 321

7.1	Temperature		321
	7.1.1	Heat and cold tolerance	321
	7.1.2	Temperature dependence of growth and reproduction	328
	7.1.3	Temperature dependence of photosynthesis and respiration	331
7.2	Salinity		332
7.3	Desiccation tolerance		337
7.4	Nutrients		342
7.5	Water motion and hydromechanical adaptations		344

8	**Biotic Factors in the Euphotic Zone. Strategies, Productivity of Seaweeds, and Commercial Uses**		347
	8.1	Competition among seaweed species and the relationships between seaweeds, herbivores, and predators	347
		8.1.1 Eulittoral zone	347
		8.1.2 Sublittoral zone	349
		8.1.3 Protection against grazing	351
	8.2	Epiphytes, endophytes, endozoans, and parasites	353
	8.3	Strategies for the survival and growth of seaweeds	356
	8.4	Special growth strategies of sublittoral, perennial seaweeds: storage, translocation, and growth in darkness	358
	8.5	Productivity and growth rates of seaweeds	360
	8.6	Constituents, commercial use, and decomposition of seaweed biomass	364

BIBLIOGRAPHY 371

TAXONOMIC OVERVIEW OF GENERA 485

INDEX 489

SEAWEEDS

PART ONE
Distribution and Structure of Seaweed Vegetation

1 Introduction to Vertical and Geographical Distribution

The world under water requires its own terminology, and this chapter will take the reader right into it. The **pelagos** comprises unattached organisms of the open water or **pelagic realm,** and the **benthos** includes organisms that live on the sea bottom or **benthic realm** (Fig. 1.1). The plankton (passively drifting organisms) and the nekton (motile organisms) are found in the pelagos. The animals in the benthos are called **zoobenthos** and the plants **phytobenthos.** The marine macroalgae or seaweeds and the seagrasses are called **macrophytobenthos,** and the microalgae the **microphytobenthos.** There about 8000 species of seaweeds along the world's coastlines, and they may extend as deep as 270 m. The seagrasses are marine flowering plants; they include about 50 species and may extend down to about 70 m. Seaweeds and seagrasses are important contributors to food webs in coastal waters (Chapter 8).

1.1 VERTICAL ZONATION

In this book the term "zone" is used for vertical divisions, whereas the term "region" is reserved for biogeographical areas. The **euphotic zone** of the benthic realm is the lighted layer inhabited by autotrophic plants. In this sunlit zone there is sufficient light to satisfy the photosynthetic requirements of plants. The lower limit of the euphotic zone (termed "compensation depth" or "extinction depth" by Sears and Cooper 1978) is at around 200 m (to a maximum of 270 m) in very clear waters (Chapter 6). This happens to coincide with the **shelf break** where the continental slope begins (Fig. 1.1) and some past glacial sea levels occurred (Seibold and Berger 1982). Oceanographers also use the term "littoral zone" for the shelf area. For that reason marine biologists should avoid this term when describing intertidal zonation and use terms such as "eulittoral zone" (see below). The importance of the shelf regions is evident from the fact that nearly 90% of the world's marine invertebrate taxa are contained in the benthic communities of the continental shelves, and the remaining 10% in the plankton or deep sea (Valentine 1973).

Subdivisions of the pelagic and benthic realms below the euphotic zone. Within the pelagic realm the euphotic zone is also termed the **epipelagic zone,** below which is the **mesopelagic** (twilight) zone and the sunless **bathypelagic zone** (Fig. 1.1). In clear waters sunlight penetrating below 200 m is insufficient to drive photosynthesis of most

4　INTRODUCTION TO VERTICAL AND GEOGRAPHICAL DISTRIBUTION

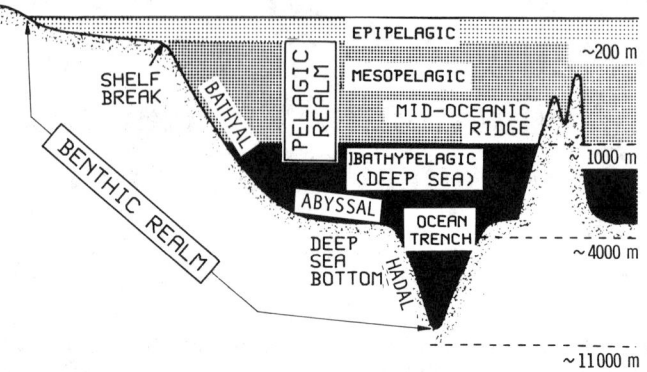

Fig. 1.1 Vertical subdivision of the benthos and the pelagos, the two realms of the sea. (After Hedgpeth 1957*b*; Tait 1972; Pérès 1982*a*; Tischler 1984.)

algae, but it can be detected by animals and humans in submersibles. Within the benthic realm, the zone below the euphotic zone is the **aphotic zone,** inhabited by heterotrophic organisms (animals, fungi, and bacteria). The aphotic zone may be further subdivided into **bathyal** (area of the continental slope), **abyssal** (ocean bottoms), and **hadal** (ocean trenches, down to the maximum depth of 11,022 m in the Mariana Trench of the northwestern Pacific). This terminology, mainly based on Hedgpeth (1957*b*) and Peres (1982*a*), differs from that of other authors (e.g., Menzies et al. 1973, who preferred the term "archibenthal" to "bathyal").

1.1.1 Subdivisions of the Euphotic Zone

The euphotic zone is subdivided in the following way (Fig. 1.2):

(a) The **supralittoral zone,** which is reached by spray water;
(b) The **eulittoral zone,** that is, the intertidal zone, which is submersed and emersed, either periodically due to tides or aperiodically due to irregularly occurring factors, as in the enclosed seas of the Baltic or the Mediterranean;
(c) The **sublittoral zone,** which is submersed with the upper part at extreme low water levels occasionally emersing; the lower limit is set by the deepest-occurring algae.

The positions of boundaries between species are fixed by a variety of physicochemical and biotic factors. Upper limits in the supra- and eulittoral zone are frequently determined by tolerance to abiotic factors such as desiccation and extreme temperatures (Chapter 7). Biotic interactions may also be involved, as experimental removal of competing species has shown (Section 8.1). Lower limits are commonly determined by the presence of competetively superior species (Section 8.1), but conversely, physicochemical

VERTICAL ZONATION 5

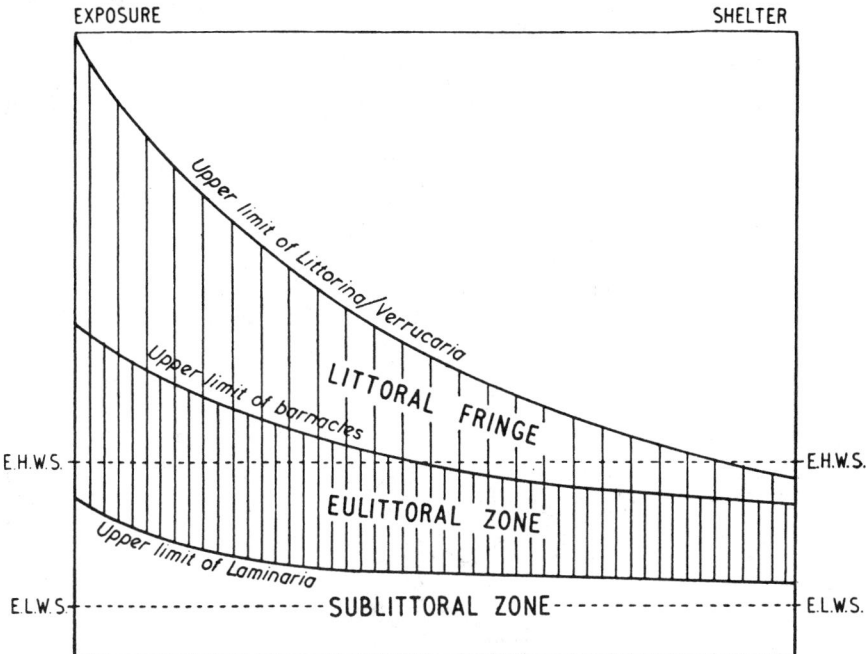

Fig. 1.2 Vertical subdivision of the euphotic zone, the upper limits rising on wave-exposed shores. The term "littoral fringe" has been replaced by "supralittoral zone" in the present book (compare with Fig. 1.4). E.H.W.S. = extreme high water of spring tides; E.L.W.S. = extreme low water of spring tides. (From J. R. Lewis 1964.)

tolerances may be involved, such as tolerance to low light levels (Chapter 6). The horizontal beltlike arrangement resulting from these and other factors is termed **zonation.**

Shifting of euphotic zone boundaries due to wave exposure. At extremely wave-exposed locations, where waves and ocean spray splash up high, the organisms of the supralittoral zone may live many meters above the high-water mark (Fig. 1.2). Under such circumstances the boundary between land and sea may appear as a broad transition zone populated by terrestrial vegetation able to withstand sea spray, and by representatives of the supralittoral zone adapted to withstand air exposure. In contrast, on a sheltered shore, the presence of the supralittoral organisms may be confined to a narrow zone between extreme and mean high-water levels. In the same way, the upper limits of occurrence of organisms in the eulittoral and the sublittoral zones rise with increasing **wave exposure** (Figs. 1.2 and 2.47). The complex action of wind and waves in connection with the topography of a coastline is as important as tidal range in regulating the absolute limits of the biological zones on any shore.

In the **sublittoral zone** light is a major limiting factor. This has caused a variety of adaptations and an algal zonation within this zone. The largest representatives of the seaweeds in the temperate regions, members of the Laminariales

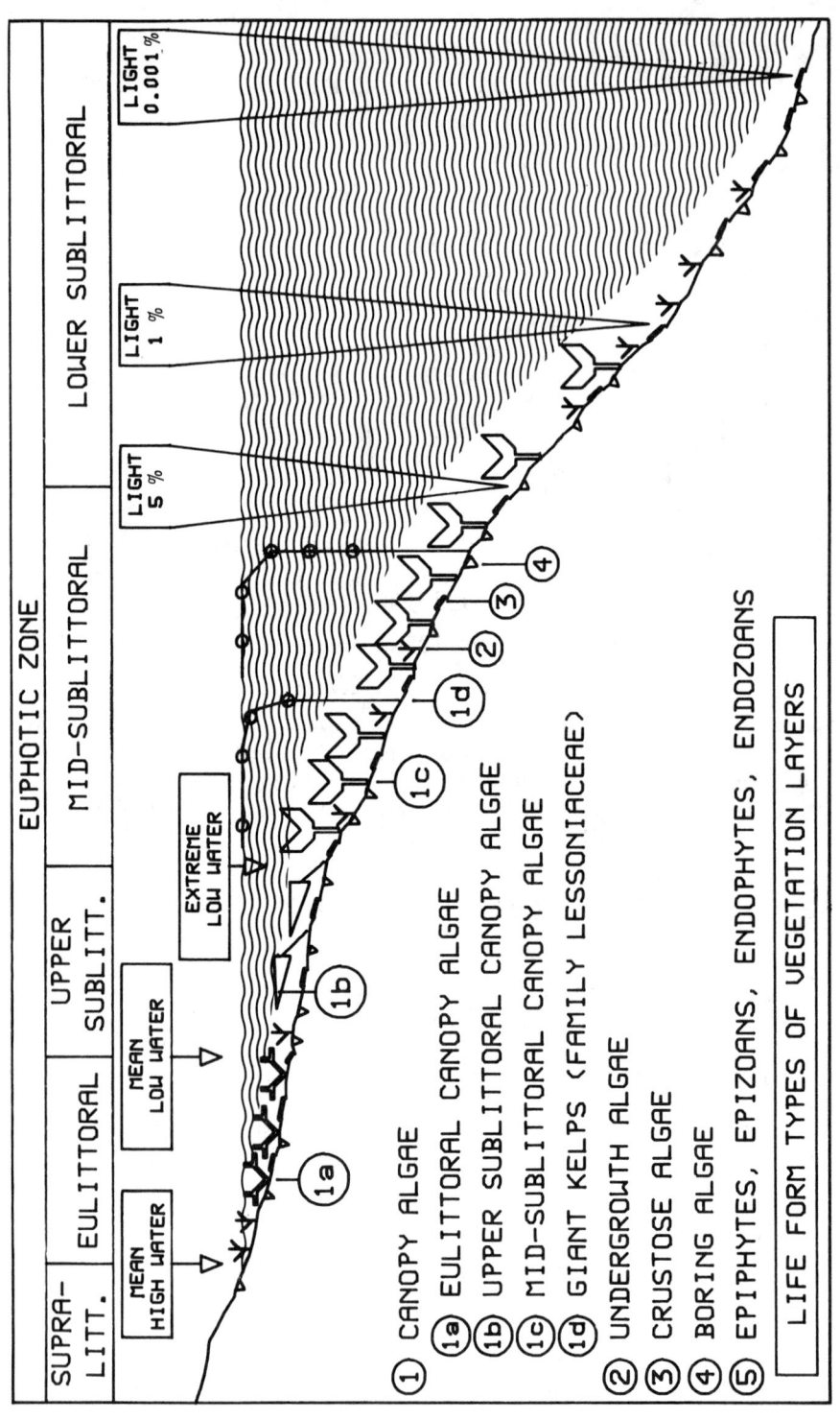

(kelps), inhabit the upper and midsublittoral zone, due to their relatively high light requirements (Section 6.2). Kelps inhabiting the **upper sublittoral zone** (from one to a few meters below mean low water) are mechanically adapted by the possession of flexible stipes (e.g., in *Laminaria digitata*) and mechanically resistant thalli to withstand the impact of the breaking waves (Section 7.5). Another important characteristic of the upper sublittoral zone is that it occasionally emerges at extreme low water (Fig. 1.3). The **midsublittoral zone** is never uncovered, and is populated by a dense vegetation of large growth forms. Struggle for light is the main problem in this subzone: each species has to find a way to spread a maximum of photosynthesizing thallus area in the underwater light field and outcompete other algae. For this purpose, blade-like thalli (phylloids) are carried by stiff stipes (e.g., in *L. hyperborea*) or held upright by air bladders (pneumatocysts) as giant branched structures, as in the case of the giant kelps (family Lessoniaceae in the order Laminariales, e.g., *Macrocystis* spp.). The impact of wave action is dampened within the midsublittoral zone, and most mechanical stress in this zone is due to currents. Down to a depth reached by about 5% of the surface light, a dense, closed vegetation canopy of kelps is still possible (Section 6.2). Below this depth, light cannot sustain a closed kelp canopy. Here the **lower sublittoral zone** begins and extends to the lower limit of algal vegetation where the light level is as low as 0.05% or 0.001% of surface irradiance in the tropics (Fig. 1.3; Section 6.2).

As in the vegetation layers of terrestrial habitats, there is a corresponding terminology for marine vegetation (Fig. 1.3). The fronds of the **canopy algae (group 1)** overtop all other algae and correspond to the tree layer of terrestrial vegetation. Among this group are canopy algae of the eulittoral zone adapted to withstand desiccation **(group 1a)** (e.g., the Fucaceae). The canopy algae of the upper sublittoral zone **(group 1b)** require, as do eulittoral algae, mechanical adaptations (e.g., flexible stipes) for their survival in a wave-beaten environment. The main adaptations of the canopy algae of the midsublittoral **(group 1c)** involve development of supporting structures for maximum exposure of the thallus area to light, through possession of either rigid stipes **(group 1c)** or pneumatocysts **(group 1d;** see above). There are some marine algal communities (e.g., in Alaska) where the kelp forests have multilayered canopies (Fig. 2.82).

Below the vegetation of the canopy is a **group 2** layer of **understory algae** (undergrowth algae, algal turf) analagous to shrubs and herbs in terrestrial vegetation. Underneath this lies the **group 3** layer of **crustose algae,** analogous to the thalloid layer in the terrestrial example (Fig. 1.3). With increasing water depth and decreasing light intensity, one vegetation layer after another

---

Fig. 1.3 Scheme of algal life-form types and their distribution within the euphotic zone. Representatives of group 5 occur throughout the whole euphotic zone and are not illustrated in the figure. The deepest crustaceous red algae were found at 268 m or at 0.001% of surface light.

disappears. A model proposed by M. E. Hay (1986), modified from one on the geometry of forest trees, predicts that monolayered algae with flattened and optically dense (highly pigmented) thalli should be successful in low-light habitats.

In limestone or in the shells of mollusks and polychaetes, the **boring algae** (**group 4**) are found. These algae penetrate into the substratum as far as possible to still receive sufficient light for photosynthesis. Within the euphotic zone many plants and animals have algae attached to them (**group 5;** Section 8.2). Algae that live on other plants or animals are said to be **epiphytic** and **epizoic,** respectively. Algae that live in other plants and animals are said to be **endophytic** and **endozoic,** respectively. Parallel to these life-form types of vegetation layering, life-form systems have been set up for thallus physiognomy and for seasonal development.

1.1.2 Comparison of Different Zonation Systems

In addition to the classic differentiation of the euphotic zone into the supra-, eu-, and sublittoral zones, several other terminology schemes have been introduced by various workers, and these should not be confused with that used here (Fig. 1.4).

Classical system. The term "supralittoral" ("Supralitoral" in German) was introduced by Lorenz (1863). The term "sublittoral" goes back to F. R. Kjellman (1877, 1878), who distinguished a "littoral region" (intertidal), a "sublittoral region" (from low-water mark down to a depth of 20 fathoms, or 37 m), and an "elittoral region" (with scarce algal vegetation). The meaning of the term "littoral" was subsequently obscured by its application to the shelf area (see above), and therefore it was replaced by the term "eulittoral."

Stephenson system. This universal system was developed on the basis of worldwide investigations of intertidal zonation (Stephenson and Stephenson 1949, 1972). There are a few weak points in this system, as addressed by Womersley and Edmonds (1952), or den Hartog (1959). The use of the term "infralittoral zone" instead of "sublittoral zone" conflicts with the use of "infralittoral" in the Genoa system (see below and Fig. 1.4). Designating the upper sublittoral zone as the "infralittoral fringe" creates a problem because the term "fringe" cannot be translated directly into other languages.

Lewis system. The supralittoral zone in this system is called the "littoral fringe" (J. R. Lewis 1955, 1964), following the suggestion of Womersley and Edmonds (1952); this term also produces a problem of translation. Lewis wanted to avoid the term "supralittoral zone," because this might evoke the impression that nonmarine organisms would inhabit this zone ("supra" = beyond the littoral).

Genoa system. This system of the "French school" was introduced in 1957 at a colloquium in Genoa by marine biologists working on zonation in the Mediterranean; it has since been used exclusively in French and Italian publications (Pérès and

Biological boundaries	CLASSICAL SYSTEM	STEPHENSON SYSTEM	LEWIS SYSTEM	GENOA SYSTEM
		SUPRALITTORAL ZONE	MARITIME ZONE	
Uppermost marine organisms				
	SUPRALITTORAL	SUPRALITTORAL FRINGE	LITTORAL FRINGE	ETAGE SUPRALITTORAL — System
Barnacles				
	EULITTORAL	MIDLITTORAL ZONE	EULITTORAL ZONE	ETAGE MEDIOLITTORAL — Littoral zone
Upper laminarians		INFRALITTORAL FRINGE		Phytal zone
	SUBLITTORAL	INFRALITTORAL ZONE	SUBLITTORAL ZONE	ETAGE INFRALITTORAL (high light zone)
				ETAGE CIRCALITTORAL (dim light zone)
Lowest algae				

Fig. 1.4 Terminology of vertical zonation according to four schemes.

Molinier 1957; Pérès 1982a). The sublittoral zone is divided into an upper zone ("étage infralittoral") and a lower zone ("étage circalittoral"). The "infralittoral zone" is inhabited by sun-loving ("photophilous") algae and the seagrasses, whereas shade-loving algae live in the "circalittoral zone" (Section 2.4.6). This system has been accepted for higher latitudes by Hiscock and Mitchell (1980), with the infralittoral subzone of the sublittoral zone harboring kelp forests and erect algae. In the circalittoral subzone animals dominate; the erect and crustaceous algae are also found but with a scattered distribution.

1.2 GEOGRAPHICAL DISTRIBUTION

The present distribution of benthic marine species is the result of their migration and of the displacement of coastlines in geological periods. In the middle of the 19th century the British marine zoologist and paleontologist Forbes was a pioneer of marine zoogeography. It was Forbes who said: "The student of history follows, with intense interest, the march of a conqueror, or the migrations of a nation . . . Yet, absurd as it may seem to those who have not thought of such things before, there is deeper interest in the march of a periwinkle, and the progress of a limpet" (cited in Hedgpeth 1957a).

Two pioneers of marine phycogeography were the Swedish phycologist F. R. Kjellman (e.g., 1883) and the North American phycologist W. A. Setchell (e.g., 1920, 1922). Kjellman was interested in the comparison of Northern Hemisphere floras, and Setchell emphasized the role of temperature in algal distributions. (For a history of phycogeography before 1900, see Setchell 1917). Van den Hoek (1975, 1982a,b, 1984), Pielou (1977, 1978, 1979), and Michanek (1979, 1983) revived interest in this subject.

Vicariance biogeography, cladistic biogeography, and dispersal biogeography. Vicariance and cladistic biogeography bring together the phylogenetic (cladistic) techniques proposed by W. Hennig (e.g., 1966) and the panbiogeography of L. Croizat (e.g., 1964). The main approach of vicariance biogeography (as developed by Rosen 1976) and cladistic biogeography (as introduced by Platnick and Nelson 1978) is to develop branching diagrams (cladograms) reflecting phyletic relationships among groups of organisms. Biohistory may then be tied to geohistory by comparing cladograms of organisms with cladograms for areas. (For examples of this see textbooks by G. Nelson and Platnick 1981, G. Nelson and Rosen 1981, Wiley 1981, Humphries and Parenti 1986, Myers and Giller 1988). Following are some of the main ideas. Most species evolve by geographical subdivision of the ranges of ancestral species (allopatric speciation). Many closely related taxa, which today live in different geographical areas (e.g., continents), evolved in one place as ancestral taxa and later became separated (e.g., by fragmentation of continents) to form the recent vicarious taxa. The resulting vicarious distributions reflect the splitting of an ancestral biota or a taxon by the development of biogeographical barriers, rather than by the dispersal of organisms across such barriers, starting from a supposed center of origin. The adherents of vicariance theory strongly criticized the adherents of dispersal biogeography, and vice versa (e.g., Briggs 1984). It seems clear to many observers that both vicariant speciation and dispersal events produced present distributions, possibly with the dominance of one or the other mechanism in certain stages of history. The establishment of the Panama land bridge was a vicariant event for marine organisms, and at the same time it allowed dispersal of land organisms. The submergence of the Bering land bridge had a contrary effect. The concepts of cladistics and cladistic biogeography were introduced into phycology by Garbary (1987) and Lindstrom (1987a), after the first cladistic analyses, for example, by Prud'homme van Reine (1982) of the European Sphacelariales.

1.2.1 Marine Biogeographical Regions

Seven groups of biogeographical regions in the world's seas (Figs. 1.5–1.7) are recognized by modern marine zoogeography (e.g., Ekman 1953; Hedgpeth 1957a; Briggs 1974; Vermeij 1978) and marine phycogeography works (e.g., van den Hoek 1975, 1984; Michanek 1979, 1983):

(1) Arctic group (one region);
(2) Cold temperate group, Northern Hemisphere (three regions);
(3) Warm temperate group, Northern Hemisphere (four regions);
(4) Tropical group (four regions);
(5) Warm temperate group, Southern Hemisphere (five regions);
(6) Cold temperate group, Southern Hemisphere (five regions);
(7) Antarctic group (one region).

The delimitation of these seven groups of regions (Figs. 1.5–1.7) follows Briggs (1974), with the modification that the whole of Iceland is regarded as cold temperate. The analyses of van den Hoek (1984) and Joosten and van den Hoek (1986) largely supported Briggs's system. It should be noted that a similar terminology is also in use in marine plankton biogeography (e.g., "polar," "subpolar," "transitional," "subtropical," and "tropical" regions or zones; van der Spoel and Heyman 1983). Variation of the scheme given above, with the Mediterranean and parts of northwestern Africa and Europe as distinct biogeographical regions, have been proposed by M. Alvarez et al. (1988) and Prud'homme van Reine and van den Hoek (1988).

The boundaries of each group of biogeographical regions are primarily recognized by drastic changes in the composition of the coastal flora and fauna. Furthermore, with some exceptions, the boundaries follow certain surface seawater isotherms. This is because water temperature is the main factor governing the geographical distribution of species, although an essential condition for the existence of most seaweed species is the existence of solid substrata along any coast.

The **isotherms** (boundaries of the same mean water temperature averaged over many years for a particular month) used for the delimitation of groups of regions, are given in Fig. 1.8. It is clear that a scheme of this sort can only be used as a rough guide, since the boundaries between regions have to be based on floral and faunal discontinuities.

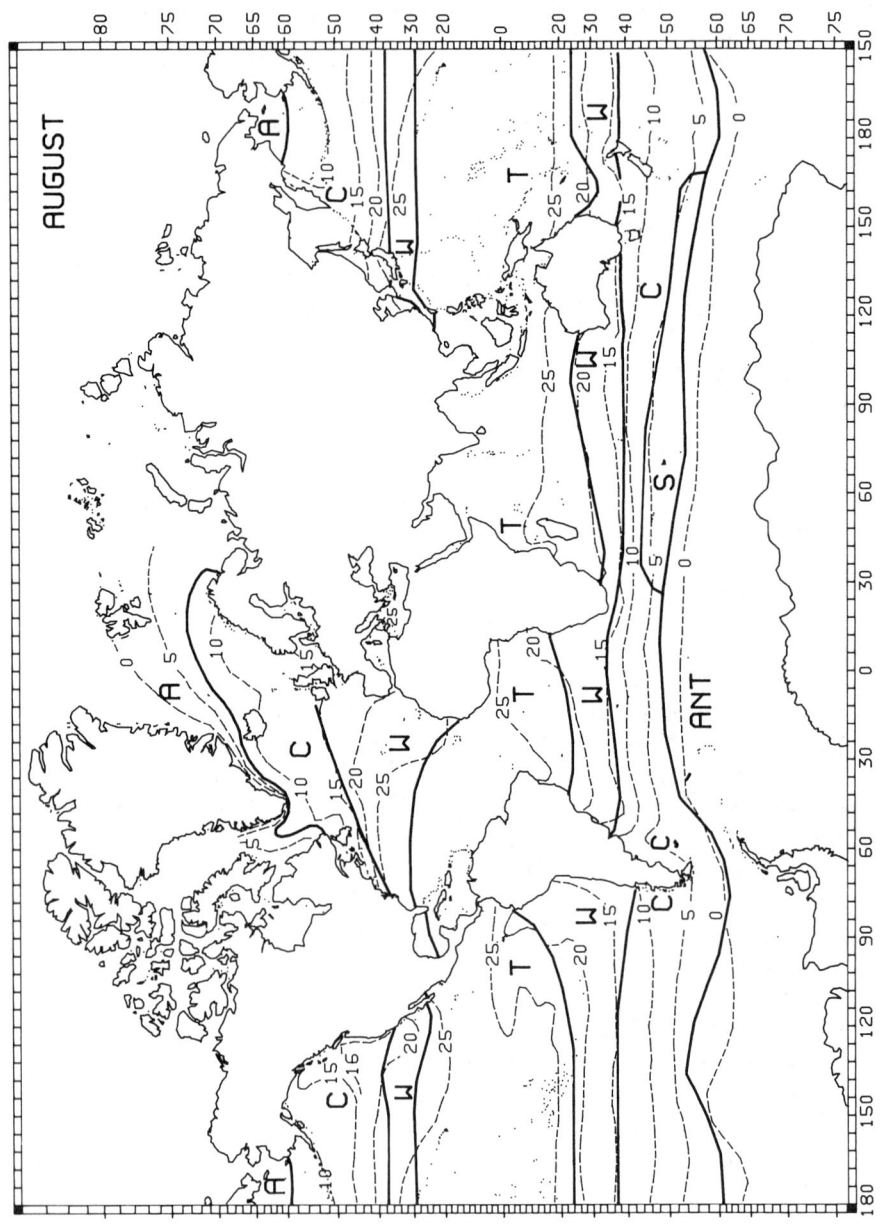

Setchell's contribution to marine biogeography. The main reason the scheme shown in Fig. 1.8 is based on intervals of 5°C and 10°C is of course the human propensity to round off numbers. It may appear, for example, only as a matter of convenience to use the 15°C-summer isotherm and the 10°C-winter isotherm as the boundaries between "warm" and "cold" temperate regions (Fig. 1.8). Such rounded temperatures often coincide with biological limits (e.g., the upper temperature limit of many members of the Laminariales is 20°C), as suggested by Setchell as early as 1893 (compare, e.g., Figs. 1.5 and 2.10), and the lower temperature limit of corals is 20°C. Setchell (1920) used a temperature classification developed by C. Hart Merriam in 1894 to define crop zones in North America. This system defined "isothere" as the mean temperature for the six warmest weeks of the year and "isocryme" as the mean temperature for the coldest six weeks. Setchell (1920) borrowed the concepts of stenothermy and eurythermy (organisms with narrow or wide temperature requirements) from M. Moebius, a German zoologist. Setchell combined this concept with that of Merriam's to provide us with the first comprehensive attempt at defining floristic regions in the sea (L. Druehl, personal communication).

1.2.2 Temperature and Boundaries of Species Distribution

With regard to the delimitation of a species' distribution by seawater isotherms in the Northern Hemisphere, one may distinguish the following cases according to Hutchins (1947), van den Hoek (1982a,b), and Breeman (1988). A species may be limited by a certain isotherm because:

(a) Lethal limits occur at higher or lower temperatures: the southern lethal limit for cold-adapted species corresponds to the August isotherm of the seawater; the northern lethal limit for warm-adapted species corresponds to the February isotherm. Lethal limits are set by the tolerance limit of the hardiest life-history stage, often the microthallus phase; near the distribution boundary the macrothallus phase occurs as a seasonal annual.

(b) Reproduction limits are set by temperature requirements for completion of the life history. The temperature requirements of a macrothallus phase often do not set the limit. The temperature requirements of other phases in the life history, such as sexual reproduction of the microthallus phase, or the vegetative initiation of macrothalli from the microthallus (e.g., a crust), may be more important.

Fig. 1.5 Seven groups of marine biogeographical regions in relation to the August isotherms (summer in Northern Hemisphere, winter in Southern Hemisphere): A = Arctic region; C = cold temperate, W = warm temperate regions of the Northern and Southern hemispheres; T = tropical regions; ANT = Antarctic region; S designates the sub-Antarctic (cold temperate) islands region. (Modified after the marine zoogeographical scheme by Briggs 1974. Deviation: all of Iceland is regarded as cold temperate.)

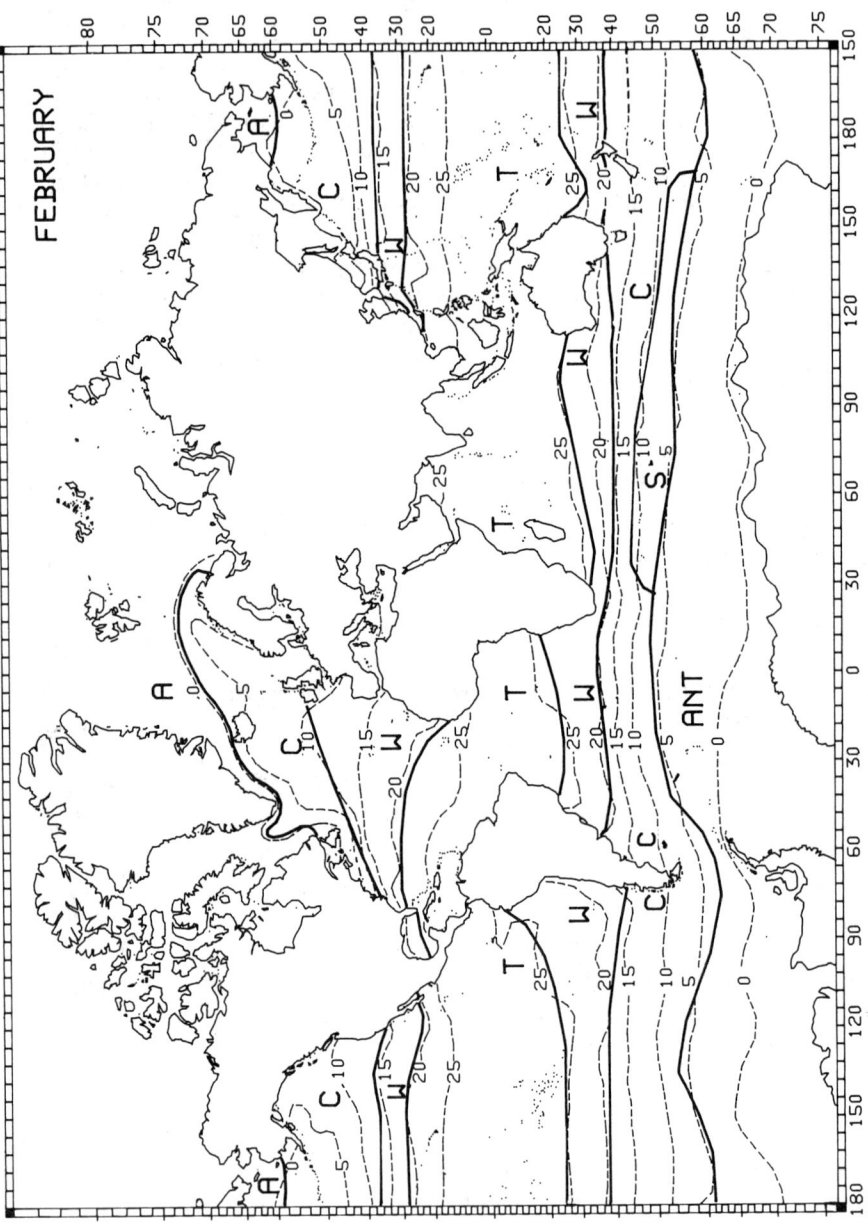

Fig. 1.6 Seven groups of marine biogeographical regions in relation to February isotherms (winter in Northern Hemisphere, summer in Southern Hemisphere). See legend to Fig. 1.5 for further details. (After Briggs 1974.)

```
Arctic region
```

ATLANTIC OCEAN	PACIFIC OCEAN
....A. Strait of Belle Isle (S. Labrador/ Newfoundland)	Barents Sea: ....E. Kola fjord ....U. (Murmansk)	....A. Bering Sea: Cape Olyutorsky	Bering Sea: ....A. Nunivak Island
....M. Cold temperate North Atlantic region Western province	....R. Eastern Province ....O.	....S. Cold temperate North- western Pacific reg.	Cold temperate North- eastern Pacific reg. ....M.
....E. North Carolina: Cape Hatteras	W. Ireland/ ....P. English Channel ....E.	....I. E. China: Wenchow SW./SE.Korea SW./SE.Japan	California: Point Conception ....E.
....R. Warm temperate Carolina region	Warm temperate Mediterranean- Atlantic region ....A.	....A. Warm temperate Japan region	Warm temperate California region ....R.
....I. Florida: Cape Kennedy	W. Africa: Senegal: ....F. Cape Verde ....R.	Hongkong, N.Taiwan/S.Taiwan, Ryukyu Isl.: Amami	Mexico, Baja Cal.: Magdalena Bay,La Paz Mexico: Topolobampo
....I. Western Atlantic tropical region	Eastern Atlantic ....I. tropical region ....C.	Indo-West Pacific tropical region	....I. Indo-West Pacific tropical region
....C. Brazil: Cabo Frio (Rio de Janeiro)	W. Africa: ....A. Angola: Mossamedes	INDIAN OCEAN	....A. ....U.
		S. Africa: Western Australia: ....S. Eastern Australia, ...E. Capeland/ Shark Bay ....I. Fraser Island: Natal ....R. Sandy Cape	S.Ecuador/N.Peru: ....C. Gulf of Guayaquil
....A. Warm temperate Eastern South American region	Warm temperate Southern African region	Warm temperate Warm temperate Southern Australian Northern New Zealand ....A. region region	Warm temperate Western South American region
.... N. Argentina: Rio de la Plata		North Island: Robe/Bermagui Manukau/East Cape	Island of Chiloe
Cold temperate Eastern South American region		Subantarctic (cold temperate) Cold temperate Cold temperate Islands region Victoria-Tasmania Southern New Zealand region region	Cold temperate Western South American region

```
                    Antarctic Convergence

                    Antarctic region
```

Fig. 1.7 Geographical localities as boundaries between marine biogeographical regions. See legend to Fig. 1.5 for further details. (After Briggs 1974.)

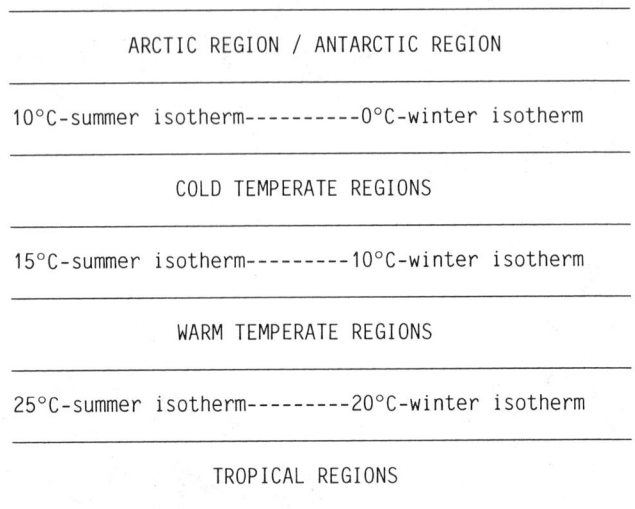

Fig. 1.8 Scheme for the delimitation of marine biogeographical regions by isotherms. (After Stephenson 1948.)

(c) **Growth limits** are operating mainly in those species persisting without (sexual) reproduction, reproduction limits being mostly narrower than are the growth limits.

Along a coast there are southern growth and/or reproductive boundaries determined by the February isotherm and northern growth and/or reproductive boundaries determined by the August isotherm. Since isotherms are based on averages of many years, and since a species may be eradicated at a particular location due to an exceptionally warm summer or cold winter (e.g., El Niño phenomenon; Section 5.5), a "safety margin" of 1–3°C is used on an isotherm for correlating the critical temperature limits of a species (van den Hoek 1982a,b).

The latitudinal intervals between the isotherms in the North Atlantic and the North Pacific are compressed on the western sides of these oceans. Warm tropical water is transported from west to east because ocean gyres in the Northern Hemisphere rotate clockwise (Fig. 2.43) due to the tendency of moving objects to veer to the right in the Northern Hemisphere. (This is the Coriolis effect: the earth rotates from west to east. A point at the equator [circumference 40,000 km] moves from west to east at a higher rate—i.e. 40,000/24 km h^{-1}—than a point situated farther north, where the circumference is smaller. A ball thrown from the equator in a northerly direction, carries with it the higher rotation rate, falls down on a more slowly moving point on the earth and hence is deflected to the right. The same happens to winds and ocean currents moving northward, such as the Gulf Stream). The compression of isotherms reduces the latitudinal distribution span of many benthic marine algal species. This is the case on the Atlantic coast of North America and on the Pacific coast of Japan (Figs. 1.5 and 1.6). Algae that flourish only within a temperature span from 0–20°C may inhabit a latitudinal interval of 35° on the European coasts (ranging from northern Norway to southern Portugal), whereas on the North American coast of the Atlantic the latitudinal interval would only be 5° (ranging from Nova Scotia to New York).

As a further consequence, the yearly temperature range at a given locality is 2 to 3 times greater on the North American coasts than on the European coasts. For example, the mean yearly temperature range may be 0–17°C at Cape Cod (−1–20°C in extreme years), 3–10°C near Tromsö in northern Norway, 8–16°C on the coast of Brittany, France, and 14–19°C on the coast of central Portugal. Therefore, **eurythermal** species (organisms with a wide range of temperature tolerance; see Section 7.1) dominate the western coasts of both oceans. These species may also occur on the eastern coasts, but there the floras are enriched by many **stenothermal** species (organisms with a narrow range of temperature tolerance).

The following examples taken from van den Hoek (1982a,b) demonstrate how one may correlate the boundaries of a species' distribution with its temperature tolerance range. The cases of many species have been analyzed, resulting in a set of temperature response types common to groups of species with similar distributions in the North Atlantic; for a review, see Breeman (1988).

Example 1: Stenothermal, warm temperate species with northern lethal and southern reproductive boundaries. Extreme cases of stenothermal algae are to be found in the group of warm temperate European algal species. The kelp *Saccorhiza polyschides*, a characteristic, European Atlantic representative of the Laminariales (Fig. 1.9), survives only at locations with a narrow temperature span. The sporophyte of this alga dies at 3°C, or at 24°C, and gametophytes become fertile at temperatures of 17°C or less (Norton 1977). This kelp may survive, therefore, in an area situated between the 4°C-February isotherm in the north (lethal boundary, allowing a 1°C safety margin for the coldest years) and the 22°C-August isotherm (lethal boundary allowing a 2°C safety margin for the warmest years) in the south. A location with these conditions does not exist on the North American coast, as the two isotherms cross over in the western North Atlantic (Fig. 1.9). In this way any species with a narrow temperature survival span is excluded on the western side of the Atlantic. In contrast, on the eastern side the two isotherms delimit a long stretch of coastline (Fig. 1.9), and here *Saccorhiza polyschides* grows from mid Norway (a yearly span of 4–12°C) to Morocco (a yearly span of 15–22°C), although only on open coasts. On the coasts of the North Sea, temperatures are lower in winter and higher in summer and the species is almost entirely absent there. In the Mediterranean the species occurs locally, where the summer temperatures just surpass 22°C.

Example 2: Tropical–warm temperate species with northern lethal boundaries in North America and northern growth and reproductive boundaries in Europe. The brown alga *Dictyota* spp. occurs with two morphologically similar species on both sides of the North Atlantic (Fig. 1.10). Schnetter et al. (1987) found that European and

18 INTRODUCTION TO VERTICAL AND GEOGRAPHICAL DISTRIBUTION

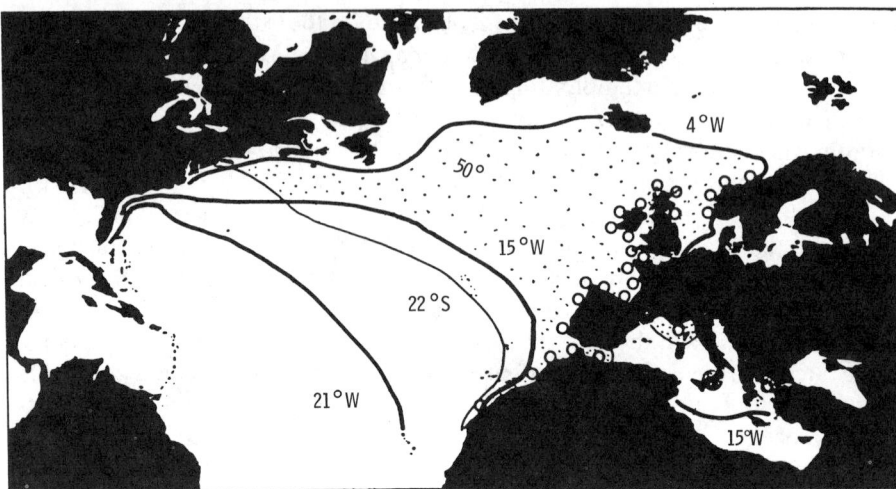

Fig. 1.9 Distribution of the kelp *Saccorhiza polyschides*. Circles represent presence (species is absent on North American coasts); distribution area is dotted (occurring, however, only along the coasts); 4°C-winter isotherm = northern lethal boundary; 15°C-winter isotherm = southern reproduction boundary; 22°C-summer isotherm = southern lethal boundary; 21°C-winter isotherm = southern growth boundary (not reached by the species). (From van den Hoek 1982*b*).

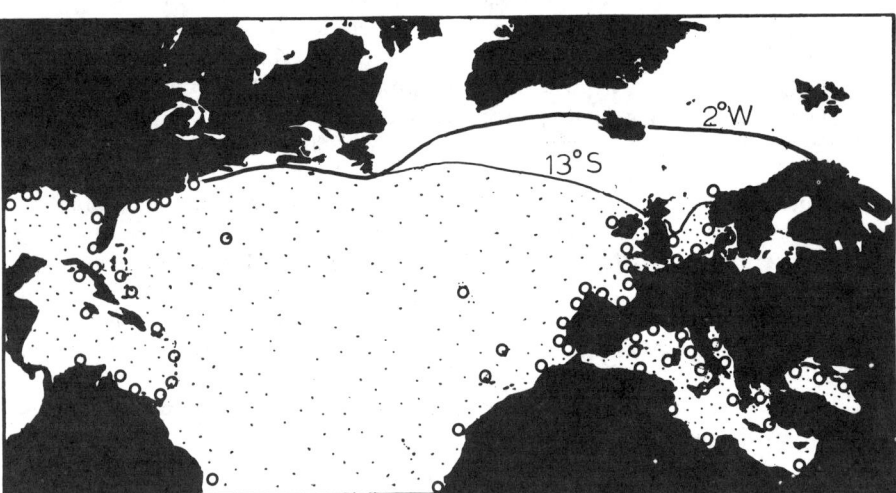

Fig. 1.10 Distribution of the brown alga *Dictyota* spp. on both sides of the North Atlantic: *D. dichotoma* in northwestern Atlantic and *D. menstrualis* in northeastern Atlantic). 2°C-winter isotherm = northern lethal boundary (limiting in North America); 13°C-summer isotherm = northern growth and reproduction boundary (limiting in Europe). (From van den Hoek 1982*a*.)

North American isolates of *Dictyota dichotoma* do not interbreed, since the American species (renamed *D. menstrualis*) has a chromosome count of $n = 24$, whereas in *D. dichotoma* $n = 16$. *D. menstrualis* occurs on the Atlantic coast of North America to the 2°C-February isotherm (Virginia). This probably represents a northern lethal boundary with a lethal limit at 1°C (safety margin of 1°C). On the European Atlantic side the 2°C-February isotherm approaches the coast in northern Norway. *D. dichotoma* penetrates only to southern Norway, where the lowest temperature is 5°C, and hence the northern distribution limit cannot be determined by lethal winter temperatures. For growth the alga requires temperatures above 12°C during a few summer months, and, in fact, its northern distribution limit follows the 13°C-summer isotherm in Europe, which therefore represents a northern growth boundary. The observation that the northern distribution limits follow a winter isotherm on one side and a summer isotherm on the other side of the Atlantic is a consequence of the fact mentioned above, that several critical isotherms cross over instead of running parallel when spanning the North Atlantic. A word of caution is required in regard to the assumption that the American *D. menstrualis* and the European *D. dichotoma* may exhibit similar temperature requirements. Biebl (1958, 1962) found that material from Brittany survived a temperature span of 3–27°C, whereas in material from Puerto Rico the range was 5–32°C. In the amphi-Atlantic green alga *Cladophora coelothrix,* an eastern Atlantic isolate survived down to 0°C, and a western Atlantic isolate to 5°C, while both isolates survived at 33°C and died at 35°C (Cambridge et al. 1987).

Example 3: Cold temperate species with southern lethal or growth boundaries and northern growth and/or reproductive boundaries. In the case of the red alga *Chondrus crispus* (Fig. 1.11) the 24°C-summer isotherm limits the southern distribution

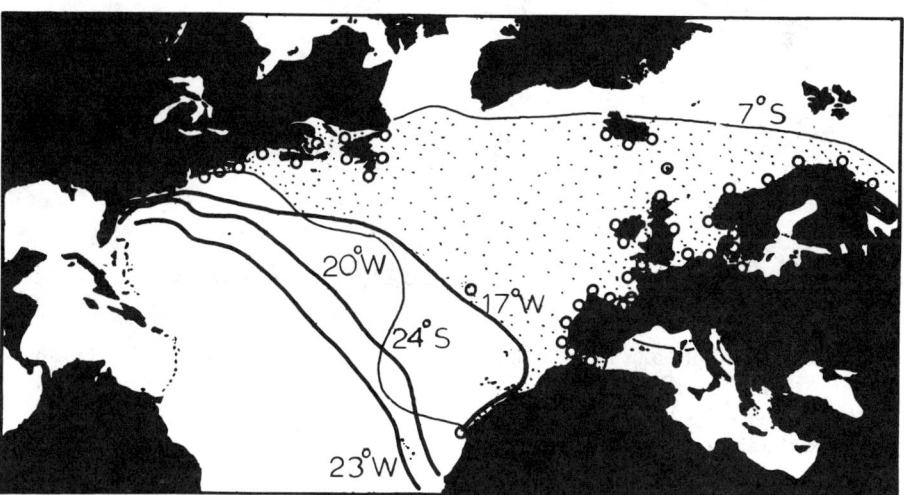

Fig. 1.11 Distribution of the red alga *Chondrus crispus* at both sides of the North Atlantic. 24°C-summer isotherm = southern lethal boundary (limiting on both sides of the Atlantic); 17°C-winter isotherm = southern reproduction boundary (additionally limiting in North Africa); 7°C-summer isotherm = northern growth boundary (limiting on both sides of the Atlantic). A southern growth and reproduction boundary (23°C-winter isotherm) is not reached by the species (From van den Hoek 1982*a*.)

20 INTRODUCTION TO VERTICAL AND GEOGRAPHICAL DISTRIBUTION

area on each side of the North Atlantic and corresponds to a southern lethal boundary on the western Atlantic coast, since the species does not survive at 28°C. On the eastern side of the North Atlantic the limit of distribution also coincides with the 17°C-winter isotherm, which may therefore represent a southern reproductive boundary. Gametophytes and tetrasporophytes become fertile and sporogenous, respectively, in the laboratory at 15°C but not at 20°C. To the north the distribution area is limited by the 7°C-summer isotherm as a northern growth boundary (minimal growth occurs in the laboratory at temperatures of 5°C and lower).

Example 4: Photoperiodically controlled, temperate species with complex boundaries. The tetrasporophyte of the red alga *Bonnemaisonia hamifera* (Fig. 1.12) forms tetraspores only in short days, at maximum photoperiods of 11 h light per day, and only in the temperature range of 11–18°C (Lüning 1980*b*, 1981*b;* Breeman et al. 1988). The **northern** distribution limit of the **gametophyte** (*Bonnemaisonia* phase), which is formed from the tetraspores, corresponds to the 13°C-October isotherm representing a northern reproductive boundary. Still farther to the north one finds the vegetatively propagating **tetrasporophyte** (Trailliella phase), which penetrates up to the 10°C-summer isotherm corresponding to a northern growth boundary. A northern

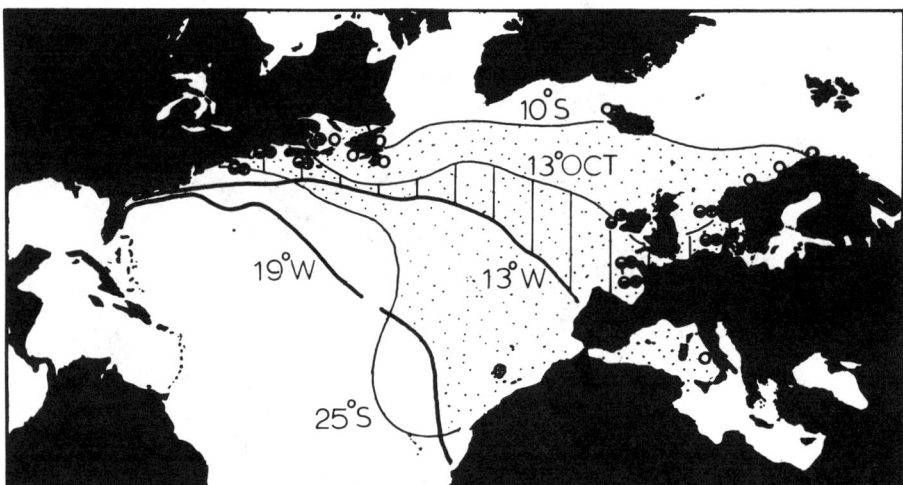

Fig. 1.12 Distribution of the red alga *Bonnemaisonia hamifera* at both sides of the North Atlantic. ● and hatched area = distribution area of the gametophyte (*Bonnemaisonia* phase); dotted area = distribution area of the tetrasporophyte (Trailliella phase) with (⊕) or without (O) tetrasporangia. 10°C-summer isotherm = northern growth boundary of the tetrasporophyte, which propagates vegetatively; 13°C-October isotherm = northern boundary for formation of tetrasporangia and gametophytes evolving from tetraspores; 13°C-winter isotherm = growth and reproduction boundary of gametophyte (limiting in Europe); 25°C-summer isotherm = southern lethal boundary of the gametophyte (limiting in North America). A southern reproduction boundary of the tetrasporophyte (19°C-winter isotherm) is not reached by the species. (From van den Hoek 1982*a*).

lethal boundary is not reached, since both generations of the species survive water temperatures just below 0°C. According to van den Hoek (1982a), the **southern** distribution limit of the **gametophyte** corresponds to a southern growth and reproductive boundary on the eastern side of the North Atlantic (13°C-winter isotherm) and to a southern lethal boundary on the western side (25°C-summer isotherm). Breeman et al. (1988) pointed out that reproduction of the gametophyte is not required for its occurrence. In southerly directions the gametophytes occur sporadically in spring and experience a progressively shorter time interval between their initiation from tetraspores in late winter and the critical temperature of 20°C in spring, which prevents further development of the gametophytes. The **tetrasporophyte** occurs to the south on both sides of the North Atlantic to the 25°C-summer isotherm, which represents a southern lethal limit (Fig. 1.12). It must be mentioned that this alga was introduced into the North Atlantic, probably from Japan, around the turn of the 20th century. In contrast to Japanese shores with basically higher seawater temperatures in autumn and winter, there is a well-known lack of synchronization in the occurrence of reproductive male and female gametophytes in various parts of the North Atlantic Ocean. Breeman et al. (1988) found that a main obstacle to such synchronization on many North Atlantic coasts is the early maturation and senescence of male plants at low winter temperatures before the female plants become reproductive. Hence there are numerous reports for North Atlantic coasts of sterile plants, isolated male gametophytes, and isolated female gametophytes with cystocarps lacking carposporangia.

2 Seaweed Vegetation of the Cold and Warm Temperate Regions of the Northern Hemisphere

2.1 PALEOGEOGRAPHY AND THE POSSIBLE HISTORY OF THE MARINE-BENTHIC FLORA

Literature: Marine biogeography and marine paleobiogeography: Briggs (1974, 1989); Ekman (1953); Haq and Boersma (1978); Hedgpeth (1957a); Joosten and van den Hoek (1986); Michanek (1979, 1983); Pérès (1982b); Pielou (1979); Prud'homme van Reine and van den Hoek (1988); Ramsay (1977); Setchell (1893, 1899, 1917, 1920, 1922); Sims et al. (1983); South (1987); South and Tittley (1986); J. W. Valentine (1973); van den Hoek (1975, 1982a,b, 1984); van der Spoel and Pierrot-Bults (1979); Vermeij (1978). **Paleogeography, paleoceanography, marine geology, and paleobiology:** Audley-Charles and Hallam (1988); Brenchley (1984); Brown and Gibson (1983); Hagevang et al. (1983); Hallam (1973); Hopkins (1967); Hurdle (1986); Kennett (1982); Kontrimavichus (1986); Meyen (1987); Nilsen (1983); Nunns (1983); Pomerol (1982); Ross and Scotese (1988); Rowley and Lottes (1988); Schopf (1980); Sclater et al. (1977); Scotese et al. (1988); Seibold and Berger (1982); Srivastava and Tapscott (1986); Stanley (1986); Summerhayes and Shackleton (1986); Thomas and Spicer (1987); Toomey and Nitecki (1985); D. H. Valentine (1972); Vogt (1986). **Paleotemperatures:** Frakes (1979). **Paleocontinental maps:** Smith et al. (1981). **Pleistocene and pre-Tertiary glaciations:** Denton and Hughes (1981); Hambrey and Harland (1981). **Global sea levels:** Haq et al. (1987); Vail and Hardenbol (1979). **Epicontinental seaways:** C. G. Adams (1981); Termier and Termier (1960); Wiedmann (1979). **Timescale:** Kent and Gradstein (1986).

Worldwide patterns in the distribution of seaweeds in particular and in benthic, shallow-water marine organisms in general, along with possible paleogeographical causes for these patterns, can be summarized as follows:

(1) The tropical regions exhibit much taxonomic similarity. **Explanation:** since Mesozoic times there has been a continuous warm-water girdle around the earth, and many tropical marine taxa have a circumglobal distribution. The warm-water girdle became interrupted by land bridges relatively recently: in the late Tertiary 17 Ma with closure of the seawater connection between the Mediterranean and the Indian Ocean (Section 2.4.6), and 3–4 Ma with the Central American land bridge (Section 4.1). (Million of years before

present will be represented by "Ma". This is the abbreviation for 1 megaannum or 10^6 yr measured from the present.)

(2) The cold temperate and polar marine-benthic floras and faunas of the Northern Hemisphere differ fundamentally from those of the Southern Hemisphere. **Explanation:** cold-water biota must have evolved in each of the three large ice ages during the Phanerozoic, that is, the last 600 million years (Fig. 2.1A). Ice ages are relatively short-lived phenomena occurring roughly every 200 million years and having a duration on the order 1 to 10 million years (Fairbridge 1973; Frakes 1979). The present cold temperate and polar marine-benthic floras and faunas are a product of the last climatic deterioration, starting in the Tertiary (which begins at 65 Ma) and culminating in the glaciation of the Miocene–Pleistocene interval; the Miocene begins at 23 Ma and the Pleistocene at 2 Ma (Figs. 2.1B,C). The cold-water biotas of both hemispheres are fundamentally different. During long evolutionary periods they were separated from each other by a tropical warm-water girdle.

(3) The cold temperate seaweed floras of the Southern Hemisphere exhibit certain similarities. **Explanation:** part of the similarities, on the species level, may be due to dispersal of marine organisms via the West Wind Drift, and part, above the species level, may be due to the common history of the southern continents as fragments of a southern supercontinent (see Chapter 5).

(4) The cold temperate benthic marine floras and faunas of the North Atlantic and North Pacific differ widely from each other, and the North Pacific is much richer in genera. **Explanation:** for most of the Tertiary, cold-water biota developed separately in both oceans because the Bering Strait was closed. More cold-water taxa could evolve in the North Pacific because it is an old ocean. Since Mesozoic times it has offered, together with the early Arctic Ocean as its northern embayment, thousands of kilometers of coastline exposed to cooler water (Fig. 2.2A). In contrast, the younger North Atlantic formed its northern coastlines only during the Tertiary (Figs. 2.2B–D). A biotic exchange between both oceans, mainly an inflow of North Pacific species into the North Atlantic via the Arctic Ocean, was possible only in the late Tertiary.

In more detail one may distinguish **five phases** that are important for the biotic exchange between the North Pacific and the North Atlantic via the Arctic Ocean.

Phase 1 (around 230–65 Ma, Mesozoic, Triassic to Cretaceous): Pacific Ocean Wide Open to the Early Arctic Ocean (Fig. 2.2A)

Plate tectonics. According to the theories of plate tectonics and continental drift, the formation of the Atlantic Ocean and the breakup of **Pangaea,** a combined landmass of all continents surrounded by Panthalassa, began in the early Mesozoic. At 200 Ma the **Atlantic Ocean** first formed its tropical parts (central Atlantic, Caribbean, and Gulf of Mexico), when Africa and South America drifted away from North America (Kennett

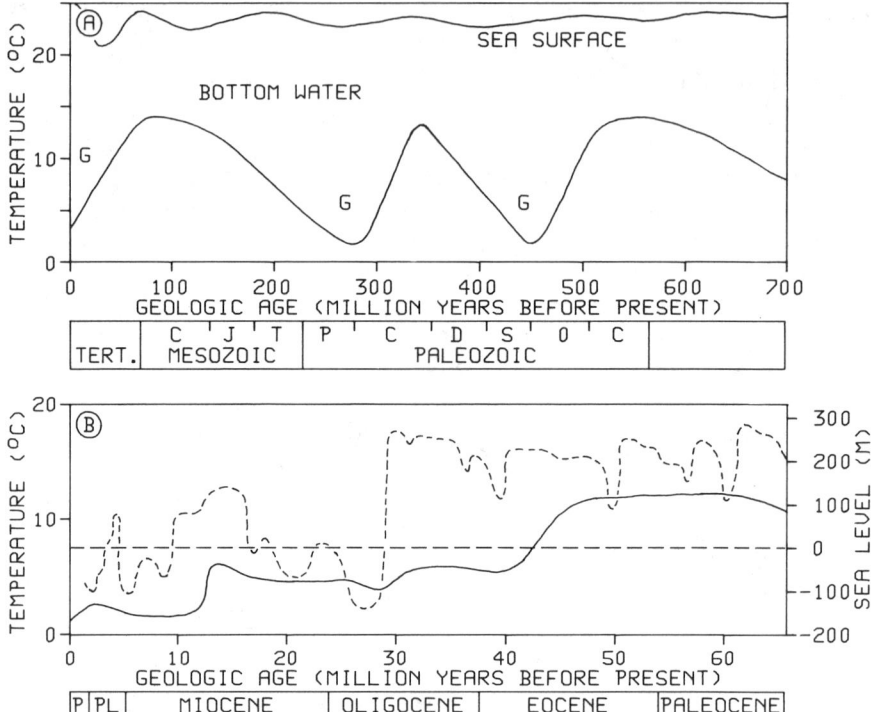

Fig. 2.1 Paleoclimate, paleotemperatures, and paleo-sea levels. **(A)** Estimated change in mean summer sea-surface temperature and in bottom water temperature for tropical latitudes over the past 700 Ma (million years before present). There were three periods of extensive glaciation (G) and cold bottom water: (1) Around 450 Ma: late Ordovician–early Silurian; (2) 300 Ma: late Carboniferous–early Permian; (3) around 2 Ma: Pleistocene. Abbreviations, Paleozoic: C = Cambrian; O = Ordovician; S = Silurian; D = Devonian; C = Carboniferous; P = Permian. Mesozoic: T = Triassic; J = Jurassic; C = Cretaceous. Tert. = Tertiary. **(B)** Tertiary and Pleistocene. Solid line = high-latitude seawater surface temperature. Major cooling phases occurred near the Eocene/Oligocene boundary (40 Ma) and in the Miocene (10 Ma). Broken line = Tertiary eustatic changes of sea levels. Meters above or below present-day sea level (= O) are tentative. Pl = Pliocene; P = Pleistocene. **(C)** Marginal eastern North Pacific: scheme of major oscillations of temperature-sensitive planktonic foraminiferal biofacies from late Miocene to Pliocene and Pleistocene. Exclusively sinistral coiling populations of *Neogloboquadrina pachyderma* indicate surface temperatures of 10°C or lower (sub-Arctic biofacies); exclusively dextral populations represent temperatures of 15°C or higher (temperate biofacies); the leading edge of the subtropical–tropical biofacies indicates temperatures of 20°C or higher. **(D)** Ice limits and approximate August seawater isotherms (in °C) 17,000 to 18,000 years ago, during the last ice-age maximum. The unfamilar-looking shapes of the continents are due to the fact that the seawater level was 85 m lower than today. (A after Schopf 1980; B after Savin et al. 1975, and Vail and Hardenbol 1979; C after Ingle 1977; D after McIntyre et al. 1976.)

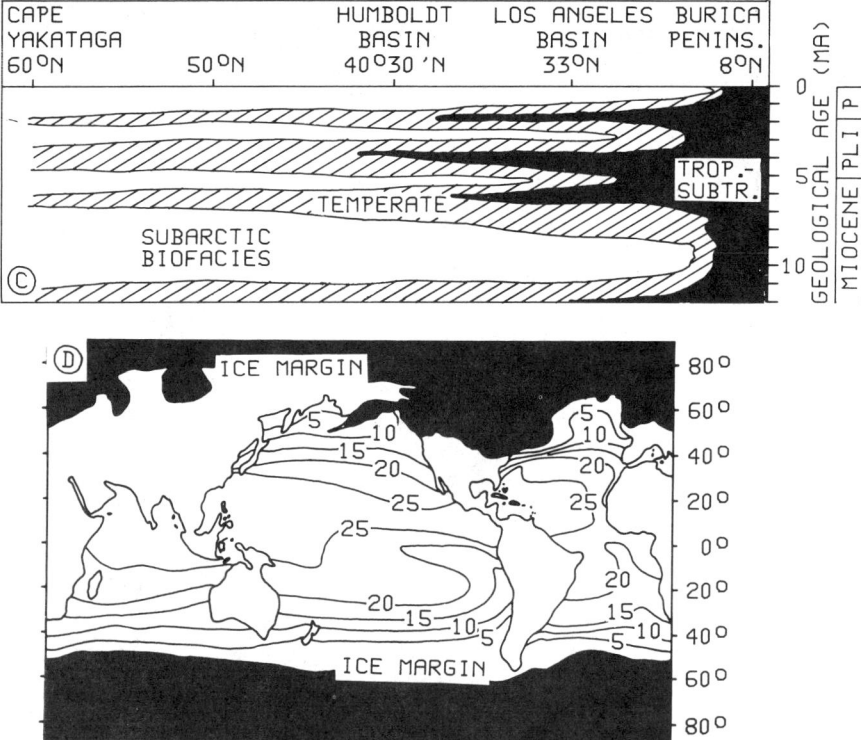

Fig. 2.1 (*Continued*)

1982). In this way a northern supercontinent, **Laurasia,** and a southern supercontinent, **Gondwana** (Gondwanaland), were formed, separated by the circumglobal tropical **Tethys Sea.** As distinct from the Atlantic, which is increasing in size, the **Pacific,** the old superocean, has steadily decreased since the early Mesozoic; Kennett (1982) has suggested that it will disappear completely in about 200 million years. The **opening of the North Atlantic** started at 165 Ma with the separation of North America from Africa. The **marine shelf biota profited** from the Mesozoic–Cenozoic **breakup of Pangaea** (Valentine 1973). More living space was created, ocean barriers were produced, and so diversity rose. Since high-latitude cool water was isolated from warm low-latitude water, the latitudinal temperature gradient rose, and the biota of the marine shelves responded by forming latitudinal provincial chains with increased endemism. These arguments support the view that **plate tectonics** was a **major driving force** for biotic revolutions in past oceans (Valentine 1973).

Evolution of the Arctic Ocean. The western Arctic Ocean, which has also been called the Boreal Gulf, may have been formed in Mesozoic times as a northern embayment of the Pacific Ocean (Valentine 1972; Herron et al. 1973; Churkin and Trexler 1981; Kitchell and Clark 1982; Fujita 1978; Sweeney 1985). Today this part is the deep **Amerasian Basin,** bounded by the continental margins of North America and Siberia, and the Lomonosov Ridge. The **Eurasian Basin,** an extension of the growing North

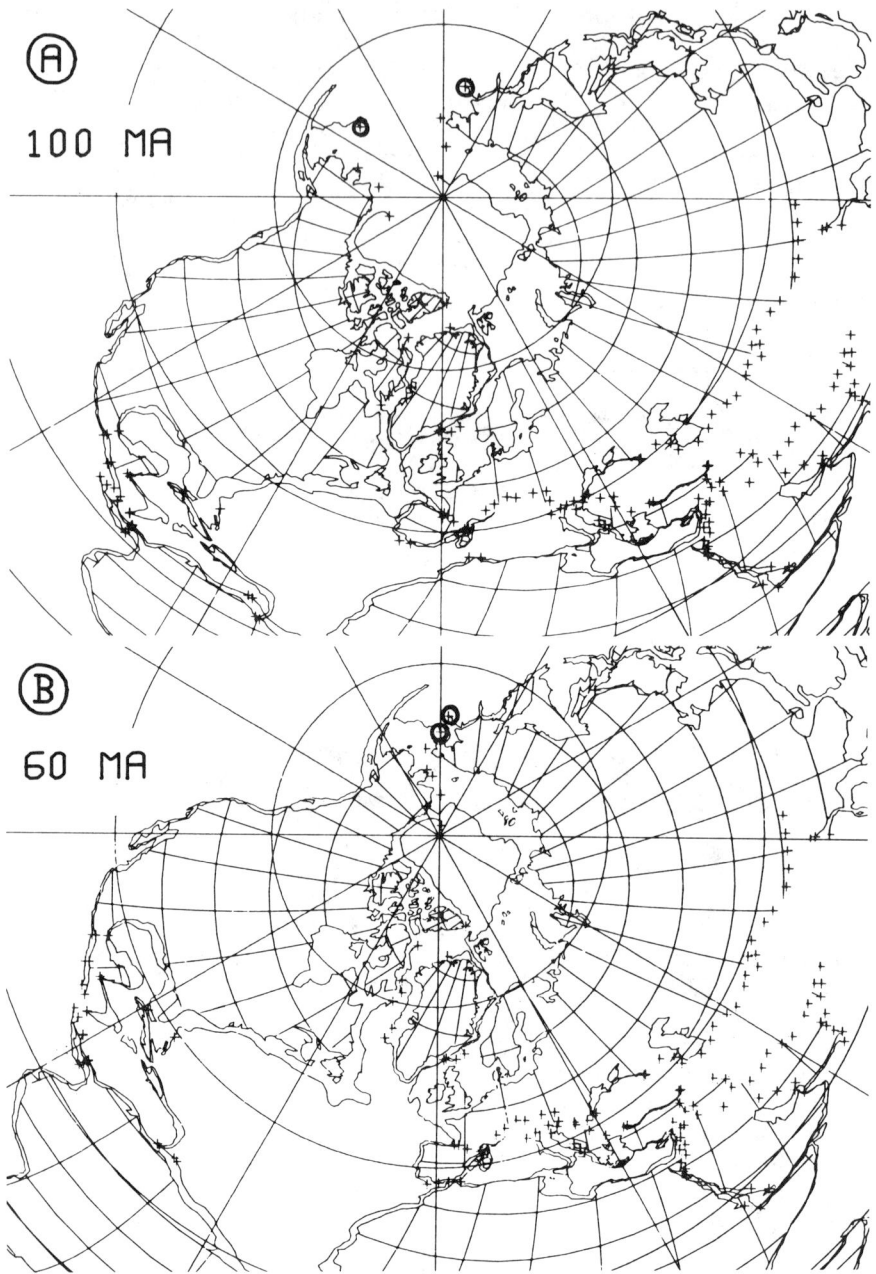

Fig. 2.2 Continental drift and formation of the Atlantic Ocean. **(A)** 100 Ma (phase 1, Cretaceous): North Pacific wide open to the early Arctic Ocean. **(B)** 60 Ma (phase 2, early Tertiary): North Pacific closed off from the early Arctic Ocean. Notes: (1) The paleocontinental maps show, in addition to the coastlines, the present-day 1000 m submarine contour, which is much more enduring than the coastlines. (2) The two endpoints of the 1000 m submarine contour in the Bering Sea area (circled crosses) approach each other in the time interval of 100–60 Ma. Note that the position of the Pole was near the Bering Strait at 60 Ma. **(C)** 55–38 Ma (phase 2, early Tertiary): North Pacific closed off from the early Arctic Ocean. Epicontinental (shallow) marine waterways connecting the Arctic Ocean to the Tethys Sea: Turgai Strait via Eurasia

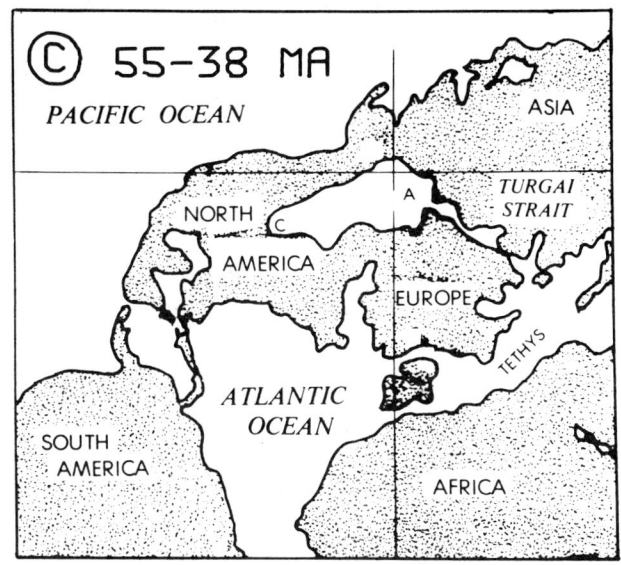

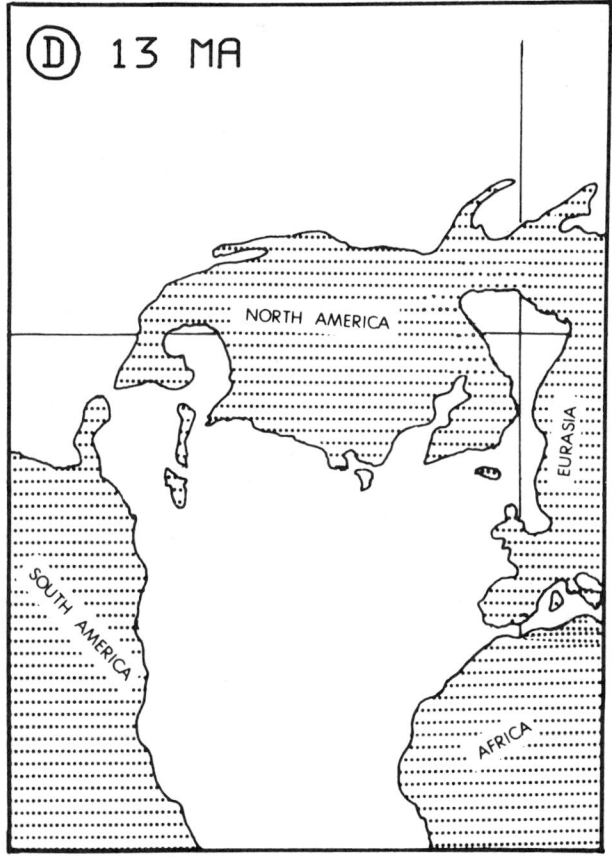

and C = Cannonball Sea (intermittently closed) via North America. Note that Beringia was formerly at a higher paleoaltitude. **(D)** 13 Ma (phase 3, mid-Tertiary): North Atlantic connected to the Arctic Ocean at around 40 Ma; epicontinental seaways dried up; the Bering Strait still closed by a land bridge. (A,B from Smith et al. 1981; C,D after Hopkins and Marincovich 1984.)

Atlantic, originated through the Tertiary seafloor spreading simultaneous to the opening of the Norwegian-Greenland Sea (Talwani and Eldholm 1977, Thiede 1979, Srivastava 1985). In Mesozoic times, tropical biota reached to much higher latitudes than today. Seawater surface temperatures in the primeval Arctic Ocean were nearly 14°C in the late Cretaceous (Emiliani 1961). As suggested for Jurassic times by Valentine (1973), there was a tropical Tethys Sea, the Pacific Ocean with intermediate temperatures, and the early Arctic Ocean with the coolest waters. One may imagine that warm temperate algae, as the forerunners of recent cool-water groups (e.g., the ancestors of the Laminariales) may have had a chance to evolve along the coasts of the Mesozoic Arctic Ocean and the adjacent northern coasts of the wide-open North Pacific (Fig. 2.2A). Besides the Pacific, the Tethys Sea may have been a donor area for further evolution in the early Arctic Ocean. Faunal migrations along the epicontinental seaways (see below) connecting the Tethyan and high-latitude regions are well documented (Casey and Rawson 1974). Some ammonites and belemnites originated in the Tethys Sea and evolved further in the primeval Arctic Ocean (Rawson 1974).

Epicontinental seas. A late Cretaceous silicoflagellate assemblage from the Arctic Ocean (Alpha Cordillera, a ridge running west from the Lomonosov Ridge) exhibited clear Pacific and Russian relations (Clark 1977), because the shallow, epicontinental **Obik Sea** on the Russian platform (a forerunner of the Turgai Strait, Fig. 2.2C) connected the Amerasian Basin to the tropical Tethys Sea (Adams 1981). Another epicontinental seaway, the **Midcontinental Sea** or **Cannonball Sea** (Fig. 2.2C) connected the early Arctic Ocean to the early Gulf of Mexico from early Cretaceous times but closed up towards the end of the Cretaceous, 70–65 Ma (Hallam 1981). Also at this time the last species of dinosaurs, which inhabited the rivers and floodplains of this seaway, became extinct (Schopf 1981).

Regressions and transgressions of the seas and biotic diversity. During the more extended of the numerous transgression periods in the Mesozoic, when epicontinental seas flooded the margins of the continents, biotic diversity (e.g., of marine mollusks) increased. New habitats and open ecological niches were created (Cooper 1977). Periods of extended marine regression were survived only by the more eurytopic species (i.e., those with wide tolerance limits). The more "continental" climate and parallel lowering of the water temperature may have been related to an increased albedo. (Land has a higher albedo, that is, reflects more radiation than the sea; see p. 32.) Hence, the rate of evolution was proportional to the rate at which new ecological opportunities occurred. Examples of these opportunities include new warm-water marine habitats on land during extended Mesozoic transgressions, and cold-water habitats developing on a large scale at higher latitudes towards the end of the Tertiary.

Phase 2 (around 65–50 Ma, early Tertiary): Arctic Ocean closed off from the North Pacific (Figs. 2.2B,C)

Near the end of the Cretaceous, possibly around 100 Ma, continental plate movements closed the deep waterway connecting the North Pacific to the primeval Arctic Ocean (Pitman and Talwani 1972). This compression was concurrent with the opening of the North Atlantic and eradicated the oceanic area between Eurasia and Alaska, bringing to an end the Pacific–Arctic biotic continuity. At the same time the **North Pacific with its vast northern coastlines** likely became the primary area for the evolution of endemic cool water taxa (Fig. 2.2B). Such taxa were highly isolated in the north by the Bering land bridge for about 60 million years and in the south by tropical water. While the **North Atlantic** offered **few northern coastlines** at this time (Fig. 2.2B), the now isolated Arctic Ocean may have become a second area of biogeographic importance, at least as a refugium.

Ocean Point. Forty kilometers south of the Arctic Ocean coastline of northern Alaska an unusual assemblage of early Tertiary (Paleocene) marine fossils (e.g., mollusks and ostracods), has been detected by Marincovich et al. (1985). Mesozoic taxa that became extinct and were replaced by modern forms in the adjacent North Pacific, survived as relicts at Ocean Point until the early Tertiary (65–50 Ma, Paleocene and early Eocene); an example is the bivalve *Eburneopecten*. Later, such relicts could migrate into the North Atlantic when a seaway formed; *Eburneopecten* is in fact recorded in Eocene and early Oligocene (around 35 Ma) deposits of the eastern United States. The Paleocene marine fauna from Ocean Point also contains endemic genera that must have evolved in the early Tertiary Arctic Ocean and later appeared in the North Atlantic after it had broken through to the Arctic Ocean (see below, phase 3).

Primitive Laminariales on Eurafrican shores and in the Arctic. There is the puzzling situation that the order Laminariales most probably evolved in the North Pacific, but the family **Phyllariaceae** occurs today exclusively on warmer Eurafrican shores (*Phyllariopsis reniformis, P. purpurascens; Saccorhiza polyschides,* Fig. 2.36), or in the cold North Atlantic and adjacent Arctic (*S. dermatodea,* Fig. 2.21). As Henry and South (1987) have pointed out, several characters of this family indicate primitiveness, such as hair tufts on young laminae, zoospores with eyespots, and no lamoxirene as the sexual pheromone. If the ancestors of the Phyllariaceae had once lived in the North Pacific, they **may have survived in the Arctic Ocean** and later spread into the North Atlantic. If they had reached the western North Atlantic, they were most likely exterminated during the ice ages (South 1987). Perhaps the warm temperate Mediterranean–Atlantic region is a refuge for marine algae that survived from the early Tertiary, just as, for example, the Canary Islands were a refuge for Tertiary higher plant relicts (Humphries 1979). Alternatively, one might suggest that the Laminariales evolved as a monophyletic group in the North Pacific, and the Phyllariaceae are relicts from an earlier time in the evolution of the autochthonous seaweed flora of the North Atlantic and adjacent Arctic. This may also be true for the **Tilopteridales,** the only brown algal order endemic to the North Atlantic, which may also represent an ancient relict group (South 1987). The order consists of the two cold-adapted species, *Haplospora globosa* and *Phaeosiphoniella cryophila* (Hooper et al. 1988), restricted to Newfoundland, and *Tilopteris mertensii* with a more southern distribution. This group has a reduced oogamous type of sexual reproduction, and trichothallic growth (South 1987).

Arctic innovations. The Arctic is not merely a passive area, but in the Tertiary a "birthplace for important biotic innovations and for major groups that later radiated to lower latitudes" (Hickey et al. 1983). This concept had been invoked by the French naturalist Buffon and was accepted by botanists such as Asa Gray, S. Gardner and A. Engler (see Krassilov 1986 for the history of the concept). Engler (1879-1882) regarded an "Arcto-Tertiary Flora" as the forerunner of the modern midlatitude flora. Obviously this innovative idea stimulated phycologists of the time, namely Kjellman (1883), Reinke (1889*a*), and Simmons (1906), to regard the Arctic as the cradle of the temperate seaweed species, which were thought to have evolved from "Tertiary-polar species" (Simmons 1906). Meanwhile, for terrestrial woody plants, the Arctic-origin hypothesis has been rejected by Wolfe (1980) and Spicer et al. (1987). They support the idea that new genera and families evolved at midlatitudes, migrated northward, and diversified at the species level in the early Tertiary Arctic, which, contrary to the views of Hickey et al. (1983), was a "dead end" for woody plants. For marine organisms and the vertebrate land fauna, it seems more likely that Arctic innovations were important, as indicated by the endemic genera of the Paleocene marine fauna from Ocean Point and the Ellesmere Eocene fauna.

Eureka Sound formation, Ellesmere Island. Early Tertiary terrestrial fossil vertebrates have been found within the present Arctic Circle (Dawson et al. 1976). They were discovered from the 3300 m thick Eureka Sound formation at about latitude 78° north on Ellesmere Island, Arctic Canada. This Eocene fauna includes taxa of crocodiles, flying lemurs, small primates, tapirs, and turtles, and it indicates a warm temperate climate with perhaps adaptations to twilight conditions. The Ellesmere Eocene fauna shows for a large part unspecific North American–European zoogeographic affinities (Estes and Hutchison 1980; McKenna 1980, Kent et al. 1984). How important was the early Tertiary Arctic Ocean for the evolution of marine taxa? One must bear in mind that at this time the Arctic Ocean was still relatively small and warm. Shallow surface temperatures near Ellesmere Island may have reached 15°C in the Eocene, when this island was located at almost the same latitude as at present (McKenna 1980). In the warm Eocene ocean surface temperatures may have been a little over 20°C in the tropics and about 10°C near the poles (Shackleton and Boersma 1981). Equable climates, characterized by moderate temperatures and a low annual range, were widespread in the Cretaceous and early Tertiary. Relict taxa such as the redwood of coastal California and the laurel forest of the Canary Islands are narrowly confined to an equable climate today (Axelrod 1984). The areas that harbor relict taxa today have to be regarded as "museums" with environmental conditions similar to early Tertiary conditions rather than as centers where such taxa originated. In the algal family Phyllariaceae, *Saccorhiza polyschides* is a relictlike "kelp" that tolerates a narrow temperature range and today lives not too far from the laurel forest of the Canary Islands (Fig. 2.36). On the other hand, there is the view that "living fossils," that is, the "survivors of various periods of extinction, are likely to be those lineages that became the best adapted to the worst of whatever conditions were eliminating species" (Valentine 1973). We may then ask, which special ecological niche was occupied by the few recent species of the Phyllariaceae or Tilopteridales to favor their survival?

Polar light regime. One may further ask: how much of the high-latitude conditions of daylength regime, with a polar night and a polar day, can still be found as an imprint in the photoperiodic behavior of seaweed taxa? These would be taxa whose ancestors originated in an ice-free, still relatively warm Arctic Ocean and later dispersed to the middle latitudes. How Tertiary and Cretaceous forest taxa and the Eocene vertebrates from Ellesmere Island coped with the two to four of months total winter darkness in polar latitudes has led to suggestions about a lower inclination of the earth's axis (e.g., Wolfe 1980) and to the more likely concept of an essentially dormant state of the plants, and hibernation of the vertebrates during winter darkness in polar regions (Axelrod 1984). Some kelps, the Arctic *Laminaria solidungula,* and the cold temperate *L. hyperborea* perform most of their seasonal length increase in the long continued darkness of the winter months (Section 8.4).

Epicontinental seas. It should be borne in mind that prior to the formation of the seaway into the North Atlantic (phase 3, see below), isolation of the Arctic Ocean was not complete because of the existence of an epicontinental seaway, the long-lasting **Turgai Strait** (Fig. 2.2C). From the Jurassic through the late Eocene (around 40 Ma) this shallow seaway connected the early Arctic Ocean to the Tethys Sea across western Siberia (McKenna 1975, Hallam 1981, Marincovich et al. 1985). At times it was a huge marine area on the Russian platform (**Obik Sea**), at other times a 100 km wide seaway, and sometimes intermittently dry. Since the marine fauna at Ocean Point contain no Tethyan taxa and the ostracods exhibit weak similarity to those of the Turgai Strait, its importance in linking the Arctic Ocean to the world ocean may have been limited.

Cretaceous/Tertiary boundary: asteroid-impact hypothesis. Dust in the stratosphere may have darkened the earth for months to several years when a huge asteroid or comet (5–10 km in diameter) struck the earth at the Cretaceous/Tertiary boundary. This darkness may have caused many extinctions, mainly of larger organisms (L. W. Alvarez et al. 1980). Anomalously high concentrations of iridium and other platinum-group elements in rocks at the Cretaceous/Tertiary boundary, which are presumed to be of extraterrestrial origin, add weight to this hypothesis. This "lights out" scenario was at first dimly regarded by some paleontologists (e.g., Schopf 1981), who pointed to a more continuous rather than a sudden extinction in many groups, such as the dinosaurs. The hypothesis seems to have gained ground lately (Vogt 1986; Thomas and Spicer 1987; Garwin 1988; Marshall 1988) and is supported by data showing dramatic vegetational changes on land following the winter of impact, with the tendency for extinction of evergreen and survival of deciduous plants (Wolfe and Upchurch 1986). In the marine fossil record, examples of mass mortality and mass extinction have been cited; examples include ammonites, cheilostomate bryozoans, brachiopods, and bivalves (W. Alvarez et al. 1984), and deep-sea ostracods (Benson et al. 1984).

Arctic freshwater injection hypotheses. The possibility has been proposed that a temporarily isolated, brackish Arctic Ocean flushed the world ocean with a low-salinity surface layer when the North Atlantic broke through to the Arctic Ocean. This injection event should have caused mass extinctions and sudden climatic change, such as the terminal Cretaceous event at 65 Ma (Gartner and Keany 1978) or at the Eocene/Oligocene boundary around 37 Ma (Thierstein and Berger 1978). These Arctic freshwater spillover hypotheses have been shortlived (Vogt 1986); they have been met with reservation for tectonic reasons (Eldholm and Thiede 1980) and because of a lack of fossil plankton evidence for a brackish or freshwater Arctic Ocean (Clark and Kitchell 1979).

Phase 3 (around 50–3 Ma): North Atlantic Connected to the Arctic Ocean, by Deep-water Seaways, with the North Pacific Still Closed Off from the Arctic Ocean by the Bering land bridge (Fig. 2.2D)

Breakthrough of the North Atlantic. In the early Tertiary (**Paleocene,** 66.4–57.8 Ma; DNAG timescale; see Kent and Gradstein 1986) there was a land bridge that reached from the North American plate over Greenland to Europe, with shelf areas. Terrestrial fossils of freshwater fishes, snakes, and crocodiles still give evidence of the common fauna (Hoch 1983). The opening of the **Norwegian–Greenland Sea** and surface exchange with the Arctic Ocean, small as it was, occurred towards the end of the **Eocene** (57.8–36.6 Ma) at around 40 Ma (Thiede 1979, 1980; Hagewang et al. 1983; Srivastava 1985). In the Eocene, the Labrador Sea also continued to evolve, and Baffin Bay opened (Gradstein and Srivastava 1980; Srivastava and Tapscott 1986). Evidence from plate tectonics is in accord with the mammalian evidence, which indicates a reduced similarity of North American and European land vertebrates at about 49 Ma in the Eocene because the northern land bridge (**De Geer Route**) over the Euramerican landmass (McKenna 1975) was severed. In the following era, the **Oligocene** (37–25 Ma), Greenland continued to separate from Svalbard (Spitsbergen and adjacent islands), and deep-water exchange was initiated between the Norwegian–Greenland Sea and the Arctic Ocean (32–34 Ma) (Eldholm and Thiede 1980; Thiede 1980; Kristoffersen and Huseby 1985).

Eocene/Oligocene boundary event. In the vicinity of the Eocene/Oligocene boundary, at around 40 Ma, a global decline in surface and bottom temperature occurred (Fig. 2.1B). Water on the high-latitude shelves became cold enough to sink and to change the warm deep sea with temperatures of 15°C in Mesozoic times into today's cold deep sea, with temperatures ranging of 0–5°C. The huge, cold deep sea called the "psychrosphere," which is overlayered by the relatively shallow warm-water layer of the "thermosphere," formed gradually after the 40 Ma event. One example of evidence for this is the development of a cool-adapted, deep-sea ostracod fauna, which became isolated from the shallow shelf regions by the development of thermal stratification and spread rapidly throughout the world in deep water (Benson et al. 1984). The Eocene/Oligocene boundary event or terminal Eocene event was the first of the two marked, rapid temperature drops during the Tertiary (Savin 1977); the second occurred in middle Miocene, about 10 Ma (see below).

Conformance between drop in sea level and drop in sea-water temperature. The global drops in seawater temperature may have been related to the global lowering of sea levels, whereby the emerging shelf areas would have produced more cooling by reflection of sunlight (albedo) than the same areas would have if covered with water (Seibold and Berger 1982). There were in fact relatively low sea levels during the Oligocene (Fig. 2.1B), possibly caused by the fact that the Atlantic Ocean had grown deep enough relative to continental areas to decrease the average depth of the seafloor, whereas the Mesozoic and early Tertiary transgression had been caused by the shallower Atlantic (Vail and Hardenbol 1979: Haq et al. 1987; Seibold and Berger 1982).

Phylogenetic size increase. The conformance between sea levels and water temperature may also be connected to Cope's rule, the evolutionary trend towards increasing body size within a phylogenetic series (Cope 1896). As Cooper (1977) pointed out, evolution in marine endotherms proceeded from small forms at times of marine transgression to larger forms during the ensuing regression and climatic deterioration. The well-known Bergmann's size rule (Bergmann 1847; Brown and Gibson 1983; Colinvaux 1986), with body size increasing towards the Poles during glacial epochs, may thus point to an extreme case of Cope's rule. Evidence to support this hypothesis may be sought in the Laminariales, which may have proceeded from small forms, maybe in the Cretaceous in the Boreal Gulf, to the present size of the giant kelps. Since long-term changes in sea level are probably tied to plate tectonics, the relationships between sea levels, water temperature, and evolutionary trends stress the view that plate tectonics is also the primary factor governing patterns of biotic evolution (Cooper 1977).

Development of cool-water taxa. The rapid temperature drop at around 40 Ma and the cooler surface water inflow into the North Atlantic from the Arctic Ocean impoverished the planktonic fauna and flora of the warm branches of the Eocene North Atlantic; for example, the planktonic foraminiferal fauna became sparse and poor in species (Berggren and Schnitker 1983). A distinct boreal faunal province developed in the Oligocene (Berggren and Hollister 1974). The temperature drop would also have enhanced the evolution of temperate seaweed groups, since seawater temperatures below 10°C occurred at high latitudes (Fig. 2.1B). The lower lethal limit of many present algae inhabiting tropical regions is around 10°C. Possibly, cold temperate algal species evolved when temperatures fell below this value along the rocky coasts of the North Pacific, northern Greenland, and the northern coasts of the landmass that later fragmented into the Canadian Arctic archipelago.

Greenland–Scotland Ridge. Later, as temperatures decreased at the middle latitudes, the temperate algae, which had developed in the Arctic Ocean and the Norwegian–Greenland Sea, invaded the more southern coasts of the North Atlantic and spread to its eastern and western shores along a series of stepping stones. These were the fragments of the Greenland–Scotland Ridge, an anomalous shallow bathymetric feature that had formed across the ocean with the opening North Atlantic (Bott et al. 1983; Nielsen 1983; Nunns 1983). This ridge starts at the **Iceland–Greenland Ridge** in Denmark Strait, crosses the active spreading center in Iceland, continues in the **Iceland–Faeroe Ridge,** the Faeroe Islands, and leads to the north Scottish shelf. The Greenland–Scotland Ridge, a modern equivalent of the **Thulean land bridge** of older biogeographers, was probably severed as a land bridge in early Eocene shortly after forming. This break possibly occurred in the Faeroe–Shetland area (McKenna 1983; Thiede and Eldholm 1983). The Greenland–Scotland Ridge did not act as an impenetrable hindrance to surface exchange of marine biota between the Arctic Ocean and the developing North Atlantic (Berggren and Schnitker 1983). The main parts of the Greenland–Scotland Ridge submerged as late as the **Miocene** (25–5 Ma), with deepwater exchange between the Arctic Ocean and the North Atlantic by the middle Miocene (Kitchell and Clark 1982). Thus a chain of coastal fragments crossing the widening gap of the eastern North Atlantic favored the colonization by seaweed species dispersing southward from the Arctic Ocean via the Norwegian–Greenland Sea on both sides of the North Atlantic (van den Hoek 1984). The submergence of the Greenland–Scotland Ridge enabled warm-water species to disperse northward, as evidenced by diatoms (Schrader and Fenner 1976).

Warmer seas in early Miocene. During the Miocene the seas again invaded the continental shelves, and in tropical waters there was no decline in surface temperatures. A temperature rise occurred in tropical waters that somewhat counteracted the overall Tertiary cooling trend (Savin 1977; Seibold and Berger 1982). Even around the shores of the Arctic Ocean the climate was still relatively mild in the middle Miocene (e.g., around 13 Ma, Fig. 2.2D). Evidence for this is given by middle Miocene forests of *Tsuga, Metasequoia, Juglans,* and *Tilia* found at latitude 74°N on northwestern Banks Island (Hopkins and Marincovich 1984).

Miocene cooling. Around 15–10 Ma in the middle Miocene the second major temperature drop during the Tertiary occurred (Fig. 2.1B) leading to global glaciation (Woodruff et al. 1981). The major ice accumulation in the Antarctic around 11 Ma correlated with a drop in global sea level (Fig. 2.1B). On land the early and mid-Tertiary boreal flora of the Northern Hemisphere with temperate affinities and broad-leaved trees and shrubs was replaced by cold-resistant conifers and small-leaved arborescent types, the forerunners of the modern Eurosiberian and North American Atlantic floras (Meyen 1987). In the sea, surface water temperatures below 5°C were reached at high latitudes (Fig. 2.1 B). Sub-Arctic planktonic Foraminifera with temperature requirements lower than 10°C progressed well into the Los Angeles Basin at this major cooling (Fig. 2.1C). Furthermore, in the Pacific, many radiolarian species indicative of the water masses of this ocean, evolved around the early or middle Miocene (Casey 1977).

Speciation in **Laminaria.** Stam et al. (1988) inferred from single-copy DNA–DNA hybridizations that North Pacific *Laminaria* species radiated in the Miocene (19–15 Ma) and suggested that the cooling of the higher latitudes would have driven speciation in potential cold-water groups. "The North Pacific with, on both sides, archipelagos with continuously changing configurations due to tectonic events and sea level fluctuations, could provide the scenario for alternating genetic isolation of subspecific populations (demes) and mixing of these populations after speciation" (Stam et al. 1988).

Pliocene cooling. In the **Pliocene** (5–2 Ma) there was a major series of warming and cooling phases. This caused migration pulses (Fig. 2.1C), which heralded the approach of the glacials and interglacials of the Pleistocene (from 2 Ma onwards).

Phase 4 (3 Ma, Pliocene): Bering Strait Open; North Pacific and North Atlantic Interconnected via the Arctic Ocean

Opening of the Bering Strait. The boundary between the American and Eurasian plates is not situated in the area of the Bering Strait but in the Verkhoyansk Mountains region of northeastern Siberia or the extension of the Lena River (Herron et al. 1973; Kitchell and Clark 1982; Cook et al. 1985). The biogeographer interested in the question of when the Bering Strait was open or closed can glean more evidence from paleontology than from plate tectonics. The Bering Strait is a shallow, epicontinental seaway that could become a narrow Bering land bridge if the sea water level were lowered by 46 m (Hopkins 1967). In the Pliocene, at around 3 Ma, a reorganization of the ocean circulation and a further sharp temperature decline occurred (Fig. 2.1B), coincident with the **opening of the Bering Strait** and the emergence of the Isthmus of Panama (Herman and Hopkins 1980; Hopkins and Marincovich 1984; McKenna 1983). Evidence for the submergence of the Bering land bridge can be found in **the sudden increase in mollusks of North Pacific ancestry** occurring in the North Atlantic. The fossil remains of these Pacific invaders have been found in the 550 m thick Plio-Pleistocene sequence of rocks at the **Tjörnes peninsula,** northern Iceland, and in the **Red Crag** of England (Hopkins 1967; Strauch 1972; McKenna 1983; Gladenkov 1979, 1986). The migrations took place through an ice-free Arctic Ocean, about 1 million years before the first widespread Pleistocene glaciation in Iceland (Einarsson et al. 1967).

Neptunea as an example for a Pacific invader. Of the 27 living species of the gastropod genus *Neptunea* (Fig. 2.3A), 23 live in the North Pacific (14 endemic on the western side), four in the North Atlantic, and two in the European Arctic (Strauch 1972). This clear preponderance of species in the North Pacific occurs in many genera of marine benthic animals (Ekman 1953), and for many seaweed genera as well. The genus *Neptunea* originated in the Eocene (50 Ma), or in the Oligocene according to Nelson (1978), with one known fossil species (*N. altispirata*) in the warm Sea of Japan. During the Tertiary in the western North Pacific an array of *Neptunea* species evolved, but many became extinct in the Pliocene, with its large temperature oscillations in the North Pacific (Fig. 2.1C). The ancestors of one warm temperate group of species (*N. arthritica* and its allies) evolved in the Oligocene; this group is still represented on the Asiatic coasts but was not able to disperse along the Aleutians to the North American Pacific coast. This dispersion was achieved by a more eurythermal, again Recent species, *N. lyrata,* which occurs from Japan (24°C in summer) to the Bering Sea (0°C in winter). This species is known from the late Miocene in the western North Pacific, and shortly after that from the eastern side. In the Pliocene at 3 Ma, *N. lyrata* dispersed through the Bering Strait, occurred "suddenly" on the northern coast of Alaska (the Gubik formation of the Colville River), and then in the Icelandic Tjörnes layers as *N. lyratodespecta*. From this ancestral species a latitudinal array of three European and two western North Atlantic species subsequently evolved. It seems remarkable that *Neptunea* has no planktonic larvae and that its large benthic larvae "had literally to crawl every inch of the way" (MacNeil 1965).

Other Pacific invaders. There are further, well-documented examples of mollusks with a Pacific origin, such as the clams *Mya*, and *Panomya*. Both have a preference for cooler water and evolved in the Oligocene, in the northern Japanese Sea and in South Alaska respectively (Figs. 2.3B,C). *Panomya* dispersed with one, and *Mya* with two species through the Bering Strait at 3 Ma. Because *Mya* has planktonic larvae, it reached the North Atlantic earlier and changed less genetically than *Neptunea* (MacNeil 1965). Generally these cases illustrate that the North Atlantic received only a few species of a North Pacific genus, and then the more cold-adapted eurythermal species. Furthermore, periods the length of the Tertiary may be required for the evolution of cold-water biota from warm-water ancestors because Holocene cold temperate taxa and Arctic–cold temperate taxa seem to have a long history of ancestors in the early Tertiary.

Atlantic invaders in the Pacific. There is also fossil evidence for North Atlantic species invading the North Pacific at 3 Ma, such as the mussel genus *Cyrtodaria,* which originated in the early Tertiary in the Norwegian–Greenland Sea (Fig. 2.3D). To make things more complicated, reimmigration has also occurred. Strauch (1972) reported that the clam *Panomya,* which came from the North Pacific, formed new species upon arrival in the North Atlantic, two of which dispersed back to the North Pacific (Fig. 2.3C). Such evidence for reimmigration makes it difficult to extrapolate plausible evolutionary scenarios for groups with little or no fossil evidence like seaweeds.

North Pacific as the main donor area. It had been suggested by Durham and MacNeil (1967) that the influx of North Pacific mollusks started earlier, in the late Miocene (around 10 Ma), but this interpretation runs into difficulties with evidence from late

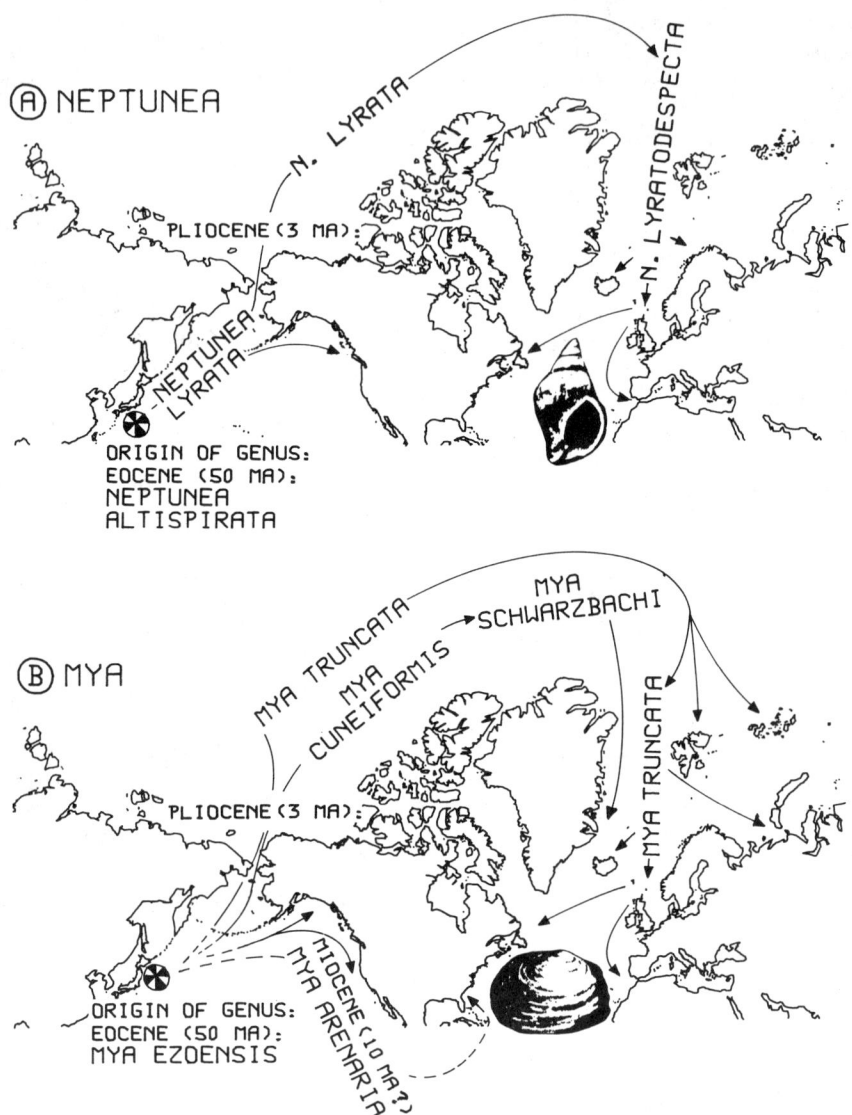

Fig. 2.3 Probable dispersal routes of one gastropod (A) and three mussels (C–D) through the opened Bering Strait, at 3 Ma. **(A)** *Neptunea:* Origin in the Eocene in the warm Sea of Japan and intensive speciation near the Pliocene/Pleistocene boundary in the North Pacific (23 recent species). Among the 14 species endemic to the northwestern Pacific are warm temperate species that were forced southward in the Late Tertiary, lost contact with the northeastern Pacific across the Aleutian Ridge, and were excluded from passing the Bering Strait. Only the eurythermal species *N. lyrata* dispersed through the Bering Strait to the North Atlantic, where it formed four recent species with different temperature demands for reproduction. **(B)** *Mya:* Origin in the Eocene near Hokkaido, northern Japan. After the opening of the Bering Strait in the

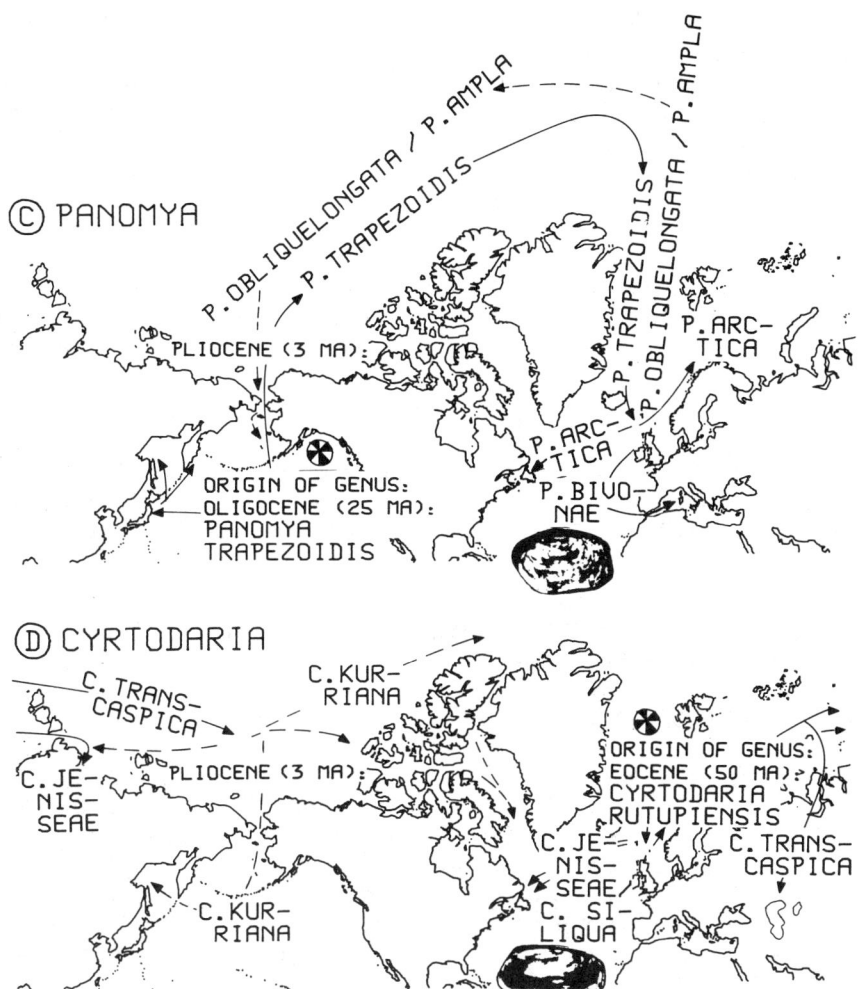

Pliocene, two species dispersed to the North Atlantic. A third species may have dispersed earlier to the North Atlantic via the Central American sector of the Tethys Sea, possibly in deep and cooler water (100–500 m depth). This would explain the appearance of *Mya arenaria* in the Yorktown formation of Virginia in Miocene times, when the Bering Strait was still closed. **(C)** *Panomya:* Origin of this cool-preferent genus in the Late Oligocene in South Alaska. One species dispersed from the North Pacific to the North Atlantic, formed a few Atlantic species, of which two species reimmigrated to northern and western Alaska. **(D)** *Cyrtodaria:* origin in the Eocene Norwegian-Greenland Sea. Of the species formed in the eastern North Atlantic, a few dispersed to the western North Atlantic, the Caspian Sea (via the Turgai Strait), Siberia, and northern Alaska. There, *C. kurriana* evolved in the Early Pleistocene, dispersed through the Canadian Arctic, and entered the North Pacific after the opening of the Bering Strait. (A–D simplified from Strauch 1972.)

Cenozoic terrestrial and marine mammal biogeography (McKenna 1983). It seems clear, however, that at least eight times more marine invertebrate species migrated from the North Pacific to the North Atlantic (at least 125 species, mostly mollusks) than vice versa (Durham and MacNeil 1967). Thus the Arctic and North Atlantic faunas became strongly imprinted by new "arrivals" from the Pacific in the Pliocene (Gladenkov 1979; 1986). Pacific invaders may have been put at an advantage by being carried by the eastward flowing longshore currents along the coasts of northern Alaska and the Canadian archipelago around northern Greenland to the Atlantic entrance. Atlantic invaders might have had to use the counterclockwise current along the inhospitable Siberian coast, as suggested by MacNeil (1965). Nevertheless, a great portion of the rich North Pacific fauna and flora never dispersed through the Bering Strait to the North Atlantic and represents the highly endemic character of today's North Pacific marine biota.

Panama seaway. The second Tethyan waterway connecting the Atlantic with the Pacific near the equator for most of the Tertiary was the Panama seaway. This was closed by the **Panama land bridge** at 6–4 Ma. It seems unlikely, however, that cold-adapted species would have used this tropical waterway instead of the Bering Strait.

Phase 5 (Pleistocene, 2 Ma-18,000 years ago): Bering Land Bridge Reemerged in Glacial Periods of the Pleistocene

Bering land bridge. During the glaciation phases of the Pleistocene the sea level fell, for example, by some 130 m about 17,000 years ago, during the last major ice age (called the Würm in Europe and the Wisconsin in North America). Thus, the Bering land bridge was submerged and reemerged repeatedly, blocking marine dispersal (Hopkins 1967). Migrations of benthic marine organisms were hampered during the glacial periods by ice blocking the Bering Strait and the coasts of the Arctic Ocean, possibly from 1 Ma onwards (Hopkins and Marincovich 1984). It should be noted that there is controversy about the beginning of perennial sea-ice cover of the Arctic Ocean, dated by Herman and Hopkins (1980) and Herman (1985) at 0.7 Ma or even at 3 Ma by Clark (1982). Moderate-scale continental glaciation and southward penetration of polar water probably began in the late Pliocene, at 3.5-3 Ma, or even at 5 Ma, the first major glaciation of the Northern Hemisphere at 2.4 Ma (Vogt 1986).

Displacements of the marine biota. In the glacial periods of the Pleistocene, the marine biota was displaced southward, and subsequent migrations to the north occurred during the warmer interglacial periods. There were about 20 glacials and an equal number of interglacials. These climatic oscillations are reflected in the migrations of Pleistocene plankton provinces in the Atlantic (Thunell and Belyea 1982). There were certainly species lost in these migrations; however, on the whole, the Pleistocene may have had a minor impact on the structure and composition of the marine biosphere. Valentine (1973) points out that the Pleistocene extinction rates among marine invertebrates did not appear exceptional. Since the tropical regions remained stable and warm, although latitudinally narrowed, there were always climatic refuge areas. No major climatic regimes were eliminated, and a few species adapted by surviving water temperatures below 0°C. The latitudinal shift of the nontropical biota resembled "variations on an established theme" (Valentine 1973).

One may also suggest that the Pleistocene did not dramatically change the ecological background of many species because they could escape to favorable conditions. For many woody plants of temperate Eurasia and North America, there is evidence that the ecological characters have remained almost unchanged since the Pliocene or even the Miocene (Kornas 1972).

Pleistocene conditions more severe in the North Atlantic than in the North Pacific. The last great ice sheets began to advance about 25,000 years ago and had their maximum extent from 21,000 to 17,000 years ago, the time differing from one area to another (Denton and Hughes 1981; Vogt 1986). Eighteen thousand years ago, at the peak of the last glaciation, the surface water temperatures of the North Atlantic, with its wide-open connection to the Arctic Ocean, were probably 10°C lower than today (McIntyre et al. 1976). Temperatures in the North Pacific were only 4°C lower, since the Bering Strait was dry land during the glacial periods, a barrier against the cold water from the Arctic Ocean. At the same time, in the North Atlantic, the 5°C-August isotherm, which now marks the southern limit of the Arctic region in southern Greenland, was situated on the coast of Brittany, and *Laminaria solidungula* could have grown there. Today's 15°C-August isotherm, on the coast of Brittany, was displaced to southern Spain during the last glaciation (compare Figs. 1.5 and 2.1D). The seaweed flora of today's cold temperate regions was probably similar, in glacial times, to the present flora in the Arctic. In general, cold temperate species with northern boundaries today at 30–50°N may have been shifted 15 to 20 degrees of latitude to the south on the European side of the North Atlantic (Fig. 2.1D). On the North American side, winter temperatures were lowered to a smaller extent than summer temperatures (McIntyre et al. 1981). Northern species with a demand of low winter temperatures for reproduction may have exhibited little shift of their southern limits on the American side of the North Atlantic (Breeman 1989). In the **North Pacific** these conditions were also not as drastic (1) because the reduction of seawater temperature was not as severe as in the North Atlantic, and (2) because the displacement migrations of the benthic organisms along the North Pacific coasts took place along a largely uninterrupted, rocky coastline well suited to the existence and migration of seaweeds. In contrast, in the **North Atlantic**, the long open-sea distances between Greenland, Spitsbergen, Iceland, and the European continental coasts created problems for the migration of benthic organisms. Additional migration barriers in the North Atlantic exist in the form of extensive soft-substratum coastlines, especially in the present warm temperate Carolina region. The coasts of western France and the coast of the North Sea also act as barriers. Thus it seems likely that, in the course of displacement migrations during the Pleistocene, more species were lost in the North Atlantic than in the North Pacific, a partial explanation of the greater richness of cool-water taxa in the North Pacific (van den Hoek 1975, 1982b, 1984).

Crossing the equator. For temperate, nontropical algae in the ice-age ocean, crossing the equator was faciliated by several factors: not only were the warm-water isotherms characterizing the tropical regions compressed latitudinally, but the ocean surface was cooler as a whole (compare Figs. 2.1D and 1.5). The fact that the ice rim was nearer to the equator and that the temperature difference between ice (less than 0°C) and tropical water (25°C) was compressed caused much stronger winds, surface currents, and equatorial and coastal upwelling (Seibold and Berger 1982). Hence, several temperate species crossed the equator faster, in a cooler and more nutrient-rich equatorial water than today.

2.2 DISTRIBUTIONS OF CHARACTERISTIC SEAWEED SPECIES OF THE NORTH ATLANTIC

Table 2.1 contains a list of many of the larger seaweeds occurring in the cold temperate regions of the North Atlantic. Many of them also inhabit the Arctic and North Pacific (amphioceanic species). They are listed in the same order as their distributions shown in Figs. 2.4–2.42. The sequence of these figures begins with the distribution of species moving from north to south. The few species endemic (restricted to a certain area) to the Arctic are represented in Figs. 3.6–3.10 and are treated in Chapter 3. The literature on which the distribution maps are based is cited in Chapters 2 and 3.

2.2.1 Arctic-Cold Temperate Amphioceanic Species

In view of the lack of fossils, it is hard to suggest which amphioceanic seaweed species (Table 2.1, group 1) and genera originated in the North Pacific and which in the North Atlantic. It may be that just prior to the opening of the Bering Strait at 3 Ma, a rich North Pacific cool-water seaweed flora, not adapted to Arctic conditions, existed south of the land bridge to serve as a source of immigrants to the North Atlantic via the coasts of the Arctic Ocean, which were probably still ice-free. To the north of the Bering land bridge, one may imagine a less diverse Arctic–North Atlantic seaweed flora whose species had adapted to the Arctic conditions of polar night and polar day over millions of years, that is, since the Norwegian–Greenland Sea was connected to the mid-Tertiary Arctic Ocean.

The rich endemic element in the North Pacific is evidence for the inability, now and in the past, of most North Pacific algae to disperse through the Arctic. This applies to the marine fauna as well (Ekman 1953; Briggs 1974). To cite an example from the group of cool-water red algae of the Northern Hemisphere: there are about 60 genera (or groups of genera) in the North Pacific, of which about 40 are endemic, whereas only four of the approximately 30 genera in the Arctic–North Atlantic group of cool-water seaweed genera are endemic: *Furcellaria, Ceratocolax, Halosacciocolax,* and *Kvaleya* (Lindstrom 1987*a*). In the brown algae, the genus *Alaria,* with about 10 species in the North Pacific, "sent" only one species, *A. esculenta,* to the North Atlantic (Fig. 2.12). Likewise, the diverse North Pacific genus *Laminaria* spread with only a few species reaching the North Atlantic, so that there are now sister taxa in both oceans (Pacific and Atlantic *L. saccharina;* tentative: *Laminaria setchellii/L. hyperborea* and *L. bongardiana/ L. digitata*). This evidence suggests that the North Pacific was the richer source of the present Arctic–cold temperate, amphioceanic species (Figs. 2.3A,B,C) (van den Hoek 1984; Lindstrom 1987*a*). In addition, genera originally endemic to the North Atlantic may have dispersed to the North Pacific (Fig. 2.3D), explaining the low number of endemic genera in the present North Atlantic.

B = brown alga, R = red alga, G = green alga, N = north, S = south, M = middle.*

		!--northern limits--!		!--------southern limits--------!	
		Atlantic Europe/Africa	North America	Atlantic Europe/Africa North America Pacific North America North Asia	Fig.

(1) Arctic-cold temperate, amphioceanic group

B:	Chordaria flagelliformis	N-Arctic	N-France	Connecticut	S-Alaska	Korea	2.4
B:	Eudesme virescens	S-Arctic	N-France	Connecticut	Brit. Columbia	N-China	2.5
B:	Dictyosiphon foeniculaceus	N-Arctic	N-France	Virginia	N-Washington	N-China	2.6
B:	Desmarestia aculeata	N-Arctic	M-Portugal	Connecticut	Oregon	Kurile Is.	2.7
B:	Desmarestia viridis	N-Arctic	N-France	Connecticut	Mexico	S-Japan	2.8
B:	Chorda filum	S-Arctic	N-Portugal	Connecticut	S-Alaska	S-Japan	2.9
B:	Agarum cribrosum	N-Arctic	---------	N-Washingt.	N-Washingt.	Korea	2.10
B:	Laminaria saccharina	M-Arctic	N-Portugal	New York	Oregon	Vladivostok	2.11
B:	Alaria esculenta	M-Arctic	N-France	New Hampshire	S-Alaska	Vladivostok	2.12
B:	Fucus distichus	N-Arctic	Ireland	Virginia	M-California	Vladivostok	2.13
R:	Ahnfeltia plicata	S-Arctic	S-Portugal	Connecticut	Mexico	Korea	2.14
R:	Dumontia contorta	Ber. Sea	N-Portugal	Connecticut	S-Alaska	Hokkaido	2.15
R:	Callophyllis cristata	M-Arctic	Sweden	New Hampshire	N-Washington	Kurile Is.	2.16
R:	Palmaria palmata	N-Arctic	M-Portugal	Connecticut	California		2.17
G:	Chaetomorpha melagonium	N-Arctic	N-France	Connecticut	Oregon	S. of Ochotsk	2.18

(2) Arctic-cold temperate, North Atlantic group

B:	Chorda tomentosa	N-Arctic	N-France	Connecticut	2.19
B:	Laminaria digitata	M-Arctic	N-France	Connecticut	2.20
B:	Saccorhiza dermatodea	S-Arctic	N-Norway	New Hampshire	2.21
B:	Ascophyllum nodosum	N-Arctic	N-Portugal	North Carolina	2.22
B:	Fucus vesiculosus	S-Arctic	N-Africa	North Carolina	2.23
B:	Fimbriofolium dichotomum	M-Arctic	M-Norway	New Hampshire	2.24
R:	Phyllophora truncata	N-Arctic	Ireland	New Jersey	2.25
R:	Ptilota serrata	M-Arctic	M-Norway	New Hampshire	2.26
R:	Membranoptera alata	S-Arctic	N-France	New Hampshire	2.27
R:	Phycodrys rubens	N-Arctic	N-France	Connecticut	2.28
R:	Odonthalia dentata	M-Arctic	Sweden	Nova Scotia	2.29
R:	Rhodomela confervoides	N-Arctic	N-France	Connecticut	2.30

(3) Cold temperate North Atlantic group

(3 a) Amphiatlantic

B:	Fucus spiralis	Newfdl., N-Norw.	N-Africa	Delaware	(N-Washington)	2.31
B:	Fucus serratus	Novaya Zemlya	N-Portugal	Gulf of St.Lawrence		2.32
R:	Chondrus crispus	Newfdl., N-Norw.	Mauritania	New Jersey		2.33
R:	Phyllophora pseudoceranoides	Newfdl., M-Norw.	N-France	Delaware		2.34

(3 b) European-North Atlantic

B:	Laminaria hyperborea	N-Norway	M-Portugal		2.35
B:	Saccorhiza polyschides	M-Norway	N-Africa		2.36
B:	Fucus ceranoides	N-Norway	S-Portugal		2.37
B:	Pelvetia canaliculata	N-Norway	M-Portugal		2.38
B:	Himanthalia elongata	N-Norway	M-Portugal		2.39
B:	Bifurcaria bifurcata	Ireland	N-Africa		2.40
R:	Halidrys siliquosa	N-Norway	N-Portugal		2.41
R:	Delesseria sanguinea	N-Norway	N-France		2.42

* This is also a survey for Figures 2.4-2.42.

Figs. 2.4–2.42 Distribution areas of important seaweeds that occur in the North Atlantic. See Table 2.1 for the arrangement of the figures (starting with the species penetrating farthest to the north). Where a continuous distribution may be assumed and many records are available, occurrence is represented by an uninterrupted line and otherwise by squares. **Figs. 2.4–2.18** Arctic cold temperate, amphioceanic species. **Figs. 2.19–2.30** Arctic cold temperate Atlantic species. **Figs. 2.31–2.34** Cold temperate Atlantic species, amphi-Atlantic. **Figs. 2.35–2.42** Cold temperate Atlantic species, European Atlantic. **Taxonomic details:** Fig. 2.7: *Desmarestia aculeata* = *D. intermedia* Postels and Ruprecht on the Pacific American coast (Lindstrom 1977). Fig. 2.8: *Desmarestia viridis* is also present in the Southern Hemisphere (van den Hoek 1982a). Fig. 2.11: the distribution of *Laminaria saccharina* in this figure includes the distribution of *L. longicruris* at the eastern North American coast (Section 2.5.1). Fig. 2.12: *Alaria esculenta* = *A. grandifolia* J. Ag = *A. pylaii* (Bory) J. Ag. (Widdowson 1971; South and Hooper 1980). Fig. 2.13: *Fucus distichus*. The distribution map is drawn without differentiating between *F. distichus* (shown as habit; grows in the upper part of the eulittoral zone and includes subspecies *anceps*), and *F. evanescens* (grows in the mid-eulittoral to sublittoral zones and includes the subspecies *edentatus;* see Rice and Chapman 1985; H. T. Powell 1957, 1981; South and Tittley 1986). Fig. 2.14: *Ahnfeltia plicata* in the northeastern Pacific, south of Alaska; it probably equals another species, *A. fastigiata* (Post et Rupr.) Makienko (Maggs et al. 1989). A new order, Ahnfeltiales, has been proposed by Maggs and Pueschel (1989). Fig. 2.15: *Dumontia contorta* = *D. incrassata* (O. F. Müller) Lamour. (Abbott 1979) = *D. filiformis* Rosenv. (R. K. S. Lee 1980). *D. contorta* does not strictly belong to the Arctic–cold temperate species because it is largely absent from the Arctic region. Fig. 2.16: *Callophyllis cristata* = *Euthora cristata* (C. Ag.) J. Ag. (Parke and Dixon 1976). Fig. 2.17: *Palmaria palmata* = *Rhodymenia palmata* (L.) Grev.; *P. palmata* forma *mollis* on the North American Pacific coast has been raised to species status as *P. mollis* (van der Meer and Bird 1985). There are a few more taxonomically uncertain *Palmaria* spp. in the western North Pacific and uncertain additional species in the eastern North Pacific (distribution map according to Guiry 1974, 1975). Fig. 2.18: *Chaetomorpha melagonium* (Web. et Mohr) Kütz. See also Blair (1983) for distribution. Fig. 2.24: *Fimbriofolium dichotomum* (Lepech.) G. Hansen = *Rhodophyllis dichotoma* (Lepech.) Gobi (see Hansen 1980). Fig. 2.25: *Phyllophora truncata* = *P. brodiaei* (Turn.) Endl. (Newroth 1971; Newroth and Taylor 1971). Fig. 2.28: *Phycodrys rubens* = *P. sinuosa* (Good. et Woodw.) Kütz. Fig. 2.30: *Rhodomela confervoides* = *R. lycopodioides* (L.) C. Ag. = *R. subfusca* (Woodw.) C. Ag. (Rueness 1977). Japanese records of *R. lycopodioides* forma *tenuissima* (Masuda 1982) are not regarded. Fig. 2.31: *Fucus spiralis* has been located for the North American Pacific coast by Norris and Conway (1974). Fig. 2.34: *Phyllophora pseudoceranoides* = *P. membranifolia* (Good. et Woodw.) J. Ag. (Newroth and Taylor 1971). (Sources for habit drawings: Dixon and Irvine 1977: 2.14; Gayral 1966: 2.20, 2.37; T. F. Lee 1977: 2.4, 2.5, 2.17–2.19, 2.21, 2.25, 2.26, 2.30; Newton 1931: 2.6, 2.12, 2.27, 2.28, 2.33, 2.38–2.42; Oltmanns 1922–1923: 2.35; Sauvageau 1918: 2.36; W. R. Taylor 1957: 2.7–2.10, 2.15, 2.16, 2.22, 2.23, 2.31, 2.32, 2.34; A. D. Zinova 1953: 2.11, 2.13; A. D. Zinova 1955: 2.24, 2.29.)

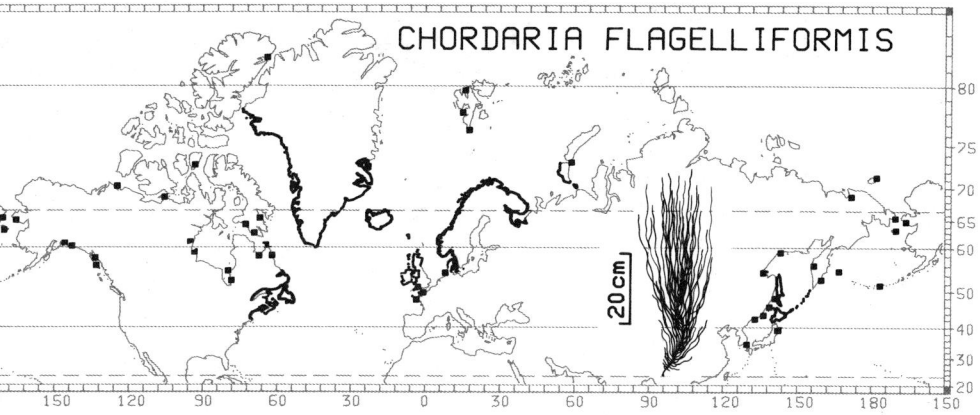

Fig. 2.4

Chances were good for most of the amphioceanic species (Table 2.1, group 1) to spread to both sides of each ocean, whether they entered the North Pacific via the Bering Strait or the North Atlantic via the Canadian Arctic archipelago. A striking exception is the kelp *Agarum cribrosum,* which has only migrated eastward in the North Atlantic to the southern coast of Greenland (Fig. 2.10).

Most of the amphioceanic species colonized the cold temperate and Arctic regions as well. This may indicate that the climate of 3 Ma, after the Bering land bridge became submerged, was so severe that both adaptation to the polar light conditions and the ability to live at low temperatures selected which species could disperse to the other ocean. On the other hand, there are amphioceanic species that now colonize only the southern Arctic region, and cold temperate, amphioceanic sister groups without a present Arctic distribu-

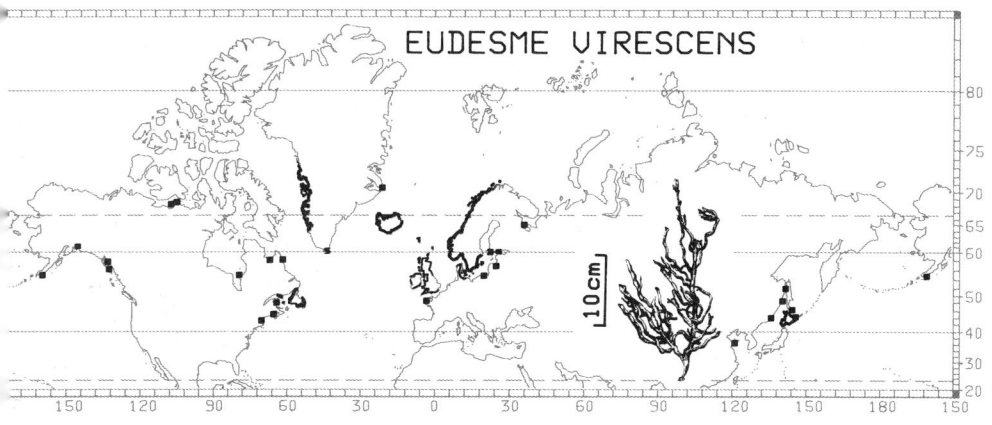

Fig. 2.5

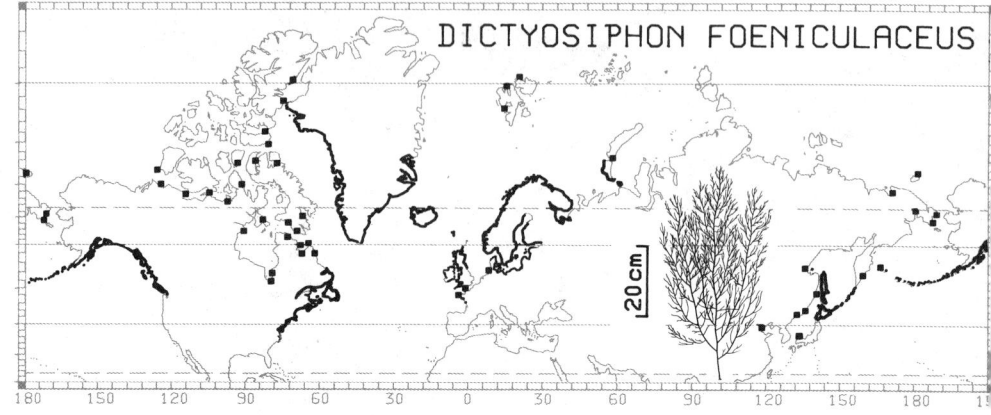

Fig. 2.6

tion. These probably left the Arctic when ice formation began. For example, the amphioceanic *Laminaria saccharina* group (Fig. 2.11) still has an Arctic–cold temperate distribution, whereas *Laminaria hyperborea* (Fig. 2.35), and *L. setchellii* (Fig. 2.82), its tentative sister taxon in the North Pacific, do not enter the Arctic region today.

The red alga *Dumontia contorta* does not strictly belong to the Arctic–cold temperate group, as this species barely enters the Arctic region (Fig. 2.15). The tribe Dumontieae of the family Dumontiaceae includes six genera that are all found in the North Pacific (Lindstrom 1985, 1987a, 1988). The tribe possibly evolved in the Pacific from a tropical or subtropical ancestor, diversified on the Asiatic side of the North Pacific, and subsequently spread to the American side. In addition there was some spillover into the North Atlantic, at first with the amphioceanic *D. contorta* (Fig. 2.15) probably expanding its range before the Arctic Ocean froze. Another spreading episode of the tribe

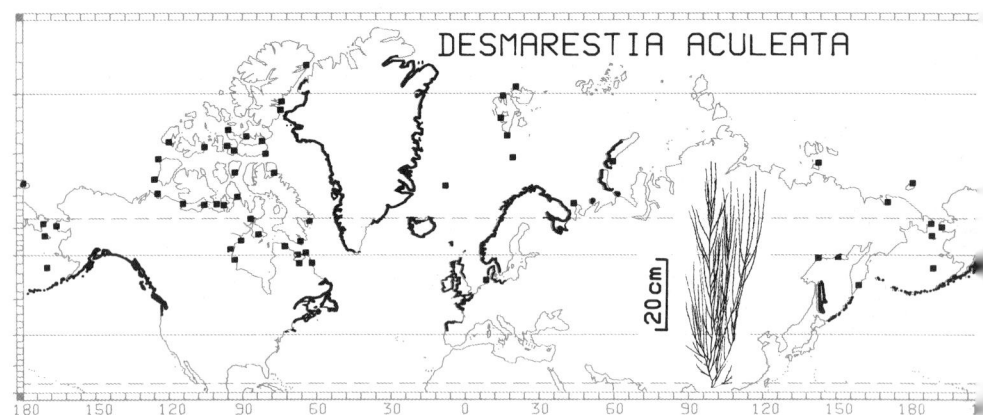

Fig. 2.7

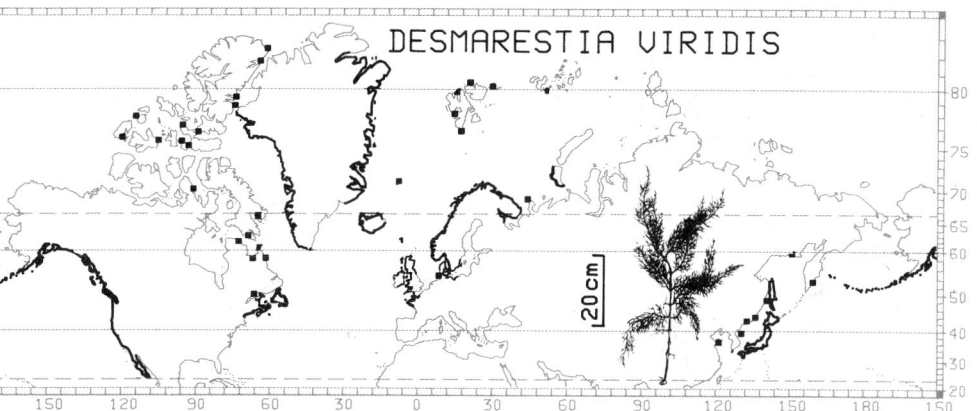

Fig. 2.8

Dumontieae may have resulted in the present distribution of the sister taxa *Neodilsea yendoana* and *D. californica* in the North Pacific, and the Arctic–western North Atlantic *Dilsea* (= *Neodilsea*) *integra* (Fig. 3.8) as well as the eastern North Atlantic *Dilsea carnosa* (Lindstrom 1988).

Another amphioceanic red algal group, this one without any distribution in the Arctic, is the genus *Mastocarpus* (= *Gigartina* pro parte), with *M. pacificus* occurring on both sides of the North Pacific, *M. papillatus* and *M. jardinii* in the eastern North Pacific, and *M. stellatus* on both sides of the North Atlantic (Guiry et al. 1984; Lindstrom 1988).

Of the Arctic–cold temperate amphioceanic algae, the following species penetrated farthest into the Arctic region after the last glaciation and have reached at least latitude 80°N along the coasts of eastern and/or western Greenland: *Chordaria flagelliformis* (Fig. 2.4), *Dictyosiphon foeniculaceus*

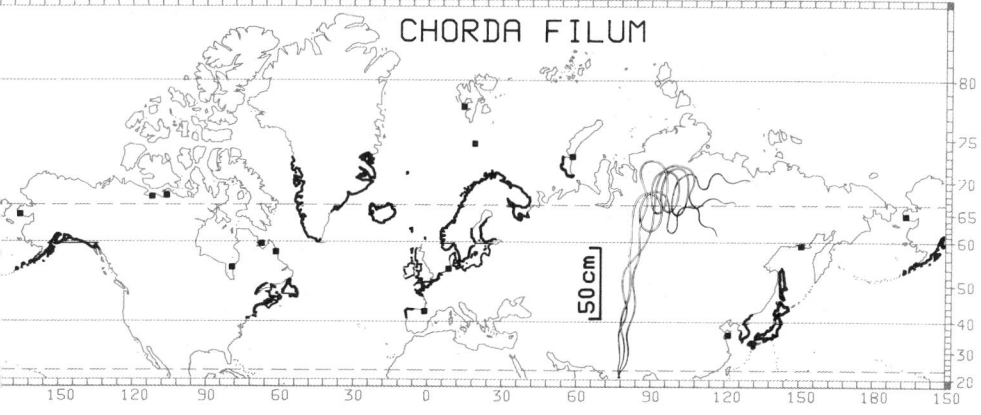

Fig. 2.9

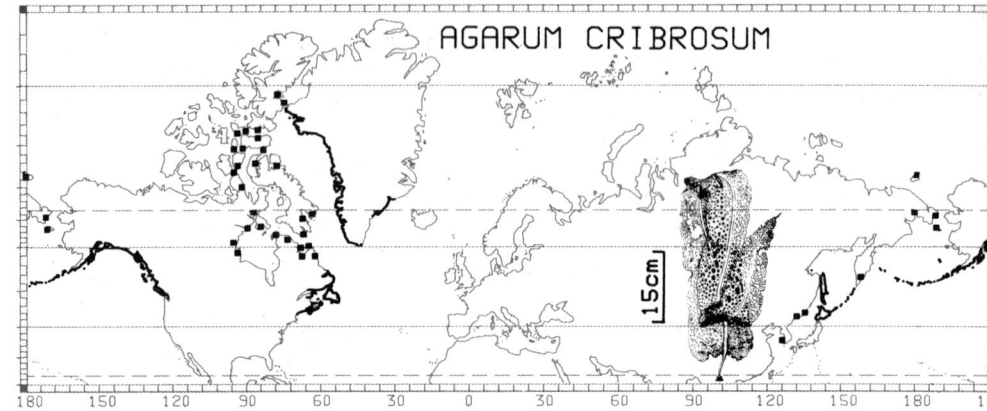

Fig. 2.10

(Fig. 2.6), *Desmarestia aculeata* (Fig. 2.7), *Desmarestia viridis* (Fig. 2.8), *Laminaria saccharina* (Fig. 2.11), *Palmaria palmata* (Fig. 2.17), and *Chaetomorpha melagonium* (Fig. 2.18). To the south, the distribution of *Desmarestia aculeata*, *Laminaria saccharina*, and *Palmaria palmata* ends along the European coasts of Portugal; the southern limit of the distribution is northern France for all the others.

Reduction of the number of amphioceanic species due to taxonomic revison. Our knowledge of the distribution of a species also depends on the changing views of those taxonomists who tend to split more and more species into vicariant (geographically isolated) sister species. These species have originated after the geographical subdivision of the ranges of ancestral species. (For surveys on speciation processes and vicariance see Brown and Gibson 1983; Myers and Giller 1988; Nelson and Rosen 1981; Pielou 1979.) For example, the red alga *Palmaria palmata* s. str. (sensu stricto)

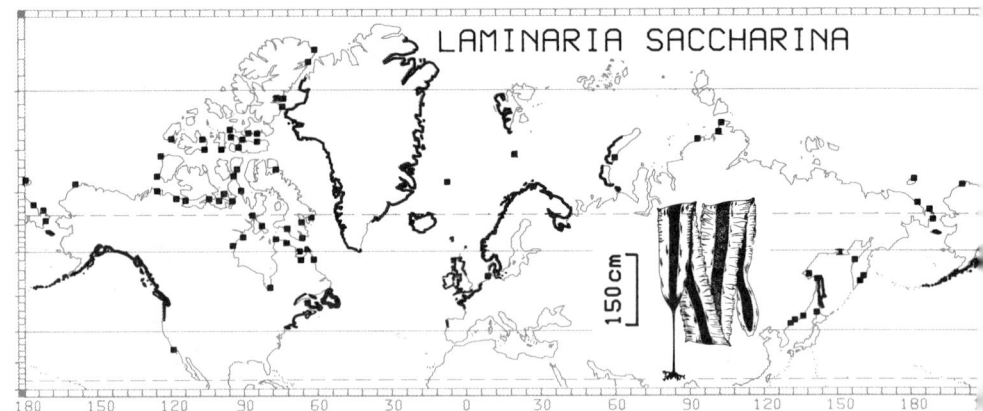

Fig. 2.11

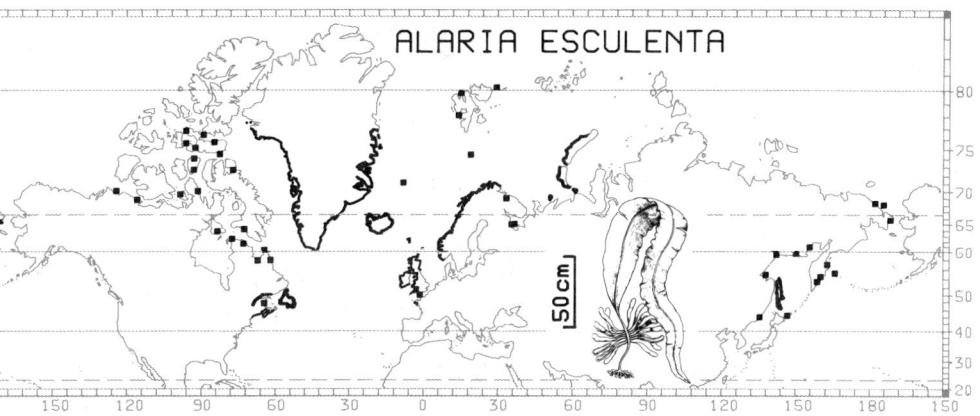

Fig. 2.12

occurs only in the North Atlantic (Fig. 2.17). It was formerly regarded as conspecific with populations on the Pacific coast of North America, subsequently named *P. palmata* forma *mollis* (Guiry 1975) and then elevated to full species status as *P. mollis,* after unsuccessful hybridization attempts (van der Meer and Bird 1985). The populations along the coasts of northern Asia probably represent other species (Guiry 1975), and there are more species besides *P. mollis* in the eastern North Pacific (Lindstrom and South 1989). Crossing experiments (van der Meer 1986) and starch gel electrophoresis (Lindstrom and South 1989) suggest incipient speciation among eastern and western Atlantic *P. palmata*. The red algae *Phyllophora truncata* (Fig. 2.25), *Ptilota serrata* (Fig. 2.26), *Rhodomela confervoides* (Fig. 2.30) and *Polysiphonia urceolata* (southern Arctic to North Portugal or North Carolina) were formerly recognized on the coasts of northern Asia; however, they have since received their own specific names (Perestenko 1980). Another example of losing an amphioceanic, and also an amphi-Atlantic distribution, is exhibited by *Gracilaria verrucosa,* for which records were given for the cold and warm temperate regions of the Northern and

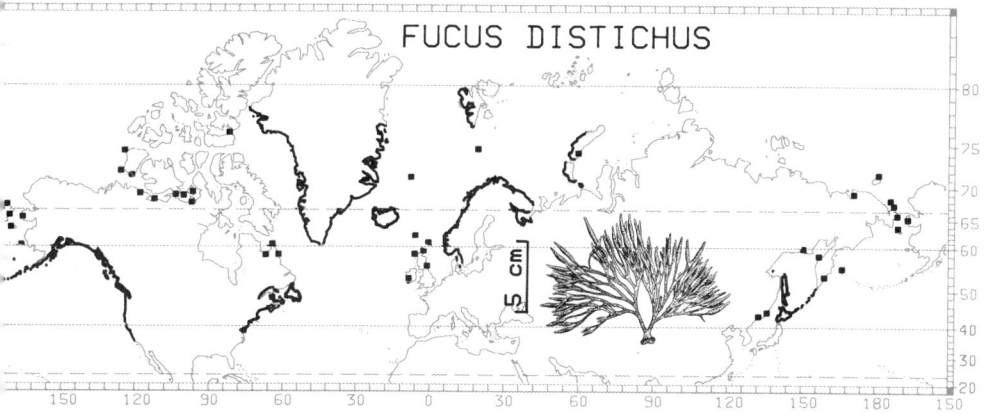

Fig. 2.13

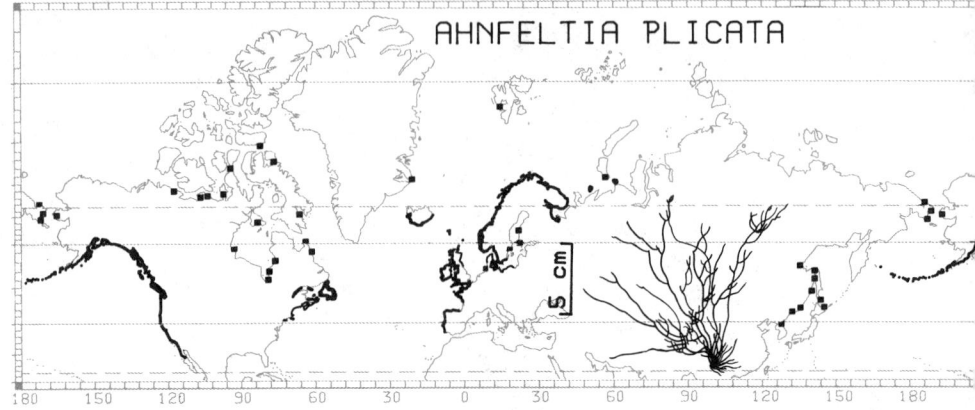

Fig. 2.14

Southern Hemispheres. Bird et al. (1982) found that material of this species from Great Britain and Vancouver had chromosome numbers of $n = 32$ and $n = 24$, respectively, and that crossing was not possible. In addition, the *Gracilaria* floras of the western and eastern Atlantic have been recognized to be distinct (McLachlan and Bird 1984), a natural consequence of the opening of the Atlantic Ocean after the Jurassic. *G. verrucosa* is confined to the Eurafrican side of the Atlantic, while *G. tikvahiae* has been identified, through hybridization experiments, as an entity on the North American side. The North Pacific populations are no longer regarded as authentic *G. verrucosa* (McLachlan 1979; Bird et al. 1982; South and Tittley 1986; van der Meer 1986). Similarly, *Ahnfeltia plicata* in the eastern North Pacific is now regarded as another species, *A. fastigiata* (Maggs et al. 1989).

Taxonomy of Alaria. Intensive speciation obviously took place in the kelp genera *Alaria* and *Laminaria,* and taxonomists are beginning to unravel the taxonomic chaos evident in the older literature. According to a revision of *Alaria* by Widdowson (1971), the North Atlantic and neighboring coasts of the Arctic seem to harbor only one

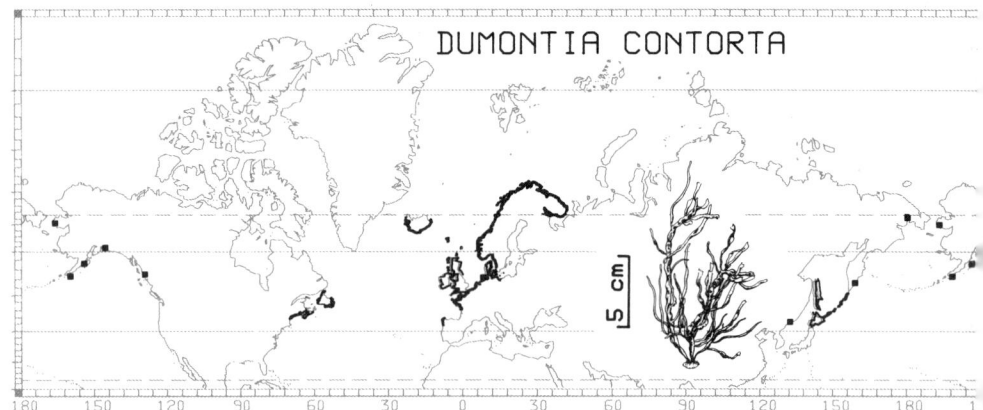

Fig. 2.15

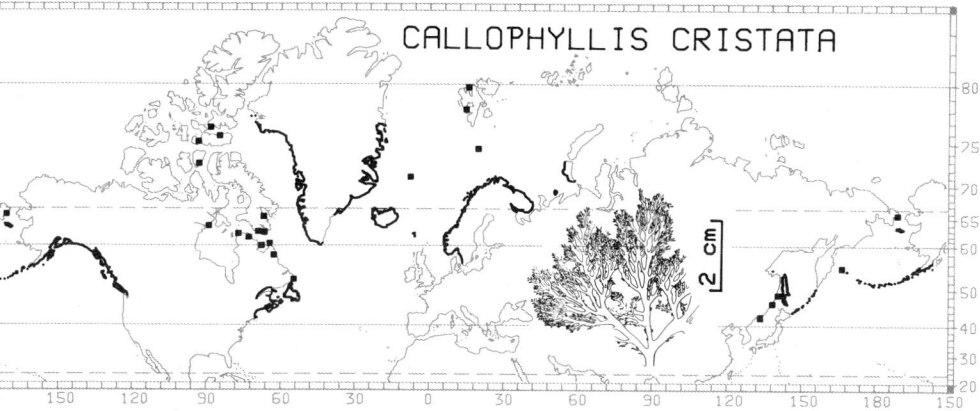

Fig. 2.16

species: *A. esculenta* with its northern forms *A. esculenta* forma *grandifolia* and forma *pylaii* (South 1984; South and Tittley 1986). In contrast, in the North Pacific 11 species of *Alaria* are distinguished on the basis of morphological criteria, as well as several other genera of the family Alariaceae. It seems likely that the family originated in the North Pacific and only *A. esculenta* migrated, via the Arctic, into the North Atlantic.

Taxonomy of* Laminaria, *section Digitatae. In the genus *Laminaria,* revised by Kain (1979), a section Simplices (undivided blade) can be distinguished from a section **Digitatae** (divided blade). The Digitatae section is represented by the North Atlantic species *L. digitata, L. hyperborea,* and *L. ochroleuca;* the North Pacific species *L. dentigera, L. setchellii, L. bongardiana* (= *L. groenlandica* sensu Druehl 1968; see Petrov 1973), and *L. yezoensis* (with discoid holdfast); the South African species *L. pallida;* and the Brazilian species *L. brasiliensis*. The possession of rigid stipes may unite a group of related species in the North Atlantic (*L. hyperborea* with a

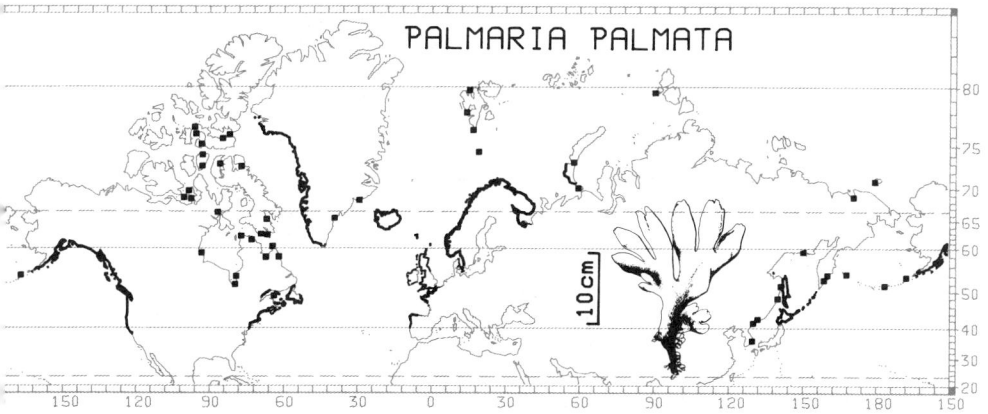

Fig. 2.17

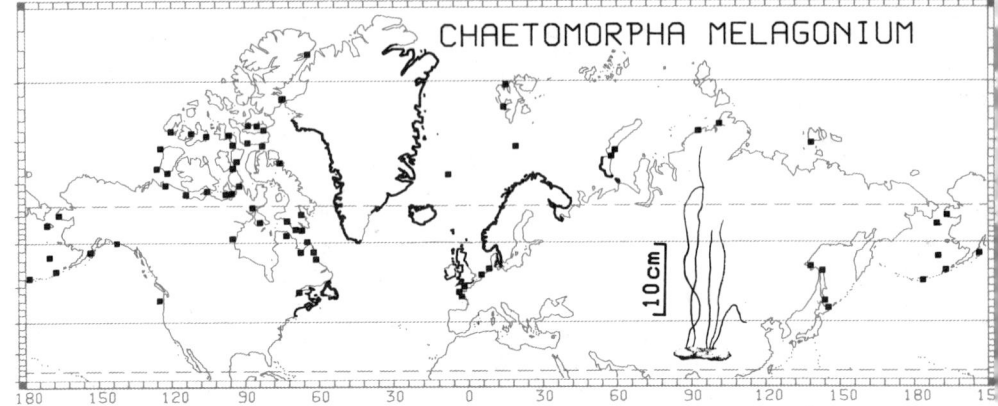

Fig. 2.18

northern and *L. ochroleuca* with a more southerly distribution) and in the eastern North Pacific (*L. dentigera,* northern distribution; *L. setchellii,* southern distribution). Similarly, *L. digitata* with a flexible stipe may be the North Atlantic sister species of the North Pacific *L. bongardiana.*

***Taxonomy of* Laminaria,** *section Simplices.* Within the section **Simplices** (undivided blade) are two distinct species with an attaching disk instead of haptera: *L. solidungula* and *L. ephemera* (combined as the subgenus *Solearia* by Ju. E. Petrov 1974). There are also three species in the section Simplices with a rhizomatous attaching system (combined as the subgenus *Rhizomaria* by Petrov 1974): the Mediterranean deep-water species *L. rodriguezii,* and the northern Pacific species *L. sinclairii* and *L. longipes* included in the section Simplices. The remaining species of the Simplices group have a branched holdfast and a morphology rather similar to *L. saccharina,* although, in the western North Pacific, species with a median thickened part (fascia) may be split off as a section **Fasciatae** (Druehl et al. 1988c). If one follows Petrov

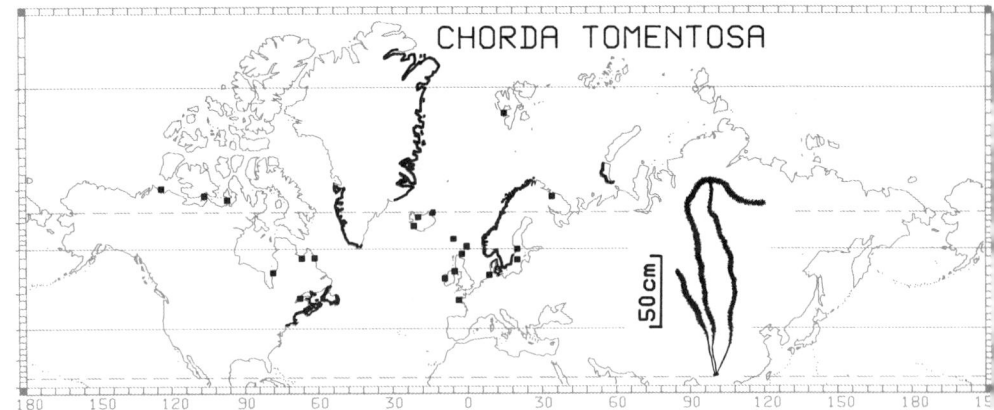

Fig. 2.19

DISTRIBUTIONS OF SEAWEED SPECIES OF THE NORTH ATLANTIC

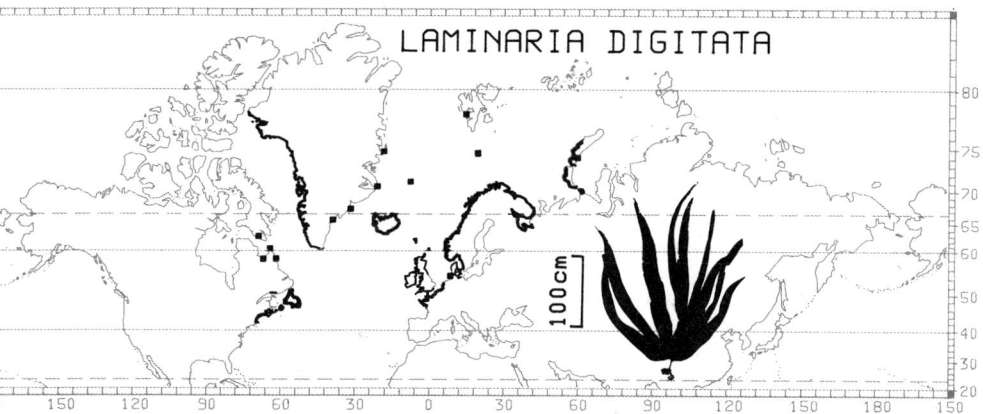

Fig. 2.20

(1974), the subgenus *Laminaria* includes all species with a branched holdfast, that is, members of the sections Digitatae, Simplices, and Fasciatae (Petrov 1974).

***Crossing and DNA–DNA hybridization experiments in* Laminaria.** Crossing was successful using the gametophytes of *L. saccharina* from Europe, of *L. longicruris* from Nova Scotia, of *L. saccharina* from British Columbia, and of *L. ochotensis* from Japan (Lüning et al. 1978; Bolton et al. 1983). Meter-long F_1 sporophytes were obtained on underwater stations in the sea. On the other hand, the onset of the formation of crossing barriers was also detected. Crossings of spermatozoids from male gametophytes of *L. longicruris* with eggs from female gametophytes of *L. ochotensis* resulted in the formation of viable macroscopic sporophytes that did not occur in the reciprocal crossing. In all these cases the hybrids were removed from the sea before they became sporogenous in order to prevent contamination of the native seaweed flora with allochthonous species. Hence, the capability of hybrids to produce

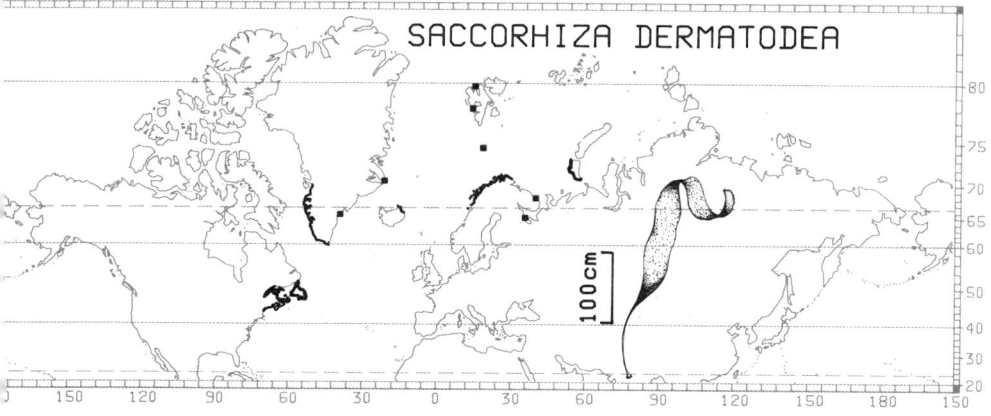

Fig. 2.21

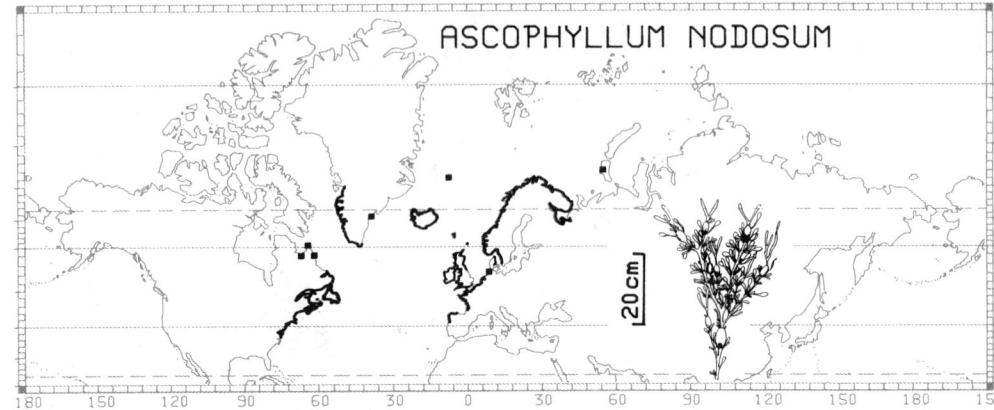

Fig. 2.22

another generation of normal plants, an important prerequisite for experimental taxonomical conclusions (van der Meer 1986), was not tested. In one case, however, viable gametophytes and an F_2 sporophyte generation were obtained in tanks in the laboratory, from the spores of the crossing of an European *L. saccharina* and *L. longicruris* from eastern Canada (characterized by long, hollow stipes and broad blades). As Lindstrom (1987a) pointed out, the ability to interbreed may be carried as a primitive (plesiomorphic) condition a long way through evolution. Hybrids between individuals of different species or even genera may be obtained. Hybridization experiments may add valuable information to hypotheses of phylogenetic relationships. Cosson and Olivari (1982) reported on interspecific hybrids of *L. digitata* with *L. saccharina* and *L. ochroleuca,* and between each of these and *Saccorhiza polyschides.* It seems that the hybrids are not capable of producing another generation of normal sporophytes (Cosson 1987). Another piece of information from experimental taxonomy has come from DNA–DNA hybridization experiments performed by Stam et al. (1988). These authors concluded that *L. digitata, L. hyperborea, L. ochroleuca,*

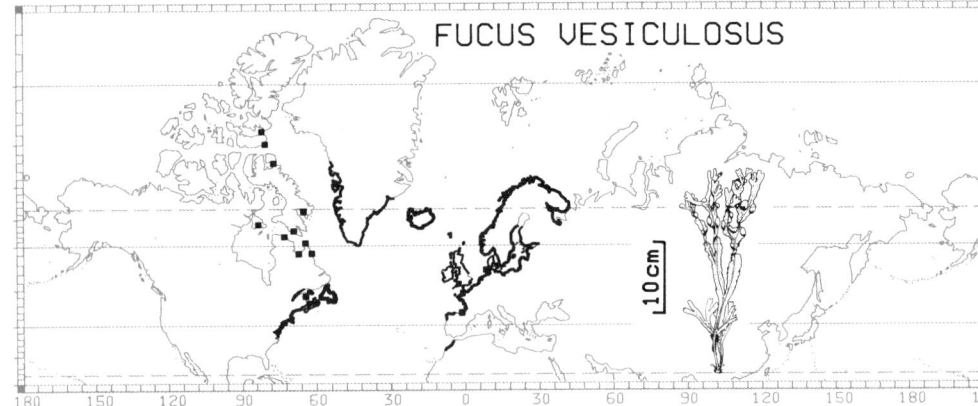

Fig. 2.23

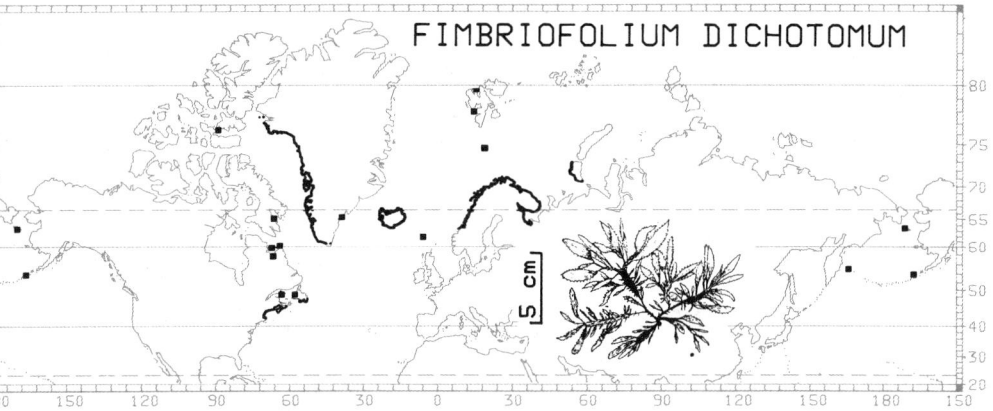

Fig. 2.24

L. saccharina, and *L. rodriguezii* probably evolved at about the same time, about 15–19 Ma.

2.2.2 Arctic–Cold Temperate North Atlantic Species

Examples of characteristic species of the North Atlantic cold temperate algal flora that do not occur in the North Pacific (Table 2.1, group 2) are *Chorda tomentosa* (Fig. 2.19), *Laminaria digitata* (Fig. 2.20), and *Phycodrys rubens* (Fig. 2.28), which also inhabit the northern part of the Arctic region. *Fucus vesiculosus* (Fig. 2.23), *Ascophyllum nodosum* (Fig. 2.22), and *Membranoptera alata* (Fig. 2.27) have penetrated less into the Arctic region. The southern limits along the European coasts are situated on the coast of Brittany (*C. tomentosa, L. digitata, P. rubens,* and *M. alata*) and northern

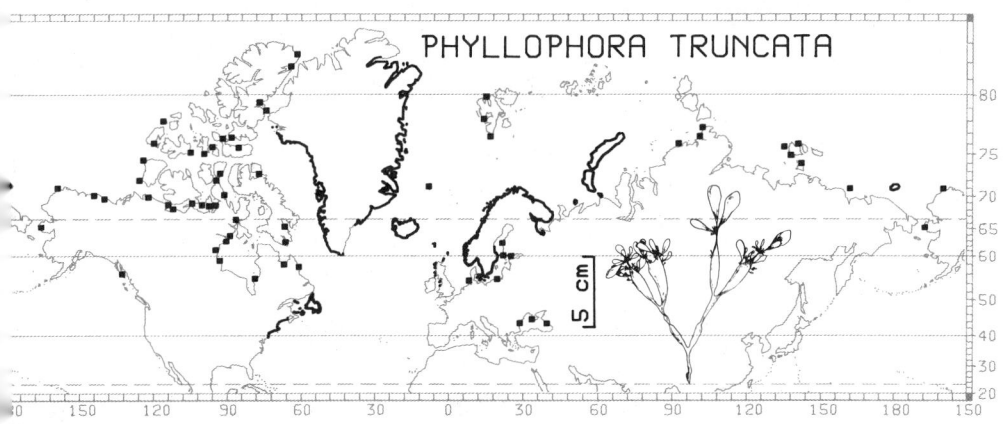

Fig. 2.25

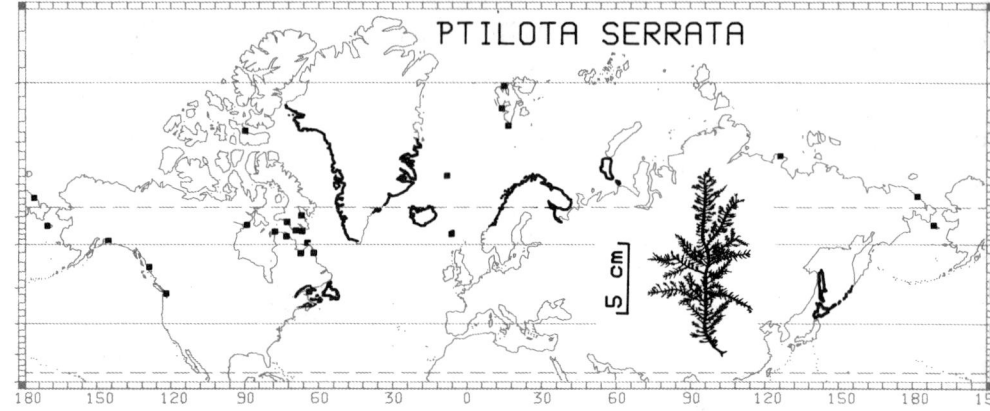

Fig. 2.26

Portugal (*A. nodosum*). In *C. tomentosa,* like in several other heteromorphic seaweed species (with a microscopic gametophyte and a macroscopic sporophyte), a requirement of low winter temperatures for reproduction of the gametophyte sets the southern distribution limit (see review by Breeman 1988). The species *Fucus vesiculosus,* which occurs on the coast of northwestern Africa, is not only eurythermal (widely tolerant to high and low water temperatures; see Chapter 7), but also euryhaline (widely tolerant of different salinities), since it penetrates far into the Baltic.

For *Ascophyllum nodosum* there is no sister taxon known from the North Pacific. In contrast, one may assume that the red algal species *Membranoptera alata* (Fig. 2.27), *Phycodrys rubens* (Fig. 2.28) and *Odonthalia dentata* (Fig. 2.29) speciated from North Pacific ancestors, since several species of *Membranoptera, Phycodrys,* and *Odonthalia* occur on the North Ameri-

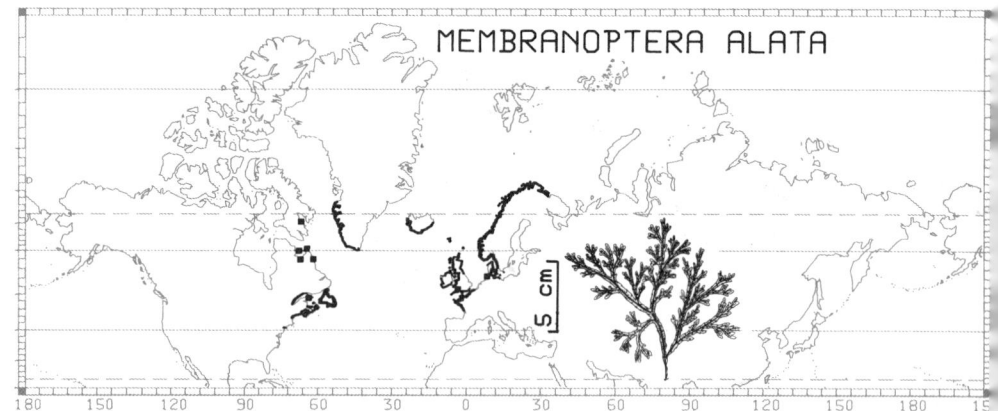

Fig. 2.27

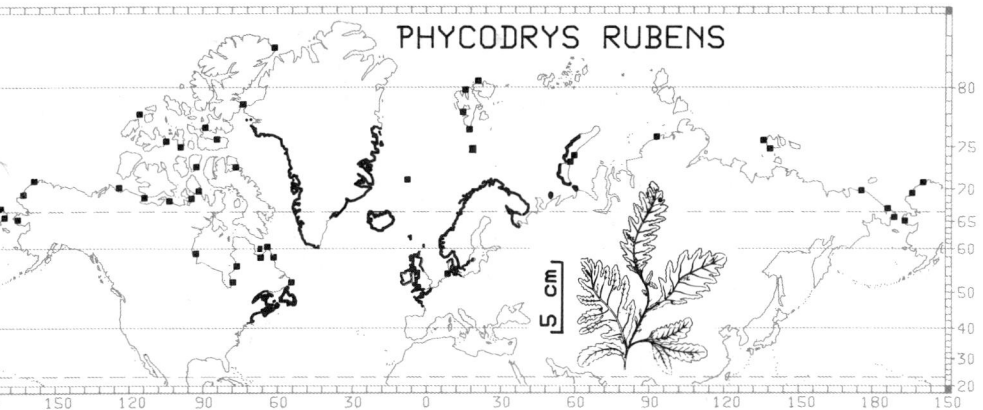

Fig. 2.28

can side of this ocean (Wynne 1970; Abbott and Hollenberg 1976; Lindstrom 1977). More examples of pairs of closely related red algal species that occur in the cool temperate waters of the North Atlantic and North Pacific, with or without an additional distribution in the Arctic, have been given by Lindstrom (1987a).

2.2.3 Cold Temperate North Atlantic Species

The genera with species endemic to the North Atlantic and not inhabiting the Arctic region (Table 2.1, group 3) either do not occur in the North Pacific or are represented with fewer species. For example, the genus *Fucus* occurs in the North Atlantic with six species, namely *F. distichus* (Fig. 2.13), *F. evanescens* (Rice and Chapman 1985; South and Tittley 1986), *F. spiralis*

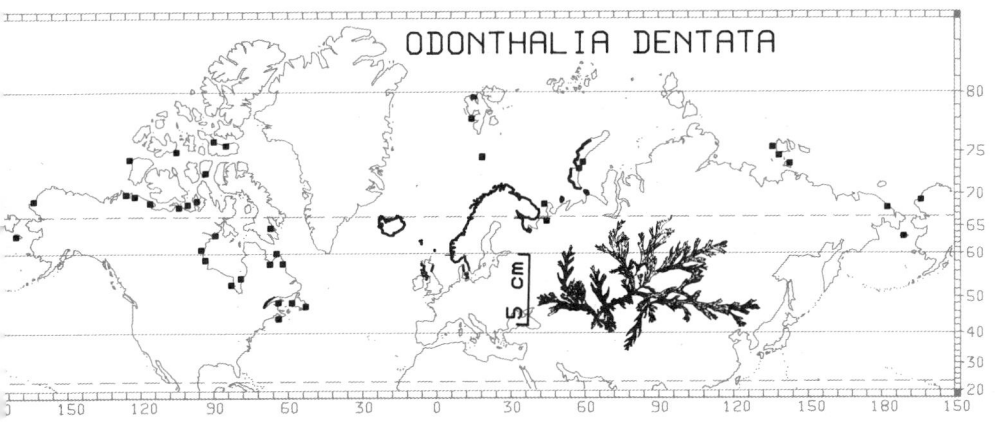

Fig. 2.29

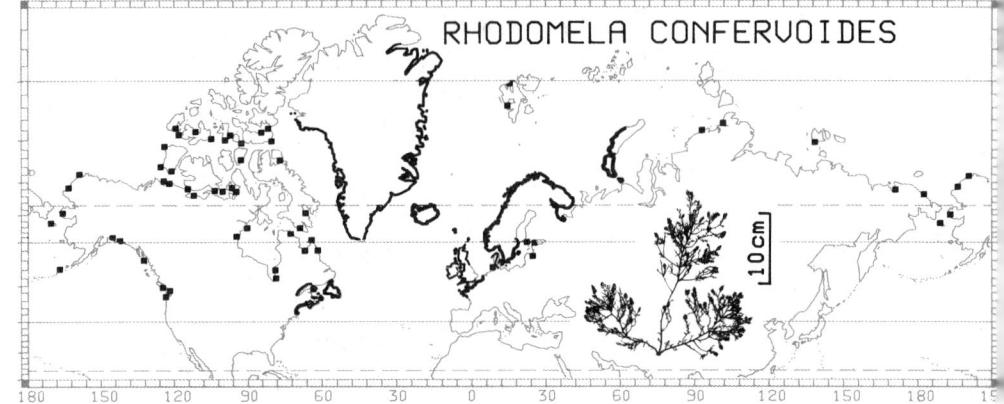

Fig. 2.30

(Fig. 2.31), *F. vesiculosus* (Fig. 2.23), *F. serratus* (Fig. 2.32), and *Fucus ceranoides* (Fig. 2.37). On the Pacific coast of North America *F. spiralis* has also been reported (Norris and Conway 1974), while another *Fucus* species, *F. gardneri,* is endemic (Silva 1979); the latter has been included in *F. distichus* by Abbott and Hollenberg 1976. According to Rice and Chapman (1985) *F. evanescens* and *F. distichus* occur on the Pacific coast of North America.

Among the Fucales, the genera *Bifurcaria* (Fig. 2.40) and *Himanthalia* (Fig. 2.39) do not occur in the North Pacific, whereas *Halidrys* (Fig. 2.41) is represented there by *H. dioica,* and *Pelvetia* (Fig. 2.38) by *P. fastigiata*. The red algal genus *Phyllophora* is represented in the North Atlantic by five species (Schotter 1968; Newroth 1971; Newroth and Taylor 1971; Parke and Dixon 1976), but only *Phyllophora truncata* (Fig. 2.25) reaches southern

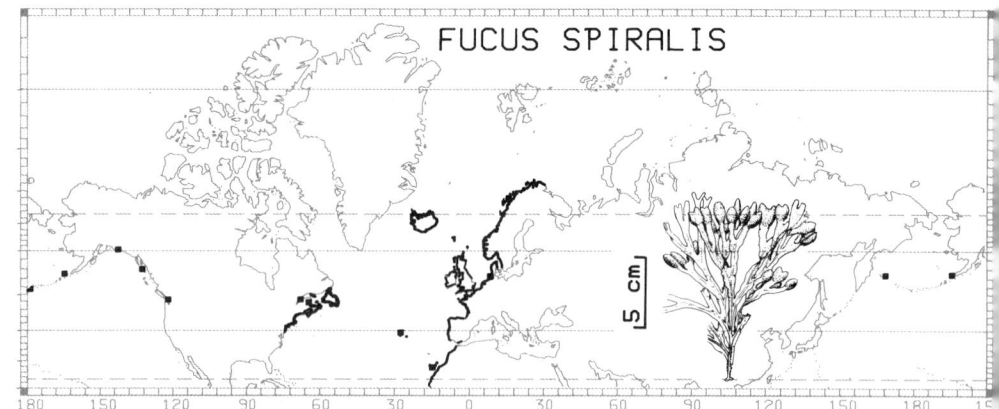

Fig. 2.31

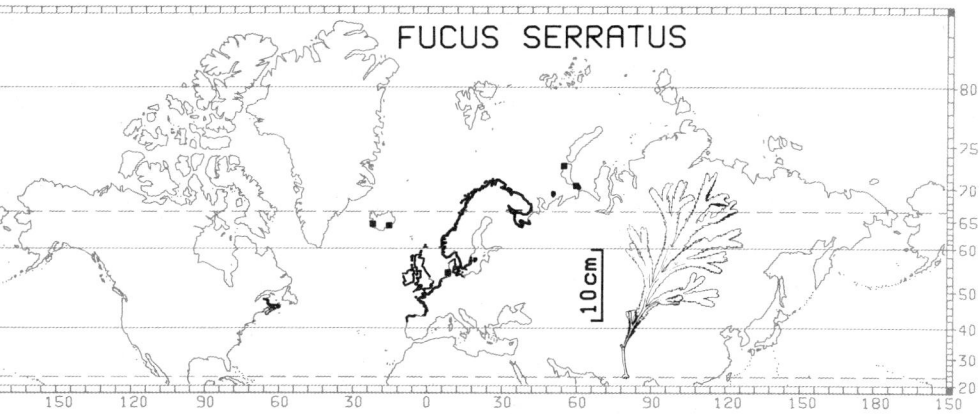

Fig. 2.32

Alaska via the Arctic region (Lindstrom and Scagel 1979), and there may be some relation between *Phyllophora* and the genus *Ozophora* in the North Pacific (Wynne, personal communication). For the eastern North Atlantic *Delesseria sanguinea* (Fig. 2.42) there is no closely related North Pacific or western North Atlantic sister taxon. The only species of *Delesseria* in the North Pacific is *D. decipiens* (Abbott and Hollenberg 1976; Hawkes et al. 1978).

An **amphi-Atlantic** distribution (Table 2.1, group 3a) is exhibited by *Fucus spiralis* (Fig. 2.31), and the two red algae *Chondrus crispus* (Figs. 1.11 and 2.33) and *Phyllophora pseudoceranoides* (Fig. 2.34). *Chondrus crispus* is definitely known only from the North Atlantic, but there are related taxa in the North Pacific, such as *C. nipponicus, C. giganteus,* and *C. pinnulatus* (Lüning et al. 1987). *F. spiralis* does not inhabit the Arctic region. Since it has

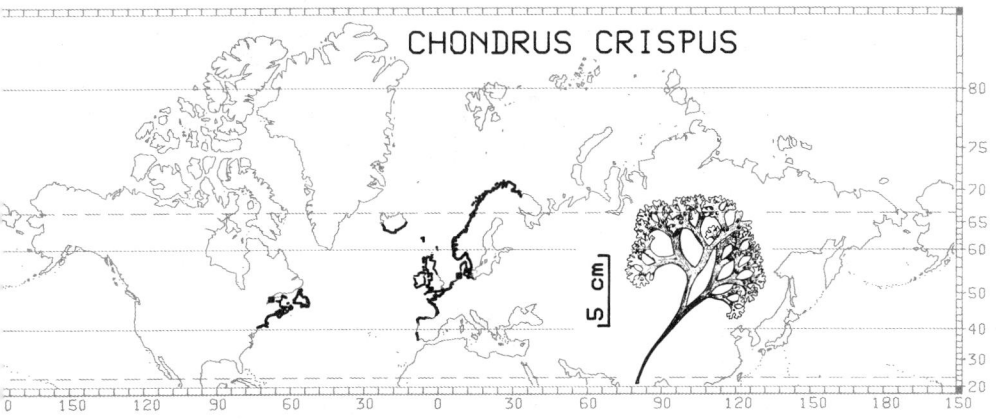

Fig. 2.33

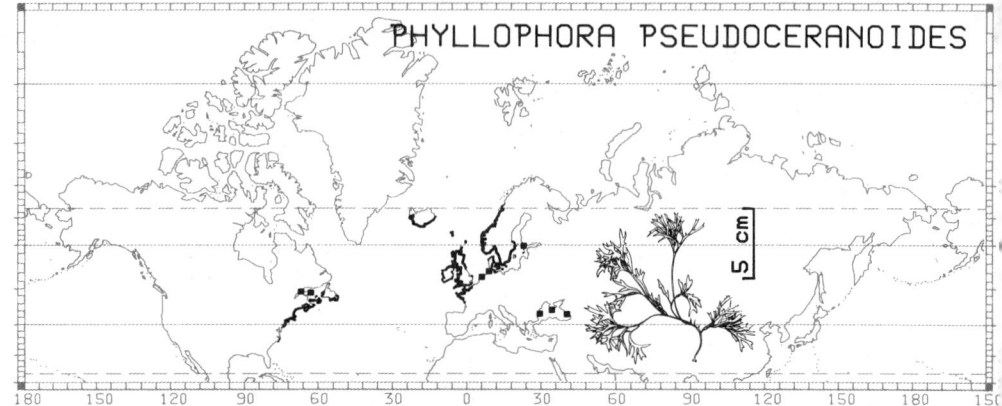

Fig. 2.34

only recently been discovered on the Pacific coast of North America (Norris and Conway 1974), the species is placed here among the North Atlantic group of algae; one should consider the possibility that it has been introduced into the North Pacific. *F. serratus* (Fig. 2.32) has a peculiar distribution pattern that, in Europe, stretches from the southern Arctic region (White Sea, Novaya Zemlya; but not Greenland) to northern Portugal. At the eastern side of the Atlantic the species is confined to the coasts of the southern Gulf of St. Lawrence and Atlantic Nova Scotia. Perhaps the species, attached to ballast stones, was taken there by ships from European coasts in the 19th century (McLachlan, personal communication).

Of the **European Atlantic** species (Table 2.1, group 3b) two characteristic species of the midsublittoral zone, *Laminaria hyperborea* (Fig. 2.35) and *Delesseria sanguinea* (Fig. 2.42), reach the northern coast of Iceland and the

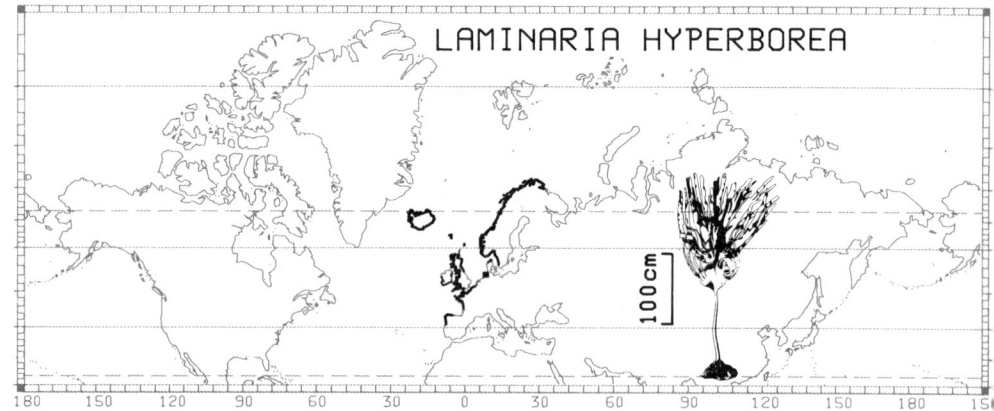

Fig. 2.35

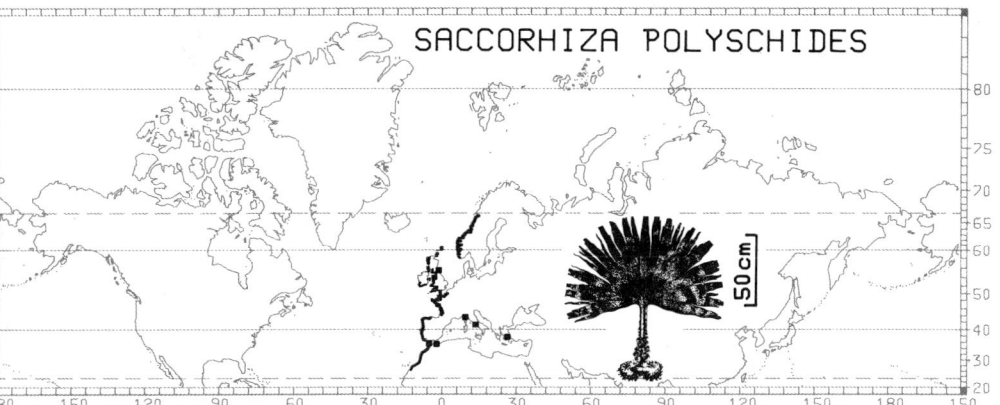

Fig. 2.36

Russian coast near Murmansk. The southern limit for *L. hyperborea* is Cape Mondego in mid-Portugal, and for *Delesseria sanguinea,* Sables d'Olonnes, a little south of the mouth of the Loire. Other characteristic European endemic species are confined to the upper sublittoral zone, namely *Himanthalia elongata* (Fucales, Himanthaliaceae; Fig. 2.39) and *Halidrys siliquosa* (Fucales, Cystoseiraceae; Fig. 2.41). Their southern limit is again on the Portuguese coast, but their northern limit is not in Iceland, as with *Laminaria hyperborea* and *Delesseria sanguinea,* but the Faeroe Islands and northern Norway. Farther south, *Saccorhiza polyschides* (Laminariaceae; Fig. 2.36), is distributed from the Shetland Islands and Rörvik (65°N) in Norway to Cape Jubi on the coast of Morocco. The species is also to be found in the western Mediterranean. Besides these sublittoral European algae there are also endemic species in the eulittoral zone, such as *Pelvetia canaliculata* (Fig. 2.38) or *Fucus ceranoides* (Fig. 2.37), the latter preferring brackish waters.

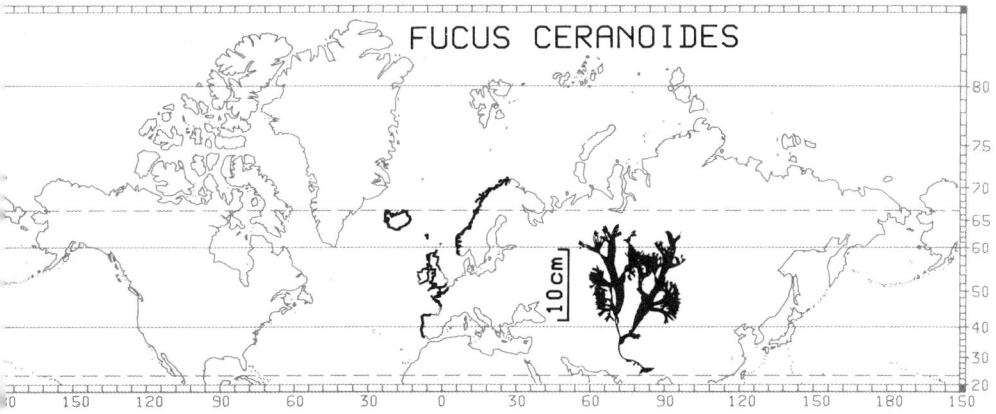

Fig. 2.37

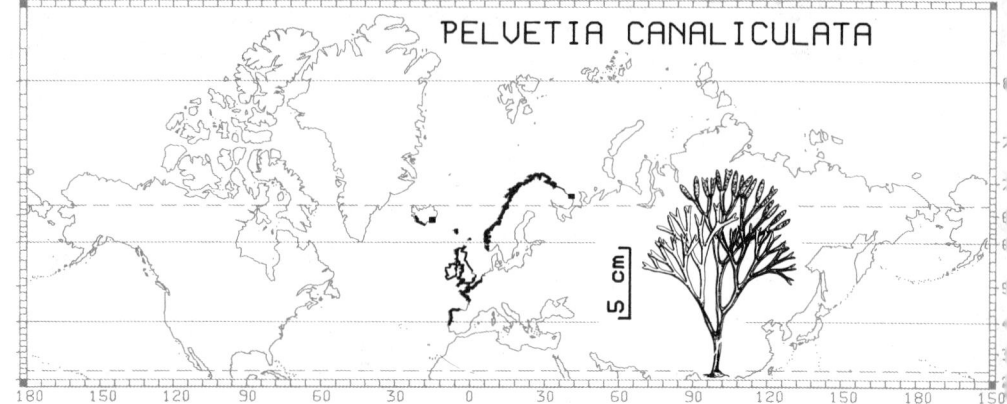

Fig. 2.38

2.3 EUROPE: COLD TEMPERATE REGION (NORTHERN NORWAY AND ICELAND TO NORTHERN FRANCE)

Literature: Of the older literature a few classic papers are cited. Modern papers preceded by an "e.g." mean that more papers pertinent to the same subject are given in the reference list of the cited paper. **Entire region:** Börgesen and Jónsson (1905); van den Hoek (1975, 1982a,b); van den Hoek and Donze (1967); Tittley et al. (1985, 1989). **Checklist, North Atlantic Ocean:** South and Tittley (1986). **Iceland:** Caram and Jónsson (1972); Munda (e.g., 1972a,b, 1975, 1980, 1987). **Faeroe Islands:** Börgesen (1903–1908); D. E. G. Irvine (1982); Price and Farnham (1982); Tittley et al. (1982). **British Isles and Ireland:** Earll and Farnham (1983); Guiry (1978); K. Hiscock (1985); K. Hiscock and Mitchell (1980); S. Hiscock (1979, 1986); Irvine et al. (e.g., 1975); Kain (1962, 1963, 1975b, 1976a, 1979); Kitching (1941, 1987); Lewis (1955, 1964); Maggs et al. (1983); Moore and Seed (1985); Newton (1931); Norton (1985a);

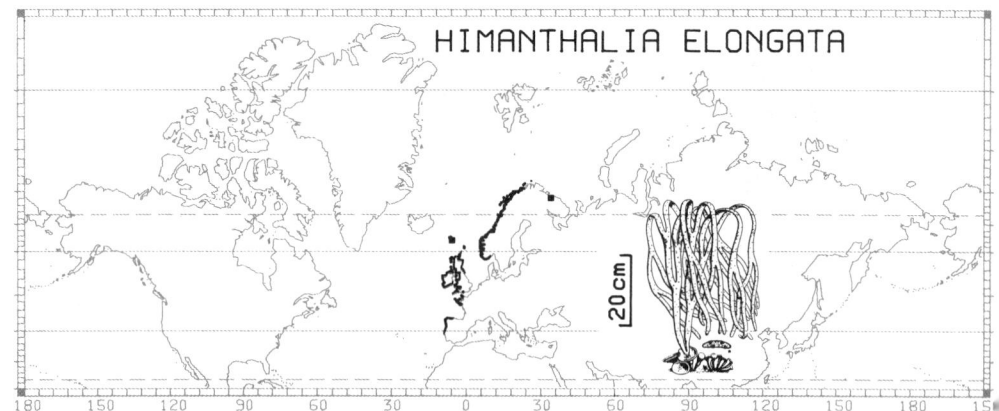

Fig. 2.39

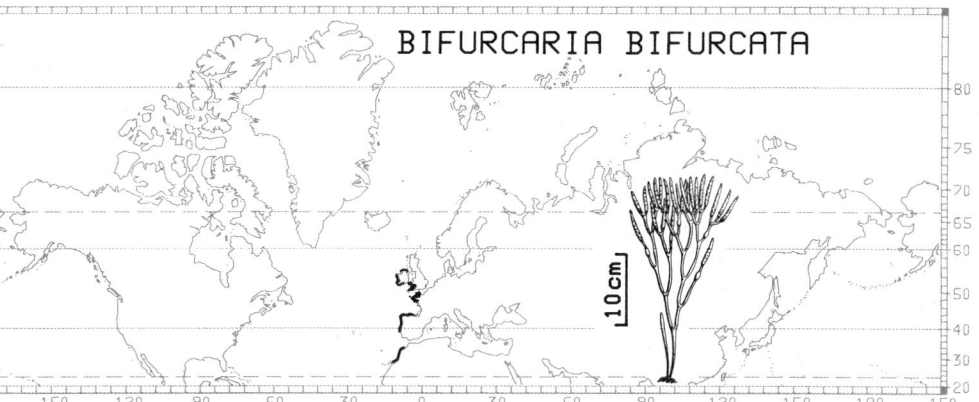

Fig. 2.40

Norton and Milburn (1972); Norton and Powell (e.g., 1979); Norton et al. (1977); Parke (1931); Parke and Dixon (1976); Price et al. (e.g., 1977); G. Russell (e.g., 1973); Stephenson and Stephenson (1972); Tittley (1986); Wilkinson (e.g., 1982). **Seaweed flora of the British Isles:** Christensen (1987); Dixon and Irvine (1977); Fletcher (1987); L. M. Irvine (1983). **Distribution maps of the marine algae of Britain and Ireland:** Norton (1985*b*). **Helgoland:** den Hartog (1959); Hagmeier (1930); Hoffmann (1940); Kornmann and Sahling (1977, 1983); Kuckuck (1894, 1897); Lüning (1970); Markham and Munda (1980); Munda and Markham (1982); Nienburg (1925); Reinke (1889*b*). **Norway:** Jaasund (1965); Jorde and Klavestad (1963); Levring (1937); Printz (1926, 1953); Rueness (1977); Sundene (1953). **Denmark:** Rosenvinge (1909–1931); Rosenvinge and Lund (1941–1950). **Checklist, Danish waters:** Christensen et al. (1985). **Sweden:** Gislén (1929–30); Kjellman (1878); Kylin (1944–1949). **Baltic:** Dietrich and Köster (1974*a*); Hällfors et al. (1981); Hoffmann (1940); Ketchum (1983); Kornas et al. (1960); Lakowitz (1929); Levring (1940); Luther (1951: hydrophytes);

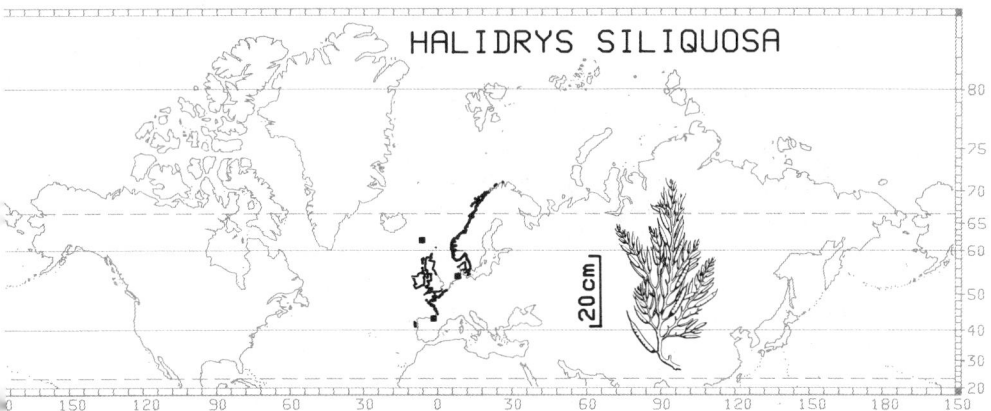

Fig. 2.41

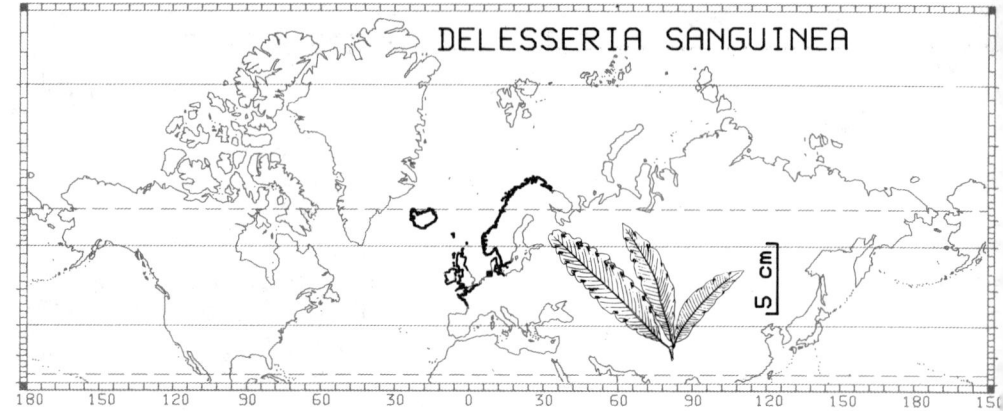

Fig. 2.42

Pankow (1971); Pankow et al. (1971); Ravanko (1968, 1972); Reinke (1889a,b); Remane (1933); Schwenke (1969, 1974); Segerstråle (1957); Voipio (1981); Waern (1952); Wallentinus (1978, 1979). **Netherlands:** den Hartog (1959); Nienhuis (1970); Stegenga and Mol (1983); van den Hoek et al. (1979); van Goor (1923). **Belgium:** Coppejans (1982a,b); Coppejans and Beeckman (1986); Coppejans and van der Ben (1980). **France:** Bouxin and Dizerbo (1971); Castric-Fey et al. (1973); Chalon (1905); Crisp and Fischer-Piette (1959); Dizerbo (1970); Ernst (1955); R. G. Evans (1957); Feldmann (1954); Feldmann and Magne (1964); Fischer-Piette (1932); Gayral (1966); Hamel (1924–1939); Lancelot (1961); L'Hardy-Halos (1972); Renoux-Meunier (1965); van den Hoek and Donze (1966); Virville (1966). **Some taxonomic groups: Chlorophyta: Ulvales:** Bliding (1963, 1968); Koeman and van den Hoek (1980, 1982a,b, 1984); Kornmann and Sahling (1978). **Prasiolales:** Kornmann and Sahling (1974). **Phaeophyta: Sphacelariaceae:** Prud'homme van Reine (1982). **Laminariales:** Kain (1979). **Rhodophyta: Corallinaceae (crustose algae):** Adey and Adey (1973). **Maritime and marine lichens:** Fletcher (1980). **Seagrasses:** den Hartog (1970); McRoy and Helfferich (1977); Phillips and McRoy (1980). **Temperature requirements of cold temperate European seaweeds:** Lüning (1984b). **Paleogeography, Europe:** Termier and Termier (1960); Thiede (1979, 1980). **Paleogeography, North Sea:** Ziegler (1977).

The North Alantic Current (North Atlantic Drift) flows with a northern branch as the "central heating" of northern Europe and turns to the south with branches of the Portugal Current and Canaries Current (Fig. 2.43). Due to the input of warm water by the North Atlantic Current, many warm-adapted species of the Mediterranean and the northwestern coast of Africa migrated, after the last glaciation, up to the coasts of Brittany and western Ireland. A plot of the latitudinal distribution spans of many algal species occurring from North Africa to Spitsbergen (Fig. 2.44) shows that there is a drastic reduction in the number of species to the north. More and more southern species drop out northwards, and there are fewer northern species to replace them (Fig. 2.44). There are clear floristic discontinuities on the coasts of Brittany and western Ireland, where the southern limits of a consid-

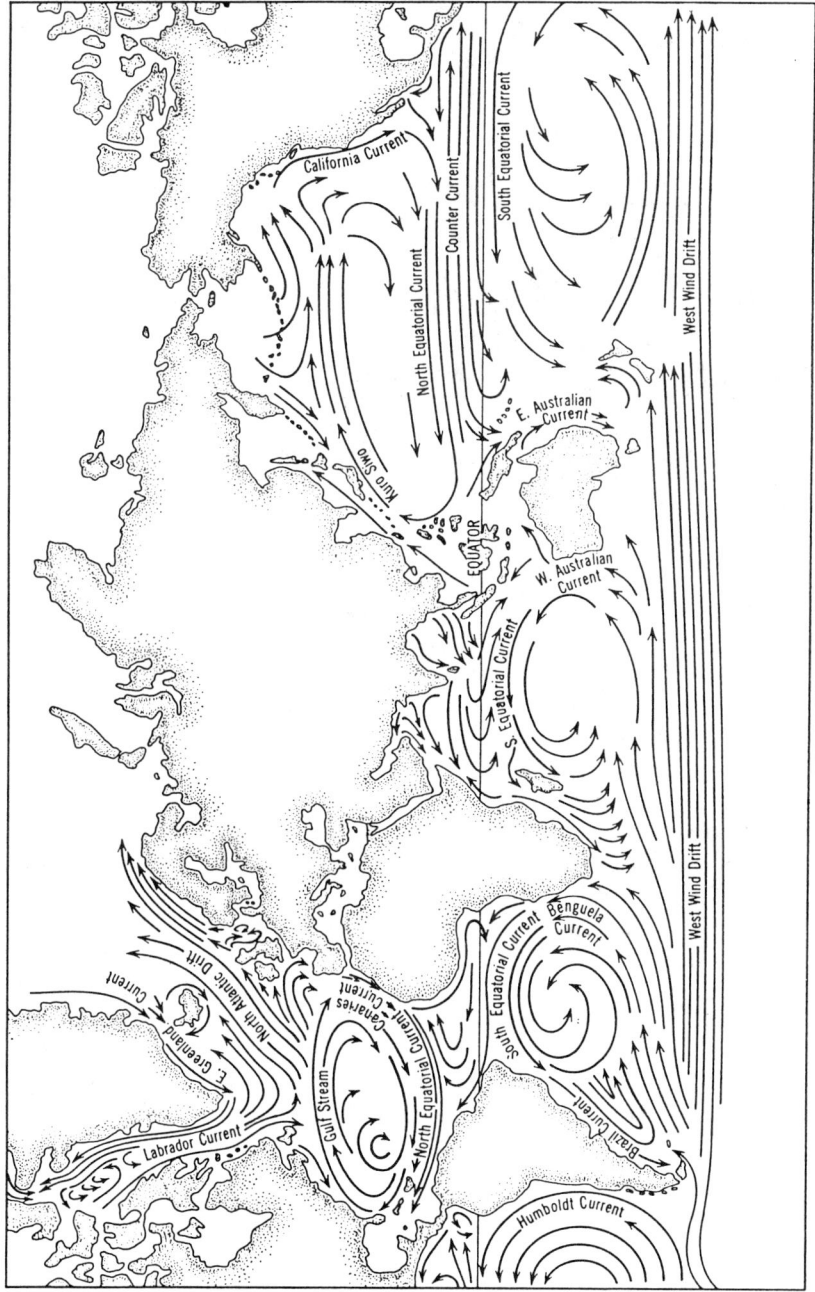

Fig. 2.43 Ocean currents. (From MacArthur and Connell 1970.)

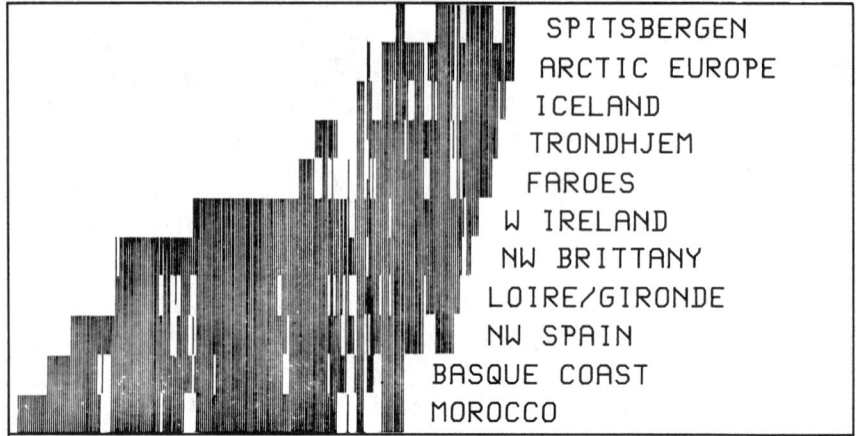

Fig. 2.44 Latitudinal distribution of seaweed species along the coasts from Morocco to Spitsbergen. Each vertical line represents the distribution span of one species. (After van den Hoek 1975.)

erable number of northern species are situated. These coasts, with a nearby 10°C-February isotherm and a 15°C-August isotherm (Figs. 1.5 and 1.6), mark the boundary between the eastern province of the **cold temperate Northern Atlantic region** and the **warm temperate Mediterranean Atlantic region**. The latter reaches to Cape Verde (Senegal, West Africa), where the eastern Atlantic tropical region begins (Figs. 1.5, 1.6, and 2.45).

Since many species of the cold temperate North Atlantic region occur on the European as well as the American coasts, one may regard these two sides as the eastern and western provinces of the same biogeographic region. The northern limit of the region roughly follows the courses of the 8°C-August isotherm and the 0°C-February isotherm, both situated between Norway and Spitsbergen, where a floristic discontinuity is again evident (Fig. 2.44; compare with Figs. 2.31–2.39, 2.41–2.42, 3.7, 3.9 and 3.11).

The basic zonation patterns within the euphotic zone of the cold temperate European coasts are illustrated by examples from Scotland (Fig. 2.46), North Wales (in relation to wave exposure; Fig. 2.47), Helgoland (Figs. 2.48 and 2.49), and Iceland (Fig. 2.50). Arctic water influences the northern and eastern coasts of Iceland, documented by the presence of the red alga *Devaleraea ramentacea*, which is distributed mainly in the Arctic but also extends to southern Nova Scotia and northern Norway (Figs. 2.50 and 3.9).

Along the low-lying coastline from northern France to Denmark and of eastern England between Flamborough Head in Yorkshire and Herne Bay in Kent, rocky substratum is limited (Dixon and Irvine 1977). Therefore many of the more prominent macroalgae exhibit distribution gaps along these coastlines (Figs. 2.4–2.42). On the other hand, Tittley (1986) has reported a total of 175 seaweed species from human-made structures in the southern

Fig. 2.45 Phytogeographical regions and provinces in the North Atlantic. A = warm temperate Carolina region; B = western Atlantic tropical region; C,D,E = warm temperate Mediterranean–Atlantic region (C = Canary province; D = Mediterranean province, E = Lusitania province); F,G = cold temperate North Atlantic region (F = eastern province; G = western province), H = Arctic region. (From van den Hoek 1975.)

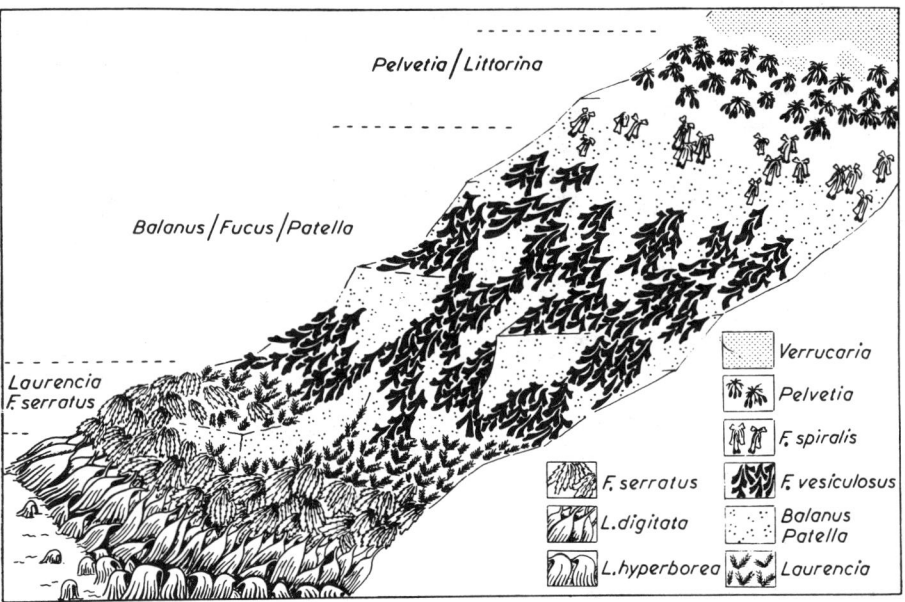

Fig. 2.46 Zonation on the medium exposed Scottish coast. (From J. R. Lewis 1964.)

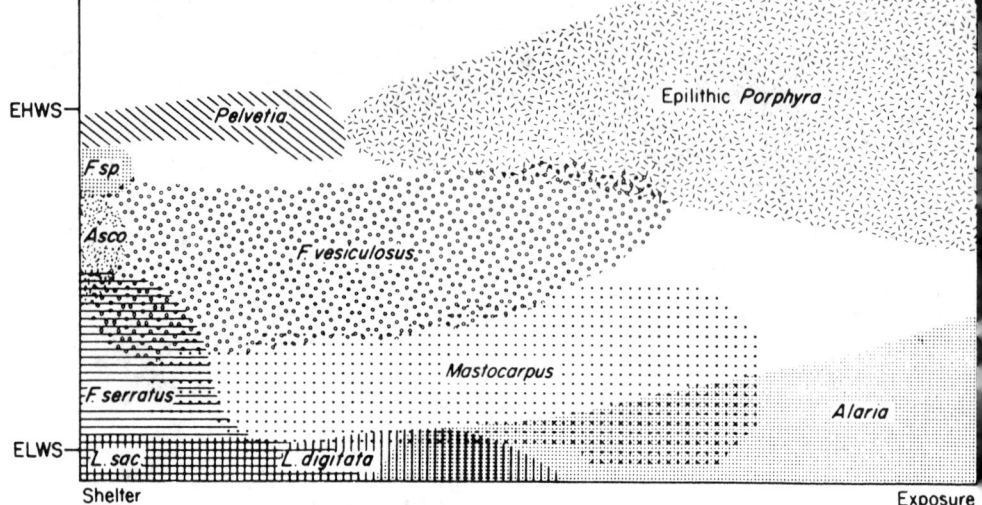

Fig. 2.47 Distribution of eulittoral seaweeds in relation to wave exposure in North Wales. EHWS = extreme high water of spring tides; ELWS = extreme low water or spring tides; *F. sp.* = *Fucus spiralis*; Asco = *Ascophyllum nodosum*; *L. sac.* = *Laminaria saccharina*; *Mastocarpus* = *M. stellatus* (= *Gigartina stellata*). (From Norton 1985*b*; modified from W. E. Jones and Demetropoulos 1968.)

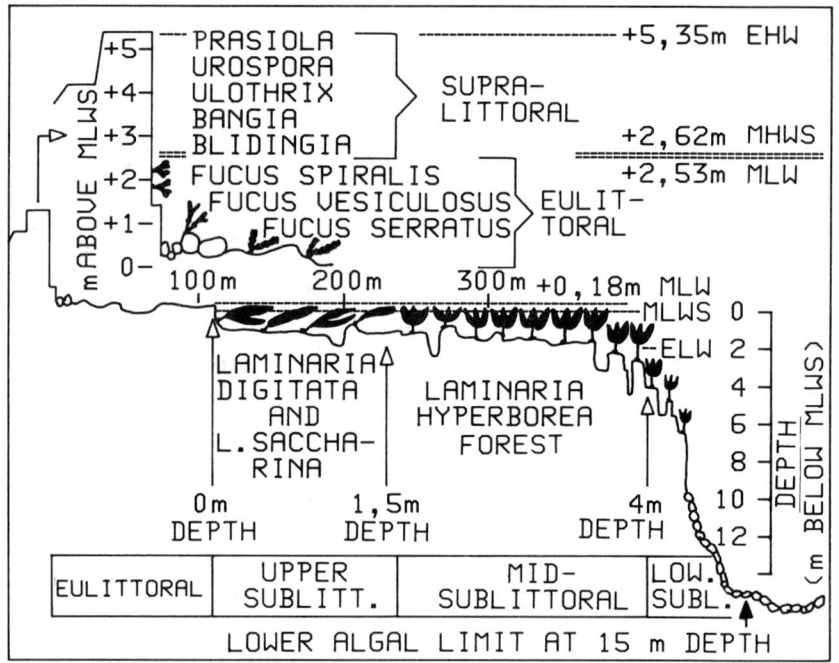

Fig. 2.48 Zonation off Helgoland. General scheme. EHW = extreme high water; MHWS = mean, MLWS = mean low water of spring tides; ELW = extreme low water.

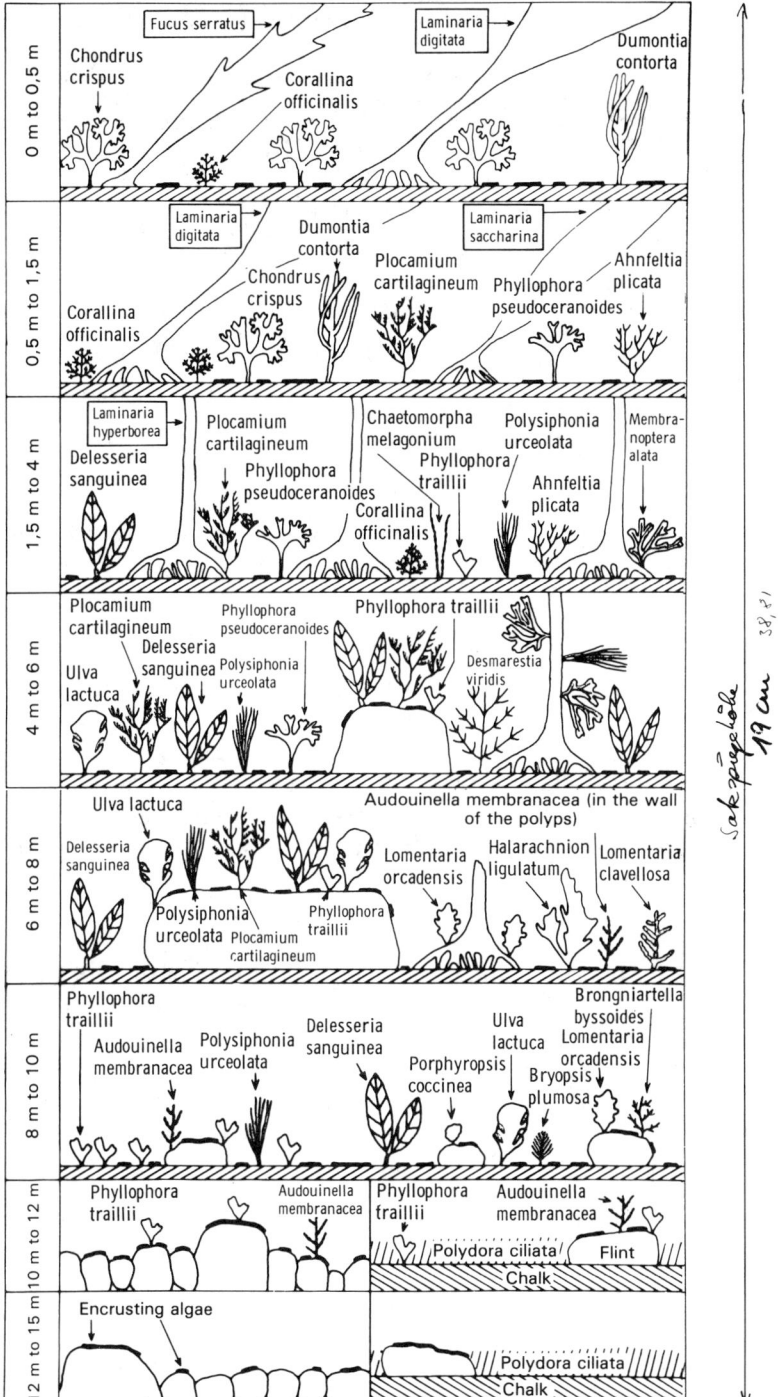

Fig. 2.49 Zonation off Helgoland. Typical understory algae at different depths in the sublittoral zone. (From Lüning 1970.)

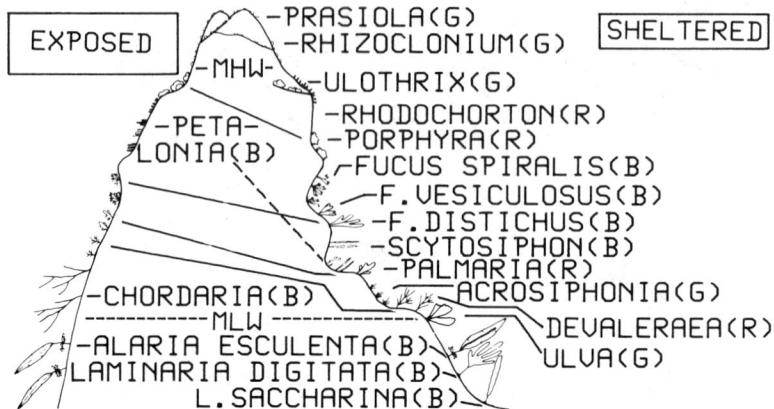

Fig. 2.50 Zonation in the eulittoral zone of Grimsey, northern Iceland. MHW = mean high water; MLW = mean low water; G = green alga; B = brown alga; R = red alga. Exposed side: "F. distichus" = F. evanescens. (From Munda 1971a.)

North Sea and the eastern English Channel; the most frequently recorded alga is the green *Blidingia minima,* followed by *Enteromorpha intestinalis, E. prolifera, Ulva lactuca,* and *Fucus spiralis* (Fig. 2.31). In the course of the 20th century the introduced new habitats have favored the spread of a few species to the southern North Sea, such as *Ascophyllum nodosum, Pelvetia canaliculata,* and *Laminaria digitata.* The last-named species, for example, was not known from Essex prior to the 1970s, from the Netherlands prior to 1871, or from the North Sea coast of Limfjord Inlet, Jutland, prior to 1939 (Tittley 1986).

The microphytobenthos dominates the soft bottoms of the Wadden Sea, which covers an area of 10,000 km^2 along the North Sea coasts of the Netherlands, West Germany, and Denmark (surveys of vegetation by Dijkema and Wolff 1983; van den Hoek et al. 1979). The seagrasses *Zostera marina* and *Z. noltii* are the main components of the comparatively insignificant macrophytobenthos of the Wadden Zee, as well as species of *Ulva* and *Enteromorpha,* which grow mainly as epiphytes on mussels and seagrasses.

Helgoland. This rocky island with its cliffs of red sandstone, chalk, and shell lime, is situated in the southern North Sea and represents an "oasis" for macroalgae. After the last glaciation and the rise of the sea level 4000 years ago, Helgoland became an island and has since been colonized by benthic marine organisms. Several characteristic species of the open Atlantic coasts, such as *Pelvetia canaliculata* and *Alaria esculenta,* are missing at Helgoland. *A. esculenta* does not survive temperatures above 16°C (Sundene 1962); the waters near Helgoland in summer regularly surpass this temperature by 1–2°C due to its position in relatively shallow water near the mainland. *A. esculenta* has been successfully cultivated in the sea near Helgoland from early spring until early summer (Munda and Lüning 1977).

History of algal investigation and marine biological stations. Pioneering the investigations of the cold temperate European algal vegetation were Kjellman (1878; Skagerrak), and Börgesen (1903–1908; Faeroe Islands). Later highlights were, for example, the phytosociological investigation of the algal vegetation of northern France, the Netherlands, and Helgoland by den Hartog (1959) and the extensive work on zonation on British shores documented by Kitching (e.g., 1941) and J. R. Lewis (e.g., 1964). Three of the larger and older marine biological stations are the Station Biologique at Roscoff (established 1872), the Plymouth Laboratory of the Marine Biological Association of the United Kingdom (established 1888), and the Biologische Anstalt Helgoland (established 1892).

2.3.1 Supralittoral Zone

Among the last representatives of the terrestrial vegetation on rocky substratum are characteristic yellow and gray lichens (Figs. 2.51B,C). Below this lichen zone the marine euphotic zone begins with the black thalli of the lichen *Verrucaria maura* and other saltwater-tolerant lichen species (Fletcher 1980) and with the uppermost specimens of the periwinkle *Littorina saxatilis* (Figs. 2.46 and 2.47). Characteristic multicellular algae in the **supralittoral zone** (**littoral fringe** according to the terminology of Lewis) are the cyanophyte *Calothrix crustacea* and, especially on dikes manured by sea birds, the green algae *Prasiola* and *Rosenvingiella*. Lower down are filamentous green algae

Fig. 2.51 Maritime zone and supralittoral zone on the coast of the Faeroe Islands (north coast of Eysturoy). **(A)** Higher plants. Left: *Plantago maritima;* right: *Armeria maritima;* in the background the grass *Festuca rubra*. **(B)** Yellow lichen *Xanthoria parietina*. **(C)** Above: gray lichen *Lecanora;* below: the lichen *Verrucaria maura*. (Photographs by the author.)

such as *Urospora* and *Ulothrix* and, during the cold season, the red alga *Bangia atropurpurea* (= *fuscopurpurea*) (Figs. 2.48 and 2.50). All these genera are distributed in temperate waters of both hemispheres, indicating that they have formed highly specialized species that withstand the extreme conditions of desiccation and osmotic stresses at the margin of the sea.

2.3.2 Eulittoral Zone

The upper biological limit of the eulittoral zone is indicated by characteristic barnacles (Fig. 1.2): *Semibalanus balanoides* and species of the more southern genus *Chthamalus*, which also occurs in the Mediterranean. These barnacles are accompanied by another, *Elminius modestus*, introduced into Europe by ships from New Zealand during World War II (den Hartog 1959; J. R. Lewis 1964; Harms 1984). Under conditions of strong wave exposure, for example, on exposed vertical cliffs or dikes, the eulittoral zone is dominated by barnacles and/or mussels (*Mytilus edulis*, Fig. 2.52A), whereas species of the Fucaceae prevail under more sheltered conditions (Figs. 2.47 and 2.52B).

Algal zonation: upper eulittoral zone. Here the fucalean species *Pelvetia canaliculata* and somewhat deeper *Fucus spiralis* and *F. distichus* subsp. *anceps* form a canopy (Fig. 2.47). (For subspecies and distribution of *F. distichus* in the North Atlantic, see Powell 1957, 1981; Rice and Chapman 1985; South and Tittley 1986). *F. distichus* has its southern limit in Scotland and western Ireland. Typical smaller algae in the upper eulittoral zone are the green alga *Blidingia minima* and the brown crustose alga *Ralfsia verrucosa*.

Algal zonation: mideulittoral zone. Here the main canopy algae are *Ascophyllum nodosum* (Fig. 2.22) and with increasing exposure *Fucus vesiculosus* (Figs. 2.23 and 2.47); they are accompanied by the bright red crustose alga *Hildenbrandia rubra* (= *prototypus*), which also reaches the upper eulittoral zone in shaded places. A luxuriant cover of the green alga *Enteromorpha* and the red alga *Porphyra umbilicalis* develops mainly in the summer and on areas not occupied by the fucalean canopy algae. This vegetation and species of *Ulothrix* and *Urospora* are typical of an early stage in the sequence of recolonization, after an initial settlement of bacteria and diatoms (den Hartog 1959). This sequence, termed **succession,** can be studied after experimental removal of the organisms (e.g., Markham and Munda 1980; see Chapter 8), on newly built breakwaters, and on newly formed islands, such as Surtsey, which emerged from the sea near Iceland in 1963 (Jónsson 1970). It may take several years for the final stage of succession when a well-established multilayered vegetation, the **climax community,** is reached again (Fig. 2.52A), a much shorter timespan than on land. The uppermost species of the crustose coralline (calcified red) algae (family Corallinaceae) is *Phymatolithon lenormandii*. The periwinkle *Littorina littorea* may be abundant on the rocky substratum in this subzone, whereas *L. obtusata* and *L. mariae* are mainly epiphytic on the fucalean canopy algae.

Algal zonation: lower eulittoral zone. This is often dominated by *Fucus serratus* (Figs. 2.32, 2.47, 2.48, and 2.52*B*), or, with increased exposure, by the red alga *Mastocarpus* (= *Gigartina*) *stellatus* (Fig. 2.47). Typical **understory** algae are the

Fig. 2.52 Aspects from the eulittoral zone of Helgoland. (**A**) Zonation on a harbor wall. Barnacles and mussels dominate in the upper and mideulittoral zone, with a belt of red algae (*Polysiphonia urceolata, Ceramium rubrum,* and *C. deslongchampsii*) in the lower eulittoral zone. Left: organisms have been scraped off, and mainly *Ulva* has appeared as a stage of succession. (**B**) Flat rocky areas covered with green algae in the mideulittoral zone with mainly *Fucus serratus* in the lower eulittoral zone. (**C**) Lower eulittoral zone with *F. serratus,* the red alga *Chondrus crispus,* the epiphytic green alga *Chaetomorpha capillaris,* and the snail *Littorina littorea.* (Photographs by the author.)

rhodophytes *Chondrus crispus, Corallina officinalis,* and the green alga *Cladophora rupestris.* Below this are the **crustose** coralline algae *Phymatolithon purpureum* (= *polymorphum;* see Woelkerling and Irvine 1986), *P. laevigatum, P. lenormandii,* and the uncalcified species *Haemascharia hennedyi.* In soft rock the **boring,** filamentous green algae *Entocladia* and *Tellamia,* and the unicellular Codiolum phase of several green algae are represented throughout the whole eulittoral zone (Wilkinson and Burrows 1972). An **opportunistic** species (*r*-strategist), one of the first to appear during succession in the mid- and lower eulittoral zone as well as in the upper sublittoral zone, is the green alga *Ulva lactuca.* The genus comprises several species, for example five on the coast of the Netherlands (Koeman and van den Hoek 1981). In spring several annuals appear, such as the red alga *Dumontia contorta,* the green alga *Acrosiphonia arcta,* and the brown algae *Scytosiphon lomentaria* and *Petalonia fascia.* Characteristic red algae in the mid- and lower eulittoral zone are *Mastocarpus stellatus* and *Laurencia pinnatifida.* Other red algae such as *Palmaria* (= *Rhodymenia*) *palmata, Lomentaria articulata,* as well as filamentous species such as

Ceramium rubrum, C. deslongchampsii, Polysiphonia urceolata, Plumaria elegans, and *Rhodomela confervoides,* are confined to the lower eulittoral zone and may reach into the sublittoral zone. The fucalean species *Himanthalia elongata* (Figs. 2.39, 2.53D, 2.54B) and *Bifurcaria bifurcata* (Fig. 2.40) also occur in the lower eulittoral zone, just above the uppermost representatives of the sublittoral Laminariales.

2.3.3 Sublittoral Zone

The vegetation structure of the sublittoral zone could only be investigated with the development of scuba diving techniques (review: Riedl 1967). Pioneers in this field include Drach (1949), Ernst (1955), Kornas et al. (1960), and Kain (1962). The first marine biologists, who descended with a diving helmet, were Berthold (1882; Gulf of Naples), Gislén (1929–1930; Gullmarfjord, Sweden), Waern (1952; Öregrund, Sweden), and Kitching (1941; Scottish coasts). Kitching was the first to describe the structure of the laminarian forests typical of European Atlantic coasts.

The **upper sublittoral zone** is, like the eulittoral zone, physically characterized by fast turbulent flow and multidirectional mechanical stress (Koehl and Wainwright 1977; Koehl 1982). The uppermost zone of the upper sublittoral, which is emergent at wave suckback, has been referred to as the "infralittoral fringe zone" (Fig. 1.4) (Stephenson and Stephenson 1949, 1972; Womersley 1981*a*). The main canopy algae at wave-exposed localities are laminarian species such as *Alaria esculenta* (Figs. 2.12, 2.47, and 2.54B), supplied with a flexible midrib, and *Laminaria digitata* (Figs. 2.20 and 2.47), which is well-adapted to this environment due to its flexible stipe and deeply divided blade. At Rockall, an isolated rock 20 m in diameter in the North Atlantic, west of the Hebrides, *A. esculenta* reaches to a 35 m depth and *L. digitata* to 20 m; *L. hyperborea* is lacking, possibly because of the extremely strong wave action (S. Hiscock, personal communication). Under more sheltered conditions these species are replaced by *L. saccharina* (Figs. 2.11 and 2.47), which is characterized by its undivided and mechanically less tolerant blade, and by the fucalean species *Halidrys siliquosa* (Figs. 2.41 and 2.53D). Both of these species have flexible stipes, which are important for life in the upper sublittoral zone where they may be exposed at extreme low tides. These canopy species can lie flat, the uppermost thalli protecting the lower ones. Typical examples of understory algae in the upper sublittoral zone are the red algae

---

Fig. 2.53 Brown algae from the eulittoral zone and upper sublittoral zone of the European Atlantic coasts. (**A**) *Pelvetia canaliculata* and *Fucus spiralis* (Roscoff, Bretagne). (**B**) *Ascophyllum nodosum* (Roscoff). (**C**) From left to right: *Ascophyllum nodosum, Laminaria saccharina* (adrift from the sublittoral zone and exhibiting a bullate pattern of the frond), and *Fucus vesiculosus* (northern Ireland). (**D**) *Himanthalia elongata* (Roscoff). (**E**) *Saccorhiza polyschides* (Roscoff). (**F**) *Chorda filum* in the upper sublittoral zone (Trondheim). (A–C,F photographs by the author; D,E courtesy of K. Jahnke.)

Fig. 2.54 Aspects from the euphotic zone of the Faeroe Islands. **(A)** Aspect of one of the islands. **(B)** Upper sublittoral zone. From the background forwards: *Himanthalia elongata, Alaria esculenta,* and the uppermost fronds of *Laminaria hyperborea*. **(C)** midsublittoral zone at 4 m deep: laminarian forest (*L. hyperborea*). **(D,E)** Understory algae at places free of *L. hyperborea* at 4 m deep. **(F)** Lower sublittoral zone at 15 m deep: red algae *Phycodrys rubens* on the holuthurian *Cucumaria* and on the mussel *Modiolus modiolus;* in the background the soft coral *Alcyonium digitatum*. **(G)** Lower sublittoral zone at 20 m deep; the uncalcified, crustose red alga *Cruoria pellita* completely covers the rocky substrate. **(H)** *Alaria esculenta* and *L. digitata,* both with the red alga *Palmaria palmata* as an epiphyte. **(I)** *L. hyperborea* with epiphytic red algae. **(J)** *L. saccharina* collected from a medium exposed locality. **(K)** *L. saccharina* forma *faeroensis* from the inner part of a fjord. **(L)** Upper sublittoral zone at the inner part of a fjord; *Chorda filum* covered with sediment. **(M)** *L. hyperborea* forma *cucullata* from the inner part of a fjord. (Photographs by the author.)

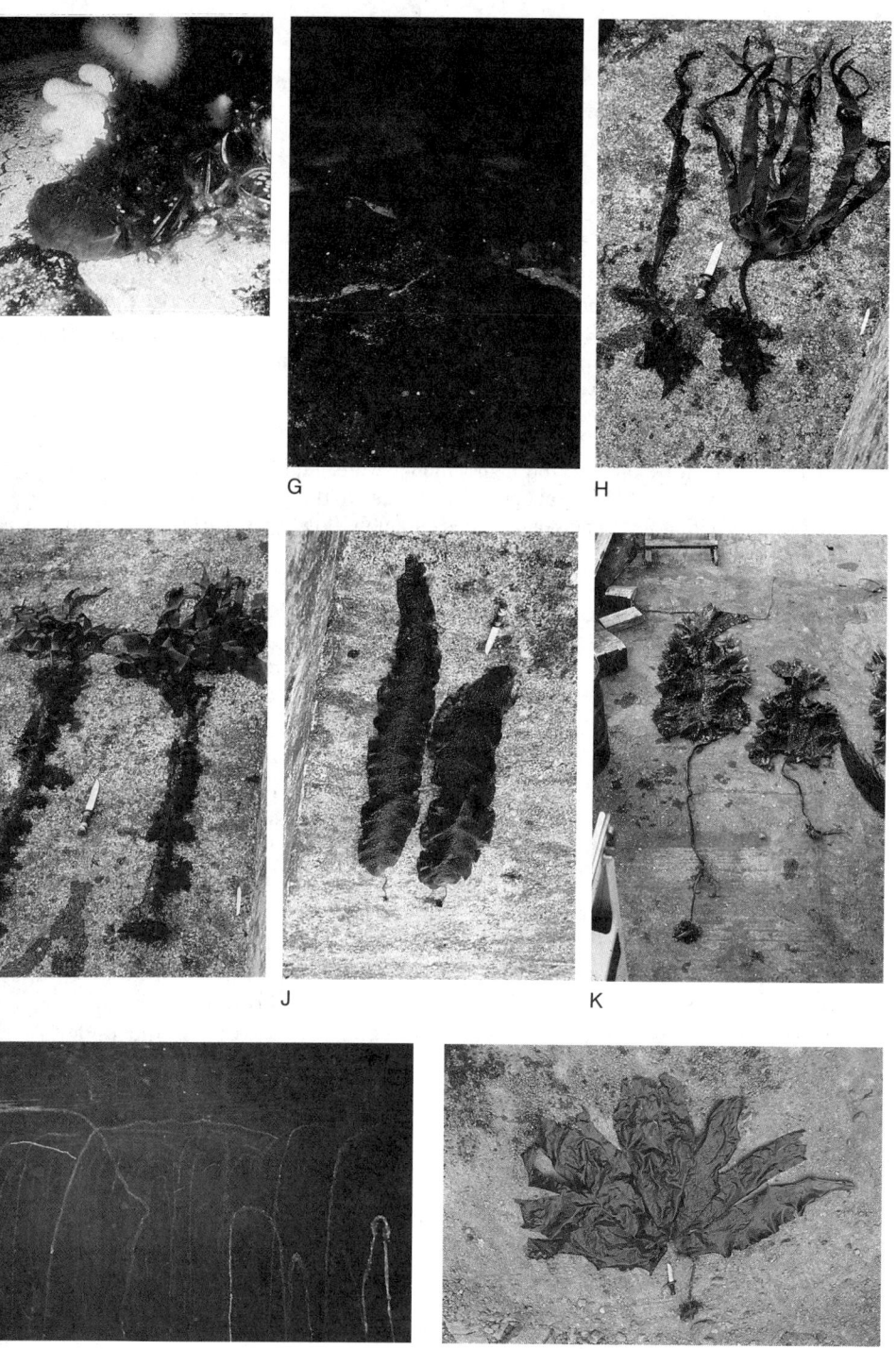

Fig. 2.54 (Continued)

Corallina officinalis, Chondrus crispus, Mastocarpus stellatus, Dilsea carnosa, and *Palmaria palmata* (Norton et al. 1977).

Seagrasses. The seagrass *Zostera marina* has a wide distribution in all temperate regions of the Northern Hemisphere and may be found on sandy and muddy substrata in the upper sublittoral zone of the European coasts (Fig. 4.3). A second species, *Z. noltii* (= *nana*), is confined to more brackish waters (den Hartog 1970). *Z. marina* is the only seagrass whose range extends as far north as the White Sea, western Greenland, and Labrador and in the Pacific to Alaska and Kamchatka. In the Atlantic the southern limit of the species occurs near Gibraltar and on the coast of North Carolina. In the North Pacific its range extends to Baja California and southern Japan. In the 1930s the populations of *Z. marina* were almost eradicated in the North Atlantic. At first this near extinction was thought to be caused by a disease—the slime mold *Labryrinthula macrocystis*. Later studies have suggested that local factors may have been involved, for example, too high summer water temperatures for *Z. marina* in Denmark or increased ecological disturbances due to the closure of the Zuyder Zee in the Dutch Wadden Sea. The seagrass beds are now slowly recovering (for details see van den Hoek et al. 1979; den Hartog 1987).

The **mid sublittoral zone** is dominated, along the European Atlantic coasts from northern Norway to mid-Portugal, by the canopy alga *Laminaria hyperborea* (Figs. 2.35, 2.54C and 2.55). The "laminarian forest" formed by this alga begins 1–2 m below mean low water. The uppermost individuals may die at intervals of several years, when the growing zone, which is situated at the lower end of the blade, is exposed to the air at extreme low water (Fig. 2.55C). In the midsublittoral zone, individuals of *L. hyperborea* may attain an age of up to 15 years, in contrast to the laminarian species of the upper sublittoral zone, which generally live no longer than three years. The haptera of the long-lived *L. hyperborea* harbor a rich epiphytic fauna (Norton et al. 1977). With its rigid stipe the alga supports a blade canopy situated 1–2 m above the rocky substratum. It is this dense canopy that absorbs most of the downwelling light (Kitching 1941; Norton et al. 1977), since there may be up to 12 m^2 of frond area on one square meter of rocky bottom (Jupp and Drew 1974). Thus, only a scarce understory vegetation is to be found under *L. hyperborea*. The lower parts of the stipes are mainly covered with hydroids and bryozoans (Fig. 2.55D). Nearer the light the stipes may carry epiphytic red algae such as *Polysiphonia urceolata, Membranoptera alata, Palmaria palmata, Cryptopleura ramosa,* and *Phycodrys rubens* (Norton et al. 1977). After experimental frond removal, the biomass of the epiphytic red algae increases rapidly (Harkin 1981). The perennial nature of the rugose *L. hyperborea* stipes seems both to attract a specialized epiphytic flora composed of a few species and to discourage opportunist epiphytes (Kain 1979).

Recolonization within the Laminaria hyperborea *forest*. The rigid stipe is the main adaptation with which *L. hyperborea* outcompetes all other canopy algae. This was demonstrated by recolonization experiments (Kain 1975*b*, 1976*b*). On cleared areas in the midsublittoral zone a mixed vegetation of *L. hyperborea*, and canopy species with

Fig. 2.55 Aspects from the sublittoral zone of Helgoland. **(A)** From left to right: *Laminaria digitata*, *L. hyperborea*, and *L. saccharina*. On Helgoland the latter species exhibits a smooth surface, without any bullations (compare with Fig. 2.53c). **(B)** Vegetation of *L. digitata* and *L. saccharina* in the upper sublittoral zone at extreme low water (1.5 m below mean low water at spring tides.) **(C)** Upper laminarian forest of *L. hyperborea*, again at extreme low water, which occurs only in intervals of several years. **(D)** Dense *L. hyperborea* forest at 3 m deep; stipes heavily covered with hydroids and bryozoans. **(E)** Smaller algae in the laminarian park at 5 m deep. From the background forwards: *Plocamium cartilagineum*, *Ulva lactuca*, and *Delesseria sanguinea*. **(F)** *L. hyperborea* at its lower limit of occurrence, at 8 m deep. The frond is totally covered by the bryozoan *Membranipora membranacea*, the stipe by other bryozoan species. (Photographs by the author.)

flexible stipes developed, for example, *L. digitata, L. saccharina, Saccorhiza polyschides, Alaria esculenta,* and *Desmarestia* species. Maximal species diversity was created at moderately low supply rates of space, and thus at moderate disturbance rates, reinforcing Tilman's (1982) view of resource competition and community structure. From the fourth year the algae were overtopped by the relatively slow growing *L. hyperborea,* whose stipes had grown sufficiently long and the blade canopy dense enough to shade out the opportunistic competitors. Therefore the existence of perennial species such as *L. saccharina, Desmarestia aculeata,* or the annual species *D. viridis* and *Saccorhiza polyschides* was confined, within the midsublittoral zone, to areas that did not support the long-lived *L. hyperborea* forest, such as bottoms covered with small stones (Kain 1962).

The dense *Laminaria hyperborea* vegetation ends at a light percentage depth of about 5%. At this depth the "laminarian forest" passes into a "laminarian park," and the **lower sublittoral zone** begins. An extensive meadowlike vegetation of smaller green, brown, and red algae (Figs. 2.48, 2.49, 2.54D,E and 2.55E), no longer shaded by the canopy of *L. hyperborea,* begins in the lower sublittoral zone. Scattered individuals of this species are also found in the lower sublittoral zone, down to a light percentage depth of 1% (Fig. 1.3). This depth, which depends on the prevailing Jerlov water type, is as low as 8 m near Helgoland (Figs. 2.48 and 2.49), at 15–25 m in the clearer waters of the open Atlantic coasts, and even at 34 m in southern Norway (Kain 1979). The deepest-occurring individuals of *L. hyperborea* grow so slowly that their blades are in danger of being completely overgrown by *Membranipora membranacea* (Fig. 2.55F). According to measurements taken from another *Membranipora* species attached to glass slides, irradiance may be reduced by about 50% due to the cover of this epiphytic bryozoan (Cancino et al. 1987). It may also impede gas and ion exchange for the alga. The snail *Helcion* (= *Patina*) *pellucidum,* which eats out cavities in the stipes and may also infest the blades, is a grazer of *L. hyperborea* (Kain and Svendsen 1969; Norton et al. 1977).

In the deeper parts of the lower sublittoral zone, at light percentage values below 1% and hence below the deepest-occurring individuals of *L. hyperborea,* shade-adapted smaller algae still flourish, such as *Delesseria sanguinea* (Fig. 2.42), *Phycodrys rubens* (Figs. 2.28 and 2.54F) and other species illustrated in Fig. 2.49. Characteristic deep-water red algae, which may occur along the British coasts down to 30 m (Norton 1968; Norton and Milburn 1972; K. Hiscock and Mitchell 1980; S. Hiscock 1986), include *Phyllophora crispa* (Gigartinales), *Kallymenia reniformis* (Cryptonemiales), *Crytopleura ramosa, Hypoglossum hypoglossoides* (= *woodwardii*), *Polyneura gmelinii* (Ceramiales, Delesseriaceae), *Polysiphonia urceolata, P. elongata,* and *Brongniartella byssoides* (Ceramiales, Rhodomelaceae). Characteristic deep-water brown algae are *Desmarestia ligulata, Dictyota dichotoma* (Fig. 2.59), and, on the coast of Brittany (Castric-Fey et al. 1973), *Carpomitra costata* (Sporochnales) and *Halopteris filicina* (Sphacelariales; Fig. 2.59).

The lower algal depth limit occurs, where about 0.05% of the surface light penetrates (Fig. 1.3). This is equivalent to a depth of 50 m, if Jerlov water type 1 prevails, that is, the clearest waters along the European Atlantic coasts. However, the algal depth limit is often situated at 20–40 m or less. Among the deepest multicellular algae are red crustose algae, such as the uncalcified species *Cruoria pellita* (Fig. 2.54G) or the calcified species *Lithothamnion sonderi,* representing the major part of the algal biomass near the lower limit of the lower sublittoral zone (Adey and Adey 1973; Clokie and Boney 1980). The record of the deepest occurence of a multicellular alga on the cold temperate European coasts is the crustose, calcified red alga *Phymatolithon rugulosum* from 90 m deep at the Rockall Bank, off northwest Ireland (Clokie et al. 1981). Crustose corallines occupy a vertical range from intertidal tide pools to the greatest depths recorded for marine algae (reviewed by Littler et al. 1985; Steneck 1986; Vadas and Steneck 1988). Of the **endozoic algae,** the minute red alga *Audouinella membranacea,* which lives in the outer layers of hydroids, may occur down to the lower algal limit (Fig. 2.49), and some minute **Conchocelis-like red algae** were detected by Clokie et al. (1981) on the shelf west of Scotland down to 78 m. The substrata of these latter, shell-boring algae are represented by mollusk shells and the calcareous tubes of the polychaete *Pomatoceros triqueter.* The Conchocelis-like phases may belong, as sporophytes, to eulittoral *Porphyra* species and *Bangia atropurpurea* (= *fuscopurpurea*), as well as more highly-evolved red algae such as *Scinaia furcellata* and *Helminthocladia calvadosii* (Kornmann and Sahling 1980).

2.3.4 Maerl Vegetation

Among the crustose coralline algae (order Corallinales) there are unattached, free-living species with irregularly branched outgrowths between 5 and 40 mm in diameter, for example, *Lithothamnion* (= *Lithothamnium*) *corallioides* and *Phymatolithon calcareum* (Fig. 2.56). Such corallines composed of loose-lying branches are called **maerl** (a Breton word) or marl (Earll and Farnham 1983; Steneck 1986). At 3–25 m deep the two species mentioned form their characteristic maerl beds on sandy bottoms, mainly near the coasts of southern Ireland, southern Cornwall, and Brittany. Drowned river valleys (rias) are the favored localities for maerl beds along the southwest coast of England (Farnham and Bishop 1985). The northern distribution limit of the two species lies on a line between the Orkney Islands and southern Norway. Their southern limit is the Mediterranean, or the Canary Islands for *Lithothamnion corallioides* (Afonso-Carrillo and Gil-Rodriguez 1982a). Further maerl-forming species in the British Isles are *Lithophyllum dentatum, L. incrustans,* and *Lithothamnion glaciale,* the last extending into the Arctic region (Farnham and Bishop 1985). Other free-living corallines may form more densely branched, spherical-to-ellipsoid thalli called "nodules" (Steneck 1986); or spherical forms known as "rhodoliths" (or "rhodolites") may result by growing around an initial "nucleus" such as a small stone. In a

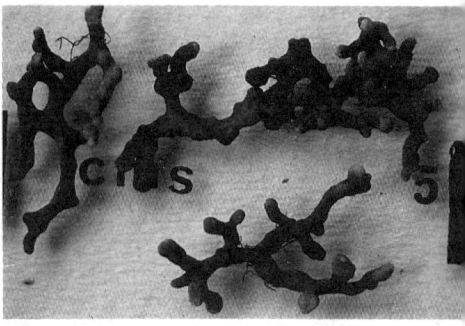

Fig. 2.56 Maerl, a sublittoral vegetation of branched, crustose and calcified red algae. **(A)** Underwater photograph of the maerl vegetation at Falmouth, southern Cornwall. **(B)** *Phymatolithon calcareum,* one of the two maerl species (Roscoff, Brittany). (Photographs courtesy of W. F. Farnham.)

rhodolith the coralline comprises more than 50% of the specimen. When this percentage is lower, one speaks of a "coralline coating" (Steneck 1986).

Numerous other small algae flourish on the branched thalli of *Lithothamnion corallioides* and *Phymatolithon calcareum* (Cabioch 1969; Adey and Adey 1973; Adey and MacIntyre 1973); the open network creates an ideal habitat for small crustacea and an abundant epifauna (Farnham and Bishop 1985). Throughout the centuries the dead, calcified thalli of the two maerl species have piled up to form layers several meters high. These layers are commercially harvested as a source of lime and trace elements for agricultural use. The yearly harvest amounts to 300,000 metric tons (Blunden et al. 1975, 1981; Wildgoose and Blunden 1981; see Table 8.5). Eighty percent of the dried maerl consists of calcium carbonate and 10% of magnesium carbonate. The maerl vegetation occurs only on moderately wave-exposed coasts. If there were too little water motion, sedimentation would be too heavy for the maerl vegetation, whereas with too strong wave action the slowly growing algae would be removed.

2.3.5 Significance of Water Motion in the Sublittoral Zone

The composition of algal communities in the eulittoral zone is strongly influenced by wave action and in the sublittoral mainly by currents (for reviews see, e.g., K. Hiscock 1983; Koehl 1986). This can be clearly seen in a comparison between the algal vegetation at the mouth of a fjord and that inside the fjord. This striking difference was stressed by Börgesen (1903–1908) for the algal vegetation of the Faeroe Islands, which have been recently reinvestigated by Irvine (1982), Tittley et al. (1982), and J. H. Price and Farnham (1982). Figure 2.54 illustrates the situation. At the fjord entrance surf-adapted vegetation, including the kelps *Alaria esculenta* (Fig. 2.54B) and

Laminaria digitata, dominates and is replaced, in the inner fjord, by the brown alga *Chorda filum* (Fig. 2.54L) and others. This almost motionless vegetation is heavily covered with sediment. *L. hyperborea* is thickly stiped at the exposed mouth of the fjord (Fig. 2.54I) but is found with a thin stipe and a large, slightly divided blade inside the fjord (forma *cucullata;* Fig. 2.54M). Since mechanical stress is reduced with increasing water depth, such still-water forms also occur with increasing depth (Fig. 2.57). Svendsen and Kain (1971) demonstrated, by transplantation experiments, that these morphological differences are not necessarily genetically fixed but may be induced by the particular environment. This result was also obtained with *L. digitata* (Sundene 1964) and *Saccorhiza polyschides* (Norton 1969). The fronds of these two species are deeply divided when growing under exposed conditions, hardly divided at more sheltered locations, whereas big, "cucullate" blades are formed under extreme shelter.

A further example is provided by *Laminaria saccharina,* which does not inhabit wave-exposed places (Fig. 2.47) but forms narrow fronds and solid stipes under conditions of medium exposure (Fig. 2.54J). In the inner parts of Faeroese fjords it occurs as a form that was described as *L. faeroensis* by Börgesen (1903–1908). This taxon is characterized by a broad, undulate frond and a hollow stipe that may attain a length of up to 3.5 m (Fig. 2.54K). As shown for the eastern North Pacific kelp *Nereocystis luetkeana* by Koehl (1986), wide undulate blades suffer greater drag than narrow flat ones, for one reason because ruffled blades flap with greater amplitude. Possibly genetic differentiation is involved in the case of *L. faeroensis,* and the striking fact that hollow-stiped *L. saccharina* has only been found along the European

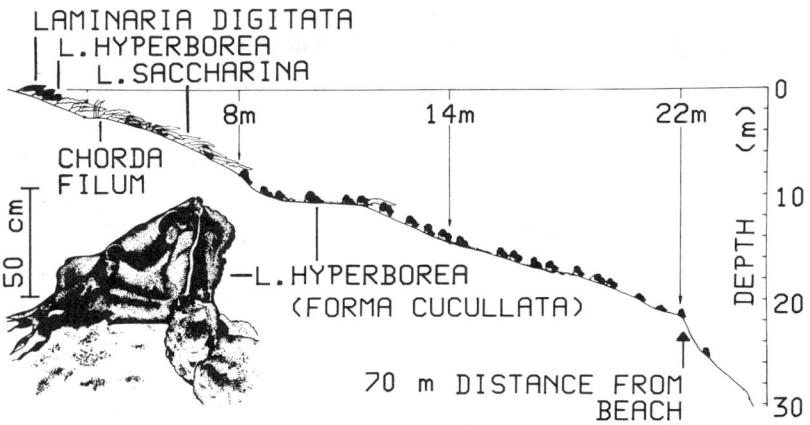

Fig. 2.57 Diving transect from a fjord at Espegrend, western Norway. Upper sublittoral zone: *Laminaria digitata*. Midsublittoral zone: in the upper part the normal form of *L. hyperborea,* deeper down *L. saccharina,* and below 8 m deep the quiet-water form *L. hyperborea* forma *cucullata*. The habit of this latter form is illustrated at the left. (From Svendsen and Kain 1971.)

coast at the Faeroe and Shetland islands may indicate some form of migration of hollow-stiped *Laminaria longicruris* from the eastern coast of North America to the eastern Atlantic (Kain 1976a; Yarish and Egan 1987; Egan and Yarish 1988; Brinkhuis et al. 1989). This may be compared to the eastward migration of *Agarum cribrosum*.

2.3.6 Brackish Water Areas: Fjords, Estuaries, and the Baltic Sea

A continuous transition from marine to freshwater (limnetic) environments takes place in **estuaries** (for a review see Ketchum 1983); examples are estuaries of the big rivers running into the North Sea along the Netherlands and German coasts, the fjords of the Norwegian and Scottish coasts, and on a bigger scale in the Baltic. Algal investigations of the Norwegian **fjords,** such as the Trondheim Fjord (Printz 1926), Hardanger Fjord (Jorde and Klavestad 1963), and Oslofjord (Sundene 1953; Klavestad 1978) have shown a "fjord effect" (Jorde and Klavestad 1963), caused mainly by two features. First, with decreasing wave action and salinity from the mouth to the inner fjord the number of species declines, and second, the lower vegetation limit is shifted upwards. The latter effect is probably due to decreased light transmission in the inner parts of the fjord, caused by suspended material brought in by the inflowing fresh water. Where freshwater is laden with organic wastes, a mass development of species of *Enteromorpha* and *Ulva* indicates high levels of eutrophication (Klavestad 1978). Likewise, a mass development of plankton may contribute to a further reduction of light transmission and enhance the fjord effect.

Biotopes with different salinities and their algal vegetation. The change of algal vegetation is particularly striking when one enters an estuary (den Hartog 1967; Wilkinson 1980). Many seaweed species are euryhaline and inhabit marine environments with normal salinity as well as the **euhaline** (salinity 40–30‰) and **polyhaline** (30–18‰) environment. The **mesohaline** environment (18–3‰) is divided into α-mesohaline (18–8‰) and β-mesohaline (8–3‰). A few species tolerate the mesohaline environment; examples in the estuaries of the Rhine, Maas, and Scheldt are *Fucus vesiculosus, Blidingia minima,* species of *Enteromorpha* and *Ulothrix,* the green alga *Rhizoclonium riparium,* and the crustose red alga *Hildenbrandia rubra.* The fact that in marine environments these species grow in the upper eulittoral or supralittoral zone (littoral fringe) indicates that they can withstand not only desiccation but also osmotic stress (den Hartog 1967, 1968). The mesohaline interval contains few freshwater algae, as the problem of colonizing brackish waters is greater for them than for marine algae. A characteristic alga of the mesohaline interval is *Fucus ceranoides* (Fig. 2.37), which occurs in brackish environments from Norway to Spain and also in the estuaries named above. In the **oligohaline** environment (3–0.5‰), the last interval before the freshwater (limnetic) realm, one finds several freshwater algal species mixed with *Blidingia minima* and *Rhizoclonium riparium.* These latter two species, regarded as true "brackish water algae" (den Hartog 1967), also occur in the algal vegetation of the **salt marshes,** where halophytes such as *Spartina townsendii* and *Salicornia herbacea* dominate. (For reviews see den Hartog 1973; Nienhuis 1970; Polderman 1979; Polderman and Polderman-Hall 1980.)

The seaweed flora of the **Baltic** is an "offshoot" of that of the North Sea, as Reinke (1889a) suggested. Most of the characteristic, cold temperate species of the European coasts, such as *Laminaria hyperborea,* have migrated via the Skagerrak only as far as the entrance of the Baltic at the Kattegat. Within the Baltic, which harbors an algal vegetation down to about a 30 m depth, the number of species and often their size (Fig. 2.58) are drastically reduced (e.g. in *Delesseria sanguinea;* Fig. 2.58), especially in the areas of the Kattegat (with salinity fluctuations of 20‰) and the Arkona Sea (10‰ salinity). Of the approximately 150 conspicuous species of the Swedish western coast only 15% reach the archipelago of the southern Finnish coast (Schwenke 1974). Figures 2.4–2.42 illustrate the different ability of the more characteristic cold temperate species to penetrate into the Baltic. Besides the reduction in salinity, the existence of seaweeds in the Baltic is also hampered by the **lack of rocky substratum** for most of the German, Polish, and Russian coasts (Dietrich and Köster 1974b). Scattered stones and erratic blocks, remnants of the glacial period, provide the only suitable substrata for seaweeds.

Brackish water submergence. Marine water from the North Sea enters the Baltic as a deep current, and less dense water of low salinity leaves the Baltic at the surface. Many organisms of the eulittoral zone of the marine environment, such as *Enteromorpha, Ulothrix,* and *Hildenbrandia rubra,* descend in the Baltic; others, such as *Fucus vesiculosus* in the inner Baltic or *Fucus serratus* in the Belt Sea, are confined to the permanently submerged substrate. As G. Russell (1985a) has pointed out, "it is the intertidal species, already pre-adapted to low salinity as part of their equipment for survival in such habitats, that have been able to exploit the desalinated Baltic."

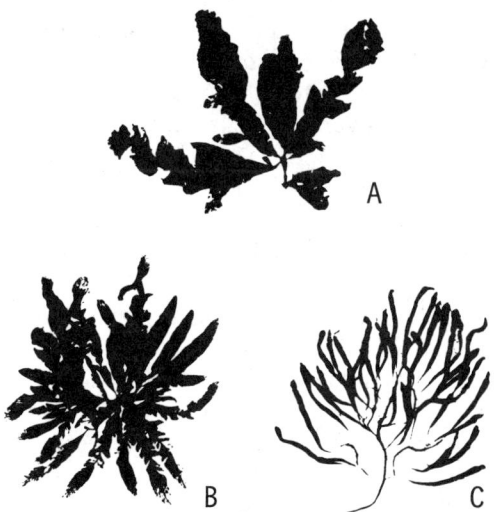

Fig. 2.58 Red alga *Delesseria sanguinea.* **(A)** North Sea form (Helgoland). **(B)** Baltic form (southern Kattegat). **(C)** Baltic form from the southern distribution limit (Blekinge, southern Sweden). (From Nellen 1966.)

Laminaria digitata and *L. saccharina* also descend from the upper sublittoral zone to greater depths on their way from the Skagerrak via the Kattegat to the Belt Sea. The descent to deeper water of higher salinity has been termed **brackish water submergence** by Remane (1955). The algae also descend because of the action of scouring ice in winter and long-lasting reductions of sea level (Hällfors et al. 1981). *Fucus vesiculosus* lives permanently submerged in the whole of the Baltic and *F. serratus* in the western part, for example, down to 15 m near Blekinge, southern Sweden (Levring 1940). This suggests that these species, confined in the marine environment to the eulittoral zone, are excluded from the sublittoral zone not for physiological reasons but due to competition, for example, with *Laminaria hyperborea*, which in the Baltic does not compete with these fucalean species. *Fucus* species exhibit the phenomenon of brackish water submergence in the Arctic, the Gulf of St. Lawrence, and the inner parts of the Norwegian fjords (Jorde and Klavestad 1963). In the Baltic, the lower boundaries of several algae, such as *Fucus* species, have moved upward considerably, possibly due to **increased turbidity** of the water caused by **eutrophication** (Wallentinus 1981; Kautsky et al. 1986; Svane and Gröndahl 1987; Breuer and Schramm 1989).

Since there are practically no tides in the Baltic, no real eulittoral zone exists. However, meteorologically caused, aperiodical fluctuations of water level may amount to 1 m in the spring. Thus a special terminology for the upper, aperiodically emerging part of the euphotic zone is required. According to Du Rietz (Waern 1952, Hällfors et al. 1981), the **geolittoral zone** occurs below the zone of the terrestrial vegetation and above the mean summer water level. The geolittoral zone is populated by marine lichens and the green alga *Prasiola* among others. Below this, the **hydrolittoral zone** reaches the lowest low-water levels ever recorded. This is inhabited by algae such as *Cladophora glomerata, Dictyosiphon foeniculaceus* (in summer), *Ceramium tenuicorne* (autumn and spring), and *Pilayella littoralis* and *Monostroma grevillei* (spring), which also occur in the **sublittoral zone.** Other algae of the hydrolittoral zone, such as *Bangia atropurpurea, Urospora penicilliformis,* and *Blidingia minima,* do not extend to the sublittoral zone. In other words, these species, along with several animals such as the barnacle *Semibalanus balanoides* (Remane and Schlieper 1971), do not exhibit the phenomenon of brackish water submergence (Waern 1952). This may mean that these species, also missing in the marine environment from the sublittoral zone, require phases of emergence. The physiological background for such a requirement is not known.

Brackish water areas are often ephemeral environments viewed on the geological time scale. The Baltic was a freshwater environment 12,000 years ago, when it was filled by melting ice after the last glaciation. Salinity was high 10,000 years ago, low 9000 years ago, and increasing from 7000 years ago, remaining brackish until now (Dietrich and Köster 1974a). This short erratic history was not suitable for the evolution of endemic species, so there are no endemic seaweeds in the Baltic (Schwenke 1974; Russell 1985a). However, the marine algae within the Baltic show significant changes of form (Fig. 2.58), as well as a reduction in size, often the loss of sexuality and attachment (Waern 1952; Prud'homme van Reine 1982). There is experimental evidence in Baltic marine algae for increased tolerance to low salinities and reduced tolerance to high salinities (Russell 1985a). Baltic *Ceramium tenuicorne* and *Rhodomela confervoides* die in seawater of normal salinity (Russell 1985a). Rueness (1978) discovered that the morphological traits of the Baltic form of the red alga *Ceramium tenuicorne* are genetically fixed, although this form can still be crossed successfully with the North Sea form. Thus the formation of genetically differentiated ecotypes has started in Baltic marine algae. In contrast to the Baltic, the Caspian Sea has existed as a brackish water environment since the late Tertiary, and a few endemic algal species are found here.

2.4 EUROPE–NORTHWEST AFRICA: WARM TEMPERATE MEDITERRANEAN–ATLANTIC REGION

This region stretches from the 10°C-winter isotherm near western Ireland, southern England, and Brittany to the 20°C-summer isotherm near Cape Verde, West Africa; (Figs. 1.5, 1.6, and 1.7). Within the region, which exhibits high endemism (40–50% in echinoderms and fish; Briggs 1974), fall the Lusitania province, the Canary province, and the Mediterranean province.

2.4.1 Western Ireland–Northern France to Senegal (Lusitania Province)

Literature: France: see Section 2.3. **Northern and northwestern Spain:** Donze (1968); Fernández and Niell (1982); Fischer-Piette (1963); John (1971); Niell (1978: **checklist**); Pérez-Cirera (1975); Weber-Peukert and Schnetter (1982). **Portugal:** André (1970–1971). **South Spain:** Fischer-Piette (1959); Seoane-Camba (1965). **Africa:** Schmidt (1957); Seoane-Camba (1969). **Morocco:** Dangeard (1949); Feldmann (1951, 1955); Gayral (1958). **Western Sahara and Mauritania:** Lawson and John (1977). **Senegal:** Sourie (1954). **Temperature requirements of warm temperate Eurafrican seaweeds:** Yarish et al. (1984, 1986, 1987).

The distributions of several cold temperate species that dominate the upper sublittoral zone, such as the kelps *Alaria esculenta* (Fig. 2.12) and *Laminaria digitata* (Fig. 2.20), end on the coasts of **West Ireland** and **Brittany,** at the southern limit of the cold temperate North Atlantic region in Europe (Figs. 1.5 and 1.6). Along the coastline that stretches to Cape Verde (Senegal), termed the "Lusitania province of the warm temperate Mediterranean–Atlantic region," one still finds many of the cold temperate seaweeds. For example, *Laminaria hyperborea* (Fig. 2.35) and *L. saccharina* (Fig. 2.11) are distributed to mid- and northern Portugal, respectively. The term "Lusitania province" was used by marine zoogeographers of the last century, such as Dana (1853, see Hazel 1970) and goes back to the Roman Empire. "Lusitania" was the name of a Roman province that covered roughly modern Portugal. "Mauretania" was another Roman province including areas now in northeastern Morocco and western Algeria.

Warm temperate, Lusitanic algae: ancestral affinities. Diversity in the cold temperate kelps is reduced southward. These kelps are replaced by others with warm temperate affinities: the ancestral kelp *Saccorhiza polyschides* (Figs. 1.9 and 2.36; synopsis by Norton and Burrows 1969), mainly in the upper sublittoral zone, and *Laminaria ochroleuca*, mainly in the midsublittoral zone. In the northern part of the Lusitania province, on the coast of Brittany, both species form a mixed vegetation with *L. hyperborea* (Ernst 1955), and both also enter the Mediterranean (Figs. 2.73 and 2.74B,D). *L. ochroleuca*, which occurs as far north as to the English Channel, west to the Azores and south to Morocco, thrives down to a depth of 25–30 m. It has a rigid stipe, is long-lived like *L. hyperborea*, and forms a dense laminarian forest. (Details about *L. ochroleuca* can be found in John 1971; Braud 1974; Sheppard et al. 1978.) On the other hand, *L. ochroleuca* is similar to *L. digitata* in having a smooth stipe and spores in summer. Thus *L. ochroleuca* may be similar to a warm temperate ancestor of both *L. digitata* and *L. hyperborea*.

Warm temperate algae crossing the equator. *Laminaria pallida* growing along the coast of southern Africa (Figs. 5.7 and 5.8) may have descended from the similar *L. ochroleuca*. Probably *L. ochroleuca* crossed the equator during one of the glaciation periods (van den Hoek 1982*b*). An opposite transequatorial migration may have brought the kelp *Ecklonia muratii* to the Northern Hemisphere. This rare deep-water species occurs on the coasts of Mauritania and Senegal (Feldmann 1937*c*). *E. maxima* is the nearest related species on the coast of southern Africa.

Other warm temperate Lusitanic algae. Examples of other southern species with distributions reaching the northern limit of the Lusitania province are the green alga *Codium bursa* (Fig. 2.69J), the brown algae *Bifurcaria bifurcata* (Fig. 2.40), *Cystoseira* species (Fig. 2.60B), *Padina pavonica* (Fig. 2.69D), and also the species illustrated in Fig. 2.65, with the exception of *Fucus virsoides*. These are the Atlantic floral element of the Mediterranean province (Fig. 2.65). Some of these also reach the cold temperate region. The long-term changes of the distribution of *P. pavonica* have been described in detail by Price et al. (1979). Figure 2.59 shows several typical smaller algae occurring in the mid- and lower sublittoral zone near Brittany, where the lower limit of the *Laminaria hyperborea* forest is situated at a 15–25 m depth and the lower algal vegetation limit at about a 45 m depth (Ernst 1955; Castric-Fey et al. 1973). Here one finds well-known Mediterranean algae, such as the brown algae *Dictyota dichotoma* (Fig. 2.69C), *Dictyopteris membranacea* (Fig. 2.59C), *Halopteris filicina*, *Carpomitra costata*, and the red alga *Schottera nicaeensis* (Fig. 2.70D). These algae flourish at the depth where the dense laminarian forest ends.

Most of the dominating algal species of the European coasts exhibit a distribution gap along the **French Atlantic coast** south of the estuaries of the Loire or the Gironde. The algae are found again on the coast of northern Spain. This may be seen from the distribution maps of, for example, *Laminaria hyperborea* (Fig. 2.35), *L. saccharina* (Fig. 2.11), *Fucus serratus* (Fig. 2.32), *Asco-*

Fig. 2.59 Typical smaller algae and animals from the lower limit of the *Laminaria hyperborea* forest at the Bay of Morlaix (Brittany). Also the haptera and stipes of *L. hyperborea* carry a rich epiphytic fauna and flora, which is not represented. B = brown alga; R = red alga; AN = anthozoan; BR = bryozoan. 1 = *Drachiella spectabilis* (R); 2 = *Phycodrys rubens* (R); 3 = *Heterosiphonia plumosa* (R); 4 = *Kallymenia reniformis* (R); 5 = *Cryptopleura ramosa* (R); 6 = *Acrosorium uncinatum* (R); 7 = *Bonnemaisonia asparagoides* (R); 8 = *Dictyota dichotoma* (B); 9 = *Halopteris filicina* (B); 10 = *Callophyllis laciniata* (R); 11 = *Meredithia* (= *Kallymenia*) *microphylla* (R); 12 = *Nitophyllum bonnemaisoni* (R); 13 = *Rhodophyllis divaricata* (R); 14 = *Delesseria sanguinea* (R); 15 = *Pterosiphonia parasitica* (R); 16 = *Polyneura gmelinii* (R); 17 = *Spondylothamnion multifidum* (R); 18 = *Dictyopteris membranacea* (B); 19 = *Sphaerococcus coronopifolius* (R); 20 = *Alcyonium glomeratum* (AN); 21 = *Eschara foliacea* (BR); 22 = *Eunicella verrucosa* (AN). (from L'Hardy-Halos 1972.)

phyllum nodosum (Fig. 2.22), *Himanthalia elongata* (Fig. 2.39), and *Chondrus crispus* (Fig. 2.33). The distribution gap is caused by the lack of rocky substratum from the estuary of the Loire to the **Basque** coast (Crisp and Fischer-Piette 1959), where locally higher water temperatures threaten the cold temperate species. Near Biarritz the surface water temperature rises to 22°C in August, and this summer temperature occurs again only on the open coast of Morocco and in the Mediterranean. Therefore the upper sublittoral

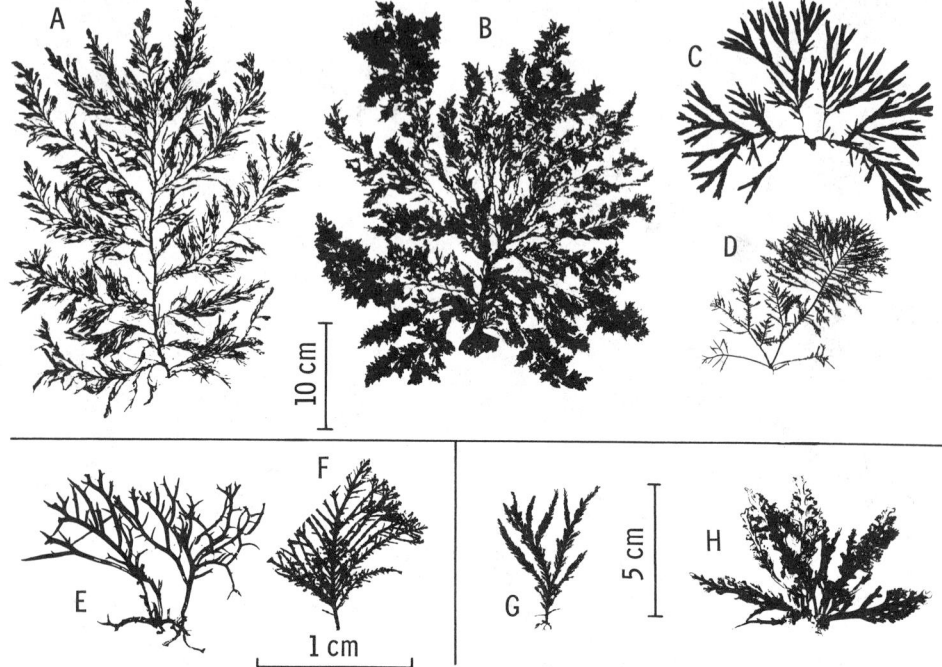

Fig. 2.60 Examples of dominating seaweed species along the coastline from southern Portugal to Morocco. **Brown algae:** (**A**) *Cystoseira baccata* (= *fibrosa;* Fucales; Ireland to Morocco). (**B**) *C. tamariscifolia* (= *ericoides;* Scotland and Ireland to Mauritania). (**C**) *Dictyopteris membranacea* (Dictyotales; Ireland to Morocco, Mediterranean, western Atlantic tropical region). **Red algae:** (**D**) *Gelidium sesquipedale* (Nemaliales; southern England to Mauritania). (**E**) *Caulacanthus ustulatus* (Gigartinales; Biarritz to Senegal). (**F**) *Corallina officinalis* (Cryptonemiales; Iceland to Mauritania, Newfoundland to South Carolina). (**G**) *Pterosiphonia complanata* (Ceramiales, Rhodomelaceae; western Ireland to Mauritania). (**H**) *Laurencia pinnatifida* (Ceramiales, Rhodomelaceae; Norway to Mauritania). (A–E,G,H from Gayral 1958; F from Kornmann and Sahling 1977.)

vegetation on the Basque coast is not dominated by the cold temperate Laminariales but by typical warm water species such as *Cystoseira tamariscifolia* (Fig. 2.60B) and the red alga *Gelidium sesquipedale* (Fig. 2.60D) and *Corallina officinalis* (Fig. 2.60F; van den Hoek and Donze 1966). In the upper part of the euphotic zone there are still species well known from the cold temperate region. In the supralittoral zone (littoral fringe) the marine lichen *Verrucaria maura* is found. In the upper eulittoral zone the algae *Blidingia*, *Enteromorpha*, and *Ralfsia verrucosa* are accompanied by the barnacle *Chthamalus stellatus*. Somewhat deeper the belts of the Fucaceae are missing, except for some scattered individuals of *Fucus spiralis*. The crustose, calcified red alga *Lithophyllum lichenoides* (= *tortuosum;* Fig. 2.68), which forms the "trottoirs" in the western Mediterranean, reaches the Basque coast as its northern geographical limit (Crisp and Fischer-Piette 1959). However, not too many warm temperate species appear on the Basque coast to compensate for the loss of cold temperate species. Thus this coastline generally exhibits a reduced algal flora (van den Hoek and Donze 1967).

Along the coast of **northern Spain,** from Santander or Gijon southwards, one again finds many well-known species from the cold temperate region last seen in northwestern France, such as *Laminaria hyperborea*. Due to temperature fluctuations over decades, the distribution limits of many algae correspondingly have fluctuated considerably along the northern Spanish coast. For example, species at Cape Penas such as *Himanthalia elongata, Fucus serratus, Laminaria hyperborea,* and *L. saccharina,* which were abundant in the 1930s, have now disappeared and been replaced by more southern species (Fernández and Niell 1982). The coast of **northwest Spain** with its characteristic drowned river valleys (in Spanish called Rias) is an upwelling area, where cold water flows to the surface and the temperature does not surpass 18–19°C in August. In this respect it might be more correct to say that the continuous distribution of many cold temperate algal species ends on the coast of Brittany and that they reappear in isolated patches within the Lusitania province. These patches are upwelling areas with somewhat colder water found from northwestern Spain to northern Portugal. Thus the northern species *Himanthalia elongata* and *Gigartina stellata* are still found in exposed locations on granite at the mouth of the Ria de Arosa (Donze 1968) in the lower eulittoral zone. At sheltered locations *Pelvetia canaliculata, Fucus spiralis, F. vesiculosus, F. serratus,* and *Chondrus crispus* are found in the lower eulittoral zone, while in the sublittoral zone *Laminaria hyperborea*, and in sheltered places *L. saccharina* flourish. The distributions of the abovementioned species (with the exception of *F. spiralis* and *F. vesiculosus*), as well as 40 other cold temperate algal species that extend to the Iberian peninsula, end just south of the 20°C-August isotherm on the coast of southern **Portugal** (Fig. 1.5). The minimum water temperature in February on the coast of southern Portugal does not drop below 14–15°C. This is important to the gametophytes of *Laminaria hyperborea* and *L. saccharina,* which require lower temperatures for optimal maturation (Lüning 1980*a*). Conversely, about 20 southern species, including the tropical green alga *Valonia utricularis* (Fig. 2.64C), reach their northern limits on the Portuguese coast. Fluctuations of the algal distribution limits in the course of years have been thoroughly investigated by Ardré (1970–1971).

Many Lusitanic species are mixed with the last algal representatives of the cold temperate region in the vegetation of the northwest Spanish rias. This vegetation, on sandy bottoms replaced by maerl, prevails southward, from **southern Portugal** and **southern Spain** to **Morocco.** Such Lusitanic species are *Lithophyllum lichenoides* (= *tortuosum*) with its thick (up to 10 cm), hemispherical thalli, and *Bifurcaria bifurcata* in the wave-exposed lower eulittoral zone. In the sublittoral zone *Saccorhiza polyschides, Laminaria* *ochroleuca,* and also the species illustrated in Fig. 2.60 are typical of the Lusitania province.

Nevertheless, on the coast of **Morocco** the vegetation belts of *Fucus spiralis* in the mideulittoral zone and the belts of marine lichens *Lichina pygmaea* and *Verrucaria maura* in the upper eulittoral zone are the remainders of the cold temperate flora. *Fucus vesiculosus,* whose southern limit is Morocco, is rarely found, being restricted to estuaries. There are **upwellings of cold water** along the coast of Morocco (Barber and Smith 1981), shifting the isotherms southward along the coast. This may be the reason some Lusitanic species (e.g., *Bifurcaria bifurcata* and *Saccorhiza polyschides*) still occur along the Moroccan coast but are absent at the same latitude on the warmer Canary Islands. A plateaulike substratum in the lower eulittoral zone is typical of the Moroccan coast. The coast of Morocco has many species in common with the Mediterranean, and in general the Moroccan seaweeds exhibit an unsual mixture of temperate and warm-adapted species (Dangeard 1949). On the other hand, a number of Lusitanic species, which migrated via the Strait of Gibraltar to the southern Spanish coast of the Mediterranean (Sea of Alborán) and the coast of Algeria, introduced a certain Atlantic influence along these coasts.

The Lusitanic members of the Laminariales, *Saccorhiza polyschides* and *Laminaria ochroleuca,* which form dense beds along the northern Moroccan coast, reach their southern limit at the 20°C-August isotherm at the boundary of southern Morocco and **Western Sahara** (formerly Spanish Sahara). Two other warm temperate members of the Laminariales, the deep-water species *Phyllariopsis brevipes* (= *Phyllaria reniformis;* Fig. 2.74A) and *Phyllariopsis* (= *Phyllaria*) *purpurascens,* also their range here. Many other warm temperate species penetrate farther south, to Cape Blanc in Western Sahara. It is surprising to find here a representative of the genus *Ecklonia,* namely *E. muratii.* Near Cape Blanc is also the northern limit of many tropical species.

South from Cape Blanc the rocky substratum passes into the sandy beaches of **Mauritania.** The last temperate species known from the Moroccan coast occur south at Cape Verde on the coast of **Senegal,** where the temperature in February is not lower than 17°C and reaches 23°C in August. Cape Verde is regarded as the boundary between the warm temperate Mediterranean–Atlantic region and the eastern Atlantic tropical region.

2.4.2 Canary Province

Literature: Feldmann (1946). **Azores:** Schmidt (1931); Prud'homme van Reine (1988). **Madeira:** Audiffred and Prud'homme van Reine (1985); Augier (1985); Levring (1974). **Salvage Islands:** Weisscher (e.g., 1983). **Canary Islands:** Afonso-Carrillo and Gil-Rodríguez (1982*a,b*); Afonso-Carrillo et al. (1984*a,b*); Börgesen (1925–1936); Gil-Rodríguez et al. (1985); Lawson and Norton (1971); Viera-Rodríguez et al. (1987). **Checklist, Canary Islands:** Gil-Rodríguez and Afonso-Carrillo (1980). **Cape**

Verde Islands: Prud'homme van Reine and Lobin (1986); Prud'homme van Reine and van den Hoek (1988).

The island groups of the Azores, Madeira, Salvage, Canary and Cape Verde Islands are collectively known as the **Macaronesian** Islands. With the exception of the Cape Verde Islands, which are situated at the boundary to the Eastern Atlantic tropical region, this group forms the **Canary province** of the warm temperate Mediterranean–Atlantic region.

Eighty percent of the algal flora of the **Canary Islands,** which are near the mainland, are common to those of the Lusitanic and Mediterranean provinces (Feldmann 1946). Several characteristic Lusitanic species that occur along the opposite Moroccan coast are absent (e.g., *Bifurcaria bifurcata* and *Saccorhiza polyschides*) or rare (e.g., *Laminaria ochroleuca,* found only in the Azores). August water temperatures rise to 23°C and do not fall below 18°C in February. Thus the tropical floral element is well represented with green algae such as *Caulerpa* and *Valonia,* with the red algae *Liagora* and *Galaxaura,* and with the brown algae *Dictyota, Padina,* and *Zonaria.* According to a zonation example given by Lawson and Norton (1971), in the upper eulittoral zone the periwinkle *Littorina neritioides* and the barnacle *Chthamalus stellatus* prevail. In the mideulittoral zone *Fucus spiralis,* the red algae *Caulacanthus ustulatus* (Fig. 2.59E) and *Centroceras clavulatum* (Fig. 4.6) and the green alga *Ulva rigida* are important. In the lower eulittoral, as well as in the upper sublittoral zone, one finds the red algae *Gelidium arbuscula* and *G. canariensis* as endemic species (Prud'homme van Reine 1988; = *G. cartilagineum* var. *canariensis* in Feldmann 1946) and the brown algae *Cystoseira abies-marina, Stypocaulon scoparium* (Fig. 2.69E), and *Padina pavonica* (Fig. 2.69D).

The algal floras of the more northerly island group of **Madeira** and particularly the northwest **Azores** have fewer tropical algae and more Lusitanic and Mediterranean species (Schmidt 1931). Of the 189 species of seaweeds of the Azores, 10 have been described as endemic, and of the warm temperate members not a single species occurs on American coasts (Prud'homme van Reine 1988). Tropical algae predominate in the algal flora of the southerly **Cape Verde Islands,** with a distinct eastern American, largely Caribbean imprint (Prud'homme van Reine and van den Hoek 1988). Scattered corals that may have migrated to the islands in the Quaternary from the Caribbean occur here.

Paleobiogeography. The Macaronesian Islands, which are of volcanic origin and have ages of 1–2 million years (Azores), and 16–36 million years (Canary Islands; Mitchell-Thomé 1976), are not continental islands but uplifted islands that have undergone primary succession (for island theory see Section 4.8). In view of the somewhat isolated position of the Macaronesian Islands, one would expect to find there a certain richness in endemic species. However, as distinct from higher plants (Humphries 1979), this does not occur for seaweed or fish species. Briggs (1974) suggests that earlier marine endemic species were eradicated due to drastic reduction in water temperatures during the glaciation periods. Nevertheless, the relative age of the Macaronesian Islands is reflected in the number of endemic seaweeds found on each island group. Madeira has 1.4% endemism in seaweeds, the Canary Islands (413 species) 2.8%, the Azores 3.2%, and the Cape Verde Islands (220 species) 4% (Prud'homme van Reine and van den Hoek 1988; Prud'homme van Reine, personal communication). The terrestrial vegetation of the Canary Islands, with its 1800 species of flowering plants, exhibits a much higher endemism (25%; Humphries 1979). The difference between endemism in the marine and terrestrial vegetation reflects the roles of the sea as a migration barrier supporting speciation for the terrestrial plants and as a potential migration route for marine algae attached to some floating material (van den Hoek 1987).

2.4.3 Seaweed Migrations and Long-distance Dispersal

During the first half of this century, seawater temperatures increased by about 1°C in the Northern Hemisphere, with a warm period in the interval 1920–1950 and a maximum in 1940 (Fig. 2.61). Several warm temperate species migrated northwards along the Norwegian coast, such as *Dictyota dichotoma, Desmarestia ligulata,* and *Gracilaria verrucosa* (Printz 1953). In 1940 *Laminaria ochroleuca,* which was confined to the northern coast of France, crossed the English Channel and settled on the coast of southwestern England. The Mediterranean brown algae *Taonia atomaria* (Fig. 2.65C) and *Dictyopteris membranacea* (Fig. 2.59C) also spread to the British coasts

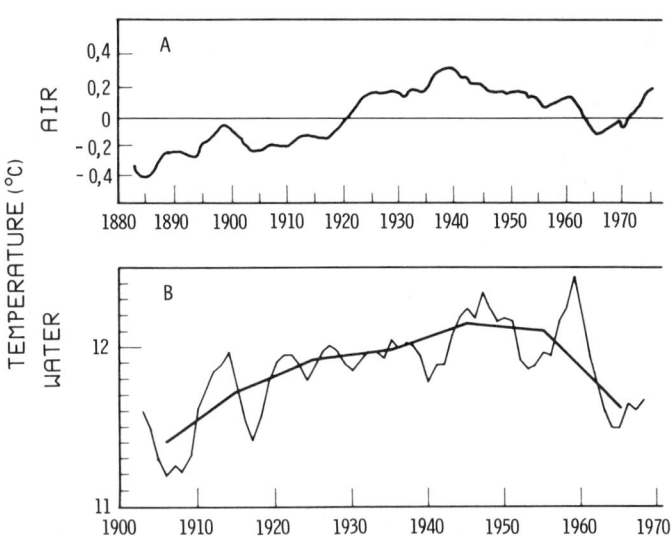

Fig. 2.61 (**A**) Changes of the mean air temperature in the Northern Hemisphere, 17.5–87.5°N. (**B**) Changes in surface seawater temperatures at Plymouth, England. The bolder, straighter line indicates a trend line drawn through the decade means. (A from Budyko and Vinnikov 1977; B from Southward and Butler 1972.)

(W. E. Jones 1974). Species such as *Dictyota dichotoma* (Fig. 2.69C) and *Laurencia pinnatifida* (Fig. 2.59H), which had flourished on Helgoland from the turn of the century (Kuckuck 1894, 1897) to the 1930s (Nienburg 1925), have now disappeared. Possibly, with decreasing water temperatures since 1950 (Fig. 2.61B), an opposite migration of northern species has set in. One of the first examples may be *Fucus evanescens* (= *F. distichus* subsp. *edentatus*), which moved southwards in the 1940s along the Scandinavian coasts to Copenhagen and Malmö (Lund 1949). This subspecies migrated from the Shetland Islands to the Scottish coast (H. T. Powell 1981). *Fucus* species and *Pelvetia canaliculata* progressed as "northern elements" southward along the southwestern coast of France from 1955 to 1970 (Fischer-Piette and Lahondère 1973).

It is not necessary to assume that striking changes in the distributions of algal species are always due to true migrations. **Cryptic microthalli,** which may be more tolerant than macrothalli, may survive for decades at a given locality, producing macrothalli as soon as temperatures surpass a critical value. This view is substantiated by Kornmann and Sahling (1980), who were able to culture the macrothalli of the warm temperate red algae *Helminthocladia calvadosii* and *Scinaia forcellata* from Conchocelis-like stages still living in shells in the sublittoral zone of Helgoland. The macrothallic form of the species had "disappeared" from this island for more than 50 years.

In general, algal spores, gametes, or zygotes drifting in the plankton seem to be only suited for short-distance transport, normally measured in meters. (For reviews see A. R. O. Chapman, 1986; van den Hoek 1987.) Evidence for this exists for species of the Laminariales, namely *Macrocystis pyrifera* (E. K. Anderson and North 1966), and *Postelsia palmaeformis* (Dayton 1973). Most red algal spores seem to be adapted for rapid sinking. There is evidence that the propagules of a few typical early colonizers (*Enteromorpha, Blidingia,* and *Ulothrix*) seem to stay long enough in the planktonic phase to reach distances of a few tens of kilometers. Amsler and Searles (1980) showed that swarmers of a coastal population of a species of *Enteromorpha* reached exposed artificial substrata on a submarine plateau 35 km away. Cultures of filtered seawater, partly from offshore areas, revealed the predominance of *Enteromorpha, Blidingia,* and *Ulothrix* (Hruby and Norton 1979; Zechman and Mathieson 1985).

In various groups of marine warm-water invertebrates such as gastropods, polychaetes, and crustaceans, there are "long-distance" larvae, which have a long planktonic larval phase and special morphological adaptations (Scheltema 1971). These "teleplanic larvae" seem to maintain reciprocal exchange and thus gene flow between, for example, the coastal biota of the eastern and western Atlantic. They also often belong to species with long-term geological persistence, still existing today from mid-Tertiary times, because widespread species are more immune to local extinction (T. A. Hansen 1978; Vermeij 1978). The hypothesis that there may be teleplanic seaweed propagules in the plankton, which consequently would be important for long-distance dispersal, has hardly been investigated (see review by Hoffmann 1987); neither has it been experimentally proved nor disproved, as emphasized by van den Hoek (1987). He also pointed to the clear indirect evidence for the existence of long-distance dispersal of seaweeds, because midoceanic islands like Tristan da Cunha established a seaweed flora of 125 species in less than 1 million years, the approximate geological age of this island group. Other striking examples are to be seen in the repopulation of the vast coastal area from the North Sea to northern Greenland, which was freed from permanent ice cover in the course of the last 18,000 years, after the maximum coverage of the last major glaciation. The Faeroes, heavily covered with ice until 10,000 years ago, may have received about 180 cold temperate seaweed species from the donor areas of Scotland and Norway, while a further 40 Arctic–cold temperate species may either have survived in place on portions of rocky coast, or had to reimmigrate from the south (van den Hoek 1987). It should be noted that there is evidence for a faster repopulation of extremely isolated offshore shoals by marine animals than by seaweed species; the example is the Virgin Rocks on the Newfoundland Grand Banks, 200 km east of Newfoundland, the nearest landmass (Hooper 1987).

An example of a recent repopulation event is the case of the volcanic island of Surtsey, which evolved in 1964, 4 km from the nearest rocks and 33 km from the mainland of Iceland. After five years the coastline of this island was colonized by 25 seaweed species, following the presence after two years of *Urospora, Ulothrix, Enteromorpha, Porphyra umbilicalis, Petalonia fascia,* and *Scytosiphon lomentaria* and after three years of *Alaria esculenta, Laminaria hyperborea,* and also various red algae, such as *Polysiphonia urceolata* or *Phycodrys rubens,* with their "immotile spores" (Jónsson 1970; Jónsson and Gunnarsson 1982).

Fouling communities on North Sea platforms provide another example of the possible capability of long-distance dispersal by seaweeds. The sequence starts with rapidly colonizing *Enteromorpha* species and ends with kelp-dominated communities, first with *L. saccharina* and then mainly with *Alaria esculenta, L. digitata,* or *L. hyperborea,* (Moss et al. 1981; Edyvean et al. 1985; Terry and Picken 1986).

Floating algae with gas-filled thalli and floating substrata such as pumice (volcanic rock) and wood are obvious mechanisms for long-range dispersal of seaweeds (van den Hoek 1987) and marine invertebrates (Vermeij 1978). Although long-range dispersal exists in seaweeds, climatic and geographical barriers seem to be more important for seaweeds than for land plants in restricting most species to particular areas. A few species, however, have "escaped" from their natural restrictions due to human activity.

2.4.4 Introduced Seaweeds

The dispersal of algae as a consequence of human activity is quite distinct from natural distribution changes brought about by long-term temperature fluctuations. Faster ships can carry alien animals and algae either attached or in ballast water (Carlton 1985). Flying boats may have provided another possible vector (Maggs and Guiry 1987a). Even more dangerous in contaminating natural floral and faunal compositions has been the shipment of marine products such as lobsters and the exchange of living material by aquaculturists, for example, the introduction of oysters wrapped in seaweeds and subsequently thrown into the sea.

Classic examples of introduced marine animals are represented by the snail *Crepidula fornicata* brought from America to Europe at the end of the last century, the barnacle *Elminius modestus* introduced to Europe from Australia after 1940, and the crab *Eriocheir sinensis* brought from China to Europe after 1912. Among seaweeds, the sudden appearance and spread of *Bonnemaisonia hamifera, Asparagopsis armata, Colpomenia peregrina, Codium fragile* and *Sargassum muticum* has been relatively well documented (see below; for reviews see W. E. Jones 1974; Farnham 1980). Recently discovered and possibly introduced red algae are *Lomentaria hakodatensis* on the coast of Brittany (Cabioch and Magne 1986); *Porphyra yezoensis* near Helgoland, North Sea (Kornmann 1986); and the North Pacific *Pikea californica* in the surge zone of the Isles of Scilly (Maggs and Guiry 1987a). The eastern Pacific brown alga *Undaria pinnatifida* was found in the northern Mediterranean and transferred to the coast of Brittany for seaweed cultivation (Pérez et al. 1984). *U. pinnatifida* is now known from both New Zealand (Hay and Luckens 1987) and Tasmania (Sandersson 1988). *Laminaria japonica* was brought to China for cultivation purposes in the 1940s (Section 2.7).

The red alga **Bonnemaisonia.** The red alga *Bonnemaisonia hamifera* probably arrived in Europe **from Japan** at the turn of the century (Figs. 2.62A,B; see Fig. 1.12 for distribution in the North Atlantic). Its heteromorphic tetrasporophyte, the Trailliella phase, was discovered in 1890 near Falmouth (southern England). In 1893 the *Bonnemaisonia* phase appeared. Around 1900 the Trailliella phase occurred on Helgoland (Koch 1951) and on the Danish coast, in 1916 near Bergen on the Norwegian coast, and in 1929 on the Orkney Islands. To the south the species spread to the Canary Islands and to isolated locations in the western Mediterranean (Feldmann and Feldmann 1942). On the eastern coast of North America, where the Trailliella phase turned up in 1927, it is distributed from Connecticut to Newfoundland (McLachlan et al. 1969; Schneider et al. 1979). On the Pacific coast of North America the species occurs in California (Abbott and Hollenberg 1976).

Asparagopsis armata (Figs. 2.62C,D). This red alga was probably carried by ships **from Australia** to the Northern Hemisphere. The species was first sighted in 1923 on the coast of Algeria, in 1925 at Biarritz, in 1926 in the northwestern part of the Mediterrananean (Feldmann and Feldmann 1942), in 1939 in Ireland, in 1959 in Scotland, and in 1973 on the Orkney and Shetland islands (Irvine et al. 1975).

The brown alga **Colpomenia.** The cushionlike brown alga *Colpomenia peregrina* (Fig. 2.62F), which today occurs in temperate latitudes of the Northern and Southern

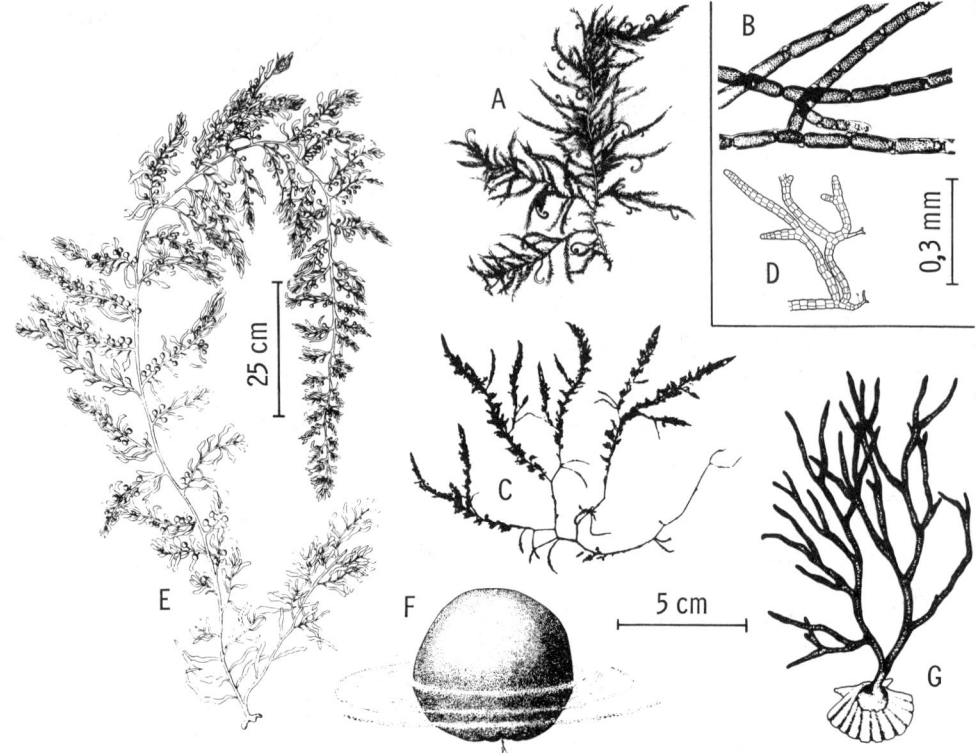

Fig. 2.62 Examples of algae that were introduced into the North Atlantic. **Red algae:** **(A)** *Bonnemaisonia hamifera* (gametophyte). **(B)** *Bonnemaisonia hamifera* (tetrasporophyte = Trailliella phase). **(C)** *Asparagopsis armata* (gametophyte). **(D)** *Asparagopsis armata* (tetrasporophyte = Falkenbergia phase). **Brown algae:** **(E)** *Sargassum muticum*. **(F)** *Colpomenia peregrina*, detached from the substrate and floating at the sea surface. **Green alga:** **(G)** *Codium fragile*. (A from W. R. Taylor 1957; B from Kornmann and Sahling 1977; C from Dixon and Irvine 1977; D from Feldmann and Feldmann 1942; E from Lund 1942; F from Dawes 1981; G from T. F. Lee 1977.)

Hemispheres, appeared in Europe for the first time on the French Atlantic coast, where it caused losses to the oyster cultures ("oyster thief"). The gas-inflated thalli fixed on the oysters, eventually floating them away. This illustrates how well this species is adapted for migration. The species, probably introduced **from Japan** by exchange of oyster cultures, subsequently spread along the British coasts and reached Denmark and southern Norway by the middle of this century (Lund 1942; Rueness 1977).

***The green alga* Codium.** The weedy green alga *Codium fragile* (Fig. 2.62G) was probably also introduced **from the Pacific** (Japan). It occurs today in both hemispheres. The subspecies *atlanticum* spread first to the European coasts from Ireland. Since 1900 the subspecies *tomentosoides* spread to the European coasts from the Netherlands and invaded the western Mediterranean from the 1940s onwards (Silva 1955; Meslin 1964). A subspecies *scandinavicum* occurred from 1919 onwards on the Danish coast and probably originated from the western Pacific Asiatic coasts (Silva 1957). In 1957 *C. fragile* subsp. *tomentosoides* was first detected in Long Island Sound, New York. It was probably introduced from western Europe around 1956 and spread north to Cape Cod by 1961 and south to New Jersey by 1966. Subsequently it moved to disjunct areas in the north (New Hampshire and Maine) and in the south (Virginia and North Carolina). Carlton and Scanlon (1985) suggested that *Codium* performed its worldwide voyage (1973, New Zealand; 1977, San Francisco Bay) perhaps as a fouling organism on ships' hulls. Whereas sexuality exists in Pacific populations of *C. fragile*, its subspecies *tomentosoides* is reported to reproduce exclusively by parthenogenetic female gametes in the North Atlantic. Although populations from North Carolina were found to be haploid, a parallel to the well-known polyploid character of weedy species might be seen in the enlarged size of the chromosomes and the increased DNA content of *C. fragile* subsp. *tomentosoides* compared to autochthonous species (Kapraun and Martin 1987). Species of *Codium* that are natural components of the warm temperate North Atlantic flora include *C. decorticatum* and *C. carolinianum* in the waters of North Carolina (Searles et al. 1984) and *C. tomentosum* and *C. vermilara* on European coasts.

The brown alga Sargassum muticum. During the 1940s this conspicuous brown alga (Fig. 2.62E) was accidentally introduced **from Japan** to the Pacific coast of North America, probably by oyster aquaculturists. This species, which is a nuisance in harbors and on beaches, spread quickly from British Columbia to California using their air bladders. Baja California was reached in the 1970s, so that a coastline of 3000 km was invaded within 30 years. During the spring of 1972, Japanese oysters from British Columbia were introduced to France, and Druehl (1973) predicted the establishment of *Sargassum muticum* in the eastern Atlantic. During the month following publication of Druehl's prediction, *Sargassum muticum* was discovered growing attached to solid substratum on the coast of southern England and soon after in northern France (Farnham et al. 1973; Critchley et al. 1983). *S. muticum* has been a permanent member of the marine flora of Roscoff, Brittany, and the Netherlands since 1980, with drift material cast upon Dutch shores since 1977 (Cabioch 1981; Critchley et al. 1987). Attached populations have been recorded in the Limfjord, northern Denmark, since 1984 (Christensen 1984). Drift specimens were recorded from 1981 onwards in the eastern Frisian Islands (i.e. Borkum and Norderney; Kremer et al. 1983), in 1984 for the Norwegian Skagerrak coast (Rueness 1985) and attached specimens in southern Norway in 1988 (Rueness 1989). With the appearance of this species along the Channel coasts, the distribution of the genus *Sargassum* has moved 1200 km north along the European coasts. The nearest naturally occurring species of the genus are *S. flavifolium* in southern Portugal and *S. vulgare* in Morocco. *S. muticum* may outcompete the local fucalean species for several reasons. The species grows fast, becomes fertile in its first year of life, and, since it is monoecious, a single plant may produce many zygotes as a result of self-fertilization. The zygotes are not suitable for long-distance transport since they sink; they have been found only up to 1.3 km from the parent plant (Deysher and Norton 1982).

2.4.5 Mediterranean Province: Introduction

Literature: Entire area: Cinelli (1985); Coppejans (1983); Feldmann (1958); Boudouresque (1971); Furnari (1984); Giaccone (1973, 1974); Ketchum (1983); Meinesz et al. (1983); Moraitou-Apostolopoulou and Kiortsis (1985); Pérès (1967a,b, 1982a,b); Pérès and Picard (1964). **Spain and Balearic Islands:** Ballesteros and Martinengo (1982); Bas (1949); Gili and Ros (1984); Gomez Garreta et al. (1982); Navarro and Bellon Uriarte (1945); J. J. Rodriguez (1889); Seoane-Camba (1969, 1975). **South France:** Belsher et al. (1976); Boudouresque (1971); Boudouresque and Cinelli (1976); Feldmann (1937a, b, 1939–1942); Ollivier (1929); Verlaque (1981). **Corsica:** Boudouresque and Perret (1977); Coppejans and Boudouresque (e.g., 1983); Molinier (1960). **Italy:** Furnari (1984); Giaccone (1969b). **Northwestern Italy:** Cinelli (1969). **Gulf of Naples:** Berthold (1882); Boudouresque and Cinelli (1971); Falkenberg (1878); Funk (1927, 1951, 1955, 1957). **Strait of Messina:** Giaccone (1972); Giaccone and Rizzi Longo (1976). **Sicily and Straits of Sicily:** Cinelli (1979, 1981); Furnari (1984); Giaccone and Sortino (1974); Giaccone et al. (1985). **Checklist, eastern Sicily:** Cormaci and Furnari (1979). **Adriatic Sea:** Ercegovic (e.g., 1948, 1957a,b, 1959, 1960); Ernst (1959); Giaccone (1978); Lorenz (1863); Munda (e.g., 1973, 1979, 1982); Pignatti (1962); Riedl (1966, 1983); Serman et al. (1981); Span (1980). **Greece and Aegean Islands:** Athasaniadis (1987); Bianchi and Morri (1983); Coppejans (1975); Diannelidis et al. (1977); Diapoulis and Haritonidis (1984); Gerloff and Geissler (1974); Haritonidis and Tsekos (1975, 1976); Giaccone (1968a,b); Moustakas (1981); Nizamuddin and Lehnberg (1970); Pérès and Picard (1958); Tsekos and Haritonidis (1977). **Black, Caspian, and Aral seas:** Ketchum (1983); K. M. Petrov (1967); A. D. Zinova (1967). **Turkey:** Güven and Ötzig (1971); Mayhoub (1976b). **Israel:** Edelstein (e.g., 1964); Lipkin and Safriel (1971). **Egypt:** Aleem (e.g., 1951), Nasr (e.g., 1940). **Suez Canal:** Lipkin (1972); Por (1978). **Libya:** Nizamuddin et al. (1978). **Tunisia:** Meñez and Mathieson (1981). **Algeria:** Boudouresque (1969); Feldmann (1931, 1937–1947, 1943). **Some taxonomic groups: Caulerpales:** Meinesz (1979, 1980). **Cystoseira:** Ercegovic (1952); Giaccone (1971); Giaccone and Bruni (1971, 1973); Huvé (1972). **Laminariales:** Drew (1972); Feldmann (1934); Huvé (1955, 1958); Giaccone (1969a, 1972); Molinier (1960). **Peyssonnelia:** Boudouresque and Denizot (e.g., 1975). **Sphacelariaceae:** Prud'homme van Reine (1982). **Paleoceanography:** Audley-Charles and Hallam (1988); Gealey (1988); Livermore and Smith (1985); Nairn et al. (1977, 1978); Rehault et al. (1985); Rögl and Steininger (1984); Sonnenfeld (1985); Stanley and Wezel (1985); Vanney and Gennesseaux (1985).

Shallow water areas in the Strait of Sicily (between Tunisia and Sicily), and the Strait of Messina (between Sicily and southern Italy) separate the western and eastern basins of the Mediterranean. The Mediterranean is almost tideless because of its narrow connection to the Atlantic. The gigantic water masses that would have to be exchanged between the Atlantic and the Mediterranean throughout each tidal rhythm cannot pass the Strait of Gibraltar, which is only 350 m deep. Instead there is a periodic "decanting" to and fro of water from the eastern and western basins so that strong currents exist in the Straits of Sicily and Messina. As deep, cold water comes to the surface in these areas, there are several seaweed species of the Atlantic floral element near the surface. The tidal range in the Mediterranean amounts to 20–40 cm. At two areas, namely in the North Adriatic Sea and in the Gulf of Gabes (Tunisia), tidal ranges are 1.5 and 2.2 m respectively. The changes in water level due to the tides cause less desiccation stress on the uppermost marine organisms than the aperiodic low-water levels caused by high atmospheric pressure changes (Baltic).

Currents and salinities. The surface current entering the Mediterranean from the Atlantic flows along the North African coast, with a branch north of Sicily turning westwards along the coasts of western Italy, southern France, and Spain. In the eastern Mediterranean the main current runs counterclockwise. Salinity increases from 36‰ to 39‰ from west to east due to high evaporation losses, but water of high salinity leaves the Mediterranean as a deep current via the Strait of Gibraltar.

Light. The water of the Mediterranean is very clear, and Secchi depths (the depth at which one can observe a lowered white 30 cm disk from the surface) reach to 50 m. The Jerlov water type may be IA or IB (see Section 6.1), and so the minimum light supply necessary for survival of the deepest algae (0.05% of the surface light) penetrates to 150–200 m.

Temperature. In the Mediterranean surface water temperatures range from 12–25°C in the west to 15–29°C in the east (Fig. 2.63A). The relatively coldest parts of the Mediterranean are situated near the Spanish–French border (Côte des Albères), in the northern Adriatic Sea (Gulf of Trieste), and in northern Greece. Along these coastlines the water temperatures rarely surpass 22°C in August and may be as low as 7–10°C in February. In the lower sublittoral zone the Mediterranean seaweeds grow at considerable depths, in water that may be up to 10°C lower than at the surface. The rapid decrease of water temperature with depth is caused by persistent stratifications of the water body. For example, the August water temperatures in the Gulf of Naples may be 25°C at the surface, 17°C at 25 m, 15°C at 50 m, and 14°C at 100 m deep (13°C in February at the latter two depths). Thus representatives of the Laminariales, which may appear at first sight to be unable to grow in the warm Mediterranean, occur at greater depths in the western part but not in the eastern part, where the temperatures at depths of 50 and 100 m range between 15°C and 20°C.

History of seaweed investigations and marine biological stations. The algal vegetation of the western Mediterranean has been intensively studied, initially by dredging near the local centers of marine biological research. The Marine Biological Station of Naples was founded in 1872 by the German zoologist Anton Dohrn. The classic investigations of algal flora in the Gulf of Naples were performed by Falkenberg (1878), Berthold (1882), and Funk (e.g., 1927), whereas Lorenz (1863), one of the pioneers of algal research in the Mediterranean, worked in the northern Adriatic Sea. The largest French marine biological station on the Mediterranean coast was founded in 1881 at Banyuls, near the Spanish border. The algal vegetation near Banyuls, and at the rocky "Côte des Albères," was studied in a classic way by Feldmann (1937*a,b*). The algal flora of the French "Côte d'Azur" was initially investigated by Ollivier (1929) and that of the Yugoslavian coast (marine biological stations at Rovinj and Split) by Ercegovic (e.g., 1948).

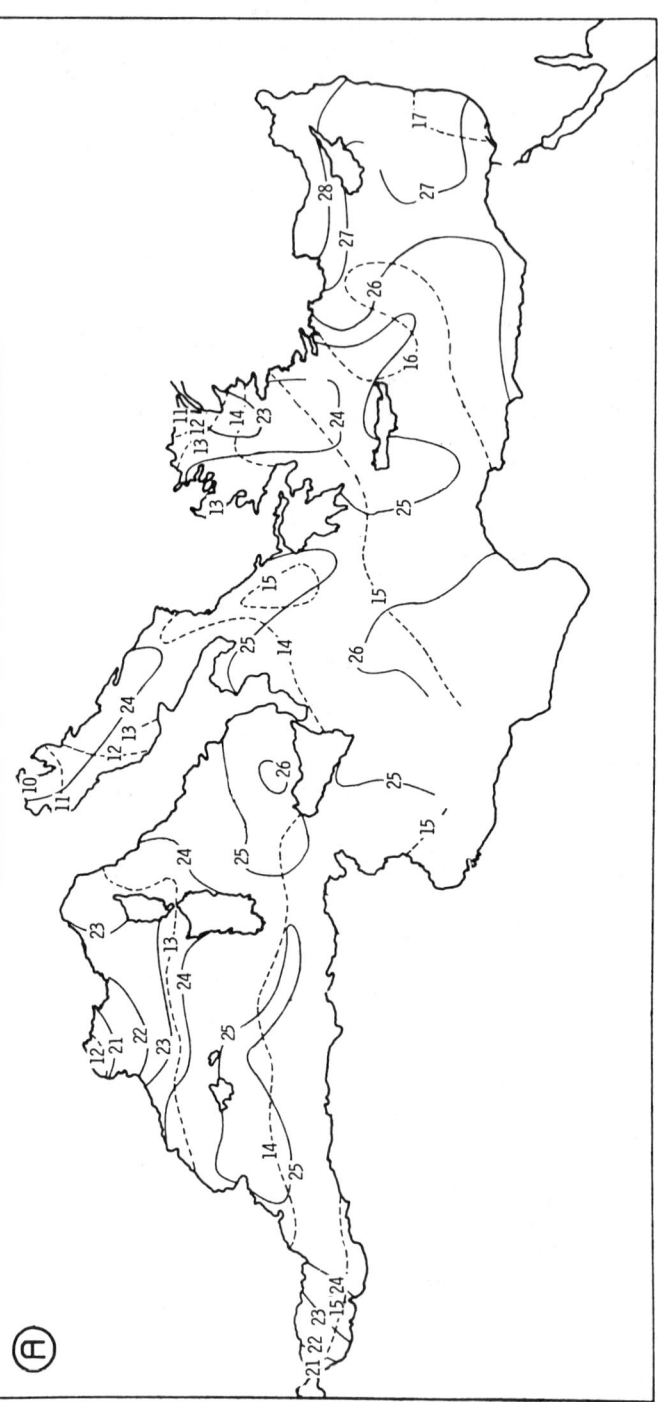

Fig. 2.63 Oceanography and paleoceanography of the Mediterranean. **(A)** Surface water temperatures (°C) in August (solid lines) and in March (dashed lines). **(B)** Oligocene (25–23 Ma): Eurasia and Africa are separated by the wide Tethys Sea. The northern arm of the Tethys Sea developed in the Oligocene and is called the Paratethys Sea. It was the forerunner of the Black, Caspian, and Aral seas. **(C)** Early Miocene (20–17 Ma): a land bridge at Suez stopped the continuity of the circumglobal Tethys Sea, permitting the first extensive mammal exchange between Africa and Eurasia (arrow). At 20–19 Ma there was a short-lived seaway from the Mediterranean to the North Sea along the Rhone Basin and the Rhine Graben. The fully marine time of the Paratethys ended around 14 Ma. **(D)** Late Miocene (6–5.5 Ma): in the Messinian event a land bridge at Gibraltar closed off the Mediterranean from the world ocean. The Mediterranean was transformed into a series of evaporitic basins. The Paratethys changed into land-locked seas with low salinity. Mammal exchange occurred between North Africa and the Iberian peninsula, as well as between Mediterranean islands. (A from Lipkin and

EUROPE–NORTHWEST AFRICA: WARM TEMPERATE ATLANTIC REGION 101

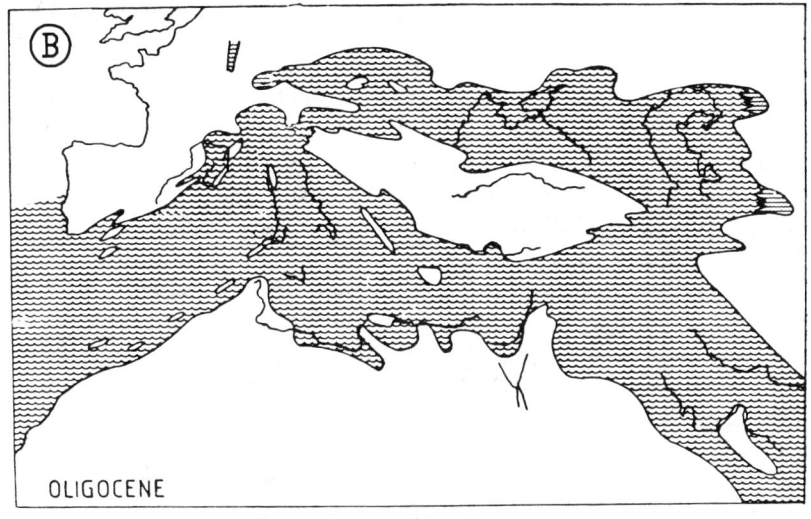

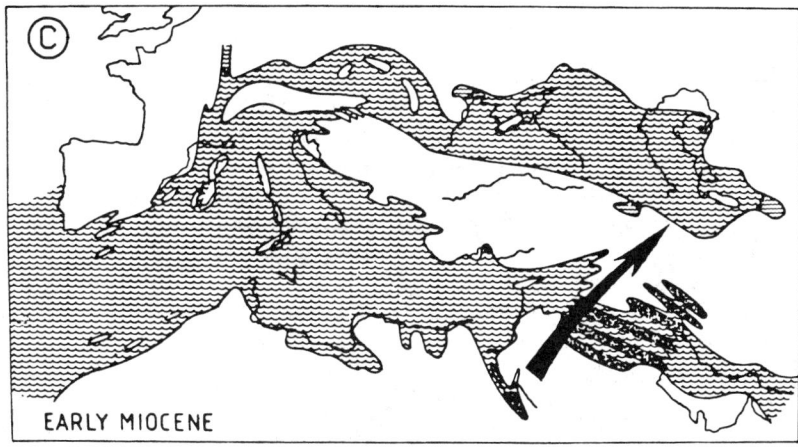

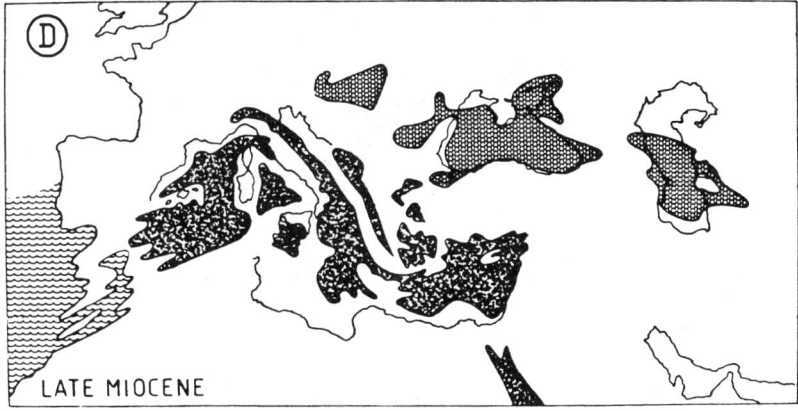

Fig. 2.63 (*Continued*)

2.4.6 Mediterranean Province: Distribution Groups and Floral History

The number of seaweed species occurring in the Mediterranean was estimated by Giaccone (1974) to be 1000, 620 of which occur along the coast of Italy (Giaccone 1969b). Of the warm temperate regions, only the Mediterranean and southern Australia exhibit a large east-west coastline paralleling the latitudes (Figs. 1.5 and 1.6). Both areas may be predicted to have a large number of species solely on the basis of length of coastline available for speciation. This argument is further expanded for faunal richness of the Indo-West Pacific (Section 4.8). In the Mediterranean, speciation is favored by compartmentation into geographically isolated parts. Of these, the Adriatic Sea, the Tyrrhenian Sea, and the Balearic basin are particularly rich in endemic species (Giaccone 1974). In the sublittoral vegetation of the Adriatic Sea (Giaccone 1978), the Strait of Messina (Giaccone and Rizzi Longo 1976), and the Strait of Sicily (Cinelli 1981), about half of the species belong to the Atlantic floral element, one-third are endemics, 10% are cosmopolitan, and 3% are recognized from the Indo-Pacific floral element.

Paleoceanography. In the Jurassic (around 180 Ma) the central part of the Tethys Sea originated by rifting, separating Africa from Europe. During the following era (the Cretaceous) the two continents changed their direction of motion and collided. The African plate subsequently slipped beneath the Eurasian plate. Most of the present-day Mediterranean Sea is not a remnant of the Jurassic Tethys Sea. The present Mediterranean is the most recent of a "series" of Mediterraneans whose evolution was controlled by the relative motions between Africa and Europe. Continental suturing and the elimination of the Mediterranean Sea may be complete in about 34 million years (Livermore and Smith 1985). In the **western Mediterranean,** the Tethys Sea, 1000 km wide in a north-south direction, became compressed to what is today the Alps, less than 100 km wide. The Western Mediterranean Basin, and its subunits the Balearic and Tyrrhenian basins, are young, small ocean basins. In the Tyrrhenian Basin rifting was initiated in the Tortonian (late Miocene) and seafloor spreading occurred since early Pliocene (Gealey 1988). The Balearic Basin started to open from Oligocene times onward, with the main oceanic accretion around 20 Ma. Subsequently, the young oceanic floor subsided to the present maximum depths of nearly 4000 m, which would be expected for a typical 20 million-year-old oceanic crust (Rehault et al. 1985). The Adriatic Sea is a shallow arm of the Mediterranean on a deeply foundered foreland, heavily laden with sediments from the Po River (Vanney and Gennesseaux 1985). The **eastern Mediterranean** is based on Middle Jurassic and younger oceanic crusts. It is thus partly a remnant of the Jurassic Tethys Sea, although it has been steadily restricted due to continued northward movement of Africa (Kennett 1982). During the **Messinian stage** (6.2–5 Ma; the final stage of the Miocene series in Italy) the Mediterranean's gate to the Atlantic was dammed for about 500,000 years. This dramatic event happened because Europe and Africa collided at Gibraltar, and/or global sea level fell by as much as 50 meters (Adams 1981, Kennett 1982, Stanley 1986). Evaporite deposits (rock salt, gypsum, anhydrite), in places 2000 m thick as revealed by core samples from the Mediterranean's seafloor, indicate that **the Mediterranean dried up repeatedly between 5 and 6 million years ago** (Hsü 1972; Hsü et al. 1977). This desiccation of the Mediterranean is called the

Messinian salinity crisis or the **Messinian event**. The time that would be required to dry up the present Mediterranean is about 1000 years (Kennett 1982). Most marine organisms would have died due to the increase in salinity prior to the deposition of the evaporites. A series of hypersaline, inland lakes may have been left on the bottom of the deep Mediterranean desert. Rivers like the Nile carved very deep canyons, until the Atlantic reflooded the Mediterranean in an awesome deluge and marine life appeared again. The Messinian event is documented by dramatic faunal changes between the late Miocene and the early Pliocene in Mediterranean fossil successions (C. G. Adams 1981). The "deep dry basin model" with its subsequent Pliocene "giant waterfall at Gibraltar" has been questioned for several reasons, one of which is that it seems unlikely that all rocks in a wall 2.5 km high and at the most 100 km across are watertight. Salt deposition may hypothetically have occurred in a relatively shallow late Miocene Mediterranean, which subsided to its present, considerable depth after the Messinian event (e.g., Sonnenfeld 1985). Whether the salts were deposited in a deep or in a shallow Mediterranean Sea, it remains clear that normal marine faunas were absent from Messinian layers and that marine biota had intermittently left the Mediterranean Sea at 5-6 Ma.

Paleobiogeography and origin of the floral elements. Due to the relatively recent recolonization of marine biota at around 5 Ma, the Mediterranean may show a greater receptivity to immigrating and variously adapted species than one would expect in an old, evolved ecosystem in which space is scarce and competition high (Almaca 1985). Today the Mediterranean Sea harbors species originating in tropical, warm, and cold temperate regions.

Tropical floral element. The Tethys Sea existed as a circumglobal ocean from Cretaceous to late Tertiary times. The **tropical floral element** of the Mediterranean is thought to date back to the algae of the circumglobal Tethys Sea (Feldmann 1937*b*, 1958; Pérès 1967*a*). This floral element includes species of pantropically (occurring in all four tropical regions) spread genera and species (Fig. 2.64). As mentioned earlier, the present Mediterranean Sea is only the most recent in a series of Mediterraneans, and the marine biota was intermittently expelled to nearby Atlantic shores during the Messinian event at 5 Ma. At this time the Mediterranean lost its coral reefs and its tropical character in general. Planktonic Foraminifera show that during the early Pliocene (5 Ma), the northern tropical limit retracted to a position south of Gibraltar, possibly close to where it is now (Cifelli 1976). Most tropical species did not return to the Mediterranean after the desiccation event.

Indo-Pacific floral element. The **connection** between the **Mediterranean and Indian Ocean** via the Red Sea was closed in the Tertiary during the Miocene (17 Ma, Fig. 4.2; 18–14 Ma according to S. M. Stanley 1986), when Africa and Arabia came into contact with Eurasia. After this time, separate tropical faunas and floras developed in the Indian Ocean on the one hand and in the Mediterranean and adjacent warm temperate Atlantic coasts on the other. After the Mediterranean lost its tropical character at 5 Ma, few representatives of the tropical **Indo-Pacific flora** were left.

Lessepsian immigrants. Much later, during historic times, some Indo-Pacific species again migrated into the Mediterranean through channels dug through the land bridge of Suez, first in Egyptian, then in Roman-Arabic times (intermittent canal from the period of Ramses II in the 13th century B.C. until the period of Calif Al Mansur in the

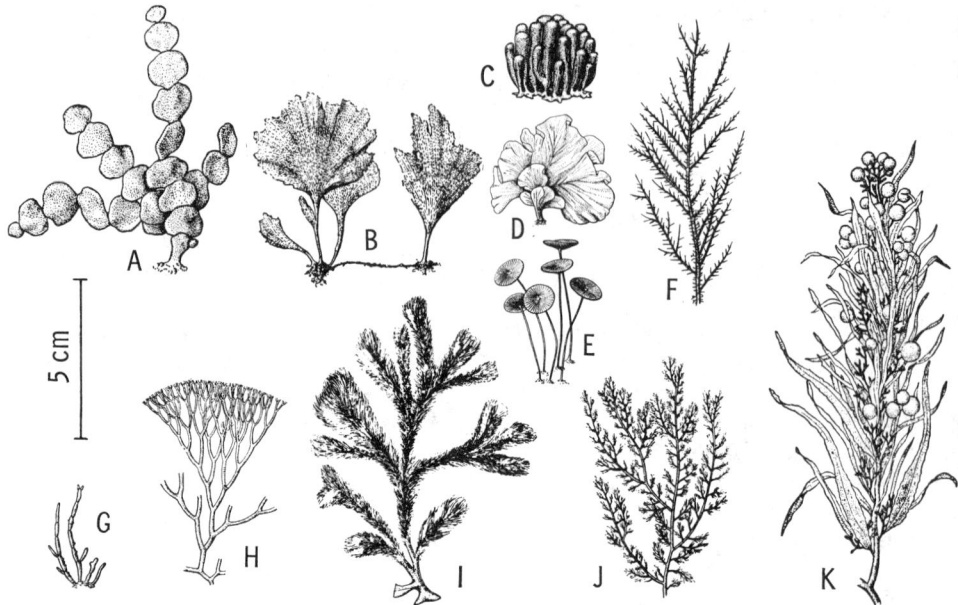

Fig. 2.64 Examples of the tropical floral element in the Mediterranean. Distribution groups: EA = eastern Atlantic (tropical); IWP = Indo–West Pacific (tropical); MED = Mediterranean (endemic); MWA Mediterranean–warm temperate–Atlantic; PAN = pantropical; WA = western Atlantic (tropical). **Green algae: (A)** *Halimeda tuna* (IWP, WA). **(B)** *Udotea petiolata* (MWA). **(C)** *Valonia utricularis* (IWP, WA). **(D)** *Anadyomene stellata* (IWP, WA). **(E)** *Acetabularia acetabulum* (= *mediterranea*, MWA). **Red algae: (F)** *Hypnea musciformis* (PAN). **(G)** *Amphiroa rigida* (EA, WA). **(H)** *Liagora viscida* (MED). **(I)** *Digenea simplex* (PAN). **(J)** *Wrangelia penicillata* (PAN). **Brown alga: (K)** *Sargassum vulgare* (PAN). (A–K from Riedl 1983.)

8th century A.D.; Por 1971), and finally after the opening of the **Suez Canal** in 1869. At least 20 Indo-Pacific algal species and one seagrass invaded the eastern, warmer part of the Mediterranean via the Red Sea; examples are the green alga *Caulerpa racemosa* and the seagrass *Halophila stipulacea* (Fig. 2.75C). The passage by marine organisms through the Suez Canal was slowed early in the history of the Suez Canal by the high salinities of the Bitter Lakes, which were initially as high as 68‰ (52‰ in 1924, today probably constant at 41‰). The Indo-Pacific immigrants have been termed **Lessepsian immigrants,** named after the builder of the Suez Canal (Pérès 1967a; Por 1978). The seagrass *H. stipulacea,* which occurs in the eastern Indian Ocean and the Red Sea, was first sighted near the Island of Rhodes in 1894, and today the species is found in the whole of the eastern Mediterranean (den Hartog 1970). At first the brackish area of the Nile delta hampered the distribution of the Lessepsian immigrants further west, but after 1966 this effect was reduced when the Assuan Dam allowed only a quarter of the water of the Nile to reach the Mediterranean (Por 1978).

Messinian species. When the Mediterranean dried up, probably a few species remained in the Messinian lagoons in evaporated marine waters with high salinities or in brackish waters diluted by rivers. These Messinian species (e.g., species groups of the ostracod genus *Cyprideis* and the cyprinodont fish genus *Aphanius*) are the oldest inhabitants of the present Mediterranean Sea and its neighboring seas. Today the Messinian survivors thrive in brackish or highly saline marginal waters and lagoons, whereas the fully marine species of the Mediterranean Sea are relative newcomers of Pliocene age (Por and Dimentman 1985).

Atlantic floral element. During the **glacial periods** the seawater temperatures at the surface were reduced by about 2–3°C in the early Pleistocene and by about 5–6°C in the late Pleistocene, as compared with present values, for example, 25°C in the Ionian Sea or Tyrrhenian Sea in summer (Thunell 1979). As a consequence, more of the remaining tropical species were eradicated and a number of species of the northern fauna and flora invaded the Mediterranean (Kosswig 1956; Briggs 1974; Fig. 2.65). Some representatives of the **Atlantic floral element** are today found as **glacial relics**; examples are *Fucus virsoides* (Fig. 2.65B) in the northern Adriatic Sea (probably a descendant of *F. spiralis*) and the red alga *Plocamium cartilagineum* on the southwestern coast of France (Fig. 2.65E). The cold-water alga *Desmarestia viridis* lives at greater depths in the Adriatic Sea (Ercegovic 1948) and may be another glacial relic (Verlaque 1981). Also, the straits of Sicily and Messina with their cooler waters harbor species with northern affinities (Cinelli 1981, 1985) such as representatives of the Laminariales (see p. 117). From the refuge of the tropical West African coast the warm temperate species reinvaded the Mediterranean in postglacial times; this can be seen today by the continuous distribution of many Lusitanic–Mediterranean species.

Cosmopolitan and endemic floral element. Besides a **cosmopolitan floral element** with species such as *Ulva lactuca* or *Ceramium rubrum* (with reservations as to further

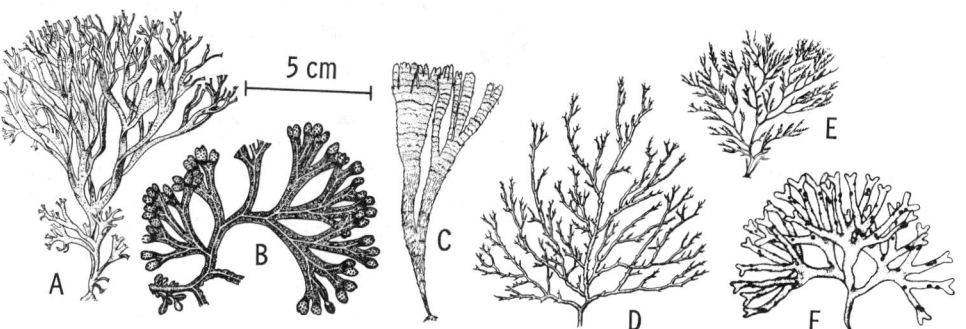

Fig. 2.65 Examples of the Atlantic floral element in the Mediterranean. **Brown algae: (A)** *Cutleria multifida.* **(B)** *Fucus virsoides.* **(C)** *Taonia atomaria.* **Red algae: (D)** *Sphaerococcus coronopifolius.* **(E)** *Plocamium cartilagineum* (= *coccineum*). **(F)** *Gymnogongrus crenulatus* (= *norvegicus*). (A from Oltmann 1972–1973; B–E from Riedl 1983; F from Schotter 1968.)

taxonomic revision) and the brown alga *Scytosiphon lomentaria*, there is an **endemic** floral element. According to Giaccone (1974) 20% of the seaweeds in the Mediterranean are endemic species, a value that must be substantiated by further taxonomic work. Probably this number is not exaggerated in view of the fact that there is 70% endemism of red and brown algae on the southern coast of the geographically more isolated continent Australia. Feldmann (1937*b*) called the early endemics (which may have a Tethyan past and be of Indo-Pacific origin) **paleoendemics;** one example is the characteristic red alga *Rissoella verruculosa* (Fig. 2.67), which is endemic to the western Mediterranean and the coast of Morocco. Mediterranean **neoendemics** (according to Feldmann 1937*b;* Cinelli 1981, 1985) evolved after the formation of the Suez land bridge 17 Ma from closely related Atlantic immigrant species. Examples are furnished by the brown algal genus *Cystoseira* (see below) and the red alga *Phyllophora nervosa*. The latter species occurs in the Mediterranean and the Black Sea. *P. nervosa* probably descended from the Atlantic species *Phyllophora crispa* (= *rubens*). In view of the Messinian desiccation event, one must assume that both paleo- and neoendemics had temporarily survived as species on the Atlantic coasts adjacent to the Mediterranean Sea.

Cystoseira *as a species- rich brown algal genus.* According to the views of Feldmann (1937*b*), Ercegovic (1959), and Giaccone and Bruni (1971, 1973), many of the nearly 30 Mediterranean species and 10 morphological variants of the brown alga *Cystoseira* are neoendemic (Giaccone 1971). This genus is represented on the Côte des Albères (near Banyuls) by 10 species (Feldmann 1937*a*), and at the island of Jabuca (in the Northern Adriatic Sea) with 11 species (Ercegovic 1957*a*). On the European North Atlantic coast this genus reaches northwards to the coast of southern England with five species, where, for example, *C. tamariscifolia* (southern British Isles to Mauritania) occurs. In the Mediterranean the latter species is replaced by other species of similar morphology (vicariant species): *C. mediterranea* in the Gulf of Lions (southwestern France) and on the Balearic Islands, *C. stricta* at the Côte d'Azur and in the Tyrrhenian Sea, *C. spicata* and *C. corniculata* in the Adriatic Sea, and *C. amentacea* in the Aegean Sea (Feldmann 1958, Giaccone and Bruni 1971). In total, two-thirds of the species of *Cystoseira* are endemic to the Mediterranean (Giaccone 1974), and the speciation process within this genus is probably still actively progressing.

Role of the Fucales in the Mediterranean. Apart from the representatives of the Laminariales occurring locally at greater depths, the largest algae in the sublittoral zone of the Mediterranean belong to the brown algal order Fucales. They play the role of canopy algae (Fig. 1.3), taken over in cold temperate regions by the Laminariales. Only three genera of the Fucales are present: *Cystoseira* with many species; *Sargassum* with three species, *S. vulgare, S. linifolium, S. hornschuchii;* and *Fucus* with one Adriatic-endemic species, *F. virsoides*.

2.4.7 Mediterranean Province: Depth Distribution and Vegetation Types

It is remarkable that phytosociological methods have been used almost exclusively by marine biologists with Mediterranean seaweed vegetation (terminology in Table 2.2). The strong interest in subdividing marine vegetation into "associations" is evident in the early phycological literature of the Mediter-

Table 2.2 Survey of biocenoses (communities) in the Mediterranean euphotic zone on rocky substrate, in relation to depth, as well as to light and wave exposure; photophilous = sun-exposed sites, sciaphilous = shaded sites; BG = biocenose-groups.

depth zones[a]	depth zones[b]	light exposure	wave exposure	BG[c]	example of biocenose
supralittoral zone	étage supralittoral			BG1	Verrucario-Melapharetum neritioides
eulittoral zone	étage mediolittoral				
upper				BG2	Nemalio-Rissoelletum verrucolosae
lower				BG3	Neogoniolitho-Lithophylletum tortuosi
sublittoral zone					
upper	étage infralittoral	photophilous	exposed	BG4	Cystoseiretum strictae
		photophilous	protected	BG5	Cystoseiretum crinitae
		sciaphilous	exposed	BG6	Lomentario-Plocamietum cartilaginei
		sciaphilous	protected	BG7	Udoteo-Aglaothamnietum tripinnati
lower	étage circalittoral			BG8	Rodriguezelletum strafforellii

[a] Classic European terminology.
[b] Terminology of the French school ("Genoa system").
[c] The numbers of the biocenose-groups are used in the text.
Compiled from the work of Boudouresque, Cinelli and Giaccone.

107

ranean. Before Möbius introduced the term "biocenosis" in 1877, Lorenz (1863) described the algal vegetation types of the North Adriatic Sea with terms such as "Ulvetum" or "Litoral-Cystosiretum" (German spelling is "Litoral" instead of "littoral"). Lorenz also made important contributions to the nomenclature of vertical zonation, his term "supralittoral zone" still being used today (Riedl 1964*a*). A set of 69 symbols characterizing the biocenoses in the Mediterranean has been proposed by Meinesz et al. (1983).

According to the extensive work on the Mediterranean marine biocenoses (communities) by Molinier, Péres, and Picard and by the more recent diving investigations of Boudouresque, Cinelli, and Giaccone, there exists a complex biocenotic system, which is basically represented in Table 2.2. Generally, this applies to the whole of the Mediterranean; however, individual biocenoses exhibit local variations. Furthermore, it should be realized that the phycological investigation of the eastern Mediterranean has not been as intensive as of the western part.

The **supralittoral zone,** reached only by spray water (**BG1** in Table 2.2, Fig. 2.66), is only poorly developed on Mediterranean shores due to the strong insolation and the generally small fluctuations of water level. Several cyanophytes and the lichen *Verrucaria symbalana* are found here. Characteristic animals are the periwinkle *Littorina* (= *Melaraphe*) *neritioides* and the isopod *Ligia italica*. The periodically or aperiodically uncovered **eulittoral zone,** exhibiting a vertical extent of only a few decimeters, is colonized in its **upper part** (**BG2** in Table 2.2) by the barnacle *Chthamalus stellatus* and by species of the limpet *Patella*. At wave-exposed locations a narrow belt of the Mediterranean-endemic, coarse red alga *Rissoella verruculosa* (Fig. 2.67) follows together with, in spring, the red algae *Bangia atropurpurea, Porphyra leucosticta,* and *Nemalion helminthoides* and the brown alga *Scytosiphon lomentaria*. At protected and eutrophic places the green alga *Enteromorpha compressa* may cover the entire upper eulittoral zone.

In the **lower part** of the eulittoral zone (**BG3** in Table 2.2) of the western Mediterranean and the Adriatic Sea the crustose coralline alga *Lithophyllum lichenoides* (= *tortuosum;* see Woelkerling 1983), with its hemispherical thalli and meandering furrows, forms continuous protruding belts called "trottoirs" (Figs. 2.66 and 2.68; for details see Blanc and Molinier 1955; Molinier 1960). In the eastern Mediterranean these belts are built by *Lithophyllum byssoides, Tenarea tortuosa* (= *undulosa;* see Woelkerling et al. 1985), *Neogoniolithon notarisii,* and by serpulid polychaetes (Lipkin and Safriel 1971; Coppejans 1975). Associated species along the coast of southern France include the red algae *Laurencia undulata, Gastroclonium clavatum,* the green alga *Bryopsis muscosa,* and the Chinese limpet *Patella aspera*.

In the northern Adriatic Sea, southward to Dubrovnik in Yugoslavia or Ancona in Italy, the only representative of *Fucus* in the Mediterranean, *F. virsoides,* also occurs in the lower eulittoral zone. The vegetation of *F. virsoides* has been described in detail by Pignatti (1962) and Munda (1973, 1982).

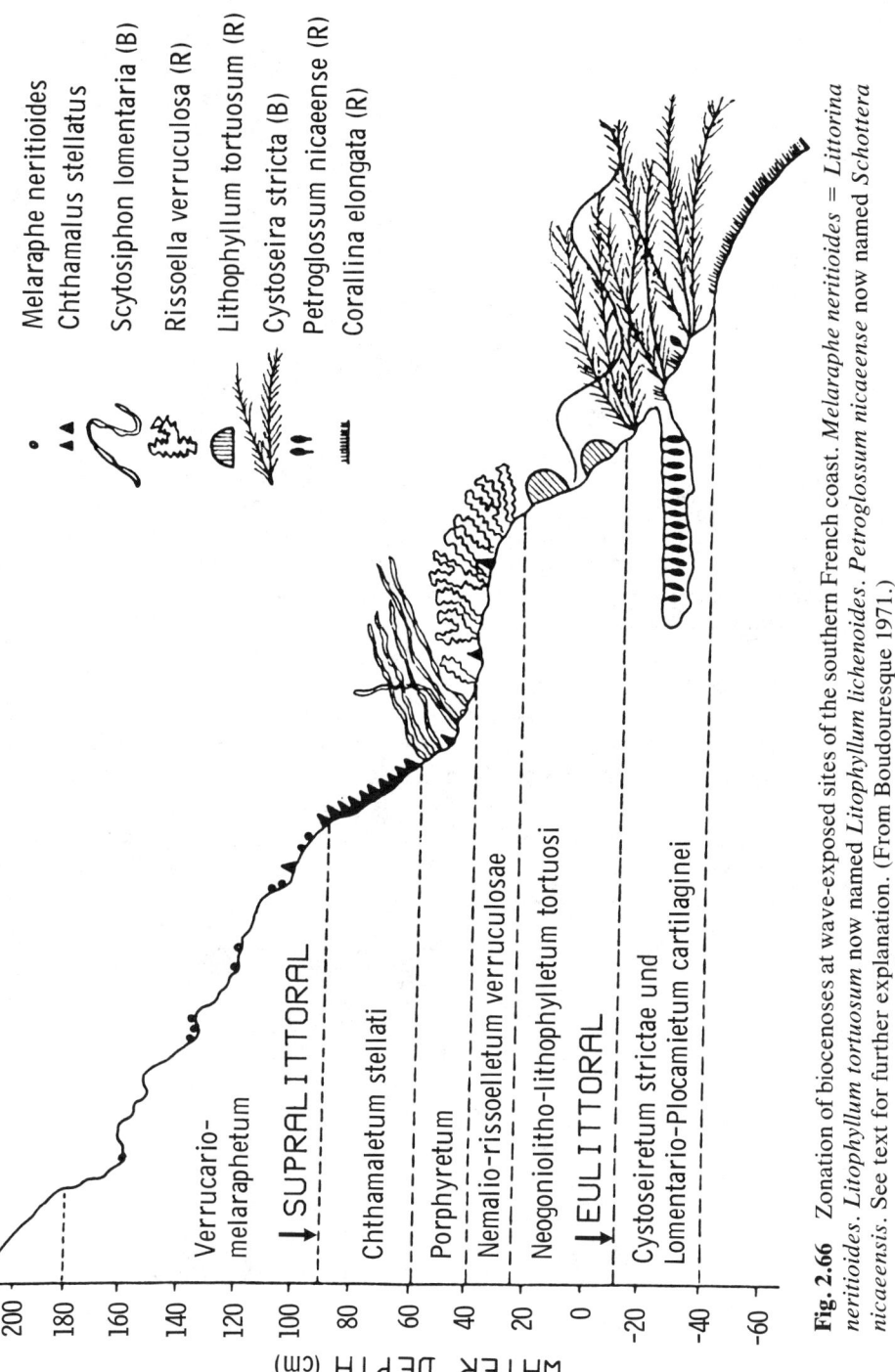

Fig. 2.66 Zonation of biocenoses at wave-exposed sites of the southern French coast. *Melaraphe neritioides* = *Littorina neritioides*. *Litophyllum tortuosum* now named *Litophyllum lichenoides*. *Petroglossum nicaeense* now named *Schottera nicaeensis*. See text for further explanation. (From Boudouresque 1971.)

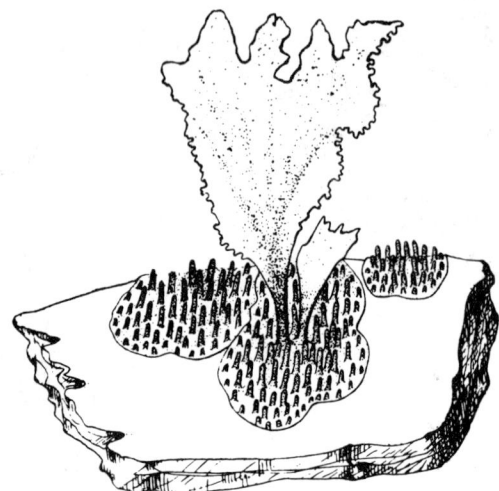

Fig. 2.67 *Rissoella verruculosa* (red alga, Gigartinales) in its autumn form. Erect fronds form from a perennating, crustose thallus. The adult frond disappears by October. Monotypic genus (containing only one species); occurs in the western Mediterranean and in the southern Adriatic Sea. (From Feldmann 1937b.)

The **upper sublittoral zone** ("etage infralittoral" according to the Genoa system; see Fig. 1.4) reaches from the level of the mean low-water mark to the deepest seagrasses at 20–45 m deep, depending on water clarity. At locations where the Jerlov water type IA prevails, the photon fluence rate at 45 m deep amounts to 10% of that at the surface (Fig. 6.3) or to 200 μmol m^{-2} s^{-1}, if one assumes a surface value of 2000 μmol m^{-2}s^{-1}, which may be realistic for many days in summer. In the upper sublittoral zone the sun-adapted or

Fig. 2.68 *Lithophyllum lichenoides* (= *tortuosum*) at Banyuls, southern France. Characteristic crustose coralline alga, which forms a protruding, solid belt ("trottoir" in French) up to 50 cm thick in the eulittoral zone of the western Mediterranean. (Photograph by the author.)

photophilous algae flourish, while at places protected from sunlight the shade-adapted or sciaphilous algae grow. It should be realized that these are descriptive terms, and experimental work to quantify them is still lacking.

Characterization of the sublittoral zone according to water motion. Riedl (1964b, 1966) showed that within the upper sublittoral zone three subzones can be recognized, differing in their hydrodynamic traits and accordingly in the mechanical adaptations of the organisms inhabiting them. There is an upper "surf zone" (0.3–2 m or to a maximum water depth of 4 m) with undirected particle movement and multidirectional mechanical stress on the organisms. It is followed by an "oscillation zone" (2–10 m or to a maximum water depth of 20 m) with harmonically circulating particles and bidirectional mechanical action on the organisms. Below 10 to 20 m deep one finds the "current zone" with uniform particle movement, mostly parallel to the coastline, and here mechanical action on the organisms is weak and unidirectional.

In sunlit water and at strong or medium wave exposures in the upper sublittoral zone (**BG4** in Table 2.2) one finds sun-adapted species of *Cystoseira* (Fucales, Cystoseiraceae).These canopy algae form the greatest part of the algal biomass in this subzone. This biocenosis has been named after *C. stricta* (Table 2.2), which occurs at the Côte d'Azur and is replaced by *C. mediterranea* as a vicariant (morphologically similar but geographically distant) species near Banyuls, the Balearic Islands, and along the coast of western Italy (Fig. 2.69A). Below the dense canopy of *Cystoseira* numerous smaller understory algae grow, for example, the red algae *Laurencia pinnatifida, Schottera* (= *Petroglossum*) *nicaeensis,* and the green alga *Valonia utricularis.* At locations where there is no light-protecting canopy of *Cystoseira,* biocenoses develop that are dominated by, for example, the green alga *Acetabularia acetabulum* and the mussel *Mytilus galloprovincialis.* In the eastern Mediterranean the crustose coralline algae *Tenarea tortuosa* (= *undulosa*) and *Lithophyllum byssoides* are dominant. In warmer parts of the Mediterranean (Corsica, Balearic Islands, Algeria, Sicily, and Lebanon) the sessile vermet snail *Vermetus cristatus* forms a second protruding belt in the upper sublittoral zone below the trottoir of *Lithophyllum lichenoides* (= *tortuosum;* Pérès 1967a).

Again in sunlit water, at habitats sheltered from wave action (**BG5** in Table 2.2), other *Cystoseira* species dominate, such as *C. crinita* and *C. barbata* in the western Mediterranean or *C. adriatica,* which is endemic in the Adriatic Sea. At these protected locations a luxurious accompanying flora and fauna understory is developed in the shadow of these canopy algae on the rocky substratum. Hundreds of algal and animal species live on the richly branched *Cystoseira* thalli. Some typical species are illustrated in Fig. 2.69. Several algae of the tropical floral element (Fig. 2.64 D–K) and some species of the Atlantic floral element (Fig. 2.65 A–C) occur as sun-adapted (photophilous) algae in the upper sublittoral zone. Where the erect algae have disappeared due to intensive grazing by the sea urchin *Paracentrotus lividus,* only crustose coralline algae (e.g. *Lithophyllum incrustans*), which may be eaten by

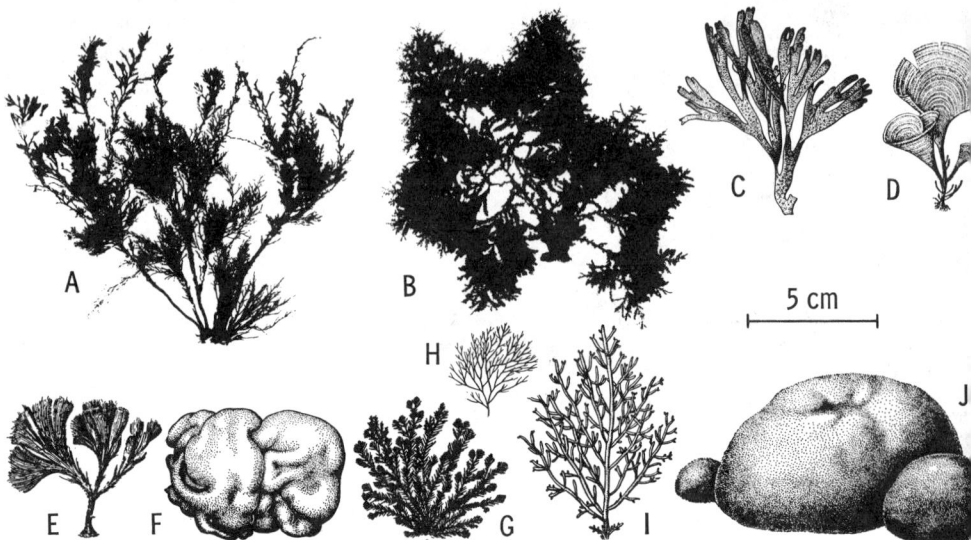

Fig. 2.69 Examples of algal species in the sun-exposed upper sublittoral zone of the Mediterranean. Abbreviations with regard to distribution groups as in legend to Fig. 2.64. **Brown algae: (A)** *Cystoseira mediterranea* (MED). **(B)** *Cystoseira crinita* (MED). **(C)** *Dictyota dichotoma* (PAN). **(D)** *Padina pavonica* (IWP, WA). **(E)** *Stypocaulon scoparium* (= *Halopteris scoparia*, PAN). **(F)** *Colpomenia sinuosa* (PAN). **Red algae: (G)** *Corallina elongata* (= *mediterranea*, WMA). **(H)** *Jania rubens* (PAN). **(I)** *Laurencia obtusa* (PAN). **Green alga: (J)** *Codium bursa* (WMA). (A–B from Feldmann 1937a; C–J from Riedl 1983.)

another sea urchin, *Arbacia lixula,* remain on the rocky substratum (Verlaque 1984; see Section 8.1).

Many different biocenoses inhabit the upper sublittoral zone in shaded locations, on northerly facing substrata, or in hollows (Fig. 2.66). The biocenoses at wave-exposed places (**BG6** in Table 2.2) are treated here first. In the northwestern Mediterranean (especially in the relatively cool Gulf of Lyon), one finds from the low water level down to a 2 or 3 m depth, a biocenosis dominated by the red algae *Plocamium cartilagineum* (Fig. 2.65E) and *Lomentaria articulata* (Fig. 2.70B). These two species belong to the Atlantic floral element, for which the northwestern Mediterranean has represented a refuge since glacial times. In this biocenosis, extensively studied by Boudouresque and Cinelli (1971, 1976), the abundance of the above two species decreases to the south. In the Gulf of Naples the biocenosis is dominated by the red alga *Botryocladia botryoides* (Fig. 2.70A). Characteristic accompanying algae in these shaded biotopes are the green alga *Valonia utricularis* (Fig. 2.64C), the Atlantic–Mediterranean red alga *Gymnogongrus crenulatus* (= *norvegicus;* Fig. 2.65F), the Mediterranean-endemic red alga *Phyllophora nervosa* (Fig. 2.70E), and the Atlantic–Mediterranean red alga *Schottera*

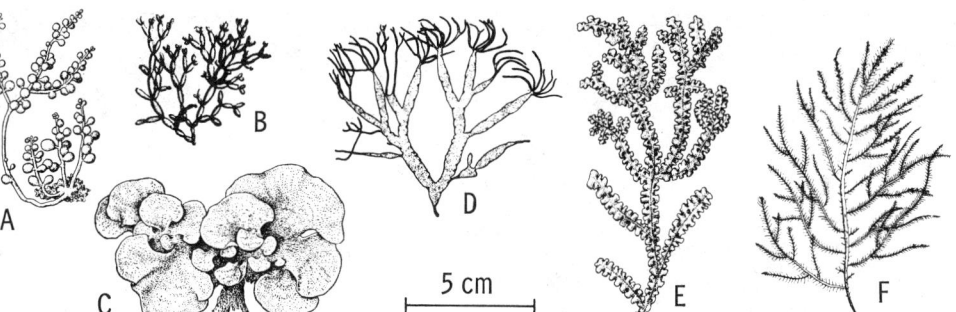

Fig. 2.70 Examples of red algae in the shaded upper sublittoral zone of the Mediterranean, at wave-exposed (A, B, D) or sheltered habitats (C, E, F). **(A)** *Botryocladia botryoides*. **(B)** *Lomentaria articulata*. **(C)** *Peyssonnelia squamaria*. **(D)** *Schottera nicaeensis*. **(E)** *Phyllophora nervosa*. **(F)** *Bonnemaisonia asparagoides*. (B from Fritsch 1959; F from Dixon and Irvine 1977a; A, C–E from Riedl 1983.)

(= *Petroglossum*) *nicaeensis* (Fig. 2.70D; see Guiry and Hollenberg 1975 for the taxonomy of this species). The last-named alga has recently been accidentally introduced into Australia (Lewis 1983).

On the shaded wave-protected substratum (**BG7** in Table 2.2) of the upper sublittoral zone, the green alga *Udotea petiolata* (Fig. 2.64B) and the red alga *Peyssonnelia squamaria* (Fig. 2.70C) dominate. Due to their weak attachment to the substratum these algae do not withstand wave exposure. These and other algae illustrated in Fig. 2.70, also penetrate into the lower sublittoral zone, where such algae encounter low irradiances and minimal mechanical stress.

The **lower sublittoral zone** or "circalittoral zone" (**BG8** in Table 2.2) extends from the deepest seagrass depths (20–45 m) to the lower algal limit (100 m to about 200 m). Note that the midsublittoral zone is not distinguished here as it is in the scheme of Fig. 1.3. This difference facilitates comparisons with the Genoa system of zonation. In terms of light penetration, the lower sublittoral zone receives 10% at its upper boundary and 0.05% at its lower limit. In summer the maximum photon fluence rate to be expected is about 200 μmol m^{-2} s^{-1} at the boundary between upper and lower sublittoral zones, and on the order of 1 μmol m^{-2} s^{-1} (or lower) at the lower algal limit. Numerous smaller red, brown, and green algae have adapted to life in the lower sublittoral zone, which is characterized by a chronic lack of light (examples in Figs. 2.71 and 2.72). However, as mentioned above, most of these algae are also found near the surface at shaded and quiet locations. Several species of the fucalean genera *Cystoseira* and *Sargassum* represent the larger algae in the lower, rocky sublittoral zone; examples are *C. spinosa* throughout the Mediterranean, *C. zosteroides* (= *opuntioides;* Fig. 2.72B) in the western part, and *S. hornschuchii* (Fig. 2.72C) throughout the Mediterranean. Furthermore, in the western Mediterranean five species of the Lami-

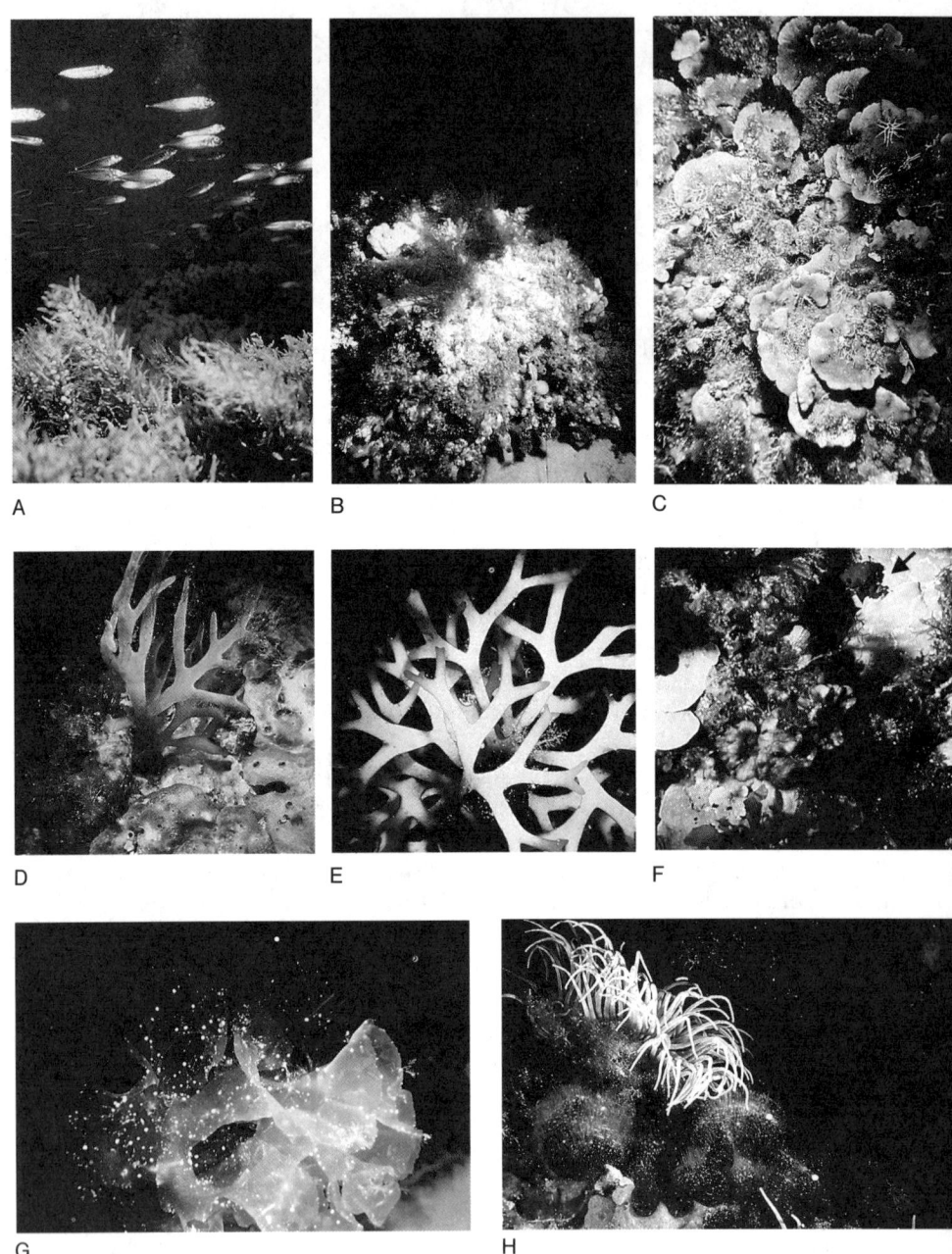

Fig. 2.71 Aspects from the lower sublittoral zone (30–45 m deep) on submarine banks between Sicily and Tunisia. **(A)** Vegetation of the brown alga *Sargassum hornschuchii*. **(B)** A block of the Coralligène. **Red algae: (C)** *Pseudolithophyllum expansum*. **(D)** *Chrysymenia ventricosa*. **(E)** *Fauchea repens*. **Green algae: (F)** *Palmophyllum crassum* (arrow points to the red alga *Pseudolithophyllum expan-*

nariales are present (see below). At the lower algal limit, only crustose coralline algae survive; typical species in the Mediterranean include *Lithothamnium philippii* and *Pseudolithophyllum expansum* (Fig. 2.71C,F).

Lower algal depth limits. At the Côte des Albères, near the French–Spanish border, the lower algal depth limit is about 40 m, because at this depth the rocky substratum passes into sandy bottom (Feldmann 1937b). Algae are found at other, deeper localities wherever a rocky substratum is present. Algae were found at 120–130 m deep in the Gulf of Naples (Berthold 1882), and even at 180 m in the Balearic Islands (Rodriguez 1889) and southern Crete (Pérès and Picard 1958). In the Adriatic Sea, Ercegovic (1960) tried to locate the deepest-growing seaweeds by dredging. He thought that the brown algae *Sargassum hornschuchii, S. vulgare, Laminaria rodriguezii,* and the red alga *Halarachnion ligulatum* might still grow at 200 to 260 m deep. However, these specimens might have drifted to these deep locations from upper depths. More reliable observations of the algal flora are made from submersibles at depths greater than 100 m. From such a diving expedition down to 130 m depth near Corsica (Jerlov water type IA; see Chapter 6) Fredj (1972) reported that at the depth range of 95–110 m the green algae *Udotea petiolata, Palmophyllum crassum,* the red alga *Pseudolithophyllum expansum,* and the kelp *Laminaria rodriguezii* still occurred, but not below 120 m. Ercegovic (1957b) and later Riedl (1964a, 1966) speculated that benthic diatoms could exist to depths of 300 m, considerably deeper than multicellular algae are found. This speculation has yet to be tested.

Green algae at greater depths. Several green algae, such as *Ulva olivascens* (Fig. 2.71G) and *Codium coralloides* (Fig. 2.71H), inhabit the lower but not the upper sublittoral zone. This fact can be taken as one of the contradictions to Engelmann's (1883) hypothesis. According to him green algae should be confined to the upper layers of the water. The green alga *Dasycladus vermicularis* is found at a 90 m depth off the coasts of Egypt and Israel (Edelstein 1964). *Caulerpa proliferà,* which is also present in the upper sublittoral zone, was found at a 150 m depth near Crete and off

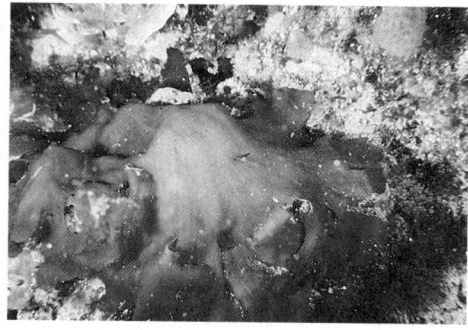

I
J

sum). **(G)** *Ulva olivascens.* **(H)** *Codium corallioides.* **Brown algae: (I)** *Arthrocladia villosa.* **(J)** *Zanardinia prototypus.* The algal species E and G–J only occur in the lower sublittoral zone; the other species ascend to shaded locations in the upper sublittoral zone. (Photographs courtesy of F. Cinelli.)

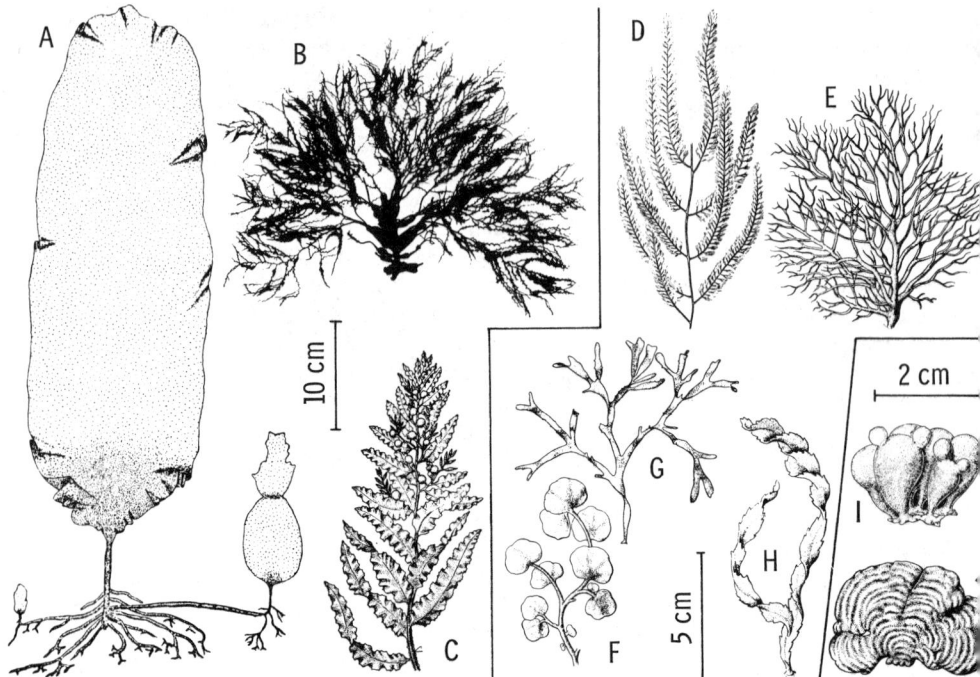

Fig. 2.72 Examples of seaweed species in the lower sublittoral zone of the Mediterranean. **Brown algae: (A)** *Laminaria rodriguezii.* **(B)** *Cystoseira zosteroides* (= *opuntioides*). **(C)** *Sargassum hornschuchii.* **(D)** *Sporochnus pedunculatus.* **Red algae: (E)** *Acrosymphyton purpuriferum.* **(F)** *Neurocaulon foliosum* (= *reniforme*). **(G)** *Rhodymenia ardissonei* (= *corallicola*). **(H)** *Vidalia volubilis.* **Green algae: (I)** *Valonia macrophysa.* **(J)** *Palmophyllum crassum.* Species A occurs below 50 m deep, C–E mostly below 10 m deep, and the remaining species in deep water and at shallower, shaded habitats. (A from Huvé 1955; B from Feldmann 1937a; C–J from Riedl 1983.)

Alexandria (Pérès and Picard 1958). The species *Palmophyllum crassum* (Fig. 2.71F), which is also found in the upper sublittoral zone at shaded places, is in fact not a multicellular alga but belongs to the order Chlorococcales. The individual cells are held together by a gelatinous matrix, and in this way a crustose, deep-greenish thallus of up to 20 cm in diameter is formed (Feldmann 1937a; Giaccone 1967). A similar genus, *Palmoclathrus,* occurs at greater depths on the southern Australian coast. *Codium corallioides* and *C. effusum* (= *difforme*) have similar morphologies and are found in the lower and shaded upper sublittoral zones of the Mediterranean (Feldmann 1937a). Other members of the genus in the Mediterranean include two dichotomously branched species: *C. vermilara* (Silva 1955; Delépine 1959) and *C. fragile.* The latter has invaded the Mediterranean since 1940.

Submarine banks and the Coralligéne. Characteristic locations for the circalittoral biocenoses are submarine slopes or deep horizontal plateaus. The communities inhabiting the numerous submarine banks between Sicily and Tunisia (Fig. 2.71) have been extensively studied by Cinelli (e.g., 1981). Another typical biotope is the **Coralligène** ("coralligène de plateau"), which consists of blocks of organic concretionary material, rising like hard substratum islands several meters from the sand-covered bottom (Fig. 2.71B). The blocks are composed of the calcareous shells and housings of mollusks, polychaetes, bryozoans and gorgonians and of the dead thalli of crustose coralline algae. The blocks are covered with the typical algae of the lower sublittoral zone. Live coralline species, mainly *Mesophyllum lichenoides,* and other species such as *Pseudolithophyllum expansum* may cover an average of 30% of the outer surface of these blocks. The recent algal reefs of the Mediterranean Coralligène are similar to Miocene coralline limestones from Tethyan areas (Bosence 1985). The precious red coral *Corallium rubrum,* from which the Coralligène was originally named, does not typically occur in this biotope but on the underside of overhanging rocks in the lower sublittoral zone (Pérès 1967a; Riedl 1983). The Coralligène communities near Banyuls were described in detail by Laubier (1966) and the algal vegetation at greater depths by Giaccone (1968b; Tyrrhenic and Aegean Sea) and Ercegovic (1960; Adriatic Sea).

Laminariales in the Mediterranean. The distribution of the representatives of this order is shown in Fig. 2.73. *Laminaria rodriguezii* is endemic to the Mediterranean and characterized by a rhizomatous holdfast system from which erect plants are vegetatively formed (Figs. 2.72A and 2.74C). It grows at a 50–120 m depth on the coasts of Algeria, Tunisia, Majorca, Corsica, Sicily, and in the Adriatic Sea (Feldmann 1934; Giaccone 1969a). At these depths, where temperatures never surpass 15°C in summer, representatives of the Laminariales find the temperature regime similar to the cold temperate region (Fig. 1.8). The phenomenon that occurs when species sensitive to higher water temperatures descend into deeper, colder water has been termed "isothermic submergence" (Briggs 1974). It may be compared to the "equatorial submergence" of pelagic animal species that also avoid the warmer surface waters towards the equator by diving down to deeper waters. Other representatives of the Laminariales in the Mediterranean are *L. ochroleuca* (with a pale blade base; Fig. 2.74B), *Phyllariopsis brevipes* (= *Phyllaria reniformis;* Fig. 2.74A; see Henry and South 1987), *Phyllariopsis* (= *Phyllaria*) *purpurascens* (both species with hairs on the frond; *P. purpurascens* with a discoid holdfast instead of haptera), and *Saccorhiza polyschides* (with cushionlike holdfast; Fig. 2.36). These species are distributed along the eastern Atlantic coast from Morocco northwards: *P. purpurascens* to the Spanish-Galician coast, *P. brevipes* to Biarritz, and *L. ochroleuca* to southern England. *Saccorhiza polyschides* is the only one that penetrates the cold temperate region as far as mid-Norway and the Shetland Islands. In the Mediterranean one finds this species on the coast of Algeria and in the Tyrrhenian Sea. In the Strait of Messina, where strong currents exist and cold water rises to the surface, *S. polyschides* and *P. brevipes* may be collected just below the surface and at moderate depths (Fig. 2.74D; Huvé 1958; Giaccone 1969a, 1972). Recently, in one of the southwestern French lagoons (Étang de Thau) the Japanese species *Undaria pinnatifida, Laminaria japonica* (Fig. 2.91B) and *Sargassum muticum* (Fig. 2.62E) were accidentally introduced by oyster aquaculturists (Pérez et al. 1981; Boudouresque et al. 1985). A few cold-water algae such as *Desmarestia viridis,* also detected in the Étang de Thau, may have been introduced (Verlaque 1981). Previously, Ercegovic (1948) had dredged a *Desmarestia* species similar to *D. viridis* from a 40–80 m depth in the Adriatic Sea.

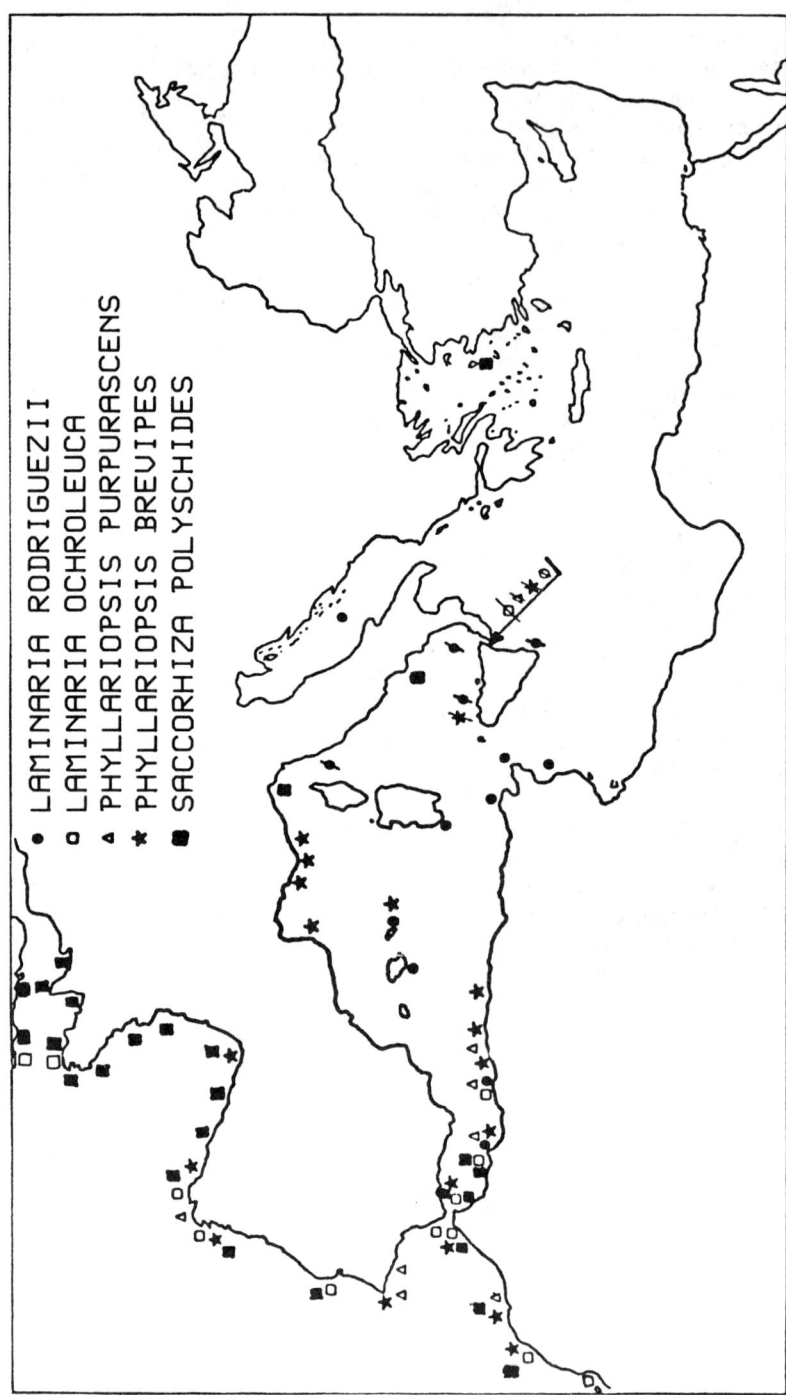

Fig. 2.73 Distribution of the Mediterranean representatives of the Laminariales: *Laminaria rodriguezii*; *L. ochroleuca*; *Phyllariopsis* (= *Phyllaria*) *purpurascens*; *Phyllariopsis brevipes* (= *Phyllaria reniformis*); and *Saccorhiza polyschides*. (From Giaccone 1969a.)

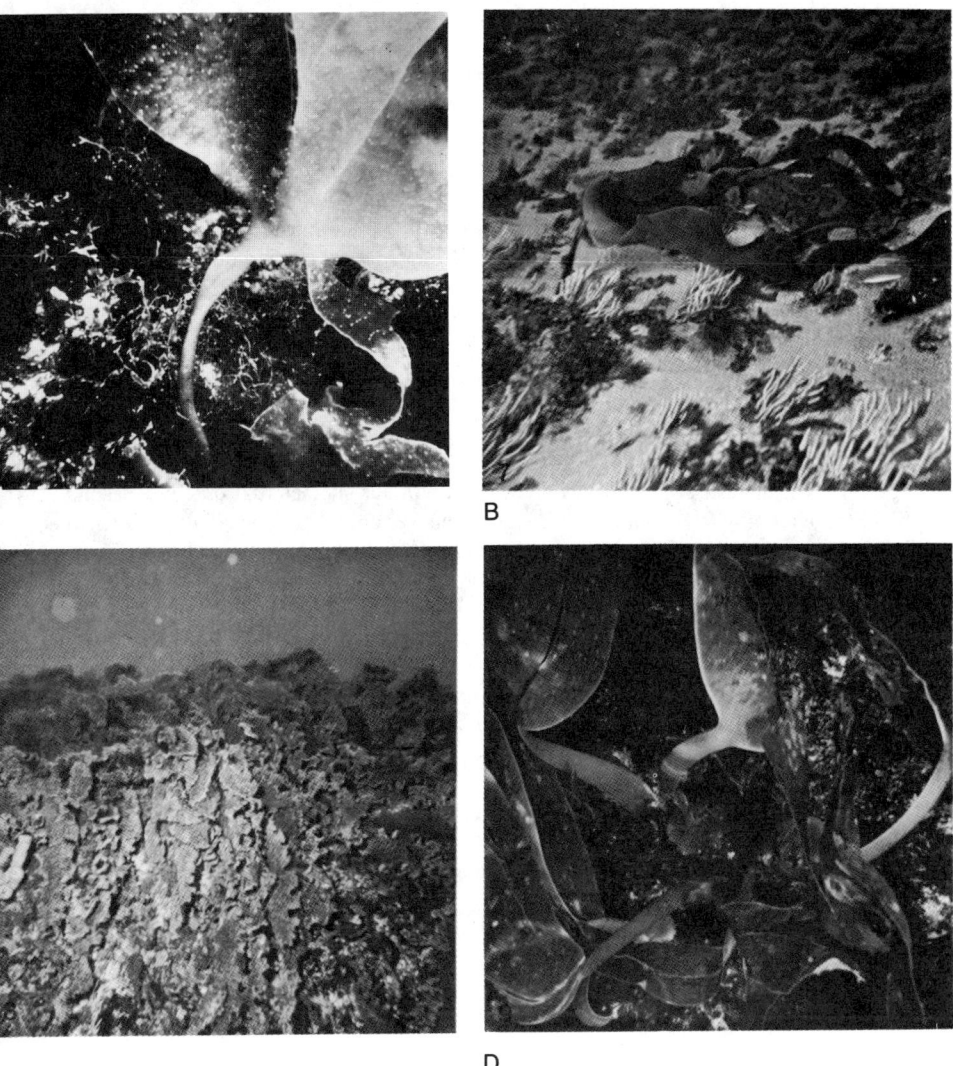

Fig. 2.74 Laminariales in the Mediterranean. **(A)** *Phyllariopsis brevipes* at 25 m deep on the "Banco Graham" south of Sicily. **(B)** *Laminaria ochroleuca* at 75 m deep in the Strait of Messina. **(C)** Dense vegetation of *Laminaria rodriguezii* at 65 m deep on the "Banco Apollo" near the island of Ustica. **(D)** *Saccorhiza polyschides* at 1 m deep in the Strait of Messina. (Photograph A courtesy of F. Cinelli; photographs B–D from Giaccone 1969a.)

Vegetation of sea caves. These habitats were extensively studied by Riedl (1966), and were also the subject of phycological studies (e.g., Larkum et al. 1967). The gradient of irradiance reaching from full sunlight to the algal limit occurs over a relatively short distance. Early investigators (e.g., Falkenberg 1878) found the same algal species at the same irradiance limit in the inner cave as at comparably great depths in the lower sublittoral zone (extinction depth). However, at the irradiance limit within a cave are found shade-adapted species of the upper sublittoral zone (Funk 1927). Why they do not occur at the extinction depth in the lower sublittoral zone remains an unanswered question.

Seagrasses. The seagrass species illustrated in Fig. 2.75 occur on **sandy and muddy bottoms**. The Mediterranean endemic species *Posidonia oceanica*, which is distributed throughout this sea down to a 40 m depth (Pérès and Picard 1958), grows on coarse sand with vertical and horizontal rhizomes. As the rhizomes are covered by sediment, they rise vertically at the rate of about 1 m per century (Pérès 1982*b*). They may attach to rocky substrata and over time enough sediment builds up to cover the rocks. Within the sediment the rhizomes form a continuous network, the lowermost part of which is made up of dead rhizomes. These *Posidonia* "mats," up to 4 m thick, are well known to the fishermen as suitable anchoring places within otherwise sandy areas. *P. oceanica* starts to grow in August and stops in May, possibly due to an internal, circannual rhythm (Section 6.4.3; Ott 1979, 1980). On more muddy bottoms one finds the seagrass *Cymodocea nodosa*, a pioneer species that on coarser sand is finally replaced by *P. oceanica*. This species also occurs along the adjacent northwest African coast, on the Canary Islands, and on the coast of southern Spain. *Zostera noltii* (= *nana*), is confined to estuaries and brackish water lagoons, or "étangs," of southwestern France. The species is also widely distributed along the European coast, north to southern Norway, in the Baltic to Kiel, and south to Mauritania. At habitats that are colonized in the Mediterranean by *Z. noltii*, one may also find *Z. marina*, which, outside the Mediterranean, also colonizes fully marine habitats (den Hartog 1970). Finally, *Halophila stipulacea* is confined to the eastern Mediterranean, and its westernmost outpost is Peloponnesos (Haritonidis and Tsekos 1976). This euryhaline species (surviving a wide salinity range) occurs in the Indian Ocean and the Red Sea. It has invaded the Mediterranean sandy bottom seagrass vegetation of *Cymodocea nodosa, Posidonia oceanica,* and the algal meadows of *Caulerpa prolifera* after the opening of the Suez Canal. There are records of *H. stipulacea* for

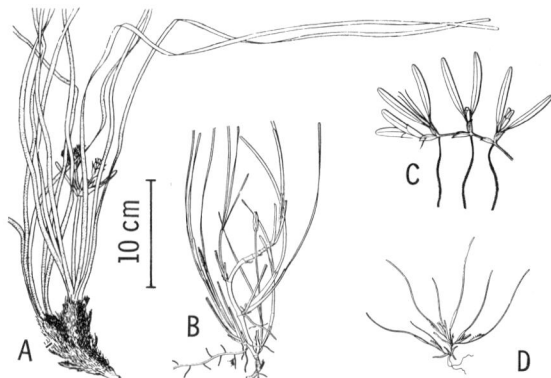

Fig. 2.75 Seagrasses of the Mediterranean. **(A)** *Posidonia oceanica.* **(B)** *Cymodocea nodosa.* **(C)** *Halophila stipulacea.* **(D)** *Zostera noltii.* (A from Riedl 1983; B–D from Hartog 1970.)

depths that are unusually great for seagrasses (e.g., at 100 m; see Pérès 1982a), and should be rechecked, since they might relate to drift material, as pointed out by den Hartog (1970) in his evaluation of some other depth records.

Seaweeds on sand. The green alga *Caulerpa prolifera* (Fig. 2.76A; order Caulerpales) grows on sand in the lower sublittoral zone. This is a tropical Mediterranean species that is absent in the colder parts of the Mediterranean such as the Adriatic or the northern Aegean seas. Additional examples of sand-inhabiting species of *Caulerpa* are *C. ollivieri* on the southern French coast and *C. racemosa* in the eastern Mediterranean. *C. prolifera* often occurs with the seagrass *Cymodocea nodosa,* which prefers somewhat muddy bottoms, and is accompanied by the green alga *Cladophora prolifera. Penicillus capitatus* (Fig. 2.76B) represents another typical seaweed species growing on sand. Two characteristic brown algae on sandy gravel bottoms, sometimes occurring on solid substrate, are *Arthrocladia villosa* (Fig. 2.71I) and *Sporochnus pedunculatus* (Fig. 2.72D). The branched, loose-lying, crustose coralline algae *Phymatolithon* (= *Lithothamnium*) *calcareum* and *Lithothamnium corallioides* (=*Lithothamnium solutum*) form the maerl vegetation on sand similar to that on the coast of Brittany. In the western Mediterranean (e.g., on the coast of Provence), the maerl vegetation is found down to a 50 m depth. In the Aegean Sea and near Crete it may extend down to a 100 m depth (Pérès and Picard 1958; Jacquotte 1962).

2.4.8 Black, Caspian, and Aral Seas

These marine and pseudomarine areas can be regarded as provinces of the warm temperate Mediterranean–Atlantic region (Briggs 1974). The **Black Sea** is a brackish appendage to the Mediterranean, characterized by a salinity of 17–18‰ and by a seasonal range of surface temperatures of -1–29°C (Caspers 1957; Zenkevitch 1963). The Black Sea is inhabited by about 130 species of red algae and 70 species of brown and green algae. This is an impoverished Mediterranean flora that consists mainly of the more cold-tolerant species (for details see Zinova 1967).

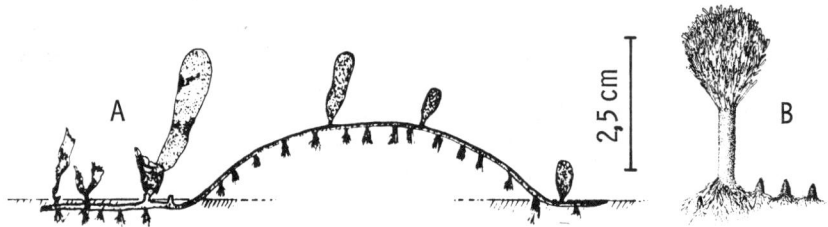

Fig. 2.76 Green algae inhabiting sandy bottoms (both species with a western Atlantic tropical distribution). **(A)** *Caulerpa prolifera:* with buried rhizomes up to 80 cm long. Bladelike thalli, formed from June onwards, at first grow towards the light and then bend down to the substrate. **(B)** *Penicillus capitatus:* also called "Neptune's shaving brush." Rhizomes are buried beneath the sand and have stipes that carry siphonous, dichotomous branches. (A from Meinesz 1979; B from Meinesz 1980.)

Phyllophora meadows. At a depth of 10–60 m in the Black Sea the red alga *Phyllophora nervosa* flourishes at the upper depths, accompanied by *P. truncata* (= *brodiaei*) mainly at the lower depths (Vasiliu and Bodeanu 1972). This vegetation is loose-lying or grows on a substratum of mussel shells. It propagates primarily vegetatively by fragmentation and makes up about 90% of the algal biomass. At the relatively shallow northwestern part of the Black Sea, on the coasts of Romania and the Ukraine, probably the largest mass of red algae in the world occurs. About 5 million metric tons fresh weight of red algae covers 15,000 km^2 of mud and shells. This area has been termed **Sernow's *Phyllophora* Sea** after its explorer (Zenkevitch 1963). The isolated occurrence of the Arctic–cold temperate species *P. truncata* (Fig. 2.25) in the Black Sea, absent in the Mediterranean, points to its origin from the glacial Mediterranean flora. This species is widely tolerant of low salinities and may even penetrate far into the Baltic (Fig. 2.25). Its euryhaline character may be as important for its success in the Arctic as it is in the Baltic and Black seas.

The seagrasses *Zostera marina* and *Z. noltii* grow on soft bottoms along the coasts of the Black Sea. The brown alga *Cystoseira barbata* penetrates into the Black Sea, forming 9% of the biomass within the euphotic zone (Zenkevitch 1963). Due to the clarity of the water (Gessner 1955–1959) the deepest crustose coralline algae may extend down to a 100 m depth. The characteristic lack of oxygen and mass production of hydrogen sulphide (due to sulphate-reducing bacteria) begins at depths below 150 m, well below the lower algal limit. Fresh water flows out from the Black Sea along the surface, while saltier marine waters from the Mediterranean enter at the bottom, sink down in the Black Sea, and become stagnant. The sill of the Bosporus was a land bridge during glacial periods because of the worldwide lowered sea levels, and is it today only 35 m below sea level. During the cold interludes with an excess input of fresh water, the Black Sea became a freshwater lake, well aerated at all depths, and the Bosporus was not a strait but a river. During the warm periods including the present, the Black Sea was made brackish by water spilling from the Mediterranean through the Bosporus. After the last glaciation, with its maximum extent around 18,000 years ago, the world sea level rose sufficiently high some 10,000 years ago for salt water to enter the Black Sea. At this time the Black Sea attained its present brackish character, and its existence as an anoxic, lifeless abyss began (Hsü 1978).

Remnants of the Paratethys Sea. When the Mediterranean was separated from the Indian Ocean by the Suez land bridge in the Miocene, 17 million years ago, vast northern extensions, comprising the Paratethys (or Sarmatic Inland) Sea still covered parts of eastern and central Europe (Figs. 2.63B,C). During the Messinian crisis, the Paratethys Sea became separated from the Mediterranean and disintegrated into a network of lakes. The Black Sea, along with the **Caspian Sea** and **Aral Sea,** is a remnant of the Paratethys. (For details see Ekman 1953; Lattin 1967; C. G. Adams 1981; Hsü 1978; Por and Dimentman 1985). During the Messinian event, the Paratethys did not dry up, but part of its waters probably drained into the desiccating Mediterranean, adding further to the disintegration of the Paratethys (Fig. 2.63D). When the Mediterranean refilled 5 million years ago, salt water and marine life briefly returned to the Black Sea and possibly to other parts of the Paratethys. The marine connection with the Mediterranean was, however, soon severed again, and the Black Sea began a long interval as a freshwater lake (Hsu 1978).

Caspian and Aral seas. These two inland seas are still inhabited by relic marine species of the Sarmatic Inland Sea, although later introductions of marine species cannot be excluded. The Caspian Sea, which has a salinity of 12–13‰, well below that of normal seawater, still harbors the seagrass *Zostera noltii* and about 100 marine benthic algal species (30 red, 10 brown, and the rest green seaweed species; Zinova 1967). Most of these have Mediterranean or Mediterranean–Atlantic affinities. Endemic to the Caspian Sea are the brown alga *Monosiphon caspicus* (Chordariaceae) and the red algae *Callithamnion kirillianum, Polysiphonia caspica, Laurencia caspica,* and *Titanoderma caspicum* (= *Dermatolithon caspicum;* see Woelkerling 1986). In the Aral Sea (salinity of 10‰ until the 1960s; since then increased due to loss of inflowing water consumed for field irrigating) there are a few marine species that remain; examples are *Polysiphonia violacea* and the seagrass *Zostera noltii.* Algal genera characteristic of brackish and freshwater habitats, such as *Ulothrix, Rhizoclonium,* and *Vaucheria,* are found here (Zinova 1967).

2.5 NORTH AMERICA: COLD AND WARM TEMPERATE REGIONS IN THE NORTH ATLANTIC

Literature. Entire area: Hillson (1977); Humm (1969); T. F. Lee (1977); Pielou (1977, 1978); Searles (1984); Steele (1983); Stephenson and Stephenson (1972); W. R. Taylor (1957). **Checklist:** South and Tittley (1986); Tittley et al. (1985, 1989). **Newfoundland:** Bolton (1981); Hooper (1987); Hooper et al. (1980); Keats et al. (1985); South (1976, 1983a,b); South and Hooper (1980); Steele (1983). **Quebec, Gulf of St. Lawrence:** Bird et al. (1983); Cardinal (1967); Cardinal and Villalard (1971); Gauthier et al. (1980); Ketchum (1983); Sève et al. (1979); South and Cardinal (1973). **Nova Scotia and Bay of Fundy:** Bell and MacFarlane (1933); C. J. Bird et al. (e.g., 1976); A. R. O. Chapman (1981); Edelstein et al. (1970); Mann (e.g., 1972); Stephenson and Stephenson (e.g., 1972); Tittley et al. (1987); Wilson et al. (1979). **South of Bay of Fundy to Cape Cod (southern New Brunswick, Maine, New Hampshire, Massachusetts):** Coleman and Mathieson (e.g., 1975); Davis and Wilce (1987a,b); Kingsbury (1969); Lamb and Zimmermann (1964); Mathieson and Hehre (e.g., 1982); Sears and Cooper (1978); Setchell (1922); South et al. (1988); Vadas and Steneck (1988). **South of Cape Cod to Cape May (Rhode Island, Connecticut, New York, New Jersey):** Schneider et al. (1979); Sears and Wilce (1975); Yarish and Edwards (1982). **Cape May to Cape Hatteras (Delaware, Maryland, Virginia, North Carolina):** Humm (1979); Kapraun (1980, 1980–1984); Orris (1980); F. D. Ott (1973); Peckol and Searles (1983); Searles (1984); Zaneveld (1972); Zaneveld and Willis (1974–1976). **Cape Hatteras to Cape Kennedy (South Carolina, Georgia, Florida):** Cheney and Dyer (1974); Dawes (1974); Mathieson and Dawes (1975); Searles (1984); Searles and Schneider (e.g., 1978, 1980); Stephenson and Stephenson (e.g., 1972). **Gulf of Mexico, northern coast (Texas, Louisiana, Alabama, Florida):** Baca et al. (1979); Conover (1964); Earle (1969); Edwards (1969, 1970); Edwards and Kapraun (1973); Hamm and Humm (1976); Hanisak and Blair (1988).

On the Atlantic coast of North America the Arctic region is followed by the **western (American) province of the cold temperate Northern Atlantic region.**

This is followed by the **warm temperate Carolina region,** which finally borders the **western Atlantic tropical region** (Figs. 1.5–1.7, 2.45). The biogeographical boundaries between these regions are the Strait of Belle Isle, Cape Hatteras, and Cape Kennedy (formerly Cape Canaveral; Fig. 2.77).

History of algal investigation and marine biological stations. The history of subdividing the North American Atlantic coast into biogeographical regions or provinces, with early work by such people as J. D. Dana (1853), has been given by Hazel (1970) and Briggs (1974). The first algal investigations were carried out in this area by European botanists who made occasional references to the northeastern American flora. The first substantive work on the Newfoundland algae was published by De la Pylaie in 1829. In 1858 the Irish phycologist W. H. Harvey, upon invitation of the Smithsonian Institute, published his classical work *Nereis Boreali-Americana.* Further floristic collections in the late 19th and early 20th centuries culminated in the works of Farlow, Collins, Holden, Setchell, and Howe. The floristic approach of these early phycologists ultimately led to the classic treatises by W. R. Taylor on the *Marine Algae of the Northeastern Coast of North America* (1957) and *Marine Algae of the Eastern Tropical and Subtropical Coasts of the Americas* (1960). (For a historical review see W. R. Taylor 1957.) Examples of marine biological stations along this coast are Logy Bay Marine Laboratory of the Memorial University of Newfoundland,

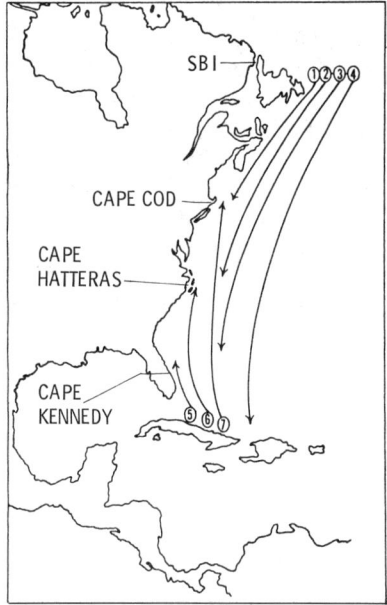

Fig. 2.77 Distribution groups of algal species on the eastern North American coast, with cold-water affinities (groups 1–4) and warm-water affinities (groups 5–7). Many species adapted to cold water reach only as far south as Cape Cod (group 1); many warm-adapted species are found only as far north as Cape Hatteras (group 6). SBI = Strait of Belle Isle (between Newfoundland and Labrador). (From Humm 1969.)

Huntsman Marine Laboratories near St. Andrews in New Brunswick, Jackson Estuarine Laboratories of New Hampshire, and the Marine Biological Laboratory at Woods Hole, Massachusetts.

2.5.1 American Province of the Cold Temperate North Atlantic Region (Newfoundland to Cape Hatteras)

In comparing the American and the European provinces of the cold temperate Northern Atlantic region, R. T. Wilce once pointed out that when a North American phycologist travels to Europe, he finds many seaweed species he had not seen before in the field, whereas the European phycologist coming to the North American Atlantic coast for the first time feels familiar with all but a few species, such as the kelp *Agarum cribrosum* (Fig. 2.10).

Paleobiogeography and Paleoceanography

Arctic–cold temperate species. The fossil origin of the **molluscan fauna** on the shallow shelf of the northwest Atlantic (to a 150 m water depth) has been compiled by Franz and Merrill (1980). The 76 investigated gastropods and bivalves are distributed mainly from the Arctic to Newfoundland, declining rapidly in number of species towards Cape Cod ("Arctic–boreal fauna"). All live today on both sides of the Atlantic (amphi-Atlantic species), and 81% have a documented fossil origin in the Pacific. This reflects the mass invasion of the North Atlantic by Pacific ("preformed") cold-water species through the Bering Strait after its opening at around 3.5 Ma. During the glacial episodes the Arctic–cold temperate algae had to move to more southern latitudes. The general fact that many of the amphi-Atlantic species also occur in the circumpolar Arctic region (Table 2.2) has been well known for many marine animal groups (Ekman 1953; Hazel 1970; Briggs 1974; Steele 1983) and points to the possibility of a still-existing, circumpolar gene flow. A small percentage (11%) of the Arctic–boreal molluscan species with a documented Pacific origin, such as the kelp *Agarum cribrosum*, did not reach the eastern North Atlantic (Franz and Merrill 1980). Another small group (8%) of the Arctic–cold temperate molluscan species is endemic to the northwest Atlantic. For the benthic algae the two brown algae *Papenfussiella callitricha* and *Omphalophyllum ulvaceum* are examples.

Cold temperate species and northwestern Atlantic endemics. The cold temperate ("boreal") group of **molluscan** fauna, with a distribution from Labrador to Cape Hatteras, is of mixed origins. It contains a small group of Pliocene trans-Arctic migrants from the Pacific invaders. There is a second group of amphi-Atlantic species with long histories in the Atlantic, and a third group of northwestern Atlantic endemics derived from American Miocene ancestors. The endemic mollusk group becomes dominant around Cape Hatteras ("Transhatteran species group"; Franz and Merrill 1980). Although it is clear that the soft-bottomed coast south of Cape Cod offered optimal conditions for the speciation of mollusks, there are some northwestern Atlantic endemics among the **seaweeds** as well such as the red algae *Grinnellia americana, Lomentaria baileyana,* and *Agardhiella subulata* (Yarish et al. 1984).

Foraminiferal evidence: biogeographical origin and species duration. Fossil origins have also been documented in the fast- evolving and rapidly dispersing **benthic Foraminifera.** For the species on the Atlantic continental margin of North America it has been shown that the first fossil occurrences in the Pleistocene or Pliocene are widespread in the Northern Hemisphere (Buzas and Culver, 1986). For example, of the 23 species occurring today from Cape Hatteras to Newfoundland and documented since the **Pleistocene,** eight occurred first on the American side of the North Atlantic, three on the European side, and the remainder in Alaska (nine species), Vancouver (two species), and California (one species). The average **species duration** for benthic Foraminifera now occurring at depths less of 200 m from Florida to Cape Hatteras is on the order of 20 million years, and only 7 million years for species restricted to north of Cape Hatteras (Buzas and Culver 1984). In this example, species restricted to the harsher and more variable environments in the north seem to have the shortest durations. This is also reflected in the fact that only 15% of the modern benthic foraminiferal fauna from Cape Hatteras to Newfoundland extend as far back as the Pliocene.

Lack of stenothermal species and the development of cold currents. As mentioned before, the lack along the North American eastern coast of many European "stenothermal" algal species, (requiring a narrow temperature range for survival and often having a sensitivity to cold water) must be primarily due to the relatively large yearly temperature there compared with European locations along the open Atlantic coast. Since the **Oligocene** (about 35 Ma), cooler water began to flow along the eastern coast of North America.and eroded the deep continental margin (Burke 1979), when the sill between Arctic Ocean and the North Atlantic was breached and Arctic water spilled over the severed Greenland–Scotland Ridge. Nevertheless, there was still tropical fauna in the Oligocene and **Miocene** on the continental shelf and slope off Nova Scotia (Cifelli 1976). Subsequently, an abrupt change from warm temperate to cold-water faunas occurred at 3 Ma in the Labrador Sea. At this time the cold **Labrador Current** in its present form was generated, mainly as a result of the development of glaciation in the Northern Hemisphere (Berggren and Hollister 1974). The Labrador Current deflected the Gulf Stream southward to a latitude of about 45°N. The juxtaposition of these two currents produced the strongest horizontal thermal gradient in the oceans and thus compression of the isotherms in the western North Atlantic (Figs. 1.5 and 1.6). There was a reduction in number of cold-sensitive species during the **Pleistocene** displacement migrations that occurred along the extensive soft-bottom coasts of the Carolina region. On the European side of the Atlantic there are shorter coastlines with unsuitable substrata for seaweeds (Loire estuary to the Basque coast and North Sea). Beginning about 5000 years ago, the algal flora in the western Atlantic may have approached the composition of the modern flora, as evidenced for the diatom flora in Late Glacial and Holocene successions in the Gulf of Maine (Schnitker and Jorgensen 1990).

The number of seaweed species increases by 35% from Labrador, the southern part of the Arctic region, to **Newfoundland** with its 270 species of marine multicellular algae (South 1983*b*). Generally the coastal vegetation of Newfoundland, as illustrated in Fig. 2.78, is somewhat characteristic of the whole region with its southern limit at Cape Cod. The exception is the eulittoral and upper sublittoral zone on the western and northern coasts of Newfoundland,

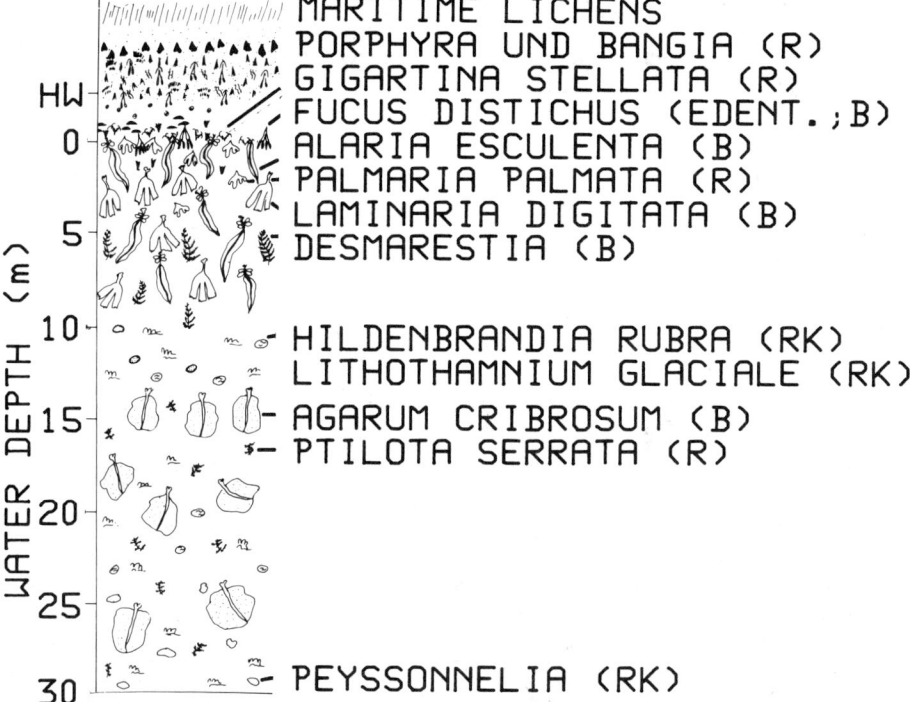

Fig. 2.78 Zonation on the southern coast of Newfoundland (wave-exposed habitat). HW = high-water level; R = red alga; RK = red crustose alga; B = brown alga. *Fucus "distichus"* occurs as the subspecies *edentatus* and is termed *F. evanescens* (Rice and Chapman 1985) in this case. *Desmarestia* = *D. aculeata*. *Gigartina stellata* now named *Mastocarpus stellatus*. (From South 1983b.)

which are scoured by drifting ice. Ice scour removes the perennial kelp canopy of *Alaria esculenta* and favors the colonization of the well-lighted rocky substratum by annual algae such as the brown algae *Chordaria flagelliformis, Dictyosiphon foeniculaceus, Scytosiphon lomentaria,* and *Saccorhiza dermatodea* and the green algae *Spongomorpha arcta* and *S. aeruginosa* (Keats et al. 1985).

Temperature conditions for the Newfoundland seaweed flora. On the eastern coast of Newfoundland facing the Atlantic (Avalon Peninsula) the winter seawater temperature sinks below zero, but in autumn it may rise to 13–14°C for short periods. This temperature is sufficiently high for the successful reproduction of various cold temperate algae that do not penetrate into the Arctic region. The cold water of the Labrador Current (Fig. 2.43) is superimposed on the western and southern coasts of Newfoundland by warmer Atlantic water. Below the thermocline separating the two water bodies at 10-20 m deep, Arctic algae at their southernmost range (e.g., *Laminaria solidungula*) find suitable life conditions down to about 40 m deep. This may be compared to the "isothermic submergence" of the representatives of the Laminariales in the Mediterranean. On the Grand Banks of Newfoundland, 200 km east of the nearest land, Hooper (1987) investigated a benthic algal community on shallow pinnacles with a predominantly Arctic species composition, *Laminaria digitata* and *Alaria esculenta* being prominent species. Few cold temperate algae seem to have reached these isolated rocks, which still mirror the impact of the last glaciation with its maximum at 18,000 years ago.

In the Strait of Belle Isle (Fig. 2.77) a large branch of the Labrador Current (Fig. 2.43) turns west and imparts a cold-water influence on the algal vegetation of the northern **Gulf of St. Lawrence** and the western coast of Newfoundland. In the western part of the Gulf of St. Lawrence the seaweed vegetation is influenced by the fresh water of the St. Lawrence river, by the long-lasting ice coverage, and by the scouring of drift ice. Ice action is still important in the phycologically well-investigated Bays of Chaleurs and Gaspé, along the red sandstone coasts of northern New Brunswick, and from Prince Edward Island to Cape Breton and northern Nova Scotia. The eulittoral zone of these areas looks rather barren because of the ice, but algal vegetation flourishes in the sublittoral zone.

Zonation near Prince Edward Island. In an example from the north shore of Prince Edward Island, red algae are the main components of the vegetation, whereas members of the Laminariales, other than *Chorda filum* and *C. tomentosa,* occur only as scattered plants (up to 60 cm long) or small patches, namely *Laminaria digitata, L. longicruris, L. saccharina* (Bird et al. 1983). The normally intertidal brown alga *Fucus vesiculosus* and the red alga *Chondrus crispus* dominate at depths of 0–2.5 m and 2.5–5 m respectively. *Fucus serratus,* possibly an import from Europe, is associated and competes with *F. vesiculosus*. Numerous, normally intertidal annuals also thrive at 0–5 m, for example, the brown algae *Chordaria flagelliformis, Dictyosiphon foeniculaceus,* and *Scytosiphon lomentaria*. The red alga *Furcellaria lumbricalis* codominates with *Chondrus crispus* at 5–7.5 m and becomes prevalent at 7.5–10 m. Both species are harvested from storm-cast deposits on the beaches. The deep-water red algae *Phyllophora truncata* and *P. pseudoceranoides* flourish at 10–15 m, sometimes to 20 m, whereas the vegetation at 20–25 m consists mainly of the deep-water kelp *Agarum cribrosum,* the filamentous red alga *Polysiphonia urceolata,* and crustose corallines. Except for *Agarum cribrosum,* all these algae are well known from European coasts, which emphasizes the common biogeographical character of the cold temperate North Atlantic region.

Warmer-water species are found in a few shallow embayments and estuaries in the southern Gulf of St. Lawrence because of relatively high (up to 28°C) surface temperatures in summer. These species, which must in addition tolerate a temperature of 0°C, are also found farther south on the Atlantic coast; examples are the red algae *Lomentaria baileyana* and *Dasya baillouviana* and the brown alga *Stilophora rhizodes*. They may be relics of a warmer period, 7000 years ago ("climatic optimum," or "mid-hypsithermal period") (Bousfield and Thomas 1975), when warm-water biota penetrated further north than today. The above red algae require the summer temperatures for reproduction and overwinter as small pads (Novaczek et al. 1987).

Taken as a whole, the southern Gulf of St. Lawrence is not a refuge for warmer-water species but harbors a cold-water flora that tolerates brief exposure to higher temperatures (Bird et al. 1983). Along the Atlantic coast of **Nova Scotia** there is a luxurious algal vegetation present in the eulittoral zone.

Atlantic coast of Nova Scotia: eulittoral zone. At slightly wave-exposed locations in the eulittoral zone (Fig. 2.79B) the amphi-Atlantic species *Fucus spiralis, F. vesiculosus, F. evanescens,* and *Ascophyllum nodosum* dominate. These locations are shared by the periwinkles *Littorina littorea* and *L. obtusata,* while the barnacle *Semibalanus balanoides* dominates at strongly exposed habitats. The red algae *Chondrus crispus* and *Devaleraea* (= *Halosaccion*) *ramentacea* form continuous belts in the lower eulittoral. Other red algae such as *Hildenbrandia rubra, Corallina officinalis,* the green alga *Cladophora rupestris,* and the brown alga *Pilayella littoralis* act as understory algae or as continuous belts at greater depths. The impressive tidal range in the **Bay of Fundy** gives the eulittoral zone a vertical extension of up to 18 m. A well-developed algal vegetation is present at the entrance; examples are *Fucus vesiculosus* and *Ascophyllum nodosum* (Fig. 2.79G; for details see Edelstein et al. 1970; J. S. Wilson et al. 1979; Tittley et al. 1987). The inner parts of the Bay of Fundy are locally muddy and not favorable for seaweed growth.

Atlantic coast of Nova Scotia: sublittoral zone. In the sublittoral zone of the American province, as in the European province, one finds *Alaria esculenta* as a canopy just below the low-water mark, followed by *Laminaria digitata*. A striking difference is the lack of *L. hyperborea*. Its role as a canopy alga in the midsublittoral zone, down to 10 m deep, is played on the eastern North American coast by *L. digitata* and the northeast American endemic *L. longicruris*. On cleared quadrats, the faster-growing *L. longicruris* is initially more abundant (C. M. Smith 1986). Due to its maximum age of only two to three years, *L. longicruris* becomes overgrown after three years by the increasing canopy of the slower-growing but more persistent *L. digitata,* which has a maximum age of up to five years. Besides *L. longicruris,* which is characterized by its long and hollow stipes, *L. saccharina* with its shorter, solid stipes occurs further to the south and also on the coast of Nova Scotia in wave-protected habitats (A. R. O. Chapman 1973a). In contrast to *L. saccharina, L. longicruris* consistently develops hollow stipes in culture tanks, although this may take two years in summer-developed sporophytes. The morphologies of the two species are similar in the first year of growth. The absence of *L. longicruris* from western mainland Europe may indicate a recently evolved taxon (Yarish and Egan 1987; Egan and Yarish 1988; Yarish et al. 1990). In the midsublittoral zone the green sea urchin *Strongylocentrotus droebachiensis* may become abundant and destroy the laminarian vegetation. This happened during the early 1970s; however, a disease-induced mass mortality of urchins followed, and algal biomass increased again (Miller 1985a,b; Chapman 1986; see Section 8.1). *Agarum cribrosum,* a deep-water Arctic–cold temperate kelp with a perforated blade, grows where the dense laminarian vegetation ends, normally dominating in the 10–30 m range (Figs. 2.10, 2.79F,G), together with the red alga *Ptilota serrata*. Both algal species seem to have a defense against predation by urchins and are least preferred by the green sea urchin (Vadas 1977, Keats et al. 1984). If the *Laminaria* vegetation has been destroyed by sea urchins, the occurrence of *A. cribrosum* shifts upwards so that one may assume that the upper depth limit of this species is due to competition (A. R. O. Chapman 1973b, 1981; Tremblay and Chapman 1980).

South of the Bay of Fundy, on the coasts of **Maine, New Hampshire,** and **Massachusetts,** the cold temperate algal vegetation influence remains, and the lower algal limit may be situated locally at 20–25 m (Lamb and Zimmermann 1964). Off Cape Ann (Massachusetts) on submarine banks in relatively clear water, an algal vegetation grows at an annual temperature range of 4–11°C (Sears and Cooper 1978). This vegetation consists mainly of the red algae *Ptilota serrata* (Fig. 2.26), *Phycodrys rubens* (Fig. 2.28), *Phyllophora truncata* (= *brodiaei;* Fig. 2.25), *Callophyllis cristata* (Fig. 2.16), and *Fimbriofolium* (= *Rhodophyllis*) *dichotomum* (Fig. 2.24). Also found here are crustose red algae such as *Lithothamnion glaciale* (Fig. 3.12) and the deep-water green alga *Derbesia marina*. Off the coast of Maine, using a submersible, Vadas and Steneck (1988) investigated another deep-water vegetation community. A *Laminaria* species, up to 2 m long and of uncertain taxonomic affinity (possibly *L. digitata*), was found to a 30 m depth in relatively clear water and growing on a deep-water rock pinnacle; *Agarum cribrosum* was found to 40 m. Foliose red algae, mainly *Ptilota serrata* and *Phycodrys rubens,* grew down to 50 m, fleshy crusts (Peyssonnelia sp.) to 55 m, and coralline crusts (*Leptophytum laeve*) to 63 m.

The cold Labrador Current flowing from the north finally turns eastward in summer at **Cape Cod** and farther south in winter near Cape Hatteras at the southern boundary of the cold temperate North Atlantic region. The Gulf Stream reaches the southern coast of Cape Cod, and consequently the distribution of most of the Arctic–cold temperate seaweed species ends either at Cape Cod or slightly further north (Table 2.1).

Between **Cape Cod** and **Long Island Sound,** where the rocky substratum of the northern coast ends, the loss of Arctic–cold temperate species from Cape Cod (Fig. 2.77) is not balanced by the gain in warm-adapted algae. This decline in diversity is also seen on the European coast south of the border area of Brittany. For many of the southern species the local and seasonal temperature gradient south of Cape Cod is probably too steep, and so they appear only at Cape Hatteras, North Carolina, about 800 km south of Cape Cod. Long Island Sound is also the southern limit of distribution of the genus *Laminaria* in shallow water in the western North Atlantic, and harbors still the three species *L. digitata, L. saccharina,* and *L. longicruris*. The southernmost outpost in the western Atlantic may be a deep-water population of *L. saccharina* that was discovered at 28–36 m deep at the edge of the continental shelf off the coast of New Jersey (Egan and Yarish 1988).

Fig. 2.79 Aspects of the seaweed vegetation on the east coast of Nova Scotia. **(A–F)** Wave-exposed coast near Halifax. **(A)** Coastal view. **(B)** Mideulittoral zone with *Fucus vesiculosus* and *Chondrus crispus*. **(C)** Upper sublittoral zone (1 m deep): *Alaria esculenta*. **(D)** Midsublittoral zone (5 m deep): *Laminaria longicruris* and *L. digitata* (at the right). **(E)** *L. digitata*. **(F)** 12 m deep: *Agarum cribrosum*. **(G)** Eulittoral zone of the outer Bay of Fundy: *Fucus* spp. and *Ascophyllum nodosum*. (Photographs by the author.)

2.5.2 Warm Temperate Carolina Region (Cape Hatteras to Cape Kennedy)

This region, starting south of Cape Hatteras, is bordered by the 10°C-winter isotherm in the north and the 20°C-winter isotherm at Cape Kennedy (= Cape Canaveral) in the south. These are the same isotherm limits as the warm temperate Mediterranean–Atlantic region (Figs. 1.5–1.7). Close relationships between the warm temperate algal floras of the European–North African region and the North American coast do not exist because both have been geographically isolated for so long (van den Hoek 1984).

The question as to whether Cape Hatteras or Cape Cod should be regarded as the northern boundary of the warm temperate region of the mid-Atlantic coast of the United States has been discussed in detail by such workers as Humm (1969), Briggs (1974), and Searles (1984). Both authors advocate Cape Hatteras as the northern boundary, mainly because farther north the cold temperate biota prevail.

Whereas on the European coasts the warm temperate region extends over 30° of latitude, this distance is reduced to almost 5° on the mid-Atlantic coast of North America. The latter coast is mainly unfavorable to seaweeds because hard substrata is replaced by sandy barrier islands and shallow sounds. The main substratum for seaweeds in shallow water is provided by shell rubble, seagrasses, and manufactured structures, such as seawalls and jetties. There is, however, a deep offshore vegetation on the continental shelf; the latter widens north and south of Cape Hatteras to 100 km or more off the New Jersey and Georgia coasts.

The diversity of endemic seaweeds within the Carolina region is relatively low. The region encompasses 320 species of seaweeds, with about 30 endemic species (Searles 1984). Most of these (about two-thirds) are restricted to the deep-water vegetation and possibly adapted to the moderate but seasonally cooler temperatures that occur on the continental shelf. The shallow and offshore floras each contain around 200 seaweed species. Approximately 100 species occur exclusively in the shallow coastal environment and another 100 only in the deep offshore flora. The latter group consists almost exclusively of tropical species, most of which occur at their northern limit and have a peak development in mid summer. In the shallow vegetation there is a eurythermal tropical element that develops mainly in the summer, whereas northern species flourish in winter and warm temperate species all year round (for details see Earle 1969; Kapraun 1980; Searles 1984). Some eurythermal, northern species such as *Fucus vesiculosus* (Fig. 2.23) or the seagrass *Zostera marina* also occur.

Chesapeake Bay, bordered by Maryland and Virginia, has extensive salt marshes dominated by halophytic higher plants such as *Spartina alterniflora*. Water temperatures of up to 30°C occur in this brackish water habitat, and only a limited number of marine algae (like species of *Enteromorpha*) are found growing on the jetties and other artificial substrata (Ott 1973). Likewise, in the turbid water near the sandy coasts of **North Carolina,** seaweeds do not find favorable conditions. However, in relatively shallow water off the coast, solid substratum is available in the form of rocky outcrops, stones and organic concretions, for example, 10–20 km off **Cape Hatteras.** Here a characteristic deep-water algal vegetation is present 15–60 m deep in relatively clear water (Peckol and Searles 1983; Searles 1984), at water temperatures (at a 35 m depth) of 11°C in February and 24°C in August (Hazel 1970).

At Cape Hatteras the Gulf Stream turns eastward into the Atlantic (Fig. 2.43). Thus many warm-adapted algae thrive here in deeper water; examples are the green algae *Caulerpa prolifera* (Fig. 2.76A), *Codium carolinianum,* the brown algae *Sargassum filipendula* (Fig. 4.1P), *Dicytota dichotoma* (Fig. 2.69C), *Zonaria tournefortii,* the red algae *Botryocladia occidentalis* and *Gracilaria mammillaris,* and many others of the approximately 800 seaweed species of the western Atlantic tropical region. Species diversity rises from about 100 species on the coasts of Maryland and Virginia to about 300 on the coast of North Carolina because of the northern extension of the sublittoral Caribbean algal flora.

The soft-bottomed coasts of South Carolina and Georgia, where extensive salt marshes prevail, are also unfavorable for seaweeds (there are only about 80 seaweed species on the coast of Georgia). This situation also applies to the sandy coasts of northern Florida, from the southern limit of the Carolina region to the beginning of the western Atlantic tropical region at **Cape Kennedy.** Here the 23°C-February isotherm and the 29°C-August isotherm approach the coast (deviating from the scheme given in Fig. 1.8), and the distributions of the last northern species (*Petalonia fascia, Ectocarpus siliculosus,* and *Porphyra leucosticta*) ends. The deep-water macroalgal community on the East Florida continental shelf is mainly of Caribbean origin, particularly diverse at a 30–40 m depth, and contains about 200 taxa, excluding crustose corallines (Hanisak and Blair 1988).

Gulf of Mexico. The northern coasts of the Gulf of Mexico, from Tampico (Mexico) to the area between Tampa Bay and Cape Romano, are not regarded as tropical but belong to the southern part of the Carolina region (Briggs 1974). This coastline includes the soft-bottomed coasts of Texas, Louisiana, Alabama, and West Florida with locally occurring limestone. Many tropical species are able to survive here due to the high summer water temperatures, which may rise to 28–30°C, uncharacteristic of a warm temperate region. The genus *Sargassum* is represented by attached species such as *S. filipendula* and by the floating *S. fluitans* and *S. natans*. These latter species also thrive in the Sargasso Sea. However, the winter temperatures in the surface water from the Mississippi estuary to West Florida may be as low as 13–15°C, well below the minimum temperature in tropical regions. In winter, northern species flourish; examples are the macrothalli of *Porphyra leucosticta* and *Bangia atropurpurea,* and *Ectocarpus siliculosus, Spongonema tomentosum,* and *Petalonia fascia*. (For details see Earle 1969; Edwards 1969; Edwards and Kapraun 1973.) These species do not occur at the southern tip of Florida, which belongs, as an isolated "island," to the western Atlantic tropical region. They appear again on the warm temperate western coast of Florida. Although today southern Florida separates the floristically similar southern parts of the Atlantic coast of the Carolina region from the northern Gulf of Mexico, one can assume that there was a continuous distribution of warm temperate species around southern Florida during the Pleistocene (Earle 1969). Furthermore, from Miocene times up to 1.8 Ma, the Suwannee Straits connected the Gulf of Mexico to the Atlantic Ocean across northern Florida (Berggren and Hollister 1974; Bert 1986).

2.6 NORTH AMERICA: COLD AND WARM TEMPERATE REGIONS IN THE NORTH PACIFIC

Literature. Entire area: Abbott and Hollenberg (1976); Carefoot (1977); Kozloff (1983); Ruprecht (1852); Setchell and Gardner (1920–1925); Waaland (1977). **Keys and synopsis, southeastern Alaska to northern Washington:** Gabrielson et al. (1987); Scagel et al. (1989). **Alaska and islands of the Bering Sea:** Calvin and Ellis (1978); Calvin and Lindstrom (e.g., 1980); Hansen et al. (1981); Johansen (1971a); Lebednik et al. (1971); Lindstrom (1977); Lindstrom and Scagel (1979); Setchell (1899); Wynne (1970, 1985–1987). **British Columbia and Washington:** Dayton (1971, 1975); Dethier (1987); Garbary et al. (1980–1982); Hawkes et al. (1978); Kylin (1925); Neushul (1965a, 1967); Rigg and Miller (1949); Scagel (e.g., 1957, 1967, 1973); Stephenson and Stephenson (e.g., 1972); Widdowson (e.g., 1973–1974). **Oregon:** Doty (1947); Markham and Celestino (1976); Phinney (1977). **California:** Abbott and Hollenberg (1976); Abbott and North (1972); Aleem (1973); Dawson and Foster (1982); Dayton (1985a); Dayton et al. (1984); Devinny and Kirkwood (1974); Foster (1975); Foster et al. (1988); Galbraith and Boehler (1974); Hommersand (1972); McLean (1962); Murray and Littler (e.g., 1981); Neushul (1965a); Quast (1971); Ricketts et al. (1985); Schiel and Foster (1986); Silva (1979); Stephenson and Stephenson (e.g., 1972); Thom (1980); Wilson et al. (1977). **Baja California:** Dawson (1941); Dawson et al. (e.g., 1960); Devinny (1978); Ketchum (1983); Littler and Littler (1981). **Taxonomic groups: Laminariales:** Dayton (1985a); Dayton and Tegner (1984); Dayton et al. (1984); Druehl (1968, 1970, 1981); Ebeling et al. (1985); Foreman (1984); Foster and Schiel (1985); Neushul (e.g., 1971, 1977); North (e.g., 1971); Sanbonsuga and Neushul (1978); Schiel and Foster (1986); Widdowson (1971); Womersley (1954). **Crustose coralline algae:** Steneck and Paine (1986).

The boundary between the Arctic region and the cold temperate northwestern Pacific region runs parallel to the mean southern limit of pack ice, about 60°N, or the latitude of Nunivak Island off the Alaskan coast (Figs. 1.5–1.7). The cold temperate region ends at Point Conception in mid-California, and from there the warm temperate California region extends to Baja California, where local upwelling occurs (see below). The North American Pacific coast harbors one of the richest seaweed floras of the world and may be compared, in number of species and endemics, with other rich algal floras such as the Mediterranean, Japanese, and southern Australian coasts.

History of algal investigation and marine biological stations. The first algal investigations were carried out in this area as a result of Captain George Vancouver's expedition between 1791 and 1795. Accompanying him was Archibald Menzies, who collected some seaweeds that were later described by European phycologists. Postels's and Ruprecht's classical work (1840) was based on a Russian expedition and illustrated among others the giant kelps of the Pacific coast of North America (Fig. 2.82G). Further collections in the late 19th and early 20th centuries culminated in the works of Harvey, Anderson, Setchell, Gardner, and G. M. Smith. (For historical reviews see Scagel 1957; Papenfuss 1976.) Major biological stations are Bamfield Marine Station on Vancouver Island; Friday Harbor Laboratories (formerly known as the Puget Sound Biological Station) in Washington; and in California Bodega Marine Station near Bodega Bay, Hopkins Marine Station on Monterey Peninsula, and Scripps Institute of Oceanography at La Jolla.

Paleoceanography and paleobiogeography. Due to the continuous northward movement of the Pacific plate, thousands of kilometers of oceanic crust slipped under the Kamchatka, Beringian, and Alaskan margins in the **Mesozoic,** and hundreds of kilometers slipped under the Aleutian Ridge during the Cenozoic. The Aleutian Ridge was formed during latest Cretaceous or earliest Tertiary times (Scholl et al. 1975; Kennett 1982). The basic paleobiogeographical traits have been mentioned in Section 2.1. It is important to remember that throughout the **Tertiary** cool water taxa had a better chance to evolve in the North Pacific than in the North Atlantic. The upper temperature tolerance of typical cold temperate kelp species is several degrees lower in the eastern North Pacific than in the North Atlantic (Lüning and Freshwater 1988). At least eight major phases of intense cooling occurred in the marginal North Pacific during the last 15 million years, with maxima in the late Miocene, late Pliocene, and Pleistocene (Ingle 1967, 1977). During the **Miocene cooling** (10–5 Ma) and glacial times (Fig. 2.1C), for example, sub-Arctic foraminiferal species with temperature reqirements below 10°C progressed to the Los Angeles Basin. During the warmer events, warm temperate taxa penetrated the Alaskan Current gyre above 50°N. The major **water masses** of the North Pacific are thought to have developed their present properties in the early or middle Miocene (Casey 1977). During the past 15 million years the cooling effect was less on the Asiatic side of the North Pacific where the high velocity Kuroshio (Kuro Siwo) Current transports warm water and tropical–subtropical plankton northward to about 41°N. It then converges with the cold Oyashio Current (Fig. 2.43). In contrast, on the American side the relatively slow **California Current** has its source at high latitudes, transporting cold water southward and converging with equatorial water about 10°N, south of southern California (Fig. 2.43). This convergence has changed more dramatically in latitude during the last 15 million years than has the more stationary convergence of the Kuroshio and Oyashio Currents on the Asiatic side. Only minor penetrations of cold-water species southward to about 30°N occurred in the most intense cooling phases on the Asiatic side (Ingle 1967, 1977). Generally, the analysis of planktonic faunas from the Miocene through the Pleistocene has revealed that the distribution of some cool-water species reaches further south on the American side of the North Pacific than on the Asiatic side (Ingle 1967). This **persistent cool-water aspect on the American side** of the North Pacific may have favored the evolution of certain cool-water taxa that do not occur on the Asiatic side. Furthermores, Estes and Steinberg (1988) have pointed out that limpets and herbivorous marine mammals (e.g., sea otters), obligately associated with North American kelps, appeared late in the Cenozoic. An unusually low intensity of herbivory in the North Pacific may have favored the evolution of large brown algae (see Section 8.1.2).

Occurrence of the Laminariales. The Laminariales provide the most noticeable assemblage of cool- and cold-water algae on the Pacific North American coast. About 75% of the genera are endemic, and the greatest species diversity is encountered in the vicinity of Vancouver Island (Druehl 1981). Particularly characteristic are the various genera of the **family Lessoniaceae,** which are all endemic (Fig. 2.81), except for *Macrocystis,* which also occurs in the Southern Hemisphere (Fig. 5.2), and *Lessonia* species, which are restricted to the Southern Hemisphere (Fig. 5.6). The largest and most highly differentiated algal species are represented by the three kelps *M. pyrifera, Nereocystis luetkeana,* and *Pelagophycus porra* (Fig. 2.80), all members of the family Lessoniaceae. They are probably young geologically and may have originated towards the end of the Tertiary. A fossil link was found in Miocene deposits from southern California that combines traits of the recent genera *Pelagophycus* and *Nereocystis* (Figs. 2.80E,F; Parker and Dawson 1965). Furthermore, it seems as if speciation in this family is still at an active stage, as intermediate hybrids between *Macrocystis pyrifera* and *Pelagophycus porra* have been found in the field (Neushul 1971; Coyer *et al.* 1982). These crosses can be produced in the laboratory (Fig. 2.80 D; Sanbonsuga and Neushul 1978). A phylogenetic analysis of kelp chloroplast DNA restriction fragments indicated that *Macrocystis, Lessonia* and *Alaria* constitute one lineage and *Nereocystis* and *Laminaria* are in another one (Fain et al. 1988). The giant kelps may thus be of polyphyletic origin, i.e. they may not represent a closely related group. The geographical distribution of the Laminariales along the Pacific coast of North America is illustrated in Fig. 2.81. Of the family **Laminariaceae,** *Laminaria saccharina* and *Agarum cribrosum* occur in the Pacific and the Atlantic. *L. yezoensis, Costaria costata, Cymathere triplicata,* and *Thalassiophyllum clathrus* inhabit the North Pacific as far as northern Japan, whereas the remaining 11 species of the Laminariaceae listed in Fig. 2.81 are endemic to Pacific North America. Half of the species of the *family Alariaceae* in the North Pacific are endemic to Pacific North America, and the other half extend into the western North Pacific. According to Widdowson (1971), these latter species reach the coast of northern Japan (*Alaria praelonga* and *Eisenia arborea*), to the Kurile Islands (*A. crispa* and *A. fistulosa*), and to the western Aleutian Islands (*A. taeniata* and *A. tenuifolia*). Almost as large as the giant kelps, *A. fistulosa,* with blades up to 18 m long and 1.7 m wide, is buoyed to the surface by an inflated midrib (Fig. 2.82, lower; Druehl 1981). The Pacific–Atlantic species *Chorda filum* (Fig. 2.9) which is a member of the primitive family **Chordaceae,** is also present. No other coast of the world harbors such a wealth of endemic species and genera of all four families of the Laminariales. The North Pacific may be regarded as a center of evolution of this order. Only a few (e.g., *Alaria, Laminaria* and *Agarum*) have migrated into the Atlantic or to the Southern Hemisphere (e.g. *Macrocystis pyrifera*). Further evidence that the possible origin of the Laminariales is the Pacific coast of North America is that the basic morphological patterns are all represented here, within the genus *Laminaria.* There are species with a digitately divided blade (section Digitatae), such as *L. dentigera* and *L. setchellii* (Druehl 1979), which have a rigid stipe like the eastern Atlantic species *L. hyperborea.* Another, *L. bongardiana,* has a flexible stipe like the Atlantic species *L. digitata* but develops slits in the frond much later than does *L. digitata. L. saccharina* and *L. farlowii* are species with an entire blade (Section Simplices). A discoid holdfast is found in *L. ephemera* and *L. yezoensis* (subgenus *Solearia,* according to Petrov 1974). This trait also occurs in the Arctic species *L. solidungula* (Figs. 3.3E and 3.6), which may be a descendant of its North Pacific allies. Finally, the subgenus *Rhizomaria* (accord-

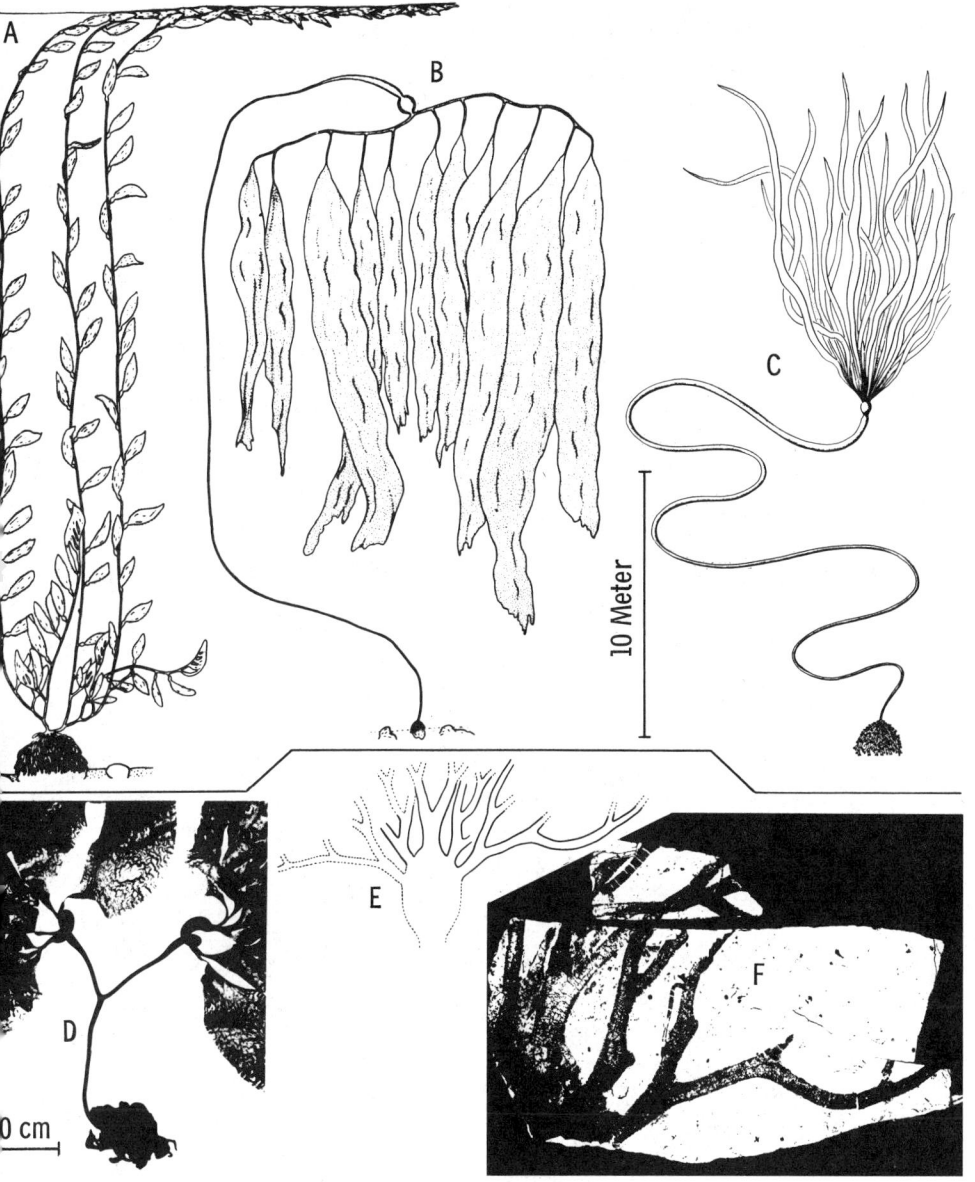

Fig. 2.80 Giant kelps of the North American Pacific coast. **(A)** *Macrocystis pyrifera*. **(B)** *Pelagophycus porra*. **(C)** *Nereocystis luetkeana*. **(D)** Generic hybrid between *M. pyrifera* (dichotomous branching) and *P. porra* (bladders in the stipe), produced by a crossing experiment in the laboratory. **(E)** Branching pattern. **(F)** Imprint of the fossil algal species *Julescraneia grandicornis* from Californian Tertiary deposits. The genus *Julescraneia* is intermediate between *Pelagophycus* (antlerlike stipes on bladders) and *Nereocystis* (multiple branching). (A from Galbraith and Boehler 1974; B from Dawson et al. 1960; C from Abbott and Hollenberg 1976; D from Sanbonsuga and Neushul 1978; E,F from Parker and Dawson 1965.)

138 SEAWEED VEGETATION OF THE COLD AND WARM TEMPERATE REGIONS

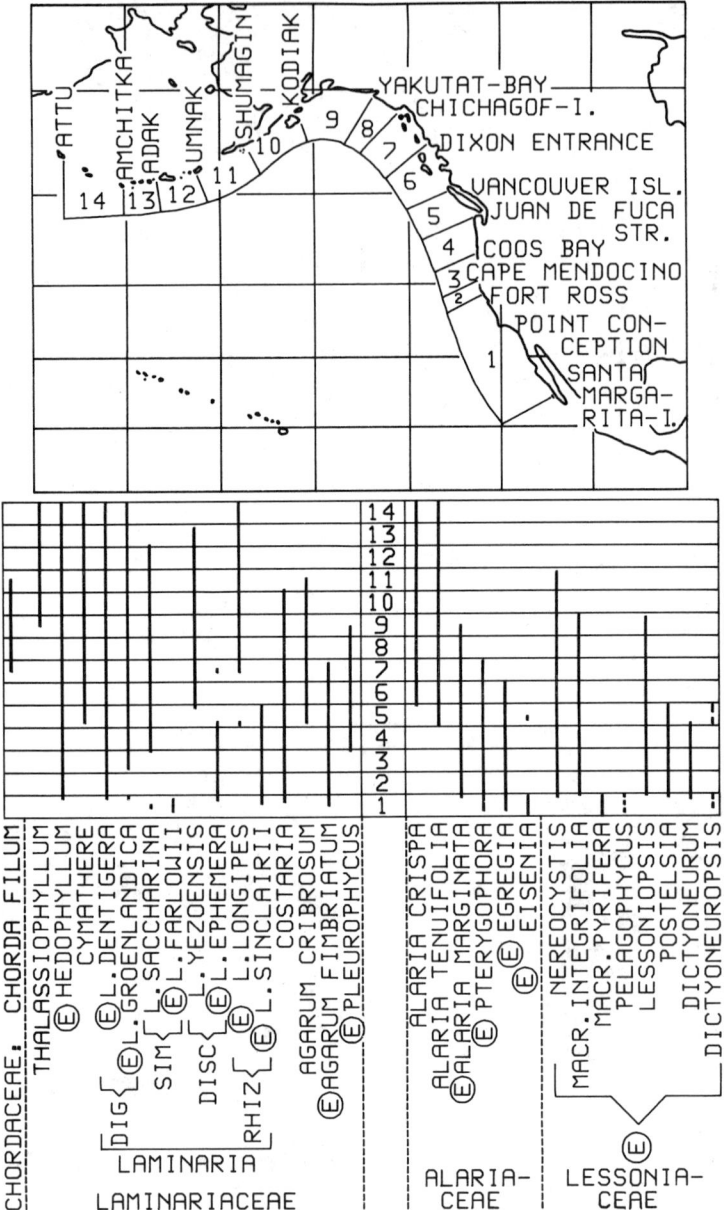

Fig. 2.81 Distribution of representatives of the Laminariales in 14 sectors along the North American Pacific coast. E = endemic to this coast. Abbreviations for species groups of *Laminaria:* DIG section Digitatae (divided blade); SIM = section Simplices (undivided blade); DISC = discoid holdfast (subgenus *Solearia* according to Ju. E. Petrov 1974); RHIZ = rhizomatous holdfast (subgenus *Rhizomaria* according to Petrov 1974). *Laminaria dentigera* and *L. setchellii* are conspecific according to

ing to Petrov 1974) with a rhizomatous attachment system is represented by the two species *L. sinclairii* and *L. longipes*. The relationship to the third member of this subgenus, *L. rodriguezii* in the Mediterranean, remains a puzzle (see Section 5.5 for Brazilian deep-water *Laminaria* spp.).

2.6.1 Cold Temperate Northeastern Pacific Region (Alaska to Central California)

Along the Pacific coast of North America, as along the European coasts and in general along the eastern sides of the oceans, the isotherms diverge (Figs. 1.5 and 1.6). Thus the cold temperate northeastern Pacific region extends over a vast latitudinal range. A relatively uniform temperature regime is affected by the North Pacific Current. It originates at the Asiatic coast from the warm Kuroshio (Kuro Siwo) Current and crosses the Pacific at approximately 50°N (Fig. 2.43). About 500 km off the coast of Oregon this current branches, sending off the Alaska Current to the north and the California Current to the south (Fig. 2.43). Both branches tend to keep the temperature regime uniform. Furthermore, upwelling of cold water along the Californian coast keeps the surface water temperatures relatively low even in the southern, central Californian part of the region. In consequence, the water temperatures along the coast from the Aleutian Islands to Point Conception (central California) range between 10°C and 17°C in August and between 5°C and 14°C in February. Local coastal temperatures can vary more in British Columbia and Southeast Alaska, which have numerous inland passages, bays, and fjords. Various seaweed species are spread over enormously long distances from California to Alaska or even to the Aleutian Islands (Figs. 2.81 and 2.82). A similar wide occurrence is also exhibited by several amphioceanic species, such as *Desmarestia viridis* (Fig. 2.8), *Scytosiphon lomentaria*, *Codium fragile* (Fig. 2.62G) and *Plocamium cartilagineum* (Fig. 2.65E).

The algal flora of the **Aleutian Islands** consists of a mixture of Arctic–cold temperate species and cold temperate species of the eastern and western North Pacific. On the coast of Amchitka Island the water temperature ranges between 10°C in summer and 1–2°C in winter. About 10% of the seaweed species of Amchitka Island are endemic to the area of the islands of the Bering Sea (Lebednik et al. 1971) and are probably adapted to the narrow temperature range of this area. In the supralittoral zone of Amchitka Island one finds the green algae *Prasiola borealis*, species of *Ulothrix*, and the red algae *Audouinella* (*Rhodochorton*) *purpurea* and *Porphyra* species. However,

---

Abbott and Hollenberg (1976). Druehl (1979) regards the populations occurring in segments 14–9 as *L. dentigera,* in segments 8–1 as *L. setchellii. L. groenlandica* = *L. bongardiana* (see legend to Fig. 2.82). The genus *Alaria* is represented by more species than indicated in the figure: *A. nana, A. praelonga* (both endemic), *A. fistulosa* (westwards to Kurile Islands and Sakhalin), and *A. taeniata* (westward to the Aleutian Islands). (After Druehl 1970.)

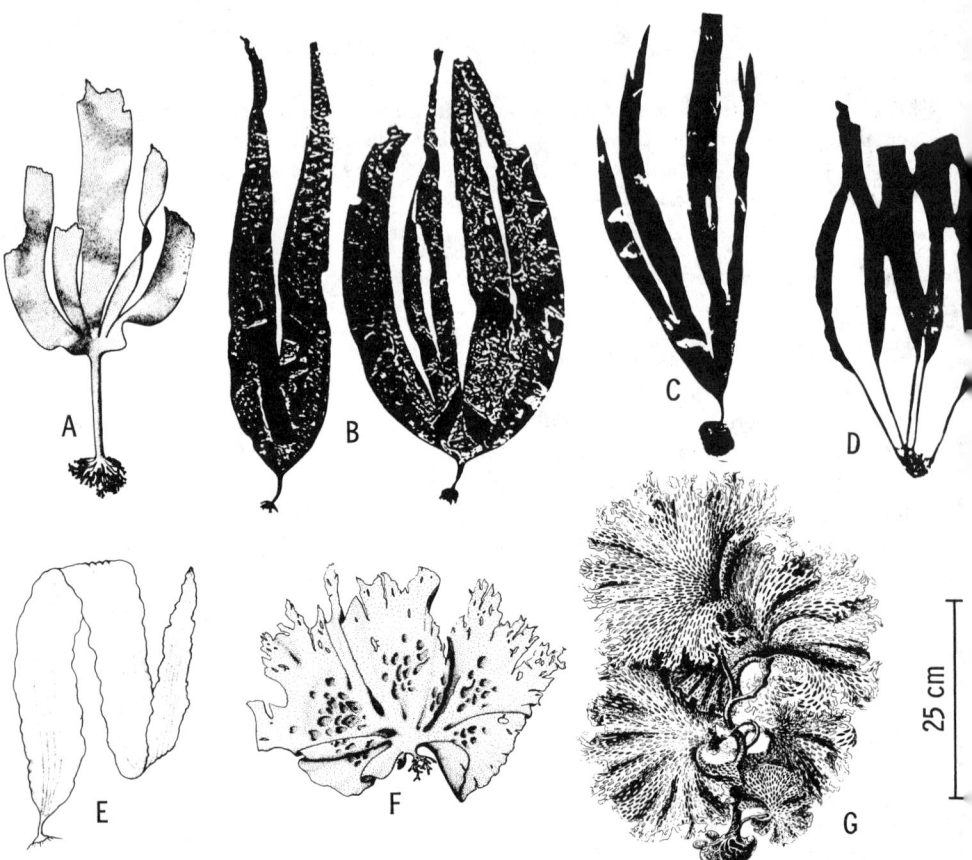

Fig. 2.82 Kelp vegetation of Alaska. **Above:** some representatives of the Laminariales (family Laminariaceae), which occur on the southern coastline of Alaska. **(A)** *Laminaria setchellii (dentigera));* **(B)** *L. bongardiana* (=*L. groenlandica*) (left: first-year plant; right: second-year plant). **(C)** *L. yezoensis* (with discoid holdfast). **(D)** *L. longipes* (with rhizomatous holdfast). **(E)** *Cymathere triplicata.* **(F)** *Hedophyllum sessile.* **(G)** *Thalassiophyllum clathrus.* Note: *L. bongardiana* Post et Rupr. (see Petrov 1973), a digitate species in the northeastern Pacific, had been called *L. cuneifolia* by Setchell and Gardner (1920–1925) and *L. groenlandica* by Druehl (1968). *L. cuneifolia* and *L. groenlandica* refer to North Atlantic representatives of the Simplices section of *Laminaria* and are included in *L. saccharina* (South 1984; South and Tittley 1986). Wilce (1959–1965), in work in the North Atlantic on the section Simplices, had included *L. cuneifolia* in *L. groenlandica;* subsequently the Pacific alga received its peculiar name. (A,F from Abbott and Hollenberg 1976; B from Druehl et al. 1987; C,D from Druehl 1968; E from Scagel 1967; G from Postels and Ruprecht 1840). **Right:** succession leading to Laminaria forest at Torch Bay (southeastern Alaska, Glacier Bay National Park; see segments 7–8 in Fig. 2.81). **(A)**Algal-barren rock (only crustose coralline algae left) with sea urchins. **(B)** One year after experimental removal of sea urchins: fast-growing annual kelps *Nereocystis luetkeana* and *Alaria fistulosa* as upper canopy algae, *Costaria costata* and *Laminaria dentigera* as lower canopy algae. **(C,D)** Second and third year: transgression to pure stands of the perennial *L. setchellii* (similar to *L. dentigera*) originally cited as *L. groenlandica.* (From Duggins 1980.)

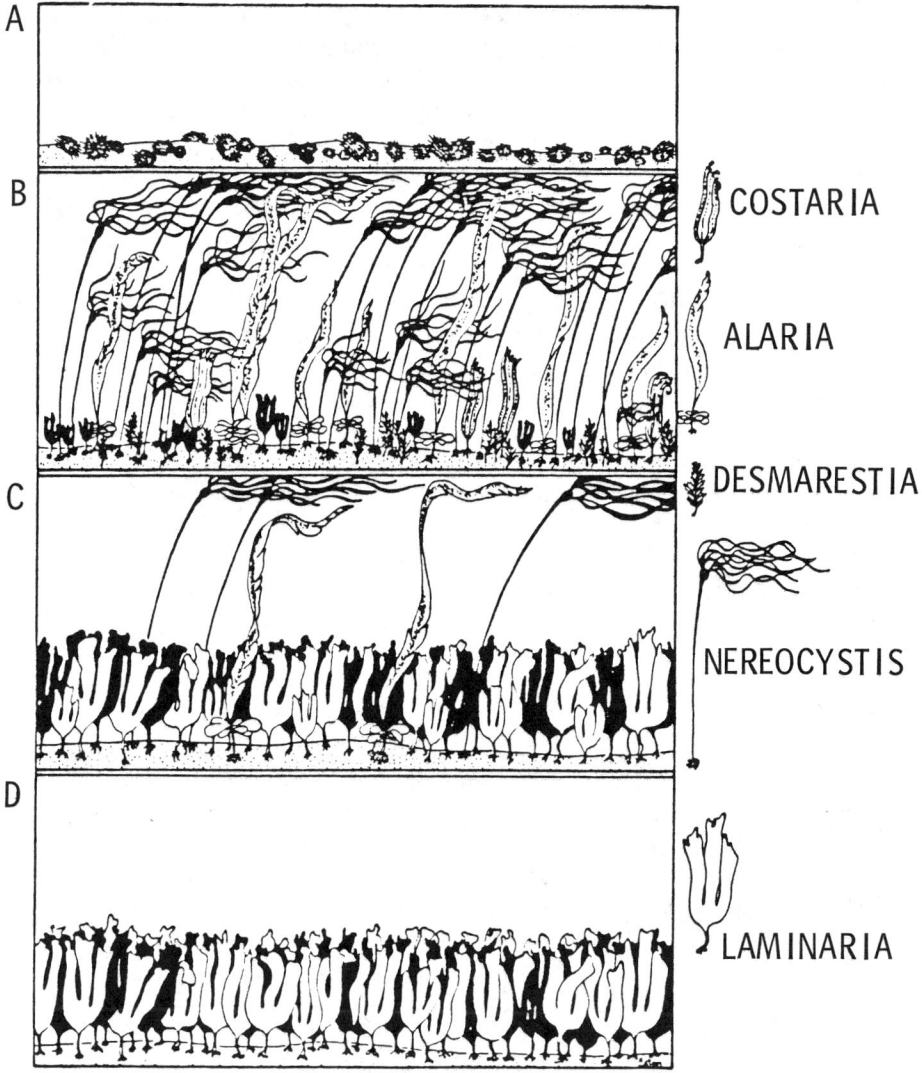

Fig. 2.82 (*Continued*)

there are few barnacles and marine lichens. The eulittoral zone is dominated by the brown algae *Fucus* cf. *distichus* (Fig. 2.13), *Hedophyllum sessile* (Fig. 2.82F), and the red alga *Halosaccion americanum* (Fig. 2.83), whereas in the upper sublittoral zone the kelps *Alaria crispa* and *Laminaria longipes* (Fig. 2.82C) prevail.

The **southern coast of Alaska** may be divided into two units, southcentral Alaska (Alaska Peninsula, Kodiak Island, Cook Inlet, and Prince William Sound), and southeastern Alaska (the Alexander Archipelago, which is geographically the northern extension British Columbia). The coastline between Kodiak Island and Prince of Wales Island (comprising segments 8 and 9 in Fig. 2.81) represents an overlap area (Druehl 1970, Lindstrom and Scagel 1979), as it is inhabited by many southern algae and by some Arctic–cold temperate species, such as *Phyllophora truncata* (Fig. 2.25). An intensive study has been conducted on the seaweed vegetation of Prince William Sound, where in 1964 an earthquake lifted the rocky substratum up to a height of 10 m above the water level leaving the remnants of dead attached seaweeds exposed to the air (Johansen 1971a). In Prince William Sound the upper sublittoral zone is dominated, at wave-exposed places, by *Laminaria bongardiana* (Fig. 2.82B, upper) and *L. saccharina* (Fig. 2.11). In the midsublittoral zone at Kodiak Island, situated in the Gulf of Alaska, the greater part of the algal biomass is represented by *Laminaria dentigera* (similar to *L. setchellii;* Fig. 2.82A, upper; compare with Fig. 2.81), which forms a dense laminarian forest in the depth range of 3–10 m and passes into a laminarian park at 18 m deep (Calvin and Ellis 1978). Other kelp species found here are *L. yezoensis* (Fig. 2.82C, upper), *Pleurophycus gardneri* and *Agarum cribrosum* (Fig. 2.10).

The perennial laminarian forest represents the endpoint of a rapid algal succession (Duggins 1980, 1983). Where urchins have overgrazed and left an algal-free substrate with only crustose corallines present, the succession moves to a diverse kelp community in the first year and to an almost pure stand of *Laminaria setchellii* (similar to *L. dentigera;* originally stated as *L. groenlandica*) in the third year (Fig. 2.82, lower). This succession takes place whenever the sea urchin populations of *Strongylocentrotus* species become decimated by the sea otter *Enhydra lutris* or the large starfish *Pycnopodia helianthiodes*. The latter, which may attain a diameter of up to 1.5 m, is the largest starfish in the world. In Torch Bay (southeastern Alaska) about two-thirds of the rocky area in the midsublittoral zone is characterized as overgrazed, barren ground; almost one-third is a kelp forest of *L. setchellii;* and about 5% has stands with early successional, annual kelp species such as *Nereocystis luetkeana, Alaria fistulosa,* and *Costaria costata*. These are outcompeted later by the dense *L. setchellii* forest.

Zonation: British Columbia to Central California

Eulittoral zone. The composition of the algal vegetation along the coasts of **British Columbia** and **Washington** (Fig. 2.83) does not change dramatically. Throughout the whole of the cold temperate northwestern Pacific region, along the coasts of **Oregon** to **central California** and also to the phycologically well-investigated coast of the Monterey peninsula (south of San Francisco; Fig. 2.84), the algal vegetation is similar. In the **upper and mideulittoral zone** the small red alga *Endocladia muricata,* which occurs from Alaska to Mexico, forms a feltlike settlement, together with barnacles and mussels (e.g., the barnacles *Chthamalus dalli, Balanus glandula* and *B. crenatus;* and somewhat deeper the mussel *Mytilus californianus*). Another widely distributed red algal species is *Mastocarpus* (= *Gigartina*) *papillatus* (see Guiry et al. 1984) with its tetrasporophytic, crustose Petrocelis phase, formerly known as *Petrocelis middendorfii* (Polanshek and West 1975). This extensive crust grows very slowly. Paine et al. (1979) determined, on the basis of growth measurements extending over seven years, that specimens with an area of 120 cm^2 may be up to 87 years old, or 50 years on average. A survey on the vertical distribution and phenology of intertidal fleshy crustose algae was provided by Dethier (1987). Particularly well represented is the red algal genus *Porphyra* with around 17 species (Garbary et al. 1980–1982). Herbivorous snail species of *Acmaea* and *Littorina* are present. At wave-beaten places in the lower eulittoral zone, at "sublittoral exclaves" on permanently spray-wet outcrops in the eulittoral zone, a peculiar light-brown representative of the Laminariales occurs, the sea palm *Postelsia palmaeformis* (family Lessoniaceae). Its unusual height of occurrence may also be due to high air humidity on the fog-bound coast. The annual alga reaches a length of up to 50 cm and remains erect even when emergent due to its rigid stipe. The alga settles higher in the eulittoral zone than any other species of the Laminariales. *P. palmaeformis* grows together with the mussel *Mytilus californianus*, which is continuously removed by the waves. The mechanically resistant alga obtains free rocky substratum for its sporophytes arising in spring (Dayton 1973; Paine 1979). At less-exposed locations in the eulittoral zone one finds *Pelvetia fastigita* and *Fucus gardneri*. In wave-protected habitats the introduced brown alga *Sargassum muticum* or extensive meadows of the red alga *Iridaea cordata* prevail. Most of the 35 described species of *Iridaea* occur in the eastern Pacific (Hannach and Waaland 1986). In *I. cordata* upright blades develop from the perennial holdfast in autumn and winter. The cuticle of the blade is formed of alternating electron-opaque and -translucent layers, and this thin layer produces a brilliant iridescence due to constructive and destructive light interference (Gerwick and Lang 1977). On wave-exposed rocks in the **lower eulittoral zone,** periodically **buried by sand,** a typical kelp is *Laminaria sinclairii*. Further to the north is found the morphologically similar species *L. longipes* (Fig. 2.82D, upper). From the rhizomatous holdfast new erect fronds regenerate, as in the Mediterranean species *L. rodriguezii* (Fig. 2.72A). After periods of prolonged sand burial, usually occurring from July to October along the coast of Oregon, the blades die back. Autumn storms remove the sand, and in March new stipes with blades grow from the rhizomatous holdfast (for details see Markham 1973).

Sublittoral zone. At the boundary between the **lower eulittoral zone** *and* **the upper sublittoral zone** (Figs. 2.83 and 2.84) one finds the 4–5 m long fronds of the canopy algae *Alaria marginata* and *Egregia menziesii* (Alariaceae). Close to these Alariaceae the following larger seaweeds are prevalent: *Laminaria bongardiana* (Laminariaceae; with a flexible stipe), *Cystoseira osmundacea* (Fucales), *Desmarestia ligulata* (Desmarestiales), and finally two representatives of the family Lessoniaceae: *Lessoniopsis littoralis* (up to 2 m long) and *Nereocystis luetkeana.* This vegetation occupies a somewhat analogous position to the European upper sublittoral vegetation of *Alaria esculenta, Laminaria digitata* (both with flexible stipes), and *Himanthalia elongata* (Figs. 2.52–2.55). *Nereocystis luetkeana* grows in the upper and midsublittoral zone. Its stipe, which may attain a length of up to 15 m, ends near the water surface in a pneumatocyst (gas-filled bladder) from which hang narrow blades up to 4 m long. The whole gigantic thallus is formed in one growing season. *N. luetkeana* has extensive sporangial sori on the blades, and the whole sorus drops out and sinks to the bottom where the zoospores are released near where they can grow. This is in contrast to the other members of the Laminariales, which directly release zoospores from the

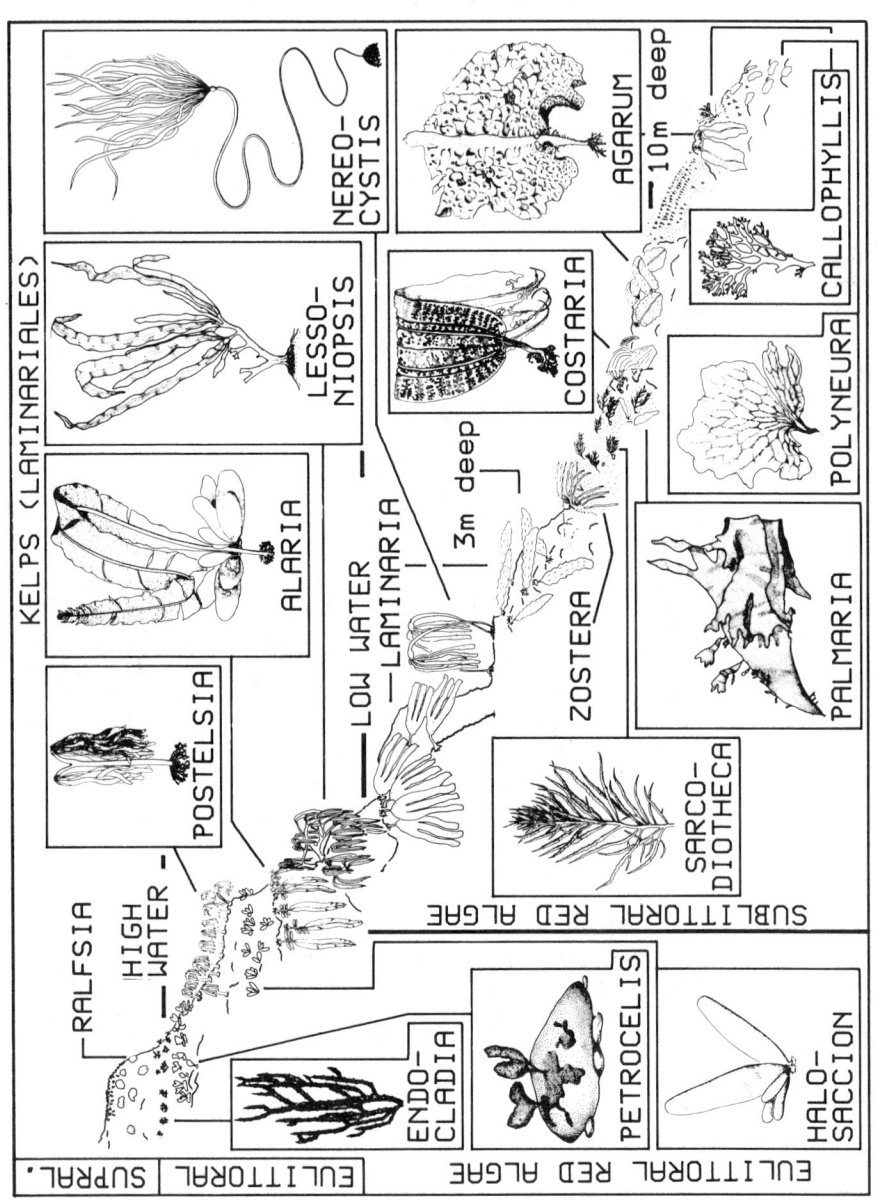

sori on the blades. As an opportunistic, annual species, *N. luetkeana* rapidly covers substrata that are cleared yearly by winter storms and forms its sori from June onwards, with a maximum in September and October (Foreman 1984). This species tolerates strong wave exposure (Dayton et al. 1984; Schiel and Foster 1986; Harrold et al. 1988). Another kelp species of the upper sublittoral zone is *Macrocystis integrifolia* (up to 6 m long; family Lessoniaceae; Fig. 2.84), which, like *N. luetkeana*, exhibits a wide latitudinal distribution along the Pacific coast of North America (Fig. 2.81). Since the giant kelp *Macrocystis pyrifera* is absent north of Monterey (Santa Cruz), the **midsublittoral zone** is dominated by *N. luetkeana* and by the comparatively smaller representatives of the family Laminariaceae (Fig. 2.83), accompanied by understory red algae such as *Opuntiella californica* and *Constantinea subulifera*. Typical algae of the **lower sublittoral zone** are the kelp *Agarum fimbriatum* and, at depths of 15–25 m, *Polyneura latissima, Fryella gardneri,* and species of *Callophyllis* (Fig. 2.83). The vegetation at the lower algal limit, 30–60 m depth (according to water clarity) is dominated by crustose coralline algae, that is, North Pacific endemic species of *Lithothamnion* and *Lithophyllum*. On **sandy bottoms** in the upper part of the sublittoral zone from Alaska to Mexico, the circumpolar seagrass *Zostera marina* flourishes. Usually on rocky substrata one finds the endemic North American Pacific seagrasses *Phyllospadix scouleri* and *P. torreyi*.

2.6.2 Warm Temperate California Region (Point Conception to Baja California)

Various cold temperate species are found at their southern limit of distribution south of Point Conception where the south-flowing California Current is deflected from the coast (Fig. 2.43). A warm current from the south enables warm temperate species to penetrate up to Point Conception, and so here major floristic and faunistic discontinuities occur (for details see Thom 1980; Murray and Littler 1981). Upwelling areas along Baja California provide cool-water niches further south (see below).

The warm temperate California region, inhabited by somewhat eurythermal northern species and by warm-adapted southern species, extends from

---

Fig. 2.83 Zonation on the coasts of British Columbia and Washington (wave-exposed habitats). **Details on species names and dimensions:** *Ralfsia verrucosa* (crustose brown alga, several centimeters in diameter). **Kelp species:** *Postelsia palmaeformis* ($\times$ 0.05); *Alaria marginata* ($\times$ 0.07); *Lessoniopsis littoralis* ($\times$ 0.1); *Nereocystis luetkeana* ($\times$ 0.007); *Costaria costata* ($\times$ 0.2); *Agarum fimbriatum* ($\times$ 0.07). **Red algae:** *Endocladia muricata* ($\times$ 0.7); Petrocelis phase of *Mastocarpus papillatus* ($\times$ 0.1); *Halosaccion americanum* ($\times$ 0.15); *Sarcodiotheca gaudichaudii* (= *Neoagardhiella baileyi*) ($\times$ 0.06); *Palmaria mollis* ($\times$ 0.08); *Polyneura latissima* ($\times$ 0.15), *Callophyllis flabellulata* ($\times$ 0.2). **Seagrass:** *Zostera marina*. The *Laminaria* species illustrated just below the low-water level is *L. bongardiana*. (Vegetation profile from Waaland 1977. Zonation schemes: for the eulittoral zone based on Rigg and Miller 1949 and for sublittoral zone based on Neushul 1965a. Habit drawings: *Endocladia* from Waaland 1977; *Lessoniopsis* from Scagel 1967; remainder from Abbott and Hollenberg 1976.)

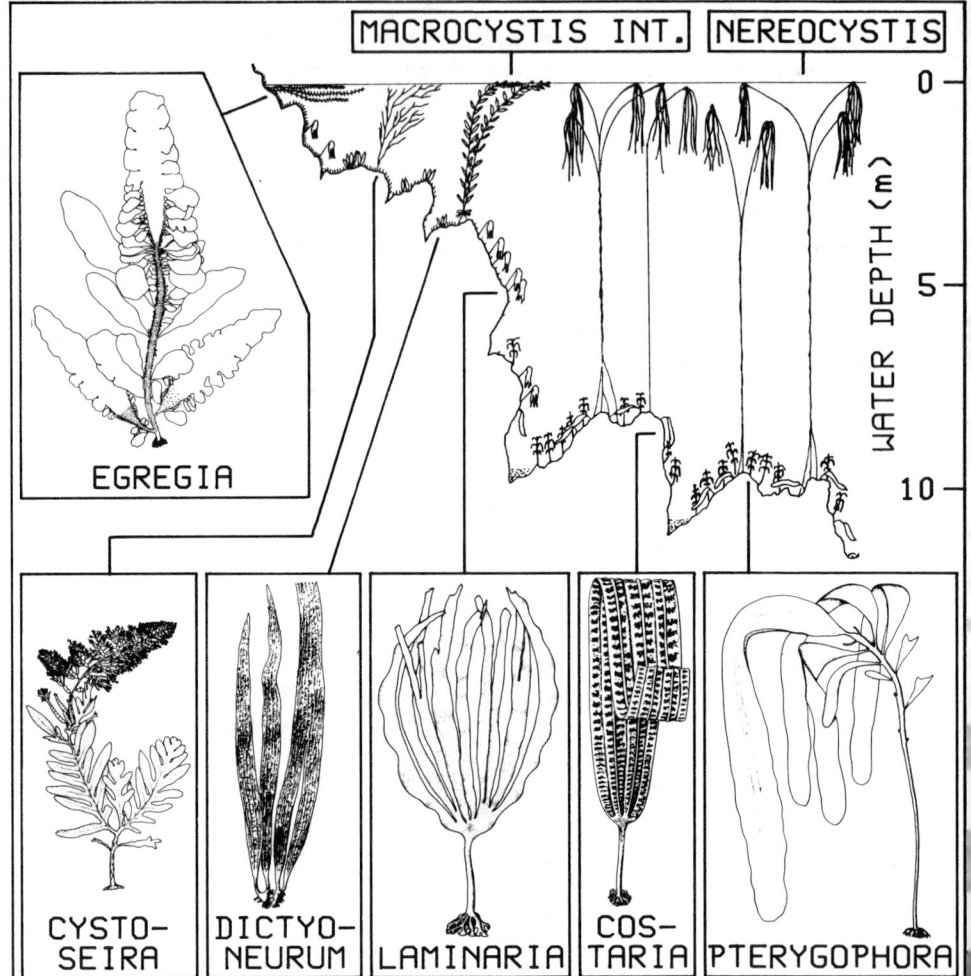

Fig. 2.84 Zonation in the sublittoral zone at Monterey and examples of typical kelp species. INT = *Macrocystis integrifolia*. The larger species, *M. pyrifera,* grows more southerly (see Fig. 2.86). **Details on species names and dimensions:** *Egregia menziesii* (× 0.3); *Cystoseira osmundacea* (× 0.1); *Dictyoneurum californicum* (× 0.3), *Laminaria setchellii* (× 0.2); *Costaria costata* (× 0.2); *Pterygophora californica* (× 0.1). (vegetation profile from McLean 1962. Habit drawings: *Dictyoneurum* from Abbott and Hollenberg 1976; *Cystoseira* and *Pterygophora* from Galbraith and Boehler 1974; remainder from Scagel 1967.)

Point Conception to Baja California, bordered by the 18°C-summer isotherm to the north and by the 20°C-winter isotherm to the south (Figs 1.5 and 1.6). Besides a few amphioceanic algal species (Figs. 2.85A–C) various Californian–Japanese sister species occur, as well as quite a number of endemic seaweeds (Figs. 2.85D–J).

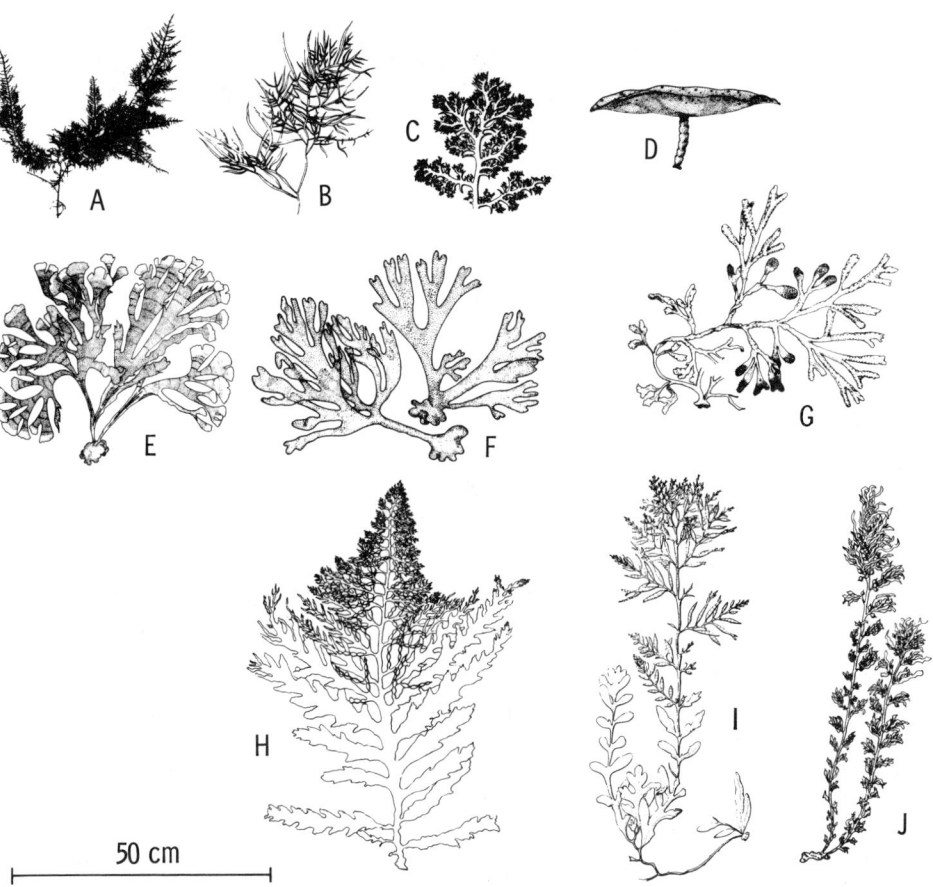

Fig. 2.85 Examples of algal species of the warm temperate California region (lower eulittoral and sublittoral zone. **A–C: Pacific–Atlantic red algae: (A)** *Pterocladia capillacea*. **(B)** *Grateloupia filicina*. **(C)** *Plocamium cartilagineum*. **D–J: endemic species: (D)** *Maripelta rotata* (red alga, Rhodymeniales; only in the lower sublittoral zone). **Dictyotales: (E)** *Zonaria farlowii*. **(F)** *Dictyota flabellata* (to Panama). **Fucales: (G)** *Hesperophycus harveyanus* (only in the eulittoral zone). **(H)** *Cystoseira setchellii*. **(I)** *Halidrys dioica*. **(J)** *Sargassum agardhianum*. (Drawings from Abbott and Hollenberg 1976.)

Zonation. **The upper sublittoral zone** of the southern Californian coast (Fig. 2.86) is dominated by *Egregia menziesii*, with fronds up to 15 m long, and locally by the smaller species *Eisenia arborea*, which does not reach as far north as the former. Instead of the typical species found north of Point Conception such as *Nereocystis luetkeana* and *Lessoniopsis littoralis* (distribution shown in Fig. 2.81), one finds, in the sublittoral zone and in the **lower eulittoral zone,** southern endemic representatives of the Fucales; examples are *Sargassum agardhianum, Halidrys dioica,* and the monotypic (genus having only one species) *Hesperophycus* (Fucaceae). The vegetation is enriched by various species of the brown algal order Dictyotales and

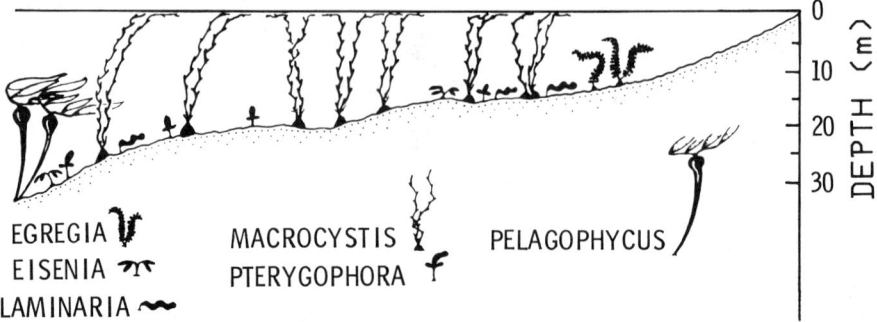

Fig. 2.86 Zonation in the sublittoral zone of the coast off La Jolla (southern California). Canopy algae: *Egregia menziesii* in the upper sublittoral zone, and *Macrocystis pyrifera* and *Pelagophycus porra* in the midsublittoral zone. Smaller representatives of the Laminariales (e.g., *Laminaria farlowii*) form an understory vegetation. (From Dawson et al. 1960.)

other warm-adapted red algal genera such as *Pterocladia*, and *Grateloupia*, (Fig. 2.85). The rarity of large brown algae in the low intertidal zone of southern California has been partly attributed to sea-urchin grazing (Sousa et al. 1981). The kelp *Egregia menziesii*, the fucalean species *Cystoseira osmundacea*, and *Halidrys dioica* successfully colonized low intertidal areas from which urchins had been removed. A further problem for the brown algae with their recruitment only from settled spores or zygotes is created by successful vegetative proliferation of the dominant perennial red turf algae *Gigartina canaliculata, Laurencia pinnatifida,* and *Gastroclonium coulteri* (Sousa et al. 1981). In the **mideulittoral** zone the northern species *Halosaccion americanum* and *Postelsia palmaeformis* are absent and replaced by red algae such as *Corallina*, again *Gigartina canaliculata*, or the fucalean alga *Pelvetia fastigiata*. In the upper eulittoral zone, as in the cold temperate region, the red alga *Endocladia muricata* occurs (Fig. 2.83). The giant kelp species of the family Lessoniaceae add another layer to the usual sequence of three vegetation layers within the sublittoral zone: (1) representatives of the Laminariaceae and Alariaceae); (2) understory algae; and (3) crustose algae (Fig. 2.87; compare with Fig. 1.3). Typical, smaller representatives of the Laminariales (Figs. 2.86 and 2.87) occur below the Californian *Macrocystis pyrifera* kelp beds, *Pterygophora californica* and *Eisenia arborea* (both belonging to the Alariaceae), and *Laminaria farlowii* (undivided blade Laminariaceae). These are perennial and long-lived kelp species (Dayton et al. 1984), with an age of up to 11 years (*P. californica* and *E. arborea*), or six years (*L. farlowii*). Below and among these species numerous smaller deep-water and shade-adapted species occur, such as erect growing members of the Corallinaceae and the characteristic, umbrellalike red algae *Maripelta rotata* (Fig. 2.85D) and *Constantinea simplex*. Off the coast near La Jolla, the deepest erect algae are to be found at about 50 m deep, and the deepest crustose red algae at the lower algal limit of 60 m (Neushul 1965*a*).

The Giant Kelps and their Morphology (Fig. 2.88).

Among the family Lessoniaceae the species *Pelagophycus porra* is endemic to southern California (segment 1 in Fig. 2.81). The occurrence of *Macrocystis pyrifera* begins

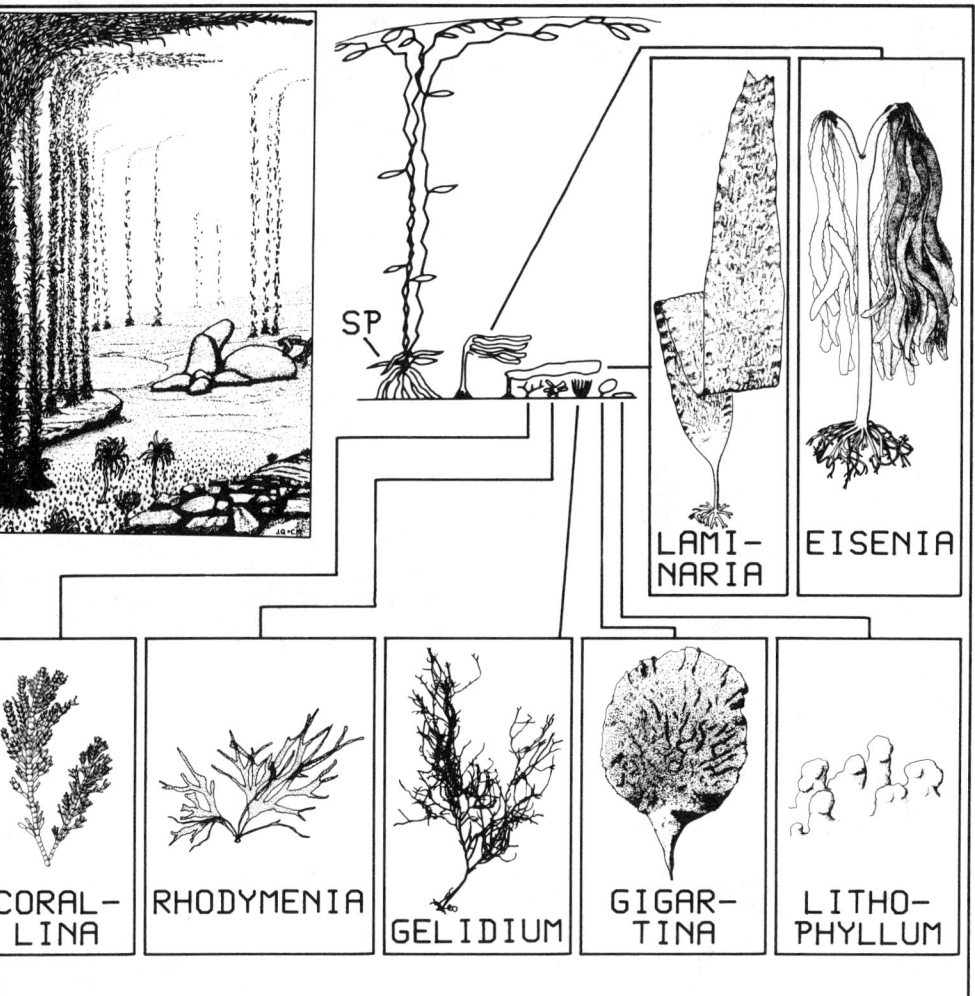

Fig. 2.87 *Macrocystis* forest on the southern Californian coast with typical understory algae. **Upper left:** panoramic view of *Macrocystis* forest with *Eisenia arborea* as understory. **Following to the right:** scheme of the four vegetation layers in the *Macrocystis* forest: *Macrocystis pyrifera*, representatives of the Laminariaceae and Alariaceae, small sheetlike and branched algae, crustose red algae. SP = sporophylls. **Details on species names and dimensions: brown algae:** *Laminaria farlowii* (× 0.08); *Eisenia arborea* (× 0.04). **Red algae:** *Corallina officinalis* var. *chilensis* (0.4); *Rhodymenia californica* var. *attenuata* (× 0.3); *Gelidium nudifrons* (× 0.5); *Gigartina corymbifera* (× 0.1); *Lithophyllum imitans* (× 2). (Panoramic view from Quast 1971; scheme of vegetation layers from Foster 1975; drawings from Abbott and Hollenberg 1976.)

Fig. 2.88 Aspects of the southern Californian giant kelps. **A–F,I:** *Macrocystis pyrifera*. **(A)** Kelp bed (aerial view). **(B)** Fronds at the water surface. **(C,D)** Underwater aspects, at about 25 m deep. **(E)** Meristematic zone at the distal frond end. **(F)** Stipe with blades. **(I)** Individual exposed on a road. **G–H:** *Pelagophycus porra*. **(G)** holdfast. **(H)** Individual exposed on a road. (A,C,D,E photographs courtesy of M. Neushul; remaining photographs by the author.)

Fig. 2.88 (*Continued*)

south of Monterey, about 200 km north of Point Conception, where the distribution of the smaller, northern species *M. integrifolia* ends. With chromosome numbers of 2n = 64 in *M. pyrifera* and 2n = 32 in *M. integrifolia* from British Columbia, the former species is a polyploid of the latter, although *M. integrifolia* from Sitka, Alaska, with 2n = 64, also seems to be polyploid (Yabu and Sanbonsuga 1985, 1987; *M. pyrifera* around Santa Barbara recorded as *M. angustifolia;* see Abbott and Hollenberg 1976). As soon as *Macrocystis pyrifera* appears, the vegetation structure within the sublittoral zone is fundamentally changed by the giant kelps up to 50 m long whose fronds may typically absorb 80% of the incident light in the uppermost 1 m of the water column (Gerard 1984*a*). The smaller algal species live on the bottom in a **shaded habitat,** occasionally lit by sunflecks as beams of light penetrate gaps in the constantly moving canopy. This subordinate role applies to the representatives of the Laminariaceae and Alariaceae, which in other temperate regions and the Arctic coasts are the dominants of the sublittoral zone. **Gas-filled bladders** give *M. pyrifera* an important strategic advantage, also present in the other two giant algae *Pelagophycus porra* and *Nereocystis luetkeana.* Organs providing buoyancy are absent in the families Laminariaceae and Alariaceae (except *Alaria fistulosa*), which must rely on other methods to expose their fronds to the light, namely by having rigid stipes or by taking advantage of water turbulence. With the latter two methods, thallus lengths of several meters may be obtained. However, the formation of thalli 10 times longer is made possible if buoyancy is used to hold the thallus in an upright position, as is shown by the habit (characteristic form) and dimensions of the three giant kelps (Figs. 2.80 and 2.88H,I).

The thalli of these species are attached to rocky substrata by correspondingly large **holdfasts**. The **fronds** are formed and reach the surface in one year and in this way take advantage of the ample light supply in the upper water layers. The **stipes** and the individual **blades** move continually between the water layers on a time scale of 1 min to 1 h, whereby photoinhibition effects are reduced and total net photosynthesis is optimized (Gerard 1986). At a 5 to 20 m depth, at some distance from the coast, *Macrocystis pyrifera* forms dense vegetation (kelp beds and forests) **from Monterey** (Santa Cruz) **to Baja California** (Fig. 2.88A). High water temperatures (above 20°C) and low nutrient concentrations associated with El Niño events may occasionally damage the kelp canopy (Dayton and Tegner 1984; Gerard 1984b). *M. pyrifera* is a major source of raw material for the Californian **alginate industry,** which consumes about 120,000 tons fresh weight of *M. pyrifera* per year and supplies one-third of the world alginate market. The perennial thallus part of *M. pyrifera* is represented by the holdfast (maximum age: 16–30 years; Neushul 1977), from which new, subdichotomously branched fronds grow to the water surface. The maximum growth rate may be 30 cm per day. This growth rate is obtained by the translocation of organic compounds from the sunlit, upper parts of the fronds to the new fronds arising from depths where irradiances are too low for any substantial photosynthesis. Fronds can grow up to 20 m through the water column in 10 months (Neushul 1977). After the new fronds have reached the water surface, they spread along the surface for several meters. Up to 100 erect fronds may arise from one holdfast. They grow from an apical meristem (Fig. 2.88E) and live on average for six months. The buoyancy of the fronds of *M. pyrifera* is not achieved by a single gas-filled bladder, as in the two other giant kelps, but by multiple small bladders at the base of each blade. Blades may be up to 80 cm long, and as many as 200 may be attached to one frond. Still farther off the coast, at the seaward margin of the kelp beds of *M. pyrifera,* the equally large thalli of *Pelagophycus porra* (stipes up to 25 m long) grow in areas deeper than 30 m with strong currents. At the distal end of the stipe of this species a gas-filled, clublike structure is formed that passes into a sphere carrying antlerlike branches with several 6-20 m long blades (Figs. 2.80 and 2.86H). Since the 16th century, Spanish sailors sailing from the Philippines to Central America used detached and drifting specimens of *P. porra* as a means of navigation and as an indication that the Californian coast was near.

Dynamics and biotic relations in Californian kelp communities. The kelp forests slow longshore currents by almost one-third (Jackson and Winant 1983). The relatively stagnant water inside the forest serves as a refuge from predators for many animals. At the periphery of the kelp forest there is much biological action, because here are concentrated the predators such as fish feeding on plankton or on animals encrusting kelp fronds; this is known as the "edge effect" (Dayton 1985a). The kelp canopy is thinned out by winter storms; severe storms may remove the complete *M. pyrifera* canopy but spare most understory kelps, mainly *Pterygophora californica* (Ebeling et al. 1985; Foster and Schiel 1985, Schiel and Foster 1986). Once established, such understory patches may persist over years and resist invasion by *M. pyrifera* (Dayton and Tegner 1984). Habitats with strong water movement have no *M. pyrifera* because the alga is susceptible to damage or removal, and because it has a high surface area, large stipe bundles, and a holdfast that degrades in the center (Dayton et al. 1984; Harrold et al. 1988). More tolerant to physical stress is the northern giant alga *Nereocystis luetkeana,* which in the small overlapping geographical area (Fig. 2.81) may outcompete *M. pyrifera* in areas where water motion is high (Dayton et al. 1984; Harrold et al. 1988). The smaller, stipitate kelp species *P. californica, Eisenia arborea,* and *Laminaria setchellii* are also tolerant to wave surge, while the stronger stress tolerators seem to be the "prostrate" species *L. farlowii, Cystoseira osmundacea,* and *Dictyoneurum californicum,* which are even more tolerant to wave stress (Dayton et al. 1984). Articulated coralline algae are more abundant at wave-exposed than protected sites, whereas fleshy red algae show the opposite pattern (Harrold et al. 1988). Grazing by sea urchins on *Macrocystis* is mainly due to the red urchin *Strongylocentrotus franciscanus,* purple urchins *S. purpuratus,* and white urchins *Lytechinus anamesus* (Leighton 1971). White urchins kill small laminarian sporophytes and gametophytes, whereas small individuals of the fucalean *Cystoseira osmundacea,* mainly growing in the upper 10 m, appear relatively unaffected so that kelp may finally be excluded by *C. osmundacea* in this way (Dean et al. 1988).

On the southern part of the warm temperate California region, along the Pacific coast of the peninsula of **Baja California, upwelling** occurs (Barber and Smith 1981). This colder water enables many temperate species to penetrate as far as the southern boundary of the California region, which is situated 300 km north of the southern tip of the peninsula, at Magdalena Bay or at the island of Santa Margarita (for details see Dawson 1960). At this boundary the temperature of the surface water surpasses 25°C in summer, and the eastern Pacific tropical region begins. The distribution of *Macrocystis pyrifera* and *Eisenia arborea* extends to this boundary, although the two species occur more and more sporadically, depending on the local presence of upwelling. These cool-water areas are important for explaining equatorial crossings during glacial times. *Egregia menziesii* and *Pelagophycus porra* reach as far as the Punta Eugenia, halfway along the 1000 km peninsula (Fig. 2.81). At the more northerly island of Guadeloupe, 250 km offshore, these kelp species are absent because there is no upwelling. In the warm temperate algal flora of Baja California one finds numerous smaller algae that also occur as identical species or as closely related sister species on the warm temperate coasts of Japan; examples are the brown alga *Pachydictyon coriaceum* (Dictyotales) and the red alga *Lomentaria hakodatensis*. This similarity, not seen in the two temperate regions bordering the North Atlantic, points to the existence of a common North Pacific warm temperate flora. This floral continuity was broken when the Bering Strait opened and cold Arctic water divided the North Pacific (van den Hoek 1984).

Gulf of California. The southern end of Baja California belongs to the eastern Pacific tropical region, whereas the coasts bordering the **Gulf of California,** north of La Paz on the coast of Baja California, and north of Topolobampo on the coast of the Mexican mainland, again exhibit a warm temperate character. The Gulf of California or **Cortez province** part of the warm temperate California region is distinguished from the **San Diego province** on the open Pacific coast from Point Conception to Magdalena Bay. (Cortez Sea was the old Spanish name for the Gulf of California.) The two provinces are quite different in respect to their marine fauna and flora. The southern part of the San Diego province, with its local upwelling areas, could still be reached by eurythermal northern species, which could not penetrate the southern end of Baja California due to its tropical character. Hence, the warm temperate algal flora of the Cortez province evolved mainly from descendants of the adjacent eastern Pacific tropical region. This isolation resulted in an endemism probably as high as 35% (Dawson 1960). The Cortez province (Gulf of California) does not harbor any representatives of the Laminariales. In the Cortez province's northern algal flora, brown algal genera such as *Sargassum* and *Padina* are typical and are accompanied by numerous western Pacific species. This observation stresses again the close alliance with the Japanese warm temperate algal flora, whereas the southern part of the Cortez province has been invaded by many tropical species, such as the red alga *Digenea simplex* and the green alga *Halimeda discoidea*.

2.7 ASIA: COLD AND WARM TEMPERATE REGIONS IN THE NORTH PACIFIC

Literature: (Note: In Russian, Japanese and Chinese language publications the algal species names are printed in the roman alphabet, so some idea of the composition of the algal flora is possible even if one cannot read the characters of the cited languages.) **USSR:** Petrov (e.g., 1974); Zenkevitch (1963). **Kamchatka and Commander Islands (Beringa):** Vozzhinskaja (e.g., 1965); E. S. Zinova (1940, 1954c). **Sea of Okhotsk:** Ketchum (1983); E. S. Zinova (1954a). **Sakhalin and Tatar Strait:** Tokida (1954); Vozzhinskaja (e.g., 1964); E. S. Zinova (1954b). **Kurile Islands:** Kussakin (1961); Nagai (1940-1941); A. D. Zinova (1959). **Russian coast of the Sea of Japan, Vladivostok:** Funahashi (1973); Makienko (1975); Perestenko (1980). **Korea:** Boo (1987); Boo and Lee (1986); Kang (1966); E.-A. Kim et al. (1986); H. S. Kim et al. (1983); Y. H. Kim and Lee (1981, 1985); H. B. Lee and Lee (e.g., 1981); Nam (1986); Song (1986). **Checklist, Korea:** I. K. Lee and Kang (1986). **Japan:** Chihara (1975); Funahashi (e.g., 1973); Hommersand (1972); Katada and Satomi (1975); Masuda (e.g., 1982); Ogawa and Machida (1976–1977); Okamura (1932); Segawa (1971); Suda (1987); Yamada and Tanaka (1944); Yoshida (1963). **Checklist, Japan:** Yoshida et al. (1985). **China:** Tseng (1983); Tseng and Chang (e.g., 1964). **Hong Kong:** B. Morton and Morton (1983); Tseng et al. (1980). **Paleoceoanography, seas neighboring China:** Wang (1985).

From the northern Bering Sea the Kurile or Oyashio Current brings cold water to the eastern coast of North Japan. The cold water descends near Cape Inubo, situated at the latitude of Tokyo, whereas warm water is transported northwards to Japan by the Kuroshio (Kuro Siwo) Current, the "Gulf Stream of the Pacific" (Fig. 2.43). Along the northeastern Asiatic coast cold water flows southward. The **cold temperate northwestern Pacific region** (Figs. 1.5–1.7) stretches from the Arctic region at Cape Olyutorsky in northern Kamchatka (10°C-summer isotherm) to mid-China (Wenchow, south of Shanghai) and approximately to mid-Japan (Cape Inubo). The 10°C-winter isotherm approaches the two coasts forming the border of the **warm temperate Japan region,** although the August temperature at this boundary is not 15°C but 25–28°C. (The usual temperature to delimit cold from warm temperate regions is 15°C in winter; Fig. 1.8.) This discrepancy follows from the compression of the isotherms and the wide annual spans of seawater temperature on the Asiatic side of the Pacific, or in general on the western sides of the oceans. If in this particular case one uses as the boundary between cold and warm temperate regions the winter isotherm of 10°C rather than the summer isotherms of 25–28°C (which are characteristic of a tropical environment), the cold requirement for reproduction of northern species is emphasized more than the effects of high summer temperatures.

The Kuroshio Current imparts a warm temperate character to the algal vegetation of southern Japan (south of a line from Hamada in southwestern Honshu to Cape Inubo), the coasts of South Korea, and southern China from Wenchow to Hong Kong. The coastlines south of Hong Kong, the eastern coast of Taiwan, and the islands of the Ryukyu archipelago belong to the Indo-West Pacific tropical region (Fig. 1.7).

The 0°C-February isotherm is shifted further to the south in the **Sea of Okhotsk** because of the adjacent Siberian cold center in the Northern Hemisphere (Fig. 1.6). In several parts the ice cover may persist for up to 10 months per year. Large distances of the coasts of **Kamchatka** are unsuitable for seaweeds because of soft substrata, but rocky areas are interspersed.

Zonation at Kamchatka. On rocky substratum in the upper eulittoral zone the 5–10 cm long red alga *Gloiopeltis furcata* (Cryptonemiales, Endocladiaceae) occurs (Fig. 2.89). The distribution of this characteristic eulittoral species extends to Japan (Fig. 2.90), southern China, and the North American Pacific coasts to Washington. In the middle and lower eulittoral zone one finds the red alga *Halosaccion glandiforme,* which is a sister species of *H. americanum* on the North American Pacific coast (Fig. 2.83; I. K. Lee 1982). The dense laminarian vegetation in the sublittoral zone of western Kamchatka consists of a mixture of amphioceanic species, such as *Alaria esculenta* (forma *dolichorachis*) and *Laminaria saccharina* (with taxonomically uncertain, local races or subspecies; see Ju. E. Petrov 1974). Other laminarian eastern Asiatic-endemic species such as *Alaria crassifolia* (Fig. 2.91E) occur in the upper sublittoral zone. The midsublittoral zone, down to 20 m deep, is inhabited by *Laminaria* species with a divided blade (section Digitatae). These are represented by the North Pacific species *L. yezoensis* (Figs. 2.82C upper, 2.91C), possibly by *L. setchellii* or *L. dentigera* (compare with Fig. 2.82A upper), and by other species such as *L. bongardiana* (synomous to *L. groenlandica* sensu Druehl 1968; see legend to Fig. 2.82). Among the smaller algae, of which some are illustrated in Fig. 2.89, the eastern Asiatic red alga *Phycodrys riggii* (= *P. serratiloba* = *P. fimbriata;* see Perestenko 1980) is a typical deep-water alga. The algal vegetation ends on the coast of western Kamchatka often at about 30 m deep (Vozzhinskaja 1965).

In the southern part of the Sea of Okhotsk, on the coasts of the island of **Sakhalin** and the **Kurile Islands,** a belt of the North Pacific brown alga *Analipus japonicus* (= *Heterochordaria abietina*) is apparent in the eulittoral zone. Many species that are endemic to the North Pacific Asiatic coasts are also found here, such as the red algae *Iridaea cornucopiae, Corallina pilulifera,* the brown alga *Pelvetia wrightii,* and several *Sargassum* species.

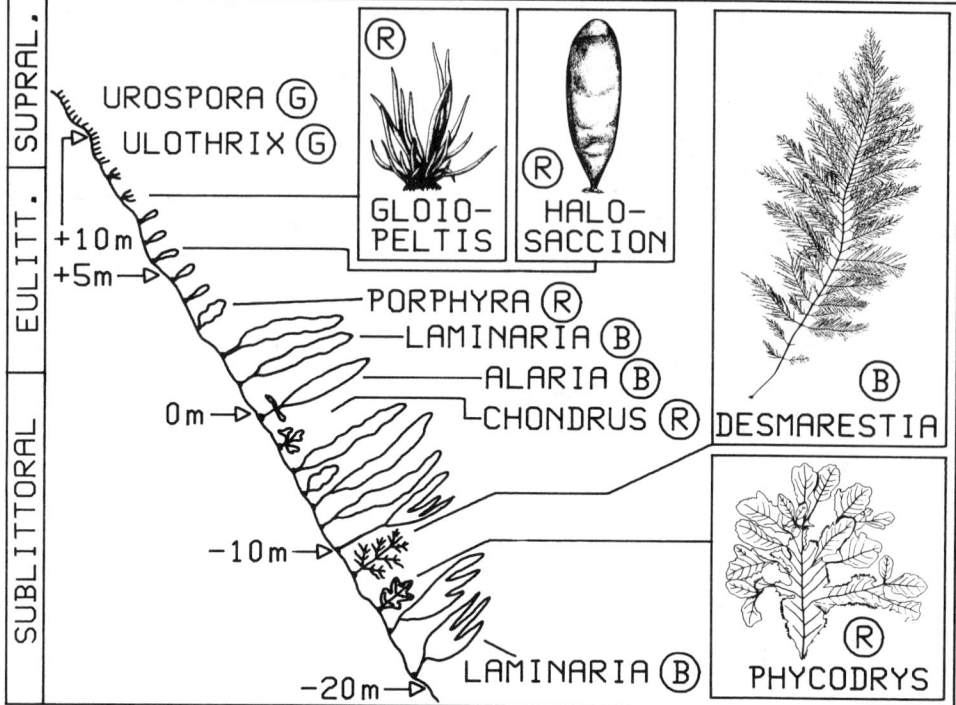

Fig. 2.89 Zonation on the west coast of Kamchatka (wave-exposed habitat). G = green alga; B = brown alga; R = red alga. **Details on species names and dimensions:** *Gloiopeltis furcata* (× 0.5); *Halosaccion glandiforme* (× 0.2); *Desmarestia ligulata* (× 0.2); *Phycodrys riggii* (× 0.2); *Urospora penicilliformis; Ulothrix pseudoflacca; Porphyra ochotensis; Laminaria saccharina* (upper) and possibly *Laminaria dentigera* (lower); *Alaria crassifolia; Chondrus armatus*. (Vegetation profile after Vozzhinskaja 1965; drawings from Perestenko 1980.)

In the sublittoral zone representatives of the Laminariaceae are present, namely, *Arthrothamnus bifidus* (possibly distributed to the Aleutian Islands) and *Kjellmaniella gyrata;* both species are somewhat similar to *L. saccharina*. All these algae belong to the rich cold temperate seaweed flora of the **Sea of Japan** (750 species according to Funahashi 1973). The Sea of Japan is bordered by the Russian continental coast from the mouth of the Amur River to Vladivostok, and by the adjacent coasts of Sakhalin, Japan, and Korea. The migration of North American Pacific species to the Asiatic coasts may have been facilitated by the counterclockwise gyre of the Alaska, Aleutian, and Oyashio currents (Fig. 2.43). This may have occurred with the cold-adapted kelp *Thalassiophyllum clathrus* (Fig. 2.82G) which exhibits a continuous distribution from northern Japan to Alaska (Fig. 2.81).

Of the 800 to 900 seaweed species of **Japan** (Abbott and Hollenberg 1976; Yoshida et al. 1985) one-third belong to the northwest Pacific endemic group and 20% are amphi-Pacific occurring also along the North American Pacific coast. The southern parts of Japan are warm temperate and a large proportion of the species here show similarities with the southern tropical algal flora. The more eurythermal species could easily migrate to Japan through the northward-flowing Kuroshio Current. Detached *Sargassum* individuals and other floating algae drift mainly along the western coast of southern Japan into the Sea of Japan (Yoshida 1963).

The seaweed richness of the Japanese coasts and the Sea of Japan may have to do with the probable stability of the high annual temperature fluctuations for long geological periods and the predominance of rocky shores, which is in contrast to the situation in the northwestern Atlantic (van den Hoek 1975). This algal richness occurs despite the compression of isotherms and the large annual temperature fluctuations on the western side of the North Pacific (Figs. 1.5 and 1.6).

Zonation. On the eastern coast of northern Honshu (Fig. 2.90) the dominating species in the upper eulittoral zone is the red alga *Gloiopeltis furcata,* mentioned earlier from the coast of Kamchatka. Endemic to the coasts of Japan, Korea, and Hong Kong in the mideulittoral zone, the monotypic fucalean genus (i.e., it contains only one species) *Hizikia* occurs as *H. fusiforme.* In southern Japan the highest settling representative of the warm-water genus *Sargassum, S. thunbergii,* is found in the eulittoral zone (Fig. 2.92F). Further representatives of the Fucales in the Japanese algal flora are *Pelvetia wrightii, Fucus distichus* subsp. *evanescens, Coccophora langsdorfii* (a monotypic genus endemic to Japan and Korea), and several species of *Cystophyllum* (a widely distributed genus in the Indo-West Pacific tropical region, Australia and New Zealand; see Table 5.1). The boundary of the sublittoral zone is marked by the uppermost representatives of the Laminariales. On the coasts of northern Honshu (Fig. 2.90A) and Hokkaido one finds the eastern Asiatic species *Alaria crassifolia* (Fig. 2.91E), *Undaria pinnatifida*(Figs. 2.91D, 2.92E,F), and the amphi-Pacific species *Costaria costata* (habit shown in Fig. 2.83). In the midsublittoral zone on the coasts of Hokkaido and northern Honshu, *Laminaria* species with an undivided blade (section Simplices) dominate; examples are *L. japonica* (Figs. 2.91A), *L. angustata* (Fig. 2.91B), and *L. longissima* (Figs. 2.92A). However, to the south along the coasts of Honshu (Fig. 2.90B) the sublittoral zone is dominated by *Sargassum* species. Other dominants in the sublittoral zone are representatives of the Alariaceae, *Undaria pinnatifida, Eisenia bicyclis* (Fig. 2.91F), and *Ecklonia cava* (Figs. 2.91G, 2.92C,D). The last two species form dense marine forests in the sublittoral zone, with *E. bicyclis* dominating in the upper sublittoral down to 5 m deep, and *E. cava* in the midsublittoral from 3 m to 25 m or more (Hayashida 1983, 1984; Maegawa et al. 1987).

There are many **warm temperate** seaweed species that occur on both sides of the North Pacific or occur as sister species. These species are probably the remains of the warm temperate Tertiary algal flora, once continuously distributed in the North Pacific. Their presence may also be a result of the prevailing current system (Fig. 2.43), inasmuch as transport via the Kuroshio and North Pacific currents to Californian shores is conceivable. Of the 110 examples of amphi-Pacific algae given by Hommersand (1972), the following pairs of sister

A
BANGIA ATROPURPUREA (R)
PORPHYRA (R)
GLOIOPELTIS FURCATA (R)
SEPTIGER (MUSSEL)
ANALIPUS JAPONICUS (B)
NEMALION VERMICULARE (R)
HIZIKIA FUSIFORME(B;FUCALES)
CHONDRUS YENDOI (R)
CORALLINA PILULIFERA (R)
CALLIARTHRON YESSOENSE (R)
ALARIA CRASSIFOLIA (B)
COSTARIA COSTATA (B)
UNDARIA PINNATIFIDA (B)
LAMINARIA JAPONICA (B)
EULITTORAL
SUBLITT.

B
Elsenia bicyclis
Grateloupla elliptica
Sargassum ringgoldlanum
Gelidlum amansli
Glgartina tenella
Sargassum nlgrlfolium
Galaxaura falcata
Sargassum serratifollum
Ecklonia cava
Depth in meter
0
2
4
6
8
10
12
14
16
18
0
200
400
600
800
1000
1200 m
Distance from the shore

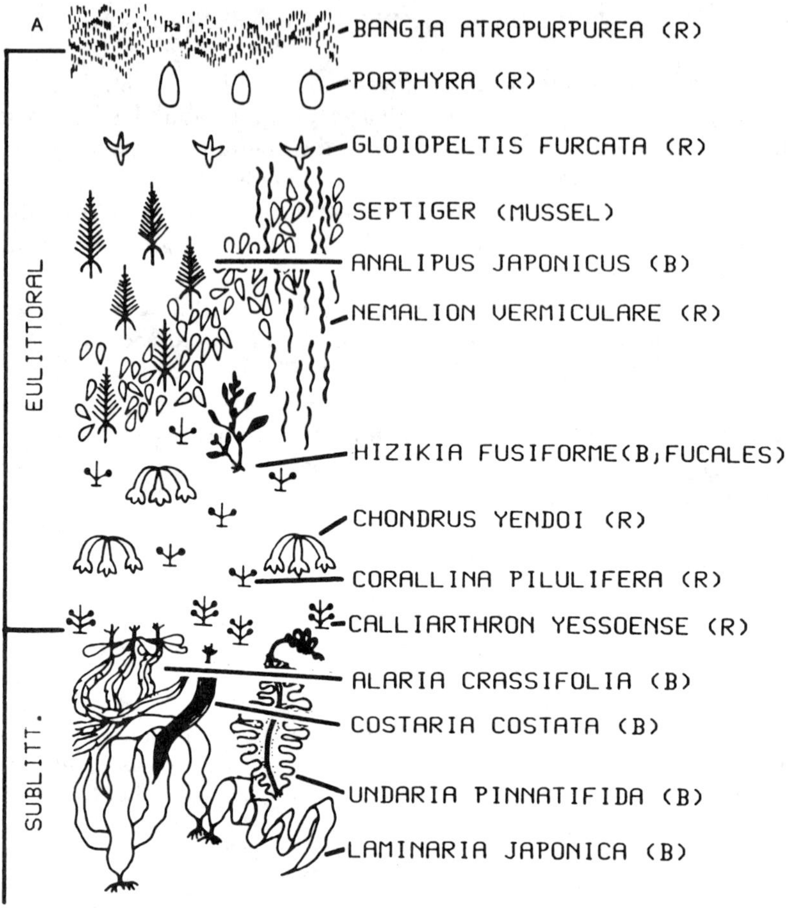

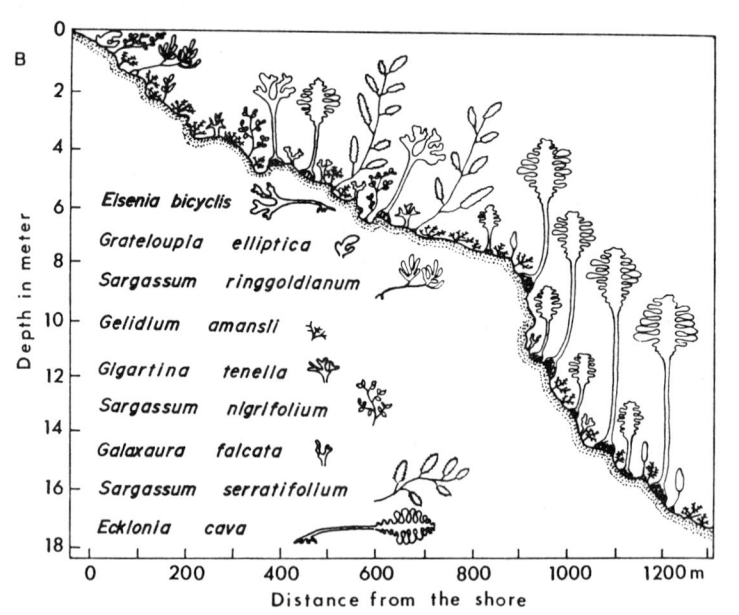

species may be cited (first-named species found on the American coasts, second-named on the Japanese coasts): *Eisenia arborea/E. bicyclis, Nemalion helminthoides/N. vermiculare,* and *Corallina pinnatifolia/C. pilulifera.*

Laminariales in the northwest Pacific. In this area about 75% of the genera of the Laminariales are endemic (as in the northeastern Pacific), and the greatest species diversity occurs in the southern Kuriles–northern Hokkaido region (Druehl 1981). Within Japan most of the representatives of the Laminariales are restricted to Hokkaido, which is bordered to the south by the 20°C-August isotherm (Fig. 1.5).

Japan: Laminariaceae. Ninety percent of the Japanese harvest of *Laminaria* for commercial use is collected on the coasts of Hokkaido, the rest on the coast of northern Honshu. Within the genus *Laminaria* about 13 species with an **undivided blade** are reported to occur on Hokkaido (Tokida et al. 1980; Druehl et al. 1988c). Not all of these represent truly distinct species, since many are interfertile (Kain 1979). On the basis of chromosome counts, crossing experiments, and clustering analysis these species may be divided into five groups (Druehl et al. 1988c): (1) the *L. angustata* group (*L. angustata, L. longissima,* forms of *L. diabolica;* Fig. 2.91B) on the colder eastern coast of Hokkaido; (2) the *L. japonica* group (*L. japonica,* and, for example, *L. religiosa, L. ochotensis,* and *L. coriacea;* Fig. 2.91A) in the warmer coasts of western and northern Hokkaido; (3) a group comprising *L. sachalinensis, L. cichorioides,* and *L. saccharina* found mainly in the north; (4) *L. yendoana* with restricted occurrence in the southeast, and (5) *L. longipedalis* with local occurrence in the northeast. The species contained in groups 1, 2, and 5 have been combined as Section Fasciatae. Their blades are usually narrow and long and have a thickened median part ("fascia"). The blades of the other groups are broader and have either an inconspicuous or no fascia (section Simplices). *L. yezoensis* is the only representative of the section Digitatae, with a **divided blade;** however, this has a discoid holdfast and may, according to Ju. E. Petrov (1974), be placed into a subgenus *Solearia* (see Section 2.2.1).

Japan: Alariaceae. *Eisenia* exhibits a discontinous worldwide distribution. *Eisenia bicyclis* (Fig. 2.91F) is found in Japan, *E. arborea* (Fig. 2.87) on the Californian coast, and *E. cokeri* (Fig. 5.6D) on the western coast of South America. Since the genus is present on both sides of the Pacific and in both hemispheres in relatively restricted areas, it must have crossed the equator. The same must be assumed to have occurred with the genus *Ecklonia,* which is represented along the Japanese coasts by *E. cava* (Figs. 2.91G, 2.92D,E), *E. kurome,* and *E. stolonifera* with its rhizomatous holdfast system. Other species of *Ecklonia* occur on the coasts of West Africa, South Africa, southern Australia, and New Zealand. The crossing of the tropical regions during the Pleistocene was possible in the Pacific along the American coasts, as illustrated by the distribution of *Macrocystis pyrifera.* Perhaps *Eisenia* has followed this route. However, the genus *Ecklonia* is absent on the American Pacific coasts, and its

---

Fig. 2.90 Algal zonation on Japanese coasts. **(A)** North Honshu (Rikuchu National Park). B = brown alga; R = red alga. *B. atropurpurea* = *Bangia fuscopurpurea; Chondrus yendoi* = *Iridaea cornucopiae.* **(B)** East Honshu (Izu Peninsula). (A after Chihara 1975; B from Hayashida 1983.)

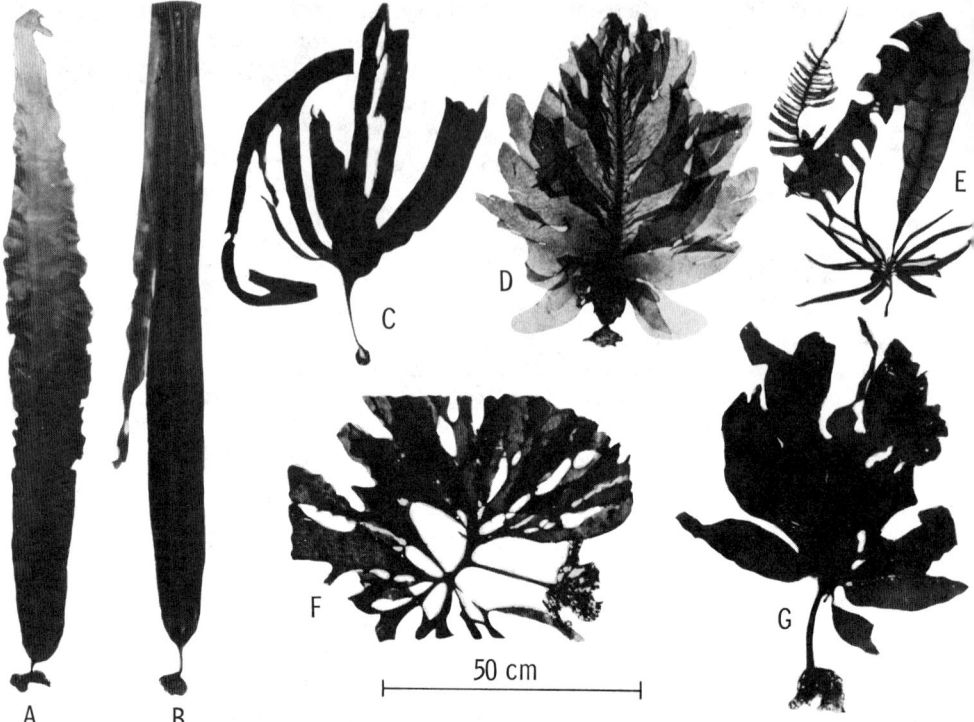

Fig. 2.91 Examples of Japanese representatives of the Laminariales. **Laminariaceae:** **(A)** *Laminaria japonica*. **(B)** *L. angustata*. **(C)** *L. yezoensis*. **Alariaceae:** **(D)** *Undaria pinnatifida*. **(E)** *Alaria crassifolia*. **(F)** *Eisenia bicyclis*. **(G)** *Ecklonia cava*. (Photographs from Segawa 1971.)

worldwide distribution presents a puzzle. To conceive of its crossing the equator in the vast Indo-West Pacific region along the Asiatic side of the Pacific is difficult. When *Eisenia bicyclis* and the three Japanese species of *Ecklonia* were crossed, F_1 sporophytes up to 40 cm long were obtained between any pair of the four species (Migita 1984).

Japan: Ancestral Laminariales. Japanese waters harbor peculiar and possibly ancient members of the Laminariales. *Pseudochorda nagaii* has a whiplike thallus up to 60 cm long and distinctive plurilocular antheridia in the male gametophyte (Kawai and Kurogi 1985). It occurs from the Aleutian Islands to Hokkaido and the Japan Sea coast of the USSR and was previously placed in the Chordariales. This species has now been recognized as a primitive member of the Laminariales (family Pseudochordaceae). *Streptophyllopsis kuroshioensis*, a small and peculiar deep-water representative of the Laminariaceae, has been discovered at a 20–50 m depth off the Izu Peninsula on the Pacific coast of Honshu (Kajimura 1981, 1987). Two short-stiped annual blades up to 20 cm long arise from a dichotomously branched perennial system up to 10 cm high. Sori develop not only on the relatively small blade but also on the annual stipe, which attains a maximum length of 4 mm. The prostrate holdfast is reminiscent of the rhizomatous *Laminaria* species on the Pacific coast of North America and in the Mediterranean, and also of *Ecklonia stolonifera* (family Alariaceae, see above). The presence of *Pseudochorda* and *Streptophyllopsis* leads to the question of whether Japanese waters should be regarded as a "museum for Laminariales" or an area in which this order may have originated, as certain mollusks did in the early Tertiary (see Section 2.1).

The western coast of Japan is influenced by the warm Tsushima Current, which constitutes a northward extension of the Kuroshio Current. The eastern Japanese coast receives cold water from the southward-flowing Oyashio Current. This is comparable to the continental coast of the Sea of Japan due to the cold, southward-bound Linan Current. In consequence many cold temperate algae mainly inhabiting the colder eastern coast of Hokkaido are to be found more to the south on the continental coast, from **Vladivostok** to the northeastern coast of **Korea**. Seventy five percent of the 400 species of Korean seaweeds are common to the islands of Japan and 50% to Hokkaido (Kang 1966). Along the western coast of Korea, at water temperatures of 20–26°C, only kelp species with warm-water affinities, namely, *Undaria pinnatifida* and *Ecklonia* species (see above), can survive. Dominant species in the intertidal zone on the western coast are the brown algae *Sargassum thunbergii*, and *Pelvetia siliquosa* and the red algae *Gelidium divaricatum, Gloiopeltis furcata,* and *Corallina pilulifera* (Y. H. Kim and Lee 1985). The distributions of *Costaria costata, Laminaria japonica, L. religiosa,* and *Agarum cribrosum* end at the somewhat colder eastern Korean coast (H. S. Kim et al. 1983; Nam 1986).

The seaweed flora of **China** contains approximately 900 species, 15% of which are reported to be endemic to China (Tseng 1983). The Yellow Sea, bordered by the Asiatic continental coast and the western coast of Korea, receives a certain warm-water influx from the south. It is characterized by large annual water-temperature fluctuations, which may sink to 0°C in winter in the north. The western part of the Yellow Sea (240 seaweed species; Tseng and Chang 1964) belongs to the cold temperate northwestern Pacific region (Briggs 1974). The shallow part of the China Sea was formed in the Tertiary, from north to south (Wang 1985). The Bohai Gulf was formed as a rift basin in the Eocene, followed by the East China Sea, and then in the Oligocene by the South China Sea. During the Pleistocene the Yellow Sea was dry land for much of its shallower area, and was subsequently invaded by mainly eurythermal species from the Sea of Japan.

Of the Laminariales *Undaria pinnatifida* (Figs. 2.91D, 2.92B,E) occurs along the Chinese coast southward to the province of Zhejiang (northern limit of the species near Vladivostok and the warmer western coast of Hokkaido). *Ecklonia kurome* is found scattered along the coasts of the provinces of Zhejiang and Fujian. *Laminaria japonica,* which is cultivated extensively on northern and middle Chinese coasts, did not occur there naturally because of

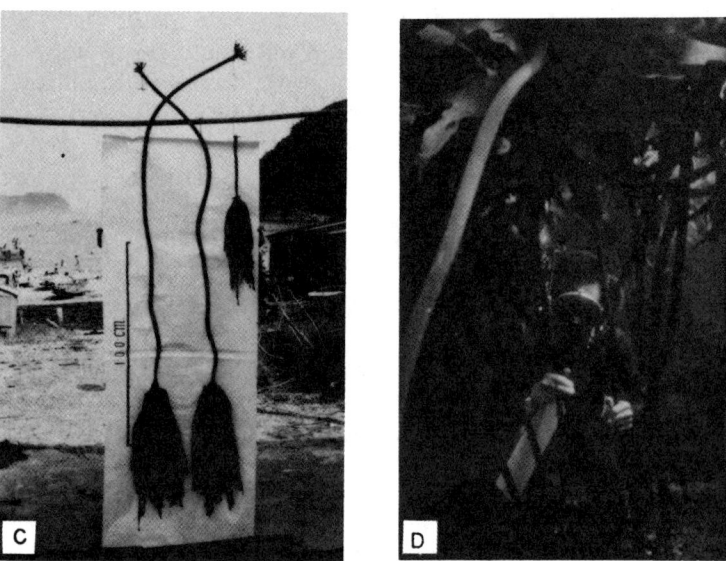

Fig. 2.92 Aspects of the Laminariales on Japanese and Chinese coasts. **(A)** *Laminaria longissima* (= *angustata* var. *longissima;* up to 15 m long) on the eastern coast of Hokkaido, exposed on the beach for drying. **(B)** Underwater view of *Undaria pinnatifida* in the upper sublittoral zone on the east coast of Hokkaido. **(C,D)** *Ecklonia cava* on the eastern coast of Honshu (Izu Peninsula), individual sporophytes in July and kelp forest at a 10 m water depth. **(E)** *U. pinnatifida* and young specimens of *L. japonica* on the shore of Qingdao (North China). **(F)** Vegetation of the lower eulittoral and upper sublittoral zone near Qingdao with *Sargassum thunbergii* and *U. pinnatifida*. (Photographs A,B,E,F by the author; C,D courtesy of F. Hayashida.)

E F

Fig. 2.92 (*Continued*)

the high summer temperatures but was accidentally introduced from Japan in 1927 (Tseng 1981*a,b*).

Along the northern Chinese coast one still finds a few seaweeds known from the North Atlantic. The southward distribution of *Desmarestia viridis* (Fig. 2.8) extends to the coasts of the Yellow Sea, while *Chorda filum* (Fig. 2.9) and *Scytosiphon lomentaria* reach as far as the South China Sea (Tseng 1983). Along the coasts of the East China Sea to the northern boundary of the Indo-West Pacific tropical region near Hong Kong, the rich algal flora of the warm temperate Japan region dominates. This flora is supplemented by various species endemic to China.

3 Seaweeds of the Arctic

The peak of the last glaciation occurred 18,000 years ago (Fig. 2.1). When the ice melted, about 150 seaweed species, capable of living at low water temperatures and surviving prolonged periods of darkness, reinvaded the Arctic from the North Atlantic rather than from the North Pacific. There are few endemic seaweed species in the Arctic region, but more in the Antarctic region due to its longer persistence as a cold-water habitat. Continental glaciation in the Arctic may have started in the Pliocene, at 3–5 Ma, while in Antarctica the first glaciers at sea level occurred in the Oligocene, at 14 Ma. The inception of perennial sea-ice cover has been dated at 3 Ma or 0.7 Ma. Ekman (1953) pointed out that the Arctic does not show the variety of marine fauna that could have evolved, given a sufficiently long time for the processes of speciation. The role of the Arctic Ocean as a bridge between the North Pacific and the North Atlantic has been pointed out in Section 2.1.

Literature: Kjellman (1883); Taylor (1954). **Paleoceanography, paleobiology:** many references on this subject have been cited in Section 2.1; only literature dealing exclusively with the Arctic region is cited here. Churkin and Trexler (1981); Clark (1977, 1982); Dawson et al. (1976); Denton and Hughes (1981); Estes and Hutchison (1980); Forbes (1975); Fujita (1978); Herman (1974, 1983); Herman and Hopkins (1980); Herron et al. (1973); Kent et al. (1984); Kitchell and Clark (1982); Kristoffersen and Huseby (1985); Marincovich et al. (1985); McKenna (1980); Nairn et al. (1981); Sweeney (1985).

Expeditions and early algal research in the Arctic. Several of the various expeditions sent out in search of navigable routes along the Arctic coasts between the Atlantic and Pacific oceans brought back seaweed samples. The phycologist W. H. Harvey, for example, worked on the samples of the unfortunate Parry Expedition. This expedition, with the vessels *Fury* and *Hecla,* reached only as far as the Foxe Basin in its quest to discover the **Northwest Passage** through the Canadian Arctic. Most disastrous was the Franklin Expedition (1845–1848), which lost the vessels *Erebus* and *Terror* between Victoria and King William Island (North-West Territories) and perished completely. Several search expeditions were sent out, and their algal collections were the basis of the early phycological work of G. Dickie on Canadian Arctic algae. (For further details see Wilce 1959 and Lee 1980.) The Northwest Passage was mastered from west to east in 1850 by Robert McClure, and vice versa by Roald Amundsen from 1903 to 1906. It was fortunate for algal research that the phycologist Kjellman was also a member of the *Vega* Expedition. This expedition, under the leadership of A. E. Nordenskjöld, mastered the **Northeast Passage** along the northern Russian coast in 1878–1879 by overwintering at Cape Jakan in the East Siberian Sea.

A. D. Zinova (1957), in her work on the seaweeds of the Siberian coast, cites Kjellman's collecting locations, and these are again incorporated in the distribution maps of the present book (Figs. 2.4–2.42; 3.7–3.11). F. R. Kjellman's book *The Algae of the Arctic Sea* (1883) was based on substantial experiences. He was a member of two earlier Swedish expeditions, to Novaya Zemlya and to the mouth of the Yenisei in 1875 and 1876, respectively. He also went on the Swedish Spitsbergen Expedition in 1872–1873.

3.1 LIMITS OF THE ARCTIC REGION

At the most northerly locations at which seaweeds have been found, there are only 4 to 6 ice-free weeks per year and none in some years. The southern limit of the Arctic region, as indicated in Figs. 1.5 and 1.6 nearly follows the 0°C-February isotherm (10°C August isotherm) or approximately the southern ice limit in winter and spring (Fig. 3.1) and extends to the coasts of

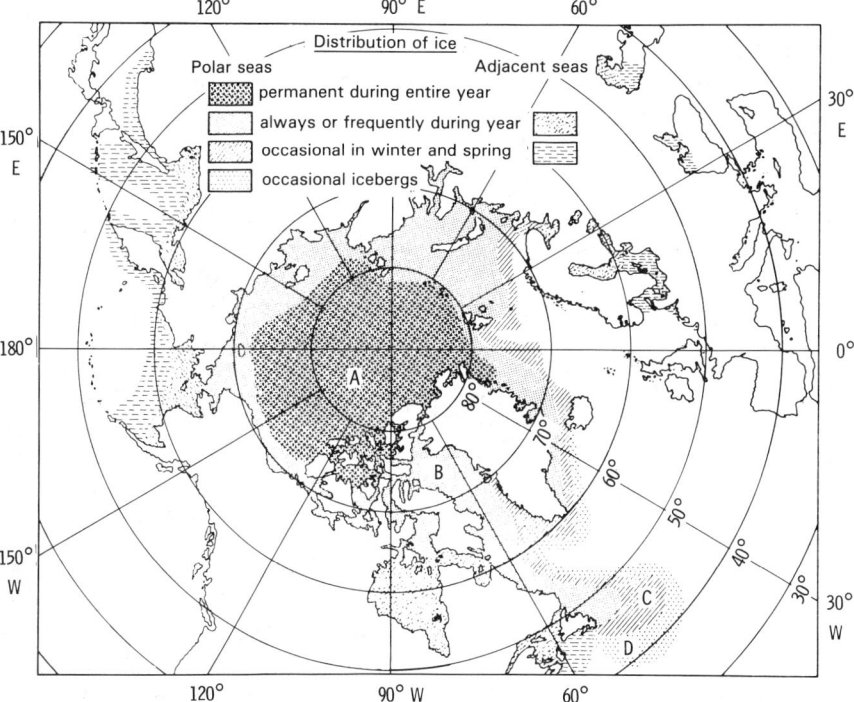

Fig. 3.1 Ice limits in the Northern Hemisphere. Four areas are distinguished. **(A)** Permanently ice-covered area (southern boundary corresponds to minimal ice extension in first half of September). **(B)** Area in which ice is found with more than 50% probability (southern boundary corresponds to minimal ice extension in March). **(C)** Area in which ice is found at a 10–50% probability (southern boundary corresponds to maximal ice extension in March). **(D)** Area in which icebergs may occur off the eastern North American coast. (From Dietrich et al. 1980.)

Greenland and Spitsbergen and the northern Russian coast from Kolafjord (Murmansk) to Cape Olyutorsky (Bering Sea). Along the North American coastline the Arctic region stretches from western Alaska north of Nunivak Island, passing the Canadian Arctic archipelago, to southern Labrador, where the Strait of Belle Isle serves as a boundary to the cold temperate North Atlantic region.

Problems with the limits of the Arctic region. In contrast to assumptions of Briggs (1974) and Michanek (1979, 1983) the northern coast of Iceland is not considered here to belong to the Arctic region. This coast is blocked by ice at irregular intervals of several years (Dietrich et al. 1980), and its seaweed vegetation exhibits, like the eastern Icelandic coast, some Arctic traits (Munda 1972a). Remmert (1980) chose the polar circle as the southern boundary of the Arctic region for the purposes of terrestrial zoogeography. On the basis of this decision, southern Baffin Island, Hudson Bay, Labrador, and southern Greenland are excluded from the Arctic region, while northern Norway becomes part of it. However, for marine organisms, distribution limits generally follow seawater isotherms. A pattern based on the polar circle is not reflected by the southern limits of seaweed species typical of the marine Arctic region (compare Figs. 3.7–3.11 with Figs. 1.5 and 1.6), or by the northern limits of cold temperate species (Figs. 2.31–2.42). A marine **sub-Arctic region,** recognized by Börgesen and Jónsson (1905) and by Michanek (1979, 1983) is not considered in the present book. The northern limits of the Arctic–cold temperate seaweeds are so scattered that it is hard to say where a benthic marine sub-Arctic region should begin or end, for example, along the eastern and western coasts of Greenland (Figs. 2.4–2.30). There are hardly any endemic seaweed species for a sub-Arctic region, and the same applies to the Arctic bryozoans and polychaetes (Powell 1968).

Cold-water currents. The coasts of Greenland and Labrador belong to the marine Arctic region because they are influenced by cold water. The coasts of Iceland and northern Norway, however, although situated farther north, receive an input of warm water from the North Atlantic Current. The **East Greenland Current** transports water southward along the east coast of Greenland (Fig. 2.43), rounds southern Greenland, and continues as the northbound **West Greenland Current.** Part of this turns west in the Davis Strait, and together with cold water from the Hudson Strait forms the southward-flowing **Labrador Current.** The latter probably originated parallel to the inception of perennial sea-ice cover in the Pliocene (Berggren and Hollister 1974) and enables many Arctic–cold temperate species today to penetrate far south along the North American coast of the Atlantic. Only the northern part of the **Bering Sea** can be regarded as belonging to the Arctic region. In the southern part, of the Bering Sea, the influence of warm water from the North Pacific causes the pack ice limit to be situated south of the Bering Strait in winter, at about 60°N latitude in the Bering Sea, and north of the Bering Strait in summer, in the Chukchi Sea (Zenkevitch 1963).

3.2 SEAWEED ENVIRONMENT IN THE ARCTIC REGION

Kjellman (1883), the pioneer of algal research in the Arctic, summarized the seaweed vegetation by writing that "the most prominent features in the general aspect of the Arctic marine flora are scarcity of individuals, monotony and luxuriancy." **Scarcity of individuals,** he proposed, was due to the lack of suitable substratum in most places. He also described dense beds where stable substratum was available. **Monotony,** he considered to be due to the dull brown color of the laminarians. The laminarian forest seemed to be similar to that in cold temperate regions. **Luxuriancy** described the growth and size of the laminarian vegetation.

The seaweed vegetation in the Arctic region may be dense in places, yet it exhibits a low productivity. This is due to the seasonal lack or scarcity of light, low nutrient concentrations, and a low photosynthetic rate because of low water temperature. In the northern Arctic region the water temperature hardly changes in the course of a year, staying a little above $-1.8°C$, the freezing temperature of seawater. The Arctic region seaweeds have not found a strategy that will enable them to grow fast at low water temperatures. The sporophytes of *Laminaria longicruris,* isolated from a location in the Canadian Arctic in the form of a gametophytic culture, grow three times faster in the laboratory at 10°C than at 0°C (Bolton and Lüning 1982). In spite of this, **large thallus size** is characteristic of many seaweed species in the Arctic. For example, the red alga *Phycodrys rubens* (= *sinuosa*) may reach a length of 30 cm (Kjellman 1883) in Greenland, whereas in the cold temperate regions half this length or less is normal for the species. In Labrador huge individuals of *Laminaria longicruris* were found. In Greenland *L. saccharina* may retain three years' growth of blade tissue (Lund 1959), resulting in a more impressive thallus size than in the cold temperate regions, where it loses more of the yearly growth at the distal end of the blade than it retains. Thus the large thallus size of seaweeds in the Arctic is not an indication of high productivity but of **longevity** of the individuals. The exact physiological and/or biotic reasons for this longevity are unknown.

Life in cold water. The growth of Arctic–cold temperate seaweed species in the Arctic, and in winter in cold temperate regions, is characterized by phenotypic adaptations to counteract the effect of low temperature on metabolism. This may be achieved by, for example, increasing the number of photosynthetic enzyme molecules at low temperature (Clarke 1983). Due to such measures almost constant photosynthetic rates are achieved in *Laminaria saccharina* at growth temperatures between 0° and 20°C (Davison 1987). The phylogenetic adaptation to life at near zero water temperatures must have occurred in the late Tertiary, when polar organisms had to keep pace with the general cooling of about 15°C in seawater over 50 million years (Fig. 2.1). Scarlato (1977), on the basis of temperature characteristics of bivalve mollusks, suggested that the first stage of the phylogenetic adaptation process was the broadening of survival and reproduction temperature towards the lower end of the temperature scale. After the species had acquired the capacity to grow and reproduce at negative temperatures, elimination of high-temperature survival and the continuous lowering of reproduction temperatures set in. This process has advanced much further in the Antarctic than in the Arctic region.

Light conditions. At latitude 80°N, where the northernmost seaweeds occur, the polar night lasts from mid-October to mid-February (Fig. 3.2). One might assume that the reduction of the yearly photosynthetic gain, due to the dark period of the polar

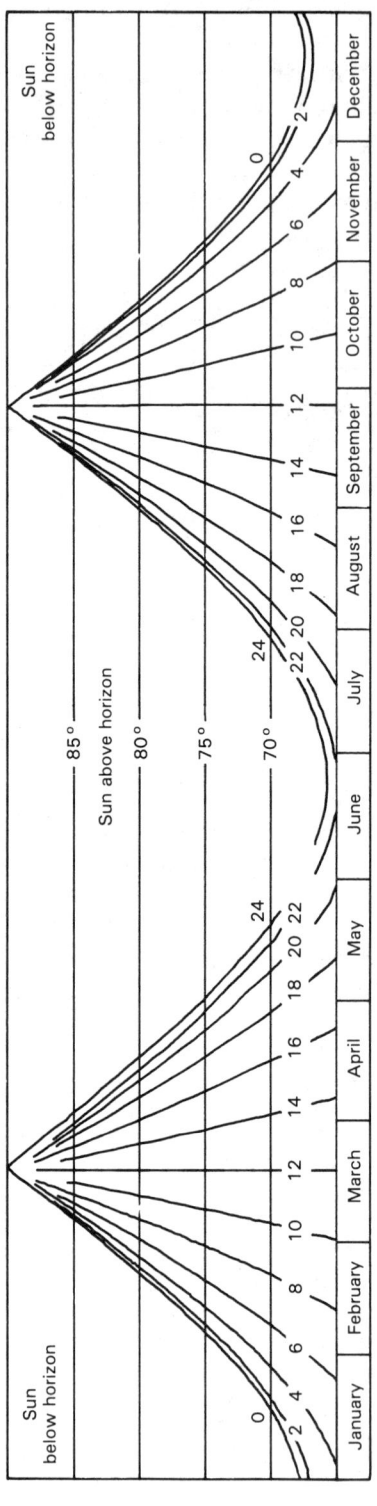

Fig. 3.2 Daylength (in hours) north of the Polar circle (66°33′N) in relation to geographical latitude. (Redrawn from Smithsonian Meteorological Tables 1951.)

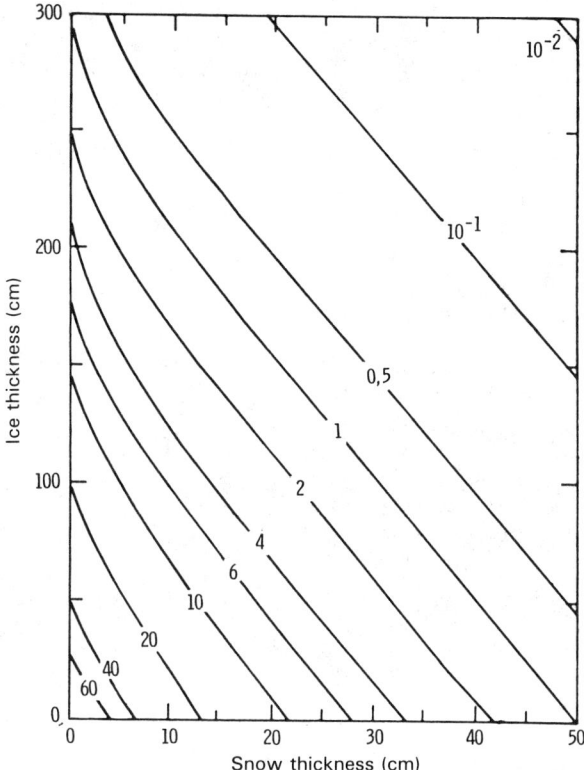

Fig. 3.3 Transmission of sea ice (newly formed) that is covered by snow (shortly before melting). Transmission values are given as percentage of irradiance above water (400–800 nm). (From Maykut and Grenfell 1975.)

winter, might be compensated for by continuous photosynthesis during the weeks or months of the polar summer. Although there may be sufficient light above water in spring and early summer, it is only in late summer when the ice layer (a few meters thick) has melted that the seaweed vegetation receives enough light to satisfy the demands of photosynthesis. Transmission of light is particularly impeded if snow is covering the ice. For example, a meter of clear ice transmits 20% of light (with maximum transmission in the wavelength range of 450–550 nm), but only 1% of light, if 30 cm of snow are present (Fig. 3.3). At Igloolik in the Canadian Arctic (69°N, 82°W), the seawater above the benthic algal vegetation is covered by an ice layer up to 1.5 m thick for most of the year, and the main photosynthetic gain of the algae is restricted to about 10 ice-free weeks from August to mid-October. Chapman and Lindley (1980) at Igloolik performed the first continuous underwater light measurement in the Arctic (Fig. 3.4F). They found that 80% of the yearly light supply reaching the kelp vegetation at 10 m deep, is received during the 10 ice-free weeks. Just after the ice has melted, the water becomes turbid due to a short-lasting plankton bloom, which consumes most of the already low nutrients in the water. Very clear water prevails after that, with transmission values of 60–80% per meter in the wavelength

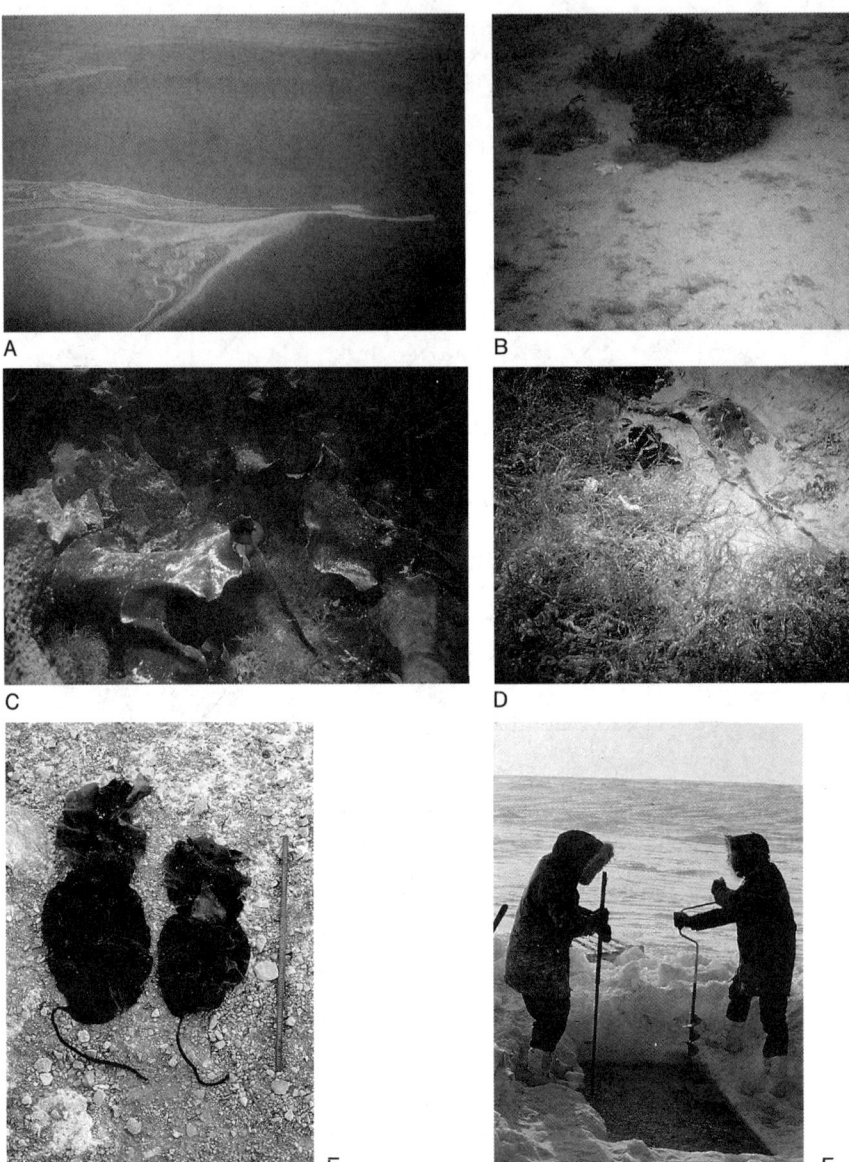

Fig. 3.4 Arctic seaweed vegetation at Igloolik, Canadian Arctic. (**A**) Aspect of the coast in September. (**B**) At 2 m deep (below low water; tidal range is about 2 m): first conspicuous seaweeds to be found (i.e., *Fucus evanescens*), sparse spots formed by the green alga *Acrosiphonia* and by benthic diatoms. (**C**) *Laminaria longicruris* at 10 m deep. The laminarian vegetation begins at 4 m deep. (**D**) *Desmarestia aculeata* in the foreground and an extremely long individual of *L. longicruris* in the background; heavy sedimentation. (**E**) Habit of *L. solidungula*. (**F**) A hole for divers who monitored a continuous light-measuring station near Igloolik (see Chapman and Lindley 1980). (Photographs A–E by the author; F courtesy of A. R. O. Chapman.)

range of 400–700 nm (Fig. 6.1). In other words, once the yearly light supply arrives, it penetrates to great depths. At a study site in the Beaufort Sea, Alaska, in August at 6–7 m deep, photon fluence rates (see Chapter 6) ranged between 100 and 200 μmol m^{-2} s^{-1} during daytime and 20–30 μmol m^{-2} s^{-1} at night (Dunton and Schell 1986). This explains why sublittoral algae (e.g. *Laminaria solidungula* or *Phyllophora truncata*) may penetrate to relatively great depths in the Arctic region.

Growth in darkness and dark tolerance. The biologists who accompanied the early expeditions to the Arctic were surprised when, on a coastline which had just become free of ice in spring or early summer, dredgings brought actively growing seaweeds to the surface. Heterotrophic growth in seaweeds has never been demonstrated (Wilce 1967; Lobban et al. 1985), although growth from reserve materials is quite common in mid- and lower sublittoral seaweeds of temperate regions (Section 8.4). The Arctic region seaweeds also grow like this, and generally a perennial growth form dominates. It has been shown that the Arctic-endemic *Laminaria solidungula* uses the annual dark period for producing its blade and the short annual light period for storage of photosynthates (A. R. O. Chapman and Lindley 1980; Dunton et al. 1982; Dunton and Schell 1986). This temporal separation of growth and photosynthesis is important since the photosynthesizing blade area must be present as soon as the small yearly light supply arrives. **Dark tolerance,** that is, the ability of vegetative tissue, algal spores, and zygotes to survive prolonged periods of darkness, is important for success under Arctic conditions. It has been demonstrated that *Laminaria* species are well adapted to survive darkness in Arctic conditions. The early, unicellular stage of the gametophyte survives at least six months in the dark in the laboratory to produce healthy gametophytes that produce sporophytes within two weeks under light (Kain 1964, Lüning 1980a). In contrast, spores of the warm temperate eulittoral Japanese species *Porphyra tenera* and *Gelidium amansii* die after a dark period of four to six weeks (Ohno and Arasaki 1969). Arctic diatoms may survive at least five months of darkness (Palmisano and Sullivan 1983). The **deciduous** habit, which slows respiration losses considerably and is an important factor in trees at high latitudes (Axelrod 1984), is another adaptation to the chronic lack of light. A typical case is the brown alga *Desmarestia aculeata* (Figs. 2.7 and 3.3D). It reaches far into the north, with only the older portions of the thallus persisting through the winter, and fresh shoots are formed early in the year.

Mechanical effects of ice. No supra- and eulittoral algal vegetation exists in the northern Arctic region because of ice scouring. Most intertidal and upper sublittoral biomass is physically removed on hundreds of kilometers of coastline by winter pack-ice (Hooper 1988). Those seaweeds completely embedded under ice may survive for a considerable length of time. This is demonstrated by healthy individuals of fucalean species in the eulittoral zone of the southern Arctic region, which is covered each year by the **"ice-foot"** (Fig. 3.5), a protruding ice belt beginning at the high-water mark (Deichmann and Rosenvinge 1908; Kanwisher 1957). Kjellman (1877), in his description of the algal vegetation of western Novaya Zemlya, gave a vivid description of the continuous crushing and grinding action of drift ice on the substratum. This leads to an often observed polished appearance of hard rocky substratum on Arctic coasts (Fig. 3.6) and to the formation of large quantities of gravel, sand, and mud as products of the everlasting crushing and grinding ice action.

Fig. 3.5 Remnants of the "ice foot" in summer, near Port Burwell, Ungava-Bay, Labrador. (Photograph from Wilce 1959.)

Little mechanical stress on sublittoral seaweeds. Where water motion is weak (with increasing depth in the sublittoral zone and especially in the inner calm parts of fjords), the algae may be heavily covered with **silt** (Fig. 3.4D). Being able to withstand such conditions in calm water is an adaptation of seaweeds in the Arctic region. In general, one may expect little mechanical stress on the seaweeds of the midsublittoral zone and lower. In winter when storms are raging, the water is covered by ice. When the ice layer is absent in late summer and autumn, the weather is often calm. A certain fragile appearance is obvious, e.g. in for example, kelp thalli that may be very old and probably not exposed to intensive mechanical stress for many years.

Fig. 3.6 Ice-scoured shore at Hebron, northern Labrador. (Photograph from Wilce 1959.)

Effects of brackish water. A further impediment of marine life in the upper sublittoral zone is the annual layer of **brackish water** after the thaw (Hooper 1988). In the inner Scoresby Sound (eastern Greenland) brackish water may be detected in summer down to 25 m and in the middle parts of the fjord down to 15 m (Lund 1959). Before the sea ice melts, fresh water from the glaciers flows beneath the fjord ice, and salinity (in a 3 m thick water layer) may be reduced for some time to 3–4‰. Temperate region eulittoral seaweed species often inhabit the upper sublittoral zone in the Arctic region, replacing the less brackish-water-tolerant representatives of the Laminariales (Fig. 3.3B).

3.3 DISTRIBUTIONS OF ARCTIC-ENDEMIC SPECIES

One may argue that there are no seaweed species endemic to the Arctic. The species denoted in Figs. 3.7–3.11 as Arctic-endemic, occurring along one coastline or another, cross the "barrier" of the Arctic region, erected with so much effort by the marine biogeographers (Figs. 1.5–1.6). Even the Arctic kelp *Laminaria solidungula* (Fig. 3.7) occurs off Newfoundland, although only in deep water. The red algae *Devaleraea* (= *Halosaccion*) *ramentacea* (Fig. 3.10; see Guiry 1982) penetrates for a long way into the cold temperate American North Atlantic region, and the red alga *Turnerella pennyi* (Fig. 3.7) into the European North Atlantic. However, such extensions, which are annoying to the biogeographer, also occur in other areas; for example, the red alga *Rissoella verruculosa,* which is a paleoendemic species in the western Mediterranean, is also found on the Atlantic coast of Morocco. Such extensions may be considered minor, yet the concept of endemism should still hold with the species illustrated in Figs. 3.7–3.11. The distribution patterns of the Arctic endemics may indicate mainly that they require low temperatures for reproduction (Breeman 1988, 1989). Their upper lethal limits are higher than in species of the older cold-water habitats of Antarctica (Section 7.1.1). Nevertheless, the few Arctic endemics seem to have physiologically adapted to life under Arctic conditions to such a degree that they can hardly compete with the Arctic–cold temperate species (groups 1 and 2 in Table 2.1) at lower latitudes.

The distribution of endemic seaweeds within the Arctic region is circumpolar with wide gaps occurring along the Russian Arctic, where soft substratum prevails. Endemics include the kelp *Laminaria solidungula* (Fig. 3.7), the red algae *Dilsea integra* (Fig. 3.9) and *Devaleraea ramentacea* (Fig. 3.10). These species penetrate far into the northern Arctic region; possibly they were among the last seaweeds to leave the Arctic when the ice advanced and among the first species to follow its northward retreat. The two red algae *Turnerella pennyi* (Fig. 3.8) and *Pantoneura baerii* (Fig. 3.11) are probably restricted to the parts of the Artic region adjacent to the North Atlantic. Species with a similar morphology are *Turnerella mertensii* and *Pantoneura juergensii* (Wynne 1970) in the North Pacific.

3.4 LOCAL SEAWEED FLORAS AND THEIR VEGETATION STRUCTURE

After Kjellman's *Algae of the Arctic Sea* appeared in 1883, it was not until 1954 that W. R. Taylor produced a survey on the seaweed vegetation of the Arctic region. Because the Arctic coasts are difficult to access, relatively few pertinent publications exist.

3.4.1 Greenland

Literature: Entire area: Rosenvinge (e.g., 1898). **East Greenland:** Lund (1951, 1959); Rosenvinge (1910). **South Greenland:** Pedersen (1976). **West Greenland:** Wilce (1963).

Most of the larger seaweed species of Greenland occur on the east and west coasts. The floristic similarity of both coasts stands at 80% (Lund 1959). A notable exception is provided by the kelp species *Agarum cribrosum* (Fig. 2.10). Relatively intensive investigations were performed along the east coast of Greenland, with the exception of 500 km of inaccessible coastline from 61°N to 65.5°N, for which no algal records exist. Information on algal vegetation is also sparse for the west coast of Greenland north of 72°N.

Arctic–cold temperate species dominate even the northernmost location where seaweeds have been investigated: the Jörgen Brönlunds Fjord (82°N) (Table 3.1). Most of the species constituting the seaweed vegetation of Greenland are known to everybody who is familiar with the algal flora of the

Fig. 3.7–3.11 Distribution areas of characteristic, Arctic-endemic seaweeds, with minor occurrence outside the Arctic region. **Special remarks: Fig. 3.7** *Laminaria solidungula* (order Laminariales). This brown alga has an attachment disc instead of haptera, forms a new blade each winter and early spring, has sori on previous years' blades, has up to six blades on one individual, has a total length of up to 2.5 m, and has a stipe length of up to 1 m. **Fig. 3.8** *Turnerella pennyi* (order Gigartinales). The thallus of this red alga is 5–25 cm long, has a heteromorphic life cycle (South et al. 1972), and a bladelike gametophyte and crustose tetrasporophyte. The tetrasporophyte was described as *Cruoria arctica* or *C. rosea*. **Fig. 3.9** *Dilsea integra* (order Cryptonemiales). This red alga has a thallus length up to 30 cm. **Fig. 3.10** *Devaleraea ramentacea* (order Palmariales). The thallus of this red alga is 10–40 cm long, has numerous forms, and an unusual life history like *Palmaria palmata* (see Fig. 2.17) (van der Meer 1980): the female gametophytes reach only microscopic size and are fertilized by spermatia of the macroscopic male gametophytes of the previous year. The similarly macroscopic tetrasporophyte develops directly from the fertilized egg cell. A carposporophyte may not have existed in the order Palmariales. **Fig. 3.11** *Pantoneura baerii* (order Ceramiales). The thallus of this red alga is 15–20 cm long. A record for Sitka, Alaska, possibly relates to a North Pacific *Pantoneura* sp. (Wynne 1970). (Sources for habit drawings: Lund 1959: 3.8 W. R. Taylor 1957: 3.9 A. D. Zinova 1953: 3.7; and A. D. Zinova 1955: 3.9, 3.10.)

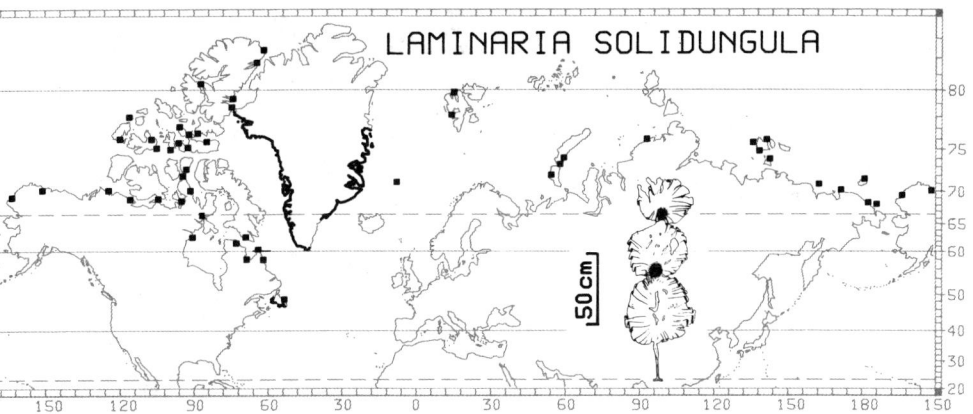

Fig. 3.7

North Atlantic. While 20 seaweed species reach as far north as Jörgen Brönlunds Fjord (82°N), 110 species occur on South Greenland (60°N) (Pedersen 1976).

Numerous algal species reach their northern limit along the eastern Greenland coast (73–77°N), between Franz Josef Fjord and Cape Bismarck: the red algae *Turnerella pennyi* (Fig. 3.8), *Callophyllis cristata* (Fig. 2.16), *Devaleraea ramentacea* (Fig. 3.10), *Scagelia pylaisei* (= *Antithamnion boreale;* Hansen and Scagel 1981), *Ptilota serrata* (Fig. 2.26), *Phycodrys rubens* (Fig. 2.28), and *Pantoneura baerii* (Fig. 3.11); the brown algae *Dictyosiphon foeniculaceus* (Fig. 2.6), *Fucus evanescens* (Fig. 2.13), *Alaria esculenta* (Fig. 2.12); and, surprisingly, the Arctic endemic kelp *Laminaria solidungula* (Fig. 3.7). Perhaps the more northerly habitat of this deeper sublittoral alga has been overlooked.

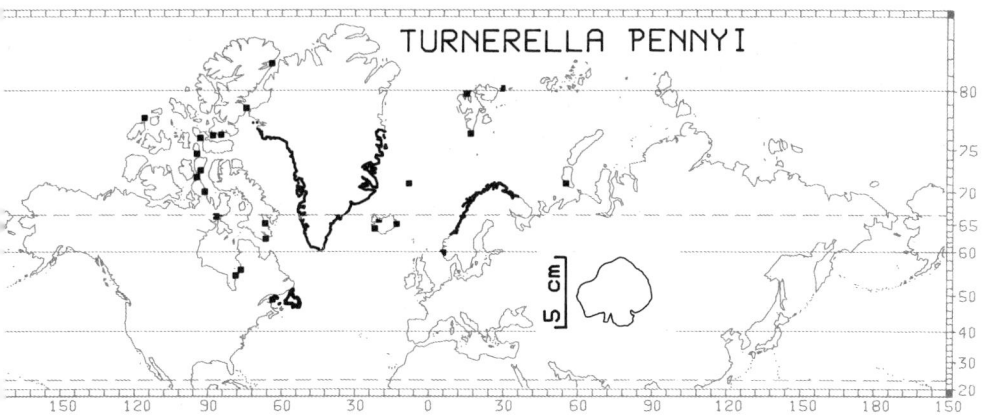

Fig. 3.8

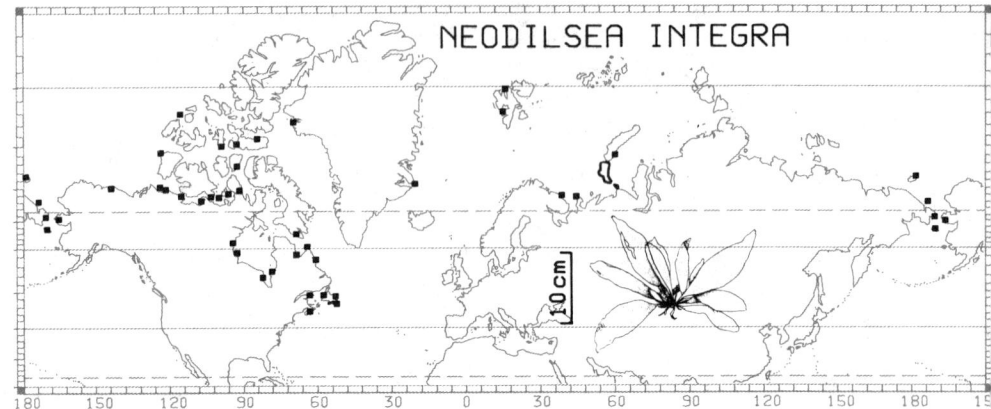

Fig. 3.9

Zonation on the east coast of Greenland. Lund (1959) studied the algal vegetation in **Scoresby Sound** (70.5°N; dark period in the polar winter of about two months; 109 seaweed species) and the more northern **Franz Josef Fjord** (73.5°N; dark period of about three months; 74 seaweed species). The seawater temperatures along these coasts range between a maximum of 3°C in summer (July) and a minimum of −1.8°C in winter. The sea is covered by a 1–2 m thick ice layer from September to May. In addition, from October to May, a 30–40 cm thick layer of snow covers the ice so that almost no light enters the water (Fig. 3.3). **Green algae** (25 species) contribute 75% of the algal species in the eulittoral zone, and one of these, *Chlorochytrium schmitzii*, an endophyte in the red crustose alga *Cruoria arctica*, is to be found with its host at the lower algal limit of 120 m deep (Fig. 3.12). The **brown algae** (53 species) dominate in the depth range of 0–30 m, where they make up more than half of the total diversity. *Sphacelaria arctica* and *Omphalophyllum ulvaceum* are the deepest-occurring brown algae. The **red algae** (31 species) reach maximum diversity at 10–20 m deep, while

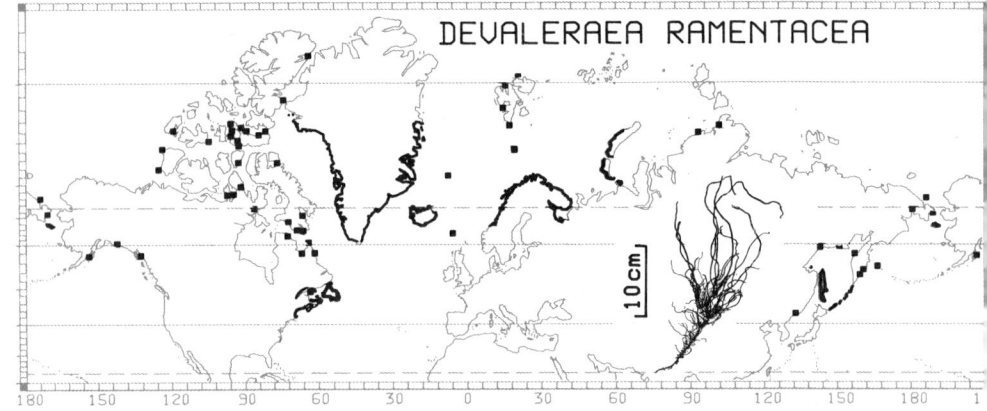

Fig. 3.10

LOCAL SEAWEED FLORAS AND THEIR VEGETATION STRUCTURE

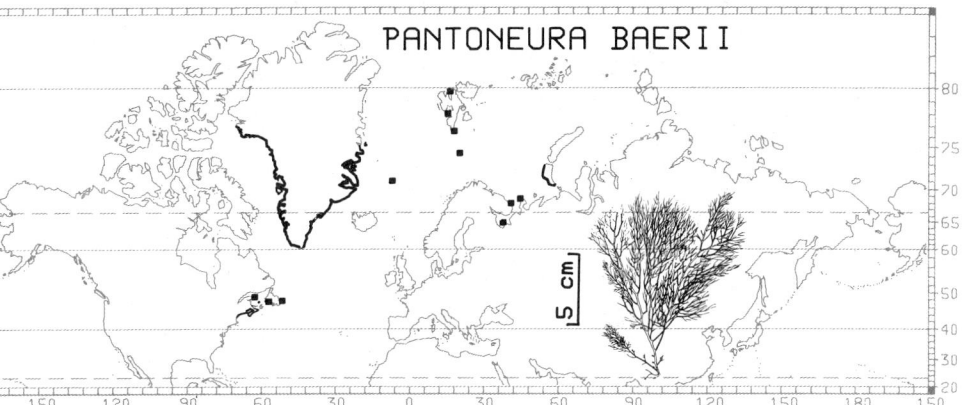

Fig. 3.11

Table 3.1 The northernmost seaweed species, collected in the Jörgen Brönlunds-Fjord, a branch of the Independence-Fjord in NE-Greenland (82°N, 31.5°W).

Green algae

Chaetomorpha melagonium   Chaetomorpha capillaris (= tortuosa)
Acrosiphonia sp.   "Chlorochytrium inclusum"[1]

Brown algae

Pilayella littoralis	Giffordia ovata	Sphacelaria arctica
Sphacelaria plumosa	Leptonematella fasciculata	
Elachista fucicola	Desmarestia viridis	Desmarestia aculeata
Stictyosiphon tortilis	Punctaria glacialis	Litosiphon groen-
Chorda tomentosa	Laminaria saccharina	landicus

Red algae

Audouinella efflorescens	Phyllophora truncata	Ceratocolax hartzii
Polysiphonia arctica	Rhodomela confervoides (= lycopodioides)	

[1] in Phyllophora truncata; and probably the Codiolum-stage of Acrosiphonia (Pedersen 1976)

After Lund 1951.

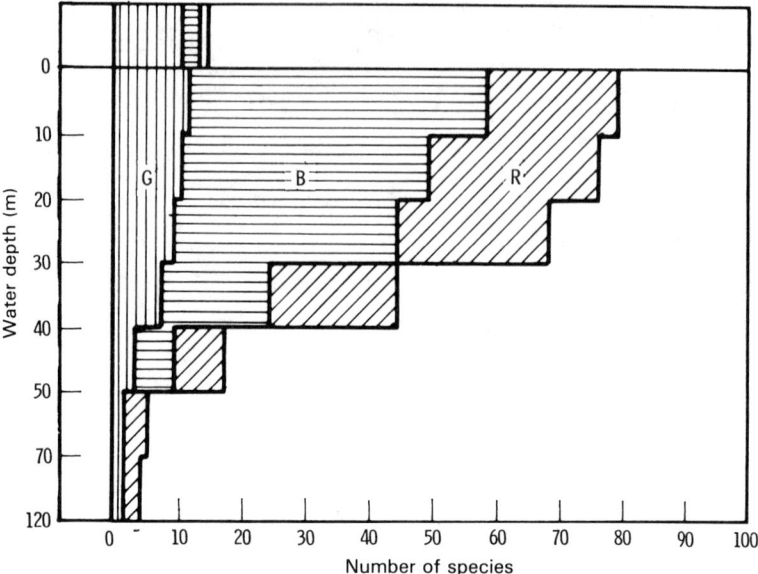

Fig. 3.12 Number of species of green algae (G), brown algae (B), and red algae (R) at different depths on the east coast of Greenland. Total species number is 109. The *x*-axis gives approximately relative percentages of each. (From Lund 1959.)

three crustose red algae are found at the lower limit of algal vegetation. At **wave-exposed habitats** the **eulittoral zone** extends about 1 m vertically, there the scarce algal vegetation consists mainly of filamentous green algae (species of *Ulothrix* and *Urospora*), the red alga *Audouinella purpurea,* and Cyanophytes (e.g., *Calothrix*). The vegetation of the upper sublittoral zone, from the low-water mark to about 3 m deep, is mainly dominated by *Fucus evanescens*. Other species that will tolerate the brackish water to a depth of 10 m include *Pilayella littoralis, Sphacelaria arctica, Stictyosiphon tortilis, Punctaria plantaginea, Scytosiphon lomentaria,* and the filamentous green alga *Spongomorpha* sp. Between 10 and 30 m, in the midsublittoral zone, a kelp vegetation flourishes, dominated by *Laminaria saccharina* and *L. solidungula,* associated with *L. digitata, Saccorhiza dermatodea,* and *Alaria esculenta* (forma *pylaii*). In the lower part of this kelp zone and further down in the *lower sublittoral zone,* to about 40 m depth, one finds a red algal vegetation with the dominating species being *Phycodrys rubens, Phyllophora truncata, Turnerella pennyi,* and *Callophyllis cristata,* which are accompanied by *Polysiphonia arctica, Rhodomela confervoides* (forma *lycopodioides*), and *Dilsea integra*. This vegetation is then followed, towards the algal limit at 120 m deep, by crustose red algae such as the calcified species *Lithothamnium glaciale* (Fig. 3.13), *Leptophytum laeve,* and *L. foecundum* and the uncalcified species *Cruoria arctica,* whose depth range is 12–120 m. Since all the cited depth limits were determined by dredgings, there is some uncertainty over the **lower algal limit of 120 m.** This is deeper than the deepest record in the European cold temperate region of 90 m on the Rockall Plateau, northwest of Ireland. Direct observations made from a submersible off the coast of northern Labrador support an algal depth limit with the deepest corallines at 130 m. On the soft bottoms of the **inner fjord**

Fig. 3.13 *Lithothamnion glaciale,* a crustose coralline alga with outgrowths (growth form of the "rhodolith"; see Section 4.2); and occurs in the Arctic region and in the cold temperate North Atlantic region. The illustrated specimen (10 cm in diameter) was collected on the eastern coast of Newfoundland. (Courtesy of Bob Hooper; photograph courtesy of J. Marschall.)

areas there are loose-lying algae or algae attached to small stones. Dominating this area are the brown algae *Desmarestia aculeata, Fucus evanescens, Stictyosiphon tortilis, Dictyosiphon foeniculaceus,* the green alga *Chaetomorpha melagonium,* and the red algae *Phyllophora truncata, Phycodrys rubens,* and *Devaleraea ramentacea.*

In **southern Greenland** at Cape Farvel (60°N), where the cold East Greenland Current mixes with warmer Atlantic water and continues its flow as the West Greenland Current, no sea ice is formed even in winter. However, drift ice and icebergs, which are brought to this coast by the East Greenland Current, block the open coast at irregular intervals. The August temperature of the surface water is about 2°C.

Zonation on the coast of southern Greenland. The 2 m wide eulittoral zone of the **inner fjords** are protected from the threat of drift ice (Pedersen 1976). This zone is reminiscent of the cold temperate North Atlantic region and supports a luxuriant vegetation of *Ascophyllum nodosum* and *Fucus vesiculosus. Fucus evanescens* also grows here but shifts downward to the upper sublittoral zone further north. At the **open and wave-exposed coast,** the fucalean species are present only in cracks and are reduced in size due to the effects of ice abrasion. In summer the **eulittoral zone** is dominated by annual green algae, (e.g., *Urospora penicilliformis, Ulothrix* spp., *Monostroma groenlandicum,* and the red alga *Porphyra miniata*). In the **upper sublittoral zone,** from the low-water mark down to 1.5 m deep, one finds a vegetation of mainly annual brown algae, as described above for the Scoresby Sound. Deeper down, the **midsublittoral** zone again harbors kelp vegetation. The vegetation structure in the sublittoral zone of southern Greenland represents a transition stage between the situation in the central and northern Arctic region. There is a gap up to 10 m deep between the low-water mark and the uppermost kelp vegetation in the central and northern Arctic, while in the cold temperate regions of the Northern Hemisphere the kelp vegetation begins in the upper sublittoral zone (Section 2.3.3). *Laminaria solidungula* (Fig. 3.7) occurs in southern Greenland, whereas *L. hyperborea* (Fig. 2.35) is absent in this southern part of the Arctic region. A further characteristic of the kelp vegetation is the presence of *Agarum cribrosum* (Fig. 2.10), which has reached southern Greenland via western Greenland from the North American coasts.

The southwestern coast of **West Greenland,** north of Cape Farvel to Disco Bay at latitude 60°N (area of the Davis Strait), is besieged by pack ice only in winter and spring (Fig. 3.1), due to the mixing of Atlantic water with the West Greenland Current. North of Disco Bay a submarine ridge marks the beginning of the colder Baffin Bay. Here the northern limits of several Arctic–cold temperate algae are to be found along the coast of western Greenland. These include *Ascophyllum nodosum* (Fig. 2.22), *Fucus vesiculosus* (Fig. 2.23), *Chorda filum* (Fig. 2.9), *C. tomentosa* (Fig. 2.19), *Membranoptera alata* (Fig. 2.27), *Polysiphonia urceolata,* and *Cladophora rupestris.*

The branched brown alga *Papenfussiella callitricha* (Ectocarpales, Chordariaceae; Hooper and South 1977; South 1987), up to 65 cm long, is restricted to the coasts of southern and western Greenland, and the adjacent coast of North America. In deeper water, over a similar distribution, one also finds the smaller (15 cm long), bladelike brown alga *Omphalophyllum ulvaceum* (Ectocarpales, Pogotrichaceae), which has an eastward distribution to the Norwegian coast. This alga was detected in the Canadian Arctic and on the coast of southeastern Alaska (Wynne et al. 1982). In view of the extensions beyond the Arctic region, previously mentioned regarding the species illustrated in Figs. 3.7–3.11, these two algae may be added to the group of Arctic-endemic species.

The seaweed vegetation of **Disco Bay** (69.5°N; 137 species) and, 1200 km to the north, of the coast of **Baffin Bay** near **Thule** (78°N; 121 species) was investigated by Wilce (1963). At locations protected from ice scouring, the eulittoral and upper sublittoral zones are inhabited by about 25 seaweed species. More species are found in the midsublittoral zone of both areas, where the dominating brown algae are *Agarum cribrosum, Desmarestia aculeata, Laminaria longicruris,* and *L. solidungula.* Since the species inventory of the Disco Bay and the area near Thule is rather similar, it does not make sense to place a major biogeographic boundary between them.

3.4.2 Spitsbergen, Franz Josef Land, Bear Island, and Jan Mayen

Literature: Spitsbergen (Svalbard): Svendsen (1959). **Jan Mayen:** Kjellman (1906); Rosenvinge (1924). **Franz Josef Land:** Marr (1927). **Bear Island:** Kjellman (1903).

In the sublittoral zone, high in the northerly archipelago of **Spitsbergen** and **Jan Mayen** (an island halfway between Spitsbergen and Iceland), one finds the same kelp species as those in the Scoresby Sound in eastern Greenland. The cold temperate species *Laminaria hyperborea*, present in Iceland, is absent from Jan Mayen, being replaced there by *L. solidungula*.

Zonation at western Spitsbergen. At the latitude of the Isfjorden (78°N), where Svendsen (1959) investigated the algal vegetation, the dark period during the polar winter lasts from the end of October to mid-February. The polar summer lasts with continuous light from the end of April to the end of August. In most years, from December to May, the Isfjorden is covered by a 1 m thick layer of sea ice. The Spitsbergen Polar Current transports cold water and drift ice from the easterly Barents Sea, passes the southern coast of Spitsbergen, and in some years may block the Isfjorden with ice until July. The water temperature at the mouth of this fjord is less than 0°C until May but may rise to 5°C by July. In comparison to the coast of eastern Greenland, the summer temperatures of the seawater are higher at Spitsbergen and Jan Mayen because of the Atlantic Spitsbergen Current. This is a branch of the relatively warm North Atlantic Current and ensures that the area west of Spitsbergen is relatively free of ice (Fig. 3.1). The less drastic Arctic conditions at western Spitsbergen are reflected in the zonation of its algal vegetation. The species *Fucus evanescens*, *Pilayella littoralis*, *Chordaria flagelliformis*, and *Devaleraea ramentacea*, which are confined to the upper and even midsublittoral zone on eastern Greenland, inhabit the eulittoral zone on western Spitsbergen (tidal range of about 1 m). These algae, together with other common species such as *Rhodomela confervoides* (forma *lycopodioides*) and *Palmaria palmata* inhabit the **upper sublittoral zone,** reaching to about 1.5 m deep. *P. palmata* has not been recorded for Scoresby Sound. The **midsublittoral zone,** reaching from 1.5 m to 20 m deep, is dominated by the kelps *Laminaria digitata*, *L. saccharina*, *Alaria esculenta* (forma *grandifolia*), and *L. solidungula*. Partly in the midsublittoral and usually in the **lower sublittoral zone,** one finds the brown algae *Desmarestia aculeata* and *D. viridis* and the red algae *Callophyllis cristata*, *Phycodrys rubens*, *Polysiphonia arctica*, and *Ptilota serrata*. With increasing depth, crustose red algae and the brown crustose alga *Pseudolithoderma* become the dominant species. The lower algal limit is attained at a 55 m depth.

Franz Josef Land. This archipelago, with some of the northernmost islands in the Arctic (80–82°N), was discovered as late as 1873 by an Austrian expedition. A British polar expedition dredging at 25 m deep near Northbrook Island (80°N), the southernmost island of the archipelago, brought *Desmarestia viridis* and young individuals of *Laminaria*, *Polysiphonia arctica*, and *Monostroma* to the surface (Marr 1927). The northern islands of the archipelago are covered by the permanent ice of the Arctic Ocean, and so the algal records near Northbrook Island belong to the few documentations of the northern limit of seaweed vegetation. The same expedition also dredged adjacent to the northeastern coast of Spitsbergen (80.5°N, 30°E), again near the permanent ice. The algae brought up from 25 m deep at this location were *Desmarestia viridis*, *Alaria esculenta*, *Phycodrys rubens*, *Turnerella pennyi* (reported as *Kallymenia rosacea*) and crustose, coralline algae. There are algal records from two other areas situated near the permanent ice, viz. the Jörgen Brönlunds Fjord in northeastern Greenland, and the coast of Brock Island in the Canadian Arctic.

Bear Island. This island lies halfway between Spitsbergen and northern Norway, Kjellman (1903) reported 19 seaweed species from Bear Island, including *Desmarestia aculeata, Laminaria saccharina, L. digitata, Alaria esculenta, Fucus evanescens, Phycodrys rubens,* and *Palmaria palmata.*

3.4.3 Russian Arctic

Literature: Entire area: Kussakin (1977); Ju. E. Petrov (e.g., 1974); Zenkevitch (1963); A. D. Zinova (1953, 1955). **Murmansk:** Zinova (e.g., 1933). **White Sea:** Gobi (1878); A. D. Zinova (1950); E. S. Zinova (e.g., 1934). **Novaya Zemlya:** Kjellman (1877); E. S. Zinova (1929, 1956). **Kara Sea:** E. S. Zinova (1925). **East Siberian Sea, Chukchi Sea:** A. D. Zinova (1957). **Northern Bering Sea:** Kjellman (1889); Vinogradova (1973).

Along the northern coast of Russia, the Arctic region extends from the **Barents Sea (Kolafjord** near Murmansk), passing the coasts of the **White, Kara, Laptev, East Siberian,** and **Chukchi seas** to **Cape Olyutorsky** in the **northern Bering Sea** (Figs. 1.5 and 1.6). Long coastlines of **soft substrata,** especially from the Kara Sea to the eastern Siberian Sea, are not suitable for seaweeds (Fig. 3.14). The large Siberian rivers transport huge quantities of sand and mud to the coast and also turn much of the area adjacent to the estuaries into a **brackish-water** environment. Salinity may decline to 10‰ on

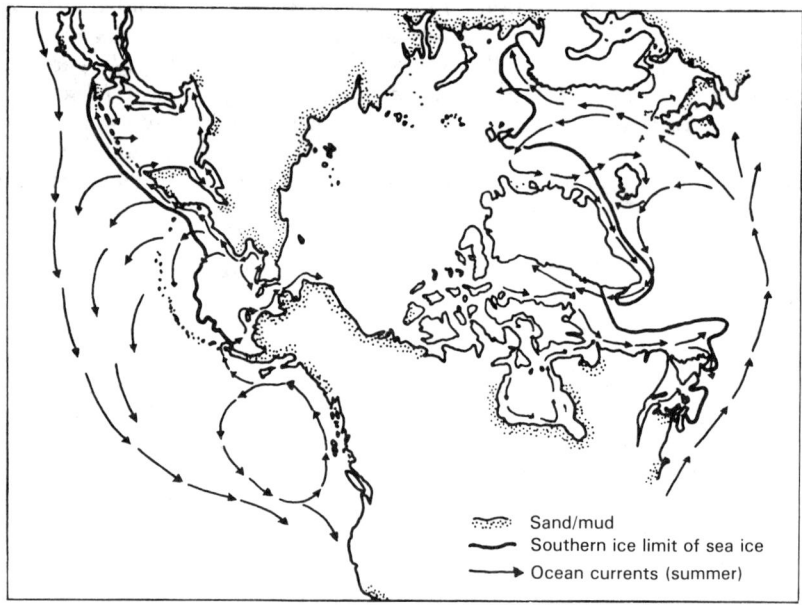

Fig. 3.14 Coastlines with soft substrata, not suitable for seaweeds. (From Widdowson 1971.)

the southern coast of the Laptev Sea due to the freshwater input of the Lena river (Zenkevitch 1963).

Rocky substratum exists for longer coastlines only from the Norwegian–Russian border to the western White Sea, and on the coasts of Novaya Zemlya (E. S. Zinova 1929). Along the Siberian coast isolated hard-substratum locations, where seaweeds can grow, exist at a few rocky promontories and islands (A. D. Zinova 1957) and on the Russian coasts of the northern Bering Sea (Kjellman 1889). For these long coastlines Nordenskjöld, the leader of several Swedish polar expeditions as well as the Vega Expedition along the whole of the north coast of the Eurasian continent, was correct in saying, "There is no rock, not even of the size of a pea."

In the western **Barents Sea** the warming influence of the North Cape Current, an extension of the North Atlantic Current, is still perceptible. However, east of Murmansk, at the Kola Peninsula situated between the Barents and White seas, the pack-ice boundary approaches the coast, on average at longitude 45°E. Eastward, the whole of the northern Russian coast is often blocked for long periods by ice (Fig. 3.1). The maximum seawater temperatures in summer reach 10°C in the western Barents Sea (for example near Murmansk), 6–10°C in the Kara Sea, 2–5°C in the East Siberian Sea, and rise again to 8°C near the southern limit of the Arctic region in the northern Bering Sea. The rise in temperature in the northern Bering Sea is due to a northbound warm current passing the Bering Strait. In winter negative water temperatures are common along the whole coastline from the eastern Barents Sea to the Chukchi Sea.

The coast near **Murmansk,** east of the Norwegian–Russian border, is the transition between the cold temperate North Atlantic region, represented by the algal flora of northern Norway, and the Arctic region with its much poorer seaweed flora. This is clearly shown by the eastern distribution limits of typical cold temperate North Atlantic species (group 3 in Table 2.1), which either do not or hardly enter the Arctic region. *Laminaria hyperborea* (Fig. 2.35), *Fucus vesiculosus* (Fig. 2.23), *Pelvetia canaliculata* (Fig. 2.38), *Chondrus crispus* (Fig. 2.33), and *Delesseria sanguinea* (Fig. 2.42) extend to about the Norwegian–Russian border. *Laminaria digitata* (Fig. 2.20), *Fucus serratus* (Fig. 2.32), and *Ascophyllum nodosum* (Fig. 2.22) reach Novaya Zemlya.

Zonation near Murmansk. In the upper sublittoral zone on the ice-free coast of the island of Kildine, near Murmansk (E. S. Zinova 1933), one finds *Himanthalia elongata, Laminaria saccharina,* and *Alaria esculenta.* The midsublittoral zone, down to 20 or even 40 m deep, is dominated by *Laminaria digitata,* which replaces *L. hyperborea.* This may indicate that the lower limit of *L. digitata* on the cold temperate European coasts is due to competition with *L. hyperborea.* Dominating red algae in the mid- and lower sublittoral zone are *Fimbriofolium dichotomum, Ahnfeltia plicata, Callophyllis cristata, Ptilota serrata, P. plumosa, Phycodrys sinuosa,* and *Odonthalia dentata,* among others.

In winter the coasts of the inner bays of the **White Sea** are extensively covered by ice. The annual range of the seawater temperature may be 20°C. In the western White Sea the soft substratum begins east of Belomorsk, although to the east a few capes and promontories offer rocky substratum to seaweeds. In the algal vegetation of such an isolated rocky location, on the coast of Lietnaia (64°N, 37°E) one finds, at 3.5–7.5 m deep, a dense vegetation of *Laminaria saccharina* and, deeper down in the range of 7.5–15 m, a vegetation dominated by *L. digitata*. At both depth ranges *Alaria esculenta* is present, a further indication that this characteristic species of the upper sublittoral zone in the cold temperate region of the European coasts is excluded from the midsublittoral zone. This is probably due to competition with *L. hyperborea*, as is the case with *L. digitata* (see above). Of the larger red algae, *Ahnfeltia plicata* is particularly common from the upper sublittoral zone to 15 m deep (E. S. Zinova 1933). This species represents a major raw material for the Russian agar industry. The red algae, found on the coast near Murmansk, again are common here, but in addition the Arctic-endemic *Pantoneura baerii* (Fig. 3.11) appears.

East of the White Sea the Arctic-endemic species *Dilsea integra* (Fig. 3.9) appears for the first time on the Kanin Peninsula, and *Laminaria solidungula* (Fig. 3.7) occurs at Novaya Zemlya.

Novaya Zemlya. The two islands of Novaya Zemlya are separated by the narrow Matotschkin Strait, which in some places is as little as 600 m wide. The seawater temperatures in the Matotschkin Strait may rise to 6°C in summer. The western coast of Novaya Zemlya faces the Barents Sea, and the eastern coast faces the colder Kara Sea. In many years the northern and eastern coasts are blocked by ice, so the main exploration of algal vegetation has been on the western coast. Common species in the sublittoral zone are the kelps *Laminaria saccharina, L. digitata, L. solidungula* (north of 73°N), *Alaria esculenta, Saccorhiza dermatodea* (with thallus lengths of up to 4 m); other brown algae *Desmarestia aculeata* and *D. viridis;* and the red algae *Dilsea integra, Palmaria palmata, Devaleraea ramentacea, Phyllophora truncata, Ptilota serrata, Pantoneura baerii, Phycodrys rubens, Polysiphonia arctica, Odonthalia dentata,* and *Rhodomela confervoides* (Kjellman 1877; E. S. Zinova 1929). Secchi depths of up to 45 m in the area of the North Cape Current indicate that the waters of the Barents Sea are relatively clear. Samples of luxurious, crustose coralline algae were dredged at depths of 50–80 m near the Gorbovy Islands (76°N) northwest of Novaya Zemlya. The dominating species were *Lithothamnium glaciale* (Fig. 3.13) and *Phymatolithon*. As elsewhere in the central and northern Arctic region, the vegetation in the eulittoral zone is poorly developed at Novaya Zemlya. Dominating in the upper eulittoral zone are the green algae *Urospora penicilliformis* and *Enteromorpha* species and in the lower eulittoral zone the brown algae *Pilayella littoralis, Fucus distichus,* and *Chordaria flagelliformis*. The 120 seaweed species of Novaya Zemlya and southern Greenland are closely related in type.

Along the coasts of the **Kara Sea** eastward **to the Chukchi Sea** there is solid substratum in isolated locations with, for example, the brown algae *Dictyosiphon foeniculaceus, Desmarestia aculeata, Laminaria saccharina, L. solid-*

ungula, Fucus evanescens, Phyllophora truncata, Devaleraea ramentacea, and *Phycodrys rubens* (A. D. Zinova 1957). These Arctic–Atlantic species continue to dominate in the northern **Bering Sea,** which has also been invaded by a few Pacific representatives. The basic changeover from the Arctic–Atlantic to the Pacific character of the seaweed flora occurs more to the south in the cold temperate regions of the North Pacific, that is, on the coasts of Kamchatka and in the Gulf of Alaska.

3.4.4. North American Arctic

Literature: Northern Bering Sea: Ketchum (1983); Kjellman (1889). **Northern Alaska:** Collins (1927); Dunton et al. (1982); Mohr et al. (1957). **Canadian Arctic:** Breton-Provencher and Cardinal (1978); R. K. S. Lee (1973, 1980). **Labrador:** Ellis and Wilce (1961); Wilce (1959).

In North America the Arctic region covers the coasts of western and northern **Alaska** (northern Bering, Chukchi, and Beaufort seas), the continental coast, and the coasts of about 70 islands of the **Canadian Arctic,** to Ellesmere Island, Baffin Island and **Labrador.** Many North Atlantic seaweeds have recolonized the Arctic region adjacent to the North Atlantic. Could a similar input from North Pacific seaweeds via the Bering Sea have occurred?

Northern Bering Sea. Only in the northern Bering Sea do the fauna and flora have an **Arctic character.** The southern pack-ice limit in winter approximately follows latitude 60°N, and the corresponding line from Cape Olyutorsky to Nunivak Island is regarded as the southern limit of the Arctic region within the Bering Sea (Figs. 1.5–1.7). The seaweed vegetation in the Arctic part of the Bering Sea was investigated by Kjellman (1889) when the Swedish *Vega* Expedition passed through this area. On the **Asiatic side** Kjellman investigated the seaweed vegetation at St. Lawrence and Konyam bays and St. Lawrence Island, situated between Eurasia and Alaska. On the American side he studied the seaweeds of Port Clarence on the coast of **western Alaska.** Kjellman's expedition, continuing along the coast north of Port Clarence and the coast of **northern Alaska** bordering the Beaufort Sea, showed that these areas were inhabited mainly by the predominantly circumpolar seaweed flora from the Arctic region adjacent to the North Atlantic. In detail, the diversity of seaweed species for the northern Bering Sea and the Beaufort Sea indicate the presence of Arctic-endemic species illustrated in Figs. 3.7, 3.9, and 3.10, and Arctic–cold temperate amphioceanic species (group 1 in Table 2.1), with the exception of *Eudesme virescens,* which has been recorded for the North Pacific. Also found on the coasts of the northern Bering Sea were *Phycodrys rubens* and *Odonthalia dentata,* which are listed as Arctic–cold temperate North Atlantic species in Table 2.1 (group 2). In these cases morphologically similar species (vicariant or sister species) exist in the North Pacific. Nevertheless, a few North Pacific seaweed species such as the red alga *Neorhodomela larix* (= *Rhodomela larix*) and the kelp *Alaria crispa,* inhabit the Arctic part of the Bering Sea.

Southern Bering Sea. When the *Vega* left the northern Bering Sea and sailed 1600 km south to Beringa Island (55°N), Kjellman was impressed to find many of the typical seaweed components of the cold temperate **northwestern Pacific region,** for example, the kelps *Cymathere triplicata, Laminaria dentigera, Thalassiophyllum clathrus,* and four species of *Alaria,* endemic to the North Pacific (according to the taxonomic revision of the genus by Widdowson 1971). These species, and many cold temperate animal species of the North Pacific as well (e.g., bryozoans; Powell 1968) have not entered the Arctic region via the Bering Sea even though a northbound current through the Bering Strait would facilitate such migrations (Coachman 1975). This may be identical to the unsolved problem of why many cold temperate European species (group 3, Table 2.1), such as *Laminaria hyperborea,* have not entered the Arctic. It is true that the bulk of the Arctic seaweed flora consists of amphioceanic species (group 1, Table 2.1). Furthermore, numerous Arctic–North Atlantic species (group 2, Table 2.1) are also represented in the North Pacific as somewhat similar, vicariant species. The North Pacific is rich in cold temperate species because it is an "old" ocean reaching far northward. It was isolated from the Arctic Ocean until late in the Tertiary by the Bering land bridge, and only the species adapting to Arctic conditions were exchanged between the North Pacific and the North Atlantic via the Arctic Ocean.

Of the Arctic–cold temperate amphioceanic species, about one-third of those listed in Table 2.1 (group 1) are present at Beringa Island, and several are distributed farther southward along the Asian Pacific coasts, for example, *Desmarestia aculeata* (Fig. 2.7) and *Devaleraea ramentacea* (Fig. 3.10) to Sakhalin and the Kurile Islands; *Dictyosiphon foeniculaceus* (Fig. 2.6), *Alaria esculenta* (Fig. 2.12), and *Fucus evanescens* (Fig. 2.13) to Manchuria (at about Vladivostok) and Hokkaido. The distributions of two species of group 1 extend to Korea and Japan and are also present on the Pacific coast of North America, namely, *Chorda filum* to the Gulf of Alaska (Fig. 2.9) and *Chordaria flagelliformis* to Vancouver Island (Fig. 2.4).

The waters of the northern coast of Alaska are navigable in summer in the western Beaufort Sea because water from the North Pacific passes through the Bering Strait. In contrast, the eastern Beaufort Sea and the ocean waters in the **Canadian Arctic archipelago** is mostly blocked by long-lasting ice, which made the search for the Northwest Passage so difficult. R. K. S. Lee (1973, 1980), incorporating the results of previous investigators, performed many helicopter expeditions combined with diving and dredging investigations, to survey the seaweed flora of the Canadian Arctic archipelago. On the basis of this catalog it becomes evident that the seaweed flora of the Canadian Arctic archipelago (175 species, incorporating many of those listed in Table 2.1 as groups 1 and 2 and illustrated in Figs. 2.4–2.30) again characterizes a cutoff of the circumpolar-distributed seaweed flora of the Arctic region and not a transgression stage between seaweed floras of the North Atlantic and North Pacific. Only three of the 175 species listed in R. K. S. Lee's catalog (1980) are confined in their distribution to the North American Arctic and the North Pacific.

Environmental conditions for seaweeds in the Canadian Arctic. The eastern coasts of Baffin and Ellesmere islands are largely rocky (e.g. the Eureka Sound formation; see Section 2.1), but westward from there the rocky substratum gives way to gravel and mud in the Canadian Arctic archipelago. The local seaweed vegetation becomes more and more isolated the farther west one progresses. However, on soft substrata one may find loose-lying algal populations (global review: Norton and Mathieson 1983), with often dwarf forms of *Fucus evanescens, Desmarestia aculeata, Sphacelaria plumosa, Devaleraea ramentacea, Phyllophora truncata,* and *Chaetomorpha melagonium.* The existence of such loose-lying algae is quite common in the Arctic region and has to do with the low mechanical stress in the deeper sublittoral zone. Along the coasts of the Hudson Strait the tidal range amounts to 3–9 m, but it may decline by 1 m or even to a few centimeters in the central and western parts of the Canadian Arctic archipelago, thereby reducing the mechanical stress on seaweeds due to tidal currents. The water temperatures in August and September may rise to just above 0°C in the eastern area, that is, from Baffin Bay to South Ellesmere Island (Dunbar 1951). In the western area negative water temperatures prevail in summer and, correspondingly, the period of annual ice cover increases to the north and west, until only a few ice-free weeks per year are left for the last seaweeds growing near permanent ice. The brackish water layer, which occurs after the thaw, may for short periods extend down to 10–20 m deep. Since environmental stress for seaweeds increases in a westerly direction, it is understandable that 95% of the algal species of the Canadian Arctic archipelago have been found on the coast of Baffin Bay and the Hudson Strait and only 40% in the western area, on the coasts of the larger islands, such as Banks, Melville, and Victoria islands. A poorly developed eulittoral vegetation exists as an extension of the Labrador algal vegetation in the warmer southeastern area, from the rocky coasts of Baffin Bay to the Hudson Strait.

Seaweeds near the permanent ice. At extreme Arctic locations, such as Brock and Mackenzie King islands (76°N, 113.5°W), the algal vegetation mainly consists of Arctic-endemic species. Lee (1973) investigated the area from a plane and discovered a 4 × 7 km area of partially ice-free water (40% drift ice) in the midst of solid ice extending over hundreds of kilometers. It probably represented a **"polyna,"** an isolated area that becomes ice-free for a few weeks every year. On the gravelly shore of Brock Island *Ulothrix*-like species were found in the sublittoral zone, along with benthic diatoms and other unicellular algae. Dredgings at 7–12 m deep yielded the brown algae *Laminaria solidungula, Desmarestia viridis, Giffordia ovata, Omphalophyllum ulvaceum, Sphacelaria arctica;* the red alga *Turnerella pennyi;* and a filamentous green alga *Spongomorpha* (or *Acrosiphonia*) with its Codiolum phase living endophytically in *Turnerella pennyi.*

The algal flora on the coasts of **Labrador** and Ungava Bay was investigated by Wilce (1959). The Labrador coast is generally rocky, but on the open coast only a reduced eulittoral algal flora exists due to ice abrasion (Fig. 3.6).

Zonation on Labrador. On partially protected shores, species known from Greenland are found in the eulittoral zone. In the mideulittoral zone *Fucus evanescens* dominates. From July to October *Chordaria flagelliformis, Petalonia fascia,* and *Scytosiphon lomentaria* form a dense brown algal belt in the lower eulittoral zone. The two latter species, along with *Ascophyllum nodosum* (Fig. 2.22) and *Fucus vesiculosus* (Fig. 2.23), which are also present on Labrador, do not penetrate far into the Arctic region. A characteristic crustose brown alga, *Ralfsia fungiformis* (Fig. 3.15), is found in tide pools and in the upper sublittoral zone. This alga occurs from northern

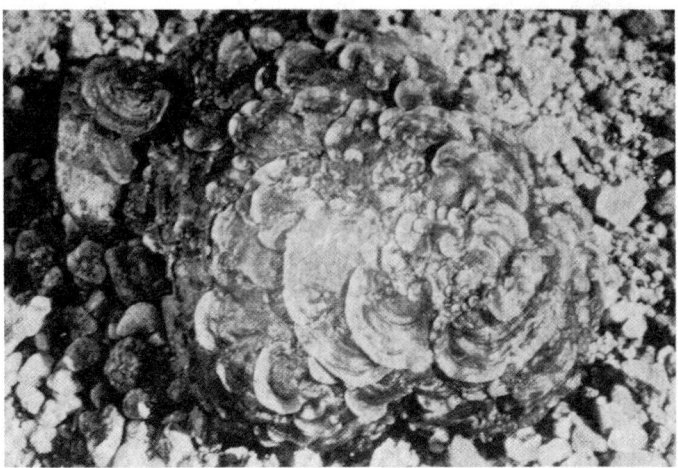

Fig. 3.15 *Ralfsia fungiformis,* a crustose brown alga in a tide pool on the coast of Labrador. (Photograph from Wilce 1959.)

Massachusetts to Hudson Bay and southern Baffin Island, but is also found at Novaya Zemlya and in the northern Bering Sea. In the midsublittoral zone *Laminaria longicruris* may attain a size of 15 m long and 1.3 m wide. Other common species of this zone are *L. digitata* forma *nigripes, L. solidungula* (mostly below a 10–20 m depth), *A. esculenta* forma *grandifolia, Agarum cribrosum, Saccorhiza dermatodea;* the red algae *Phyllophora truncata* forma *interrupta, Kallymenia schmitzii;* and the usual Arctic-endemic and Arctic–cold temperate species (Table 2.1). As observed from a submersible off northern Labrador, the **lower depth limits** are 130 m for crustose coralline algae, 110 m for *Turnerella pennyi,* and 100 m for *Ptilota serrata* (Hooper, personal communication). A dominant red algal, noncoralline crust at depths of 3–46 m in the Canadian Arctic and in Greenland is *Haemascharia* (= *Petrocelis*) *hennedyi* (Wilce and Maggs 1989).

On the southern coast of Labrador the **Strait of Belle Isle** (Fig. 2.77) functions as a biogeographical boundary between the Arctic region and the cold temperate North Atlantic region. Many southern species reach as far as Newfoundland but not Labrador (South 1983b). This applies, for example, to *Chondrus crispus* (Fig. 2.33), *Phyllophora pseudoceranoides* (Fig. 2.34), and *Fucus spiralis* (Fig. 2.31). The Arctic-endemic algae (Figs. 3.7–3.11) reach as far as Newfoundland, where they tend to be shifted into the deeper water of the southbound Labrador Current.

4 Seaweed Vegetation of the Tropical Regions

The tropical regions are bordered to the north and south by the 20°C-winter isotherm (February isotherm in the Northern Hemisphere, August isotherm in the Southern Hemisphere; Figs. 1.5, 1.6). Water temperatures may rise to 30°C in the central areas and to 25°C in the marginal areas. The boundary of the tropical regions coincides with the distribution belt of hermatypic corals (Fig. 4.1). Marine biogeographers (Ekman 1953; Briggs 1974) distinguish **four tropical regions** (Figs. 1.5-1.7, 4.2):

(1) Eastern Atlantic tropical region;
(2) Western Atlantic tropical region;
(3) Indo-West Pacific tropical region;
(4) Eastern Pacific tropical region.

Coral reefs, seagrasses, and mangroves characterize many tropical coasts. The Indo-West Pacific region exhibits maximum diversity of all three groups, with secondary diversity centers in the Caribbean (McCoy and Heck 1976).

4.1 DOMINANT SEAWEEDS AND SEAWEED PALEOBIOGEOGRAPHY

The tropical regions represent the oldest marine habitats. Even in the Pleistocene they did not lose their warm-water character, although their latitudinal extension was constricted. At the maximum of the last glaciation 18,000 years ago, the drop of surface temperature in the Gulf of Mexico, Caribbean Sea, and western equatorial Atlantic was less than 2°C, and approximately 5°C in the eastern equatorial region of the Atlantic (McIntyre et al. 1976). On the other hand, surface seawater isotherms in the tropical regions have probably not surpassed 33°C since the late Precambrian, 700 Ma (Schopf 1980). This is supported by the fact that the upper survival limits of tropical marine algae and animals are uniformly situated at 35–40°C. A rather drastic reduction of tropical surface temperatures occurred around the Eocene/Oligocene boundary at around 40 Ma, when temperatures in tropical waters dropped from an

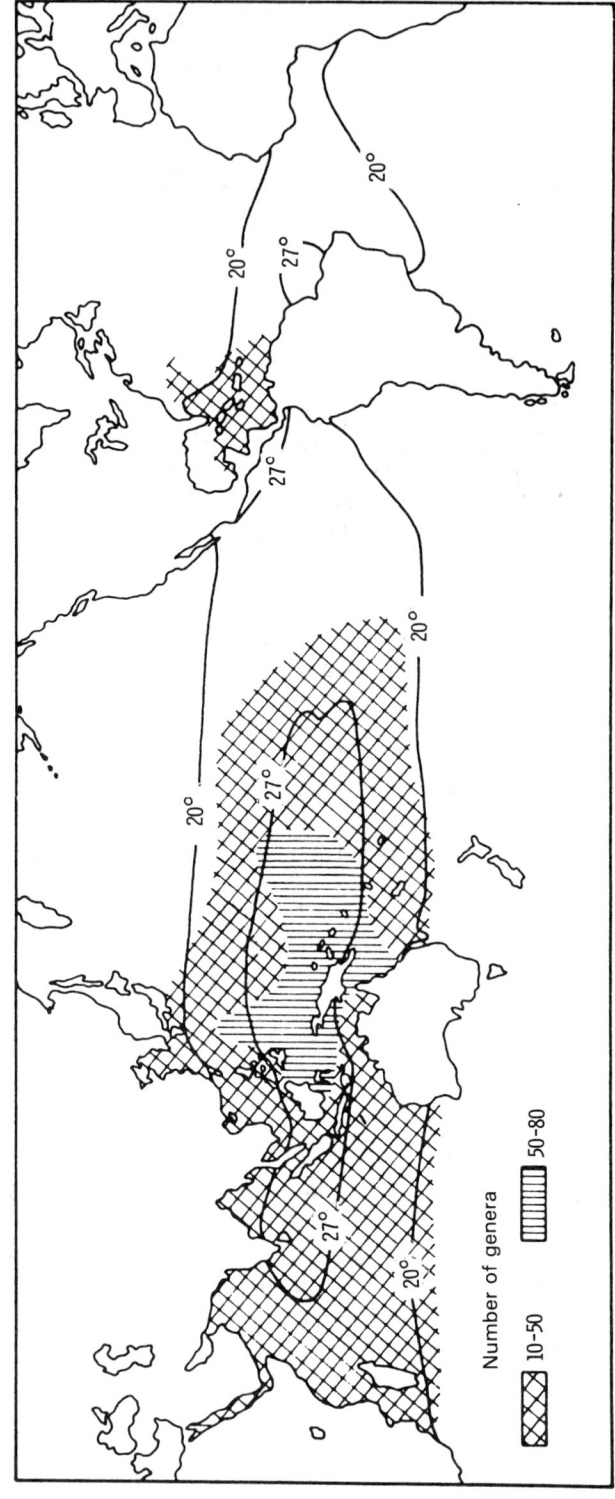

Fig. 4.1 Distribution of coral reefs and abundance of genera of hermatypic corals. Isotherms relate to winter. (From Levinton 1982.)

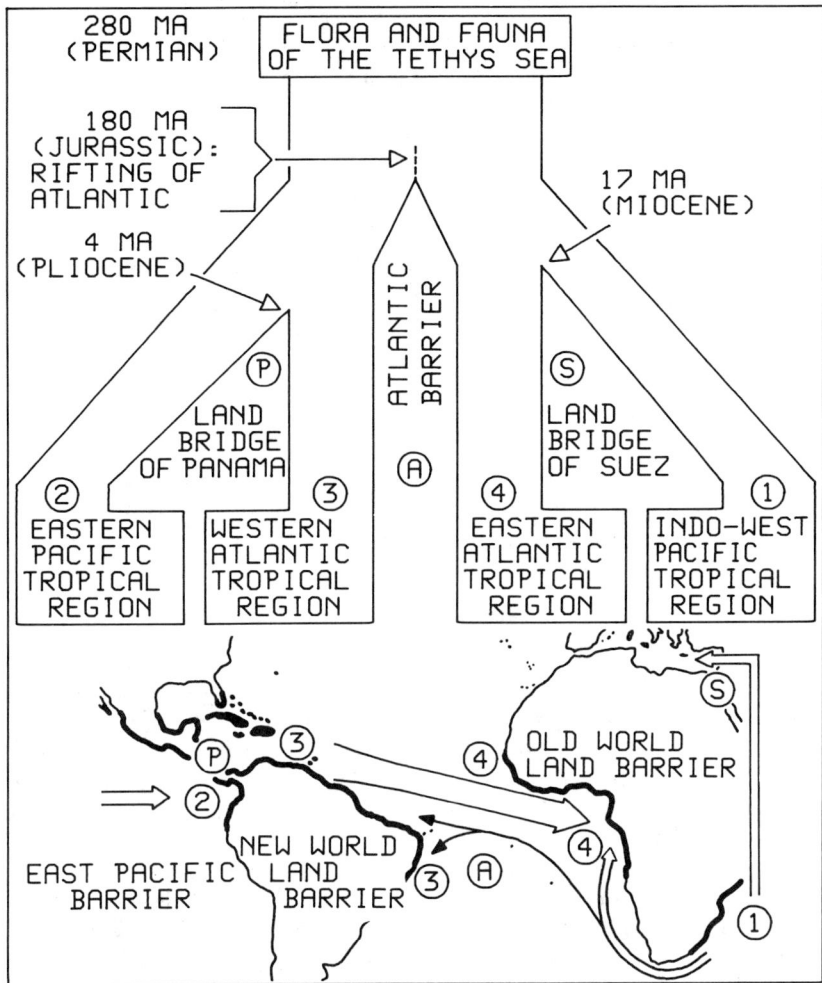

Fig. 4.2 Formation of the recent four tropical regions (thick lines) by splitting up the fauna and flora of the Tethys Sea. Migration routes of tropical species are indicated by arrows, their diameter indicating the relative significance of the routes. (Splitting of Tethys fauna after Menzies et al. 1973; map after Briggs 1974.)

average of 27°C (early Tertiary and today) to about 20°C (Grant-Mackie 1979).

From the cold temperate to the tropical regions the diversity of brown algae is drastically reduced and red and green algae begin to dominate. The **R : P index** (number of red algal species to the number of brown algal species in a given flora), introduced by Feldmann (1937b), rises from 1.1 in the cold temperate regions to a maximum of 4.3 in the tropical regions (Table 4.1).

Table 4.1 Number of seaweed species (N) in different geographical areas. R = red algae, B = brown algae, G = green algae (after Womersley 1981). Southern Africa: N is probably too low, the figure for R:P ratio is probably nearer to 4 than 3.

coast	region	N	R (%)	B (%)	G (%)	R:P	authors
Canadian Arctic	arctic	168	32	38	30	0.9	Lee 1980
Newfoundland	cold temperate	209	38	34	28	1.1	South and Hooper 1980
British Isles	cold temperate	604	48	33	19	1.5	Parke and Dixon 1976
Maryland, Virginia	cold temperate	115	44	23	33	2.0	Ott 1973
California	cold to warm temp.	666	69	20	11	3.4	Abbott and Hollenberg 1976
Southern Australia	cold to warm temp.	1100	73	18	9	4.0	Womersley 1981
Southern Africa	warm temperate	539	59	20	21	3.0	Simons 1976
Western Florida	warm temperate	261	52	16	32	3.2	Dawes 1974
Trop. West Atlantic	trop. to warm temp.	752	56	13	31	4.3	Taylor 1960
Malaysia, Indonesia	tropical	629	63	15	22	4.2	Weber van Bosse 1928

J. Bolton, pers. comm.

Species diversity. The diversity of marine fauna reaches a maximum in tropical seas. It has been estimated that an Indo-Pacific coral reef harbors 3000 animal species from protozoans to fish. The richness in diverse habitats in a huge circumglobal tropical belt, with its ever stable temperatures, is thought to cause this high species diversity (H. B. Moore 1972; Briggs 1974). The acceleration of life processes by higher temperatures may also have speeded up the processes involved in speciation (Vermeij 1978). In contrast to the marine fauna in tropical regions, the seaweed flora does not show the highest species diversity. Coasts that are "rich in algae" and situated in cold and warm temperate regions, or in tropical regions all have about 600–800 species from 200–300 genera of seaweeds (Table 4.1). Largest seaweed species diversity occurs, for example, on the warm temperate coast of southern Australia and in the Mediterranean. The hermatypic corals, whose abundance and diversity covers the hard substrata in the marine tropical regions, leave little settling space for seaweeds. This may primarily explain why algae do not attain great species diversity in the tropics.

Seaweeds with pantropical distribution. The early investigators of tropical seaweed floras were impressed by the fact that many genera and even species occur in the Atlantic and Indo-Pacific, and at both sides of each of the two oceans (pantropical distribution; examples in Fig. 4.3). This wide distribution does not occur for most temperate algae. One may assume, nevertheless, that widespread tropical species differ genetically. Strains of the green alga *Dictyosphaeria cavernosa* from Hawaii and the Caribbean may have been separated for up to 55 Ma, as suggested by single-copy nuclear DNA–DNA hybridization measurements (Olsen et al. 1987).

Pantropical green algae: Caulerpales and Cladophorales. The order **Caulerpales** comprises pantropical, species-rich genera such as *Codium* (about 80 species; Silva

1962), or *Caulerpa* (about 70 species; Calvert et al. 1976), and many species are endemic in a restricted area. This is not the case in the phylogenetically old **Cladophorales** complex (Cladophorales and Siphonocladales) with 15 or so genera. These contain few species per genus or are monotypic (one species per genus), or should be split into several genera, as is the case in *Cladophora* (Olsen-Stojkovich et al. 1986; van den Hoek 1988). Some morphological forms persist through geological time; this the principle of **homeostasis** resulting in "**living fossils.**" Genomic change may proceed independently, the conservative morphology being the combined effect of constant ecological pressure, developmental constraints, and complex genetic regulation (Thorpe 1982; Olsen-Stojkovich et al. 1986). If unexploited niches suddenly become available, there may be rapid morphological change and speciation whereas the short time permits little biochemical speciation (Thorpe 1982).

Pantropical green algal order Dasycladales. Particularly interesting, phytogeographically, is the order **Dasycladales** since it contains genera with **calcified thalli,** such as the pantropically distributed *Acetabularia* and *Neomeris* (Fig. 4.4D; absent in West Africa). The fossilized thalli of these calcified algae may be traced back as genera to the early Tertiary or even to the Mesozoic (Table 4.2). Incidentally, *Acetabularia* became famous after Hämmerling's experiments regarding the role of morphogenetic substances in cap formation (see Bold and Wynne 1985 for details).

Pantropical brown algal orders Dictyotales and Fucales. Among the brown algae, mainly tropical orders are the **Dictyotales** (review in Papenfuss 1977) with the pantropical genera *Dictyota, Dictyopteris,* and *Padina* (27 species); and the **Fucales** (review in Nizamuddin 1970) with the genera *Sargassum* (about 150 species) and *Turbinaria*.

Pantropical red algae. There is a wealth of red algae in the tropics. A few pantropically distributed genera, each containing many species, are cited as examples: *Gelidium, Pterocladia,* and the calcified erect genera *Galaxaura* and *Liagora* (order Nemaliales); *Halymenia* and *Grateloupia* (order Cryptonemiales); the erect Corallines *Jania* and *Amphiroa* (order Corallinales); *Gracilaria* (order Gigartinales); and *Laurencia* with about 80 species (order Ceramiales). Particularly in the warm temperate regions, *Gelidium, Pterocladia,* and *Gracilaria* are important sources of agar (Section 8.6).

At first glance the common occurrence of pantropical seaweed genera was surprising. Today the four marine tropical regions are separated by the land bridges of Central America and the Suez and by two vast oceans, the Atlantic barrier and the eastern Pacific barrier (Fig. 4.2) with few interspersed islands. However, the two land bridges were formed as late as the late Tertiary. Before this the Tethys Sea, now a "lost ocean," girdled the world as an uninterrupted warm-water belt. A westward circumglobal current (Gordon 1974) facilitated widespread migrations of marine tropical species. These migrations were similar to that in the cold temperate species caused later by the Antarctic Circumpolar Current. The separation of the fauna and flora of the Tethys Sea started when the coastlines of the Mediterranean and of the Caribbean separated with the opening of the Atlantic. In the course of the late

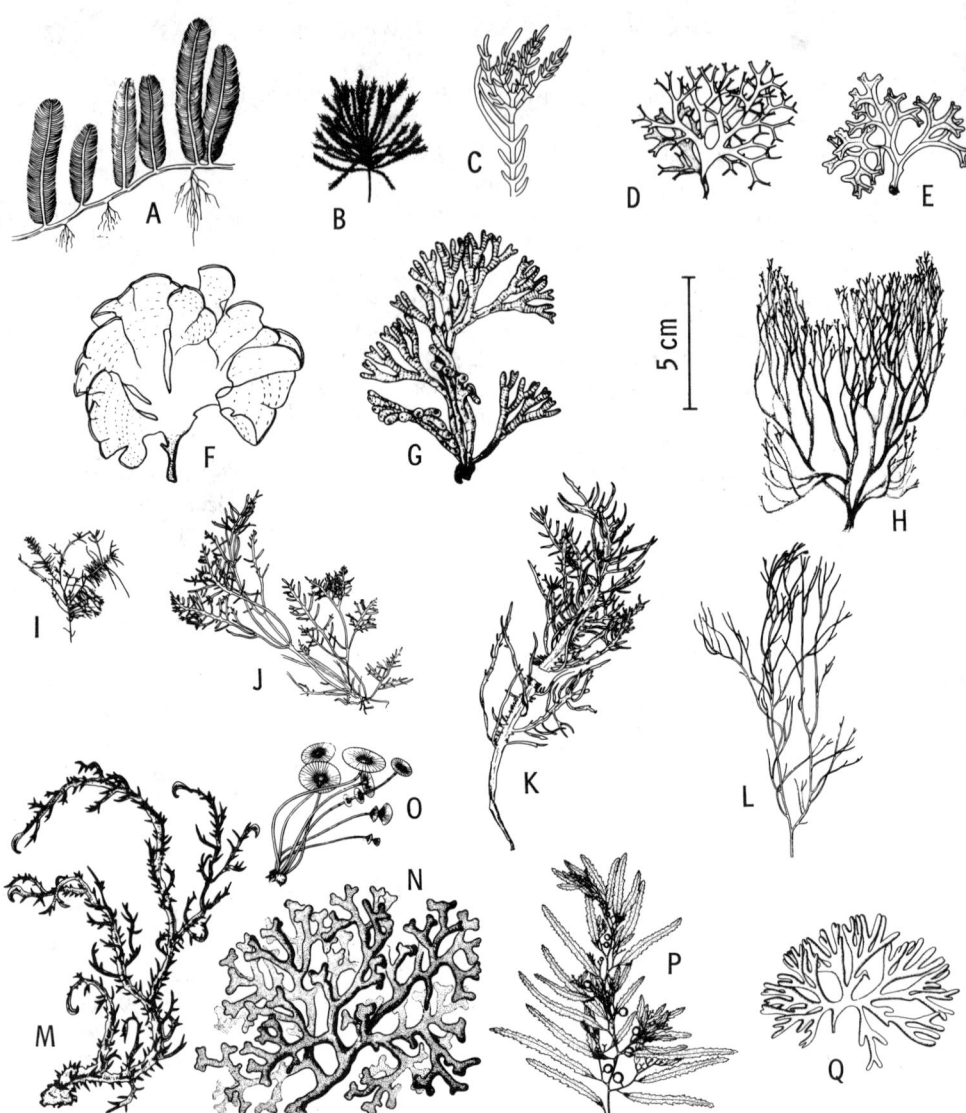

Fig. 4.3 Examples of pantropical seaweeds. EZ = eulittoral zone; SZ = sublittoral zone. **Pantropical species: Green algae: Caulerpales: (A)** *Caulerpa sertularioides* (SZ). **Siphonocladales: (B)** *Siphonocladus tropicus* (upper SZ). **(C)** *Boodlea composita* (spongelike habit, lower EZ). **Brown algae: (D)** *Dictyota pulchella* (= *divaricata,* according to Hörnig and Schnetter 1988; SZ; properly no longer a pantropical species). **(E)** *Dictyopteris delicatula* (SZ). **(F)** *Padina australis* (= *gymnospora;* upper SZ). **Red algae: (G)** *Galaxaura obtusata* (lower EZ to middle SZ; still other pantropical species of *Galaxaura* exist). **(H)** *Liagora farinosa* (calcified, but only at the tips; lower EZ to upper SZ). **(I)** *Gelidium crinale* (lower EZ; also in the Mediterranean). **(J)** *Pterocladia capillacea* (EZ and upper SZ; also in the Mediterranean). **(K)** *Grateloupia filicina* (lower EZ to upper SZ; also in the Mediterranean). **(L)** *Gracilaria verrucosa* (lower EZ to middle SZ). **(M)** *Hypnea musciformis* (lower EZ to upper SZ; also in the Mediterranean; turns green in strong light). **Representa-**

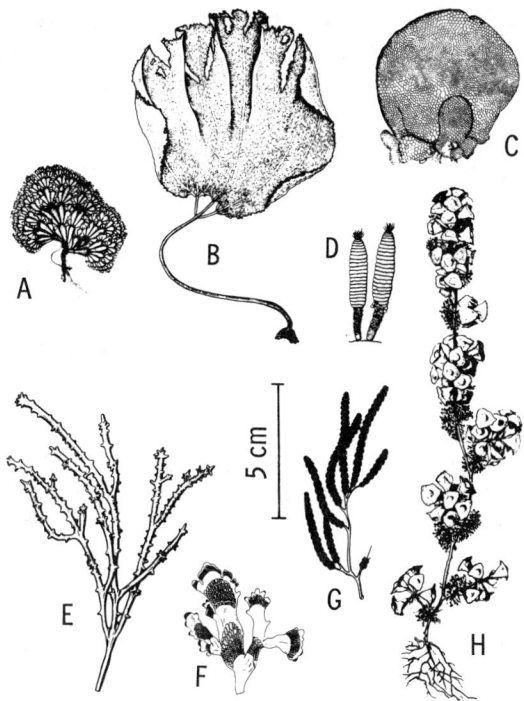

Fig. 4.4 Tropical seaweeds with discontinuous distribution, that is, occurring in the Indo-Pacific and in the western Atlantic, but not at West Africa. EZ = eulittoral zone; SZ = sublittoral zone; IWP = Indo-West Pacific, ATL = Atlantic. **Green algae: Siphonocladales:** (**A**) *Anadyomene stellata* (SZ; found in the Mediterranean; genus also contains IWP species and deep-water species; up to 45 cm in length; *A. menziesii* dredged in the Gulf of Mexico from 200 m deep). (**B**) *Struvea pulcherrima* (middle SZ; genus also contains western ATL and IWP species). (**C**) *Dictyosphaeria cavernosa* (lower EZ and SZ; genus also contains IWP and western ATL species). **Dasycladales:** (**D**) *Neomeris annulata* (calcified; genus contains both IWP and ATL species). **Red algae:** (**E**) *Eucheuma isiforme* (Gigartinales; cartilaginous, SZ; genus contains several western ATL and IWP species). (**F**) *Martensia pavonia* (Ceramiales, Delesseriaceae; SZ; genus contains a few IWP species). (**G**) *Dictyurus occidentalis* (Ceramiales, Dasyaceae; with anastomosing network at the tips, SZ; western ATL species and still another IWP species). **Brown alga:** (**H**) *Turbinaria turbinata* (Fucales, Sargassaceae; lower EZ and typical on coral reefs in upper SZ; this species found in western ATL; several other species of this genus found in IWP). (A from Bold and Wynne 1985; B from Dawes 1981; C–H from Taylor 1960.)

---

tives of pantropical genera: **Green algae:** (**N**) *Codium taylori* (amphi-Atlantic species, upper SZ). (**O**) *Acetabularia crenulata* (western Atlantic species, upper SZ). **Brown alga:** (**P**) *Sargassum filipendula* (amphi-Atlantic species, upper SZ). **Red alga:** (**Q**) *Halymenia agardhii* (amphi-Atlantic species, upper to middle SZ). (C,F,I,Q from Lawson and John 1982; G,M from Dawes 1981; remainder from W. R. Taylor 1960.)

Mesozoic and the Tertiary an Atlantic barrier of steadily increasing width separated the western from the eastern Atlantic tropical biota (regions 3 and 4 in Fig. 4.2).

New World land barrier. The Central American land bridge eliminated the tropical marine connections between the Atlantic and the Pacific. The gradual emergence of the Isthmus of Panama culminated 3–4 million years ago. Evidence for this includes the separate evolution of planktonic Foraminifera on both sides of the land bridge since that time and the exchange of continental mammals between North and South America (Berggren and Hollister 1974, Keigwin 1978; Marshall et al. 1979). South American higher plant taxa that moved to the Central American lowlands after the emergence of the isthmus have hardly differentiated; further northward migration may continue (Gentry 1982).

Effects on the marine fauna and flora: Indo-Pacific–Atlantic distributions. The effects of the formation of the land bridge were considerable. Today the similarity of coastal fauna on both sides at the species level is only 1% for shore fishes, 2% for echinoderms, and 11% for sponges (Briggs 1974). Of the four genera of **seagrasses** found today in the Indo-Pacific and the tropical western Atlantic, only *Halophila decipiens* is represented in both oceans (den Hartog 1970). However, one finds numerous pairs of morphologically similar seagrass species (sister species) (e.g., *Thalassia testudinum* in the western Atlantic and *T. hemprichii* in the Indo-Pacific). In the **brown algae,** the genus *Sargassum* has many species and seems to be in an active phase of speciation. Apart from *S. vulgare* (Fig. 2.64K) no amphioceanic species seem to be left. This also applies to *Padina,* whereas for *Dictyota* several Indo-Pacific–Atlantic species are known. Many *Dictyota* species are endemic to some areas. The **crustose coralline algae** (order Corallinales) may represent a slower-developing group. All tropical genera of the Corallinales show an Indo-Pacific–Atlantic distribution and more than 50% of the species are represented as pairs of sister species or are difficult to separate at all (Adey 1976). One of the classical investigations on tropical marine seaweeds was performed by Börgesen (1913–1920) from the **Danish West-Indies** in the eastern Caribbean. Of the 330 known species 30% occur in the Indo-Pacific, 50% are endemic to the western Atlantic, and the remaining species show an amphi-Atlantic distribution.

Panama Canal. This canal is not at the same level as the sea and, due to its freshwater sections, represents a migration barrier to many marine species (for details see Briggs 1974; Vermeij 1978). One would expect several shallow-water seaweeds from the Caribbean coast (e.g., the red algae *Laurencia papillosa* and *Hypnea musciformis*) to be found on the Pacific coast of Central America, but they are not. Hay and Gaines (1984) have suggested that these algae might survive the transport through the canal on ship hulls but not the stronger herbivore activity on the Pacific coast with its lack of reef-generated refuge. The tidal amplitude on the Pacific coast is about 7 m, making all but the highest intertidal habitats available to herbivorous fishes on a daily basis, whereas algae on the Caribbean reef flats, where the tidal amplitude is only about 0.4 m, are not as available to predation.

Old World and Atlantic as barriers. Numerous seaweed genera and species are found in the tropical Indo-Pacific and the Caribbean. Many of these are absent along the tropical coast of West Africa. This is illustrated by the discontinuous distribution pattern of *Neomeris* (Dasycladales; Fig. 4.5). The surprising situation exists that "many taxa present in the Caribbean are more likely to occur on the east coast of Africa or elsewhere in the Indo-Pacific region than on the much less distant west coast of Africa" (Papenfuss 1972). This applies to seaweeds (examples in Fig. 4.4), seagrasses, and marine animals (Ekman 1953). The present low algal diversity of the tropical West African coast was possibly caused by a considerable loss of species when the tropical region was constricted during the Pleistocene. The richer diversity of the other three tropical regions may be mainly due to a better conservation of the Tethyan stock of genera and species. The **Atlantic barrier** may have been crossed from America to Africa (arrow in Fig. 4.2) by numerous species with a present-day amphi-Atlantic distribution making use of the eastbound Equatorial Countercurrent (Fig. 2.43). The westbound North Equatorial Current was probably not a major migration route, since West Africa "had little to send" (Ekman 1953). As indicated by arrows in Fig. 4.2, a certain number of Indo-Pacific species migrated into the Atlantic via the Cape of Good Hope or into the Mediterranean after the opening of the Suez Canal.

4.2 FOSSIL CALCIFIED ALGAE AND THE EVOLUTION OF ALGAE AND SEAGRASSES

There are few fossil records of uncalcified algae. Only 10% of the recent 8000 species of benthic multicellular algae are calcified (Wray 1977). There are about 100 genera of Recent calcareous algae known, about 90 in the **red algae,**

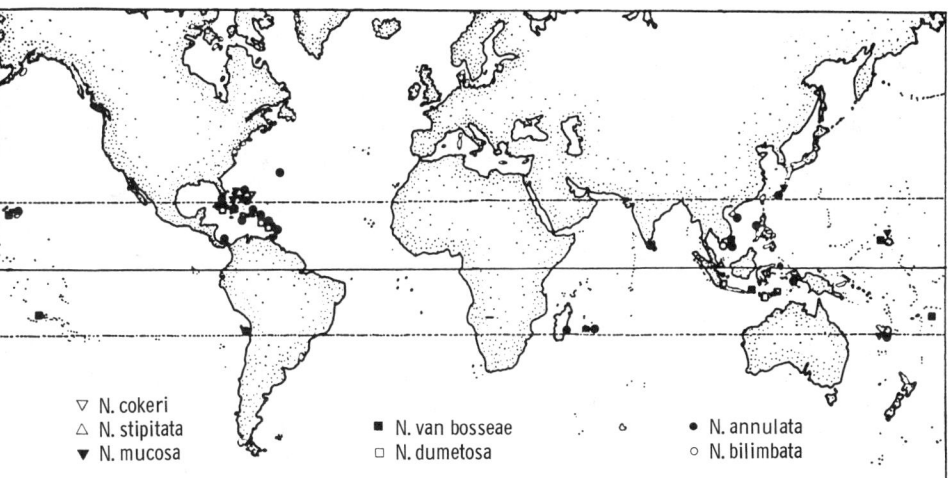

Fig. 4.5 Distribution of the tropical green algal genus *Neomeris* (Dasycladales) as an example of a discontinuous distribution. *Neomeris* does not occur on the eastern coast of Africa. Two species of *Neomeris* (not on the map) have been found at the Cape Verde Islands, *N. annulata* and *N. mucosa* (Prud'homme van Reine 1984). (From Valet 1969.)

10 in the **green algae,** and one in the **brown algae.** There are also some marine **cyanophytes** (Nostocales) capable of calcium deposition (Littler and Littler 1984). **Stromatolites** (biogenic sedimentary structures of a laminated structure) were formed by microbial mats (mainly cyanophytes) trapping sediment and promoting precipitation of carbonate since Precambrium times (see Bold and Wynne 1985).

Recognizing the first appearance of a particular algal group, with no safe fossil records, is more a matter of guesswork. **Red algae** emerged in Precambrian times, 1300–1400 Ma, while fungi and green algae came later but probably long before the brown algae. One assumes that among the **green algae** the unicellular **Chlorococcales** and the filamentous **Ulotrichales** evolved at 900 Ma in the late Precambrian and, for example, the **Cladophorales** and the **Siphonocladales** as late as the Jurassic, at 160 Ma (review chapters on fossil algae in T. N. Taylor 1981; Bold and Wynne 1985; Raven 1986; South and Whittick 1987).

From the analysis of 5S ribosomal sequences of five **brown algae,** Lim et al. (1986) concluded that the diversification between brown algal genera such as *Acinetospora* (Ectocarpales), *Chordaria* (Chordariales), *Eisenia* (Laminariales), and *Sargassum* (Fucales) occurred as late as the Tertiary. On the basis of this analysis the brown algae separated from an ancestor common to diatoms in the Mesozoic, about 200 Ma.

Reconstruction of the history of **calcified algae** is much safer. The processes involved in the deposition of carbonate within algal thalli (reviews: Borowitzka 1982, Johansen 1981, Littler and Littler 1984) have been studied intensively in the green alga *Halimeda.* Carbonate deposition is enhanced in the light due to removal of carbon dioxide and bicarbonate in photosynthesis, an increase in intracellular concentrations of carbonate, and a pH rise. But why do not all algae deposit carbonate? It is assumed that uncalcified algae use organic inhibitors to actively prevent carbonate deposition. Such inhibitors might also secure the formation of uncalcified photosynthetic joints or flex points in otherwise calcified algae.

Calcified green algae: Dasycladales. The calcified representatives of the green algae deposit calcium carbonate outside the cell in the intercellular spaces as the crystal form of aragonite (Borowitzka 1977). Representatives of the order Dasycladales were present in the early Paleozoic (Table 4.2). A large number of fossil, extinct genera of this order followed each other in the course of geological time. Their geological lifespan was in the range of 50–150 million years. For example, the genera *Rhabdoporella* and *Macroporella* existed from the Ordovician to the Silurian and from the late Carboniferous to the Jurassic, respectively (see Table 4.2 for time scale). Empirical data on mean duration of species and genera in brachiopods, bryozoans, bivalves, ammonites, graptolites, and mammals are in the range of 1–25 million years (Schopf 1977). From the fossil record of Foraminifera and mollusks it has been argued that evolution proceeds faster and genera are more short-lived in the tropics than at higher latitudes (Pielou 1979). The Recent genera of the Dasycladales existed only from the late Cretaceous or Tertiary.

Table 4.2 Calcified seaweeds and their geological age (after Wray 1977).

	Paleozoic	Mesozoic	Cenozoic
	PRC CAM ORD SIL DEV CAR PER	TRI JUR CRE	TER QUA PRESENT
million years ago	570 500 440 395 345 280 225	195 135 65	2

Green algae	
Order Dasycladales	
Diverse extinct genera (about 110)	-!---!---!---!---!---!---!---!---!---!---!
Recent: Neomeris, Cymopolia	-!--------!---------
Recent: Acetabularia	---!---!---------
Order Caulerpales (family Codiaceae)	
Diverse extinct genera (about 20)	---!---!---!---!---!---!---!---!---!---!
Recent: Halimeda	-!---!---!---------
Red algae:	
Order Cryptonemiales	
Family Solenoporaceae (crustose thalli)	--!---!---!---!---!---!---!---!---!---!
Family Gymnocodiaceae (erect thalli)	!---!---!---!---!---
Family Peyssonneliaceae (crustose thalli)	-!---!---!---!---!---!---!---------
Recent: Ethalia	--!---!---!---------
Recent: Peyssonnelia	!---!---!---------
Order Corallinales	
Family Corallinaceae, crustose thalli	
Recent: Archaeolithothamnion, Lithothamnion, Lithoporella, Lithophyllum	-!---!---!---!---------
Recent: Mesophyllum, Tenarea, Melobesia	!---!---!---------
Recent: Porolithon, Neogoniolithon	-!---!---------
Family Corallinaceae, erect thalli	
Recent: Amphiroa, Arthrocladia, Jania	--!---!---!---------
Recent: Corallina, Calliarthron	! --!---!---------

Calcified green algae: Caulerpales. In the order Caulerpales (Bryopsidales), also represented by fossilized, extinct genera from the early Paleozoic, one-quarter of the extant taxa have calcareous or only weakly calcified thalli and belong to the six genera *Halimeda, Penicillus, Udotea, Pedobesia, Rhipocephalus,* and *Tydemania.* The strongly calcified genus *Halimeda* belonged to the far-ranging Cretaceous Tethyan flora with fossil remains in limestone facies, for example, of the Mediterranean, Persian Gulf, Mexico, Texas, Thailand (Hillis-Colinvaux 1986a). An extinct, codiacean phylloid alga with calcified blades several centimeters long and 1 mm thick was *Eugonophyllum,* which formed algal meadows in the Carboniferous and Permian (Gray et al. 1981, Toomey and Nitecki 1985).

Calcified extinct red algae: Solenoporaceae and Squamariaceae. In the order **Cryptonemiales,** crustose calcified members have been known since the Cambrian. The family **Solenoporaceae,** with no preserved reproductive structures in the geologically older members, had its maximum development during the Jurassic and became extinct in the Paleocene, possibly because of a herbivore-susceptible anatomy (Steneck 1983). The virtual absence of information on reproduction precludes unequivocal assignment of the Solenoporaceae to the Corallinales (see below), but *incertae sedis* placement within this order was recommended by Woelkerling (1988) to emphasize the possibility that the solenoporoid algae represent the ancestral stock from which taxa of the Corallinales evolved. The first erect, calcified thalli occurred in the Carboniferous (extinct family **Gymnocodiaceae**) (Table 4.2).

Calcified red algae: Squamariaceae. This family, also called **Peyssonneliaceae** and belonging to the order Cryptonemiales (Table 4.2), has calcified crustose species. The Squamariaceae first appeared in the late Carboniferous, and calcium carbonate is deposited in its members within the cell wall as **aragonite**.

Calcified red algae: Corallinales. In the order Cryptonemiales, which comprises numerous other families with noncalcified species, half of the genera and species formerly belonged to the family Corallinaceae (reviews in Adey and MacIntyre 1973; Johansen 1981), which is now regarded as a separate order, **Corallinales** (Silva and Johansen 1986; review in Woelkerling 1988). The Corallinales deposit calcium carbonate as **calcite,** and in addition the calcified cell walls contain magnesium carbonate. The Recent members of the Corallinales date back to the late Mesozoic or early Tertiary and may have descended from the Solenoporaceae, or ancestral corallines in the late Paleozoic (Wray 1977). The ancestral *Archaeolithophyllum* had thin thalli and strongly raised conceptacles. As herbivore groups with increased grazing activity evolved in Mesozoic times, the Corallinaceae formed thick and herbivore-resistant thalli with sunken (and thus protected) conceptacles (Steneck 1983, 1986). Whereas in colder waters the dominating genera are *Lithothamnium, Clathromorphum, Phymatolithon, Mesophyllum,* and *Pseudolithophyllum* (the last two genera centered in the Southern Hemisphere), warm-water reef-building genera include *Neogoniolithon, Porolithon,* and *Lithophyllum.* Most of the crustose species of the Corallinaceae exist as flat growth forms on rocky substratum, thriving under and often requiring intensive herbivory (Steneck 1983). A layer of cells, the epithallus, overlies the region of growth in corallines and protects some crusts from excavating herbivores.

Rhodoliths and coralliths. Unattached "free-living" thalli are formed by the maerl species and by others that grow into nodular structures. These latter are called **rhodoliths** and may live on sand; examples are species of *Neogoniolithon, Lithophyllum, Tenarea,* and *Porolithon* in the tropics and *Lithothamnium glaciale* in the Arctic and the northern part of the cold temperate Northern Atlantic region (Fig. 3.12). Rhodoliths are found in tropical regions down to 200 m deep and may attain a diameter of 30 cm in 500–800 years (Adey and MacIntyre 1973). On shallow reef flats, rhodoliths may remain static for periods up to several months. They continue to grow around their entire perimeter because water flushes through the upper millimeter of the sediment substratum and moves the sand and gravel grains that laterally support the rhodolith. Free-living spheroidal growths of corals called **coralliths** behave similarly (Scoffin et al. 1985).

Calcified thalli in other red algal orders and uncalcified, fossil red algae. Several members of the order **Nemaliales** (e.g.,, *Liagora* and *Galaxaura*) and of the order **Ceramiales** (Crouanieae) also calcify. *Titanophora* is the sole calcifying genus of the **Gigartinales** (Littler and Littler 1984). Fossil **uncalcified** red algal genera from Californian (late Miocene) deposits are *Delesserites, Paleosiphonia,* and *Chondrides* (Parker and Dawson 1965).

Padina, a calcified brown alga. *Padina* provides the only example of calcification in the brown algae; it has a partially calcified thallus in approximately one half of its 20 or so species. Calcium carbonate is deposited as aragonite in extracellular concentric zones, beginning in the intercellular space formed by the infolded apical margin of the thallus (Okazaki et al. 1986).

Fossil brown algae. The fossil record of the uncalcified Phaeophyta is doubtful. Fossil, Paleozoic, and dichotomizing genera (e.g., *Hungerfordia, Protosalvinia,* and *Fucoides*) thought to bear similarity to the brown algal orders Dictyotales or Fucales may well belong to other dichotomizing groups, such as the red algae, liverworts, or invertebrate animals (Clayton 1984). Reliable fossil records of uncalcified brown algae come from Californian (late Miocene) deposits, and are related to the giant kelps and the Cystoseiraceae, (e.g., *Paleohalidrys, Cystoseirites, Cystoseira,* and *Paleocystophora*) (Parker and Dawson 1965). Fossil fucalean algae resembling *Sargassum* species were discovered in Carpathian Tertiary shales and siltstones (Section 4.7). Late glacial marine algae including *Desmarestia aculeata* and a *Laminaria* species were described from the Clyde estuary, Scotland, by Brett and Norton (1969); *Sphacelaria cirrosa*, mentioned in this paper is probably *Tilopteris mertensii* (Prud-'homme van Reine, personal communication).

The enigma Prototaxites. The affinity of the uncalcified, early Devonian genus *Prototaxites* remains enigmatic. The plant was a large, lacunose (full of cavities), pseudoparenchymatous mass (interwoven, contiguous filaments) and a behemoth; some specimens were up to 4 m long and 1 m thick. The internal structure consisted of hyphae with perforate septa and large tubes with no counterpart in the plant kingdom and bearing only superficial resemblance to the sieve elements of the Laminariales (Schmid 1976). Jonker (1979) viewed it as a gigantic red alga with cystocarp-like structures (fossil "genus" *Pachytheca*). Niklas (1976) isolated and identified cutin and suberin derivatives from *Prototaxites* and suggested it was a terrestrial plant or an aquatic plant experiencing periods of desiccation. Schmid (1976) and Schweitzer (1983) pointed to the possibility that *Prototaxites* belongs to a separate, bizarre order or class of plants that grew in sublittoral, marine habitats and represented one of several trials to colonize the land during the Devonian (T. N. Taylor 1981). The attempt was unsuccessful since no land plant is known with a similar internal construction. If *Prototaxites* was an alga, it would have been the only large alga that would have been fossilized in the Paleozoic in great number and quantity, and this during a relative short period of geological time.

Indo-Pacific as a refugium for "living fossils." From the fossil record it can be seen that the green algal order Dasycladales (Figs. 4.4D, 4.6B–D, and 4.7A), with a distribution center in the Indo-Pacific, contains several geologically old genera. Of these genera few species remain today, which may be taken as a further indication of great geological age. In contrast, *Acetabularia* (Fig. 4.3O), with its many recent species and also belonging to the order Dasycladales, appeared relatively late in geological time (Table 4.2). The more primitive of the 70 species of the pantropically distributed genus *Caulerpa* (order Caulerpales) live in the Indo-Pacific (Calvert et al. 1976), where most of the *Caulerpa* species are found, contrasting with the 17 species found in the tropical western Atlantic (W. R. Taylor 1960). Svedelius (1924; biography

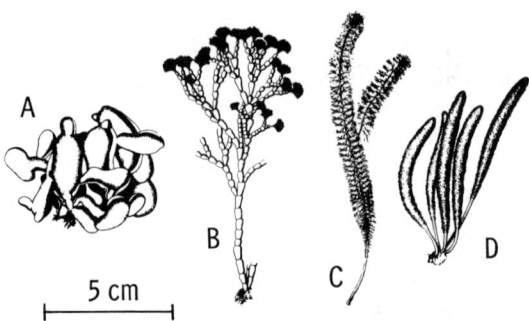

Fig. 4.6 Tropical green algae with discontinuous distribution (i.e., occurring only in the western Atlantic and partially in the Mediterranean). EZ = eulittoral zone; SZ = sublittoral zone. **Green algae: Siphonocladales: (A)** *Valonia macrophysa* (lower EZ and upper SZ; in the Mediterranean; genus also contains Indo-Pacific–Atlantic and Indo-Pacific species). **Dasycladales: (B)** *Cymopolia barbata* (calcified, upper SZ; genus found only in western Atlantic). **(C)** *Batophora oerstedi* (upper SZ). **(D)** *Dasycladus vermicularis* (SZ; only one species, also in the Mediterranean). (Habit drawings from W. R. Taylor 1960.)

in Papenfuss 1961), a pioneer of algal research in the tropics and a student of Kjellman in Uppsala, used these facts to put forward the view that the Indo-Pacific conserved many geologically old seaweeds. He assumed that the "old genera" of the Tethys Sea spread from Indo-Pacific centers to the younger Atlantic. Another interpretation may be that the central area of the Atlantic has existed since the early Cretaceous (Fig. 2.2A–D) and was possibly inhabited by the present old genera due to the Tethyan connection. These old genera may have become extinct in the Atlantic because of its open connection to the Arctic Ocean and consequently the more drastic decline of seawater temperatures during the Pleistocene. This possibility has been discussed in detail by authors like Ekman (1953) and McCoy and Heck (1976). According to this

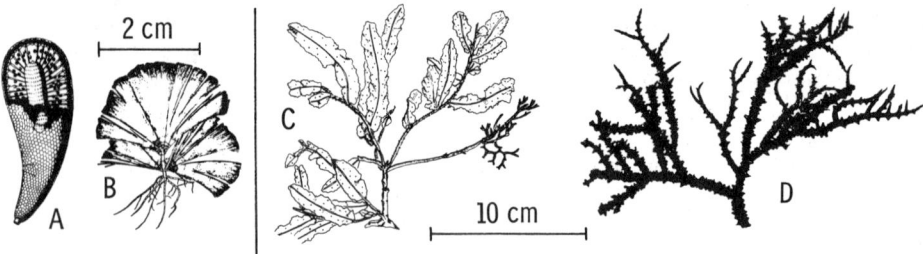

Fig. 4.7 Tropical seaweeds with discontinuous distribution (i.e., occurring only in the Indo-Pacific). **Green algae: (A)** *Bornetella oligospora* (Dasycladales, calcified). **(B)** *Tydemania expeditionis* (Caulerpales, deep-water alga). **Brown alga: (C)** *Sargassum hawaiiensis* (deep-water alga, down to 200 m deep). **Red alga: (D)** *Eucheuma denticulatum* (A from Fritsch 1959–1961; B from Gilmartin 1966; C from De Wreede and Jones 1973; D from Weber van Bosse 1913–1928.)

view, one need not regard the Indo-Pacific as a center of origin whence such "old genera" spread but rather where they were preserved until today.

Floral history of the seagrasses. Of the 12 existing genera of seagrasses, seven occur in the tropical Indo-Pacific (Fig. 4.8D–J). Of the four seagrass genera found in the tropical western Atlantic, only *Halodule* is found on the tropical coast of West Africa. The fossil evidence indicates that angiosperms evolved from gymnosperms in the Jurassic, with seagrasses first appearing in the Cretaceous. The majority of the relatively old group occur in the Indo-Pacific (review in Larkum et al. 1989). The Recent tropical genus *Cymodocea* can be traced back to the Eocene, 55 Ma. *Zostera,* a cold-water genus with fossil ancestors in Japan, probably originated in the western Pacific, dispersed to North America, and, after inundation of the Bering land bridge in the late Tertiary, to the North Atlantic (McRoy 1968). The main adaptions of the seagrasses are to salinity, underwater pollination, and flower formation (Sculthorpe 1967). Vegetative propagation is important in the majority of species (Tomlinson 1974). One genus with only one species represented in the tropics, *Thalassodendron,* produces seeds that germinate on the plant (a process called vivipary, seen in mangrove plants, and *Amphibolis* spp.) before becoming detached and sinking to the bottom.

4.3 EASTERN ATLANTIC TROPICAL REGION (TROPICAL WEST AFRICA)

Literature: Entire area: Lawson (1966, 1978); Lawson and John (1982); Schmidt (1957). **Senegal to Liberia:** Aleem (1978a); De May et al. (1977); Sourie (1954). **Gulf of Guinea:** John et al. (e.g., 1977); Lawson (1956). **Gabon to Angola:** John and Lawson (1974); Lawson et al. (1975). **Ascension:** Price and John (1980). **St. Helena:** Colman (1946).

Of the four tropical marine regions, the eastern Atlantic tropical region, situated on the West African coast between Cape Verde (Senegal) and Mossâmedes (Angola), exhibits the smallest latitudinal extension (Figs. 1.5 and 1.6). This region becomes constricted by two cool currents, one from the north (the Canaries Current) and the other from the south (the Benguela Current) (Fig. 2.43). The Canaries and Benguela currents continue westward as the North and South Equatorial currents, respectively. Both cross the Atlantic to reach South America. The warm-water character of the tropical West African coast in the lee of the two currents is caused by local warming and by an inflow of warm water transported to this region by the eastbound Equatorial Countercurrent, which, as the Guinea Current, reaches the Gulf of Guinea. The eastern Atlantic tropical region has been further subdivided into a tropical African portion in the north and a tropical transitional African portion from Gabon to Angola in the south (Lawson 1978, 1988, John 1986).

Temperate seaweed species that crossed the equator. The relatively small extension of warm water along the tropical West African coast extends only down to about 50 m

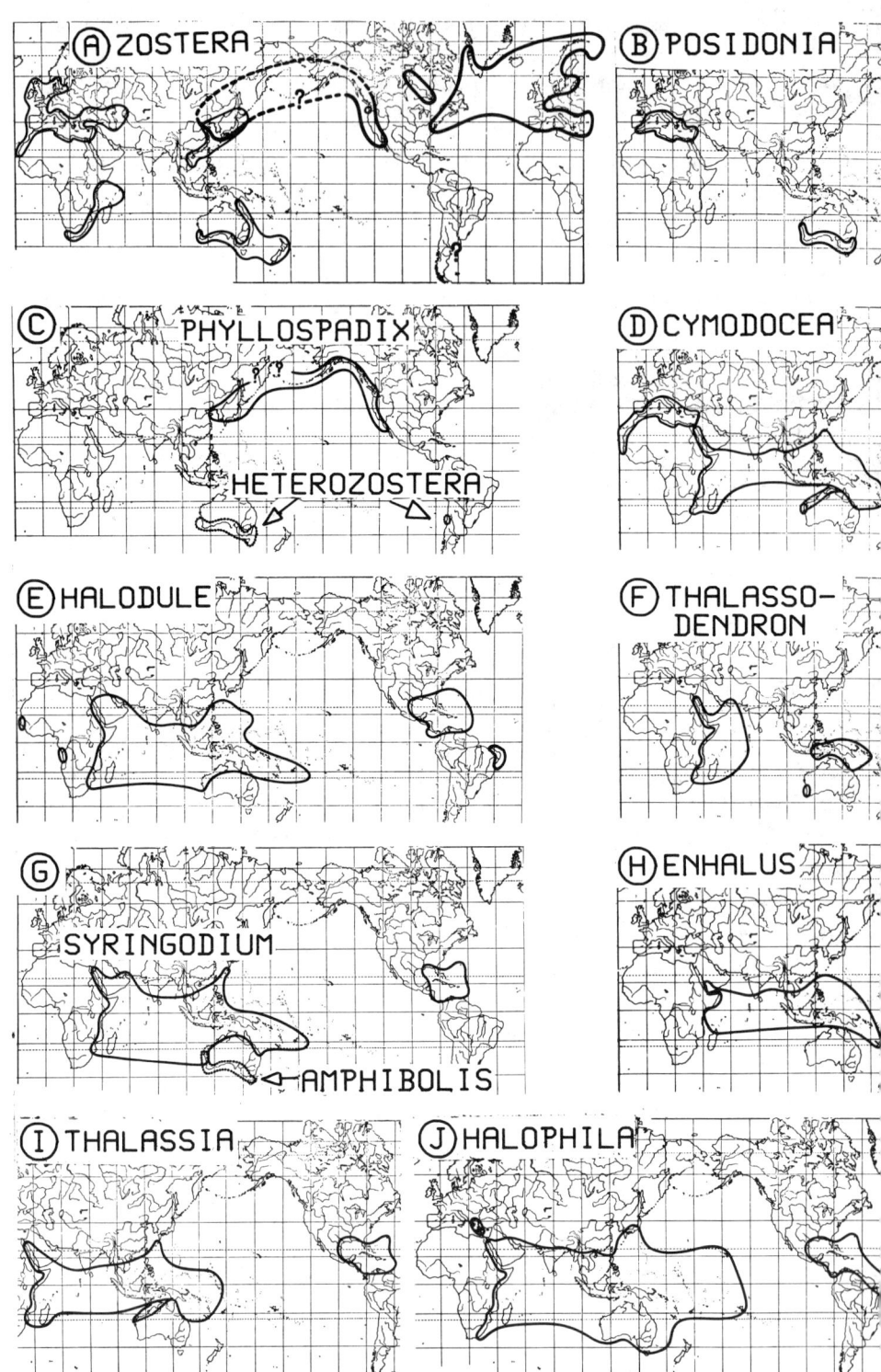

deep. This constricted warm-water region probably facilitated the crossing of the equator by certain temperate species on the African rather than the American side of the Atlantic (van den Hoek 1982b). From microfossil evidence of bottom sediments at the height of the last glaciation (18,000 years ago), the water temperatures on the tropical West African coast were 2–8°C lower than today (Frakes 1979; Sarnthein et al. 1982). The southern limit of *Laminaria ochroleuca* is now the coast of Morocco at 30 m deep, and an ancestor of it may have crossed the equator in one of the cold periods of the Pleistocene or Miocene, by spreading southward in deep water along the West African coast to South Africa, where its morphology changed to that of *L. pallida*.

The tropical West African coast has one-third the number of seaweed species of the tropical Atlantic coast of America (Lawson and John 1982). Of the approximately 300 seaweed species of tropical West Africa, two-thirds are amphi-Atlantic and one-half occur in the Indo-Pacific. Many of the tropical West African seaweed species extend far into the adjacent warm temperate regions, giving a certain eurythermal character to these species and pointing to the nonoptimal tropical character of the region. Tropical species are impeded along the West African coast by upwellings that may reduce the surface-water temperature to as low as 19°C (Ekman 1953; John et al. 1977).

The proportion of seaweeds endemic in this tropical region is 7%, several times lower than that for algae in the Caribbean. An example of an endemic West African alga is the netlike red alga *Dictyurus fenestratus*, from an 8–14 m depth (Fig. 4.9). Endemism is also low for numerous animal groups (Briggs 1974). As mentioned previously, the genus *Halodule* is the only representative of the tropical seagrasses.

Hermatypic corals are scarcely represented, forming no continuous coral reefs (John 1986). There is fossil evidence that coral reefs existed along the West African coast and in the Mediterranean, and became extinct in the course of the Pliocene (last epoch of the Tertiary). In general, one attributes the current depauperate character of the eastern Atlantic tropical region to the relatively high losses of tropical species when temperatures declined at the end of the Tertiary and Pleistocene. The seabed, largely covered by sand, gravel, or mud, and the poor underwater light along the West African coast together make an environment here that is not ideal for seaweeds (John 1986). Rocky substrata are scattered, and the coastal water is often turbid due to large sediment transport by the big rivers during the summer rainy season.

Fig. 4.8 Distribution areas of the 12 recent seagrass genera. **(A–C)** Genera in temperate regions. **(D–J)** Genera in tropical–warm temperate regions. **(A–G)** Family Potamogetonaceae. **(H–J)** Family Hydrocharitaceae. (After den Hartog 1970. Changes: *Cymodocea* and *Thalassia* according to H. Kirkman, personal communication; *Halodule* and *Halophila* according to Oliveira Filho et al. 1983.)

Long coastlines of tropical West Africa are represented by steep, sandy beaches, where green algae such as *Caulerpa* and less frequently the seagrass *Halodule wrightii* determine the floral community. The distribution of another seagrass species, *Cymodocea nodosa*, only extends southward from the adjacent warm temperate Mediterranean–Atlantic region to Senegal. In the lagoons and extensive estuaries (e.g., of Niger), one finds the **mangroves** *Rhizophora racemosa*, *Avicennia germinans*, and *Laguncularia racemosa* (John 1986). *Avicennia germinans* extends southward to about 150 km south of Cape Blanc in Mauritania (Prud'homme van Reine, personal observation).

Zonation on the Gold Coast. The infrequent rocky areas may exhibit a zonation, as illustrated for the Gold Coast in Fig. 4.9. The surface-water temperature ranges from 25–29°C but may decline occasionally during July to September to 19°C due to changes in the direction of coastal currents and subsequent upwelling.

Because of strong insolation, the **supralittoral** zone harbors few marine organisms, a common effect in the tropical regions. In the supralittoral zone of the Gold Coast, cyanophytes and snails such as *Littorina punctata* are found. Marine lichens have not been reported (Lawson 1966).

For the organisms in the **eulittoral zone** the relatively small tidal range of 0.6–1.8 m at spring tides is not as important as the strong surge. On the coast of Senegal the barnacle *Chthamalus stellatus* dominates at wave-exposed locations in the **upper eulittoral zone.** This species is well known from the warm temperate Mediterranean–Atlantic region and further south is replaced by *C. dentatus*. A belt of the crustose brown algae *Basispora africana* and *Ralfsia africana* typically follows the barnacles deeper down (Lawson and John 1982). Other members of the upper eulittoral zone flora, although not abundant, are the green algae *Enteromorpha*, *Ulva fasciata*, and *Chaetomorpha antennina;* the red algae *Bangia fuscopurpurea* and *Porphyra;* and brown algae such as *Ectocarpus breviarticulatus* and *Bachelotia antillarum*, which is morphologically similar to *Pilayella*.

In the **lower eulittoral zone,** continually wetted by breaking waves at low water, one finds a dense vegetation of crustose coralline algae and small erect red algae (e.g., *Laurencia* spp., *Gelidium* spp. and *Centroceras clavulatum*). Several of these species are known from the warm temperate Mediterranean–Atlantic region. Sessile animals associated with the pinkish-red layer of the crustose coralline algae are the mussel *Perna perna* (= *Mytilus perna*), the limpet *Patella safiana*, the carnivorous snail *Thais nodosa*, and, in stronger wave action, the large barnacle *Balanus tintinnabulum* (John 1986). Along protected locations species of *Bryopsis*, *Padina*, *Colpomenia*, *Hypnea*, and *Gracilaria* occur as well as erect-growing representatives of the Corallinaceae, namely *Jania*, *Corallina*, and *Amphiroa*.

In the **upper sublittoral zone** *Sargassum vulgare* and *Dictyopteris delicatula* (Fig. 4.9) are typical components of rocky habitats along the whole tropical West African coast, where there is moderate wave action (John 1986). The southern extent of the range of the warm temperate genus *Cystoseira* occurs at Senegal.

The **sublittoral zone** harbors about 100 seaweed species, which may grow, for example, on submarine banks consisting of solidified coastal sediment (John et al. 1977, John 1986). Most of these species, several with pantropical distribution such as *Dictyota dichotoma* or *Jania rubens*, range in size from a few millimeters to 10 cm. The largest seaweeds here are the brown alga *Sargassum filipendula* (up to 1 m long), the red alga *Halymenia actinophysa* (up to 30 cm long), and the endemic *Dictyurus*

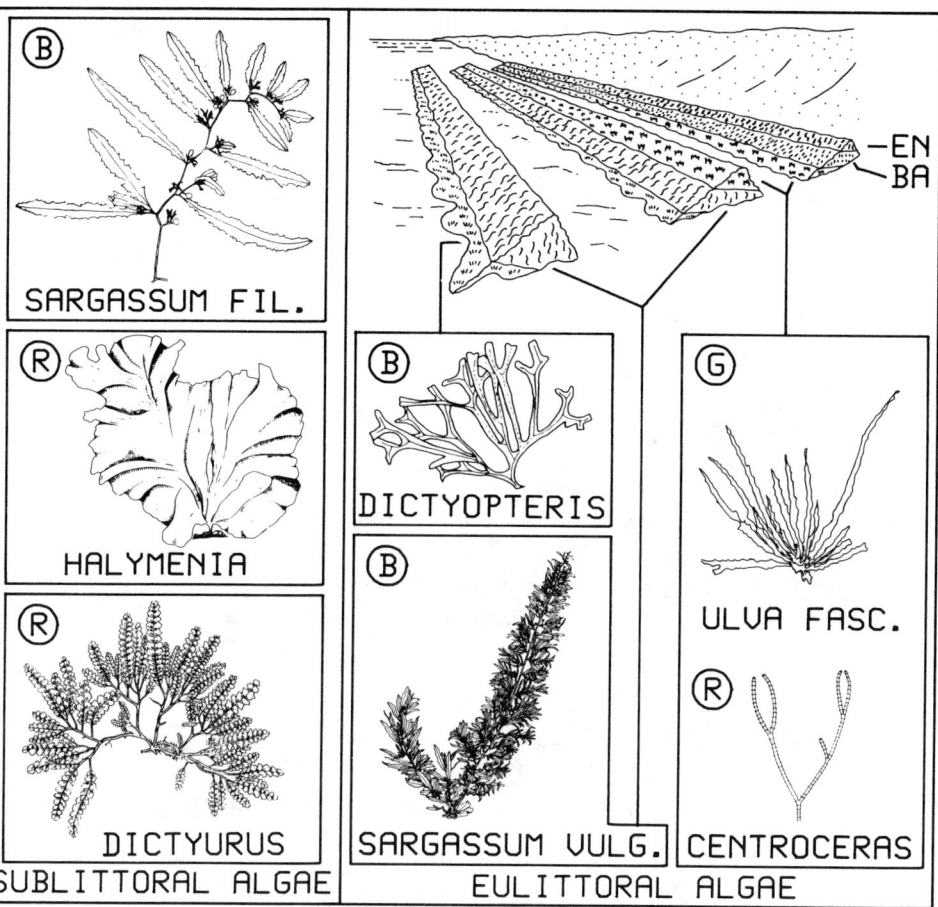

Fig. 4.9 Seaweed zonation on rocky cliffs off the sandy Gold Coast (Gulf of Guinea, southern coast of Ghana). G = green alga; B = brown alga; R = red alga; EN = *Enteromorpha* spp. BA = *Bachelotia antillarum* (morphologically similar to the filamentous brown alga *Pilayella*). **Details on species names and dimensions: eulittoral zone:** *Dictyopteris delicatula* (× 0.4); *Sargassum vulgare* (× 0.2); *Ulva fasciata* (× 0.1); *Centroceras clavulatum* (× 2.5; Ceramiaceae). **Midsublittoral zone:** the following more conspicuous species occur: *Sargassum filipendula* (× 0.2; amphi-Atlantic); *Halymenia actinophysa* (× 0.2; Cryptonemiales); *Dictyurus fenestratus* (× 0.2; Ceramiales, Dasyaceae; with a network of anastomosing branches; eastern Atlantic–endemic, tropical and warm temperate). All cited species, excluding *S. filipendula* and *D. fenestratus*, are pantropically distributed. (Zonation scheme from Lawson 1956; habit drawings from Lawson and John 1982.)

fenestratus (up to 15 cm long; Fig. 4.9). Typical **grazers** are the nocturnal sea urchin *Diadema antillarum* and fish such as *Acanthurus monroviae* or *Abudefduf saxatilis*. The territorial and aggressive damselfish *Microspathodon frontatus* frightens off other fish and allows the development of an **algal farm** in its own territory (John 1986). Since the water is very turbid during the rainy season, the sublittoral algae have to obtain their main annual photosynthetic gain in the period of clearer water from November to March (Lawson and John 1982). Sandy seabeds in many places along the West African coast are often strewn with **rhodoliths** formed by calcareous red algae. Such cobble areas have been misleadingly called "coral banks" (John 1986).

Ascension and St. Helena. These islands are a long way from the continents and are among the most isolated islands in tropical regions. **St. Helena** originated by volcanic action in mid-Tertiary. The shore-fish fauna exhibit an endemism of 22%, and one-third of the remaining shore-fish species occur on both sides of the Atlantic (Briggs 1974). **Ascension,** situated about 2000 km from West Africa and South America, is younger than St. Helena, lying only 145 km away from the Mid-Atlantic Ridge, the actively spreading center of the Atlantic bottom. Ascension probably arose some 1.5 Ma, and correspondingly the endemism of its shore-fish fauna stands at only 4%. The coast consists of hard substratum of volcanic origin thus precluding the existence of seagrasses. However, coral reefs are also absent, although the seawater temperatures are in the range 22–26°C and would be optimal for hermatypic corals. Probably enough geological time has not occurred to allow for establishing coral reefs. Another possible reason for their absence may be the extremely heavy surf (Price and John 1980). The tidal range is 0.9 m. Small tidal ranges are characteristic of all islands situated a long way from continents. The seaweed flora is dominated by species that are widely distributed, such as *Sargassum vulgare, Ulva* species and *Gelidium pusillum.*

4.4 TIDAL MANGROVE FORESTS AND THEIR SEAWEEDS

Literature: V. J. Chapman 1977; Gessner 1955–1959; King 1981; Por and Dor 1984; Tomlinson 1986.

In calm water along the tropical coasts and in brackish estuaries one finds mangrove vegetation, "trees that grow in the sea." The main components are trees up to 30 m high and bushes to 2 m in height, some of which become almost completely submerged at high tide.

The species composition of mangroves (about 90 species worldwide belonging to 20 angiosperm families) is distinguished by the **Eastern Mangrove Region** (East Africa to Polynesia) and the **Western Mangrove Region** (tropical Atlantic and Pacific American coasts), the latter being poorer in species number. Both types of mangrove share a few genera (*Rhizophora, Avicennia,* and *Xylocarpus*) but no mutual species. At both sides of the land bridges of Central America and the Suez many pairs of sister species of mangroves have evolved, a further indication of the relatively late formation of these land bridges. The northernmost mangrove vegetation is found in the Red Sea (southeastern end of the Sinai Peninsula), in the Persian (Arabian) Gulf, in Mauritania, and at Bermuda. The southernmost mangrove vegetation is represented by one species (*Avicennia marina*) in Westernport Bay, Victoria, Australia.

Mangrove plants show special adaptations to life in brackish or saline water (e.g., filter effects in the roots to avoid excessive uptake of salt, and salt removal by accumulation of salt in salt glands in leaves) and in tidal, muddy habitats. Extensive prop roots anchor the plants, and pneumatophores facilitate gas exchange by growing upright into the air from the buried roots. Many mangrove genera have floatable seeds and fruits, and in others (e.g., *Rhizophora*) germination of the seed occurs while still on the parent plant (vivipary) so that the pointed germling penetrates into the soft bottom when detached from the parent plant.

On and among the tangled prop roots, which function as sediment traps, are the common red algae *Bostrychia, Catenella,* and *Caloglossa,* whereas the muddy bottom is colonized mainly by green algae such as *Caulerpa* or *Halimeda* (reviews in Post 1963; van den Hoek et al. 1972; King 1981). The hanging roots of the red mangrove, *Rhizophora mangle,* are colonized by fleshy algae (e.g., *Acanthophora spicifera* and *Spyridia filamentosa*), which escape from herbivory, particularly by sea urchins, by living on the hanging roots. The dominant algae of the mud-embedded root habitat (e.g., *Halimeda opuntia*) show reduced edibility as a consequence of heavy calcification and potential herbivore-deterring secondary metabolites (P. R. Taylor et al. 1986).

4.5 CORAL REEFS AND THE ROLE OF ALGAE

Literature: Adey (1976); Adey and Goertemiller (1987); Adey and McIntyre (1973); Adey and Vassar (1975); Bakus (1969); D. J. Barnes (1983); Colin (1978); Dahl (1973b); E. A. Drew (1986); Goff (1983a); Goreau and Goreau (1973); Hay et al. (1988a,b); Hatcher and Larkum (1983); Hillis-Colinvaux (1980; 1986a,b); O. A. Jones and Endean (1973–1976); Kühlmann (1984); Law and Lewis (1983); J. B. Lewis (1981); Littler and Littler (1984, 1988); Longhurst and Pauly (1987); Raven (1986); B. R. Rosen (1982, 1984, 1988); Schuhmacher and Zibrowius (1985); Steneck (1982, 1983, 1986); Stoddart (1969); Toomey (1981); van der Land (1989); Wanders (1976–1977); Wells (1957); Wray (1971, 1977).

Coral reefs occur in the latitudinal range of 30°N to 30°S, are common in three of the four tropical regions, and absent only on the tropical coast of West Africa (Fig. 4.1). The **reef-building organisms** are primarily represented by hermatypic (reef-building) corals, accompanied by crustose coralline algae and erect calcified green algae such as *Halimeda*.

Hermatypic corals. The hermatypic corals (Fig. 4.10) belong to the order Madreporaria (Scleractinia) and comprise **500–700 Recent species** in the Indo-Pacific, and 60–80 Recent species in the Atlantic. The Atlantic and Indo-Pacific have almost completely different coral faunas with only about 7% overlap at generic level (Rosen 1984). About **5000 extinct species** are known. The hermatypic corals thrive at water temperatures higher than 20–22°C, at an optimum of 23–29°C. The food of hermatypic corals consists mainly of zooplankton, but organic material derived from endosymbiotic

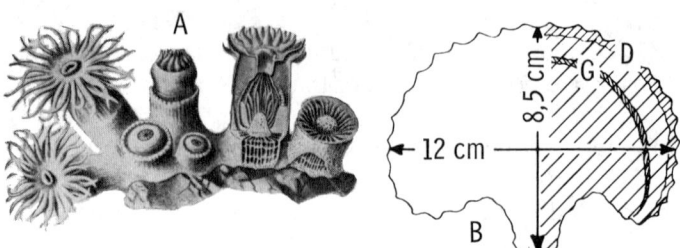

Fig. 4.10 Hermatypic corals and their endosymbiontic algae. (**A**) Coral *Astroides calycularis*. The live tissue of the polyps covers the calcified skeleton. (**B**) Cross section of the coral *Favia* with endosymbiontic algae. D = brown layer of dinoflagellates; G = green layer of the siphonous, chalk-boring green alga *Ostreobium*. (A from Kaestner 1984; B from Shibata and Haxo 1969.)

dinoflagellates (zooxanthellae) probably represents an additional food source. Most hermatypic corals are **zooxanthellate,** but there are also **azooxanthellate** reef-building corals and nonreef-building corals with zooxanthellae (Schuhmacher and Zibrowius 1985).

The coral reef consists of the calcified skeletons of reef builders forming a conglomerate, which may grow vertically 5–10 mm per year and over millions of years may reach a vertical extension of 1400 m. At this depth the Eocene basaltic basement was reached in a drill core on Enewetak Atoll (Marshall Islands). Many coral reefs are hundreds of meters thick. Darwin explained this by assuming that the deep coral reefs were formed on a continuously submerging substrate, while the reef builders added material to keep the reef's upper limit just below the water surface. The theory of plate tectonics supports Darwin's idea (Rosen 1982), since the seafloor subsides as it spreads from midoceanic ridges.

There are three different coral reef morphologies:

(1) Atolls (ring-shaped coral islands nearly or completely surrounding a lagoon);
(2) Barrier reefs (ridges of coral 300–500 m wide, up to 25 km long, and close to and parallel to the coastline);
(3) Fringing reefs (growing from the shore with little or no open water between them and the shore).

Paleozoic and Mesozoic mounds and reefs. Organic carbonate buildups (mounds) have existed since the Paleozoic; however, the animals and algae building them were taxonomically quite different from eon to eon (Wray 1971, 1977; Toomey 1981). **Stromatolitic algal mounds** prevailed from the Precambrian to the Ordovician, parallel to algal-sponge mounds in the Ordovician. Late Paleozoic buildups (e.g., in the Permian) are dominated by small calcified **phylloid green algae,** together with encrusting animals. The first **reefs** were formed in the **Paleozoic,** from the Ordovician to the

Permian, by **tabulate** and **rugose corals,** stromatopores, and other animal groups that all became extinct except for the bryozoans. Dominating algae were filamentous encrusting Cyanophyta and Chlorophyta. The modern-day coral algal reefs, with hermatypic, **scleractinian corals,** symbiotic with **zooxanthellae,** are of **Mesozoic** origin and appeared in the **Triassic** in the newly formed Tethys Sea, with intensive diversification in the Jurassic (Kühlmann 1984). At that time, reefs occurred in England, Japan, and in the epicontinental seas in central Europe, that is, 20° farther north than today. The hermatypic scleractinian families changed until the modern predominating families (Pocilloporidae, Poritidae, and Dendrophylliidae) appeared in the Cretaceous. Predominating calcareous algae on the Mesozoic reefs were the **Solenoporaceae,** while the **Corallinaceae** took over from the Cretaceous onwards (Wray 1971).

Symbiosis and depth relationships. The Paleozoic corals may not have been symbiotic, but evidence of massive exoskeletons indicative of symbiosis dates from the Triassic, when scleractinians appeared. Critical evidence for symbiosis exists from the mid-Jurassic (Law and Lewis 1983; Raven 1986). Due to their endosymbiotic algae the "modern" scleractinian (hermatypic) corals have a **light requirement** of at least 3% of the irradiance at the water surface (Stoddart 1969). For this reason, live hermatypic corals with endosymbiotic algae are only found to a maximum depth of 145 m (Fricke and Schuhmacher 1983). Ahermatypic scleractinians with no symbiotic algae (see below) and no light requirement also developed in the Mesozoic and colonized deeper zones (Kühlmann 1984). Many species of hermatypic coral occur at depths shallower than 25 m and grow optimally only above a 10 m depth, indicating higher light requirements.

Tertiary and Pleistocene temperature crises. Many coral genera originated in the Tertiary, and mainly in the Indo-Pacific. Of the living reef coral genera, 37% appear to have originated since the late Miocene in the Indo-Pacific (Rosen 1984). In the late Tertiary the continuing deterioration of the climate and the fragmentation of the Tethys Sea resulted largely in the extinction of the European coral reefs (Kühlmann 1984). The Caribbean became isolated from the Indo-Pacific by the Panamanian land bridge 3–4 Ma. In the Pleistocene several coral genera (e.g., *Fungia* and *Pocillopora*) became extinct in the Caribbean but not in the Indo-Pacific (Kühlmann 1984). The near-equatorial **reduction of sea-surface temperatures** during the last glaciation was probably less than 2°C in the western Pacific, western equatorial Atlantic, and Caribbean (McIntyre et al. 1976; Broecker 1986); however the land-locked Caribbean may have lost more species during the last glaciation. The Indo-West Pacific harbors many geologically old coral genera that were once more widespread, and, in addition, many new genera appeared from the Miocene onward in the Indo-West Pacific (Rosen 1984, 1988). With its wealth of low-lying islands, the Indo-West Pacific supported a "vicariance-and-refuge-diversity pump," whereas the Caribbean remained a modest refuge with few new originating species (Rosen 1984).

Pleistocene sea-level crises. Vast areas of **emergent coral reefs** occurred when water froze near the Poles and the global seawater level fell by 70–200 m during the glacial periods of the Pleistocene. During the interglacial periods sea levels increased as the ice melted. For example, during the peak of the last glaciation, about 18,000 years ago, the global seawater level was lowered by some 130 m; it increased after that at a rate of about 1 cm per year. About 6000 years ago the Recent seawater level was reached. The increase in sea level occurred relatively suddenly, since the major transgressions were more rapid throughout the Pleistocene than the regressions (Seibold and Berger 1982). The coral reefs that are found alive today were able to grow vertically sufficiently fast to follow this increase in seawater level, but many other reefs could not keep up and died, leaving **submerged reefs** with their upper surface below 100 m.

Algae exist on the coral reef as:

(1) Endosymbiotic dinoflagellates (*Symbiodinium*) in the outer layers of the polyps of the hermatypic corals;
(2) Boring, minute, filamentous thalli of cyanophytes and green algae in the reef corals and in hard substrate;
(3) Crustose coralline algae;
(4) Erect-growing, relatively small macroalgae.

About two-thirds of the primary production on the coral reef may be due to the endosymbiotic algae of the corals, and one-third to the other three algal groups (Wanders 1976–1977). The coral reef seems to exist as a highly productive ecosystem, like an oasis in the desert of the nutrient-poor and unproductive tropical pelagic realm. This is partly possible because the stock of nutrients is largely kept within the coral reef ecosystem and is rapidly turned over (Lewis 1981), although the concept of coral reefs having nutrient-recycling properties is not always applicable (van der Land 1989).

In addition, cyanophytes growing epiphytically in algal turfs, (e.g., *Calothrix* spp.) fix atmospheric nitrogen and add nitrogenous compounds to the coral reef community so that exported nitrogen is continuously replaced (Webb et al. 1975; Wiebe et al. 1975; Littler and Littler 1984). Another input of nutrients may be due to trade-wind-driven currents providing a low but definite nutrient source (Adey and Goertemiller 1987). The domain of the coral reef remains, however, the nutrient-poor environment. Higher concentrations of phosphate favor not only the growth of macroalgae but also act as crystal poisons of calcification, which may partly cause the lack of coral reefs in upwelling areas (Kinsey and Davies 1979, Raven 1986).

(1) Endosymbiotic dinoflagellates. Symbiotic associations between animals and photosynthesizing plants are much more common in the tropics than in the temperate zones, probably a consequence of lower nutrients, high predation in reef areas (Vermeij 1983), and the great geological age of the coral reef ecosystem. The polyps of most species of reef-building hermatypic coral species contain the dinoflagellate *Gymnodinium microadriaticum* as an endosymbiont, giving a brownish-green color to the corals. The zooxanthellate corals can be regarded as living greenhouses or photosynthetic animals (Rosen 1988). Experimentally darkened hermatypic corals grow very slowly or die. It is thought that:

(1) The endosymbiotic algae facilitate the deposition of carbonate and thus the formation of the skeleton of the coral by CO_2 consumption during photosynthesis;

(2) The coral polyps receive some photosynthates from the algae, as shown by tracer studies using ^{14}C;

(3) The algae receive nutrients like nitrate and phosphate from the coral (reviews in Borowitzka 1977; Muscatine 1980; Roth et al. 1982).

An extremely deep-growing scleractinian coral, *Leptoseris fragilis,* has only been found at a depth of 100–145 m in the Red Sea, still harboring zooxanthellae (Fricke and Schuhmacher 1983). The platelike solitary coral seems to provide the zooxanthellae with additional light. Chromatophores containing pigment granules of an unknown chemical nature are situated like a mirror behind the monolayer of zooxanthellae in the oral gastrodermis of the coral. The coral pigment absorbs the prevailing violet and short-wave blue light (380–420 nm) at depth and emits it as additional suitable light for the photosynthesis of the algal symbionts, mainly as blue light at 420–455 nm (Schlichter et al. 1985, 1986). In addition, the coral pigment displays a reddish autofluorescence.

(2) Boring, filamentous algae. The green alga *Ostreobium,* several species of cyanophytes such as *Plectonema,* and also a few red algal species bore holes of 5–10 μm in diameter in the whole vertical range of the coral reef within the euphotic zone (van den Hoek et al. 1975). Such ubiquitous boring algae are able to penetrate into corals, limestone, or shells by means of their apical cell, which probably excretes organic acids for dissolving lime; examples are the Conchocelis phases of red algae. The boring algae, which are obviously able to survive at low irradiances, are found in the somewhat transparent limestone below the vegetation layers of macroalgae and crustose algae. They are also present in the limestone skeleton of live corals, for example, as a 2 mm wide, blue-green layer 13 mm inside the coral *Favia* (Fig. 4.10B). There is evidence that cyanophytes living within the skeletons of corals fix nitrogen and enrich the nitrogen milieu of the coral reef ecosystem (Crossland and Barnes 1976). The photon flow rate at which the algae live in this peculiar habitat is on the order of 2–3 μmol m^{-2} s^{-1} (Halldal 1968; Shibata and Haxo 1969), similar to the photon flow rate present at the lower limit of the euphotic zone.

(3) Crustose coralline algae. Numerous genera of the order Corallinales, such as *Porolithon, Lithophyllum,* and *Neogoniolithon,* contribute to the vertical increase of the coral reef and may form a distinct algal ridge on the seaward, wave-exposed side of the reef. The increase in thallus thickness is on the order of 1–5 mm per year, and the increase in thallus area may amount to 1–2 mm per month (Adey and Vassar 1975; see also Table 8.4). The crustose, calcified, and uncalcified algae are present throughout the euphotic range of the coral reef, except at locations that are heavily covered with sediment.

(4a) Erect-growing, calcified macroalgae. The **calcified green alga** *Halimeda* is an important reef-building erect alga in sheltered and mainly sandy areas of coral reefs (see monograph by Hillis-Colinvaux 1980). This alga deposits calcium carbonate as aragonite on the outside of its thallus. Half the sediment contributing to the height increase of a coral reef may consist of fragments of erect calcified algae, such as *Halimeda* or *Penicillus,* or of calcified crustose algae so that this may be referred to as

an "algal" rather than a "coral" reef (Hillis-Colinvaux 1986b). The branches of *Halimeda* may produce a new segment every three or four days, and the biomass of this alga can double in 15 days, adding 7 g dry weight per day per square meter of solid substratum (Drew 1986). *Halimeda* populations grow, primarily on the sands and pinnacles of the back reef or lagoon, secondarily as an *H. opuntia* zone behind the coralline algal ridge, and least frequently at greater depths on the fore reef (Hillis-Colinvaux 1986b).

(4b) Erect-growing, noncalcified macroalgae. These are not, at first sight, striking in appearance on coral reefs. In shallow water most of the space is occupied by the massive corals, and only 10% of the space may be inhabited by macroalgae (Wanders 1976–1977). The multispecific turf or lawnlike algal community has considerable potential for photosynthesis and growth, and blue-green algae, including heterocysted and nitrogen-fixing genera, are an important turf component (Adey and Goertemiller 1987). In **deeper water** fast-growing macroalgae may become a strong factor in controlling coral growth as observed along the outer slopes of the Bermuda atoll. Here coral growth stops at 50–70 m deep, where downward irradiance reaches 5–20% of surface light and the deeper coral sites are heavily overgrown by species of *Dictyota, Padina, Sargassum,* and *Udotea* (Fricke and Meischner 1985).

What little macroalgae exist at shallower depths are usually found in crevices and hollows protected from **herbivorous animals.** Another refuge from predation for macroalgae is the substratum close to predation-resistant animals, such as the fire coral *Millepora alcicornis* (Littler et al. 1987). Many of the more prominent algae produce **toxic, secondary metabolites** to deter herbivores. Green, caulerpalean algae possess terpenoid substances (e.g., *Halimeda* spp. produce halimedatrial) as well as the mechanical protection of calcification (Paul and Fenical 1983, 1987). The long evolutionary history of tropical ecosystems enabled grazers and grazed algae to develop sophisticated strategies. *Halimeda* species produce more herbivore-susceptible tissue at night when herbivorous fishes are less active (Hay et al. 1988a). Furthermore, the concentration of halimedatrial is maximal in the young, soft, nutritionally valuable tissue. When the tissue ages, the morphological defenses increase due to calcification, and chemical defenses decrease as well.

Overall the herbivorous animals, predominantly represented by herbivorous fish species and sea urchins, seem to play a key role in securing the existence of the corals. Species of *Caulerpa, Sargassum, Turbinaria,* and *Padina* do not seem to be as resistant to grazers as are *Halimeda* species (S. M. Lewis 1985).

The herbivores keep the **undefended algae** consistently cropped. This has been convincingly demonstrated by using cages to exclude all herbivores (Fig. 4.11). Under the cages a luxuriant vegetation of macroalgae develops that not only shades the underlying corals but also functions as a sediment trap, thus endangering the existence of these corals. Van den Hoek et al. (1975) argued that no coral reef could exist without herbivorous animals and that without them the hard-substratum areas in the tropics might be largely covered with seaweeds instead of corals.

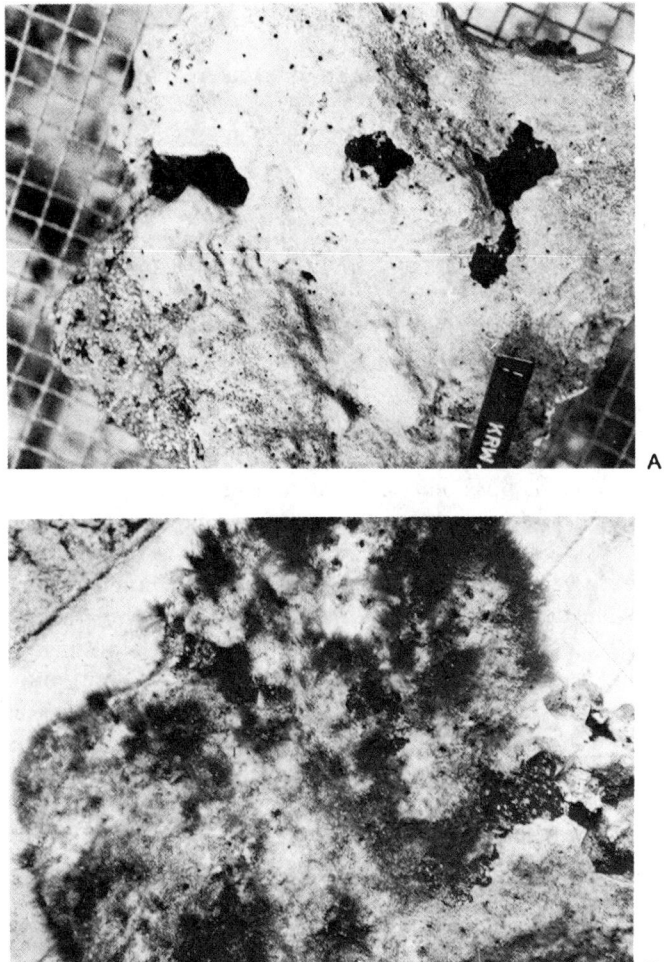

Fig. 4.11 (A) Limestone with the crustose red alga *Porolithon pachydermum*. (B) same limestone after one month in a cage excluding grazers; dense vegetation, mainly of the filamentous brown alga *Giffordia duchassaingiana*. (From Wanders 1976–1977.)

In comparison, the major part of the seaweed biomass in the temperate regions is degraded through detritus chains, and the significance of herbivores increases towards the equator (Vermeij 1978; Gaines and Lubchenco 1982). This is emphasized by the absence of any **herbivorous fish species** in the cold temperate regions. On coral reefs, 20% of the fish species may be herbivorous, 10% omnivorous, and the remainder carnivorous (Bakus 1969). The fish order Perciformes, which includes the four most important herbivore families, evolved in the late Cretaceous, and the herbivorous reef fishes evolved not before the Eocene (Steneck 1983). Two fish families (Scaridae and Acanthuridae) are able to denude primary substrata, and three families (Scaridae, Balistidae, and Monacanthidae) are capable of excavating calcareous substrata.

Besides the corals, another group of reef builders may take advantage of the activity of herbivores on the coral reef, namely, the **crustose coralline algae** (Adey and Vassar 1975). If the herbivores did not graze the erect algae, the role of the underlying crustose, red algae would probably be reduced to the comparatively low significance of crustose algae in temperate regions, as producers of rhodoliths and maerl.

With the high activity of herbivores in mind, it is difficult to imagine how the relatively long-lived kelp thalli might survive in the tropics. The hypothesis that kelps are excluded from tropical regions by high water temperatures can be opposed by the more rhetorical question: why should kelps not have evolved species to survive tropical seawater temperatures?

There remains the possibility that the nutrient levels in coral reef areas are too low for macroalgae to successfully compete with the corals, as suggested from work in the Houtman Abrolhos Islands (western Australia), where the kelp *Ecklonia radiata* meets the southernmost coral communities (Johannes et al. 1983). The dearth of corals in tropical upwelling regions may be caused not only by water temperatures too low for the corals but also by too high nutrients that favor extensive macroalgal growth. A nutrient-enrichment experiment of Hatcher and Larkum (1983) on the Great Barrier Reef revealed that, in the subtidal lagoon, nitrogen and grazing alternated seasonally in controlling the standing crop of the epilithic algal community.

4.6 WESTERN ATLANTIC TROPICAL REGION (TROPICAL-ATLANTIC AMERICA)

Literature: Entire area: W. R. Taylor (1960). **Checklist:** Wynne (1986a). **South Florida:** Dawes (1974); Dawes et al. (1967); Eiseman (1978, 1979); Hanisak and Blair (1988); Stephenson and Stephenson (e.g., 1972). **Bermuda:** Collins and Hervey (1917); Searles and Schneider (1987); Sterrer (1986); W. R. Taylor and Bernatowicz (1969); Thomas (1985). **Caribbean:** Diaz-Piferrer (1969); Littler et al. (1989); W. R. Taylor (1969); Vroman and Stegenga (1988). **Greater Antilles:** Almodóvar and Ballantine (1983); Dahl (1973a); Kusel (1972); V. J. Chapman (1961–1963); Goreau and Goreau (1973). **Lesser Antilles:** Börgesen (1913–1920); John and Price (1979); Price and John (1979); Richardson (1975); W. R. Taylor (1969, 1970); van den Hoek (1969); van den Hoek et al. (1972, 1975, 1978); Vroman (1968); Wanders (1976–1977). **Central America:** Bird and McIntosh (1979); Earle (1972); Norris and Bucher (1982); Phillips et al. (1982); Rützler and Macintyre (1982); W. R. A. Taylor (1935). **Colombia:** Schnetter (1976–1978). **Venezuela:** Diaz-Piferrer (1981); Gessner and Hammer (1967); Rodriguez (1959). **Brazil:** Eston et al. (1986); Guimarães et al. (1981); Joly (1965); Oliveira Filho (1976); Oliveira Filho and Ugadim (1976); Oliveira Filho et al. (1983); W. R. A. Taylor (1930); Ugadim (1973–1976); Ugadim and Pereira (1978).

The western Atlantic tropical region extends from southern Florida to Cape Frio near Rio de Janeiro and covers the islands of the Caribbean, the eastern continental coasts of Central America, and the coasts of Venezuela and Brazil (Figs. 1.5–1.7).

The warm South Equatorial Current branches off the Brazilian coast, and the southbound branch (the **Brazil Current;** Fig. 2.43), enables tropical organisms to survive to Cape Frio. The northbound branch (the **Guayana Current**) joins the North Equatorial Current, then flows into the Caribbean Sea, and finally reaches the Gulf of Mexico. The northward transport of warm water continues as the **Florida Current** leaving the Northeast American coast as the **Gulf Stream** and imparting a tropical character to Bermuda. The Gulf Stream crosses the North Atlantic to Europe as the **North Atlantic Drift.**

The Florida Current transports floatable, tropical-benthic organisms and their reproductive stages, including drifting thalli of *Sargassum* species (Section 4.7), to the soft-bottom coasts of the warm temperate Carolina region. These tropical drift organisms do not find appropiate conditions for survival on the North American coast because of low water temperatures and the absence of rocky substratum (Briggs 1974). The remains of tranported turtle grass, *Thalassia testudinum,* have been observed at a 3000 m water depth off North Carolina (Menzies et al. 1967).

History of algal investigation and species diversity. A classic study of marine phycology in the tropical western Atlantic is the work by Börgesen (1913–1920) on the seaweed flora of the Danish West Indies, nowadays comprising the U.S. Virgin Islands of St. Croix, St. John, and St. Thomas (northern Lesser Antilles). A further milestone was the work by W. R. Taylor (1960), which described the seaweed floras ranging from North Carolina to southern Brazil, including portions of the adjacent warm temperate regions. According to these and other sources (Diaz-Piferrer 1969), the area studied by Taylor contained 800 seaweed species, 80% of these occurring in the Caribbean, to which 15% are endemic (examples in Fig. 4.6). Börgesen had reported 327 species from the Danish West Indies (58% red algae, 28% green algae, and 14% brown algae). The seaweed flora of the eastern coasts of Central America and southern Florida (Phillips et al. 1982) is similar to that of the Caribbean. This widespread flora mainly determines the communities on hard-substratum locations of the tropical east coast of Brazil (W. R. A. Taylor 1930; Diaz-Piferrer 1969).

In contrast to the tropical West African coast, **coral reefs** are a main component in the western Atlantic tropical region. This area has about one-tenth the number of the species of hermatypic corals of the Indo-Pacific, probably due to two main factors. First, the Indo-Pacific has an area 20 times larger in which speciation could take place. Second, there were possibly higher losses of tropical species in the Atlantic in the late Tertiary and Pleistocene, when seawater temperatures declined more drastically and the tropical regions were more constricted than in the Pacific.

Zonation on coral reefs. The following example from Curaçao (Fig. 4.12) is used to illustrate the vertical distribution of benthic algae (van den Hoek et al. 1975, 1981). A submarine platform declines gradually to an 8–10 m depth, ending about 100 m from

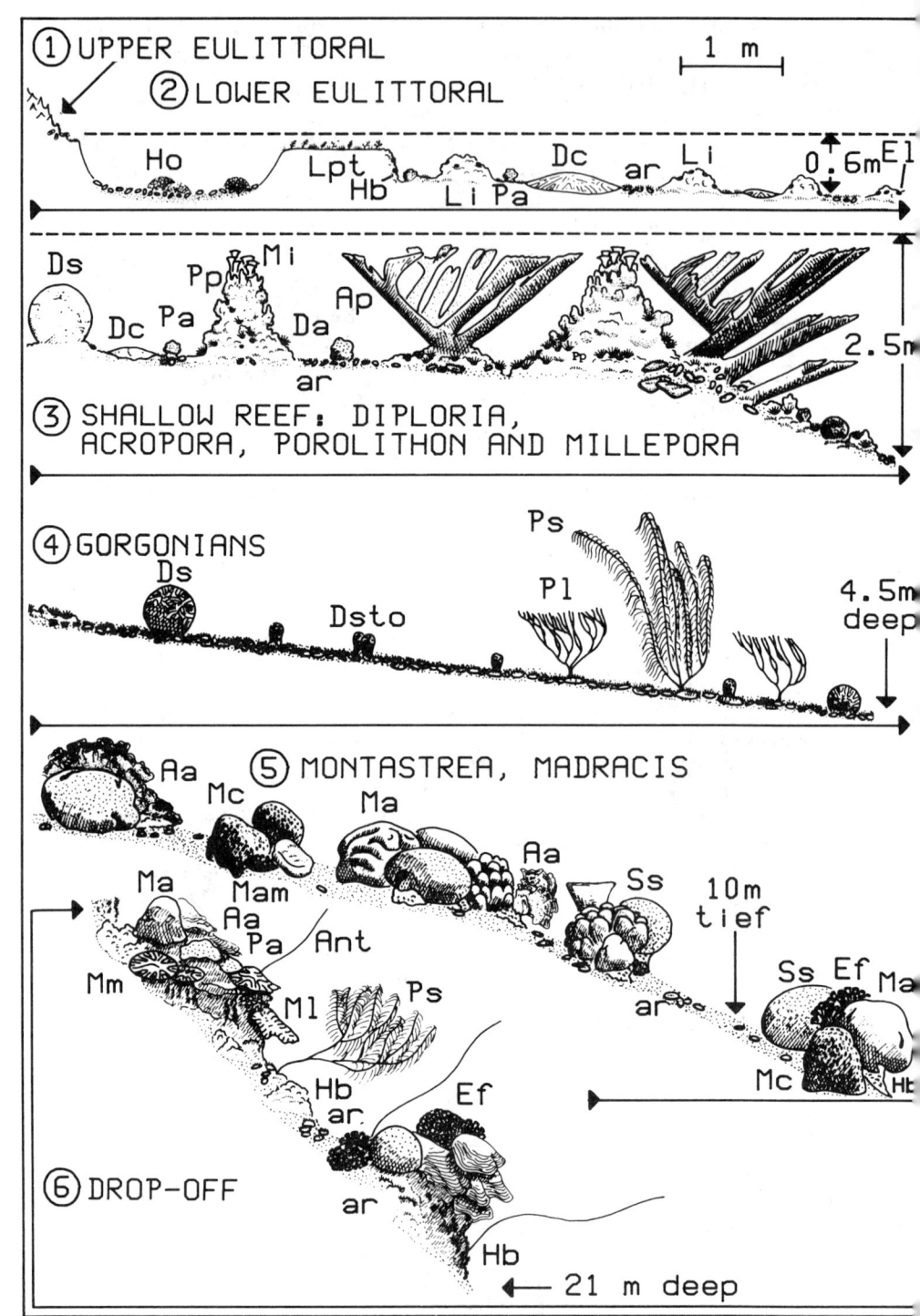

the coast with an abrupt dropoff. The seawater temperature ranges from 26°C in winter to 29°C in summer, salinity is 35‰, and there is a tidal range of about 30 cm. Seven communities are distinguished containing a total of 142 species of multicellular algae (including boring blue-green and green algae); see the description in the legend to Fig. 4.12 (the seventh group is not illustrated).

The narrow **eulittoral zone** is colonized in its upper part, by cyanophytes and in the lower part by turfs of smaller macroalgae 1–8 cm high (**communities 1 and 2** respectively in Fig. 4.12). Species of *Enteromorpha*, the brown algae *Giffordia duchassaigniana* and *Sphacelaria tribuloides*, and in the deepest part the brown alga *Sargassum polyceratium* and the red algae *Hypnea musciformis* and species of *Laurencia* dominate.

In the **sublittoral zone** on the platform, from the surf area down to 3 m deep, is a shallow reef (**community 3**), dominated by the corals *Diploria clivosa* (up to 1 m in diameter) and *Acropora* species (up to 2.5 m wide and 1.5 m high). Another reef builder is the crustose red alga *Porolithon pachydermum*, which forms elevations on which the hydrocoral *Millepora* species thrive. Due to the irregular and bizarre growth forms of such corals, the surface area of the substratum is greatly increased on coral reefs (Dahl 1973*b*). The sea urchin *Diadema antillarum* occurs abundantly, keeping the scattered algal turf short, but it may also graze on the surface of the corals (van den Hoek et al. 1978). After mass mortality of *D. antillarum* in 1983 in Caribbean cryptic reef environments, mortality of encrusting organisms decreased by half, although the community composition hardly changed (Jackson and Kaufmann 1987). Fragmented remains of the reef builders are turned into coral sand, which is slowly transported by wave action along the platform until it runs down the dropoff.

Denser macroalgal turfs about 2 cm high are found at 3–5 m deep in **community 4** (Fig. 4.12), which is dominated by the gorgonians *Pseudopterogorgia* and *Plexaura*. The substratum for the algal turf is formed mainly by coral debris and sand that becomes partly solidified by algal rhizoids. Predominating species are the brown alga *Sphacelaria tribuloides;* red algae such as *Chondria polyrhiza, Pterocladia americana, Jania* species, and *Ceramium leutzelburgii;* and the cyanophyte *Lyngbya*.

At 5–10 m deep the hermatypic corals again predominate as **community 5** (Fig. 4.12), this time with *Madracis mirabilis* (1 m high and 5 m wide) and smaller *Montastrea* species. Herbivorous fish are common here and may, like the pomacentrid *Eupomacentrus planifrons*, defend the scarce, erect algae

---

Fig. 4.12 Zonation on a fringing reef on the coast of Curaçao. **Corals:** Aa = *Agaricia agaricites;* Ant = *Antipatharian* sp.; Ap = *Acropora palmata;* Dc = *Diploria clivosa;* Ds = *Diploria strigosa;* Dsto = *Dichocoenia stokesii;* Ef = *Eusmilia fastigiata;* Ma = *Montastrea annularis;* Mam = *Madracis mirabilis;* Mc = *Montastrea cavernosa;* Mi = *Millepora* sp.; Ml = *Mycetophyllia lamarckana;* Mm = *Meandrina meandrites;* Pa = *Porites astreoides;* Pl = *Plexaura* spp.; Ps = *Pseudopterogorgia* spp.; Ss = *Siderastrea siderea*. **Sea urchins:** Da = *Diadema antillarum;* El = *Echinometra lacunter*. **Algae:** ar = algal turf. **Crustose, calcified red algae:** Hb = *Hydrolithon boergesenii;* Li = *Lithophyllum intermedium;* Pp = *Porolithon pachydermum*. **Erect, uncalcified red alga:** Lpt = *Laurencia papillosa* turf. **Green, calcified alga:** Ho = *Halimeda opuntia*. Communities 1–6 (along with a seventh not illustrated here) are explained in the text. (From van den Hoek et al. 1975.)

growing in their territory against rivals. Damaged tips and regenerating branches illustrate heavy grazing pressure. Boring cyanophytes in coral limestone and fragments of corals are important components of the vegetation.

At the upper rim of the dropoff (**community 6** in Fig. 4.12), at about 10 m deep, the coral sand runs down in cascades. Flat growth forms of diverse corals, which occur down to 30 m deep, are attached to limestone pillars. Crustose red algae such as *Hydrolithon boergesenii* are common, and seaweeds predominate at a 25–60 m depth (Fig. 4.13) as a diverse, shade-adapted **community 7**. The foliose brown alga *Lobophora variegata* forms a girdlelike vegetation at 30–38 m deep. Its blades have a half-life of only 2–3 weeks, so the seemingly constant biomass of the alga is in fact the result of an equilibrium between high grazing losses and moderately high growth rates (Ruyter van Steveninck and Breeman 1987). Season does not significantly influence abundance, size, growth, or reproduction of this alga (Ruyter van Steveninck et al. 1987). *L. variegata* is consumed by herbivores and increases in cover and blade size after moss mortality of the sea urchin *Diadema antillarum* (Ruyter van Steveninck and Breeman 1987). Of the shade-adapted community 40 species are still present at 55–60 m deep, including several representatives of the green algal order Caulerpales (e.g., *Bryobesia cylindrocarpa, Udotea,* and *Caulerpa*), the brown algae *Dictyopteris delicatula* and *Dictyota dichotoma,* and numerous red algae. Finally, crustose red algae reach 80–90 m deep where a sandy plateau starts.

Fig. 4.13 Dropoff at the southwestern coast of Curaçao at about 40 m deep showing the turf of various predominating deep-water algae. The flat coral (right) is *Agaracia lamarckii,* and the whiplike structures belong to the colonies of the antipatharian coral *Stichopathes*. (From van den Hoek et al. 1978.)

Deep-water algae. On other coasts, at which the rocky substratum extends deeper than at Curaçao, the deepest crustose red algae were found at 175 m or 200 m deep. These observations were made from submersibles near the coast of British Honduras and at the Bahamas (Adey and MacIntyre 1973; Lang 1974). Near the Bahamas, at a 60–150 m depth (with water temperatures of 20–24°C), a new green algal genus *Johnson-sea-linkia profunda* was discovered and named after the submersible used. The thallus of this alga, which resembles a radar screen, seems to be optimally adapted for collecting the feeble light in deep water (Fig. 4.14). Flat growth forms are also common in the deep water algal flora off the coast of southern Florida (Eiseman 1978). Typical deep-water algal species at 60–150 m deep off the Bahamas belong to the green algal genera *Halimeda, Caulerpa* (Caulerpales), *Anadyomene, Struvea,* and *Microdictyon* (Siphonocladales). The ubiquitous phaeophyte *Lobophora variegata* (Dictyotales) forms golden-brown, erect blades down to at least 90 m deep and dark-brown crusts in the intertidal zone (Ruyter van Steveninck et al. 1988). Littler et al. (1985, 1986) recorded the deepest-growing algae near San Salvador (Bahamas), again with a submersible. At a 520–268 m depth crustose sponges predominated. At successively shallower depths the following species appeared: at 268 m (0.001% of surface light; maximal photon flow rate of 0.03 μmol m^{-2} s^{-1}) the deepest crustose coralline algae; at 210 m (0.01% of surface light) the boring green alga *Ostreobium;* at 189 m the noncalcified crustose red alga *Peyssonnelia;* and finally at 157 m (0.1% of surface light) the deepest erect alga, *Johnson-sea-linkia profunda*. Water temperature was 29°C at the surface and 19°C at 268 m deep.

Seaweed vegetation at locations where coral reefs are absent. Hard substratum that is not covered by coral reefs may be covered with a dense *Sargassum* vegetation (Fig. 4.15), for example, at depths to 10 m on the northeastern coast of Curaçao, where eastern winds possibly prevent development of corals developing. According to Wanders (1976–1977) the productivity of this *Sargassum* vegetation is two-thirds the value that is typical for coral reefs.

Coral reefs are absent in bays and lagoons covered with soft substratum, and generally along the 3000 km, muddy coastline from the estuaries of the **Orinoco River** in Venezuela to the **Amazon** and to Fortaleza in northwestern

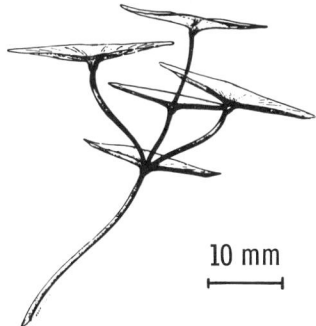

Fig. 4.14 *Johnson-sea-linkia profunda* (Caulerpales), a deep-water green alga from the Bahamas. (From Eiseman and Earle 1983.)

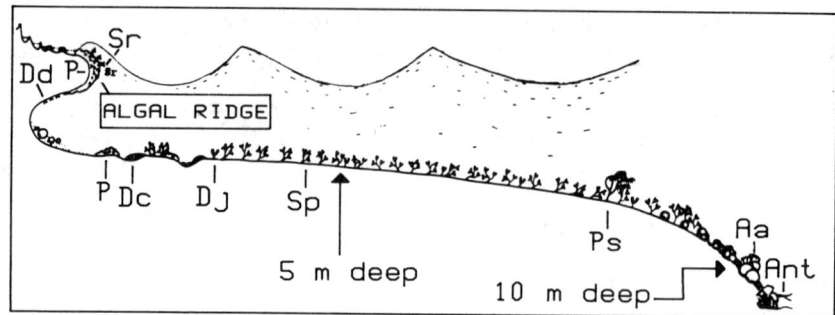

Fig. 4.15 Vegetation transect on the northeastern coast of Curaçao, where extensive coral reefs are lacking. Predominating is the brown alga *Sargassum platycarpum* (Sp). P = *Porolithon pachydermum* (crustose coralline alga); Dc = *Diploria clivosa* (hermatypic coral); Dj = *Dictyopteris justii* (brown alga); Ps = *Pseudopterogorgia acerosa* (Gorgonia); Aa = *Acropora palmata* (hermatypic coral); Ant = antipatharian coral. (From van den Hoek et al. 1981.)

Brazil. From here to the southern limit of the eastern Atlantic tropical region at Rio de Janeiro, rocky substratum with coral reefs occurs again (Briggs 1974). On soft-bottom coasts one finds a **mangrove** vegetation dominated by *Rhizophora mangle* and *Avicennia nitida*. On their prop roots in brackish water are found the green alga *Caulerpa* and the red algae *Acanthophora* and *Catenella* (details in Gessner and Hammer 1967; Vroman 1968; van den Hoek et al. 1972). The red alga *Bostrychia radicans* grows in the most shallow water and tolerates temporary emergence. The mangrove vegetation is followed by **seagrasses**, which are also present near coral reefs on locally occurring sandy areas. In the western Atlantic tropical region the seagrasses are represented by *Thalassia testudinum* and the genera *Halophila, Halodule,* and *Syringodium* (Fig. 4.8).

Brazilian deep-water algae. On the continental shelf of the northeastern and southeastern Brazilian coasts, away from the muddy Amazonian area, crustose coralline algae prevail. The most common brown alga is *Lobophora variegata,* at a 6–100 m depth, with its red algal epiphyte *Polysiphonia scopulorum* (Guimarães et al. 1981). Other frequent brown algae are *Dictyopteris justii, Stypopodium zonale,* and *Padina sanctae-crucis,* and the red alga *Hypoglossum tenuifolium* is a common epiphyte on several species. The numbers of algal species found were 24 in brown algae (Guimarães et al. 1981) and 36 in green algae (Ugadim and Pereira 1978). Near the southern limit of the western Atlantic tropical region at Rio de Janeiro, 100 km off the coast, two *Laminaria* species, *L. brasiliensis* and *L. abyssalis,* were found at greater depths.

4.7 SARGASSO SEA

Literature: Butler et al. (1983); Carpenter and Cox (1974); Gessner (1955–1959); Howard and Menzies (1969); Parr (1939); Pérès (1982b); Sisson (1976); Teal and Teal (1975); Winge (1923); Woelkerling (1972).

In 1492 Christopher Columbus first came across *sargaco* ("floating herbs" in Portuguese) at 35°W. Earlier reports of the supposed discovery of the Sargasso Sea in ancient times are dubious (Krümmel 1891). The two floating *Sargassum* species of the Sargasso Sea, *S. fluitans* and *S. natans* (Fig. 4.16B), may have originated (1) from attached species of the western Atlantic tropical region as drift algae transported northeastward by the prevailing currents or (2) from Tertiary ancestors in the European part of the Tethys Sea (see below).

The great geological age of the vegetation of the Sargasso Sea is demonstrated by the presence of a typical accompanying fauna that is well adapted with regard to color and morphological traits (Butler et al. 1983; Stoner and Greening 1984). Besides an invertebrate fauna, a "displaced benthos" (Hedgpeth 1957a) with many endemic species and several species without the usual planktonic larvae, there are fish species with bizarre, algallike appendages; examples are *Histrio histrio* (Fig. 4.16C) and *Antennarius marmoratus* with its yellow-brown striped "*Sargassum*"-uniform, in which white dots imitate the calcified epiphytic fauna of *Sargassum* (Parr 1939; Cott 1957; Pérès 1982b; Butler et al. 1983; Wicksten 1983).

Paleobiology: suggested origin of the Sargasso assemblage in the Carpathian sector of the Tethys. Fossil fucalean algae with air bladders and without holdfasts, accompanied by a rich ichthyofaunal assemblage resembling that of the present-day Sargasso Sea, were discovered in Carpathian shales and siltstones of late Eocene to early Miocene age, roughly 40–20 Ma (Jerzmánska and Kotlarczyk 1976). The mid-Tertiary Carpathian basin and its cordilleras may have provided favorable conditions

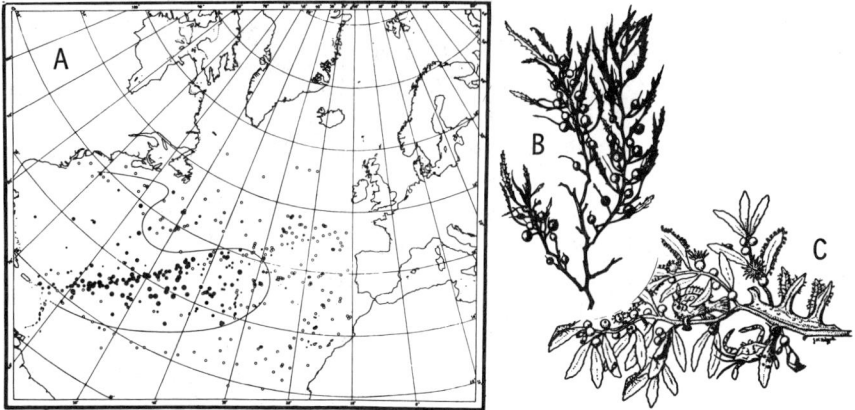

Fig. 4.16 (A) Boundaries of the Sargasso Sea, as determined by the *Dana* Expedition in 1920–1922. Black circles: collection sites of floating *Sargassum;* abundance increases with diameter of circles. Open circles: no *Sargassum* present. (B) *Sargassum natans* from the Sargasso Sea. (C) Invertebrate and fish fauna of floating *Sargassum* in the Gulf of Mexico. The Sargasso fish *Histrio histrio* hovers motionless among the *Sargassum* blades, then rapidly attacks and swallows a nearby animal. (A,B from Winge 1923; C after Hedgpeth 1957a.)

for the evolution of a pelagic *Sargassum* community from algae torn loose and kept permanently as a pelagic system. An important fish species in the Carpathian remains is *Syngnathus incompletus*, similar to the present-day *S. pelagicus* associated with the floating *Sargassum* assemblage. It is possible that the fossil remains are evidence of a quasi Sargasso Sea in the Carpathian section of the Tethys that migrated to the Atlantic Ocean after closure of the Tethys at 17 Ma. The Tethys was closed by the Suez land bridge and by the final folding of the Alps which destroyed the original habitat at the end of the Miocene. The *Sargassum* assemblage continued to evolve in the center of the North Atlantic circular current system. Another fossil trace of the assemblage of ancient algae and Syngnathidae comes from Sicily, whereas no such fossils have been found along the Central American coast (Jerzmánska and Kotlarczyk 1976).

Because of the absence of receptacles and holdfasts in the two *Sargassum* species, it is not possible to relate the two floating species to any of the Recent attached *Sargassum* species of the tropical western Atlantic region. Sexual reproduction as a means of enhancing gene recombination is obviously not important in view of the almost constant environmental conditions in the Sargasso Sea. Even if *Sargassum* were to reproduce sexually, developing zygotes would sink to unfathomable depths and die. Thus the floating *Sargassum* vegetation has probably maintained itself by vegetative growth for many millions of years. The fragmentation of larger and older thallus clumps occurs generally during the winter and is followed by growth and propagation during the spring (Butler et al. 1983).

The Sargasso Sea, as delimited by the two floating *Sargassum* species, exhibits a diameter of 2500 miles (Fig. 4.16A) and is encircled by a gyre formed by the Gulf Stream, the North Atlantic Drift, and the counter-flowing North Equatorial Current (Fig. 2.43). One might say, as Gessner (1955–1959) put it, that the advantage of a stationary life mode, which is offered to attached seaweeds by a hard substratum, is rendered to the floating *Sargassum* vegetation by these currents.

The floating *Sargassum* vegetation occurs scattered, or in windrows, aligned by the prevailing winds (Faller and Woodcock 1964). With light and variable winds, *Sargassum* disperses over the surface (Butler et al. 1983). The approximate distribution extends from the Lesser Antilles to 35°W; in the western Atlantic it covers the latitudinal range 20–30°N and in the eastern Atlantic the range of 30–40°N (Fig. 4.16A). This area is bordered by the 20°C-February isotherm to the north and by the 25°C-February isotherm to the south (Fig 1.6). The Sargasso Sea is thus situated in the northern part of the western Atlantic tropical region. *S. fluitans* and *S. natans* also occur near the coasts of the western Atlantic tropical region and in the Gulf of Mexico (Taylor 1960). However, the quantities present in these areas are too small to continually deliver the stock required for the whole of the *Sargassum* vegetation in the Sargasso Sea.

As invaders along the northwestern fringe of the Sargasso Sea, numerous individuals of *Ascophyllum nodosum* and *Fucus vesiculosus* drifted out from eastern North America (Woelkerling 1972). Accompanying these species were 14 smaller epiphytic seaweed species (e.g., *Pilayella littoralis* and *Polysiphonia lanosa*).

The **biomass** of the floating *Sargassum* vegetation is in the range of 800–2000 kg fresh weight per square kilometer, and 400 kg per square kilometer in the Gulf of Mexico. For the whole of the Sargasso Sea with an area of 3 million square kilometers, the algal biomass may amount to 4–11 million metric tons fresh weight (Parr 1939; Gessner 1955–1959; Butler et al. 1983). Although these numbers sound impressive, the biomass per square meter is as low as 0.9–2.5 g fresh weight or 75–225 mg of carbon and in this respect is smaller than the planktonic biomass in the Sargasso Sea (Pérès 1982b). These values are extremely small compared with the seaweed biomass attached to 1 m^2 of rocky substratum. The primary productivity and the growth rates of the floating *Sargassum* species are low, since the surface water in the Sargasso Sea is poor in nutrients, as is true for all tropical pelagic realms. The Jerlov water type indicates extremely clear water associated with low nutrient regimes in the Sargasso Sea.

4.8 INDO-WEST PACIFIC TROPICAL REGION

Literature: Tropical East Africa: Lawson (1980); Mshigeni (1983); Schmidt (1957). **Natal:** Farnham and Lambert (1981). **Mozambique:** Isaac and Chamberlain (1958). **Tanzania:** Jaasund (1969–1977, 1976). **Kenya:** Isaac (1971); Moorjani (1977). **Somalia:** Sartoni (1986). **Islands in the West Indian Ocean: Aldabra:** J. H. Price (1971). **Mauritius:** Baissac et al. (1962); Börgesen (1940–1957). **Seychelles:** Untawale and Jagtap (1989). **Red Sea:** Aleem (1978b); Ketchum (1983); Khoja (1987); Lipkin (1975); Mergner (1979); Mergner and Svoboda (1977); Natour et al. (1979); Papenfuss (1968); Rayss and Dor (e.g., 1963); Simonsen (1968). **Persian (Arabian) Gulf:** Basson (1979); Basson et al. (1977); Nizamuddin and Gessner (1970). **Pakistan:** Anand (1940–1943); Saifullah (1973); Shameel et al. (1989). **India:** Börgesen (1934, 1937–1938); Krishnamurthy and Yoshi (1970); Murthy et al. (1978); Ohno and Mairh (1982); Umaheswara Rao and Sreeramulu (1964, 1970). **Sri Lanka (Ceylon):** Durairatnam (1961); Svedelius (1906). **Maldive Islands:** Hackett (1969, 1977). **Bangladesh:** Nurul Islam (1976). **Thailand:** Egerod (1974); Kamura and Choonhabandit (1986). **Malaysia:** Sivalingam (1977). **Vietnam:** Dawson (1954); Pham-Hoang (1962). **South China:** Morton and Morton (1983); Tseng (1983–1984). **Taiwan:** Chiang (1960–1962, 1973). **Checklist, Taiwan:** Lewis and Norris (1987). **Philippines, Indonesia:** Cordero (1976–1979); W. R. Taylor (1966); Meñez and Calumpong (1981); Trono and Ganzon-Fortes (1980); Velasquez et al. (1975); Weber van Bosse (1913–1928). **Catalog, Philippines:** Silva et al. (1987). **Papua New Guinea:** Heijs (1987). **Checklist, northern Australia:** Lewis (1984–1987). **Northern and northwestern Australia:** Cribb (1973, 1981); Fuhrer et al. (1981); Knox (1963); Morrissey (1980); I. R. Price et al. (1976); Womersley (1958, 1981a,b). **Great Barrier Reef:** Talbot and Steene (1987). **Western Pacific islands:** Dahl (1979). **Micronesia:** Tsuda and Wray (1977). **Mariana Islands:** Tsuda and Tobias (1977). **Caroline Islands:** Trono (1968–1969). **Marshall Islands:** Dawson (1957b); Gilmartin (1960, 1966); W. R. Taylor (1950). **Melanesia: Solomon Islands:** Morton (1973); Womersley and Bailey (1969,

1970). **Fiji Islands:** Chapman (1971). **Central Pacific islands: French Polynesia:** Payri (1987). **Hawaii:** Abbott (1989); Adey et al. (1982); De Wreede and Jones (1973); Doty et al. (e.g., 1974); Grigg (1983). **Easter Island:** Börgesen (1924); Santelices and Abbott (1987).

In the Indo-Pacific Ocean the marine tropical area spans half the globe and 60° of latitude. The **Indo-West Pacific region** reaches from East Africa to Central America and South America and covers the coasts of the Indian Ocean, including Indonesia and northern Australia, and the Pacific archipelagoes of Micronesia, Melanesia, Hawaii, and Polynesia. Probably the main reasons the Indo-West Pacific tropical region harbors more marine-benthic species than any other of the marine tropical regions are the huge extent of this region (22,000 km from East Africa to the last Polynesian group of islands, the Tuamotu Archipelago), the effective geographical isolation, and the minor reduction in species diversity during the Pleistocene relative to the Atlantic (Ekman 1953; Briggs 1974). There are in this region 10 times more species of hermatypic corals and two to three times more species of mollusks or shore fishes than in the tropical western Atlantic. Since the greatest diversity of marine animal species occurs in the triangle bounded by the Philippines, Malaya, and New Guinea, this area may have either served as a distribution center, as advocated by Briggs (1974, 1984), or it may be a refugium where long-term stable temperatures preserved more older species than in other areas (Vermeij 1978). Rosen (1988) advocates a geotectonic explanation in that the northward movement of Australia towards Asia might have increased the coral diversity by convergence of Australian and Australasian tropical biota. The number of species (not of genera) of zooxanthellate corals in the Indo-Pacific is in fact not highest within the Philippines/Malaya/New Guinea triangle but in western Australia and in the Red Sea (Sheppard 1987).

The Indo-Pacific region, with its island chains and continental shelf segments forming a wide scattering of disjunct but rather similar shallow habitats, may be viewed as one of the few examples of a physically two-dimensional shelf-depth region (Valentine and Jablonsky 1983). Most of the other marine shallow water regions are linear shelf-depth regions because they are associated with relatively narrow shelves along the present continental margins. The Mediterranean region provides another example of a somewhat two-dimensional shelf-depth region and is characterized by a high number of species.

The long coastlines of **East Africa** are characterized by mangrove swamps or sandy beaches. The South Equatorial Current, (Fig. 2.43) bringing warm water to the southern part of East Africa, continues southward as the Agulhas Current and almost extends the tropical region to the southern end of Africa (Figs. 1.5 and 1.6). The algal flora of Tanzania was extensively investigated by Jaasund (1969–1977), who found 310 seaweed species. Little is known of the algal flora of Madagascar. However, the seaweed flora of the more easterly island, **Mauritius,** was investigated in a classical way by Börgesen

(1940–1957). The 13 species of *Sargassum* at Mauritius are all restricted in their occurrence to the Indo-Pacific, and nine of these species are found only in the western Indian Ocean. Further representatives of the Fucales on the East African coast and in general on the coasts of the Indian Ocean are *Cystoseira myrica, Hormophysa triquetra,* and several species of *Turbinaria* (Table 5.1).

At the entrances to the **Red Sea** and the **Persian (Arabian) Gulf** the highest oceanic isotherms (32–35°C) and oceanic salinities (42–46‰) occur. The high salinities are due to excessive evaporation and little rain. The nearest coral reefs to Europe are the northern reefs of the Gulf of Aqaba (29°N) in the Red Sea, now being threatened by pollution. In the Pacific the most northern coral reefs exist in the Ryukyu Islands (southern Japan) and Hawaii. An example of the algal zonation in the Red Sea is given in Fig. 4.17.

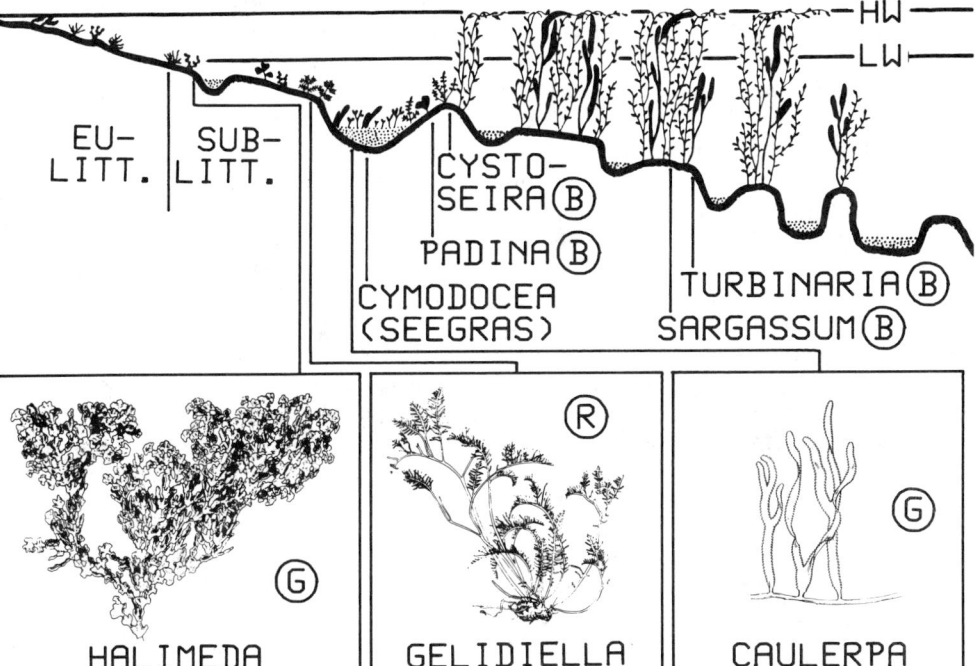

Fig. 4.17 Seaweed zonation in shallow water at the island of Sarso in the Red Sea (Farasan Archipelago, Saudi Arabia). HW = high water; LW = low water; G = green alga; B = brown alga; R = red alga. Abbreviations in legend: IWP = Indo–West Pacific; EP = Eastern Pacific; WA = Western Atlantic; ME = Mediterranean. **Details on species names and dimensions:** *Halimeda opuntia* (× 0.15, IWP, WA); *Gelidiella acerosa* (× 0.3, pantropical and ME); *Caulerpa serrulata* (× 0.5, IWP, WA); *Cystoseira myrica* (IWP, WA); *Padina pavonica* (IWP, WA, ME); *Turbinaria decurrens* (IWP); *Sargassum latifolium* (IWP); *Cymodocea serrulata* (IWP). (Vegetation transect from Simonsen 1968; habit drawings from W. R. Taylor 1960.)

***Ecklonia* species off Oman.** Quite unexpectedly, an *Ecklonia* species was found in the upwelling area off the coast of Oman at 17°N in the Arabian Sea; it was below a thermocline at a 6 m depth, together with an unidentified fucalean alga (Hiscock et al. 1984). It probably bears a relationship to the southern African *E. biruncinata* and the southern Australian *E. radiata*. The existence of a kelp species in a tropical region is to be seen here as an exception caused by upwelling cool, nutrient-rich water. The upwelling area off the Somali and Arabian coast represents the world's fifth major upwelling area. Upwelling is exhibited in this case when the southwest monsoon directly transports surface water away from the coast (Barber and Smith 1981).

Along the sedimentary coasts of **India mangroves** dominate, and extensive coral reefs occur only on the southern Indian coast. On one of the few rocky locations in northwestern India, near Okha (22.5°N), 137 seaweed species were collected by Börgesen (1934). Of these, 15% are endemic to the northern Arabian Sea, and 50% have been recorded in the seaweed floras of Japan, Indonesia, and Australia. This demonstrates the wide distribution of many seaweed species within the Indo-West Pacific tropical region.

Monsoon-caused seasonal periodicities in the seaweed flora. Svedelius (1906) recorded the seasonal fluctuations of seaweeds on a coral reef near Galle (6°N), on the south coast of **Sri Lanka (Ceylon)**. Since the seawater temperature deviates little from 26°C and daylength does not vary greatly, the usual causes of seasonality do not apply. The northern Indian Ocean (Arabian Sea and Bay of Bengal) is characterized by seasonally directional changes of wind and current from which it gets its name the "**Monsoon Sea.**" From November to April **dry winter monsoons,** blowing from the northeast, direct the prevailing current. The intertidal zone of Sri Lanka, the northern Arabian Sea, and the coast of northwestern India dries up, with the result that most seaweed vegetation disappears by the end of the monsoon season (Umaheswara Rao and Sreeramulu 1964; Murthy et al. 1978). From May to October the **wet summer monsoons** cause a reversal of the prevailing current by blowing from the ocean to the land. During this period an immense surf stirs up the sediments, which, together with river sediments, make the coastal waters very turbid. Some algal species are found only in the rainy monsoon season in the shallow waters of Ceylon, e.g., the red alga *Dermonema dichotomum* (Fig. 4.18E). Several bushy, perennating (lasting for several or many years) seaweed species and species of *Sargassum* resume growth and form new thallus branches (Fig. 4.18), whereas old thallus parts are torn off by heavy wave action. Numerous species only become fertile during the wet monsoon season. The shallow-water coast of northwestern India in July, at a seawater temperature of 33°C, is repopulated with abundant algal germlings, for example, the brown algae *Sargassum* and *Cystoseira,* the red alga *Gracilaria foliifera,* and the green algae *Ulva lactuca, Caulerpa,* and *Codium dwarkense*. This occurrence in the wet monsoon season, in summer in northern latitudes, is reminiscent of the new growth that occurs at temperate latitudes in winter and early spring. The seasonal lack of light is the common denominator for the summer-winter periodicity, and for the wet-dry monsoon periodicity, particularly in sublittoral habitats. In both cases seaweeds have adapted and use environmental signals (Chapter 6) to start their growth during or by the end of the annual low-light period, so that the light-receiving thallus area is expanded as soon as the annual high-light period begins.

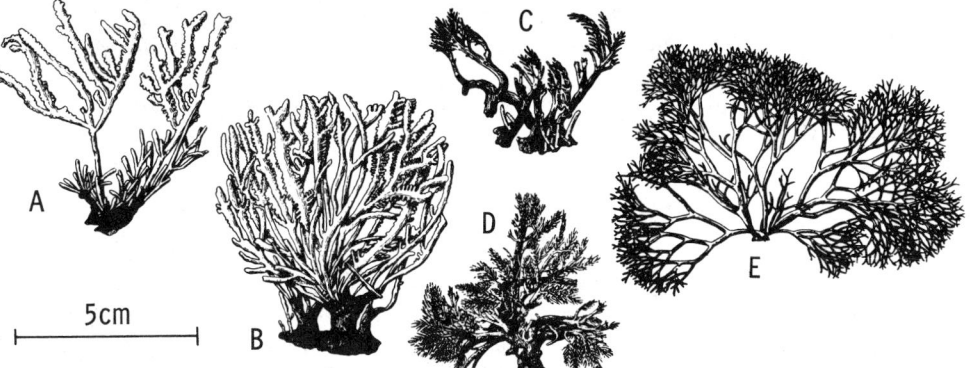

Fig. 4.18 Tropical red algae from Sri Lanka that have cast off parts of their thallus by August (period of southwestern monsoon; A, C), and are fully grown by November to March (period of northeastern monsoon; B, D). **(A,B)** *Laurencia ceylanica*. **(C,D)** *Chondria armata* (= *Rhodomela crassicaulis*). **(E)** *Dermonema dichotomum* (Helminthocladiaceae) (present only during the period of the southwestern monsoon). (From Svedelius 1906.)

West of the Indian subcontinent a submarine ridge, thought to be a submerged part of the Indian landmass, rises from a 3000–4000 m depth. On this ridge in the Arabian Sea the extensive **coral reef formations** of the **Laccadive, Maldive,** and **Chagos Islands** appear. The Maldive Islands, situated just north of the equator, consist of a double row of about 2000 islands, mainly in the form of atolls. The term "atoll" comes from the Maldivian "atolu."

Similarity of seaweed floras of distantly spaced islands: Maldive and Marshall Islands. Quite remarkably the seaweed flora of the Maldive Islands bears more resemblance to that of the Marshall Islands 8000 km away in the Pacific than to the seaweed floras of the neighboring coasts of southern India and Sri Lanka, only 400–600 km away (Hackett 1969, 1977). Among the 264 described seaweed species of the Maldive Islands (83 green, 18 brown, and 163 red algae), the numerous *Sargassum* species from southern India and Sri Lanka are absent. The coasts of southern India and Sri Lanka harbor four times as many species of brown algae as do the Maldives. A similarity of only 30% exists among the red algae from these neighboring coasts, compared with a similarity of 60% with the Marshall Islands. Many of the Maldive Islands are so short-lived that native folklore still preserves the knowledge of large islands that have disappeared below the water's surface. It is not surprising, therefore, that the coasts of such short-lived islands become colonized mainly by those seaweed species with the ability to be easily distributed over wide water distances (see theory on oceanic islands later in this section). Obviously there is also a basic difference in the type of hard substratum that has caused the discrepancy existing between the seaweed floras of the Maldive Islands and the southern Indian coast. The Maldives are almost exclusively dominated by coral reefs, while the southern Indian coast offers other kinds of hard substrata to seaweeds.

Similar to the coasts of India, Asian continental coasts from **Burma, Thailand,** and **Vietnam** to the northern limit of the Indo–West Pacific tropical region at **Hong Kong** are mainly characterized by soft substrata. Coral reefs and a more abundant seaweed vegetation exist only at interspersed hard-substratum areas.

Zonation on the coast of Vietnam. Rocky substratum is more common along the coast of this country. For the supralittoral zone cyanophytes, the red algae *Bangia atropurpurea, Porphyra crispata, P. vietnamense,* and the green alga *Enteromorpha clathrata* have been reported by Pham-Hoang (1962). The upper eulittoral zone is dominated by the barnacle *Chthamalus* and the red algae *Bostrychia, Dermonema frappieri,* and *Jania rubens.* Occurring in the lower eulittoral zone are the oyster *Ostrea;* the red algae *Gelidiella acerosa, Gelidium pusillum,* and *Centroceras clavulatum* (Fig. 4.9); the green algae *Chaetomorpha antennina;* species of *Enteromorpha* and *Ulva;* and the crustose brown alga *Lobophora* (= *Pocockiella*) *variegata.* The predominating brown algae in the upper sublittoral zone are *Padina commersonii, Turbinaria ornata,* and other *Turbinaria* species, *Hormophysa triquetra* (with a similar habit to the warm temperate eastern Atlantic–Mediterranean genus *Cystoseira*), and species of *Sargassum.* These are accompanied by representatives of the green algal order Dasycladales, such as *Boodlea composita* (Fig. 4.3C), *Neomeris annulata* (Figs. 4.4D and 4.5), and *Bornetella oligospora* (Fig. 4.7A).

Near the eastern limit of the Indian Ocean, along the western coasts of **Sumatra** and **Java** extending to **western Australia,** one again finds **coral reefs** (Fig. 4.1). A classical work on the approximately 600 species of the Indonesian seaweed flora was produced as a result of the Netherlands Siboga Expedition (1899–1900) by Weber van Bosse (1913–1928). A few examples of Indo-Pacific seaweeds occurring on the coasts of **Indonesia** are illustrated in Fig. 4.7. The red algal genus *Eucheuma* has a few species in the tropical western Atlantic but about 20 Indo-Pacific species, most of them occurring in the Malay Archipelago (the large group of islands between Southeast Asia and Australia, including Indonesia, the Philippines, and New Guinea). This genus is an important source of carrageenan.

The sea off the coast of **northern Australia,** with about 200 seaweed species (Womersley 1981*a*), and the **Malay Archipelago** can be characterized as a rather shallow shelf area bordered by largely soft-bottom substrata not as suitable for coral reefs or seaweeds as are the westerly archipelagoes of **Micronesia, Melanesia,** and **Polynesia.**

Seasonality of seaweeds on a Polynesian reef. On Moorea Island, French Polynesia, situated 17° south of the equator, the ephemeral brown algae *Hydroclathrus clathratus* and *Chnoospora implexa* (Scytosiphonales) occur from July to November, whereas in the Northern Hemisphere (Micronesia, Red Sea) they are present from February to June (Payri 1987). This shows the strong influence of the astronomical year on seasonal behavior even at low latitudes. Development of perennial algae such as *Padina tenuis, Turbinaria ornata, Halimeda incrassata,* and *Sargassum* species start in March, and growth continues until September and October. The phase of maximum growth in the austral winter coincides with the wet season, similar to Sri Lanka, which is situated a little north of the equator. On the reefs of Moorea Island, strong insolation and desiccation from October to February (the dry season) may threaten the perennial species in the austral summer (Payri 1987).

Along the tropical coast of **eastern Australia** the **Great Barrier Reef** is the most extensive reef system in the world. Stretching for 2000 km from Torres Strait (which separates northeastern Australia from New Guinea) to the Tropic of Capricorn, it is made up of over 2100 barrier and fringing reefs that emerge from the outer rim of the shelf. The individual barrier reefs consist of ridges up to 500 m wide and 3–24 km long. In 1770 on the *Endeavour,* Captain Cook entered the lagoon between the barrier reefs and the coast and was unable to reach the open sea for several weeks.

Since the Australian continent drifted slowly from polar to tropical latitudes, its coral reefs are relatively young, the Great Barrier Reef probably being not older than 18 million years, and many parts little more than 1 million years old (Talbot and Steene 1987).

Seaweed flora of the Great Barrier Reef. About 300 seaweed species with a low endemism of only 2% have been recorded (Cribb 1973). The reef flat, facing the coast and covered with 10–80 cm of water at low tide, appears like a huge tidal pool. This environment is not very suitable for corals and instead harbors a great variety of seaweeds. Examples of the latter are the green algae *Halimeda opuntia, H. tuna* (both important as reef builders), *Caulerpa racemosa, Chlorodesmis fastigiata,* and *Boodlea composita;* the brown algae *Padina gymnospora, Cystoseira trinodis, Turbinaria ornata, Sargassum crassifolium,* and *S. polycystum;* and the red algae *Amphiroa foliacea, Gelidiella acerosa,* and *Laurencia obtusa* (Morrissey 1980; Cribb 1981). On the more mobile substrata of the reef flat are seagrasses, of which the more important species are *Thalassia hemprichii* and *Halophila ovalis.* On the **seaward platform,** corals and crustose coralline algae (Corallinaceae) exhibit a luxurious development; few erect-growing seaweed species are to be seen here, the green tufts of *Chlorodesmis* species being an exception.

Islands in the central Pacific: "Island hopping." The central Pacific islands occur basically as chains of geologically short-lived atolls, which are arranged in a southeast-northwest direction. The theory is that, as the Pacific plate moves northwest and finally slides under the Asian continent, successively new islands are formed above fixed "hot spots" in the lithosphere so that an island chain is formed. This island chain moves with the Pacific plate, the oldest islands submerging to the northwest. For example, the oceanic island chains of Hawaii and the Tuamotu Archipelago/Equatorial Islands system are becoming geologically younger to the southeast and are drifting northeast with a velocity of a few centimeters each year. The youngest islands at the southeastern end of the island chain of Hawaii still have active volcanoes, while the islands at the northwestern end, which originated 5 million years ago, are now submerged and carry atolls. The marine organisms, that inhabit the coasts of these moving islands are "island hopping." This means that endemics, which have evolved on the coast of a particular island, are doomed to lose their particular set of environmental conditions and will colonize another island arising to the southeast.

Therefore, the geological age of an endemic species may be much greater than the island that it inhabits (Pielou 1979).

Island theory: oceanic islands. The central Pacific islands provide examples of oceanic islands that have never been part of a mainland. Such islands, located at the boundaries of tectonic plates, are of volcanic origin. They always have to be colonized (i.e., undergo primary succession), and their existence often lasts no more than 2–10 million years (review in Carlquist 1974). The island theory, or equilibrium theory of insular biogeography, put forward by MacArthur and Wilson (1967) and confirmed by experimental approaches (reviews in Rey 1984; Schoener 1988; Williamson 1988), states that on an island there is an equilibrium between the number of immigrating species and the number of species becoming extinct. For oceanic islands, three principles are important:

(1) The number of immigrating species increases with island size;
(2) The number of immigrating species is reduced with increasing distance from the next mainland;
(3) The number of endemic species that are evolved on an island depends on the island's geological longevity and again on its size and distance from the mainland.

Since successively new immigrants arrive and species present on the island become extinct, a certain "taxon cycle" exists for each island.

Continental islands. These have been part of the mainland and have, when separated, carried with them the original number of species of the mainland fauna and flora. After an initial high extinction rate, a number of species is finally reached that is appropiate to the size of the continental island. The islands of the **Malay Archipelago** with approximately 600 species of seaweeds are continental islands, since they are remnants of a land bridge that connected Australia to Asia.

Oceanic islands and seaweed distribution. On the oceanic islands of the central Pacific area, island theory explains why there is a reduction in number of species (e.g., of corals) to the east, where distances between the islands are vastly increasing (Grigg 1983). In general, the stock of seaweed species of these central Pacific oceanic islands is relatively small and consists mainly of immigrated, **widely distributed** species accompanied by a **few endemics.** For example, the **Solomon Islands** in Melanesia, or the **Marshall Islands** in Micronesia including the atolls of Bikini and Enewetak (Eniwetok), where atomic bomb tests were performed, have a stock of only 240 seaweed species (Womersley and Bailey 1969, 1970). More than half of the species occurring on the Solomon Islands are found in the whole Indo-Pacific tropical region and one-third even have a pantropical distribution. The seaweed flora of **Easter Island,** the most isolated island in the Indo–West Pacific tropical region, has an endemism of only 14%, and 80% of its approximately 170 seaweed species have a wide distribution either in warm or temperate waters (Börgesen 1924; Santelices and Abbott 1987). There is a general geographic affinity with the western Pacific, probably due to the isolating effect of the low temperature of the Peru Current. The similarity between the seaweed floras of Easter Island and those of other islands in the Indo–West Pacific tropical region decreases with increasing distance (Santelices and Abbott 1987).

Crustose coralline algae in Hawaii. Another example of the reduction in algal numbers eastward relates to the crustose coralline algae. Of the 27 species of Corallinaceae on the islands of Hawaii, 11 are endemic while 11 others are distributed throughout the whole Indo–West Pacific tropical region (Adey et al. 1982). Only five species reach the eastern Pacific tropical coasts of Central America, and one species is restricted in its occurrence to the Galapagos Islands, Hawaii, and Panama. The fact that none of the 11 Indo–West Pacific species is represented as identical to, but rather morphologically similar to species in the western Atlantic tropical region, points to the importance of the recently formed land bridge of Central America.

Deep-water algae. On the Marshall Islands and Hawaii the deep-water vegetation has been systematically investigated by dredging (Gilmartin 1960; Doty et al. 1974). About 100 seaweed species are present off Hawaii at a 10–165 m depth. The red algae contribute three times more species to this number than do green or brown algae, as one would expect from the $R:P$ ratio typical for the tropics. Most abundant are the green algae *Halimeda discoidea* and *Microdictyon setchellianum,* the brown algae *Dictyopteris plagiogramma* and *Dictyota acutiloba,* and the red algae *Amansia glomerata, Spyridia filamentosa, Dotyella hawaiiensis,* and *Polysiphonia apiculata.* At a 50 m depth 22 species of red algae, 10 species of green algae, and five species of brown algae are found, which does not fit Engelmann's ideas about depth distribution of the algal groups differing in pigment composition. Typical deep-water algae of Hawaii, absent in shallow water, are, for example, the brown alga *Sargassum hawaiiensis* (Fig. 4.7C) and the green alga *Codium mamillosum.* Of the crustose coralline algae the dominant representatives down to 30 m deep, as in the tropical western Atlantic, are the genera *Porolithon* and *Neogoniolithon,* while *Lithothamnium, Mesophyllum,* and *Archaeolithothamnium* are mainly found at depths below 50 m (Adey et al. 1982). Compared to the Marshall Islands, the deep-water algal vegetation of Hawaii is not as rich in species and is poorly represented by the green algal order Caulerpales. For example, the characteristic deep-water alga *Tydemania expeditionis* (Fig. 4.7B) is absent at Hawaii.

4.9 EASTERN PACIFIC TROPICAL REGION

Literature: Central America: Dawson (1962*a*). **Mexico:** Dawson (1953–1963). **Guatemala to Panama:** Bird and McIntosh (1979); Dawson (1961*b*, 1962*b*), Earle (1972). **South America: Colombia to Ecuador:** Oliveira Filho (1981); Schnetter et al. (1982); Taylor (1945). **Islands: Galápagos, Clipperton, San Benedicto:** Cinelli and Colantoni (1982); Dawson (1957*a*, 1959, 1963); Silva (1966); Taylor (1945).

The Pacific area between Polynesia and America separates the Indo-West Pacific tropical region from the eastern Pacific tropical region. This Pacific area is known as the "eastern Pacific barrier" (Fig. 4.2). The few and distant islands in this area are depauperate with regard to the number of seaweed species and genera. Many genera of the hermatypic corals are absent, for example, *Acropora,* an important reef builder in the Indo-Pacific and western Atlantic. Of the five species of the Indo-Pacific green algal genus *Chlorodesmis* (Caulerpales), only one, *C. caespitosa,* is present on the Pacific coast of Central America (Ducker 1967).

The latitudinal extent of the eastern Pacific tropical region is relatively small (southern Baja California to the Gulf of Guayaquil; Figs. 1.5–1.7) and contains no more than 300 seaweed species (Dawson 1962a). The region is latitudinally constricted due to cold-water currents from the north (California Current; Fig. 2.43) and the south (Peru or Humboldt Current), similar to the situation in the eastern Atlantic tropical region. Few coral reefs exist as fringing reefs off the coast of Panama and the southern end of Baja California. There are large yearly variations in water temperature on the Pacific side of Panama, namely 15–32°C, compared with 24–30°C in the Caribbean. Coral recruitment and growth are hindered by turbid, cold water, as well as by fast-growing fouling organisms, predators, and boring animals. Another obstacle to coral growth on the Pacific coast of Panama is the large tidal amplitude of 7 m, which prevents the formation of shallow reef flats as in the Caribbean (Hay and Gaines 1984).

At the generic level, the seaweed floras on both sides of the Central American land bridge are rather similar, as exemplified by Earle (1972) for the Pacific and Atlantic coasts of Panama. In contrast, at the species level the originally uniform seaweed flora has split into many pairs of morphologically similar species since the land bridge was formed 3.5–4 Ma.

Galápagos Islands. These islands became famous among biologists after Darwin's visit in 1835 as a part of his journey around the world on the ship *Beagle*. The insights he gained from investigating the finches of these islands (which came to be called the Darwin finches) were important steps in developing his theory on evolution. The isolated position of the Galápagos Islands, situated 900 km from the mainland of Ecuador, has resulted in a pronounced endemism. This has been reported at about 40% for the taxa of the higher, terrestrial plants (Bramwell 1979) and hard corals (although only 32 species are present here; Briggs 1974), and is almost as high for the 310 seaweed species (Silva 1966). The islands emerged from the sea relatively recently, and all evolution of its terrestrial and shallow-water biota occurred within the past 3 to 4 million years (Hickman and Lipps 1985). According to Cinelli and Colantoni (1982) the composition of the sublittoral seaweed flora of the Galápagos Islands shows a more warm temperate than tropical character, and so only with reservations, can these islands be regarded as belonging to the eastern Pacific tropical region. The seaweed flora of the small **Clipperton Island,** the only atoll in the eastern Pacific and situated 1200 km from the mainland, mainly harbors species that are widely distributed in the Indo-Pacific (Dawson 1959).

5 Temperate and Polar Seaweed Vegetation of the Southern Hemisphere

The coasts of the Southern Hemisphere can be classified into:

- **Five warm temperate regions** (western South America, eastern South America, southern Africa, southern Australia—further divided into southwestern and southeastern provinces—and northern New Zealand);
- **Five cold temperate regions** (southwestern South America, southeastern South America, Victoria–Tasmania, southern New Zealand, and the sub-Antarctic islands);
- The **Antarctic region** (Fig. 1.7).

At the generic or even family level the temperate seaweed floras of the Northern Hemisphere are basically different from those of the Southern Hemisphere. This can be explained by the hypothesis that the cold-adapted seaweeds evolved parallel to the Tertiary decline of seawater temperatures in temperate and polar regions. The developing cold-water floras of the Northern and Southern Hemispheres were permanently separated by the tropical, circumglobal Tethys Sea. Only a few temperate species may have crossed the equator during the cold periods of the Pleistocene, and today exhibit an amphiequatorial distribution (Section 5.2).

5.1 DELIMITATION OF THE REGIONS AND PALEOBIOGEOGRAPHY

5.1.1 Convergences, Currents, and Regions

The most important hydrographical and biogeographical boundaries in the southern ocean are represented by two narrow, circumpolar convergence zones (AC and SC in Fig. 5.1). Here water masses converge, cold water sinks, and the surface water temperature increases abruptly in a northerly direction. At the **Antarctic Convergence** (Antarctic Polar Front; average northern extent of sea ice in winter) the cold surface water (AS in Fig. 5.1 *right*) surrounding Antarctica sinks, and surface water temperatures at this

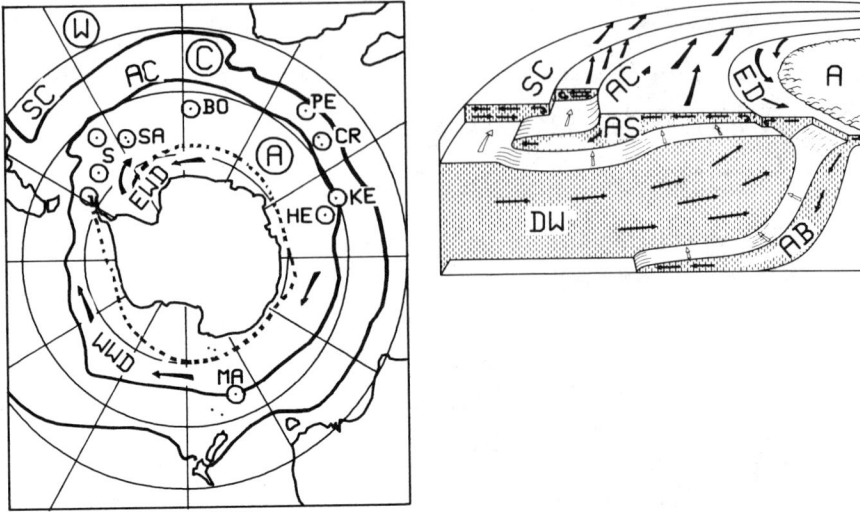

Fig. 5.1 Oceanography of the southern ocean regions **(left)**, biogeographical boundaries, and islands. A = Antarctic region; AC = Antarctic Convergence or Polar Front; C = cold temperate regions; SC = Subtropical Convergence; W = warm temperate regions; dotted line = approximate boundary of the Antarctic Divergence between the East Wind Drift (EWD) and the West Wind Drift (WWD). **Islands within the Antarctic region:** S = South Orkney and South Shetland Islands; SA = South Sandwich Islands; B0 = Bouvet Island. **Sub-antarctic Islands region:** PE = Prince Edward Islands; CR = Crozet Islands; KE = Kerguélen Islands; HE = Heard Island; MA = Macquarie Island. **Right:** structure of the water masses in the southern ocean. AC and SC as above; A = Antarctica; AB = Antarctic bottom water; AS = Antarctic surface water; DW = circumpolar deep water; ED = East Wind Drift. (left after Deacon 1964; right from Hedgpeth 1969.)

convergence are in the annual range of 3–5°C. The surface water of the area that extends northward of the Antarctic Convergence for about 10° of latitude sinks at the poorly defined **Subtropical Convergence,** where the surface water temperatures range between 14°C in summer and 5°C in winter. The two convergences approximately mark the southern and northern boundaries of the **cold temperate regions** of the Southern Hemisphere (Figs. 1.5, 1.6, and 5.1 *left*). The **Antarctic region** covers the area extending south of the Antarctic Convergence to the coast of Antarctica, and the surface water temperatures within this region range annually between − 1.3 and 3°C.

Antarctic Circumpolar Current. Antarctica is surrounded by two concentric rings of water that move in opposite directions: (1) the **East Wind Drift** (ED in Fig. 5.1 *right*), which represents the westward-flowing water near the coast; and (2) the huge eastward-moving **West Wind Drift** (Figs. 2.43 and 5.1 *right*). The boundary between the two water rings is situated within the biogeographical Antarctic region, namely, at the **Antarctic Divergence** (dotted line in Fig. 5.1 *left*), near 65°S latitude, where deep, nutrient-rich water comes to the surface and the average northerly extent of sea ice is situated in summer. The West Wind Drift, also called the **Antarctic Circumpolar**

Current, originated 30–25 Ma (see below). The West Wind Drift takes four years to travel around Antarctica and covers the vast area from the Antarctic Divergence to the Subtropical Convergence, extending over 20° of latitude (Figs. 2.43, 5.1 *right*). The West Wind Drift serves as a migration route for floatable seaweed species (e.g., *Macrocystis pyrifera;* see below) and has created the similarities that exist in seaweed floras between the five cold temperate regions of the Southern Hemisphere and the Antarctic region. The Peru Current (Humboldt Current) branches from the Antarctic Circumpolar Current at the narrow Drake Passage, which separates South America from the Antarctic Peninsula. This northward-flowing current allows cold-water species to extend along the western coast of South America almost as far as the equator.

5.1.2 Paleoceanography and Paleobiogeography of the Southern Hemisphere

Literature: Bergh (1987); Brundin (1966, 1972, 1981); Cande and Mutter (1982); Cande et al. (1988); Darlington (1965); Edgar (1986); Grant-Mackie (1979); Johnson et al. (1980); Keast (1973, 1983); Kennett (1977, 1982); Knox (1960, 1963, 1970, 1975, 1980); LaBrecque and Barker (1981); Lawver et al. (1985); Mercer (1983); Norton and Sclater (1979); Owen (1983); Pickard and Emery (1982); Powell et al. (1988); Rouland et al. (1985); Sclater et al. (1977); Scotese et al. (1988); Stanley (1986); Stock and Molnar (1987); Stevens (1980); Vogel (1984); Zinsmeister (1982).

Antarctica was part of the southern continent **Gondwana** (also called Gondwanaland; Fig. 5.2A). This is evidenced by fossil remains of the Paleozoic "Gondwana flora," with the pteridophyte *Glossopteris* as a characteristic component. This strongly seasonal flora occurred during the Permian Great Ice Age around the world, in the far south and in India (Darlington 1965; Schopf 1970; Brown and Gibson 1983; Stanley 1986). In the late Mesozoic, Gondwana split into East Gondwana (Antarctica, Australia, and New Zealand), West Gondwana (South America and Africa), and India (Fig. 5.2B). The East Gondwana landmass constituted the southern Pacific margin (Figs. 5.2B–D), on whose isolated coast highly endemic marine faunas developed (Zinsmeister 1982). This Mesozoic southern circum-Pacific coast resembled the present-day North Pacific surrounded by an almost continuous landmass, interrupted only at the Bering Strait. Note that the eastern edges of today's Australia and New Zealand were exposed to the Pacific Ocean, and the western edge of Australia to the southern Indian Ocean (Fig. 5.2D). Africa separated from Gondwana in the middle Jurassic (LaBreque and Barker 1981), India in the early Cretaceous (Johnson et al. 1980, Owen 1983) and New Zealand in the late Cretaceous (Owen 1983). There were forests in Antarctica when Australia separated from it in the late Cretaceous (Cande and Mutter 1982, Mutter et al. 1985, Rouland et al. 1985).

Trans-Antarctic relationships. The fossil remains of the southern beech *Nothofagus,* the only southern genus of the mainly northern family Fagaceae, indicate a Tertiary distribution of this genus reaching from southern South America to Antarctica, Australia, and New Zealand. There are many more examples of exclusively southern groups of organisms shared by these landmasses. The southern African terrestrial flora and fauna have relatively few trans-Antarctic examples, and these are generally

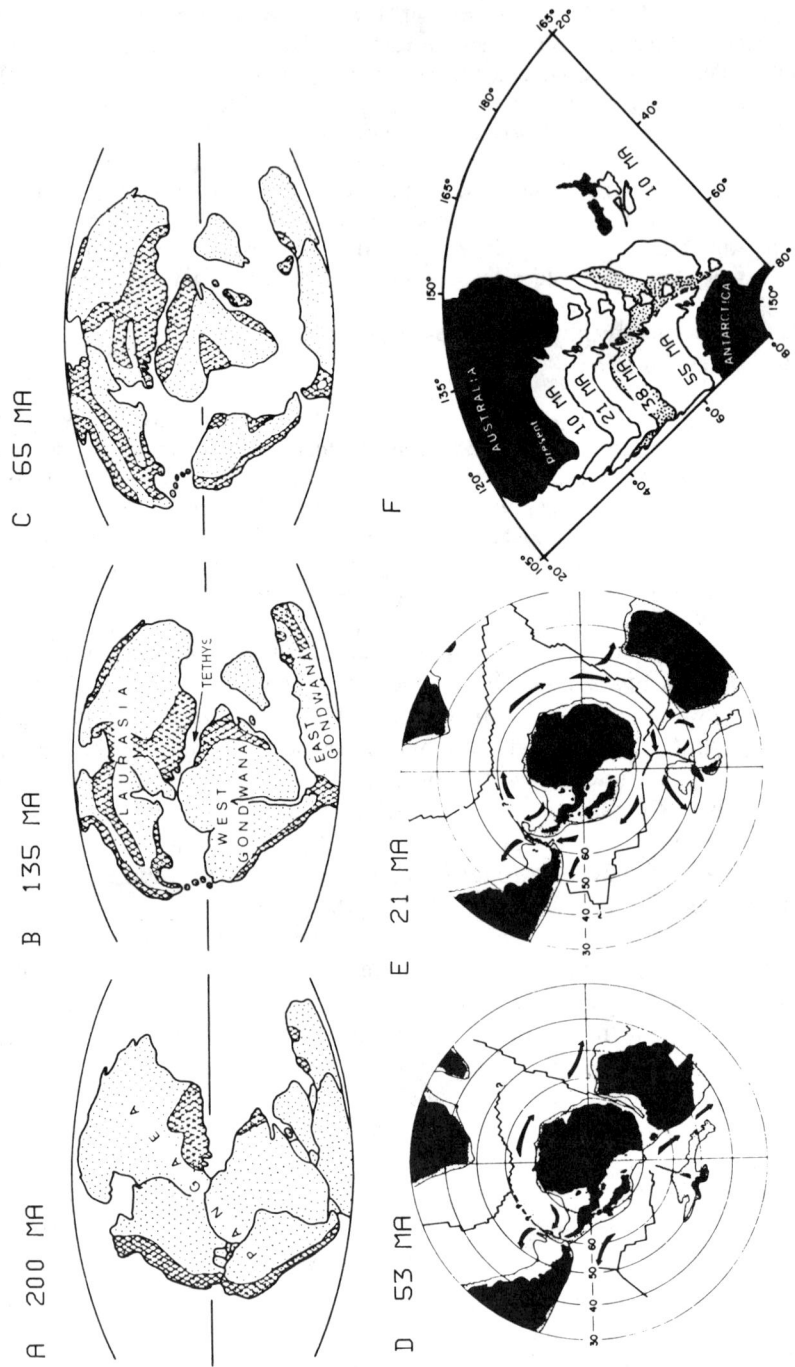

the more archaic elements (Keast 1973). The trans-Antarctic relationships have been stressed by early workers such as **Joseph Hooker,** who discovered a **circumpolar Antarctic flora** in the 19th century, and **Carl Skottsberg,** who was convinced of the existence of an **Antarctic Tertiary flora** as a counterpart to the Arctic Tertiary flora. Trans-Antarctic relationships are present in many groups and developed when the Gondwana fragments were directly connected with each other. The history of research on trans-Antarctic relationships with direct reference to **continental drift** has been reported by Brundin (1966, 1972, 1981) and Keast (1973, 1983). Through his work on Southern Hemispheric chironomid midges, Brundin developed the concept of **phylogenetic biogeography** (review in Brundin 1988). "Among the problems raised by the distribution of plants and animals in the southern hemisphere there is none which takes a more central position and is more stimulating to the imagination than the problem of transantarctic relationships. We have before us the broken circle of southern lands—southern South America, South Africa, Tasmania– Australia, New Zealand—separated by wide stretches of ocean, but populated by a biota containing numerous groups whose strongly disjunct elements are more closely related to one another than to any other group, and in the centre of the scene we are faced with the dormant Antarctic continent, hiding its secrets beneath a mighty ice cap" (Brundin 1966, p. 46).

After the fragmentation, Australia and New Zealand drifted northward. Hommersand (1986) pointed out that the main biogeographic effect of the separation of Australia from Antarctica was to bring together the high-latitude, temperate Pacific Ocean, and southern Indian Ocean marine algal floras that had evolved on the east and west coasts of Australia, respectively (Figs. 5.2D,E). Australia arrived at its present position, adjacent to Asia, as recently as 10 Ma (Fig. 5.2F). Up to 35 Ma, a land bridge connected western Antarctica to South America. It was only when this land bridge was turned into an island chain in the newly created **Drake Passage** and a deep seaway separated Tasmania–New Zealand from eastern Antarctica (26 Ma) that the **Antarctic Circumpolar Current** began to flow (Fig. 5.2E).

The taxa distributed by this cool-water current have been termed **neoaustral** migrants (Stevens 1980), while **paleoaustral** marine faunas refer to the Paleozoic Carbonian/Permian glaciation of the Southern Hemispheric continents, the second Great Ice Age before the Tertiary/Pleistocene Great Ice Age (Fig. 2.1A).

Fig. 5.2 Paleoceanography of the Southern Hemisphere. **(A– C)** Fragmentation of Pangaea, and later of Laurasia and Gondwana from the Mesozoic to the early Tertiary. **(D)** Early Eocene (53 Ma): New Zealand well separated from Antarctica; beginning of northward drift of Australia. Spreading ridges and connecting fracture zones shown as jagged lines. Arrows indicate suggested bottom-water circulation. **(E)** Earliest Miocene (21 Ma): Australia well separated from Antarctica, and the Drake passage open between South America and Antarctica. This established an open corridor for the Antarctic Circumpolar Current and thus the birth of the Antarctic ecosystem. **(F)** Successive positions of Australia during the Cenozoic. (A–C from Pielou 1979; D–F from Kennett 1982.)

Substantial Antarctic **sea ice** probably began to form during the global temperature decline near the **Eocene/Oligocene boundary,** around 36 Ma, at the same time as the deep sea cooled (Kennett 1977). With the earth cooling and glaciers forming on Antarctica during the Oligocene (36–23 Ma), a first **maximum glaciation** with glaciers at sea level occurred on the **eastern Antarctic mainland** in the **Miocene,** around 14 Ma (Kennett 1977; Grant-Mackie 1979; Mercer 1983).

The **ice cap on the eastern Antarctic** mainland formed in the Miocene and remained a semipermanent feature; at 5 Ma it had an ice volume greater than that of today. For about 10 million years between the middle and the early Pliocene, there was a period of **unipolar glaciation,** during which Antarctica was glaciated while the Arctic was essentially ice-free (Flohn 1984). Ice-sheet development on the Northern Hemisphere began about 2.5–3 Ma. Thus the marine Antarctic has a much longer history as a cold-water habitat than does the marine Arctic.

In the Miocene, the mountainous **western Antarctic archipelago** was covered by sea ice which, in the terminal Miocene event 6.5–5 Ma, built up suddenly and became grounded as an ice sheet on the shallow sea floor (Brain 1984). This coincided with a global sea-level drop and the Messinian crisis in the Mediterranean. **Western Antarctica** may have largely melted during the last interglacial age, 125,000 years ago (Mercer 1983).

Today 98% of Antarctica is covered by an ice cap up to 4.5 km thick, originating from precipitation, and representing 80% of the world's freshwater reserves. The input due to snowfall is balanced by losses along the Antarctic coasts, to which the ice slowly moves. Almost half of the circum-Antarctic coastline is bordered by floating shelf ice up to 400 m thick that eventually breaks up to create flat icebergs typical of the Antarctic.

Antarctic innovations in the early Tertiary. Early relatives of modern cool-water taxa at mid-latitudes originated in the early Tertiary at high latitudes, similar to Northern Hemispheric taxa that came from near-polar latitudes. These high-latitude origins can be seen in the discovery of fossil remains of 11 marine genera of Mollusca, Echinodermata, and Arthropoda in upper Eocene rocks on Seymour Island near the tip of the Palmer Archipelago, Antarctica (Zinsmeister and Feldmann 1984). In some cases early relatives in the Tertiary "holding tank" around Antarctica preceded their mid-latitude descendants by as much as 40 million years. The newly evolved taxa, the so-called "Weddellian species", dispersed to lower latitudes when these became cooler. Tierra del Fuego in southern South America was the appropriate first receiving area for Antarctic cool-water taxa in the Miocene. New Zealand and Australia, on their movement to the north, lost many of their Weddellian taxa, although some species adapted to the warmer conditions.

5.2 COMPARISON OF THE TWO POLAR REGIONS AND BIPOLAR DISTRIBUTED SEAWEEDS

The Circumpolar Antarctic Current dominates the area covered by the marine-benthic biogeographical regions of the Southern Hemisphere at cold temperate and polar latitudes. Two points contrasting to the Northern Hemisphere emerge:

(1) The prevailing **ocean current** has greater significance for the distribution of seaweeds;
(2) The seaweed floras inhabiting the temperate and polar regions show a more pronounced geographical isolation.

Referring to point 1, the **West Wind Drift** favors the distribution of algae suited for **long-distance transport;** examples are the floatable *Macrocystis pyrifera* and *Durvillaea antarctica* and the few plant and animal species attached to them. Holdfast-inhabiting animal species of cold temperate (Sub-Antarctic) Macquarie Island also occur 5000 km west at Kerguélen Island (Edgar 1987). The substantial percentage of southern South American cold temperate seaweed species with a wide Sub-Antarctic distribution points to *M. pyrifera* or *D. antarctica* as the floatable carrier. In the **Northern Hemisphere,** migrations at higher latitudes could mainly proceed along coastlines bridging oceanic gaps. For example, the rocky coasts of Iceland, Greenland, Labrador, and Newfoundland could be used as **stepping stones** for seaweed migrations in the North Atlantic. Such coastal migrations might have occurred as the sum of **short-distance transports.**

Referring to point 2, in the Northern Hemisphere the polar, temperate, and tropical seaweed floras have a better longitudinal coastal contact (Figs. 1.5 and 1.6) and could mix better when shifted during the Pleistocene. In contrast, corresponding regions of the Southern Hemisphere have a more island-like (Antarctica and the sub-Antarctic islands) or peninsular (southern ends of South America, South Africa, Australia, and New Zealand) character resulting in a higher endemism in these areas.

Amphiequatorial seaweed genera and species. In view of the basic differences between the cold temperate and polar seaweed floras of the Northern and Southern hemispheres, the algal taxa with an **amphiequatorial** distribution pattern represent the exceptions. Such a pattern may have arisen for two reasons. First, warm-water ancestors may have sent off cold-adapted species early in the Tertiary to the Northern and Southern hemispheres. This may well explain the cosmopolitan nature of genera such as *Ulva, Enteromorpha,* and *Ceramium,* which have a continuous distribution from the Arctic region through the tropics to the Antarctic. Alternatively, a species that evolved in a temperate region of one of the hemispheres may have crossed the equator in one of the cold periods of the Pleistocene. This probably occurred with *Macrocystis pyrifera* (Fig. 5.3), *Scytosiphon lomentaria, Petalonia fascia,* and several other species now exhibiting disjunct distributions (van den Hoek 1982a). These transequatorial migrations occurred at the east sides of the oceans, where the extension of the tropical regions was minimal (Figs. 1.5 and 1.6). *Macrocystis pyrifera* probably crossed the equator along the Pacific coast of Central America (present surface-water temperature about 28°C), and migrated southward against the northward-flowing Peru Current by using local gyres and eddies (Nicholson 1979). *Laminaria ochroleuca* crossed the equator on the West African coast (now 25°C

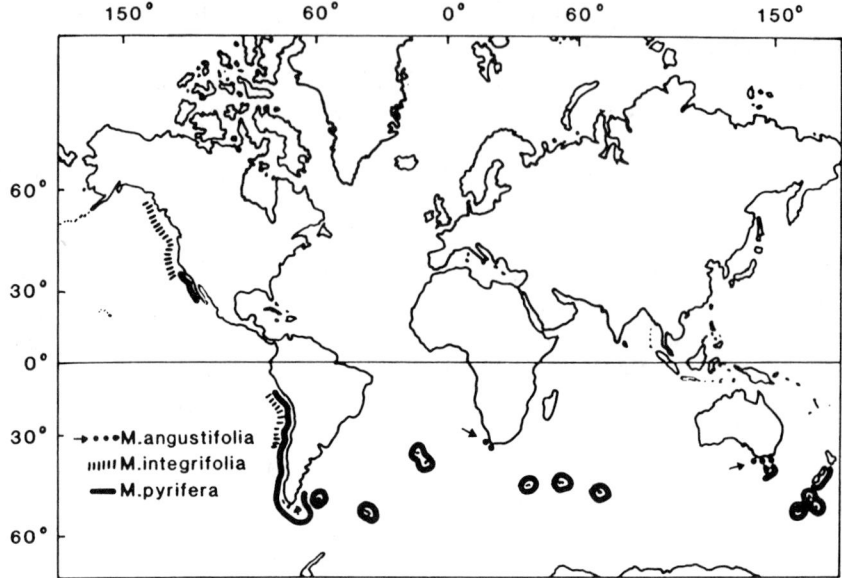

Fig. 5.3 Amphiequatorial distribution of the kelp genus *Macrocystis*. (From Foster and Schiel 1985; after Womersley 1954.)

surface water). Since the families to which these two species belong, the Lessoniaceae and Laminariaceae, respectively, are better represented in the Northern than in the Southern Hemisphere, it seems likely that the migration occurred from north to south in both cases. The only genera of the **Laminariales in the Southern Hemisphere** are *Macrocystis, Lessonia, Ecklonia, Eisenia,* and *Laminaria*. Crossing the equator has been a rare event and was facilitated by the presence of "stepping stones" in the form of local upwelling areas of cold water along Baja California and West Africa. It was also facilitated by the further constriction of the tropical regions during the Pleistocene and by glacial temperature drops as high as 8°C.

Gondwanan origin of the Fucales. Most of the genera of the Fucales are found in the Southern Hemisphere (Table 5.1). In the **Northern Hemisphere** there are only four endemic genera in the Fucaceae (all intertidal), five in the Cystoseiraceae, and one genus each in both the Sargassaceae and Himanthaliaceae (Table 5.1). Possibly, the first fucalean algae evolved on southern Australasian shores (Clayton 1984). The Northern Hemisphere may have received its few temperate fucoids during Pleistocene glaciations, when a land bridge was established between Australia and Indonesia and seawater temperatures dropped by 6–8°C in tropical New Guinea. According to this view, the speciation within the Fucales must be regarded as a recent and rapid process.

***Primitive traits in* Hormosira banksii.** This alga (Fig. 5.11A) exhibits several primitive characters: (1) the **oogonium** contains four instead of one gamete, as in all other families of the Fucales except for the Fucaceae; (2) the **conceptacles** (cavities containing reproductive structures) are not confined to special parts of the thallus (**recepta-**

Table 5.1 Distribution of the genera of the Fucales, Notheiales, Ascoseirales and Durvillaeales. ## and genus name in brackets: genus occurs on northern hemisphere only; # genus occurs on the northern and southern hemispheres; X occurrence of genus (partially only in limited areas). Northern hemisphere locations in brackets.

ORDER/Family	Genus	Number of species	Southern Australia	New Zealand	Other distribution
FUCALES					
1. Hormosiraceae	Hormosira	1	X	X	-
2. Seirococcaceae	Axillariella	1	-	-	southern Africa
	Cystosphaera	1	-	-	Antarctic
	Marginariella	2	-	X	-
	Phyllospora	1	X	-	-
	Scytothalia	1	X	-	-
	Seirococcus	1	X	-	-
3. Fucaceae	##(Ascophyllum)	1	-	-	(N Atl.)
	##(Fucus)	7	-	-	(N Pac., N Atl.)
	##(Hesperophycus)	1	-	-	(NE Pac.)
	##(Pelvetia)	3	-	-	(N Atl., N Pac.)
	##(Pelvetiopsis)	2	-	-	(NE Pac.)
	Xiphophora	2	X	X	-
4. Himanthaliaceae	##(Himanthalia)	1	-	-	(NE Atl.)
5. Sargassaceae	Antophycus	1	-	-	southern Africa
	Carpophyllum	4	-	X	southern Africa
	##(Hizikia)	1	-	-	(Japan)
	Oerstedtia	1	-	-	southern Africa
	#Sargassum	150	X	X	tropical-warm temperate
	#Turbinaria	20	-	X	tropical-warm temperate
6. Cystoseiraceae	##(Acystis)	1	-	-	(Arabian Sea)
	Acrocarpia	2	X	-	-
	#Bifurcaria	3	-	-	south. Afr., Galapagos Is.
	Bifurcariopsis	1	-	-	southern Africa
	Carpoglossum	1	X	-	-
	Caulocystis	2	X	-	-
	##(Coccophora)	1	-	-	(Japan)
	Cystophora	25	X	X	-
	#Cystoseira	40	X	-	warm temperate-tropical
	##(Halidrys)	2	-	-	(N.Pac., N.Atl.)
	Hormophysa	1	X	-	Indian Ocean
	Landsburgia	2	-	X	-
	##(Myagropsis)	1	-	-	(Japan, Korea)
	Myriodesma	8	X	-	-
	Platythalia	2	X	-	-
	Scaberia	1	X	-	-
	##(Stolonophora)	1	-	-	(Guadalupe I.)
NOTHEIALES	Notheia	1	X	X	-
DURVILLAEALES	Durvillaea	3	X	X	Subantarctic, South Am.
ASCOSEIRALES	Ascoseira	1	-	-	Antarctic-Subantarctic

Compiled after Nizamuddin 1962, 1970, Jensen 1974, Abbott and Hollenberg 1976, Clayton 1984, Yoshida et al. 1985, South and Tittley 1986, Womersley 1987.

cles); and (3) growth is brought about by a group of three-sided **apical cells** instead of a three- or four-sided apical cell as in all other fucalean genera (Clayton 1984; Clayton et al. 1985).

Durvillaeales. The genus *Durvillaea,* although displaying laminarian characteristics, has been placed in the order Fucales to which it bears resemblances that probably represent convergent traits (Clayton 1988). It has **conceptacles** containing antheridia and oogonia but a different mode of development, it has **active growth in apical regions** of the fronds, although an apical cell is absent and growth is basically intercalary (Clayton 1984; Clayton et al. 1987). K. M. Petrov (1965, 1967) placed the genus in a separate order, the **Durvillaeales** (Nizamuddin 1968; Hay 1979*a*). *D. antarctica* has a circumpolar distribution, is found in the sub-Antarctic region, and also occurs at South Georgia in the Antarctic region (Hay 1988). *D. antarctica* can float vast distances due to honeycomblike structures inside its blades (Hay 1979*b*). There are three other, nonfloatable species in the genus (p. 274).

Ascoseirales. This order also contains one genus, this time with only one species, *Ascoseira mirabilis* (Fig. 5.4). The large alga with **intercalary growth** has laminarian similarities. Since *Ascoseira* forms its reproductive cells in **conceptacles,** it was initially assigned to the Fucales but was later placed in the separate order **Ascoseirales** (Ju. E. Petrov 1963; Moe and Henry 1982). It was found that the conceptacles contain unilocular (i.e. with one chamber) sporangia, with similarities to the Desmarestiales. In the sporangia, eight sterile cells are produced, possibly remains of the reduced gametophyte, and eight biflagellate isogametes (Moe and Henry 1982; Clayton and Wiencke 1986; Clayton 1987, 1988; Wiencke 1988).

Fucalean traits in other brown algal orders. The Southern Hemisphere harbors a few seaweed taxa that may reflect the early "experimental" phase leading to the evolution of the modern Fucales; they are difficult to attach to particular brown algal orders. *Notheia anomala,* a branched and somewhat bushy alga growing almost exclusively on intertidal *Hormosira banksii* and thus in southern Australia and New Zealand, and *Splachnidium rugosum,* a cartilaginous, Southern Hemispheric intertidal alga, have **conceptacles** containing unilocular sporangia. In each of the conceptacles of *Notheia anomala* there are male gametangia with 64 spermatozoids and female gametangia with eight larger, motile gametes; sexual attraction and fertilization were observed (Clayton 1984; Gibson and Clayton 1987). It has been suggested that *Notheia* should be classified in its own family in the Fucales (Gibson and Clayton 1987) or in a separate order Notheiales (Womersley 1984–1987).The genera *Splachnidium* and *Scytothamnus* have been classified in their own families in the Chordariales (Womersley 1984-1987). The sterile counterparts of conceptacles and their possible phylogenetic forerunners, **cryptostomata** (cavities containing sterile hairs), are present in *Scytothamnus* species, *Adenocystis utricularis* (Dictyosiphonales), and the Ectocarpales *sensu lato* (Clayton 1985). These species also exhibit apical growth, and a central apical cell is present in *Splachnidium rugosum* (Clayton and Shankly 1987). *Scytothamnus australis* occurs in southern Australia, New Zealand, and Chile, whereas the region of *S. fasciculatus* is more southerly, including southeastern Tasmania, the southern part of New Zealand, Chile, Patagonia, and several sub-Antarctic islands; *S. hirsutus* is found on the Falkland Islands (= *Dictyosiphon hirsutus* according to Pedersen 1984). The life history of *A. utricularis* and *Scytothamnus* species is of a primitive phaeophyte pattern, with a macroscopic sporophyte forming unilocular sporangia and alternating with microthalloid, dioecious gametophytes (Müller 1984; Clayton 1986).

Possible evolution and radiation of the Fucales. The oogamous Fucales probably descended from **anisogamous ancestors** with apical growth and conceptacles containing sporangia. The early fucoids would have had several eggs containing an equal number of small vegetative cells per oogonium. Possibly, evolution took place in the eulittoral zone where the conceptacle provided protection for the gametophyte, which subsequently became more and more reduced, as in the "protofucalean," possibly convergent genera *Ascoseira* and *Notheia* (Nizamuddin and Womersley 1960; Clayton 1984). The sublittoral zone may have been colonized by the Fucales at a later stage. The general impression is that too many species have died out to leave a clear picture of the evolutionary links between the Fucales and genera related to the ancestors of that order.

5.3 ANTARCTIC REGION

Literature: Clayton (1987, 1988); Clayton and Wiencke (1986); DeLaca and Lipps (1976); Delépine (1966); Delépine et al. (1966); Dell (1972); Furmanczyk and Zielinski (1982); Hedgpeth (1970); Heywood and Whitaker (1984); Knox (1960, 1970); Lamb and Zimmermann (1977); Laws (1984); Moe (1985); Moe and Henry (1982); Moe and Silva (1977, 1981); Neushul (1965b, 1968); Papenfuss (1964a,b); Prescott (1979); Skottsberg (1907, 1941a, 1964); South (1979); Tranter (1982); Wiencke (1988); Wiencke and tom Dieck (1989, 1990); Wynne (1982); Zaneveld (1966a,b, 1968, 1969).

The Antarctic region covers the continental coast of Antarctica and the islands situated south of the Antarctic Convergence, namely, the South Shetland, South Orkney, South Sandwich, South Georgia, and Bouvet Islands (Figs. 1.5, 1.6, and 5.1 *left;* see Section 5.4 in regard to Heard Island). The water temperatures range annually between $-1.8°C$ and $1°C$ rising to a maximum of $3.7°C$ on the coast of South Georgia. South Georgia has also been included in the sub-Antarctic region on the basis of its terrestrial flora (Pickard and Seppelt 1984).

History of seaweed investigation. C. Skottsberg was a pioneer of phycological research in the Antarctic. He participated in the Swedish Antarctic Expedition of 1901–1903 but lost the major part of his collections when the expedition ship went down in the ice. More of the early history of Antarctic biological research has been reported by Dell (1972). Neushul (1965b) and Delépine (1966) investigated seaweed vegetation by diving in western Antarctica. Most seaweed collections have come from the coasts of the Antarctic Peninsula (western Antarctica), south of South America. Zaneveld worked on three expeditions in the mid-1960s in the Ross Sea area, mostly camping along the coast of Victoria Land and on most of the Ross Sea islands. He collected the algae primarily by diving under ice 4 m thick and less frequently by trawling from an icebreaker (Zaneveld, personal communication). **Circum-Antarctic distribution** applies to a species that occurs around the Antarctic continent. No seaweed records are available for the coastline between 72°W and 173°E, or between

50°W and 40°E. Perhaps no benthic seaweed vegetation exists here because the coast reaches into the sea as far as the shelf ice (Neushul 1968). Recent collections from western Antarctica in the 1980s by Moe, Wiencke, Clayton, and D. Müller supplied phycological laboratories throughout the world with unialgal cultures of various Antarctic seaweeds.

The seaweed flora of the Antarctic region consists of about 100 species, of which about one-third are endemic to the Antarctic region, about one-half endemic to the Antarctic Peninsula, and 20–40% have been reported for eastern Antarctica, mainly in the Ross Sea (Skottsberg 1964; Zaneveld 1966a,b; Neushul 1968; South 1978; Moe 1985). Thus there are about as many seaweed species on the Antarctic Peninsula as on the coast of eastern Greenland (109 species) or on Novaya Zemlya (120 species). However, the endemism is much higher in the Antarctic than in the Arctic (5% in the latter) due to its longer history as a cold-water habitat, and to its lack of coastal connections to the cold temperate regions. Several Antarctic endemic seaweeds grow optimally up to 5°C but do not survive at 13–15°C in the laboratory.

It is a characteristic trait of the Antarctic sublittoral seaweed vegetation that the **Laminariales** are absent and that the dominating canopy algae have been taken over by the **Desmarestiales.** The genus *Desmarestia* exhibits a circum-Antarctic distribution pattern and colonizes, with several species, the sublittoral zone down to a 40 m depth.

A giant representative of the Desmarestiales, with fronds up to 10 m long and 1 m wide, is the circum-Antarctic *Himantothallus grandifolius* (Fig. 5.4); it grows below a 5 m depth and together with species of *Desmarestia,* forms a dense vegetation down to 35 m deep. This alga was formerly known as *Phyllogigas* or *Phaeoglossum* and was regarded as a representative of the Laminariales. However, the early growth of the sporophyte of the monotypic genus *Himantothallus* was shown to occur by trichothallic growth of a uniseriate thallus axis (Moe and Silva 1977, 1981), a feature of the Desmarestiales. The gametophytes are also typical of this order (Wiencke and Clayton 1990). The flat thallus, similar to laminarian blades, grows from branches formed by the uniseriate axis (Fig. 5.4). As a consequence of this discovery, the "last member of the Laminariales in the Antarctic region" was lost. Obviously, a thallus consisting of holdfast, stipe, and blade represents an optimal, basic growth form for a sublittoral canopy alga. This form most likely evolved independently several times, as in the several species of *Durvillaea* (order **Durvillaeales**), and in the monotypic genus *Ascoseira mirabilis* (order **Ascoseirales**).

A further representative of the **Desmarestiales** endemic to the Antarctic, besides *Himantothallus,* is the much smaller *Phaeurus antarcticus* (Fig. 5.4), which occurs from tide pools to 10 m deep in the sublittoral zone; the development in culture has been studied by Clayton and Wiencke (1990). The hypothesis that the Desmarestiales originated in the Antarctic or at least the

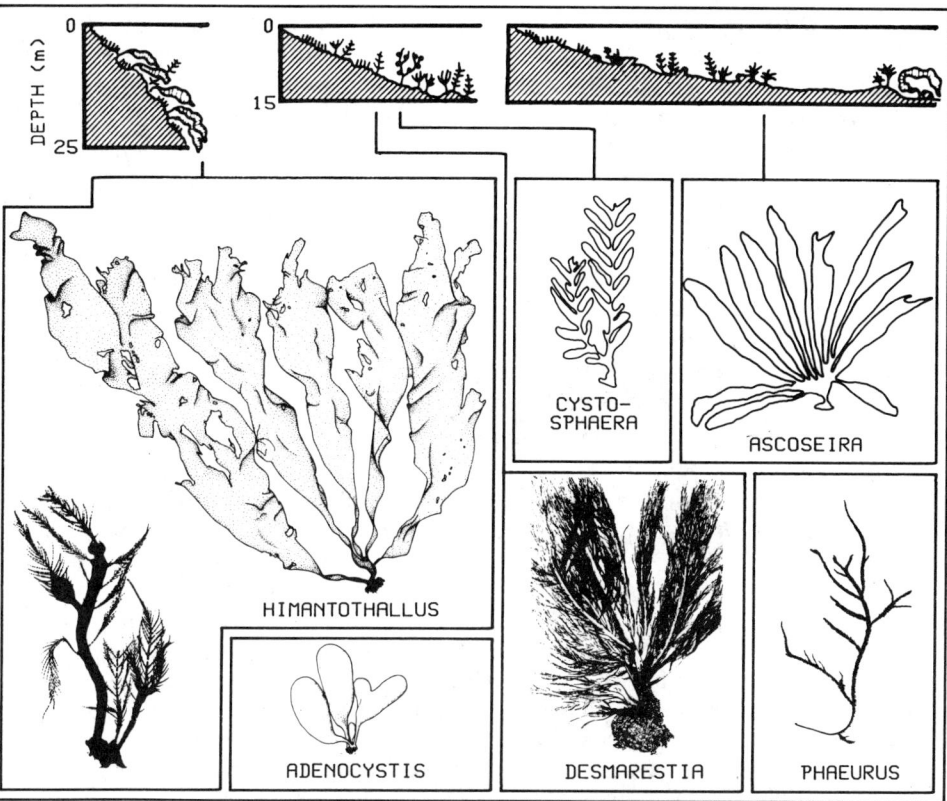

Fig. 5.4 Brown algae of western Antarctica and their vertical distribution on diving transects at the South Shetland Islands. Abbreviation in legend: Ant. P. = Antarctic Peninsula. **Details on species names and dimensions: Desmarestiales:** *Himantothallus grandifolius*(× 0.02; up to 10 m long; left: a 3 mm long, young sporophyte; circum-Antarctic, South Georgia); *Desmarestia anceps* (× 0.05; only Ant. P.; *Phaeurus antarcticus* (× 0.8, only Ant. P.). **Fucales:** *Cystosphaera jacquinotii* (× 0.08; only Ant. P.). **Ascoseirales:** *Ascoseira mirabilis* (× 0.03, Ant. P., South Georgia, South Sandwich Islands). **Punctariales:** *Adenocystis utricularis* (× 0.8, Ant. P., South Shetland Islands, southern South America, New Zealand). (Vegetation transects from Neushul 1965b; drawings from Lamb and Zimmermann 1977; *Ascoseira* and *Cystosphaera* simplified; young sporophyte of *H. grandifolius* from Moe and Silva 1977.)

Southern Hemisphere is supported by the fact that they are well represented and replace the Laminariales as canopy algae in the Antarctic region. Drops in seawater temperature during the Pleistocene were only sufficient to allow the warm temperate kelps such as *Macrocystis pyrifera* and *Laminaria ochroleuca* to cross the equator. On the other hand, due to their warm temperate affinities they could not penetrate the Arctic or Antarctic regions. The typical cold temperate species of the Laminariales of the Northern Hemisphere that are able to live in the Arctic, could not cross the equator, nor could the reverse crossing be achieved by the typical Antarctic species of Desmarestiales. This example explains why the two polar regions harbor fundamentally different seaweed floras.

The **Fucales** is represented by the Antarctic Peninsula endemic species *Cystosphaera jacquinotii* (Fig. 5.4). The alga grows at a 5–10 m depth and has, due to its air bladders, been found mainly as drift. *Ascoseira mirabilis* (Fig. 5.4), another large seaweed species, is endemic to western Antarctica and grows from the upper sublittoral downwards (C. Wiencke, personal communication). Its massive holdfast, divided blades, and especially its intercalary growth zone resemble the features of the Laminariales or the Durvillaeales. *Ascoseira mirabilis* bears possibly convergent relationships to the Fucales and constitutes the order **Ascoseirales** (see Section 5.2).

Several species of Antarctic **red algae** (examples in Fig. 5.5) are large (e.g., *Iridaea obovata*, which may attain a length of 1 m). The **green algae** are represented, for example, by widespread or Antarctic-endemic species of the cosmopolitan genera *Urospora*, *Monostroma*, or *Enteromorpha*, and by *Lambia antarctica*, a representative of the order Caulerpales in Antarctica (Delépine 1967).

Zonation on the Antarctic Peninsula. As in the northern Arctic region, a supra- and eulittoral algal vegetation is sparse in the Antarctic because of periodic ice cover and the effects of abrasion by drift ice. At King George Island the green alga *Prasiola* grows in the supralittoral zone. In the eulittoral, particularly in crevices and protected tide pools, one finds the green algae *Urospora, Ulothrix, Enteromorpha, Spongomorpha,* and *Monostroma;* the red alga *Porphyra;* and the brown alga *Adenocystis utricularis* (Fig. 5.4). Occurring in the lower eulittoral are the red algae *Palmaria decipiens* (= *Leptosomia simplex;* Fig. 5.5F) and *Iridaea obovata* (C. Wiencke and A. Peters, personal communications). Wave-exposed localities and the upper sublittoral down to 5 m deep may be devoid of erect seaweeds because of ice scouring. These areas are covered with crustose coralline algae (species of *Lithophyllum* and *Lithothamnion*). A luxuriant sublittoral vegetation of brown and red algae (Figs. 5.4 and 5.5) begins mainly below a 5 m depth. Divers have noticed the scouring effects of icebergs and their fragments on the rock surface down to 20 m deep. Algal thalli that were scoured from the substratum by ice may gather on the shore, freeze, and be transported by floating ice for long distances in the open sea. It seems unlikely that seaweeds that were collected by dredging at depths of several hundred meters had grown at these sites. These seaweeds had been transported by drift ice that melted so that the enclosed seaweeds sank to the bottom, where they possibly retained a fresh appearance for a long time due to the low water temperature (Skottsberg 1964; Neushul 1965*b*). As in other parts of the world in the sublittoral zone, perennial algae predominate, storing reserve carbohydrates during the high-light season. For example, one might expect storage to occur in the midribs of the red alga *Neuroglossum ligulatum* (= *Myriogramme mangini;* Fig. 5.5I) so that it can survive the low-light season and resume growth, possibly even in darkness.

The seaweed vegetation of **eastern Antarctica,** on the coast of Victoria Land and north of the Ross Ice Shelf, which represents the permanently frozen

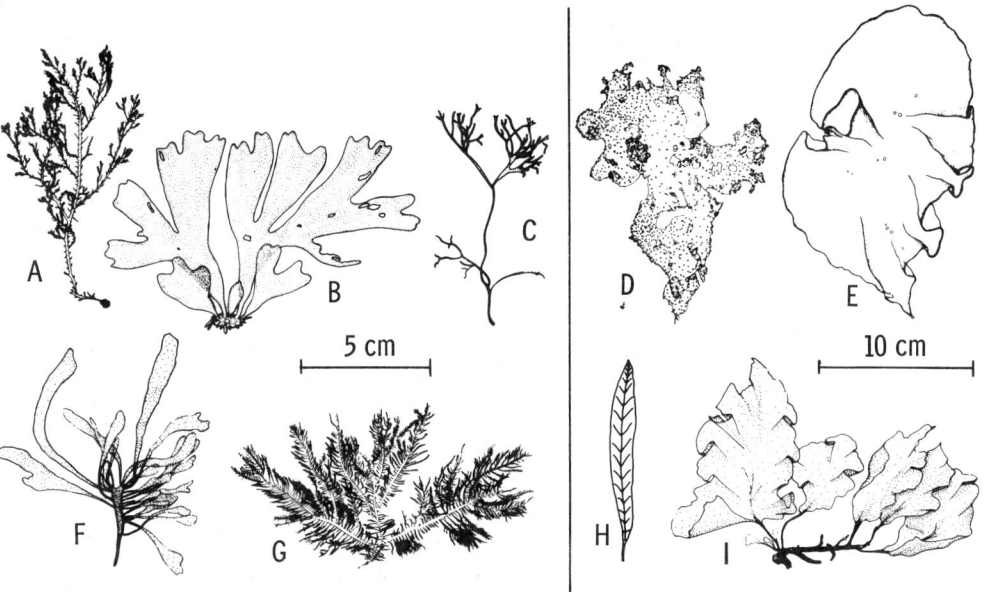

Fig. 5.5 Red algae of western Antarctica. Abbreviation in legend: Ant. P. = Antarctic Peninsula. **Nemaliales: (A)** *Delisea pulchra* (15–40 m deep; Ant. P., South Georgia, Kerguélen, Heard and Macquarie Islands, southern Australia). **Cryptonemiales: (B)** *Kallymenia antarctica* (5–33 m deep; Ant. P., Falkland Islands, Adélie coast of eastern Antarctica). **Gigartinales: (C)** *Gymnogongrus antarcticus* (in tide pools and the upper sublittoral zone; circum-Antarctic). **(D)** *Gigartina papillosa* (0-20 m deep; Ant. P., Falkland Islands, Tierra del Fuego, Kerguelen Islands, Tristan da Cunha). **(E)** *Iridaea obovata* (0–20 m deep; circum-Antarctic, sub-Antarctic islands). **Rhodymeniales: (F)** *Palmaria decipiens* (= *Leptosomia simplex*, see Ricker 1987; in tide pools and in the upper sublittoral zone; circum-Antarctic). **Ceramiales: (G)** *Georgiella confluens* (0–10 m deep; Ant. P., South Orkney, South Georgia, and Kerguélen Islands). **(H)** *Delesseria lancifolia* –(upper sublittoral zone; Ant. P., Terra del Fuego, Falkland Islands, sub-Antarctic islands south of New Zealand). **(I)** *Neuroglossum ligulatum* (= *Myriogramme mangini*, see Wynne 1982; 0–33m deep; Ant. P. and South Shetland Islands). (Drawings from Lamb and Zimmermann 1977.)

southern part of the Ross Sea, does not seem to harbor as many species as western Antarctica. The diving investigations of Zaneveld (1966*a,b*) indicate that larger brown algae such as *Himantothallus grandifolius* reach their southern limit at 76°S. The southernmost seaweed species, still present on Ross Island near the Ross Ice Shelf, are the red algae *Iridaea obovata, Phyllophora antarctica,* and *Hildenbrandia lecannellieri* and the green alga *Monostroma hariotii*. At this southern limit of seaweed vegetation the sea is annually covered by 2 m of ice for 10 months. These conditions are similar to those at the northern limit of the seaweed vegetation in the Arctic region. Zaneveld (1968) found by direct subice observations that at least *P. Antarc-*

tica and *I. obovata* continue to grow from reserves during the period of total darkness.

5.4 SUB-ANTARCTIC (COLD TEMPERATE) ISLANDS REGION

Literature: Knox (1960); Papenfuss (1964a); Skottsberg (1941a). **Kerguélen:** Délepine (1963, 1966). **Snares, Auckland, Campbell, Bounties, Antipodes:** C. H. Hay et al. (1985); Knox (1975). **Macquarie Island:** Kenny and Haysom (1962); Knox (1975); Ricker (1987).

The sub-Antarctic islands represent a fourth cold temperate region of the Southern Hemisphere (Figs. 1.5-1.7). The region encloses, from west to east in Fig. 5.1 (*left*), the island groups of Prince Edward (Prince Edward and Marion Islands), Crozet, Kerguélen, Heard (because of its position south of the Antarctic Convergence, also regarded as part of the Antarctic region), and island groups south of New Zealand including Macquarie, Auckland, Campbell, the Bounties, and the Antipodes. Briggs (1974) put the island groups south of New Zealand, except for Macquarie Island, in the cold temperate southern New Zealand region because of their rather close faunistic relationships to this latter region. One should realize that the term "sub-Antarctic region" has been used in different ways, for example, by Hedgpeth (1969) to cover the sub-Antarctic islands listed above, and Tierra del Fuego and Tristan da Cunha. South Georgia has been included in the sub-Antarctic region by terrestrial plant phytogeographers (Pickard and Seppelt 1984)

The surface water temperatures along the coasts of the sub-Antarctic islands amount to 3–11°C in winter and 5–14°C in summer, depending on the particular geographical position. According to a catalog by Papenfuss (1964a), the seaweed flora of the Antarctic and the sub-Antarctic islands regions contains 550 species and 190 genera, half of the genera being endemic to the area covered by the two regions.

Some of the characteristic seaweed species of the Antarctic region are found in the sub-Antarctic islands region; examples are the brown alga *Adenocystis utricularis* (which extends to southern South America, New Zealand, and Tasmania) and the red alga *Iridaea obovata*. Absent are the brown algae *Himantothallus grandifolius*, *Phaeurus antarcticus*, *Ascoseira mirabilis*, and the red alga *Palmaria decipiens*. Instead the Circumpolar Antarctic Current has stranded numerous seaweed species on the various islands around the southern ocean. As a group these seaweeds are responsible for showing the similarities existing among the seaweed floras throughout the cold temperate regions of the Southern Hemisphere.

Representatives of the **Laminariales**, *Macrocystis pyrifera* (Figs. 2.80A, and 5.3; South America to New Zealand, and California), and *Lessonia flavicans* (Fig. 5.6C; genus represented in southern Australia and New Zealand by other species), are found on the sub-Antarctic islands. A new species of *Macrocystis, M. laevis,* is a case of island endemism; it is found on Marion Island, one of the two entities forming the Prince Edwards Islands 1000 km west of Crozet Island and 2100 km southeast of Cape Town (C. H. Hay 1986). *M. laevis* has fleshy, smooth blades, sori on the blades forming the canopy; and it is otherwise similar to *M. pyrifera* in such areas as holdfast morphology. Marion Island is probably no older than 250,000 years and may have received its *Macrocystis* via the Antarctic Circumpolar Current from the much older (Precambrian to Lower Cretaceous) Falkland Islands, where smooth-bladed populations of *Macrocystis* were recorded on an outlying island (Skottsberg 1921–1923).

Durvillaea antarctica (Fig. 5.6F) is a widespread, large brown alga that occurs on the sub-Antarctic islands. It is found on the west coast of southern South America and southern New Zealand but not, as the name suggests, in the Antarctic region, except for South Georgia (C. H. Hay 1988). This alga grows, up to 10 m long, in a narrow, intermittent fringe along the lower eulittoral zone, at Kerguélen with a maximum standing crop of 220 kg fresh weight per square meter formed by 470 individuals (Lawrence 1986).

Zonation. In contrast to the Antarctic region, the intertidal zone of the sub-Antarctic islands (e.g., **Kerguélen** or **Crozet**) is well populated with seaweeds because ice damage is exceptional (Delépine 1963, 1966). Above the intertidal, in the supralittoral zone, the lichen *Verrucaria* and the green alga *Prasiola* are present. The upper eulittoral zone is dominated by *Porphyra* species, the green filamentous genera *Ulothrix* and *Urospora,* and the crustose red alga *Hildenbrandia lecannellieri.* In the mideulittoral zone one finds the brown alga *Adenocystis utricularis,* the red alga *Iridaea,* the green alga *Acrosiphonia pacifica,* and crustose coralline algae. In the lower eulittoral zone in such areas as the southern coasts of Chile and New Zealand, *Durvillaea antarctica* plays the same dominant role as a canopy alga. The sublittoral zone is occupied by a dense vegetation of *Macrocystis pyrifera.*

Sub-Antarctic island groups near New Zealand. To the south of New Zealand there are six wind-swept sub-Antarctic island groups: the **Snares** (48°S, 166°34E), **Auckland,** and **Campbell islands,** the **Bounties,** the **Antipodes,** and **Macquarie Island,** whose seaweeds are listed by C. H. Hay et al. (1985), except for Macquarie Island, which is covered by Ricker (1987). *Macrocystis pyrifera* and *Durvillaea antarctica* are present in all these island groups. *Lessonia* is also common, *L. flavicans* having been recorded for Auckland, Campbell, the Bounties and the Antipodes. A fucalean species growing in all these island groups except for the Bounties is *Xiphophora chondrophylla* (C. H. Hay et al. 1985). The Antipodes are of recent volcanic origin (0.2–0.5 Ma) and may harbor an endemic *Durvillaea* species somewhat similar in shape to *D. chathamensis.* In general, there is little evidence of island endemism in the New Zealand sub-Antarctic islands (C. H. Hay et al. 1985). Along the wave-lashed coasts of **Macquarie Island,** situated about halfway between New Zealand and Antarctica, the larger brown algae *Durvillaea antarctica* and *Macrocystis pyrifera* are widely distributed (Kenny and Haysom 1962; Ricker 1987). *Lessonia variegata* occurs here and in New Zealand.

5.5 SOUTH AMERICA

Literature: Entire area: Oliveira Filho (1981); Papenfuss (1964b); Santelices (1980). **Peru:** Acleto (1973); Dawson et al. (1964); Howe (1914). **Chile:** Dayton (1985b); Etcheverry (1960, 1986); Guiler (1959); Levring (1960); Santelices (1989); Santelices and Abbott (1978); Santelices et al. (1980); Skottsberg (1921–1923); Villouta and Santelices (1984); Westermeier and Rivera (1989); Wiencke and tom Dieck (1990). **Argentina, Uruguay:** Barrales and Lobban (1974); Kühnemann (1972); Skottsberg (1921–1923); Taylor (1938). **Brazil:** Baptista (1977); Coutinho and Seeliger (1984, 1986); Guimarães et al. (1981); Joly (1965); Kühnemann (1969); Ugadim (1973–1976); Yoneshigue (1985). **Pacific islands: Juan Fernandez Islands:** Etcheverry (1960); Levring (1941); Skottsberg (1941b). **South Atlantic islands: Falkland Islands:** Cotton (1915); Taylor (1938). **Tristan da Cunha:** Baardseth (1941). **Gough Island:** Chamberlain (1965), Chamberlain et al. (1985). **Ascension:** Price and John (1980). **Special groups:** *Lessonia:* Searles (1978); Villouta and Santelices (1986). *Macrocystis:* Barrales and Lobban (1974); Dayton (1985b); Neushul (1971).

The biogeographical boundaries of southwestern South America (and southern Africa) are determined primarily by the extent of the **northbound coldwater currents,** which are enforced by branches of the West Wind Drift. Along the west coast of South America, the **Peru Current** (Humboldt Current; Fig. 2.43) shifts the **tropical/warm temperate** boundary almost to the equator at the Gulf of Guayaquil.

The coast of Peru belongs to one of the major areas of **coastal upwelling** along the eastern boundaries of the oceans, where predominant equatorward winds and currents (Fig. 2.43) are part of the more or less stationary midoceanic atmospheric high-pressure systems (Barber and Smith 1981; Longhurst 1981). The other four major upwelling areas are off the coasts of California and Oregon (California current), southwestern Africa (Benguela current), northwestern Africa (Canaries Current), and Somali and Arabia. The southwest monsoon causes upwelling in the last case.

The tropical–warm temperate boundary is situated on the east coast of South America much farther to the south than on the west coast (Figs. 1.5–1.7). The **warm temperate/cold temperate** boundary has been suggested to occur on the west coast at Chiloé Island, latitude 42°S (Briggs 1974), although cold-water algae like *Durvillaea antarctica* and *Adenocystis utricularis* occur to central or northern Chile, respectively (A. Peters, personal communication).

One-third of the approximately 400 seaweed species inhabiting the warm and cold temperate regions of the Pacific coast of South America are **endemic** to these regions (Santelices 1980). Particularly rich in endemics are the coast of Peru near Callao and the southern end of South America within the latitudinal range of 53–55°S. Another one-third of the total species occur in the cold temperate South American sub-Antarctic islands regions. Examples include *Lessonia,* which represents the only genus of the Laminariales present in the Southern (but not in the Northern) Hemisphere, *Durvillaea* *antarctica* (also in New Zealand), and *Adenocystis utricularis* (also in New Zealand and in the sub-Antarctic and Antarctic regions).

Distribution of Lessonia. Since most of the Lessoniaceae are found in the North Pacific, *Lessonia* probably migrated to the Southern Hemisphere during the last ice age, or maybe earlier during the Miocene cooling. The related genus *Lessionopsis*, with the only species *L. littoralis* is found in the North Pacific. *Lessonia variegata* occurs in New Zealand, *L. corrugata* in Tasmania, and there are four *Lessonia* species in South America (Searles 1978; Villouta and Santelices 1986). The intertidal Pacific *L. nigrescens* (Fig. 5.6B), with its massive holdfast, occurs from central Peru to southern Chile, while the subtidal *L. trabeculata* (0.5–20 m depth; Fig. 5.6A), again with a massive holdfast, occurs from central Peru to central Chile. On the southern tip of South America, in the Pacific from 49°S, south to Cape Horn and on the Atlantic coast to 47–48°S, two species with unfused haptera occur, the deep-water *L. flavicans* (Fig. 5.6C; 2–20 m depth) and the shallow subtidal *L. vadosa* (0.5–2 m depth).

Of the total of 130 seaweed species with a **western South American**–sub-Antarctic distribution pattern, 60 species reach the Kerguélen Islands, and 28 species reach as far west as the Auckland Islands (Santelices 1980) situated south of New Zealand. This exemplifies the important role of the **West Wind Drift** as a carrier. Only 3% of the 400 seaweed species of temperate Pacific South America exhibit a **warm temperate–tropical** distribution. The cold **Peru Current** largely prevents tropical algae from migrating to the south, while more and more cold-water species drop out toward the north. Thus the **warm temperate western South American region** appears somewhat depauperate in seaweed diversity (van den Hoek 1984). In addition, the **El Niño** phenomenon (the unpredictable change of the prevailing wind direction), which results in the sudden absence of upwelling, enhances the somewhat unstable character of the region's temperature regime.

About 10% of the above-mentioned 400 temperate Pacific–South American species exhibit an **amphiequatorial** distribution pattern with a distribution gap in the tropics; examples are the brown algae *Macrocystis pyrifera* and *Macrocystis integrifolia* (Fig. 5.3). Since the two North Pacific species of *Macrocystis* both occur on the Pacific coast of South America, the migration to the south possibly occurred in the Pleistocene, and not earlier. *M. integrifolia* is found from northern Peru to central Chile, followed by *M. pyrifera* around the southern tip of South America (Neushul 1971). It is an open question as to why, along the western coasts of the Americas, the species nearest to the equator is *M. pyrifera* in the north but *M. integrifolia* in the south (Barrales and Lobban 1974, Dayton 1985*b*).

Interestingly, another member of the Laminariales, besides *Macrocystis* and *Lessonia*, must also have crossed the equator, namely, the genus *Eisenia* (Alariaceae). This is known from Pacific North America as *E. arborea*, which is very similar to the Peruvian species *E. cokeri*. A further species belonging to the genus *Eisenia* occurs in Japan.

Fig. 5.6 Brown algae on the Pacific coast of South America. **(A)** *Lessonia trabeculata* (sublittoral, central Chile). **(B)** Belt of *L. nigrescens* in the lower eulittoral zone on the coast of central Chile. Middle: *Durvillaea antarctica*. Top: *Iridaea laminariodes*. **(C)** *L. flavicans* in the sublittoral zone down to 20 m deep; southern tip of South America; also sub-Antarctic islands. **(D)** *Lessonia trabeculata,* underwater view. **(E)** *Durvillaea antarctica* in the lower eulittoral zone on the coast of central Chile. **(F)** *D. antarctica* (also sub-Antarctic islands and New Zealand). (Photographs A,B,D,E courtesy of Renato Westermeier; C,F from Fritsch 1959–1961.)

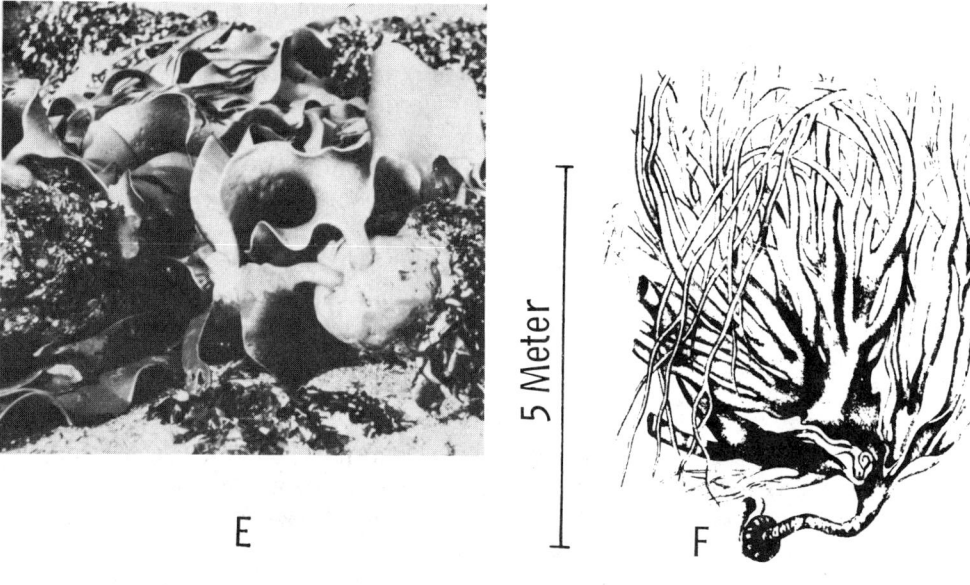

Fig. 5.6 (*Continued*)

Juan Fernandez Islands. These islands became famous because Alexander Selkirk, the real-life model for Robinson Crusoe, lived here from 1704 to 1709. The islands are situated 650 km from the midcoast of Chile, and the seaweed flora exhibits the remarkably high endemism of 32% (Silva 1966). It is assumed that the archipelago originated in the early Tertiary, but it may not be older than 4.5 million years (Stuessy et al. 1984). There are similarities between the seaweed flora here and that of the midcoast of Chile, although *Macrocystis, Lessonia,* and *Durvillaea* are absent on the Juan Fernandez Islands (Levring 1941). Few similarities exist between the seaweed flora of these islands and that of Easter Island 4000 km west of the coast of Chile or with the islands of the central Pacific.

Wave-exposed sites along the coast of central **Chile** are dominated in the **upper eulittoral** zone by the barnacle *Chthamalus cirratus* and the periwinkle *Littorina peruviana*. In the **mid eulittoral zone,** there are the erect corallines *Corallina officinalis* var. *chilensis* and *Serraticardia* spp., crustose corallines, and uncalcified red algae, e.g., *Gelidium chilense, Iridaea laminarioides* and *Centroceras clavulatum*. Also growing here are the green algae *Codium dimorphum, Ulva rigida* and *Enteromorpha compressa* and the brown alga *Ralfsia verrucosa*.

The lower eulittoral zone, from approximately 1 m above mean low water, and the upper sublittoral zone are dominated by a belt of the kelp *Lessonia nigrescens,* which is joined in more sheltered conditions by the kelplike *Durvillaea antarctica,* scattered at different heights among the *L. nigrescens* plants (Fig. 5.6) (Santelices et al. 1980; Villouta and Santelices 1984). Both algae have different strategies to resist the forces imposed by heavy seas.

L. nigrescens is a surf-adapted kelp, in which several stipes carrying narrow blades emerge from one holdfast.

In the upper sublittoral zone, there are, among the holdfasts of the two canopy species *L. nigrescens* and *D. antarctica,* red algae such as *Ahnfeltia durvillaei, Ceramium rubrum, Ballia scoparia* (also in the Antarctic region), and several *Desmarestia* species. An almost continuous cover of crustose coralline algae (e.g., *Mesophyllum* spp.) extends from the lower to the upper sublittoral zones. The midsublittoral zone is dominated by *Macrocystis pyrifera* in localities that are not too wave-exposed, while deep-water kelps are represented by *Lessonia trabeculata.* The Chilean *M. pyrifera* communities are simpler than their northeastern Pacific counterparts, with less algal and animal diversity, a possible consequence of heavy Pleistocene losses and recolonization only in recent times (Dayton 1985b).

In more sheltered places, the agarophyte *Gracilaria* is extensively harvested on Chilean coasts and may have to be artificially regrown. Cultivation experiments indicate a possible annual yield of 100 tons fresh weight per hectare (Westermeier, personal communication). Dominating in estuaries of southern Chile are the red algae *Porphyra columbina* and *Iridaea laminarioides* and the green alga *Enteromorpha intestinalis* (Westermeier and Rivera 1989). Extensive human intervention in the Chilean intertidal zone, specifically from collecting intertidal gastropods and thus potential algal grazers, favors the growth of *I. laminarioides,* which is extensively harvested and exported (Moreno et al. 1984; Westermeier 1987).

The coast of **Argentina** harbors approximately 400 species of seaweeds (Kühnemann 1972). The representatives of the Laminariales, *Macrocystis pyrifera, L. flavicans,* and *L. vadosa,* have a continuous distribution from the Pacific to the Atlantic coast of southern South America around Tierra del Fuego. The northern limit of *M. pyrifera* on the coast of **Patagonia** (tidal range of 5–6 m, maximum of 14 m) is situated on the Valdés Peninsula, at latitude 42°S, with the limit of the two *Lessonia* species at 47–48°N. The diversity of the Argentinian *Macrocystis* forest is as low as that of the Chilean (see above); no consistent grazers, like urchins, were found by Barrales and Lobban (1974), pointing to a probably geologically young ecosystem. *Durvillaea antarctica* extends only a short distance along the Argentinian side of Tierra del Fuego to about Bahía Thetis (C. H. Hay 1979b).

The **warm temperate eastern South American region** has its southern boundary at the estuary of the Rio de la Plata (at the border of Argentina and Uruguay), where the warm, southbound Brazil Current (Fig. 2.43) meets the cold, northbound Falkland Current. Kühnemann (1972) placed the boundary to the Valdés Peninsula, where the northern distribution of *Macrocystis pyrifera* ends (see above).

Warm temperate species such as the brown algae *Dictyota dichotoma, Padina gymnospora,* and *Sargassum cymosum* and the red algae *Jania rubens* and *Centroceras clavulatum* are found at rocky sites along the coast of southern **Brazil** (e.g., at Rio Grande Do Sul, 30°S; Baptista 1977). Several coastal lagoons are landlocked on the southern Brazilian coast, the largest being Patos Lagoon. The seaweeds in this lagoon, for example, *Enteromorpha* species, *Gelidium crinale,* and *Petalonia fascia,* have been investigated with respect to their seasonality and horizontal distribution (Coutinho and Seeliger 1984, 1986). The southern limit of the western Atlantic tropical region is reached at **Cape Frio** in northwestern Brazil.

Is **Jolyna laminarioides** *related to the primitive Atlantic Laminariales?* East of Rio de Janeiro in the wave-exposed eulittoral zone near Cape Frio, *Jolyna laminarioides,* an unusual, brown alga was detected; it is up to 90 cm long one to several erect blades arise from a discoid holdfast (Guimarães et al. 1986). The new genus has been named in honor of A. B. Joly, the founder of Brazilian phycology. *J. laminarioides* has some features of the Scytosiphonales, such as a single chloroplast with a conspicuous pyrenoid (differentiated region of the plastid) in each cell and plurilocular organs (with many small chambers or locules). On the other hand, highly differentiated medullary filaments are reminiscent of the primitive kelp *Saccorhiza polyschides*. The filaments look as if they may function in long-distance translocation, just as sieve elements do in the Laminariales. This indicates either that parallel evolution has occurred or that *J. laminarioides* may be a relic of the primitive Laminariales, otherwise represented on Eurafrican shores by the genera *Saccorhiza* and *Phyllariopsis,* the only members of the family Phyllariaceae (order Laminariales). The hypothesis has been put forward (p. 29) that the Phyllariaceae are a primitive group of the Laminariales of Pacific origin that survived in the early Tertiary Arctic Ocean to spread into the still small North Atlantic after it opened at around 50 Ma. At that time Cape Frio, in South America, was still not far from West Africa (Figs. 2.2B,C), where *Saccorhiza polyschides* and two *Phyllariopsis* species are still found today. The near-tropical locations of the Phyllariaceae and *J. laminarioides* point to the origin of the Laminariales from tropical–warm temperate ancestors. These relic locations are possibly the only places in the world left for these living fossils where they may not be outcompeted by the Laminariaceae, Alariaceae, and Lessoniaceae. These three families of the Laminariales form probably a higher-evolved group and have common traits; for example, they all possess the pheromone lamoxirene (Section 6.4.3).

Deep-water **Laminaria** *species off the Brazilian coast.* Near the northern limit of the warm temperate eastern South American region at Rio de Janeiro, 100 km off the coast and in an upwelling area, two *Laminaria* species were found at a 70 m depth. These are *L. brasiliensis,* with a divided frond, and *L. abyssalis,* with an undivided frond (Joly and Oliveira Filho 1967). The digitate *L. brasiliensis* may bear relationships to the North Atlantic digitate species *L. digitata* and *L. ochroleuca* and the southern African species *L. pallida* and *L. schinzii* (tom Dieck 1989). The two kelp species can exist at 40–100 m because of the lower water temperature range of 14–26°C, while temperatures at the surface are 19–27°C (Quége 1988). This situation is comparable to that for the Laminariales in the Mediterranean. As a hypothesis one may assume that an ancestor similar to *L. ochroleuca* migrated, during the cold periods of the Pleistocene from the Northern Hemisphere along West Africa, from there to southern Africa along the coast, and then possibly to South America via the South Equatorial Current. Alternatively, the ancestors of the warm-water group, consisting of *L. ochroleuca, L. pallida, L. schinzii,* and *L. brasiliensis,* did not belong to the Pacific invaders after the opening of the Bering Strait at 3 Ma but to an ancient Pacific stock of the Laminariales isolated in the early Tertiary Arctic Ocean at 65 Ma (see Section 2.1). This group may then have entered the North Atlantic at 50 Ma in a way similar to that hypothesized above for the Phyllariaceae.

Islands in the South Atlantic. The seaweed floras of the **Falkland Islands** exhibit a cold temperate character, while the four **Tristan da Cunha** islands and **Gough Island** tend more to the warm temperate side. *Macrocystis pyrifera* occurs on all of these islands, *Durvillaea antarctica* at the Falklands and on Gough Island, and *Splachnidium rugosum* and the South African species *Laminaria pallida,* at Tristan. Two of the 40 seaweed species reported by Chamberlain (1965) at Gough Island are endemic to this island and eight to Gough Island and Tristan. The latter island group has an age of 18 million years, which together with its isolated position may account for the exceptionally high endemism of 40% for seaweed species (Baardseth 1941). Gough may be quite young, of Pleistocene age, since it has a low diversity (Chamberlain et al. 1985).

5.6 SOUTHERN AFRICA

Literature: Anderson and Bolton (1985); Bolton (1986); Bolton and Anderson (1987); Bolton and Levitt (1987); Bolton and Stegenga (1987); Branch and Branch (1981); Dieckmann (1980); Field et al. (1980); Hommersand (1986); Norris and Aken (1985); Papenfuss (1940, 1942); Schmidt (1957); Seagrief (1984, 1988); Simons (1976); Stephenson (1948); Stephenson and Stephenson (1972); Stegenga (1986); Velimirov et al. (1977); Wynne (1986*b*).

By the early Cretaceous, around 130–120 Ma, Africa began its drift from Antarctica, and the western coast of Africa became separated by shallow seas from South America (Figs. 5.2A,B). The South and North Atlantic oceans had started to connect at 90 Ma, with a deep-water connection occurring around 65 Ma (Sclater et al. 1977, I. O. Norton and Sclater 1979, Kennett 1982, Bergh 1987, Cande et al. 1988, Scotese et al. 1988).

The isolated position of the **warm temperate southern African region** has resulted in an endemism of 20% for the shore fish species (Briggs 1974), and endemism is also high for the seaweeds. Within this warm temperate region one has to distinguish two provinces, the cool southwestern Africa province, which harbors several representatives of the Laminariales, and the warmer Agulhas province.

The **southwestern Africa province** reaches to the eastern Atlantic tropical region at Mossamedes in Angola (Figs. 1.5–1.7) and stretches south to Cape Agulhas, the most southerly point of the continent, southeast of the Cape of Good Hope. There are about 270 seaweed species in this province, about one-quarter not occurring on the south and west coast of southern Africa (Bolton 1986).

The west coast of southern Africa is characterized by local **upwelling** (Barber and Smith 1981; Longhurst 1981) and is influenced by the cold **Benguela Current,** which, as a part of the South Atlantic gyre, flows northward and also receives cold water from a branch of the West Wind Drift (Fig. 2.43). Consequently, the origins of the west coast seaweed flora appear to be from temperate and polar regions of the South Pacific (Bolton and Levitt 1987).

There is a remarkable paucity of brown seaweed species on the west coast, the ratio of Rhodophyta and Chlorophyta to Phaeophyta species being 5.9 (Bolton 1986). This is similar to tropical values (6 and more). In the Northern Hemisphere the corresponding values are 3–4 and 2 for warm and cold temperate regions, respectively, and 1 for the Arctic region (Kapraun 1980).

Microfossil evidence suggests that a warm flow toward the equator along the west coast of southern Africa has existed for at least 50 Ma, while the cool Benguela system developed only in the late Miocene, since about 10 Ma (Shannon 1985). This led to the evolution of the southwestern African endemics such as *Ecklonia maxima* (see below). The annual mean inshore seawater temperatures are in the range of 12–16°C, with a mean in the warmest month in the range of 13–19°C and in the coldest month 11–14°C (Bolton 1986; Bolton and Anderson 1987; Dieckmann 1980).

The warmer **Agulhas province** stretches from Cape Agulhas along the southern and eastern coasts of South Africa to southern Natal, where it meets the Indo–West Pacific region. This province is under the influence of the **Agulhas Current,** which transports warm water to the south. The east coast seaweed flora is closely related to the tropical Indian Ocean seaweed floras (Bolton and Levitt 1987). The annual mean seawater temperatures are in the range of 17–19°C, with ranges for the means of 19–21°C in the warmest month and 13–17°C in the coldest month (Bolton and Anderson 1987).

Laminariales of southern Africa. As endemics there are two *Ecklonia* species and two *Laminaria* species in southern Africa. In addition, *Macrocystis angustifolia* occurs sporadically in the Cape area (Velimirov et al. 1977; Bolton 1986).

Ecklonia maxima. This kelp, belonging to the family Alariaceae, is the prominent kelp in the cool Benguela upwelling region on the west coast of southern African. Its hollow stipe, which becomes inflated towards the upper end, may attain a length of up to 10 m and carries a multiple divided frond, the marginal portions branching off from a flat medium portion (Fig. 5.7). *E. maxima* occurs from Swakopmund, Namibia, to Aasfontein, 15 km west of Cape Agulhas (Bolton and Anderson 1987). The kelp usually carries the red alga *Suhria vittata* as an epiphyte on its stipes (Anderson and Bolton 1985). After the cool Benguela system came into being, *E. maxima* may have evolved from a smaller and more warmth-tolerant plant that may have been similar to *E. biruncinata* (Bolton and Anderson 1987).

Ecklonia biruncinata. The second *Ecklonia* species, *E. biruncinata,* is less than 50 cm long and represents the only member of the Laminariales on the warmer south coast, that is, in the Agulhas province. The species occurs from 60 km east of Cape Agulhas to the border of the Indo–West Pacific tropical region in southern Natal, at Port Edward, although in deeper water it's range may extend farther northeast. *E. biruncinata* has spiny outgrowths on the sporophylls and usually also the primary

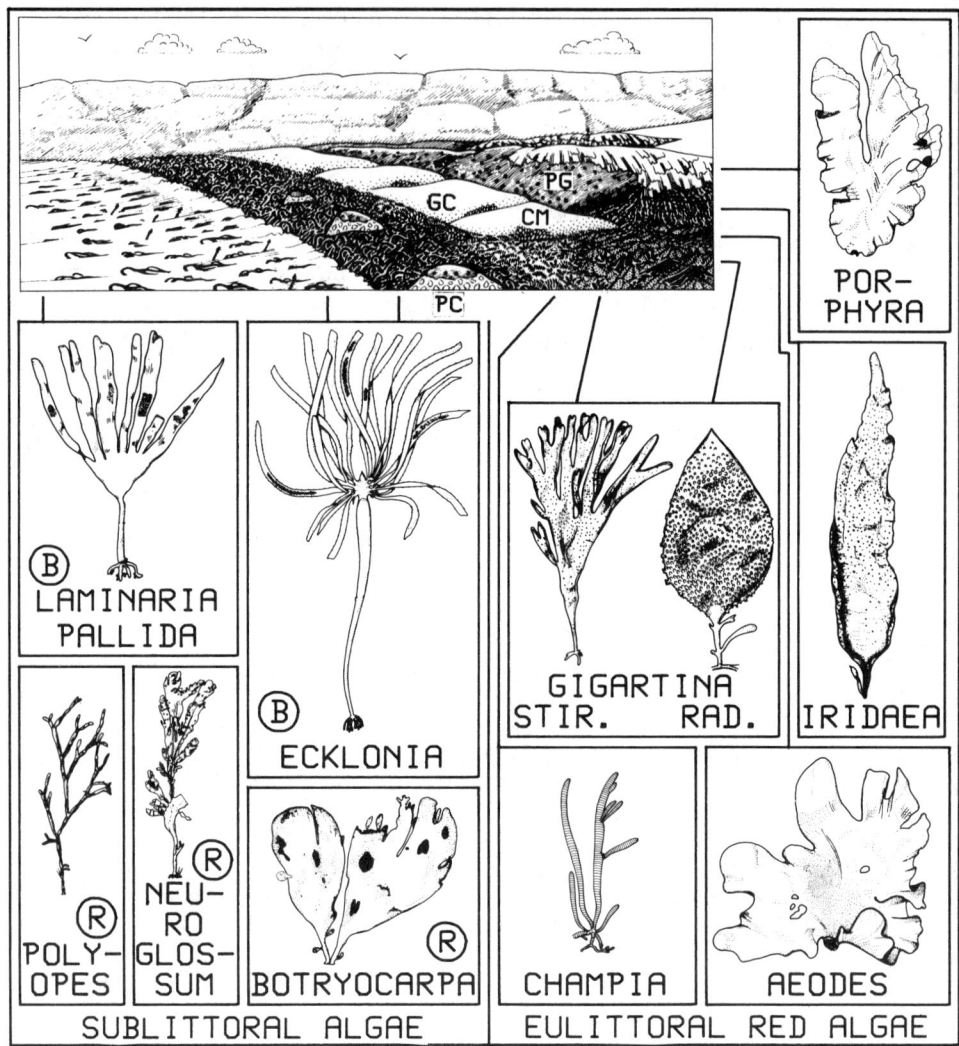

Fig. 5.7 Zonation on the western coast of southern Africa. **Details on species names and dimensions: eulittoral red algae:** *Porphyra capensis* (× 0.3); *Gigartina stiriata* (× 0.2); *G. radula* (× 0.2); *Iridaea capensis* (× 0.15); *Champia lumbricalis* (× 0.7); *Aeodes orbitosa* (× 0.1). **Sublittoral algae:** *Laminaria pallida* (× 0.03); *Ecklonia maxima* (× 0.013); *Polyopes constrictus* (× 0.3); *Neuroglossum binderianum* (× 0.4); *Botryocarpa prolifera* (× 0.4). B = brown alga; R = red alga; CM = *Chloromytilus meridionalis* (= *Mytilus crenatus*); GC = *Gunnarea capensis* (polychaete); PC = *Patella cochlear;* PG = *Patella granularis*. (Vegetation scheme from Stephenson and Stephenson 1972; drawings from Simons 1976.)

blade, and a short and solid stipe. The southern African species is related and may have to be included in *E. radiata* in southern Australia and in New Zealand (Womersley 1967; Bolton and Anderson 1987). The genus *Ecklonia* is also represented by three species in Japan, by *E. muratii* off the northwestern African coast, and by an unknown species off Oman.

Laminaria pallida. The kelp *Laminaria pallida* (Fig. 5.8) is characterized by a divided blade and by a rigid stipe up to 3 m long; the species ranges from the Cape area between the Cape of Good Hope and Cape Agulhas to Namibia (30°S). Apart from southern Africa, *L. pallida* has been found at Tristan da Cunha, and halfway between southern Africa and Australia on the islands of New Amsterdam and St. Paul (Delépine 1963). The species possibly represents the southern African descendent of the warm temperate Mediterranean and Atlantic species *L. ochroleuca,* which crossed the equator in the Pleistocene.

Laminaria schinzii. The hollow-stiped *L. schinzii* is otherwise similar to *L. pallida*. It ranges to 18°50'S, further north than *L. pallida* (Anderson and Bolton 1985; Bolton 1986).

Macrocystis angustifolia (family Lessoniaceae). This representative of the genus, up to 10 m long, is similar in size to the eastern Pacific *M. integrifolia*. It occurs mainly along the southeastern Australian coast (Figs. 5.3 and 5.11) and is to be found in southern Africa only in protected bays in the Cape area (33–34°S). The giant kelp *M. pyrifera,* which inhabits the temperate Pacific coasts of the Americas and the coasts of Tasmania and New Zealand, is absent in southern Africa. Whereas *M. pyrifera* has upright basal stipes in a conical holdfast, *M. angustifolia* has prostrate, rhizomatous branches. In this way it is again similar to the relatively small *M. integrifolia* of the American Pacific coasts (Womersley 1954; Neushul 1971).

Fig. 5.8 *Laminaria pallida* at about 5 m deep, southwest of Cape Town. (From Velimirov et al. 1977.)

Arguments for a warm temperate southern African region. The southern African members of the Laminariales exhibit more relationships to warm temperate than to cold temperate representatives of the order in other regions. For example, *Laminaria pallida* is a typical warm temperate member of the Laminariales with an elevated heat tolerance for its gametophytes (Table 7.3). Typical cold-water species, such as *Durvillaea antarctica,* are absent along the southern African coasts, as are the cold-water shore-fish species (Briggs 1974). One may therefore follow Briggs (1974) and Ekman (1953) and call the whole of the southern African region warm temperate and not cold temperate. In contrast, Stephenson (1948), the classical investigator of the marine biota along southern African coasts, had proposed to regard the cooler southwestern African province as cold temperate. One of his arguments, namely that the Laminariales are well represented in this province, cannot be accepted since this order contains characteristic warm temperate and cold temperate species. Regarding the whole of the southern African region as warm temperate seems to fit better the temperature scheme given in Fig. 1.8, and corresponds better to the delimitation of warm and cold temperate regions in the Northern Hemisphere. That is to say, the southwestern African province covers the coastline bordered to the north approximately by the 25°C-summer isotherm (Fig. 1.6, February) and to the south by the 15°C-isotherm in winter (Fig. 1.5, August; local and temporary reductions to 8°C in upwelling areas).

In the **southwestern African province** one finds as a dominant species in the **supralittoral zone** (littoral fringe) the red alga *Porphyra capensis,* abundant on the west coast, but also common on the south coast (Agulhas province). The upper **eulittoral zone** is characterized by limpets and barnacles. The mideulittoral zone is again dominated by endemic red algae (Fig. 5.7), accompanied by two members of the brown algal order Chordariales, *Chordariopsis capensis* and *Splachnidium rugosum,* the latter species exhibiting fucalean traits (see Section 5.2). The polychaete *Gunnarea capensis* may form extensive fields consisting of a layer of tubes up to 30 cm high. In the lower eulittoral zone appear crustose, calcified red algae and a belt of the red alga *Champia lumbricalis,* followed by mixed vegetation of the representatives of the Laminariales *Ecklonia maxima* and *Laminaria pallida* in the upper **sublittoral zone.**

Within the depth range of 0–4 m, *E. maxima* contributes two-thirds of the total kelp biomass and at 4–8 m deep one-half. From 8 m to 20 or 30 m deep one finds that the kelp vegetation is entirely made up of *L. pallida* (Velimirov et al. 1977; Dieckmann 1980). The endemic *Desmarestia firma* is an important understory alga and the only representative of the genus in southern Africa; it belongs to the species group with ligulate fronds. Another species, with a morphological overlap with the southern African– endemic *D. firma,* occurs in New Zealand, South America, Pacific North America and Europe. Their inclusion in the single, highly variable taxon *D. ligulata* has been questioned (Anderson 1985, Peters and Müller 1986; Ramirez et al. 1986).

The red algae of the middle and lower sublittoral zone, illustrated in Fig. 5.7, are endemic representatives of the family Delesseriaceae (order Ceramiales), with the exception of *Polyopes constrictus* (order Cryptonemiales), which also occurs in southern Australia.

Along the warmer **southern coast** of southern Africa (Fig. 5.9) in western part of the Agulhas province, one still finds numerous animal and algal species of the western coast; examples in the **eulittoral zone** are the belt-forming limpet *Patella cochlear* and the red algae *Caulacanthus ustulatus* (Fig. 2.59E) and *Centroceras clavulatum* (Fig. 4.9). The red alga *Gelidium pristoides* is an abundant agarophyte in the eulittoral of the south coast, and, together with *Gracilaria verrucosa,* a seaweed source for the South African agar industry (Carter and Anderson 1985; J. Bolton, personal communication). Species with a **pantropical** or **Indo–West Pacific** distribution pattern begin to dominate quantitatively along the eastern coast, for example, the green alga *Caulerpa racemosa* (Figs. 5.9 and 5.10I). In the **sublittoral zone** of the southeastern coast of southern Africa the members of the Laminariales

Fig. 5.9 Zonation on the southern coast of southern Africa. A = start of *Littorina* zone; B = start of barnacle zone; C = start of *Patella cochlear* zone; D = start of upper sublittoral zone G = green alga; L = snail; M = mussel; N = *Patella;* P = polychaete; R = red alga; S = ascidian. (From Stephenson and Stephenson 1972.)

from the western coast are absent and are replaced as canopy algae by *Sargassum heterophyllum* and *Anthophycus longifolius* (= *Sargassum longifolium*). Two endemic and monotypic species, *Axillariella constricta* and *Bifurcariopsis capensis*, and the endemic species *Bifurcaria brassicaeformis*, *Carpophyllum scalare* (= *Oerstedtia scalaris*), and *Cystophora fibrosa* are other members of the Fucales found in the sublittoral zone (Table 5.1). *C. fibrosa*, the only species of *Cystophora* from outside Australasia, requires detailed studies and may appear better placed in *Cystoseira* (Womersley 1984–1987).

The fucalean species mentioned have the following, restricted distributions (J. Bolton, personal communication). *Oerstedtia scalaris* is a south-coast endemic. *Cystophora fibrosa* occurs in a very small area around Cape Agulhas, from the extreme west of the southern coast a small way into the west coast–south coast overlap. *Bifurcariopsis capensis* occurs from the western end of the south coast around to the southern portion of the west coast. *Bifurcara brassicaeformis* dominates many upper sublittoral (subtidal fringe) zones in the west coast–south coast overlap and on the southern west coast. *Axillariella constricta* grows on the southern west coast.

Near Port Elizabeth, at Bird Island, the sublittoral communities are dominated by red algae, *Gelidium pteridifolium* in wave-exposed sites, *Plocamium corallorhiza* and *P. rigidum* in sites with less water movement, and *Peyssonnelia capensis* in deep water (e.g., at 22 m) (Anderson and Stegenga 1989).

The warm temperate kelp *Ecklonia biruncinata* can be found sporadically as an endemic of the Agulhas province, where it occurs in the sublittoral zone as far as southern **Natal.** There may be a relationship between *E. biruncinata* and *E. radiata* in Australia and New Zealand, to the disjunct *E. muratii* off West Africa, and to the *Ecklonia* species detected quite unexpectedly in the upwelling area off the coast of Oman. This last species occurs at 17°N in the Arabian Sea below a thermocline at a depth of 6 m (Hiscock et al. 1984).

The relationship of the Natal seaweed flora to that of Australia has been emphasized by Norris and Aken (1985). They report that 15 out of 20 previously unrecorded red algae from Natal also occur in Australia and suggest that this reflects the common Gondwana past, when the African and Australian continents had a continuous coastline (Fig. 5.2B).

Hommersand (1986), from analysis of the southern African red algal flora, has discussed the alternative possibility that western and southern Australian species migrated via the North Equatorial Current to Natal during the major worldwide cooling periods from late Miocene to the present. Long-range dispersal of red algae as epiphytes on floating brown algae or as thallus fragments containing viable or developing spores is thought by Hommersand to have occurred along the boundaries of convergent currents. Southern African species with affinities to Japan and China may have crossed the equator in a similar fashion during the Miocene to reach southern Africa. During the same time, they may have migrated along the West African coast to Europe and via the Caribbean to western North America (Hommersand 1986).

Plate tectonics may still be important to explain the modern distributions of the fucalean family Seirococcaceae, with the occurrence of the genera *Axillariella* in southern Africa, *Cystosphaera* in Antarctica, and four genera in Australasia (Table 5.1). Clayton (1984) has suggested that the Seirococcaceae originated before South Africa was effectively isolated from Antarctica, sometime later than 90 Ma.

In southern Natal the first coral reefs appear (Stephenson and Stephenson 1972) and the **Indo–West Pacific tropical region** begins (Figs. 1.5–1.7). A major biogeographical discontinuity occurs in the 130 km coastline between Hluleka and the mouth of the Kei River, probably due to a steep temperature gradient (2°C in annual mean; Bolton and Stegenga 1987).

5.7 SOUTHERN AUSTRALIA AND NEW ZEALAND

Literature: Entire area: Clayton and King (1981); Edgar (1986); Knox (1963); Stephenson and Stephenson (1972); Womersley (1954, 1959, 1981a,b, 1984, 1987). **Southwestern Australia and Abrolhos Islands:** Hatcher et al. (1987); Kirkman (1981); Wilson and Marsh (1980). **Southern coast of Australia:** Bennett and Pope (1953); Fuhrer et al. (1981); Lewis (1983); Shepherd and Womersley (1970, 1971, 1976, 1981); Womersley (1967); Womersley and Edmonds (1952, 1958). **Tasmania:** Bennett and Pope (1960); Edgar (1984). **Marine benthic flora of southern Australia:** Womersley (1984–1987). **New South Wales:** Kennelly (1987a,b); May and Larkum (1981). **Queensland:** Young and Kirkman (1975). **Lord Howe Island:** Allender and Kraft (1983); Gabrielson and Kraft (1984); R. Jones and Kraft (1984); Kraft and Olsen-Stojkovich (1985). **New Zealand:** N. M. Adams and Nelson (1985); V. J. Chapman (1956–1969); Chapman and Dromgoole (1970); Chapman and Parkinson (1974); Choat and Schiel (1982); Dellow (1955); Fleming (1979); Knox (1975); Lindauer et al. (1961); L. B. Moore (1961); J. Morton and Miller (1968); W. A. Nelson and Adams (1987); Parsons (1985a,b); Parsons and Fenwick (1984); South and Adams (1976). **Kermadec Islands:** W. A. Nelson and Adams (1984). **Special groups: Delesseriaceae (world key of genera):** Wynne (1983). **Seagrasses:** Larkum et al. (1989).

After a first rifting in the Cretaceous (95 Ma), Australia separated from Antarctica in the Tertiary, around 44 Ma, and approached its present position adjacent to Asia as recently as 10 Ma (Cande and Mutter 1982, Stock and Molnar 1987, Scotese et al. 1988). The late arrival of Australia at its present position mainly explains the biogeographical uniqueness of the Australian region and the striking biotic differences between the Australian and Asian regions (Keast 1983). The southern coast of Australia was originally joined to eastern Antarctica (Fig. 5.2 D); after fragmentation, the west and east coasts of Australia became colonized by the Indian Ocean and Pacific marine algal floras, respectively (Fig. 5.2D,E; Hommersand 1986).

New Zealand, of Pacific origin and hence mainly with a Pacific flora (Fig. 5.2D), parted from Antarctica and Australia even earlier, in the late Cretaceous (60–80 Ma). The Tasman Sea between Australia and New Zealand formed as a wide gap in the early Tertiary (65 Ma). The prevailing direction of ocean currents, particularly the West Wind Drift, impeded the exchange of species from New Zealand to Australia even more, as evidenced by the various taxa endemic to New Zealand. Only some of the vesiculate species of the fucalean genus *Cystophora* with a probable Australian origin made their way to New Zealand (Edgar 1986). Only similarities on the genus level remind us of the common past of these East Gondwanan fragments that drifted to contrasting temperature zones.

Fossil remains from the Tertiary and Pleistocene indicate that coral reefs occurred as far south as Tasmania. The warm-water marine seaweed species of the southern Australian coast could have easily moved north during progressing glaciations (Knox 1963; van den Hoek 1982*b*). New Zealand, with its restricted latitudinal range, has probably suffered large-scale extinction of warm temperate species (Edgar 1986).

Regions and water temperatures. The northern limit of the two warm temperate Australian regions (Fig. 1.7) follows the course of the 20°C-winter isotherm (August), which marks the southern limit of coral reefs and approaches the eastern and western coasts of the continent at about latitude 25°S (Fig. 1.5). The annual range of the surface water temperature on the coast of southern Australia is mostly in the range of 14–19°C but is reduced to a range of 12–14°C on the **cooler coasts of Victoria,** with main upwelling in the southeast of South Australia (Womersley 1981*a*, 1984–1987). In **Tasmania,** surface-water temperatures of 12–17°C prevail, and in shallower bays temperatures usually vary from 8°C in winter to 22°C in summer (Edgar 1984). If one places the 10°C-winter isotherm as a boundary between cold and warm temperate regions (Fig. 1.8), the whole of the southern Australian and Tasmanian coasts may be regarded as warm temperate. However, in accordance with Briggs (1974), Womersley (1981*a*), and Michanek (1983), it should be noted that along the coasts of Victoria and Tasmania the water temperatures in summer hardly exceed 15°C (Edwards 1979), the temperature at which warm and cold temperate regions are separated (Fig. 1.8). Thus a **cold temperate Victoria–Tasmania region** is recognized (Figs. 1.5–1.7). As emphasized by Womersley (1981*a*), the seawater temperatures prevailing in southern Australia and Tasmania are transitional between cold and warm temperate regimes, and this is where the nomenclature fails. Knox (1960) had avoided the problem by regarding the 12°C-winter isotherm, instead of 10°C (Fig. 1.8), as a general boundary to separate cold and warm temperate regions. Correspondingly, Knox designated the coasts of Victoria and Tasmania as cold temperate, whereas Bennett and Pope (1953, 1960) spoke of a cool (not cold) "Maugean province." The **warm temperate northern New Zealand region,** with water temperatures in the range of 12–20°C, occupies the northern part of North Island and the Kermadec Islands, situated to the northeast. The southern part of North Island, South Island, and Stewart Island form the **cold temperate southern New Zealand** region, with a temperature range of 6–14°C. The latter region includes the Auckland Islands, the Antipodes, and Campbell Island to the south as well as the Chatham and Bounty Islands to the east.

The **temperate regions** of Australia join the Indo–West Pacific tropical region at Shark Bay on the western coast of Australia and at Fraser Island on the eastern coast (Figs. 1.5–1.7). There is no upwelling on the coast of **western Australia,** as on the western coasts of South America and southern Africa, and so the tropical region extends more to the south on the western coast of Australia (Figs. 1.5.–1.7). At the **Abrolhos Islands** at 28–29°S occur the world's southernmost and yet highly diverse coral reefs; 37 hermatypic genera grow there together with a kelp vegetation of *Ecklonia radiata* and fucoid genera such as *Sargassum, Turbinaria, Cystophyllum, Caulocystis* and *Hormophysa* (B. R. Wilson and Marsh 1980; Kirkman 1981; Johannes et al. 1983; Hatcher et al. 1987).

The 5500 km long coast of **southern Australia** is exceptionally rich in seaweeds with 1100 species (Table 4.1). Endemism is high: about 70% of the 800 species of red algae and 131 of the 231 species of brown algae (Womersley 1984–1987) are endemics, as are slightly less than half of the species of green algae (Womersley 1984–1987). On the generic level endemism is about 30%, 20%, and 11% for red, brown, and green algae, respectively.

Particularly striking is the wealth of **red algae** along the coast of southern Australia (examples in Fig. 5.10). Of the over 4000 Recent species and 660 genera in the world, southern Australia harbors 20% and 43%, respectively. The family Ceramiales is especially well represented (Womersley 1981*a;* see also the dichotomous world key of the 83 genera recognized in the family Delesseriaceae by Wynne 1983). The richness in red algae is indicated by an *R:P* index (ratio of red to brown algal species) of 4. Elsewhere this value is typical for tropical regions (Table 4.1). Among the *green algae* the genera *Caulerpa* with 19 species and *Codium* with 16 species (Fig. 5.10) may be cited as examples of species richness (Womersley 1984–1987).

Of the **brown algae** the *Laminariales* are represented by three genera, *Ecklonia, Macrocystis,* and *Lessonia. Ecklonia radiata* occurs along the whole of the southern Australian coast (and in New Zealand). *E. radiata* (Fig. 5.11E) extends northwards on both the eastern and western coasts of Australia and may be considered a warm temperate representative of the Laminariales. On the eastern coast *E. radiata* occurs to Caloundra in Queensland (Womersley 1984–1987), while in Western Australia it is found to Kalbarri, at 27°S (Womersley 1981*b;* Hatcher et al. 1987), north of the Abrolhos Islands. Other species of the genus are present in southern Africa and Japan. *Macrocystis angustifolia* (Fig. 5.11F), also occurring in southern Africa, inhabits the coasts of southeastern South Australia, Victoria, and northern Tasmania, while *M. pyrifera* is restricted to Tasmanian coasts. *Lessonia corrugata* occurs in southern Tasmania (see below).

Of the approximately 35 genera of **Fucales** found throughout the world, 14 are endemic to Australia and New Zealand (Table 5.1). Among the endemic genera, *Cystophora* is the richest in species. (see Section 5.6 for the southern African *C. fibrosa,* which may belong to another genus). The order **Dictyotales** is better represented in southern Australia than in the tropical regions.

A major cause of the great species diversity and high endemism along the southern Australian coast probably results from few species being lost during

Fig. 5.10 Red and green algae of the southern Australian coast. **Red algae: Nemaliales: (A)** *Delisea pulchra* (× 0.35). **Cryptonemiales: (B)** *Callophyllis lambertii* (× 0.35). **Gigartinales: (C)** *Nizymenia australis* (× 0.35). **(D)** *Phacelocarpus labillardieri* (× 1.0). **Rhodymeniales: (E)** *Champia viridis* (× 0.5). **(F)** *Botryocladia obovata* (× 0.8). **Ceramiales: (G)** *Thuretia quercifolia* (× 0.35; Dasyaceae). **Green algae: (H)** *Caulerpa vesiculifera* (× 0.5). **(I)** *C. racemosa* var. *clavifera* (× 1.0). **(J)** *Codium duthiae* (× 0.7). **(K)** *Palmoclathrus stipitatus* (× 0.23). (K from Womersley 1971; remaining photographs from Fuhrer et al. 1981, originally colored.)

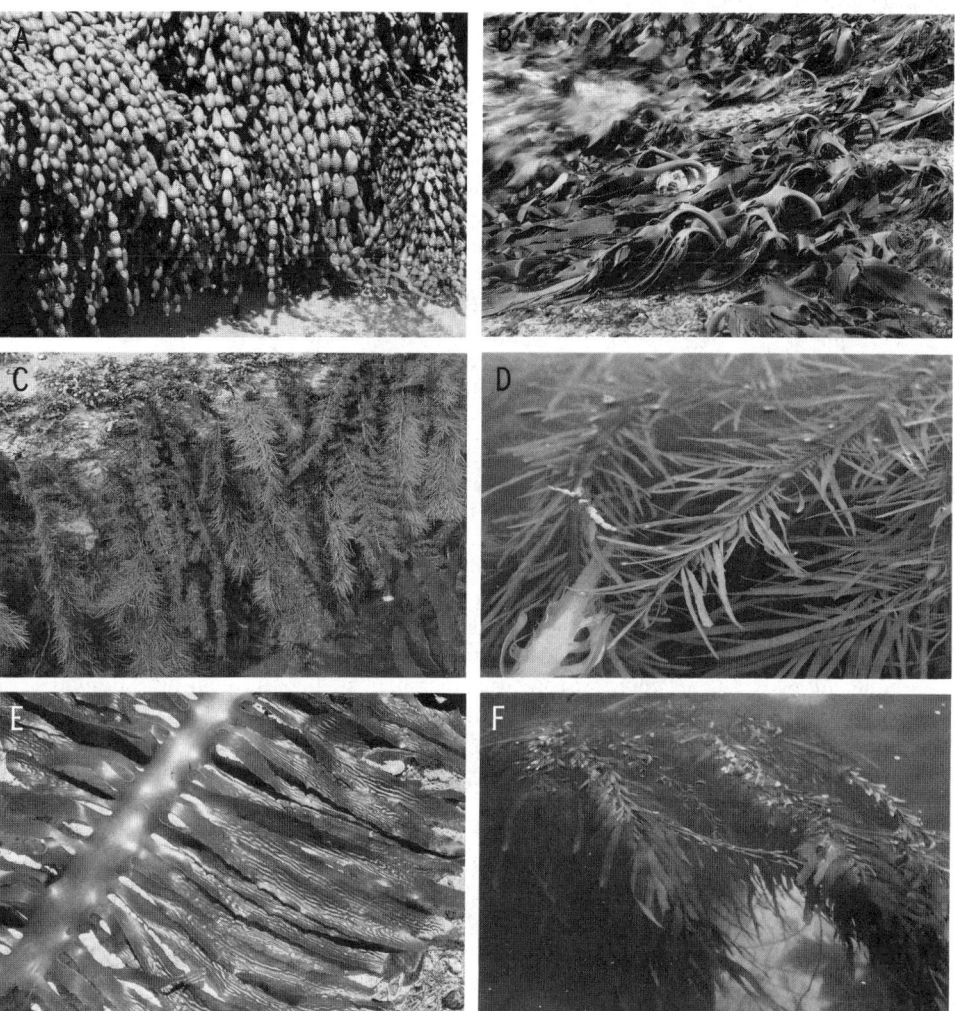

Fig. 5.11 Brown algae of the southern Australian coast. **Fucales:** (A) *Hormosira banksii* (×0.1). (B) *Durvillaea potatorum* (× 0.05). (C) *Cystophora* spp. (× 0.05). (D) *Phyllospora comosa* (× 0.08). **Laminariales:** (E) *Ecklonia radiata* (× 0.2). (F) *Macrocystis angustifolia* (× 0.02). (Photographs from Fuhrer et al. 1981, originally colored.)

the Pleistocene. Equally important may be the vast extension of the southern Australian coast parallel to the latitudes, similar to the situation in the Mediterranean. According to the biogeographic island theory, large size, richness in microhabitats, and geographical isolation all represent important prerequisites for the evolution of high species diversity and endemism. These characters are features of the southern Australian coast. As in the tropics, the relatively small temperature range seems not to be an important prerequisite for species richness.

Zonation on the southern Australian coast. In the **supralittoral zone** (littoral fringe) one finds littorinid snails (e.g., Littorina *unifasciata*) as dominants, and in the **upper eulittoral** zone small barnacles (e.g., *Chthamalus antennatus* and *Chamaesipho columna;* Womersley 1981*a*). The **mideulittoral zone** is dominated by barnacles such as *Chthamalus antennatus,* the lower eulittoral zone by small red algae such as *Gelidium* or *Centroceras clavulatum,* and the brown alga *Hormosira banksii* (Fig. 5.11A), which is the only representative of the Fucales within the eulittoral zone. This latter rather characteristic species occurs from King George Sound in Western Australia to Port Macquarie, New South Wales, and also in New Zealand. The dominant role of members of the **Fucales in the sublittoral zone** is a typical trait for the warm temperate coastlines of southern Australia and northern New Zealand. The following zonation example (Fig. 5.12) relates to the very wave-exposed coast between Cape Jaffa and the Great Australian Bight in South Australia (Shepherd and Womersley 1976). In the turbulent, **upper sublittoral** surf zone (sublittoral fringe) one finds the fucalean species *Cystophora intermedia* (mostly 20–40 cm long) as a dominant. Red algae such as *Haliptylon roseum* (= *Corallina cuvieri*) and *Jania fastigiata* grow underneath. The **midsublittoral** zone, covering the approximate depth range of 5–35 m, is dominated by *Ecklonia radiata* (Laminariales; Fig. 5.11E) and *Scytothalia dorycarpa* (Fucales), which both form a canopy layer rising to about 1 m above the bottom. In less turbulent situations, *E. radiata* dominates the upper sublittoral. *Sargassum* species, *Acrocarpia paniculata,* and *Cystophora* species (all members of the Fucales) are additional examples of canopy algae of the midsublittoral zone, although they appear more in its upper part. As understory algae with thalli lengths of 5–25 cm, one finds brown algae such as *Dictyota* and *Zonaria,* red algae such as *Plocamium,* and other genera, examples of which are illustrated in Fig. 5.10A–F. About 10 species of the green alga *Caulerpa* (Fig. 5.10H,I) grow here. Below a 35 m depth, where the dense vegetation dominated by *Ecklonia radiata* ends, the **lower sublittoral zone** extends down to as much as 60 m deep. Here the deepest crustose red algae occur. The lower sublittoral zone harbors such species as *Plocamium cartilagineum* (Fig. 2.65E), *Callophyllis lambertii* (Fig. 5.10B), genera of the Ceramiales such as *Antithamnion* and *Ballia,* and the green algae *Caulerpa hedleyi* and *Palmoclathrus stipitatus.* The latter (Fig. 5.10K) has a thallus up to 20 cm long, composed of cells immersed without order in a firm gelatinous matrix; thus the species belongs to the order Chlorococcales, such as the Mediterranean species *Palmophyllum crassum* and another species of *Palmophyllum* at the Kermadec Islands. Fucalean species such as *Myriodesma quercifolium, Scytothalia dorycarpa, Cystophora platylobium,* and *Sargassum bracteolosum* may still be found at a 50 m depth (Shepherd and Womersley 1976).

Seagrasses. These are particularly well represented in southern Australia (Fig. 4.8; review in Larkum et al. 1989) There are at least 17 species in southern Western Australia with this richness decreasing eastward across the Great Australian Bight (Kirkman 1985). Important species as far as biomass is concerned include *Amphibolis antarctica, A. griffithii, Heterozostera tasmanica, Halophila ovalis,* and especially the species of *Posidonia.* In the last genus only one species, *P. australis* occurs on the eastern Australian coast, while four species are found on the western coast and four on the southern coast (Kirkman 1985). The seagrasses, particularly *Posidonia* species, form enormous beds along the southwestern coast from Shark Bay in the north to New South Wales.

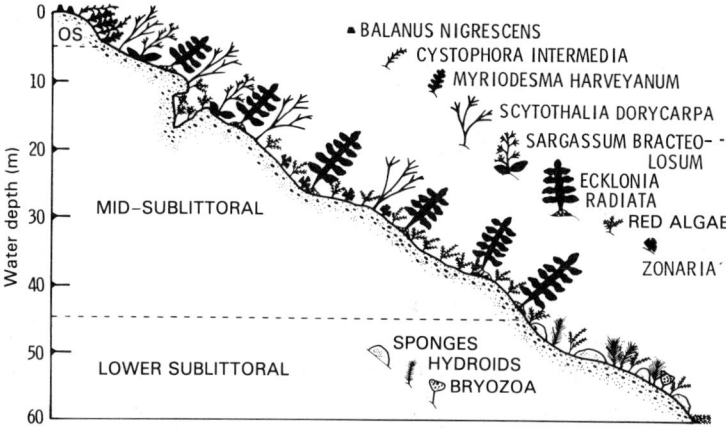

Fig. 5.12 Zonation in the sublittoral zone of the southern Australian coast at a wave-exposed site (St. Francis Island, 133 ° E). OS = upper sublittoral zone. (From Shepherd and Womersley 1976.)

In the **western portion of the southern coast** of Australia, where cliffs and sandy beaches prevail (Great Australian Bight), the upper sublittoral zone is no longer dominated by *Cystophora intermedia* but by other fucalean species, *Caulocystis uvifera, Acrocarpia robusta, Platythalia angustifolia,* and *Cystophora* species, which are again followed in the sublittoral zone by *Ecklonia radiata.* The last-named species also occurs on the cooler coasts of Victoria (see below) and Tasmania and becomes dominant again on the warm temperate part of the **eastern coast** of Australia (New South Wales), where it is accompanied by the fucalean species *Phyllospora comosa* (Fig. 5.11D), *Sargassum,* and *Cystophora. P. comosa* is common on Victorian coasts and also around Tasmania.

The dominant seaweed species change considerably in the **cold temperate Victoria–Tasmania** region. *Durvillaea potatorum,* a solid-bladed representative of the genus, dominates the upper sublittoral zone at wave-exposed sites (Fig. 5.11B). *D. potatorum* is harvested from the drift on King Island, Bass Strait, for alginate production. At medium-exposed sites the fucalean species *Phyllospora comosa* (up to 2.5 m long; Fig. 5.11D) and *Macrocystis angustifolia* (Fig. 5.11F; also in southern Africa) are important. Only *P. comosa,* and *Hormosira banksii* continue along the warm temperate eastern coast of Australia as far north as Port Macquarie (32°S). In the eulittoral zone on Tasmanian and Victorian shores dominant fucoid species is *Cystophora torulosa,* found just below the band of *H. banksii* (Womersley 1981a; Edgar 1984). *D. potatorum* only penetrates to latitude 36°S and *M. angustifolia* to the northern border of Victoria.

Southern Tasmania, where in some localities the giant kelp *Macrocystis pyrifera* forms dense forests in the sublittoral zone, represents one of the stepping stones that facilitated the circumglobal migration of this species.

Other stepping stones include South America, the sub-Antarctic islands, and southern New Zealand but not southern Africa. In the upper sublittoral zone of southern Tasmania, *Durvillaea potatorum* and an endemic member of the Laminariales, *Lessonia corrugata,* are common.

Along the 4200 km of eastern coastline the temperate flora of **New South Wales,** not extensively investigated but with a number of endemic species, grades into the tropical Queensland flora (Allender and Kraft 1983). At Jervis Bay (35°S) *Ecklonia radiata* and the fucoid species *Phyllospora comosa* dominate on rocky sublittoral substrata, while the fucoids *Sargassum* species and *Caulocystis cephalornithos* and the green alga *Caulerpa cactoides* grow on shale, sand, and shells. The seagrass *Posidonia australis* grows on sand (May and Larkum 1981). The **Indo-West Pacific tropical region** begins at Sandy Cape, Fraser Island, at 25°S (Figs. 1.5–1.7).

Lord Howe Island. At 31.5°S, 600 km east of New South Wales, and 1300 km northwest of the North Island of New Zealand, Lord Howe Island, together with the Abrolhos Islands, has the world's southernmost coral reefs. Endemism among terrestrial plants and animals is high in this isolated island of volcanic origin, with no clear affinities to either Australia or New Zealand. Endemism is lower in the marine fauna and seaweed floras, both with mainly tropical affinities (Allender and Kraft 1983; Gabrielson and Kraft 1984). Dominant groups in the seaweed flora are members of the brown algal order Dictyotales, with a few endemics and widespread species such as *Lobophora variegata* or *Padina australis.* Second in importance, especially at depths of less than 2–3 m, are the green algal genus *Codium,* with five nonendemic species, and representatives of the green algal genera *Caulerpa* and *Chlorodesmis.* There are also endemics in the red algal family Solieriaceae from Lord Howe Island. The relatively great seaweed biomass on Lord Howe Island may be due to the fact that herbivorous fish are relatively rare, compared to most Great Barrier Reef localities (Kraft and Olsen-Stojkovich 1985).

New Zealand harbors about 650 seaweed species, only 28% of which are shared with southern Australia and Tasmania (Parsons 1985*a,b*). As many as 43% are endemic; examples are the fucalean genera *Landsburgia* and *Marginariella* (Table 5.1). All this points to the early separation of New Zealand from Australia. Furthermore, dispersal from Australia to New Zealand occurs, while dispersal from New Zealand to Australia rarely or never occurs (Edgar 1986). Only 7% of the New Zealand flora circle the South Pole, examples are the brown algae *Adenocystis utricularis, Scytothamnus fasciculatus, Durvillaea antarctica, Desmarestia willii,* and *D. rossii* (Parsons 1985*a*). This low percentage of circumpolar species indicates that one should not overemphasize the role of the West Wind Drift.

Macrocystis pyrifera *as a carrier.* Drifting *M. pyrifera* does not seem to have great potential for transporting associated algae and animals from Tasmania to New Zealand (Edgar 1987). The associated fauna and algal flora of the holdfast, basically different on the species level in both geographical areas, may survive more than six months, but boring isopods may destroy the holdfast during the drift voyage of 3–6 months. Furthermore, the drifting kelp may become negatively buoyant in the Tasman Sea because of low nutrients and little growth. More to the south in the area of the sub-Antarctic islands, one may assume that the lower temperatures lessen the destruction of laminarian holdfasts by boring isopods and enhances the growth of the drifting algae and their importance as carriers for dispersal of other marine organisms (Edgar 1987).

The upper sublittoral zone in the warm temperate northern New Zealand region is dominated by the fucalean members *Carpophyllum, Cystophora torulosa, Xiphophora,* and *Sargassum* and by *Lessonia variegata* (Laminariales). *Ecklonia radiata* may form a dense canopy down to 17 m deep reaching its lower depth limit at 50–60 m (Choat and Schiel 1982).

In the cold temperate southern New Zealand region, which covers the southern part of North Island, South Island, and Stewart Island, *Durvillaea antarctica* colonizes the lower eulittoral zone. Here also are such red algae as *Gigartina, Laurencia, Polysiphonia, Caulacanthus, Gelidium,* and *Nemalion;* brown algae such as *Hormosira banksii* and *Colpomenia sinusosa;* and the green algae *Codium* and *Caulerpa.* The upper sublittoral zone is inhabited by *Lessonia variegata* (Fig. 5.13), as well as *Durvillaea willana,* the latter being endemic to New Zealand. Seaward, *Macrocystis pyrifera* forms beds. Both *D. willana* and *M. pyrifera* penetrate northward to a latitude of 40°S along the western coast and to 44°S along the eastern coast, whereas *Durvillaea antarctica* and *L. variegata* reach as far as northern New Zealand.

Fig. 5.13 Upper sublittoral zone on the coast of southern New Zealand, Otago Peninsula. Vegetation of *Lessonia variegata* (Laminariales, digitate fronds) and *Xiphophora chondrophylla* (Fucales, above *Lessonia*). (From Stephenson and Stephenson 1972.)

The belts of these two species and *Macrocystis pyrifera* remind one of the corresponding vegetation formed by *D. antarctica* and *Lessonia nigrescens* on the Pacific coast of South America. This similarity probably indicates that an important component of the seaweed flora of New Zealand is due to far-reaching migrations using the West Wind Drift.

Recently several algae have been introduced unintentionally to New Zealand and to southern Australia from Europe, such as the brown alga *Asperococcus turneri* (= *bullosus*) (Adams 1983), and *Arthrocladia villosa* (Skinner and Womersley 1983).

Kermadec and Chatham Islands. At the **Kermadec Islands,** situated to the north of New Zealand, the seaweed flora of at least 165 species has a warm temperate–tropical character (W. A. Nelson and Adams 1984). *Palmophyllum umbracola,* a green pseudo macroalga belonging to the Chlorococcales such as *P. crassum* in the Mediterranean and Bermuda and such as *Palmoclathrus stipitatus,* has been found near the Kermadec and Poor Knights Islands (W. A. Nelson and Ryan 1986). At the **Chatham Islands,** east of New Zealand (with a water temperature range of 5–16°C) the kelps *Lessonia variegata* and *Macrocystis pyrifera* are present, as they are in New Zealand.

***Island endemism in* Durvillaea.** The species *Durvillaea chathamensis* has been reported as an endemic to the Chatham Islands. The fact that the three solid-bladed species of *Durvillaea* (*D. potatorum, D. willana* and *D. chathamensis*) are endemic to the southwestern Pacific indicates the possibility that the genus originated here. Only *D. antarctica* with its buoyant blades circumnavigated the southern ocean on the West Wind Drift. For successful establishment of a population both sexes must have been washed ashore simultaneously. This seems possible, because several individuals are often attached to a composite, fused holdfast (Hay 1979*b*).

PART TWO
Ecophysiology of Seaweeds

The patterns of vertical and geographical seaweed distribution described in the first part of this book lead us to consider the environmental factors causing these patterns and thus the relationships of seaweeds to their abiotic and biotic environments. This is the field of ecophysiology.

Broadly speaking, the worldwide distribution patterns of seaweeds are mainly determined by global temperature gradients. Deeply imprinted temperature demands, evolved to mirror the geological cycle of cooling and heating of the earth at higher latitudes, keep species of algae apart. Probably millions of years were required for cold-water algae to obtain the ability to thrive at near-zero temperatures and to lose the ability for surviving at, for example, 20–30°C. This temperature range is optimal for tropical algae, which may die below 10°C (Chapter 7). Light, nutrient concentrations, water movement, and salinity are important, but biotic factors like grazing and competition are most important for local patterns. Light, as well as limiting the depth to which algae can grow, also plays a part in being an environmental signal for seasonal development. In reality all these factors are intimately interwoven in producing Recent distribution patterns, and so in addition to all this we have to take into account the scale of geological time.

6 Light

The phycologist is confronted with the following fundamental questions for a better understanding of the impact of light on distribution and seasonal occurrence:

(1) To what extent are the processes of photosynthesis and growth influenced by the spectral distribution of underwater light (Sections 6.1 and 6.3)?
(2) Which irradiances are limiting, saturating, or inhibiting these processes (Sections 6.2 and 6.3)?
(3) Which spectral ranges and irradiance levels are used by seaweeds as environmental signals in photomorphogenesis, photoperiodism, and other light-induced phenomena (Section 6.4)?

Light measurement. One measures either the energy or the number of quanta (called "photons" in the visible and ultraviolet portion of the spectrum) impinging on a unit surface area; the corresponding variables are **irradiance** (units: W m^{-2}) and **photon fluence rate,** respectively (units: μmol photon m^{-2} s^{-1}, or μE m^{-2} s^{-1} where 1 E = 1 einstein = 1 mol of photons = 6.02 × 10^{23} photons). The term "photon fluence rate" is used mainly by plant morphogeneticists (Mohr and Schäfer 1979; Mohr et al. 1983). Other workers prefer the terms "photon irradiance" or just "irradiance," instead of "photon fluence rate" (e.g., Jerlov 1976; Bell and Rose 1981; Kirk 1983; Dring 1984). **Photon fluence** (or **light exposure**) designates the sum of photons over a longer period than just seconds (e.g., moles of photons per square meter per day, per month, or per year). The portion of light exposure that is actually absorbed by an organism is rarely measured but is called the "light dose." Measurements of **natural underwater light** (reviews in Jerlov 1951, 1974, 1976, 1977, 1978; Morel and Smith 1974; Tyler 1977; Lüning 1981a; E. A. Drew 1983; Holmes and Klein 1987) under various environmental conditions exhibited a fairly constant relationship between photon fluence rate Q and irradiance I. The ratio of Q to I for visible radiation (400–700 nm) under water varies by no more than ±10%. One may therefore use the following conversions:

$$1 \text{ W m}^{-2} \sim 2.50 \times 10^{18} \text{ photons m}^{-2} \text{ s}^{-1} \sim 4.2 \text{ } \mu\text{mol m}^{-2} \text{ s}^{-1}.$$

6.1 SPECTRAL DISTRIBUTION IN THE EUPHOTIC ZONE

About one-half the solar radiation arriving at the water surface consists of visible light (range 390–760 nm) and ultraviolet (290–390 nm; reviews in

Robinson 1966; Gates 1979; Campbell 1981; Smith and Morgan 1981; Drew 1983; Ramus 1985). The other half is infrared radiation (760–3000 nm). The spectral composition of global irradiation above water (sum of the direct sunlight and the indirect light from the sky) depends on several factors, such as the sun's elevation and cloud cover. The latter factor reduces the short-wave part of the spectrum (Morel and Smith 1974).

At the surface of the water a small portion of the light is **reflected,** for example, 2–6% at solar elevations above 30° (Jerlov 1976). After the losses due to reflection the light entering the water, is further reduced by **absorption** (conversion of radiant energy into other energy forms, such as heat or photosynthetically fixed energy) and **scattering** (deviation of light). The combined effect of this reduction is called vertical **attenuation** (reviews in Jerlov 1976; Drew 1983; Kirk 1983; Ramus 1985). **Transmittance** in seawater is used in practical work. The relation between transmittance and vertical attenuation is given by

$$K_d = -\log_e T,$$

where K_d = attenuation coefficient for downward irradiance,
and T = transmittance.

Transmittance is determined by the following formula.

$$T = I_2/I_1,$$

where I_1 = irradiance at depth 1,
and I_2 = irradiance at depth 1 m lower.

Seawater has been optically classified according to its spectral transmittance by Jerlov (1951, 1976); he first described his **optical water types** on the basis of numerous measurements that were performed on the Swedish *Albatross* Deep Sea Expedition in 1947–1948. In clear (colorless) or **"oceanic" water** characterized by the Jerlov oceanic water types I–III (Fig. 6.1), the transmittance maximum is determined primarily by the absorption characteristics of the water and is located in the blue part of the spectrum ("blue" waters). Correspondingly, blue light prevails at greater depths, with minimal absorption at 465 nm (98% transmittance per meter; 1% of surface irradiance still present at 140 m; Figs. 6.1 and 6.2). Waters like this are found in the nutrient-poor open ocean or near the coast in the tropics or part of the Mediterranean.

In contrast, the yellowish **"coastal" waters** are rich in particles and yellow substance (see below) and are characterized by the Jerlov coastal water types 1–9 (Fig. 6.1). There is a tendency for green light to dominate at increasing depths, whereas the blue light and ultraviolet part of the spectrum becomes rapidly attenuated with the increasing number of the water type (Fig. 6.1).

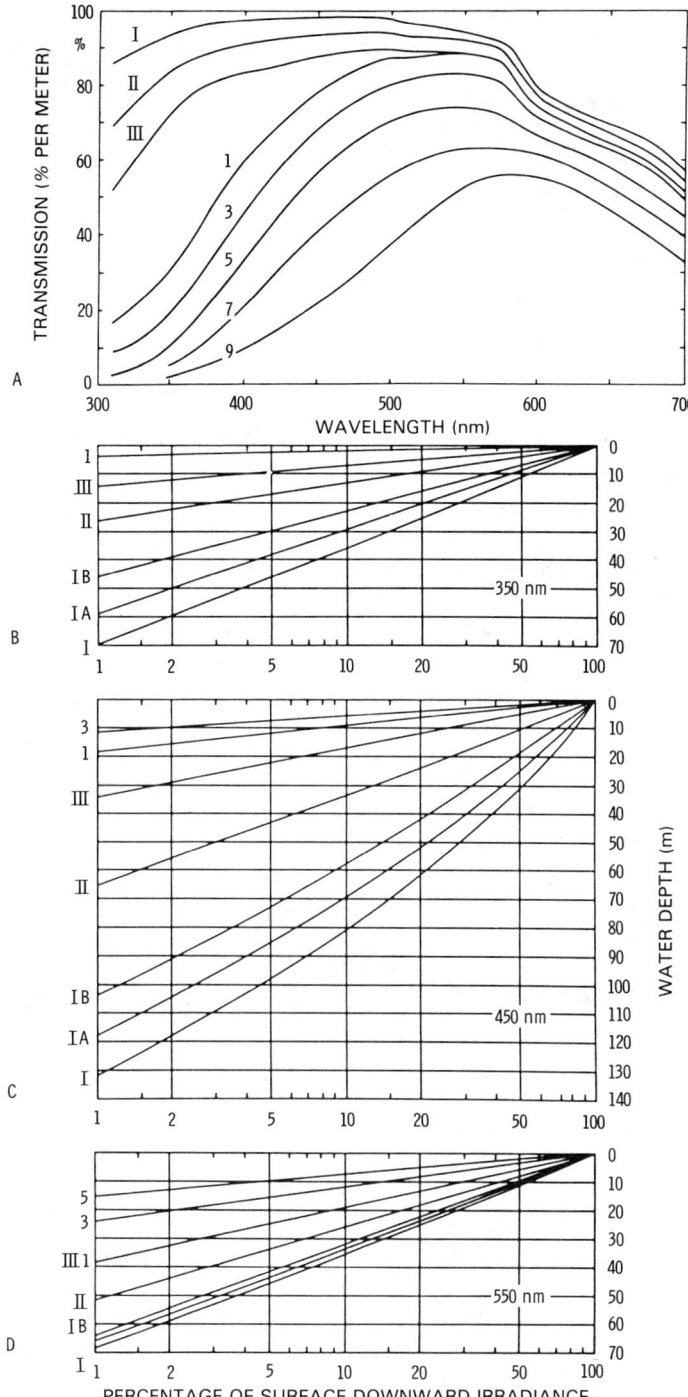

Fig. 6.1 (A) Jerlov system of optical water types. Transmission of daylight (350–700 nm) per meter at oceanic water types (I–III) and coastal water types (1–9). (B–D) Irradiance at different depths for selected wavelengths, in relation to Jerlov water types. (A from Jerlov 1976; B–J from Jerlov 1978.)

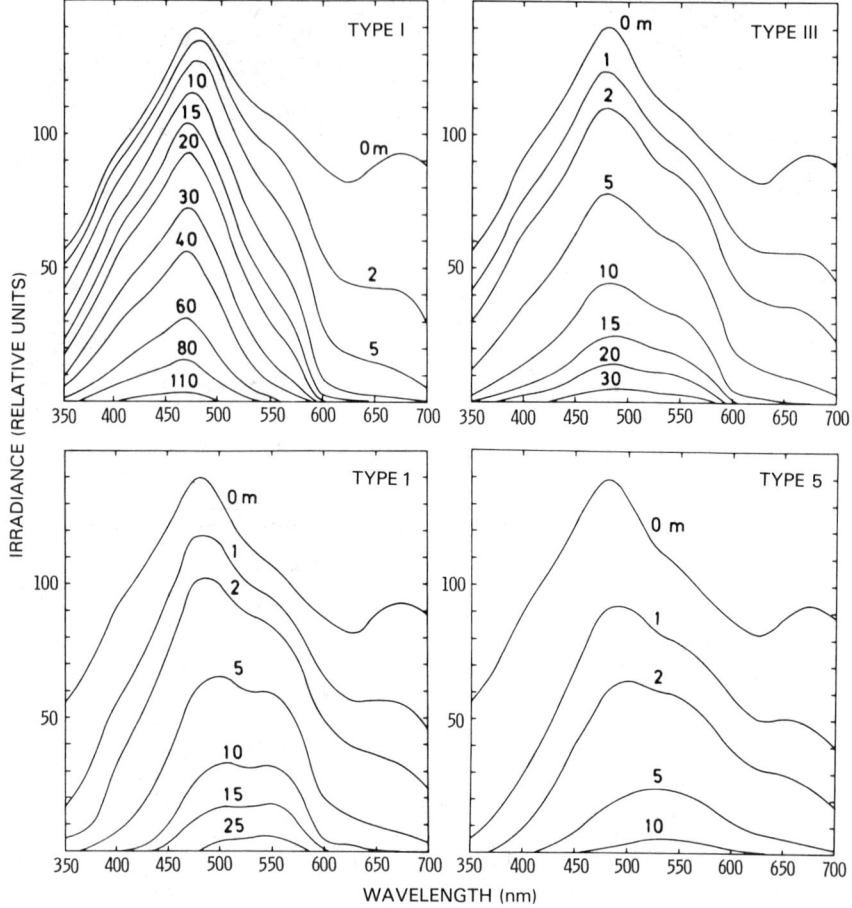

Fig. 6.2 Spectral irradiance of underwater daylight in the sea at four Jerlov water types. (From Jerlov 1978.)

Overall transmittance of the light is reduced. At 550 nm, which is the maximum transmittance in water type 5, 1% of the light penetrates to only 15 m of water depth (Fig. 6.1D). In such waters the position of the transmittance maximum is shifted to the green part of the spectrum, since blue and ultraviolet are strongly attenuated due to **scattering** by particles and **selective absorption** by **yellow substances**. The scattered light is deflected, and its optical pathlength is increased. This enhances the chance of photon absorption. The yellow substances, also termed **gelbstoff** or **gilvin** (reviews in Bricaud et al. 1981; Kirk 1983), are represented by dissolved organic and humic substances that have been brought into the sea by rivers. Phenolic substances released by seaweeds and phytoplankton may also contribute to these yellow substances, but on a minor scale (Højerslev 1982). Jerlov water types 1–3 are typical on the Atlantic coasts of Europe, where the euphotic depth may be 45 m, e.g. Brittany (Table 6.1).

The Jerlov water type may be estimated from a single measurement of transmittance in a narrow waveband, for example, at 465 nm as proposed by Jerlov (1974). A modification of Jerlov's system has been presented by Baker and Smith (1982), who related the vertical attenuation coefficient to the chlorophyll content and dissolved organic matter in the sea.

6.2 ALGAL DEPTH LIMITS IN RELATION TO NATURALLY OCCURRING LIGHT LEVELS

Above the atmosphere the irradiance of solar radiation, also termed the **solar constant,** amounts to 1361–1365 W m^{-2} (= J m^{-2} s^{-1}; 1 W = 1 J s^{-1}). From the seasonal course of insolation outside the atmosphere (Fig. 6.3B), it is clear that the maximum solar radiation is around 40 MJ m^{-2} day^{-1} (where 1 MJ = 1 megajoule = 10^6 J) all year round at the equator and in midsummer at the North Pole. Ninety-nine percent of the quanta are distributed within the waveband 200–4000 nm. After the passage of solar radiation through the atmosphere, this spectral range is narrowed to 290–3000 nm at the earth's surface. It is reduced to a maximum of 70% of the value of the solar constant in tropical regions, that is, to a global irradiance of about 1000 W m^{-2} (reviews in Gates 1979; Campbell 1981; Smith and Morgan 1981). The "photosynthetic active radiation" (PAR) of 400–700 nm (McCree 1981) is estimated to contain 50% of the global irradiation. This amounts to a maximum of 500 W m^{-2} in the tropics, corresponding approximately to a photon fluence rate of 2500 μmol m^{-2} s^{-1}. The yearly amount of insolation on the surface of the earth, after passage through the atmosphere, varies by a factor of 2–3 from the Poles to the tropics (Fig. 6.3D).

Which **annual amount** of underwater light, that is, annual light exposure or photon fluence, is available to seaweeds at different water depths? Tables 6.2 and 6.3 show the results of continuous measurements of underwater light performed for more than a year in the sublittoral zone of Helgoland, North Sea. According to this, an annual light exposure of 6 mol photons m^{-2} (= E m^{-2}), equivalent to 1 MJ m^{-2} yr^{-1}, reaches the lower algal limit of 15 m. The algae at this 15 m depth are crustose coralline algae a few millimeters in diameter, such as *Lithothamnion sonderi*. Below this the rocky substratum near Helgoland is devoid of algae. One may therefore regard these algae as representatives of the perennial, multicellular algae that have **minimum light requirements.** More light, in fact a minimum annual light exposure of about 70 mol photons m^{-2}, is needed to build up a larger thallus, like that of the kelp *Laminaria hyperborea* (Table 6.2). Chapman and Lindley (1980) measured light continuously in the sublittoral zone of Igloolik in the Canadian Arctic and found that an annual light exposure of 50 mol photons m^{-2} is available for the deepest specimens of *L. solidungula*. An annual total

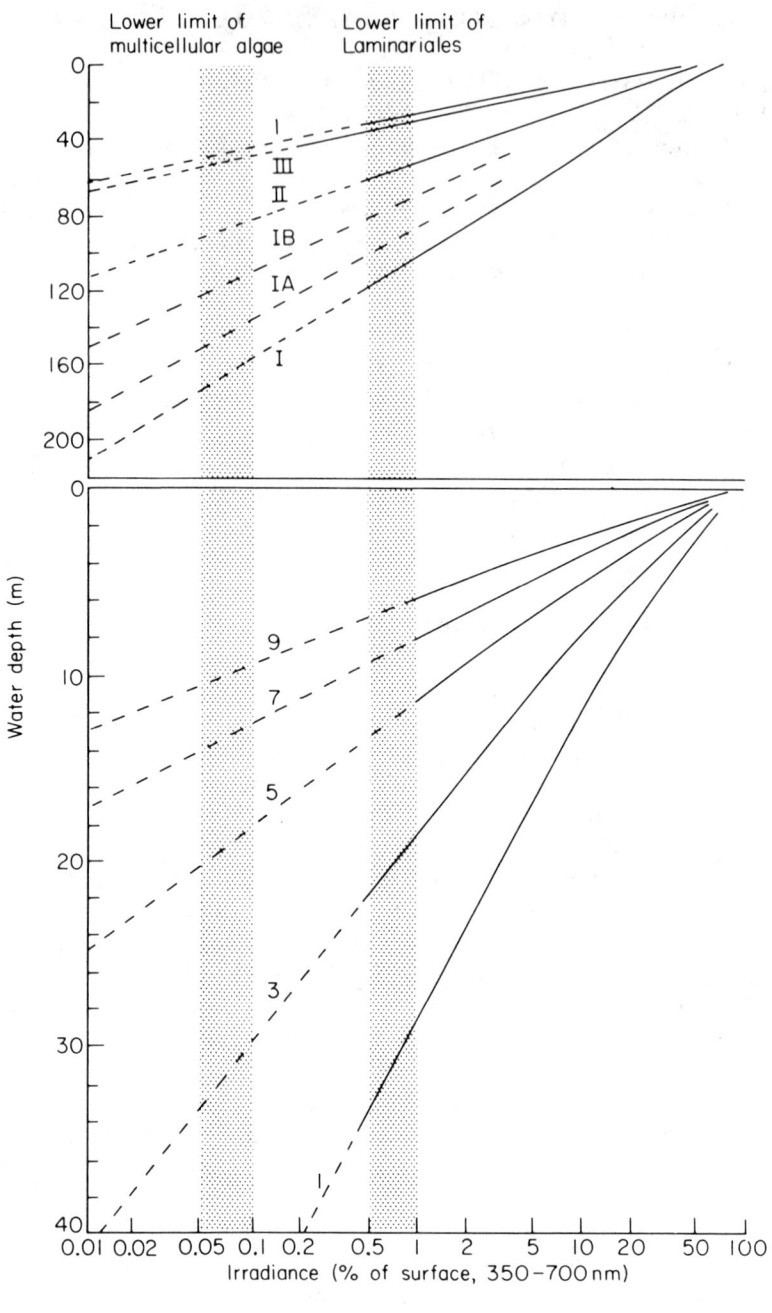

Fig. 6.3 (A) Underwater irradiance as a percentage of irradiance above water (350–700 nm), and characteristic algal depth limits in relation to Jerlov water types. (B) Latitudinal effect on calculated daily insolation (ignoring the influence of the atmosphere) throughout the year in the Northern Hemisphere. (C) Variation of lower algal limit (extinction depth) with latitude. (D) Annual average sum of photosynthet-

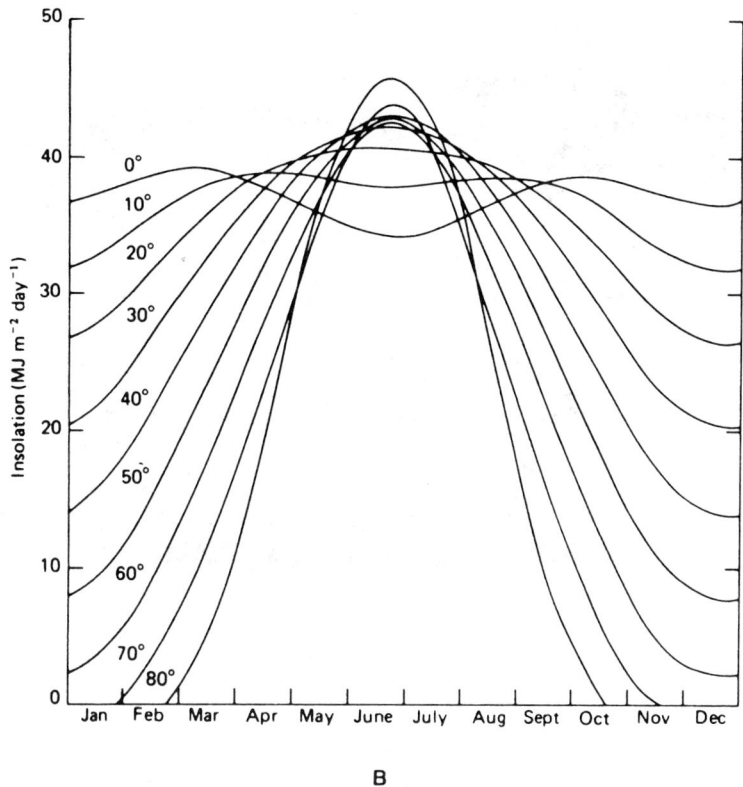

B

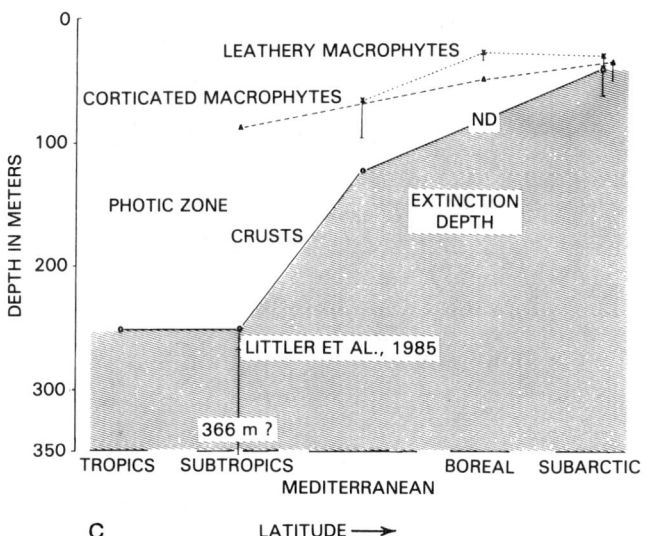

C

ically active radiation (in M J m^{-2} yr^{-1}) at the surface of the earth, after passage of sun radiation through the atmosphere. (A from Lüning 1971, modified after Lüning and Dring 1979 and Jerlov 1976); B from Kirk 1983, based on data of Kondratyev 1954; C from Vadas and Stenek 1988; D after Geiger 1965.)

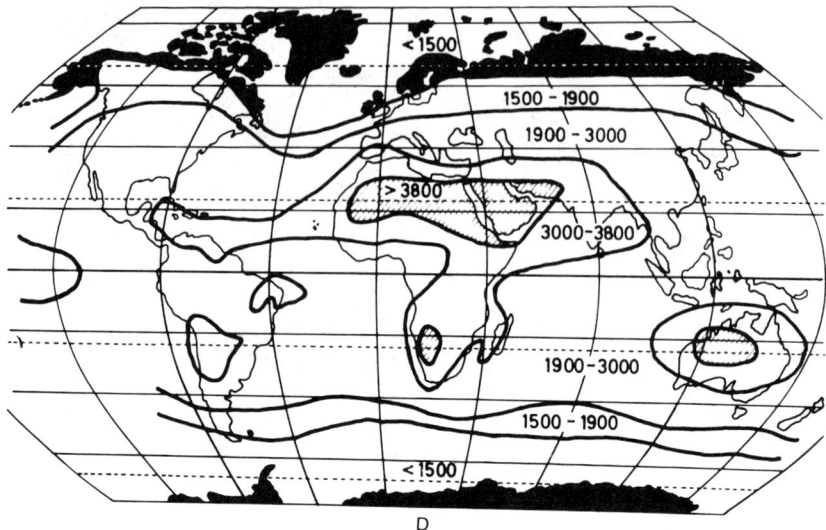

Fig. 6.3 (*Continued*)

Table 6.1 Lower depth limits of crustose coralline algae (= lower limit of algal vegetation), and representatives of the Laminariales, in relation to Jerlov-water types and irradiance P (as a percentage of light, 350–700 nm, above water) on different coasts

		Jerlov-water type	P (%)
	Deepest crustose coralline algae		
Helgoland	15 m	7	0.05
Brittany	45 m	1	0.05
Corsica	120 m	IA	0.1
Bahamas	268 m	I	0.001
	Deepest Laminariales		
Laminaria hyperborea:			
Helgoland	8 m	7	0.7
Roscoff (France)	25 m	III	1.2
off Aran Islands (west coast of Ireland)	32 m	III	1.2
Laminaria rodriguezii:			
Corsica	95 m	IA	0.6

After Lüning and Dring 1979; Guiry, pers. comm.

Table 6.2 Yearly integrals of irradiance (I) and of photon fluence rate (Q; both 400–700 nm) at different water depths in the euphotic zone at Helgoland. P indicates the percentage of irradiance at a certain water depth in relation to irradiance above water. Water depths are given in m below Chart Datum (mean low water at spring tides); MJ = Megajoule (1 MJ = 10^6 J); 1 mol photon =1 Einstein

Lower depth limits	depth m	I MJ m^{-2}	Q mol photon m^{-2}	P %
deepest individuals of Laminaria digitata	2	227.2	1037.2	11
dense forest of Laminaria hyperborea	4	84.6	387.7	4
deepest individuals of Laminaria hyperborea	8	15.6	71.2	0.7
deepest erect red algae like Delesseria sanguinea	10	7.3	33.4	0.3
algal depth limit (deepest crustose coralline algae)	15	1.3	6.0	0.05

After Lüning and Dring 1979.

Table 6.3 Euphotic zone at Helgoland: mean monthly values of photon fluence rate (Q; μmol $m^{-2}s^{-1}$; 400–700 nm) at five characteristic algal depths below mean low water springs. See Figures 2.48 and 2.49 for lower algal depth limits. Mean values of Jerlov-water type, and of daylength above and at 2.5 m below water

	depth	lower algal limits	Jan.	Feb.	March	Apr.	May	June	July	Aug.	Sep.	Oct.	Nov.	Dec.
Q	2 m	upper sublittoral zone	10.3	36.1	39.9	77.6	64.6	120.0	142.7	100.2	58.4	25.9	19.5	6.6
Q	4 m	laminarian forest	2.2	14.0	14.8	32.2	19.9	42.9	55.7	42.4	18.5	8.5	6.5	1.3
Q	8 m	deepest L. hyperborea	0.1	2.6	2.7	7.2	2.7	7.2	10.3	9.2	2.4	1.3	1.0	0.0
Q	10 m	deepest Delesseria	0.0	1.2	1.3	3.6	1.1	3.2	4.7	4.6	0.9	0.5	0.4	0.0
Q	15 m	deepest crustose algae	0.0	0.2	0.2	0.7	0.1	0.5	0.7	0.9	0.1	0.0	0.0	0.0
Jerlov-water type			9.0	6.5	6.5	6.5	7.5	7.0	6.0	5.5	7.0	8.0	8.0	9.0
Daylength above water (h)			8.0	9.8	11.8	14.0	15.9	17.0	16.7	14.8	12.7	10.6	8.7	7.5
Daylength at 2.5 m depth (h)			1.1	8.7	9.0	11.7	14.7	16.7	16.5	14.7	12.0	8.2	6.6	3.5

From Lüning and Dring 1979.

on the order of 25 mol photons m^{-2} has been calculated for the deepest-growing population of the green alga *Halimeda* at a 140 m depth at Enewetak Atoll, Marshall Islands (Hillis-Colinvaux 1986*a*).

Higher up in the sublittoral zone, annual light exposure rises to values of about 1000 mol photons m^{-2} yr^{-1}, or about 200 MJ m^{-2}, at a 2 m water depth near Helgoland (Table 6.2), but the algal vegetation becomes denser, self-shading increases, and an individual alga (e.g., a *Laminaria* thallus) may possibly receive no more light per year than deeper down. These underwater values may be compared with the much higher terrestrial light exposures. An entire temperate deciduous forest ecosystem at medium latitudes receives an annual light exposure of about 5000 MJ m^{-2}, of which 2000 MJ m^{-2} are used in the few months that the trees are in leaf, mainly from June through September (Gosz et al. 1978).

One may also express the light available at a certain water depth as a percentage of light above water, that is, as the **light percentage depth** (100% = photosynthetically active radiation, PhAR, above surface; 400–700 nm or 350–700 nm, according to the meter used; measured as mol photons m^{-2} yr^{-1}). The deepest-occurring multicellular algae receive 0.05–0.001% of the surface light, calculated from the lower algal depth limits (crustose algae) at various locations (Table 6.1). The light percentage depth for the deepest kelp individuals is situated at 0.6–1.2% (Table 6.1), and the lower limit of a closed kelp vegetation, like the lower depth limit of the *Laminaria hyperborea* forest near Helgoland, may be characterized by a light percentage depth of 4% (Table 6.2). Although the lower depth limits, expressed in meters, may vary by an order of magnitude due to differences in water clarity (e.g., the deepest *Laminaria* near Helgoland are found at 8 m deep, whereas in the Mediterranean it is 95 m deep), such limits are similar, if expressed as light percentage depth (Table 6.1). Therefore, if one has determined the Jerlov water type at a particular location, one may predict roughly the depth of the deepest-occurring algae or kelp species (Table 6.1, Fig. 6.3A). The geographical variation in solar radiation (Figs. 6.3B,D) varies the picture and leads to lower percentage depths towards the equator (Fig. 6.3C). It should be noted that in very turbid water, for example, at sites in the upper Bristol Channel, England, the 1% light depth may occur in the eulittoral zone or even above (Dring 1987*c*).

Strategy for life of deep-water algae. The shade-adapted deep-water algae use the same strategy as terrestrial shade plants (reviews in Boardman 1977; Björkman 1981; Larkum and Barrett 1983; J. M. Anderson 1986). Such shade-plant characteristics include slow growth (which minimizes respiration), and disproportionate storage of fixed carbon as reserve materials. An important prerequisite for such a long-term strategy is longevity of the thallus (Section 8.4), and mechanisms to resist grazing pressure (Section 8.1.3). The growth form of the **crustose coralline algae** is represented in all regions as the only conspicuous algal growth form left at the lower algal limit and in the inner parts of sea caves. These algae seem to be ideally adapted to a deep-water environment since they are well protected against grazing and survive in spite of slow growth. Their structure also enhances light absorption because it represents a horizontal light receiver with none of the self-shading branches that occur in erect algae.

Lower occurrence limits of photoautotrophic life. Phytoplankton ecologists in general regard the light percentage depth of 1% to represent the lower limit of the euphotic zone (Steemann-Nielsen 1975). This value has been derived from many measurements of photosynthesis of phytoplankton enclosed in bottles. These measurements on phytoplankton show this depth to be the **compensation depth,** that is, the point at which photosynthesis balances respiration. However, in special cases it has also been found that unicellular, pelagic green algae and diatoms still exhibit a net gain of photosynthesis in bottles submerged to a 200 m depth in clear oceanic waters, which corresponds to a range of light percentage depths of 0.01–0.05% (Jeffrey 1981; Kirk 1983). This range is similar to that of benthic multicellular algae (Table 6.1) and unicellular algae, such as pennate diatoms found near the lower limit of seaweeds (Sears and Cooper 1978). The range of 0.01–0.05% corresponds to photon fluence rates of 0.3–1.3 μmol m^{-2} s^{-1} or irradiances of 0.06–0.3 W m^{-2}, when maximum values of 2500 μmol m^{-2} s^{-1} or 500 W m^{-2} are recorded at the surface. Algae in terrestrial caves in southern France grow at minimal photon fluence rates of around 0.1 μmol m^{-2} s^{-1} supplied over a few hours per day (Leclerc et al. 1983). Minimum values for growth of unicellular algal species in culture are in the range 0.2–4 μmol m^{-2} s^{-1}, and near 1 μmol m^{-2} s^{-1} for terrestrial ferns with minimum light requirements (Geider et al. 1985; Raven 1986). Measured photon fluence rates in tropical habitats with clear water were, for example, 4-30 μmol photon m^{-2} s^{-1} at 80–100 m deep, from which one may calculate 0.5-5 μmol photons m^{-2} s^{-1} at 140 m deep, where the deepest populations of the green alga *Halimeda* have been found (Hillis-Colinvaux 1986a). The deepest-growing zooxanthellate coral in the Red Sea, *Leptoseris fragilis,* receives 0.5–10 μmol photons m^{-2} s^{-1} at noon, or 0.15–1.7% of surface light (Schlichter et al. 1985, 1986). Ambient photon fluence rate for benthic diatoms beneath Antarctic sea ice at 20–30 m deep was less than 0.6 μmol m^{-2} s^{-1}, that is, 0.04% of surface light (Palmisano et al. 1985). The extraordinary find of crustose coralline algae at 268 m near the Bahamas (Littler et al. 1985) shifts the minimum light requirement to 0.03 μmol m^{-2} s^{-1} (= 0.001% of 2500 μmol photon m^{-2} s^{-1}; see Fig. 6.3A) and is the deepest recorded algal life.

6.3 LIGHT DEMANDS FOR PHOTOSYNTHESIS AND GROWTH

6.3.1 Photosynthetic Pigments, Action Spectra, and Vertical Distribution of Seaweeds

The structural formulae of the photosynthetic pigments of green, brown, and red algae are represented in Fig. 6.4, absorption spectra in Fig. 6.5, and typical absorption peaks are listed in Table 6.4 (reviews: Jeffrey 1981; Ragan 1981; Ramus 1981; Larkum and Barrett 1983). All three algal groups have **chlorophyll a,** which exhibits in vivo (in the living cell) absorption peaks at about 440 nm in the blue and at about 675 nm in the red. As a consequence, all algal groups are able to absorb the prevailing blue light at greater depths in clear, oceanic water.

The typical color differences of the green, brown, and red algae are due to

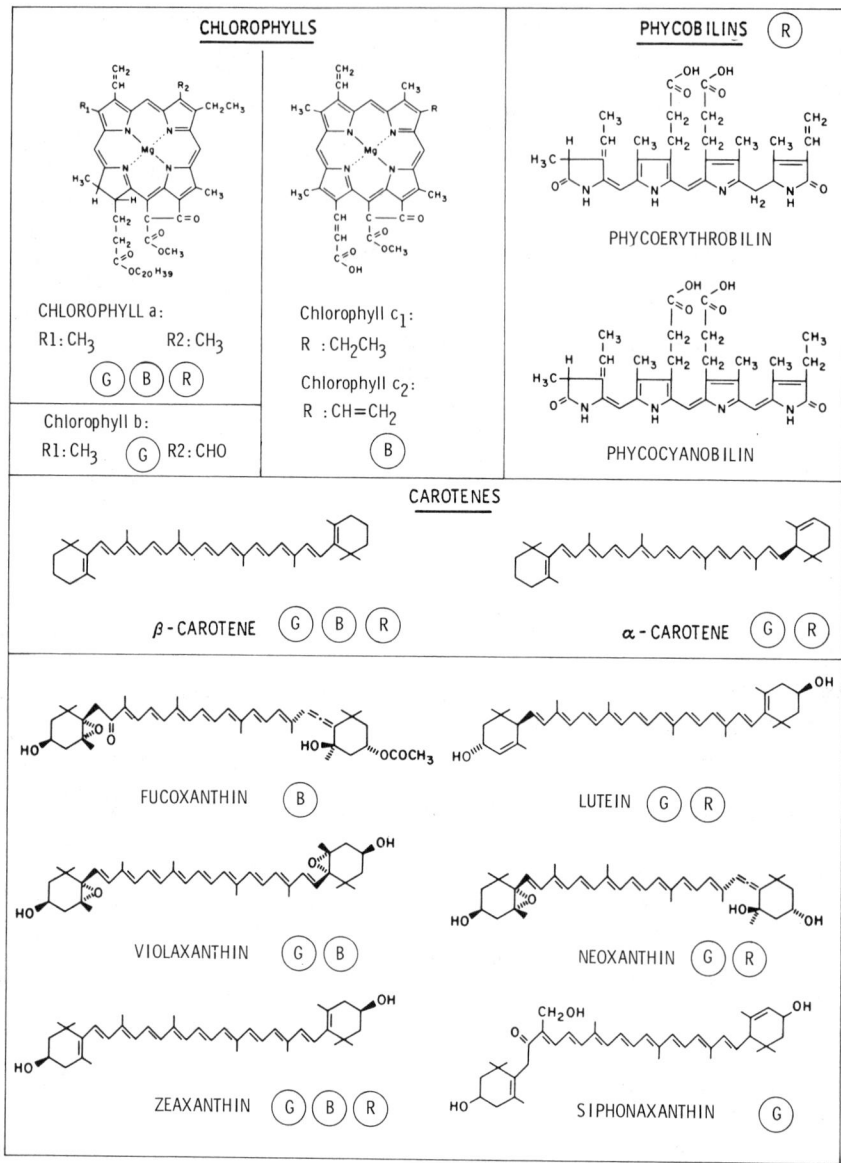

Fig. 6.4 Structural formulae of chlorophylls, chromophores of phycobilins, and carotenoids of seaweeds, G = green algae; B = brown algae; R = red algae. (Structural formulae from Ragan 1981.)

accessory pigments with maximum and differential absorption in the green part of the spectrum. These accessory pigments include the carotenoid **fucoxanthin** (reviews of carotenoids in Goodwin 1974; Goedheer 1979) in the brown algae, the reddish or bluish **phycobiliproteins** in the red algae, and the carotenoids **siphonaxanthin** and **siphonein** (review in Yokohama 1981) in the order Caulerpales and in some other groups of green algae.

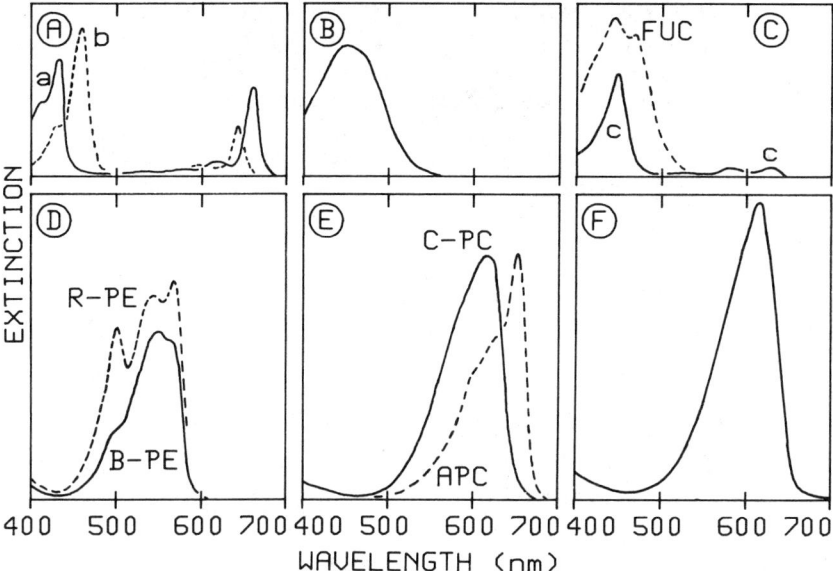

Fig. 6.5 Absorption spectra of photosynthetic pigments. **Green algae:** (A) Chlorophylls a and b in ether. (B) Siphonaxanthin in ethanol. **Brown algae:** (C) Chlorophyll c and fucoxanthin (FUC) in acetone. **Red algae:** (D) R-phycoerythrin (R-PE), B-phycoerythrin (B-PE). (E) C-phycocyanin (C-PC) and allophycocyanin (APC). (F) R-phycocyanin. (A after Nultsch 1986; B from Yokohama and Kageyama 1977; C from Goedheer 1970; D–F from Goodwin 1974.)

Green algae have, in addition to chlorophyll a, **chlorophyll b**, which exhibits in vivo absorption peaks at 470 nm and 650 nm. As in chlorophyll a, these absorption peaks are in the blue and red parts of the spectrum but slightly shifted toward the middle part of the spectrum (Table 6.4). In green algae the **action spectrum** of photosynthesis (rate of photosynthesis that occurs in monochromatic light of low, constant photon fluence rate at different wavelengths) largely follows the absorption spectrum (Fig. 6.6A), so that photosynthesis proceeds optimally in blue and red light. Green algae are abundant at greater depths in clear water, which is rich in blue light. The fact that red light is absent here is of no importance with regard to survival. In the laboratory one can rear green algae in blue light without difficulties, but not always in red light.

Thin green algae from shallow water show the so-called "**green window**" in the in vivo absorption spectrum of their chlorophyll a (Fig. 6.6A). In green algae from greater depths or from shaded areas, and in thick green algae particularly, this green window is almost completely filled by **enhancing the chlorophyll content** (Fig. 6.7). This is possible because, by increasing the chlorophyll content, the absorption in the green window part of the spectrum rises faster than in the blue and red parts, where absorption is high anyway. Thick algae may be regarded as optically "**black.**" Their many cell layers are filled with chloroplasts, there is much internal scattering, and the chance of a

Table 6.4 Absorption peaks of photosynthetic pigments. Wavelengths in nm; **main peaks** printed in **boldface**. Chlorophylls and fucoxanthin: measured *in vitro* (after extraction) or *in vivo* (in live thallus). Phycobiliproteins: absorption peaks are identical *in vivo* and *in vitro*. All *in vivo*-values may differ slightly among species and samples due to differences in the protein part of the chromoproteids

Pigment								
Chlorophyll a								
in vitro (90 % acetone)	380	410	**430**		580	615		**663**
in vivo	385	418	**438**		590	625		**675**
Chlorophyll b								
in vitro (90 % acetone)			**455**				645	
in vivo			**470**				650	
Chlorophyll c$_1$								
in vitro (diethylether)			**444**		578	630		
in vivo						634		
Chlorophyll c$_2$								
in vitro (diethylether)			**449**		582	631		
Fucoxanthin								
in vitro (90 % acetone)			**449**					
in vivo			**545**					
Phycobiliproteins (watery extr.)								
R-Phycoerythrin				498	**542**	**565**		
B-Phycoerythrin				498	**545**	**563**		
R-Phycocyanin					553		**615**	
C-Phycocyanin							**620**	
Allophycocyanin								**650**

Compiled after French 1960; Goedheer 1970; Jeffrey 1968, 1981; Jeffrey and Humphrey 1975; Jensen 1978; Meeks 1974.

photon of any wavelength being absorbed is thus extremely high (Ramus 1978, 1981; Lüning and Dring 1985).

The chloroplasts of many green algal species growing in deep water contain **siphonaxanthin** and its ester **siphonein** (Fig. 6.4; for in vitro absorption see Fig. 6.5B). Siphonaxanthin in vivo absorbs maximally in the green range, at 540 nm (Yokohama 1981; O'Kelly 1982a). Other common carotenoids of seaweeds (Fig. 6.4) absorb towards the green end near 510 nm. Both siphonaxanthin and siphonein are typical of many species of the order **Caulerpales** (formerly called **Siphonales**), which in the tropical and warm temperate regions occur as dominants in the deeper sublittoral zone. Siphonaxanthin has also been detected in members of the Chaetophorales, Ulvales, Cladophorales, and Siphonocladales, all from deep water (Yokohama 1981).

J. M. Anderson (1983) has suggested that the possession of siphonaxanthin-siphonein-chlorophyll a/b proteins in the Caulerpales may be an evolutionary relic of taxa of the early Paleozoic, which lived in deeper water when no ozone layer shielded harmful UV radiation from the surface of the earth. Those few green algal species that later moved to shallower depths retained their siphonaxanthin and siphonein.

Some deep-water green algae (e.g., species of *Cladophora* and *Microdic-*

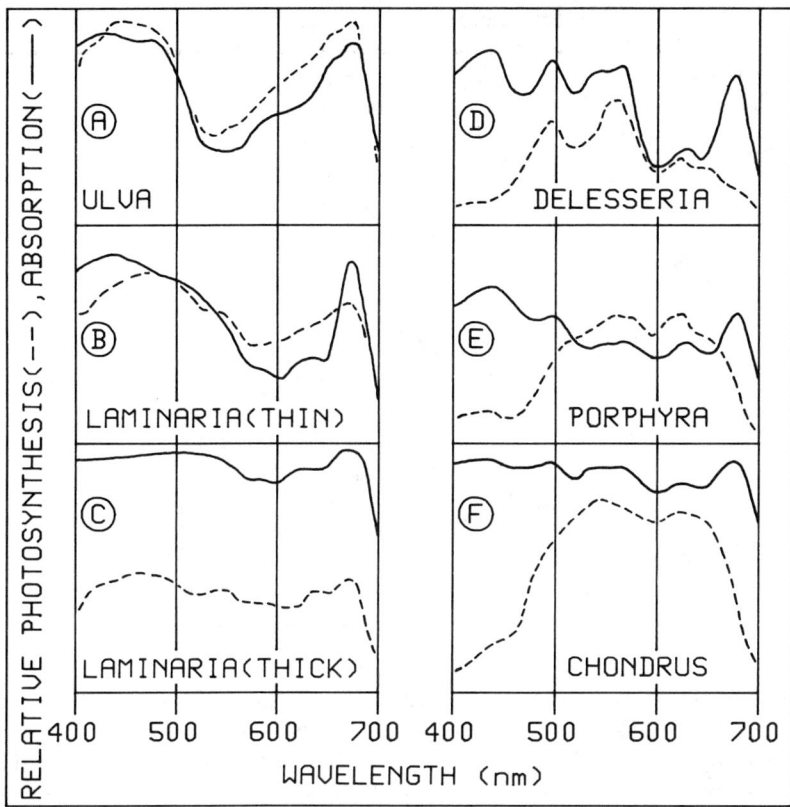

Fig. 6.6 Action spectra of gross photosynthesis (broken lines) and absorption spectra of several seaweeds. Species: **(A)** *Ulva lactuca*. **(B,C)** *Laminaria saccharina*, **(D)** *Delesseria sanguinea*, **(E)** *Porphyra umbilicalis*, and **(F)** *Chondrus crispus*. (After Lüning and Dring 1985.)

tyon) have **loroxanthin,** the precursor of siphonaxanthin, which does not absorb green light as efficiently as siphonaxanthin (Yokohama 1983). With increasing depth and shade in the sea, green algae may also accumulate other carotenoids such as β-carotene, lutein, zeaxanthin, anteraxanthin, violaxanthin, and neoxanthin (Titlyanov and Lee 1978).

The color of **brown algae** is mainly the result of **fucoxanthin,** a xanthophyll (Figs. 6.4 and 6.5C). Compared to β-carotene, which is present in all algal groups (Fig. 6.4), fucoxanthin occurs at a concentration five to eight times higher in brown algal thalli. It causes a relatively high in vivo absorption in the spectral range of 500–560 nm, ending only at 590 nm (Goedheer 1970; Kirk 1977). This is also evident from the photosynthetic **action spectrum** of a thin piece of *Laminaria saccharina* (Fig. 6.6).

In addition to chlorophyll a, the brown algae have **chlorophyll c.** This represents a mixture of two structurally different pigments, chlorophyll c_1 and c_2, which strictly speaking should be designated as chlorophyllides, since the phytol chain is absent (Jeffrey 1981; see Jeffrey and Humphrey 1975 for

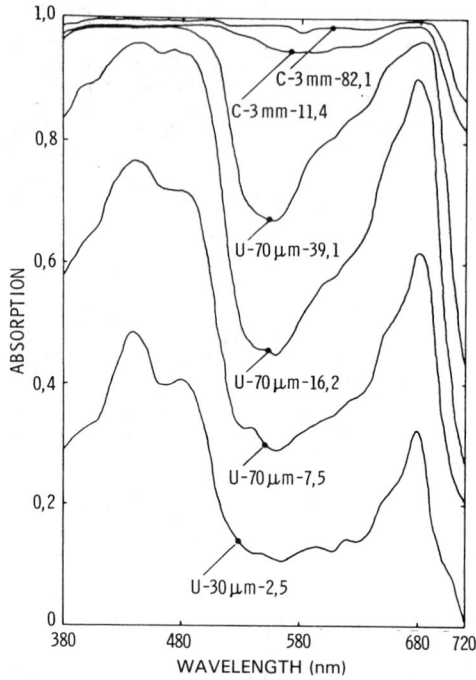

Fig. 6.7 Absorption spectra of green algae. U = *Ulva lactuca:* four thalli with differences in thallus thickness (30–70 μm), and with different pigment contents, ranging from 2.5 to 39.1 n mol cm^{-2}. C = *Codium fragile:* two thalli of 3 mm thickness with pigment contents of 11.4 and 82.1 n mol cm^{-2}, respectively. (From Ramus 1981.)

quantitative determinations). The presence of chlorophyll c may be shown in the in vivo absorption spectrum by its main absorption peak in the red at 630–638 nm and by a minor peak at 590 nm. The main absorption peaks in the blue (Table 6.4) are hidden under the main absorption peaks of chlorophyll a. The quantitative proportion of chlorophyll a : c in brown algae varies from 1 : 1 to 5 : 1. The older and **optically thick thalli,** for example, of the Fucales and Laminariales, may also be regarded as black, since their absorption spectra exhibit only small differences throughout the visible spectrum (Fig. 6.6C). However in this case the strategy has been to produce a pigment, fucoxanthin, which effectively absorbs green light.

In the **red algae** the green window in the absorption spectrum of chlorophyll a is filled by the water-soluble **phycobiliproteins,** which have high absorption in the range of 500–650 nm (Table 6.4). These pigments of the red and blue-green algae absorb optimally the green underwater light of greater depths in "coastal" waters. It has been suggested that both groups evolved during the Precambrian when the UV impact of the sun forced them to avoid shallower waters (Section 6.3.2). Both green and brown algae have equal ability to photosynthesize at depth in "oceanic" water (Larkum et al. 1967, Dring 1981). Thus it has been suggested that these algae evolved a strategy to survive below other algae in the water column above them (Larkum and Barrett 1983).

Unlike the chlorophylls and carotenoids, which are localized in the thylakoids of the chloroplasts, the phycobiliproteins are found in the **phycobilisomes.** These are particles of about 35 nm in diameter attached to the thylakoid membranes (reviews in Gantt 1981; Glazer 1982; Larkum and Barrett 1983). A phycobiliprotein consists of a chromophoric group, the **phycobilin** (formulae in Fig. 6.4), and a proteinaceous part. Among the phycobiliproteins are the **phycoerythrins** (main absorption in the green, at 500–570 nm), of which the B-phycoerythrin contains the reddish phycoerythrobilin as a chromophoric group, **phycocyanins** (main absorption at 615–620 nm), and **allophycocyanin** (main absorption in the near red, at 650 nm); the last two have the bluish phycocyanobilin as a chromophoric group (Figs. 6.4, 6.5 and Table 6.4). The deep-red algae of the deeper sublittoral zone, such as *Delesseria sanguinea* (absorption spectrum in Fig. 6.6D), are characterized by having a higher content of phycoerythrin than of phycocyanin pigments. In contrast, the red algal species of the eulittoral and the upper sublittoral zones have a higher content of phycocyanins. Hence species such as *Porphyra umbilicalis* (Fig. 6.6E) and *Chondrus crispus* (Fig. 6.6F) appear more violet. In darkness, *Delesseria sanguinea* forms phycobilins, but chlorophyll a is absent or formed in only trace amounts (Lüning and Schmitz 1988).

Another chromophoric group is **phycourobilin,** which together with phycoerythrobilin is a constituent of R- and B-phycoerythrin in red algae (Beale 1984) and some phycoerythrins of cyanophytes. At lower light levels, the red alga *Callithamnion roseum* increases the phycourobilin–phycoerythrobilin ratio within the phycoerythrin chromophore and thus obtains a stronger light absorption over a wide range of the spectrum because phycourobilin has a higher molar extinction coefficient than phycoerythrobilin (Yu et al. 1981). Furthermore, phycourobilin, with its absorption peak near 500 nm, is suited for absorption of bluish light in **deep water** and has been preferentially detected in deep-water prokaryotic picoplanktonic organisms (Glover et al. 1986). Phycourobilin and phycoerythrobilin were also found as the chromophores of a phycoerythrin in symbiotic, unicellular, and filamentous cyanophytes living in sponges and ascidians on coral reefs (Larkum et al. 1987). **Phycoerythrocyanin,** another phycobiliprotein, has been discovered in cyanophytes and contains phycocyanobilin and phycobiliviolin as chromophoric groups (Beale 1984).

The photosynthetic **action spectra of the red algae** (Figs. 6.6D–F) follow the absorption spectra in the green portion of the spectrum but not in the blue and red parts, which are the main areas of absorption of chlorophyll a. This "inactivity" of chorophyll a in red algae, first described by Haxo and Blinks (1950) as the "red drop" and "blue drop," is lessened by adding narrow-band green light to the narrow-band red (Fig. 6.8). The resulting photosynthetic rate is higher than it would be if both wavebands were applied separately and the photosynthetic rates were added together. This effect, known as the **Emerson enhancement effect,** occurs because it is only that both photosystem

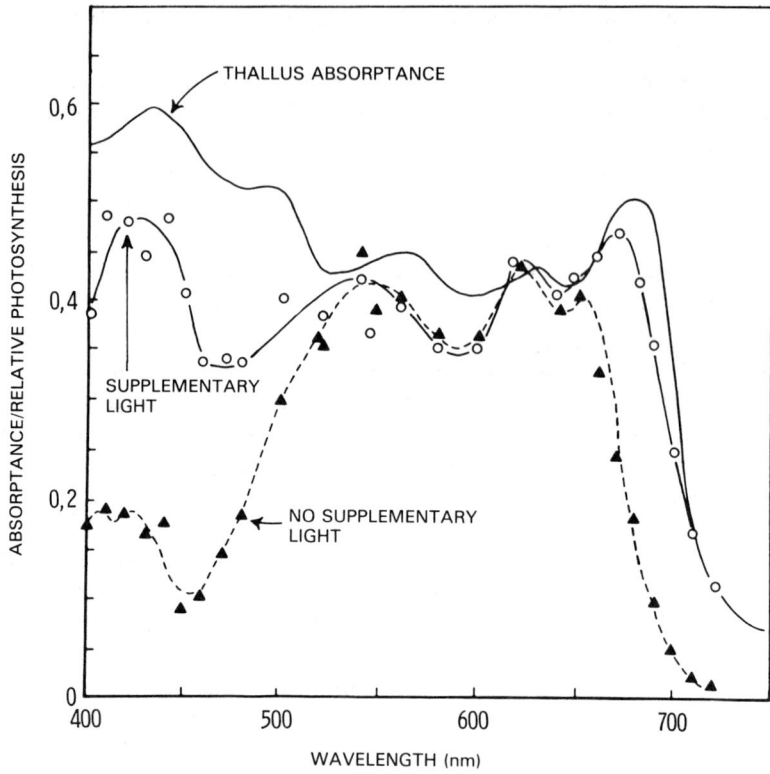

Fig. 6.8 Absorption and photosynthetic action spectra of the red alga *Porphyra perforata*. "No supplementary light": a normal action spectrum was measured in narrow-band light at 26 wavelengths. "Supplementary light": at each of the 26 wavelengths, green light with a center at 546 nm (excitation of phycobilins) was given in addition to narrow-band light. (From Ramus 1981, according to original work by Fork 1963.)

I (main absorption in the red) and photosystem II (main absorption of red algae in the green due to phycobiliproteins) are optimally supplied with photons. The accessory pigments and shorter-wavelength forms of chlorophyll a are found in photosystem II. Photosystem I consists almost entirely of chlorophyll a, with a higher proportion of long-wavelength forms of chlorophyll a than photosystem II. In **brown** and **green algae,** the red drop is evident only at wavelengths longer than 680 nm. In these algal groups there is a good overlap between the absorption ranges of photosystem I and II due to the absorption of chlorophyll c and b in the red, up to 680 nm.

Underwater light of a relatively **narrow spectral width** (although broader than the one used in the laboratory for measuring photosynthetic action spectra) occurs in the lower sublittoral zone where green light prevails in coastal, yellowish water and blue light in clear, colorless waters (Fig. 6.2). From the similarities existing between photosynthetic action spectra (Fig.

6.6) and underwater spectral distributions (Fig. 6.2), the advantage of red algae in **coastal** (yellowish) water with prevailing green light at depth is evident. As calculated by Dring (1981) for Jerlov coastal water types 3 and 9, the photosynthetic effectiveness decreases with depth in the green alga *Ulva*, increases in red algae rich in phycoerythrin, and shows little change in brown algae such as *Laminaria saccharina*.

It might be thought that in **oceanic** (colorless) waters, with prevailing blue-green and blue light at greater depths, the green and brown algae would grow deeper than the red algae because the latter have three to five times lower rate of photosynthesis in narrow blue light ("blue drop") than in narrow green light (Figs. 6.6D–F). However the light which penetrates to considerable depths in oceanic waters is blue-green (480–510 nm; Fig. 6.2) and in this region R- and B-phycoerythrin can absorb efficiently due to the presence of the PUB chromophore. Although Harder and Bederke (1957) showed in laboratory experiments that growth rates of red algae are highest in green light, the red algae may thus not be at great disadvantage in the blue-green light at the bottom of the euphotic zone.

In addition to having photosynthetic action spectra somewhat similar to underwater spectral distributions, it is important for algae at greater depths to have adaptations to **low irradiances** and to resist grazing pressure.

Engelmann in 1884 published the first accurate photosynthetic action spectra of such algae as the green alga *Cladophora*, the red alga *Callithamnion*, and the diatom *Melosira*. He used aerophilous bacteria which accumulated at thallus parts irradiated with red or blue light. Haxo and Blinks (1950) were the next to produce better photosynthetic action spectra of seaweeds, by means of an oxygen electrode. Engelmann's hypothesis of **phylogenetic chromatic adaptation** was that, because of coincidences between action spectra and spectral light distributions, red algae should prevail at greater depths, brown algae at medium depths, and green algae in shallow water (historical review in Rabinowitch 1945–1956). T. W. Engelmann interpreted his fundamental results exclusively for deep waters with prevailing green light, where the photosynthetic action spectrum of the red algae optimally matches the spectral distribution of underwater daylight.

Engelmann's view has to be modified in three respects:

(1) Members of all three algal groups occur at all water depths, and the green window in the absorption spectrum of chlorophyll is filled by different pigments in all algal groups, although in a particularly effective way by red algae.

(2) In red algae the similarity between the photosynthetic action spectrum and the spectrum of underwater light exists at greater depths in coastal (yellow) waters (Engelmann's case). In oceanic (colorless) waters R- and B-phycoerythrins are still efficient in harvesting the available light, but no more efficient than green or brown algae.

(3) The effective light absorption by optically thick, "black" thalli found

in green, brown, and red algae can hardly be used to reject Engelmann's view. Better absorption of predominantly green light by increasing the chlorophyll content (e.g., in a thick green alga) results in a high protein cost. Absorption by an optimally green-absorbing pigment such as phycoerythrin is more efficient and cost-saving in terms of biochemical economy, as Kirk (1983) has pointed out.

The discussion of Engelmann's view continues, with authors regarding seaweeds primarily as "intensity" adapters (Berthold 1882; Oltmanns 1922–1923; Dring 1981, 1982; Ramus 1981; Ramus and van der Meer 1983) with partial consent to seaweeds as "chromatic" adapters (Kirk 1983).

Ontogenetic chromatic adaptation. This term refers to the conspicuous color changes exhibited by several members of the **cyanophytes** (e.g., several species of *Scytonema* = *Tolypothrix*), cultivated in either green or red light (reviews in Bogorad 1975; Björn 1979; Jeffrey 1981; Prézelin and Boczar 1986). "Complementary chromatic adaptation" is the predominant synthesis of phycocyanins in red light and of phycoerythrins in green light. The phycocyanins make the blue-green algae appear green, while the phycoerythrins make them appear red. Some species photoregulate only the rate of phycoerythrin synthesis. Others modulate synthesis of both phycoerythrin and phycocyanin, and in this group the synthesis of phycoerythrin may cease completely in red light. Interestingly, there may exist photoreversible sensor pigments, the "phycochromes" (Björn 1979), by which blue-green algae "sense" the part of the spectrum with which they are being irradiated. Phycochromes a, c and d may be aggregation states of allophycocyanin and phycocyanin (Murakami and Fijita 1983), while the α-subunit of phycoerythrocyanin exhibits the photoreversibility characterizing phycochrome b (Kufer and Björn 1989). In **seaweeds** one cannot achieve such conspicuous color changes, even after prolonged cultivation in red, green, or blue light. Young plants of *Laminaria hyperborea* grown for up to four weeks in blue, green, red, and white light showed little pigment variation at 5 and 20 μmol photon $m^{-2} s^{-1}$. At 100 μmol photon $m^{-2} s^{-1}$, a decrease of all pigment concentrations in red and blue light was observed, and, apart from an accentuation of the red absorption peak of chlorophyll a at the high photon fluence rate, few differences in the photosynthetic action spectra occurred (Dring 1986). The greenish color that many red algal species (e.g., *Gracilaria* species, *Palmaria palmata,* and *Chondrus crispus*) may attain in the eulittoral zone under intense irradiance is due to a selective reduction of their phycobilin content. Engelmann's hypothesis has also been decribed in the literature as the "theory of complementary adaptation of marine algae," even though he was referring to phylogenetic adaptation. This has led to confusion with the experimentally testable phenomena described above, which can be detected only in cyanophytes.

6.3.2 Quantitative Light Demands for Photosynthesis and Growth

The algae inhabiting the upper part of the euphotic zone are exposed to high irradiances and must be adapted as "sun plants," whereas in the deeper euphotic zone the chronic lack of light determines photosynthetic behavior. Earlier in this text it was mentioned that many algal species are confined to either the upper or the lower part of the euphotic zone. Opponents of Engelmann's hypothesis pointed to the possibility that light quantity, instead of light quality, might have brought about such distribution patterns (Berthold 1882; Oltmanns 1892, 1922–1923; Harder 1923). To aid the following discussion, Table 6.3 (p. 285) gives some representative values for magnitudes of photon fluence rate at different depths in the euphotic zone near Helgoland throughout the year.

Photosynthesis-versus-light-intensity curves (**P vs. I curves**) result from the measurement of the rate of photosynthesis at increasing photon fluence rates (Fig. 6.9A). The rate of photosynthesis depends on the velocity at which photons are delivered during the **linear increase** in the photosynthetic rate ("linear range" LR in P vs. I curve illustrated in Fig. 6.9A).

Photorespiration. Although for practical reasons gross photosynthesis is regarded as the sum of net photosynthesis and mitochondrial respiration (Fig. 6.9A), the matter is complicated by another oxygen-consuming process, namely, **photorespiration,** which may be simultaneous to respiration in the light. Little is known about photorespiration in seaweeds, but it could be detected, particularly under high O_2 concentration (Bidwell and McLachlan 1985; Lobban et al. 1985; Reiskind et al. 1989).

The **photosynthetic light compensation point** is the photon fluence rate at which oxygen consumption due to mitochondrial respiration ("dark respiration") consumes the same amount of oxygen as that produced by photosynthesis with time. Typical values for shade plants, as derived from P vs. I curves (Fig. 6.9A), are 1–2 μmol photon $m^{-2}\ s^{-1}$ in deep-water red algae (review in Lüning 1981a) and extreme-shade plants growing on the floor of rain forests (Boardman 1977).

The photosynthetic light compensation point **increases** in cells or thalli

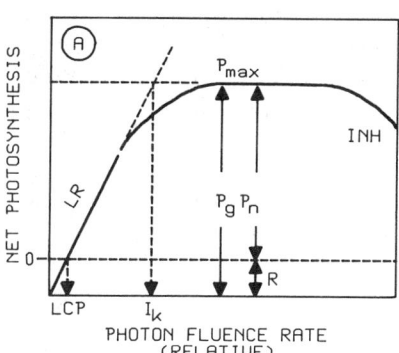 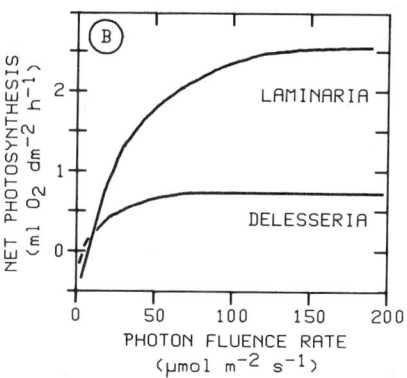

Fig. 6.9 (A) P vs. I curve (photosynthesis vs. light intensity). LR = linear range; P_{max} = maximal rate of photosynthesis; INH = initial photoinhibition at high photon fluence rates; P_g = rate of gross photosynthesis; P_n = rate of net photosynthesis; R = rate of dark (mitochondrial) respiration; LCP = light compensation point; I_k = photon fluence rate, at which the linear branch of the P vs. I curve and saturation plateau intersect. (B) P vs. I curves of *Laminaria* spp. and of *Delesseria sanguinea*, measured in summer. (A from Ramus 1981; B after Lüning 1979.)

grown **at higher light levels** because carbon-specific dark respiration increases and/or the pigment content becomes reduced; that is, the chlorophyll–carbon ratio decreases (Geider et al. 1986; Raven 1986). In *Ulva lactuca*, dark respiration was linearly related to growth rate and doubled in plants grown at 55 μmol m^{-2} s^{-1} (doubling time of thallus in carbon circa 11 days), compared to plants grown at 2.5 μmol m^{-2} s^{-1}, which was the light compensation point for growth, and close to that for photosynthesis (1.5–2.3 μmol m^{-2} s^{-1}) in this alga at 7°C (Sand-Jensen 1988*a,b*).

Algae inhabiting **low-light habitats,** for example, in the sublittoral zone, have to live for long periods each year at photon fluence rates at or below the light compensation point (Table 6.3). As mentioned previously, an essential low-light adaptation and a prerequisite for algae that live under conditions of chronic lack of light is to have a **low dark respiration.** This is generally achieved in **shade plants** through the formation of thin leaves or thalli (reduction of nonphotosynthesizing tissue) and by reduced synthesis of **photosynthetic enzymes** (e.g., RuBP carboxylase) and components of the photosynthetic electron transport chains (reviews in Boardman 1977; Björkman 1981; Larkum and Barrett 1983; J. M. Anderson 1986).

Dark respiration and growth rate must be reduced to almost zero in the deepest crustose coralline algae surviving at 0.03 μmol m^{-2} s^{-1} (see Section 6.2). The **maintenance light levels** in low-light acclimated algae, at photon fluence rates below 1 μmol m^{-2} s^{-1}, and the **maintenance respiration rates** at zero growth rate must have some finite values just above zero, but possibly they are undetectable with the presently used methods for measuring growth, photosynthesis, and respiration (Pirt 1982; Geider et al. 1985, 1986; Raven 1986). In the cyanophyte *Oscillatoria redekei* the normally observed linear decrease of respiration rate with decreasing light exposure (mol photons m^{-2} day^{-1}) is no longer valid at extremely slow growth rates (C. E. Gibson 1987). In *Ulva lactuca* kept in the dark or at subgrowth light levels, respiratory maintenance costs derived from the carbon growth balance of nongrowing thalli are extremely low, on the order of 0.01 mol of carbon lost per mole of cellular carbon per day (Sand-Jensen 1988*a*).

Relation between photon fluence rate and growth rate. The following formula, which may be used to calculate maintenance light requirements and specific growth rates at given light levels (see below), has been presented by Raven (1984*a*, 1986).

$$\mu = JAPQDFR - M$$

where μ = relative growth rate (moles of carbon assimilated per mole of plant carbon per second);

J = incident photon fluence rate (mol photons m^{-2} s^{-1});

A = fraction of the incident photon fluence rate that is absorbed by the cell or thallus;

P = factor relating moles of carbon in photosynthetic organs to area exposed to incident photons (m^2 projected cell area per mole of carbon);

Q = quantum yield of photosynthesis (value of 0.1 for 10 mol photon absorbed per mole of CO_2 fixed in photosynthesis);

$JAPQ$ = rate of photosynthesis in moles of carbon fixed per mole of carbon in the light period, not corrected for losses via "dark" respiratory losses, but corrected for photorespiratory or carbon-pumping costs and losses;

D = fraction of the 24 h light–dark cycle for which the mean photon fluence rate J occurs (e.g., value of 0.5 for 12 h light per day);

F = fraction of photosynthate carbon that is incorporated into plant material (e.g., value of 0.7, if 70% of C is incorporated);

R = fraction of plant carbon that is associated with the photosynthetic organs (e.g., values of 1 for a unicellular alga or a thin thallus with little nonphotosynthetic tissue, and 0.3 for a thick *Laminaria* thallus);

DFR = factor that converts the photosynthetic carbon assimilation rate per unit of carbon in photosynthetic organs into the rate of accumulation of carbon per unit carbon in the whole plant;

M = maintenance respiration rate at zero growth rate (moles of carbon used per mole of carbon in the plant per second; a typical value would be 10^{-8} s^{-1}).

Maintenance light requirements. From the above equation the **minimum photon fluence rate for maintenance** of existing plant material can be computed by setting $\mu = 0$ (no growth) and $F = 1$ (no growth means no growth-associated respiratory losses) and by inserting standard values, as given above:

$$J_{maint} = \frac{M}{APQDFR} = \frac{10^{-8}}{AP \times 0.1 \times 0.5 \times 0.7 \times R}$$

Calculated results for the minimum photon fluence rate for maintenance, J_{maint} (with J_{maint} given in units of mol photon m^{-2} s^{-1} = 10^6 μmol photon m^{-2} s^{-1}) are as follows:

(1) For a spherical plankton alga of 10 μm diameter ($A = 0.137$, $P = 15$ m^2 projected cell area per mole of cell carbon, $R = 1$),

$$J_{maint} = \frac{10^{-8}}{0.137 \times 15 \times 0.1 \times 0.5 \times 0.7 \times 1}$$

$$= 0.136 \; \mu\text{mol photon } m^{-2} \; s^{-1};$$

(2) For a multicellular algal thallus with little nonphotosynthetic tissue, such as a flat thalli one cell layer thick (e.g., *Monostroma* or *Porphyra*) or thin filamentous thalli, and juveniles of many macroalgae ($A = 0.60$, $P = 2$, $R = 1$),

$$J_{maint} = \frac{10^{-8}}{0.6 \times 2 \times 0.1 \times 0.5 \times 0.7 \times 1}$$

$$= 0.238 \; \mu\text{mol photon } m^{-2} \; s^{-1};$$

(3) For a *Laminaria* thallus ($A = 0.95, P = 0.2, R = 0.3$),

$$J_{maint} = \frac{10^{-8}}{0.95 \times 0.2 \times 0.1 \times 0.5 \times 0.7 \times 0.3}$$
$$= 5.01 \ \mu\text{mol photon m}^{-2} \text{ s}^{-1}.$$

These calculated values are of the right order to agree with the observed minimum thresholds of photon fluence rate for growth in various unicellular and multicellular algae. The maintenance metabolic rates refer to a state of no growth, or to the first, unobservable hints of growth. The multicellular thallus (equation 2) requires more light for maintenance than the model plankton alga (equation 1) because the higher A value in (2) does not quite offset the lower P value of the 70-μm-thick thallus. A thick *Laminaria* thallus (equation 3) requires more light because of the lower P value, 10 times lower than in the thin multicellular thallus (2), and the lower R value, with little offset due to the somewhat higher value for A. **Measured** compensation light levels for growth are, for example, 0.2 μmol photon m^{-2} s^{-1} at 15°C for the delicate, filamentous deep-water red alga *Atractophora hypnoides* (Maggs and Guiry 1987b), 2.5 μmol photon m^{-2} s^{-1} for the not so deep-growing *Ulva lactuca* at 7°C (Sand-Jensen 1988a), and 10-20 μmol photon m^{-2} s^{-1} for a benthic *Sargassum* species at 18-24°C (Hanisak and Samuel 1987).

Calculated specific growth rates at maximum and minimum light levels. From $\mu = JAPQDFR - M$ one may calculate that the specific growth rate of a planktophyte ($A = 0.137, P = 15$) at **maximum** values of photon fluence rate around 2000 μmol m^{-2} s^{-1} should be $1.4 \ 10^{-4}$ mol C mol plant C^{-1} s^{-1} (mol carbon per moles plant carbon per second), corresponding to a generation time of 1.375 h. The relation between the generation time t_d and the specific growth rate μ is $t_d = (\ln 2)/\mu$. Observed generation times for algae are not shorter than 8 h, possibly because it is difficult to fit sufficient catalysts into the cell to handle all absorbed light energy expected (Raven 1984a, 1986). At the other extreme of the range of photon fluence rates one may refer to the **minimum** value of 0.01 μmol m^{-2} s^{-1}, at the maximum depth of 270 m for the **deepest coralline crusts** near the Bahamas. Inserting the values used for thin multicellular algae (*Ulva* type) and neglecting any maintenance respiration rate, one obtains a specific growth rate of $4.2 \ 10^{-10}$ mol C mol plant C^{-1} s^{-1}, corresponding to a doubling time of 52.3 yr. One way of reducing the doubling time would be to let A, the fractional absorption of incident photons, exceed 0.6, although A cannot exceed 1.0. An increase in A might be justified, even with no more pigment per unit area, due to more scattering and large changes of refractive index caused by the calcium carbonate of the coralline crust (Raven, personal communication).

Calculation of specific growth rates at nonsaturating photon fluence rates. For a given specific growth rate the photon fluence rate required is:

$$J = \frac{\mu + M}{APQDFR}$$

Using the values stated above, the calculated results are as follows:

for *Ulva*: $\mu - 10^{-6}$ mol C mol plant C^{-1} s^{-1} (i.e., a doubling time of eight days):

$$J = \frac{10^{-6} + 10^{-8}}{0.6 \times 2 \times 0.1 \times 0.5 \times 0.7 \times 1} = 24 \text{ } \mu\text{mol photon m}^{-2} \text{ s}^{-1}$$

for a thick *Laminaria* thallus: $\mu = 4.4 \text{ } 10^{-8}$ mol C mol plant C^{-1} s^{-1} (i.e., a doubling time of six months, calculated for one side of the thallus):

$$J = \frac{4.4 \times 10^{-8} + 10^{-8}}{0.95 \times 0.2 \times 0.1 \times 0.5 \times 0.7 \times 0.3} = 27 \text{ } \mu\text{mol photon m}^{-2} \text{ s}^{-1}$$

for a thick *Laminaria* thallus in a dense *Laminaria* bed with much self-shading, (e.g., with a thallus area index of 10 m² per square meter of rocky substratum; $A = 1.0, P = 0.02$): μ as above:

$$J = \frac{4.4 \times 10^{-8} + 10^{-8}}{1.00 \times 0.02 \times 0.1 \times 0.5 \times 0.7 \times 0.3} = 257 \text{ } \mu\text{mol photon m}^{-2} \text{ s}^{-1}.$$

Looking at the last two examples, actual photon fluence rates under water in summer, as given in Table 6.3, are in fact above 100 μmol photon m^{-2} s^{-1} in the upper sublittoral zone, where there is a dense *Laminaria* vegetation, and about on the order of tens of μmol photon m^{-2} s^{-1} deeper down in the *Laminaria* "park," with scattered kelp individuals and practically no self-shading.

When the photosynthetic rate has reached a plateau, that is, **light saturation** (P_{max} in Fig. 6.9A), the enzymatic reactions limit the extent of photosynthesis. Both genetic adaptation and ontogenetic acclimation determine a plant's capability to use low or high light levels (Ramus 1981). There are genetically fixed sun and shade species, and in certain species one may shift the character by cultivation at the appropriate light level (reviews in Boardmann 1977; Larkum and Barrett 1983; J. M. Anderson 1986).

Sun plants have a relatively high content of photosynthetic enzymes and electron chain components. Typical sun algae become photosynthetically light-saturated at photon fluence rates around 500 μmol m^{-2} s^{-1} and shade algae at 60-150 μmol m^{-2} s^{-1} (Fig. 6.9B; Table 6.5). It can be seen from Table 6.3 that photosynthesis of some sublittoral algae has to proceed for long periods during the year at photon fluence rates below light saturation.

The **light saturation of growth** in seaweeds occurs at values of photon fluence rate that are lower than those for photosynthesis (Table 6.5). The possible cause for the inability of algal growth rates to keep up with high photosynthetic rates may be due to suboptimal nutrient levels in the water and to inhibition of nonphotosynthetic, light-dependent enzyme reactions at photon fluence rates below those inhibiting photosynthesis.

Table 6.5 Photon fluence rates at which photosynthesis (PS) or growth (GR) are light-saturated. Unit: μmol photon m^{-2}s^{-1}. Values in brackets: original values in W m^{-2} or lux, approximately converted to μmol photon m^{-2}s^{-1}. G = green alga, B = brown alga, R = red alga

		PS	GR	Source
Eulittoral zone				
B	Fucus vesiculosus	(600)	(300-350)	PS: Kanwisher 1966; GR: Strömgren and Nielsen 1986
B	Fucus serratus	500	(150-200)	PS: Lüning 1979; GR: Strömgren and Nielsen 1986
G	Codium fragile	(500)	28	PS: Ramus 1978; GR: Hanisak 1979
R	Mastocarpus stellatus	(460)		Mathieson and Burns 1971
R	Gelidium spp., sporelings		50-75	Correa et al. 1985
R	Devaleraea ramentaceum		50	Rueness and Tanager 1984
R	Porphyra spp.	(400)	30	PS: Ogata and Matsui 1965; GR: Fortes and Lüning 1980
	surface phytoplankton	(500)		Steemann Nielsen 1975
	terrestrial sun plants	(500)		Boardman 1977
Sublittoral zone				
B	Sargassum polycystum	80-120		Titlyanov et al. 1983
B	Sargassum cymosum (benthic)		150	Hanisak and Samuel 1987
B	Sargassum natans (pelagic)		200-300	Hanisak and Samuel 1987
B	Laminaria saccharina	150	70	PS: Lüning 1979; GR: Fortes and Lüning 1980
G	Ulva fenestrata	80-150		Titlyanov et al. 1987
R	Chondrus crispus	(180)	(50-100)	PS: Mathieson and Norall 1975a; GR: Strömgren and Nielsen 1986
R	Grateloupia turuturu	80-150		Titlyanov et al. 1987
	phytoplankton from 1%-light depth	(200)		Steemann Nielsen 1975
	terrestrial shade plants	(60-200)		Boardman 1977
R	Ahnfeltia tobuchiensis	60-100		Titlyanov et al. 1987
R	Delesseria sanguinea	60		Lüning 1979
R	Ptilota serrata	70		Mathieson and Norall 1975b
B	Macrocystis pyrifera, gametophytes and young sporophytes	70		Fain and Murray 1982
R	Atractophora hypnoides		10	Maggs and Guiry 1987b
B	Laminaria spp., early sporophytes		(15)	Kain 1969
B	Himanthothallus grandifolius, young sporophytes		5	Wiencke 1988

Within a species there may occur ecotypic differentiation. Juvenile sporophytes of *Laminaria saccharina* collected from a 12 m depth grew better at very low light levels than plants from a 5 m depth (Gerard 1988). The deep-water plants exhibited reduced growth rates in the laboratory consistently at lower light levels than did the shallow-water plants, and the photosynthetic capacity P_{max} was lowest for the deep plants.

Finally, at photon fluence rates above 200 μmol m^{-2} s^{-1} in marine phytoplankton (Harris and Piccinin 1977; Harris 1978), unicellular green algae (Lidholm et al. 1987), and in seaweeds collected from deeper water, such as *Delesseria sanguinea* or *Laminaria hyperborea* (Drew 1983), the rate of photosynthesis begins to decline continuously, due to **photoinhibition** (reviews in Powles 1984; J. M. Anderson 1986; Kyle et al. 1987; Krause 1988). In algae from shallower depths, photoinhibiton may set in at photon fluence rates above 300–500 μmol m^{-2} s^{-1} (Titlyanov et al. 1987). Pioneer work on photoinhibition in seaweeds was done by Montfort (e.g., Montfort 1933; Montfort et al. 1952).

As with light saturation in photosynthesis, ontogenetic **acclimation** determines to a great extent at which light level photoinhibition sets in. In *Macrocystis pyrifera,* canopy blades maintained their maximum photosynthetic rate during 1 h incubations at photon fluence rates of up to 2160 μmol m^{-2} s^{-1}, blades at 3–5 m deep up to 1000 μmol m^{-2} s^{-1}, and at 7–9 m deep up to 500 μmol m^{-2} s^{-1} (Gerard 1986). Kelp blades from the surface canopy exhibited an **afternoon depression** in photosynthesis. This depression was only 17% because of mutual shading and vertical displacement of blades, but it rose to 70% in blades removed from the canopy and held fixed at a 0–1 m depth in open water. In tropical waters photosynthesis can be depressed to 20% in the upper parts of the thalli of some *Sargassum* species (Titlyanov et al. 1983).

One of the major **causes** of photosynthesis decline at higher light levels may be the excessive delivery of energy at the reaction center of photosystem II, where a protein is damaged (reviews in Ohad et al. 1984; Matoo et al. 1984; J. M. Anderson 1986; Lidholm et al. 1987; Krause 1988). Raven and Sprent (1989) speculated that this deleterious property of the photosynthetic apparatus became entrenched early in the evolution of algae, when plants were forced to develop at greater depths under low-light conditions because of the UV impact of the sun. Later, under a changed environment and high-light conditions at shallow water depths and for green plants on land, the deleterious property could not be disposed of because it was too deeply entrenched in the photosynthetic apparatus (p. 305).

The first **action spectrum** of photoinhibition in a seaweed species has been measured by Nultsch et al. (1987). The photosynthetic O$_2$ production of *Dictyota dichotoma* shows photoinhibition maxima in green (542 nm) and red (672 nm) light; these ranges of wavelengths are absorbed by fucoxanthin and the red band of chlorophyll a, respectively. No significant photoinhibition occurrs at wavelengths above 700 nm. These findings support the hypothesis that the effective radiation is mainly absorbed by photosystem II and that photosystem I seems not to be involved. Fucoxanthin participates particularly in the perception of photoinhibiting radiation. In blue light, carotenoids other than fucoxanthin ensure minimum photoinhibition due to shading and/or photoprotection (see below). The detrimental effects caused by blue, green, and red light, with increasing effect in this order, fit results on survival of laminarian gametophytes. Fifty per cent of the gametophytes of *Laminaria hyperborea* cultivated at 750 μmol photons m^{-2} s^{-1} died after 1.6 h in red, 3.1 h in green, and 6.7 h in blue light (Lüning 1980*a*). The detrimental effects of excessive red and green light is also evident from cultivation experiments on other brown algae and a green alga (see Section 6.4.2).

The extent of photoinhibition in photosynthesis depends on the efficacy of **repair mechanisms** and **protective measures** against high levels of light and ultraviolet radiation. If one transfers the experimental algae to environments with low photon fluence rates or to darkness, **recovery** of photosynthetic O$_2$ production occurs if no consistent photodamage has preceded. This has been shown for a cyanophyte (Barlow and Alberte 1987) and in seaweeds for *Dictyota* (Nultsch et al. 1987). In the latter case, recovery of photosynthesis occurred within about 1 h in low light, faster than in darkness. This fits the hypothesis that the protein that is destroyed during high-intensity irradiation is synthesized again during recovery in a light-dependent reaction (Matoo et al. 1984; Ohad et al. 1984).

Carotenoids, whose concentration may increase relative to chlorophyll a in high light, probably exert a light-protecting function, since excessive energy may be conducted from excited chlorophyll to carotenoid molecules (Titlyanov et al. 1977; B. D. Lee and Titlyanov 1978; Björkman 1981; Ramus 1983, 1985; Raven 1984a; Siefermann-Harms 1985; J. M. Anderson 1986; Henley and Ramus 1989). After a few hours at photoinhibiting light levels, **photodamage** occurs, and the oxidative **destruction of pigments** begins (e.g., Titlyanov et al. 1977).

Apart from shifts of P_{max} due to high light levels, which may represent a regular and daily phenomenon in the upper part of the euphotic zone, P_{max} may also change due to an endogenous, circadian rhythm, unrelated to chloroplast position per se, as demonstrated for *Ulva* (Britz and Briggs 1976; Titlyanov et al. 1978; Mishkind et al. 1979; Nultsch et al. 1981a).

Photoinhibition of growth occurs at lower light levels than short-term measurements of photosynthetic photoinhibition may indicate. In the deepwater red alga *Atractophora hypnoides* growth becomes inhibited at 30 μmol m^{-2} s^{-1}, and 50 μmol m^{-2} s^{-1} is lethal after a few weeks (Maggs and Guiry 1987b). Light levels as low as 10-80 μmol m^{-2} s^{-1} may inhibit photosynthesis, and growth may be inhibited at 10-20 μmol m^{-2} s^{-1} in certain Antarctic **unicellular algae** in brine pockets and channels in forming ice and in the interstitial water of the platelet ice layer, as well as in Antarctic benthic unicellular algae from 18–20 m deep (Rivkin and Putt 1987).

Diurnal changes of growth rate. The afternoon depression of photosynthesis in shallow water may be partly connected to another diurnal event: namely the **maximum morning growth rate** and a continuous decline throughout the rest of the day in *Fucus* species, *Ascophyllum nodosum,* and *Chondrus crispus.* This was observed by Strömgren and Nielsen (1986) by laser diffraction, the algae being exposed to air for about 2 min and to laser light (632.8 nm) of only 0.005 μmol photon m^{-2} s^{-1} for 2-3 s. With this method significant growth increments of 5–9 μm length increase can be recorded in very brief intervals (Strömgren 1975). Average **growth rates at night** ranged from 33% of average day growth in *Ascophyllum nodosum* to 63% in *C. crispus,* respectively (Strömgren and Nielsen 1986).

The **ultraviolet** part of the spectrum is particularly harmful. UV light is divided into three bands, of which the first does not naturally occur on the surface of the earth: UV-C (200–280 nm), UV-B (280–320 nm), and UV-A (320–400 nm; Caldwell 1984). Photoinhibition occurs just below the water surface in **marine phytoplankton.** Seventy-five percent of the decline of their photosynthesis is thought to be caused by UV (290–400 nm) and only 25% by visible light (400–700 nm; Smith and Baker 1980). A scleractinian **coral** with symbiotic zooxanthellate algae transplanted from 25 to 5 m deep dies within 20 days but survives if shielded by 4 mm of plexiglas that transmits 90% of the light and absorbs 95% of the UV (Vareschi and Fricke 1986).

A similar experiment, using polycarbonate screens opaque to UV but transparent to visible light, was conducted in outdoor tanks on young, subcanopy **kelp sporophytes** of *Ecklonia radiata* and showed that algal pigment destruction and tissue necrosis occurred without the UV filter and not where UV was excluded (Wood 1987). In the field at shallow water depths, experimental removal of the kelp canopy in summer may damage the younger, subcanopy kelp sporophytes (Wood 1987) and understory **red algae** (Kain 1987a), UV and excessive light being the probable cause.

UV-B (290–320 nm) may be a major cause of inhibition of photosynthesis and growth in algae growing in the upper 10 m of the sea. It is absorbed by pigments, DNA and proteins which can all be damaged. The present very low photon fluence rate of UV-B, which is mainly absorbed by ozone and oxygen in the outer atmosphere, may increase to dangerous levels at the surface of the earth when the ozone content in the atmosphere is reduced due to such human influences as aerosol sprays and jet traffic. (For reviews of UV-B see: Caldwell 1981, 1984; Harm 1981; Calkins 1982; Worrest 1983). Due to increasing UV-B levels lethal effects on organisms down to 6 m in clear "blue" waters and down to 2.5 m in more turbid "green" waters may now occur (Smith and Baker 1979). A reduction of the ozone layer above the atmosphere by 16% may decrease the productivity of marine phytoplankton by 5% (Smith and Baker 1982).

Some organisms have evolved protective measures against UV-B. Mycosporinelike amino acids with maximum absorption at 217 nm decrease with depth in the scleractian coral *Acropora* and may well represent UV-B-photoprotective substances (Dunlap et al. 1986). UV-absorbing substances in canopy sporophytes of the kelp *Ecklonia radiata* increase towards late spring (Wood 1987). In terrestrial plants the epidermis absorbs 95–99% of the UV-B (Robberecht and Caldwell 1978).

Early evolution of algae at greater depths due to UV impact: low-light aspects. As a hypothesis (Raven 1986; Raven and Sprent 1989) there may be a strong evolutionary pressure for causing photoinhibitive-sensitive processes to become entrenched in the photosynthetic apparatus. Today the oxygen and ozone in the atmosphere shield the earth from most of the harmful UV emitted by the sun, including all of the UV-C (with wavelengths shorter than 290 nm), and most of the UV-B (290–320 nm). The early algae (i.e., those in the Precambrian and early Paleozoic) may have been forced to live at greater depths in the sea because of the sun's UV impact in a low oxygen and ozone atmosphere. Here they were safe from UV, but they also evolved their photosynthetic apparatus at low light levels with little selection pressure to evolve mechanisms to counter photoinhibition. As oxygen and ozone increased in the atmosphere due to the developing plant life, algae could colonize shallower depths, obtain more light, and evolve into larger forms, since larger algae require higher light levels. What seems not to have been lost was the sensitivity for photoinhibition.

6.3.3 Photoacclimation of Pigment Content

The **pigment content** per unit leaf or thallus area is often higher in shade plants than in sun plants. (For reviews see Björkman 1981; Harris 1978; Prézelin 1981; Larkum and Barrett 1983; Anderson 1986). Under conditions of **low**

light a higher pigment content is required to enhance the chance of photons being absorbed by the antenna molecules of the photosystems. The cell pigment content of phytoplankton cells cultivated at a photon fluence rate of 1 μmol m^{-2} s^{-1} accounts for half of the cell dry weight and enables the cells to double about every week (Raven 1984b).

In **strong light** the pigment content should be lower in order to avoid photodamage (irreversible inhibition of the photosynthetic apparatus) and photoinhibition (reversible inhibition). A greater portion of the synthetic capacity of a plant is devoted in low light to the synthesis of light-collecting pigments and in strong light to the synthesis of photosynthetic enzymes and electron chain components. There are, however, different regulation types, for example, the *Ulva* type, in which pigment content, photosynthetic enzymes, and electron chain components increase under low light conditions (Titlyanov et al. 1987).

Photoacclimation of pigment content is well known from terrestrial plants, phytoplankton, and seaweeds (Lee and Titlyanov 1978; Ramus 1981; Ramus et al. 1976, 1977; Titlyanov et al. 1987). It may occur relatively quickly in seaweeds and was demonstrated by Ramus and his coworkers, who submerged seaweeds from the eulittoral zone to 1 m and 10 m water depths. After one week, the experimental algae at 10 m exhibited a dramatic increase in the concentration of photosynthetic pigments (Table 6.6).

A **selective increase of accessory pigments** had occurred, seen from the proportions of chlorophyll b to chlorophyll a in green algae, or from the proportions of the in vivo absorption values of phycobiliproteins (565 and 620 nm) to absorption of chlorophyll a (678 nm) in red algae (Table 6.6). A selective increase in accessory pigments was also noted when algae from sunny and shaded sites were compared (Table 6.6). More evidence of selective, enhanced formation of accessory pigments at low white or green light, of chlorophyll b in *Ulva curvata* and *Codium decorticatum,* of fucoxanthin in *Dictyota dichotoma* and *Fucus vesiculosus,* and of phycoerythrin in several red algae was obtained by Ramus (1983) and Murase et al. (1989), for *Ulva* species by Henley and Ramus (1989), and for various species of seaweeds by Lee and Titlyanov (1978), Titlyanov and Lee (1978), and Titlyanov et al. (1987).

Similarly, low-light cultured and deep-water plants of *Caulerpa racemosa* had higher levels of siphonaxanthin than high-light cultured and shallow-water plants (Riechert and Dawes 1986). Small sporophytes of *Laminaria hyperborea* grown at 100 μmol photon m^{-2} s^{-1} in blue or red light exhibited a greater relative decrease of chlorophyll a than of fucoxanthin or chlorophyll c, if compared to lower light levels (Dring 1986).

There is a general tendency for **green algae,** with increasing **depth** and **shade** in the sea, to form not only more chlorophyll, but selectively **more chlorophyll b;** that is, the chlorophyll a:b ratio decreases (Yokohama and Misonou 1980). A similar decrease of the ratio in several unicellular green algal species cultured at 40 or 6 μmol photon m^{-2} s^{-1} in fluorescent white

Table 6.6 Pigment concentrations and pigment proportions in the thalli of several seaweeds, after a 1-week transfer to different water depths, or at sunny and shaded sites in the eulittoral zone. Measuring units for pigment concentrations: n mol cm^{-2} thallus area (*Ulva, Porphyra*); n mol g^{-1} fresh weight (*Codium, Chondrus*); μg g^{-1} fresh weight (*Ascophyllum, Fucus*); Chl. = chlorophyll; Fuc. = fucoxanthin; A = absorption at indicated wavelengths (nm)

Green algae	at different depths					sunny or shaded sites			
	Chl.a	Chl.b	Chl.b/Chl.a			Chl.a	Chl.b	Chl.b/Chl.a	
Ulva lactuca									
1 m depth	2.24	1.00	0.44		sunny	1.29	0.68	0.53	
10 m depth	7.68	5.11	0.67		shade	9.98	5.94	0.62	
1 m : 10 m	0.29	0.20			sunny/shade	0.13	0.11		
Codium fragile									
1 m depth	120	74	0.63		sunny	111	72	0.68	
10 m depth	275	175	0.67		shade	302	200	0.72	
1 m : 10 m	0.44	0.42			sunny/shade	0.37	0.36		

Brown algae	Chl.a	Chl.c	Fuc.	Chl.c/Chl.a	Fuc./Chl.a		Chl.a	Chl.c	Fuc.	Chl.c/Chl.a	Fuc./Chl.a
Ascophyllum nodosum											
0 m depth	431	98	178	0.23	0.42	sunny	448	94	189	0.21	0.42
4 m depth	841	191	280	0.23	0.34	shade	1155	214	304	0.19	0.27
0 m : 4 m	0.51	0.51	0.64			sunny/shade	0.39	0.44	0.62		
Fucus vesiculosus											
0 m depth	482	107	202	0.22	0.42	sunny	575	118	247	0.20	0.43
4 m depth	1201	234	396	0.20	0.34	shade	2341	378	751	0.16	0.32
0 m : 4 m	0.40	0.46	0.51			sunny/shade	0.25	0.31	0.33		

Red algae	Chl.a	A 565/A 678	A 620/A 678		Chl.a	A 565/A 678	A 620/A 678
Porphyra umbilicalis							
1 m depth	1.76	0.47	0.50	sunny	2.30	0.60	0.56
10 m depth	2.49	0.76	0.66	shade	3.55	0.73	0.61
1 m : 10 m	0.71			sunny/shade	0.65		
Chondrus crispus							
1 m depth	178	0.59	0.42				
10 m depth	254	0.89	0.51				
1 m : 10 m	0.70						

From Ramus et al. 1976, 1977.

light may reflect an increase in the light-harvesting protein complexes (LHCs). These LHCs contain most if not all of the chlorophyll b and are associated with photosystem II in these microalgal species (Osborne and Raven 1986).

More chlorophyll and selectively **more chlorophyll b** are formed **in blue light** in *Chlorella vulgaris* than if cultured in red light, resulting in chlorophyll a : b ratios of 3 : 1 and 5 : 1, respectively (Kowallik and Schürmann 1984). The same selective increase in chlorophyll b and in LHC in *Scenedesmus obliquus* thylakoids shows the importance of blue light for the regulation of chlorophyll biosynthesis in green algae (Humbeck et al. 1984). Thus the blue-light-adapted cells mimicked shade adapation, and the red-light-grown cells mimicked high-light adaptation. The contrary view based on older experiments in certain terrestrial plants has been abandoned (J. M. Anderson 1986).

6.4 LIGHT AS AN ENVIRONMENTAL SIGNAL

Light has a dual function in plants, firstly as a source for energy (absorption by pigments of the photosynthetic apparatus) and secondly as an environmental signal for the regulation of development. In the latter case light is absorbed by small concentrations of highly sensitive **sensor pigments,** for example, phytochrome in higher plants (reviews in Morgan and Smith 1981; Shropshire and Mohr 1983). As an environmental signal, light may induce developmental and morphological changes in the phenomena of **photoperiodism** (dependence on daylength), of **photomorphogenesis** (dependence on spectral range), and in other phenomena like phototropism and induction of chloroplast movements. In all these cases the signal character of light is evident from the low light requirement for the induction of a reaction, which is the energy required for the execution of the induced reactions being derived from photosynthesis.

6.4.1 Photoperiodism

Daylength is used by many plants and animals as an indication of season, in the sense of an "early warning system." (For reviews see Salisbury 1981; Vince-Prue 1975, 1983; seaweeds: Lüning 1981a,b; Dring and Lüning 1983; Dring 1984). Deciduous trees start to prepare for leaf fall and form winter buds as soon as a certain critical daylength is reached in autumn, long before the first snow falls and the winter buds are actually required. In this example the snowfall represents the primary ecological factor for which the plant has to prepare, and autumnal daylength represents an ecological factor with a signal character. For plant groups in which photoperiodism is unknown, as in the unicellular algae, one assumes that their seasonal behavior is completely governed by the primary ecological factors, such as temperature, irradiance, and nutrient levels.

The photoperiodic reaction is induced in **short-day plants** as soon as daylength falls below a certain value (critical daylength) and in **long-day plants** as soon as a certain critical daylength is surpassed. In day-neutral plants, developmental reactions occur irrespective of daylength.

Many seasonally occurring reactions are known for seaweeds. For example, along the European coasts *Laminaria hyperborea* forms a new frond early in the winter (Fig. 6.10). In spring the erect, siphonous thalli of the brown alga *Scytosiphon lomentaria* and the small, flat thalli of the green alga *Monostroma* appear in the eulittoral zone (Fig. 6.11). In late autumn the filamentous Trailliella phase of the red alga *Bonnemaisonia hamifera* forms tetraspores, and from these the erect-growing gametophyte generation evolves (Fig. 6.11). All these cases represent photoperiodic **short-day reactions** and can be induced in the laboratory at any time of the year under short-day conditions (e.g., 8 h of light per day), whereas long-day conditions (e.g., 16 h of light per day) prevent the reactions. How sensitively *Scytosiphon lomentaria* from Helgoland reacts to changes in daylength of 15 min difference may be seen from Fig. 6.12. Further short-day reactions of red and

Fig. 6.10 Photoperiodism in the kelp *Laminaria hyperborea*. Experimental specimens were cultivated in the laboratory at different regimes of daylength from September (no new blade) until May. **(A)** 8 h of light per day. **(B)** 16 h of light per day. **(C)** Night-break regime (8 h light per day and 1 h of light in the middle of the long night. **(D)** Continuous darkness from September to May. (From Lüning 1986.)

brown algae are listed in Table 6.7. In green algae, photoperiodism occurs, for example, in the previously mentioned *Monostroma grevillei* (Table 6.7). The Codiolum phase of this alga, living as a chalk-boring, filamentous growth form in the eulittoral zone, becomes sporogenous under short-day conditions, and from the zoospores the bladelike *Monostroma* phase arises (Fig. 6.11).

Long-day reactions have been found less often among seaweeds. Under long-day conditions gametangia are formed by the brown alga *Sphacelaria rigidula* (ten Hoopen et al. 1983), and gametophytes in the Chantransia stage by the freshwater red alga *Batrachospermum moniliforme* (Huth 1979). So far this latter case is one of the few examples of photoperiodism in **freshwater algae** (Dring 1984). Photoperiodic reactions are common in higher plants (reviews in Vince-Prue 1975, 1983, 1985) and animals (review, e.g., in Brady 1982). As an example, the marine calanoid copepod *Labidocera aestiva* produces subitaneous eggs (thin-shelled parthenogenetic eggs produced during summer) in long days and mostly diapause eggs in short days, in order to prepare in autumn for winter conditions (Marcus 1980).

Many of the examples of photoperiodism in seaweeds, detected so far, occur in those exhibiting a **heteromorphic life cycle** (marked by "H" in Table 6.7). Others switch seasonally between a **macroscopic** life stage (e.g., the erect thallus of *Scytosiphon*) and a small, **cryptic,** perennating stage (e.g., the crustose stage of *Scytosiphon*).

The eulittoral red algae *Porphyra tenera* and *Bangia fuscopurpurea* have a heteromorphic life cycle with a cryptic Conchocelis phase producing conchosporangia under short-day conditions (Fig. 6.11). The **ecological significance** of the short-day reaction of these two algae may be seen in the necessity to produce the macroscopic *Porphyra* and *Bangia* phases (gametophytes) at the

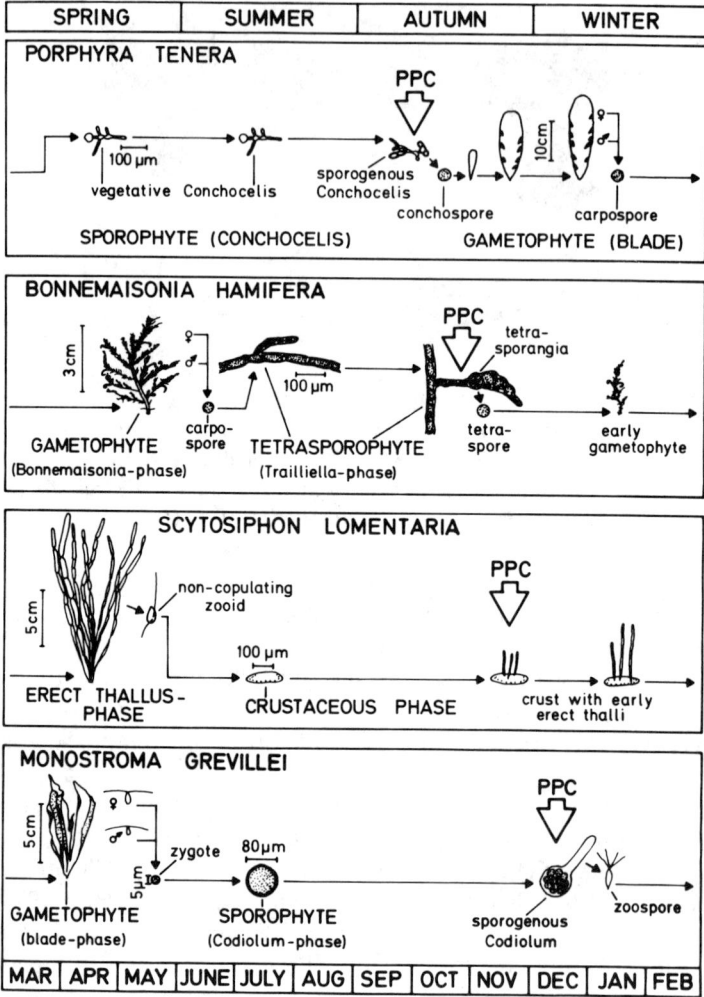

Fig. 6.11 Scheme of seasonal development of four marine macroalgae governed by photoperiodic control (arrow PPC, indicating a short-day signal). Red alga *Porphyra tenera:* formation of conchosporangia. Red alga *Bonnemaisonia hamifera:* formation of tetrasporangia. Brown alga *Scytosiphon lomentaria:* formation of erect thalli. Green alga *Monostroma grevillei:* formation of zoospores. Night-break reaction is effective in all cases. See text for further explanation. (From Dring and Lüning 1983.)

beginning of the cold season. This may be due to a cold requirement for these macroscopic phases.

The 5–10 cm long gametophytes of the red alga *Bonnemaisonia hamifera,* which arise from the tetraspores of a cryptic filamentous sporophyte, the Trailliella phase (Fig. 6.11; Table 6.7), are produced in winter. This is ensured by the fact that tetrasporangia are only formed under short-day conditions

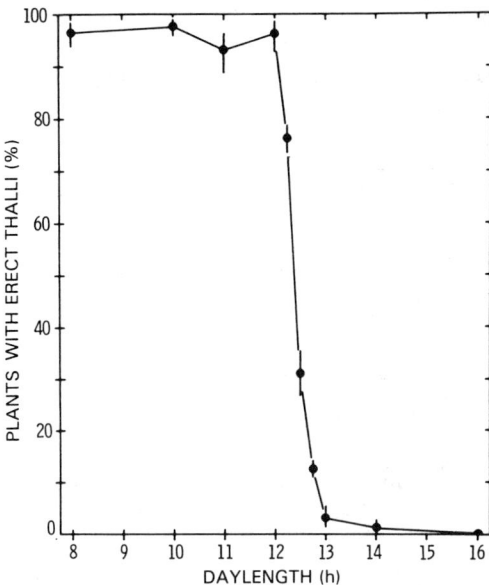

Fig. 6.12 Photoperiodism in the brown alga *Scytosiphon lomentaria*. Formation of erect thalli in relation to daylength. Experimental algae were cultivated from zoospores in petri dishes, and after 35 days the number of thalli that had formed uprights was determined ($N = 500$). Vertical bars indicate 95% confidence limits. (From Dring and Lüning 1975a.)

and only in the rather narrow temperature range of 13–19°C. Thus only in autumn and not in spring, when daylength is also short but temperatures are lower than 13°C, is the "window" open for tetrasporangia formation in the Trailliella phase (Fig. 6.13). The alga originated in Japan, where environmental conditions allow a reasonable width of the seasonal "window," and was introduced to European coasts. Here it populated coasts with autumnal temperature conditions at which only the Trailliella phase can be formed (see Sections 1.2.2 and 2.4.4). In most years at Helgoland, the seasonal window becomes very narrow (Fig. 6.13) as a result of a rapid drop in water temperatures in autumn. The Trailliella phase is thus common, but gametophytes are only occasionally found in the wild (Kornmann and Sahling 1977).

Photoperiodic ecotypes. Geographical populations of the brown alga *Scytosiphon lomentaria* (Lüning 1980b) and the red alga *Audouinella purpurea* (Dring and West 1983) differ with regard to the critical daylength at which their photoperiodic short-day reactions occur, that is, erect thallus formation and tetrasporangium formation, respectively. Similar cases reported for higher plants (review in Salisbury 1981) indicate that the critical daylength increases towards higher latitudes. In the case of *Scytosiphon lomentaria* from the Adriatic Sea, 50% of the crusts form erect thalli at a daylength of 10 h. This value rises to 12.5 h (5–15°C) in material from southern

Table 6.7 Photoperiodic responses in seaweeds. Night-break regime effective, if not otherwise stated. R = red alga; B = brown alga, G = green alga, (H) = heteromorphic life history in red and green algae, TS = tetrasporophyte

reaction		species	reacting phase	authors
(A) Short-day responses				
formation of conchosporangia	R(H)	Porphyra tenera	Conchocelis (sporophyte)	Dring 1967a,b; Rentschler 1967
	R(H)	Bangia atropurpurea	Conchocelis (sporophyte)	Richardson 1970
release of conchospores	R(H)	Porphyra torta[2]	Conchocelis (sporophyte)	Waaland et al. 1987
formation of tetrasporangia	R	Audouinella asparagopsis	tetrasporophyte	Abdel-Rahman 1982
	R	Audouinella botryocarpa	tetrasporophyte	Guiry et al. 1987a
	R(H)	Audouinella pectinata[2]	tetrasporophyte	West 1968
	R(H)	Audouinella purpurea	tetrasporophyte	Dring and West 1983
	R(H)	Bonnemaisonia hamifera	Trailliella-phase (TS)	Lüning 1980b, 1981b
	R(H)	Bonnemaisonia asparagoides	Hymenoclonium-phase (TS)	Rueness and Åsen 1982
	R(H)	Asparagopsis armata	Falkenbergia-phase (TS)	Oza 1976; Lüning 1981b
	R(H)	Atractophora hypnoides	Rhododiscus-phase (TS)	Maggs and Guiry 1987b
	R(H)	Calosiphonia vermicularis	Hymenoclonium-phase (TS)	Mayhoub 1976a
	R(H)	Meredithia microphylla	Hymenoclonium-phase (TS)	Guiry and Maggs 1985
	R(H)	Schmitzia hiscockiana[1]	crustose tetrasporophyte	Maggs and Guiry 1985
	R(H)	Acrosymphyton purpuriferum[2]	crustose tetrasporophyte	Breeman and ten Hoopen 1987
	R(H)	Farlowia spp.[1]	crustose tetrasporopyhte	DeCew and West 1981
	R	Halymenia latifolia	tetrasporophyte	Maggs and Guiry 1982
	R	Gigartina acicularis	gametophyte	Guiry 1984
	R(H)	Mastocarpus stellatus	Petrocelis-phase (TS)	Guiry and West 1983
	R	Cordylecladia erecta[2]	tetrasporophyte	Brodie and Guiry 1987, 1988
	R	Delesseria sanguinea	tetrasporophyte	Kain 1987b
formation of sporangia	B(H)	Phyllariopsis brevipes[1]	sporophyte	Henry 1987a
	B(H)	Laminaria saccharina	sporophyte	Lüning 1988
	G(H)	Monostroma grevillei, M. undulatum	Codiolum-phase	Lüning 1980b
formation of gametangia	R	Gigartina acicularis	gametophyte	Guiry and Cunningham 1984
	R	Halymenia latifolia	gametophyte	Maggs and Guiry 1987b
	R	Cordylecladia erecta[2]	gametophyte	Brodie and Guiry 1987
	R	Delesseria sanguinea	gametophyte	Kain 1987b
	B(H)	Scytothamnus spp.[1]	gametophyte	Clayton 1986
	B	Saccorhiza dermatodea[1]	gametophyte	Henry 1987b
formation of receptacles	B	Ascophyllum nodosum	macrothallus	Terry and Moss 1980
formation of erect thalli	R	Dumontia contorta	crustose stage	Rietema and Klein 1981
	R(H)	Schmitzia hiscockiana[1]	gametophyte	Maggs and Guiry 1985
	B	Scytosiphon lomentaria	crustose stage	Dring and Lüning 1975a
	B	Petalonia fascia, P. zosterifolia	crustose stage	Lüning 1980b
formation of a new blade	R	Constantinea subulifera	gametophyte and TS	Powell 1986
	B(H)	Laminaria hyperborea	sporophyte	Lüning 1986
(B) Long-day responses				
formation of gametangia	R(H)	Atractophora hypnoides	Rhododiscus-phase (TS)	Maggs and Guiry 1987b
	B	Sphacelaria rigidula	gametophyte	10 Hoopen et al. 1983
	B	Sphaerotrichia divaricata[1]	gametophyte	Novaczek and McLachlan 1987
	B	Myriotrichia clavaeformis[1]	gametophyte	Peters 1988
formation of gametophytes	R(H)	Batrachospermum moniliforme[1]	Pseudochantransia-phase	Huth 1979
formation of erect thalli	B	Bachelotia antillarum	prostrate filaments	Shanab and Abdel Rahman 1988

[1]effect of night-break regime not investigated
[2]night-break regime does not function as a long day

Norway, and to 14 h (5–10°C) in material from Iceland. In nature one finds the erect thalli of *Scytosiphon* arising at water temperatures around 10°C, that is, on the Mediterranean coasts in winter, on the cold temperate European coasts in early spring, and at northern latitudes in summer. One may assume that this is the temperature requirement of the erect thallus. The erect thallus represents the main stage for propagation of the species by zoospores, which has forced the species to evolve genetically different, photoperiodic ecotypes so that the seasonal behavior is adapted

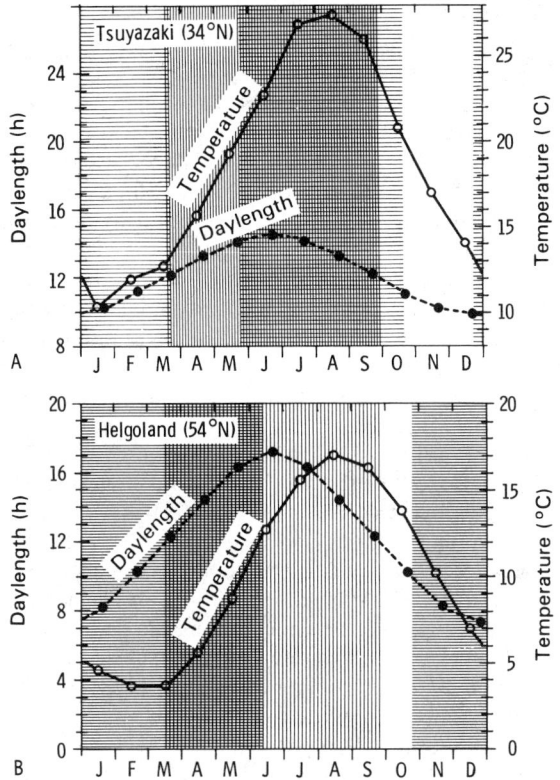

Fig. 6.13 Photoperiodism in the Trailliella phase of the red alga *Bonnemaisonia hamifera*. Restriction of tetrasporangia to a narrow "seasonal window" in autumn, due to unsuitable daylength (vertically hatched area) or to unsuitable water temperatures (horizontally hatched area). On the coast of southern Japan (**A**) the "seasonal window" is broader than near Helgoland (**B**). (From Lüning 1981*b*.)

to the latitude. There are also **daylength-neutral strains** in *S. lomentaria* that seem to have "escaped" from photoperiodic control and use temperature as an environmental trigger instead, high temperatures preventing the formation of upright fronds (tom Dieck 1987).

Night-break regime. The environmental signal for a short-day alga is generally the long night and not the short day. This may be demonstrated if the alga is cultivated in an 8 h light regime. If the 16 h long night is interrupted by 1 h or even 1 min of white light, the night is broken and development that normally occurs during long days takes place. Thus, instead of one long night (simulating winter), two short nights (simulating summer) are present per 24 h. A **night break** inhibits the response to short days in most of the responses listed in Table 6.7, which resemble most higher plants, in that they measure the length of the night, not of the day. For example, no new blades are formed by *Laminaria hyperborea* if it is subjected to a night-break regime (Fig. 6.10C). There are, however, exceptions where the night break does not work, for example, in the red alga *Acrosymphyton purpuriferum* (Table 6.7). Such "light-dominant" plants respond to the length of the day rather than to the shortness of the night (Breeman and ten Hoopen 1987; Dring 1988).

Possible sensor pigments. The **action spectrum** of the night-break effect indicates the presence of a **phytochromelike** sensor pigment in the Conchocelis phase of *Porphyra tenera* (Dring 1967b), with optimal suppression of conchosporangium formation by red (660 nm) light and reversion of this effect by subsequent far red (730 nm) light. The presence and activity of phytochrome outside the Chlorophyta cannot be regarded as proven before detailed action spectra are available and phytochrome has been extracted from nongreen algae (Dring 1984, 1988). Lopez-Figueroa et al. (1989) detected by spectrophotometry, gel electrophoresis and immunoblotting a phytochrome-like protein in extracts from the red algae *Corallina elongata* and *Gelidium,* from the brown algae *Cystoseira abies-marina* and *C. tamariscifolia,* and from the green algae *Ulva rigida, Enteromorpha compressa* and *Chara hispida.* There is also evidence for stimulation of chlorophyll synthesis by red light, with far-red reversibility, in the red algae *Corallina elongata, Porphyra umbilicalis* and the green alga *Ulva rigida* (Lopez-Figueroa and Niell 1989; Lopez-Figueroa et al. 1990). In *Scytosiphon lomentaria,* only **blue light,** supplied as a night break for 10 successive nights, inhibits the formation of erect thalli (Dring and Lüning 1975a; action spectrum similar to that illustrated in Fig. 6.14B). Probably a flavoprotein or carotenoprotein represents the sensor pigment, for example, in the numerous blue–UV-mediated photomorphogenetic, and phototropic reactions in many plant groups. The true chemical nature of the blue-UV-light photoreceptor, operationally termed **cryptochrome,** is not known (reviews in Senger 1980, 1984, 1987; Gressel and Rau 1983; Kendrick and Kronenberg 1986). The sensor pigment of *Scytosiphon,* like phytochrome, is extremely sensitive, since 50% of its thalli were prevented from forming the erect thalli after having been irradiated on each of 10 successive nights for 10 sec at a photon fluence rate of 2 μmol m^{-2} s^{-1}.

Life form types, classified according to seasonal occurrence. Annual and perennial algae were further subdivided by Sears and Wilce (1975) in the following way:

(a) **Aseasonal annuals** (macrothalli present all through the year; several short-lived generations follow each other in the course of the year, e.g., in *Ulva*);
(b) **Seasonal annuals** (present during part of the year as a cryptic stage, e.g., in *Scytosiphon lomentaria* or *Monostroma* spp.);
(c) **Perennials** (thallus survives several years without considerable losses, e.g., in *Ahnfeltia plicata* or crustose algae);
(d) **Pseudoperennials** (parts of the thallus are cast off every year, e.g., as in *Laminaria hyperborea*).

A more detailed but basically similar system has been put forward by Feldmann (1937b), derived from the life-form system of C. Raunkiaer for terrestrial plants. Garbary (1976), in a somewhat simplified manner, distinguished among the annuals the **Ephemerophyceae** *(aseasonal annuals)* and the **Hypnophyceae** (seasonal annuals) and among the perennials the **Phanerophyceae** (erect perennials), **Chamaephyceae** (crustose perennials), and **Hemi-**

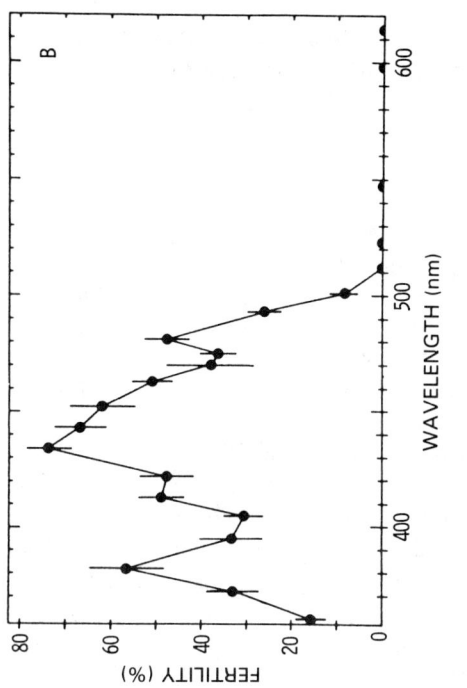
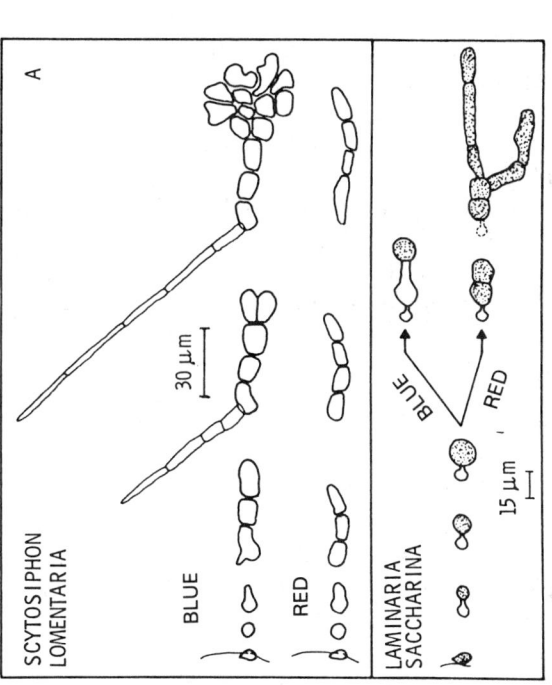

Fig. 6.14 Blue-light-induced photomorphogenetic effects in brown algae. **(A) Upper:** *Scytosiphon lomentaria*. Formation of hairs and two-dimensional, crustose thalli in blue light; filamentous thalli, without hairs in red or green light. **(A) Lower:** gametophytes of *Laminaria saccharina*. Fertilization in blue light; vegetative growth in red or green light. **(B)** Gametophytes of *Laminaria saccharina*. Action spectrum for fertilization. Experimental gametophytes were cultivated in petri dishes for 2 wk from zoospores in red light, irradiated for 48 h at different wavelengths at 15 μmol m^{-2} s^{-1} (induction of fertility by blue light and UV) and subsequently transferred for eight days to red light in order to give time for the formation of oogonia and antheridia. Fertility was determined by counting gametophytes with or without oogonia ($N = 500$). Vertical bars indicate 95% confidence limits. (A from Dring and Lüning 1983; B from Lüning and Dring 1975.)

phanerophyceae (pseudoperennials). The Feldmann system was seen in a "sophisticated" way by V. J. Chapman and Chapman (1976) to include 16 life-form types; for example, the crustose coralline algae have been designated as "chamaecalciphykes."

6.4.2 Photomorphogenetic Reactions in Seaweeds

Since the work of Voskresenskaya in higher plants (e.g., 1952, 1972) it has been found that red light favors the accumulation of carbohydrates and blue light enhances protein synthesis, respiration, and enzyme activation (reviews in Kowallik 1982; Senger 1980, 1984, 1987; Kendrick and Kronenberg 1986). Similar findings have been made for algae (reviews in Dring 1987a, 1988).

Pure **red** or **green** light has an **adverse effect** on the growth of the green alga *Acetabularia mediterranea* (Clauss 1970; Schmid 1984; Wennicke and Schmid 1987). The alga dies in red light after 2–3 wk but can be kept alive by daily pulses of blue light. The brown alga *Dictyota dichotoma* also grows optimally only in blue light (Müller and Clauss 1976). In this alga, and in *Alaria esculenta, Fucus* species, and *Laminaria* species, the light-saturated rate of photosynthesis in blue light is 50–100% higher than in red light (Dring 1987b,). Short pulses of blue light (e.g., 2 min and 20 μmol m^{-2} s^{-1}) stimulate the subsequent photosynthesis of *Laminaria* species for approximately one hour at saturating light levels in red light. Small plants of *Fucus* species died within 4–6 wk in light with a wavelength range about 575–625 nm, at a photon fluence rate of the order of 100 μmol m^{-2} s^{-1}; the addition of longer- or shorter-wavelength light reversed the deleterious effects (McLachlan and Bidwell 1983). The effects of high red and green light may be caused by the sensitivity of photosystem I to excessive radiation (see Section 6.3.2, photoinhibition of photosynthesis).

In several members of the brown algae, **blue-UV-mediated photomorphogenetic effects** were detected. Germlings of *Scytosiphon lomentaria* grow in a filamentous manner in red or green light (Fig. 6.14A). In blue or white light a two-dimensional **crustose thallus** forms (Fig. 6.14A; Dring and Lüning 1975b). A similar effect is known in fern gametopyhtes (Furuya 1983). The brown alga *Bachelotia antillarum* (Ectocarpales) grows as an unbranched filament in red light, while branches are formed in blue, green, and yellow light (Shanab et al. 1988).

A second photomorphenetic effect in *Scytosiphon lomentaria* relates to the formation of **hairs** (Fig. 6.14A), which are formed only if blue or UV light is present. The same spectral range induces the formation of lateral **hair whorls** and **reproductive caps** in *Acetabularia* species (Schmid 1984; Schmid et al. 1987), and fertility in the microscopic **gametophytes** of the **Laminariales** (Lüning and Dring 1975; Lüning and Neushul 1978). In red or green light at temperatures around 15°C, the gametophytes grow vegetatively and may finally form macroscopic balls of up to several centimeters in diameter. Blue light induces the formation of an oogonium in a unicellular female gametophyte (Fig. 6.14A) and the formation of antheridia in male gametophytes that are a few cells in size. According to the action spectrum (Fig. 6.14B), which is similar in all these photomorphoses, the sensor pigment may again be cryptochrome, as in the photoperiodic response of *Scytosiphon*.

6.4.3 Further Signal Effects of Light

Phototropic reactions, which are mediated by blue and UV light throughout the plant kingdom, have been observed in numerous algae (review in Buggeln 1981*a*). The erect thalli of filamentous algae bend towards the light, and haptera of the Laminariales or rhizoids of other algae exhibit a negative phototropism.

The zygotes of the Fucales have been classical objects for studies on the induction of **polarity** (reviews in Buggeln 1981*a;* L. V. Evans et al. 1982). The *Fucus* zygote is characterized in an early stage by a latent polarity and may subsequently use various environmental signal gradients, such as light gradient, ion gradients or contact with the substratum, for the final fixation of its polarity axis so that the rhizoid emerges at the "proper side" of the zygote. In the light (only effective in blue and UV light) the rhizoid is formed 8–14 h after fertilization on the shaded side of the zygote. A little later the germinating thallus grows out towards the light. In darkness the zygote may orient by means of chemically unidentified signal substances that are excreted in all directions but the signal substances become concentrated near the substratum so that a gradient suitable for orientation arises.

Light-induced **movements of chloroplasts** occur in various plant groups and in algae. These are mediated by blue-UV-sensitive pigments (reviews in Britz 1979; Haupt 1983) and by phytochrome in the freshwater green algae *Mougeotia* and *Mesotaenium*. Such plastid movements have been detected in **brown algae,** such as, *Dictyota dichotoma,* several members of the Scytosiphonales, Laminariales, and Fucales (Rüffer et al. 1978; Nultsch et al. 1979, 1981; Nultsch and Pfau 1979). Basically, at higher photon fluence rates (e.g., at 200 μmol m^{-2} s^{-1}) the chloroplasts move to the cell walls perpendicular to the light-exposed algal surface. They return to the horizontally oriented cell walls at a photon fluence rate 10 times lower. The **action spectrum** for moving the phaeoplasts of *D. dichotoma* into the high-intensity arrangement displays two maxima at 360 nm and 466 nms, and no action above 509 nm (Pfau et al. 1988). This suggests that a blue-UV-light photoreceptor is involved. Chloroplast movements have not been found in red algae. In the green algae *Ulva lactuca* and *Acetabularia acetabulum,* the chloroplast movements are governed by a circadian rhythm (Britz 1979; Schmid and Koop 1983; Schmid 1988). Circadian chloroplast migration also occurs in *Dictyota dichotoma* (Nultsch et al. 1984)

In several seaweed species a **light-induced release of gametes** is observed, provided that the algae are cultivated in a light–dark regime. Under such conditions a **circadian rhythm** is apparent (reviews in Bünning 1973; Sweeney

1977). If algae are cultivated in continuous light, the circadian clock is not set. In the green alga *Derbesia tenuissima,* the gametophyte (Halicystis phase) releases its gametes by an explosionlike reaction as soon as the **light cycle** starts (Ziegler Page and Kingsbury 1968).

In the brown alga *Dictyota dichotoma* egg release may be induced at the end of the dark cycle by a light pulse with a duration of 1 min at the extremely low photon fluence rate of 0.02 μmol m^{-2} s^{-1} (white light; Kumke 1973). According to the action spectrum of this effect, the sensor pigment is again a blue-UV-sensitive photoreceptor. Such a pigment also controls egg release in *Laminaria* gametophytes, which in this case occurs at the beginning of the **dark cycle,** since blue or UV light prevent the eggs from being extruded during the whole of the light cycle (Lüning 1981c). Some red algae shed spores with a maximum during the day and some during the night (Umaheswara Rao and Kaliaperumal 1987). It has not been investigated whether a circadian rhythm is involved in the diurnal periodicities of spore shedding.

Sexual attractants in brown algae. As soon as a female gametophyte of *Laminaria* releases its egg, a **pheromone (lamoxirene** in this case), is released, causing the male gametophytes growing nearby to release their spermatozoids. These are subsequently attracted by the pheromone (Lüning and Müller 1978; Maier and Müller 1981; Müller et al. 1985). In several brown algae pheromones have been found to be unsaturated hydrocarbons (Fig. 6.15). All of these pheromones serve as attractants for spermatozoids, while the additional spermatozoid-releasing function is characteristic for the Laminariales, Desmarestiales, and Sporochnales. (For reviews see Müller 1981; Maier and Müller 1986; for suggestions on laboratory experiments see Müller 1988.) The pheromone pattern of a particular alga may reflect phylogenetic relationships; for example, all investigated members of the Laminariaceae, Alariaceae, and Lessoniaceae have lamoxirene. **Ectocarpene,** the main sexual-attractant substance of the primitive *Ectocarpus siliculosus,* is accompanied in this alga by minor amounts of hormosirene, multifidene and dictyotene (Müller and Schmid 1988), major sperm-attractants in other brown algae (Fig. 6.15).

Synchronization of fertilization by a combination of environmental signals. The ecological significance of the reactions preceding fertilization in laminarian gametophytes is to greatly increase the **probability of fertilization.** These reactions include the synchronization of egg release by the light–dark regime, the subsequent release of spermatozoids induced by lamoxirene, and then attraction. In nature, as followed in situ in the sublittoral zone near Helgoland (Lüning 1980a), the zoospores of the *Laminaria* species are released in autumn and winter by the sporophytes. From the settled zoospore the first cell of the gametophyte is formed, and this overwinters in the sublittoral zone until February. Before that time light is too low for photosynthesis and growth, even at moderate water depths, due to long-lasting storms stirring up the sediment (Table 6.3). As soon as light enters the water in February, the gametophytes finish their vegetative growth phase in 1–2 wk and become fertile, provided they receive a sufficient dose of blue light while they are growing. On the basis of the

LIGHT AS AN ENVIRONMENTAL SIGNAL

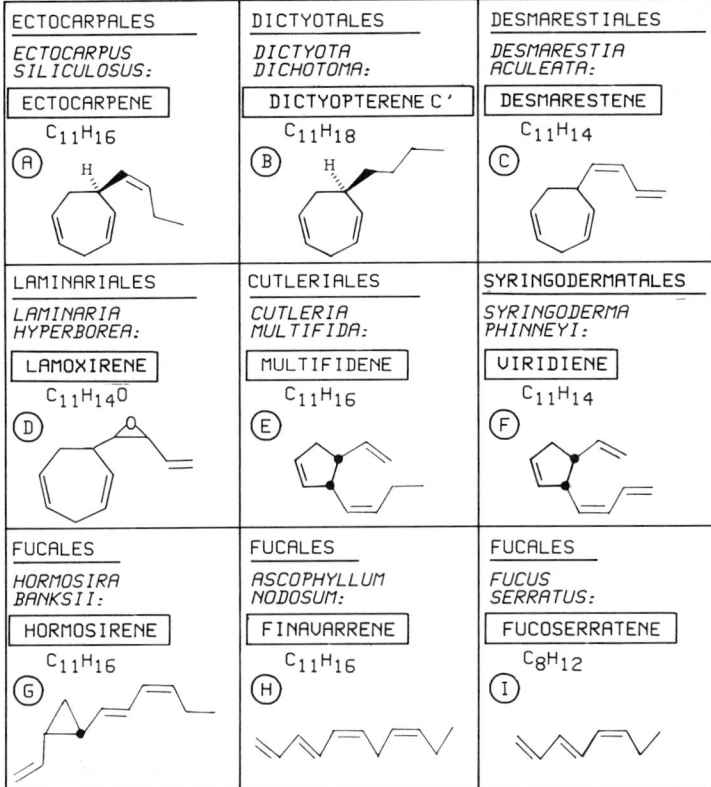

Fig. 6.15 Selected number of pheromones (sexual attractants) in brown algae. *Syringoderma*, formerly of the order Dictyotales, now constitutes a new order, Syringodermatales (Henry and Müller 1983; Henry 1984). (After Müller 1981; Maier and Müller 1986.)

reactions described above, one may then expect that, by the end of February or the beginning of March, fertilization in laminarian gametophytes occurs within the first hour of darkness during a few successive nights. Thus the timespan during which fertilization actually occurs has been narrowed from an order of months for the release of zoospores to an order of minutes as a result of the gametophytes making use of a combination of environmental signals. In this sense one would regard a pheromone acting on an antheridium or a spermatozoid as a chemical, environmental signal.

Lunar periodicity. Another environmental signal used by several marine organisms for the synchronization of reproductive development, in order to enhance the probability of fertilization, is lunar periodicity. This has been demonstrated by Müller (1962) for the brown alga *Dictyota dichotoma*, in which gamete formation and release occurs with a "semilunar rhythmicity," that is, twice per month at an interval of about 15 days. In the laboratory such a rhythm can be imitated by cultivating the alga in a light–dark regime and replacing one night per month (intervals of 28 days) with a "full

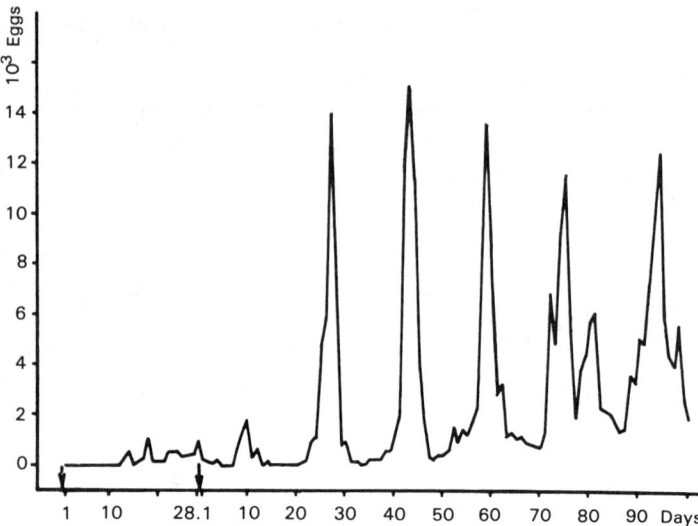

Fig. 6.16 Lunar periodism in the brown alga *Dictyota dichotoma*. Experimental female gametophytes were cultivated at 14 h of light per day. Number of eggs released was counted daily for more than three months. Continuous light was given on days 1 and 28 (arrows). Thereafter a semilunar rhythm of egg release was apparent. (From Müller 1962.)

moon," that is, with continuous light given over 24 h (Fig. 6.16). If one repeats this procedure for a few months, the semilunar rhythm with regard to egg release starts (Fig. 6.16). Possibly a long-period clock is set into motion.

Endogenous clocks. There are four kinds of endogenous clocks, the circadian, circatidal, circalunar and the circannual, with free-running periods of about 24 h, 14 days, 28 days, and one year, respectively (Sweeney 1969, 1977; Gwinner 1986). There is good evidence for the existence of **circalunar** and **circatidal** clocks in marine animal species (reviews in Neumann 1981; Brady 1982). As for seaweeds, it was noticed early that fertile thallus margins of *Ulva* (G. M. Smith 1947) and *Enteromorpha intestinalis* (Christie and Evans 1962, but see Pringle 1986) containing gametes are found twice a month, at spring tides. These observations and similar ones with regard to oogonium release in *Sargassum muticum* (Fletcher 1980) may be taken as indications that several seaweeds may be using endogenous oscillators mimicking lunar or tidal rhythms for the intrapopulation synchrony necessary to enhance the probability of fertilization. A **circannual clock** with a free-running period of a few months less than one year exists in various animal groups, such as migrating birds and molting animals, in the kelp *Pterygophora californica* (Lüning and tom Dieck 1990) and possibly in the seagrass *Posidonia oceanica* (Ott 1979). There are few other examples for plants so far (reviews in Sweeney 1969; Gwinner 1986). The annual cycle of daylength synchronizes the circannual rhythm, for example, the formation of the terminal blade in *P. californica,* to a period of 12 months, or 6, or even 3 months, if exposed to seasonal daylength cycles completed after 6 or 3 months (Gwinner 1986, Lüning and tom Dieck 1990).

7 Temperature, Salinity, and Other Abiotic Factors

Previously in this book it has been mentioned that geographical and/or vertical seaweed distribution is governed by temperature (e.g., Section 1.2.2), salinity (e.g., Sections 2.3.6 and 2.4.7), water movement (e.g., Sections 1.1.1 and 2.3.5), and light (e.g., Sections 6.2 and 6.4.1). The present chapter treats seaweed adaptations to these and a few other abiotic parameters, apart from light, in some more detail with emphasis on differential tolerances and adaptations of seaweed groups. Species with a wide tolerance range to temperature (salinity) are termed **eurythermal** (**euryhaline**), species with a narrow tolerance range **stenothermal** (**stenohaline**). The distribution of seaweeds should not be discussed exclusively in terms of their dependencies on abiotic environments. Examples of the importance of biotic factors will be given in Chapter 8.

7.1 TEMPERATURE

7.1.1 Heat and Cold Tolerance

The possible causes of algal death at higher temperatures may include processes such as denaturation of proteins and damage to heat-labile enzymes or membranes. At low temperatures, lipids and proteins of the cellular membranes are destroyed due to intracellular formation of ice crystals. (For reviews on temperature tolerance see Larcher 1980; Levitt 1980; Berry and Raison 1981; on heat tolerance see N. C. Turner and Kramer 1980; Kappen 1981; on cold tolerance see Larcher and Bauer 1981; Graham and Patterson 1982; Clarke 1983).

Eurythermal algae survive large temporal and irregular fluctuations of water temperature (Fig. 7.1A), whereas **stenothermal** species cannot exist under these conditions (Fig. 7.1B). Eurythermal species have the advantage of a wide geographical distribution; however, they are not as efficient as the stenothermal species, which may develop optimally adapted enzymes for a narrow temperature range. They may also combine the advantages of being **generalists** (eurythermal species) and **specialists** (stenothermal species) through **seasonal adaptation** to temperature tolerance (Fig. 7.1C), or by hav-

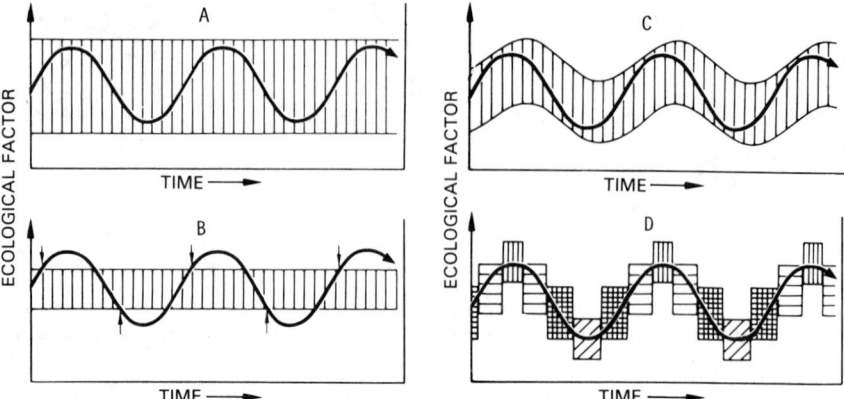

Fig. 7.1 Differential strategies of species with regard to cyclical changing environmental factors (temperature in this example). The sinusoidal curves represent the seasonal course of temperature from year to year, the vertically hatched areas the tolerance range of a species. **(A) Generalist** (eurythermal species): wide tolerance range. **(B) Specialist** (stenothermal species): narrow tolerance range. In this example the population dies at points indicated by the arrows. **(C) Seasonal adapation:** the tolerance range is seasonally adapted, with a slight time lag with regard to the course of temperature. **(D) Generations or stages with different tolerance ranges:** the species covers the whole seasonal range of temperatures by switching from one generation or stage to the next. (From Halbach 1980.)

ing **phases following each other** over the course of the year, each equipped with different tolerances (Fig. 7.1D).

The last adaptation, also termed **stress evasion** or **stress escape** (Levitt 1980), is quite common in seaweeds with a heteromorphic life history. Often a macroscopic phase completes its development before exposure to the seasonal stress, for example, high temperature in summer, which is tolerated by a microscopic phase but not the macroscopic one.

All these strategies exist together. This may be seen from the **heat tolerance** of the major components of the Helgoland seaweed flora (Table 7.1). The perennial red alga *Phyllophora pseudoceranoides* (dead at 33°C; geographical distribution in Fig. 2.34) exists in the sublittoral zone as a eurythermal species together with the stenothermal, perennial red alga *Membranoptera alata* (dead at 23°C; distribution in Fig. 2.27). A seasonal adaptation in relation to the seasonal range of water temperature (Fig. 7.1C) is exhibited by *Laminaria* species and *Desmarestia aculeata* (Table 7.1). An example of successive generations with different heat tolerances (Fig. 7.1D) is provided by the red alga *Bonnemaisonia hamifera,* in which the tetrasporophyte (Trailliella phase) dies at 28°C; however, the gametophyte (*Bonnemaisonia* phase), which grows on Japanese coasts mainly in winter and spring, dies at 25°C (Lüning 1981b).

Cold and freezing (frost or cryo-) **tolerance** is exhibited in the eulittoral seaweed species of the Arctic and temperate coasts (Table 7.2) that easily

Table 7.1 Temperature range for survival (---) and for optimal growth (+++) in seaweeds from Helgoland. Exposure in seawater for 1 week at the indicated temperatures; parameter for vitality: photosynthetic oxygen production. The values refer to experimental thalli collected in the field. *Bonnemaisonia hamifera* (gametophyte) was cultivated in the laboratory from the Trailliella-phase. Temperatures below 0°C were not tested.

Species	0	5	10	15	18	20	23	25	28	30	33
Green algae											
Monostroma undulatum	----	----	----								
Acrosiphonia arcta	----	++++	++++	----	----	----					
Blidingia minima	----	----	----	----	----	----	----				
Chaetomorpha melagonium	----	----	----	----	----	----	----	----			
Cladophora rupestris	----	----	----	----	----	----	----	----	----		
Ulva lactuca	----	----	++++	++++	----	----	----	----	----		
Codium fragile	----	----	----	----	----	----	----	----	----	----	
Enteromorpha prolifera	----	----	++++	++++	++++	++++	----	----	----	----	
Bryopsis hypnoides	----	----	----	----	----	----	----	----	----	----	
Brown algae											
Chorda tomentosa	----	----	----	----	----						
Chorda filum	----	----	----	----	----	----					
Laminaria digitata (February)	----	----	----	----	----						
Laminaria digitata (July)	----	----	----	----	----						
Laminaria saccharina (February)	----	----	++++	++++	----						
Laminaria saccharina (July)	----	----	++++	++++	----						
Laminaria hyperborea (new;Febr.)	----	----	----	----	----						
Laminaria hyperborea (new;April)	----	----	----	----	----						
Laminaria hyperborea (July)	----	----	----	----	----						
Desmarestia aculeata (new,Apr.)	----	++++	++++	----	----						
Desmarestia aculeata (new;June)	----	++++	++++	----	----						
Desmarestia aculeata (Sept.)	----	++++	++++	----	----	----					
Desmarestia viridis	----	----	----	----	----	----					
Petalonia fascia	----	----	++++	++++	----	----					
Scytosiphon lomentaria	----	----	----	----	----	----	----				
Halidrys siliquosa	----	----	----	----	----	----	----				
Fucus serratus	----	----	----	++++	----	----	----				
Fucus vesiculosus	----	----	----	++++	----	----	----	----			
Ascophyllum nodosum	----	----	----	++++	----	----	----	----			
Fucus spiralis	----	----	----	++++	----	----	----	----			
Red algae											
Phycodrys rubens (April)	----	----	----	----	----						
Phycodrys rubens (July)	----	----	----	----	----						
Membranoptera alata	----	----	++++	++++	----	----					
Rhodomela confervoides (Jan.)	----	----	----	----	----						
Rhodomela confervoides (July)	----	----	----	----	----	----					
Delesseria sanguinea	----	----	++++	++++	----	----					
Cystoclonium purpureum	----	----	++++	++++	++++	----	----				
Dumontia contorta	----	++++	++++	++++	----	----	----				
Porphyra umbilicalis	----	----	++++	----	----	----	----				
Ceramium rubrum	----	----	----	++++	----	----	----				
Polysiphonia urceolata	----	----	++++	----	----	----	----				
Corallina officinalis	----	----	----	----	----	----	----				
Chondrus crispus	----	----	++++	++++	----	----	----	----			
Ahnfeltia plicata	----	----	----	----	----	----	----	----			
Phyllophora truncata	----	----	----	----	----	----					
Phyllophora pseudoceranoides	----	----	++++	----	----	----	----	----			
Polyides rotundus	----	----	++++	----	----	----	----	----			

Compiled after Fortes and Lüning 1980; Lüning 1984b

survive temperatures below the freezing temperature of seawater (at 35‰ salinity, −1.91°C; at 30‰, −1.63°C; and at 25‰, −1.34°C). In the Arctic region, *Fucus distichus* may even survive temperatures of −40°C for several months (Kanwisher 1957), covered by an "ice foot." On the North American east coast the photosynthetic rate of *Chondrus crispus* recovered after 3 h at −20°C but not after 6 h (Dudgeon et al. 1989). The photosynthesis of *Masto-*

Table 7.2 Temperature survival ranges of eulittoral and sublittoral species from different biogeographical regions after a 12 h exposure in seawater.

Type of region	coast	annual temperature span (°C)	temperature survival range eulittoral species (°C)	sublittoral species (°C)
Antarctic	King George I.	-1.8 to +1.2		-1.8 to 11 (18)
Arctic	W-Greenland	0 to 6	-10 to 28	-1 to 22 (24)
cold/warm temp.	Brittany	10 to 16	-8 to 30 (35)	-1 (0) to 25 (30)
warm temperate	Naples	14 to 24	-7 to 35	1 (2) to 27 (30)
tropical	Puerto Rico	26 to 28	-2 to 35 (40)	+14 (5) to 35 (32)

Compiled after Biebl 1958, 1962, 1968; data on Antarctic algae from Wiencke and tom Dieck 1989.

carpus stellatus, growing higher on the shore, fully recovered from 24 h at −20°C.

Freezing tolerance, as determined in southern New Zealand seaweed species, varies with the electrical conductivity of aqueous plant extracts, a standard method for assessing frost damage in vascular plants (Frazer et al. 1988). Freezing tolerance, for a 10 h exposure time during an experiment, varied from values around −20°C in upper eulittoral species (e.g., −22°C in the brown alga *Scytothamnus australis*) to −14°C in the lower eulittoral fucalean alga *Hormosira banksii*. In the upper sublittoral zone, values above −10°C were found in some species, for example, −6°C in the fucalean species *Cystophora torulosa* and the kelp *Macrocystis pyrifera,* and −4°C in the red alga *Corallina officinalis*.

The freezing tolerance of *Fucus vesiculosus* may vary seasonally along the North American east coast, from −30°C in summer to −60°C in winter (J. Parker 1960). **Frost hardening** may occur in seaweeds in a similar way to terrestrial higher plants from cold temperate and polar regions. Bird and McLachlan (1974) report that an increase in freezing tolerance may be acquired by germlings or adult thalli of *Fucus* species after they have been exposed to temperatures around 0°C for several days.

During the freezing process **extracellular ice** is formed at an early stage in the intercellular layers and in the space between the cell wall and the protoplast. Extracellular ice exerts a desiccating effect because of the low water-vapor pressure above the ice, but the effect is not lethal. The lethal effect of freezing sets in after this stage, due to the formation of **intracellular ice,** and is associated with damage to the plasma membrane (Davison et al. 1989; Dudgeon et al. 1989). The formation of intracellular ice occurs very fast in sublittoral zone algae with large vacuoles (see below). In contrast, **freezing-tolerant** species, which exist in the supra- and eulittoral zones, tend to avoid the formation of intracellular ice crystals. They probably remain in a persistent supercooled state because of antifreeze substances in their cells (Larcher and Bauer 1981). Antarctic eulittoral green algae contain significantly higher amounts of the supposed antifreeze agent DMSP (β-dimethylsulphoniopro-prionate) than do temperate seaweed species (Karsten et al. 1989). Differential freezing tolerance may be important in structuring cold temperate communities; for example, three hr at $-20°C$ result in an immediate reduction of photosynthesis of most intertidal seaweeds, with the degree of inhibition corresponding to zonation on the shore (Davison et al. 1989).

Temperature tolerance of seaweeds from different geographical regions: exposure time. Biebl (e.g., 1958) obtained wider temperature tolerances in exposure times of 12 h (Table 7.2) than those listed in Table 7.1 where exposure time was for one week. In a study on the red alga *Polyneura hilliae* (Yarish at al. 1987), the upper survival temperature moved stepwise by 1°C, from 28°C to 24°C, after exposure times of 6 h, and 1, 5, 10, and 14 days but remained constant after further five weeks' exposure.

In general, one may recognize the following patterns.

Eulittoral species. These exhibit a wider tolerance than sublittoral species. Arctic and cold temperate species of the eulittoral zone survive temperatures below the freezing temperature of seawater in contrast to tropical, eulittoral algae. The latter species may survive the relatively low temperature of $-2°C$ for 12 h, but this temperature never occurs in their habitat. Such a "**luxuriously**" wide tolerance interval is also evident from the cold tolerance of eulittoral algae from Naples, and from the heat tolerance of eulittoral algae from Greenland (Table 7.2), suggesting that geographically widely distributed species must be eurythermal. Biebl (1962) has emphasized that the eurythermal responses of eulittoral species may be related to a cytoplasmatic constitution that allows them to lose much of their water content without damage. In spores and seeds of terrestrial plants it is well known that considerably higher temperatures are survived by tissues that have lost part of their water content (Kappen 1981). The marine red alga *Bangia fuscopurpurea* from Naples survives for 12 h in an air-dried state at 42°C, but only at 30°C if it is soaked with water (Biebl 1939). The heat tolerance of *Fucus vesiculosus* increases by 5°C if 30% of its water is lost (Schramm 1968).

Sublittoral species: cold tolerance. Sublittoral algae of the Arctic and cold temperate regions do not survive freezing in seawater (Table 7.2). The reason for this is obvious. They are never exposed to this extreme condition in their habitat. This does not even happen in the Arctic region where an ice layer 1–2 m thick covers the surface of the sublittoral habitat. Tropical sublittoral algae may exhibit only limited **cold tolerance** to temperatures below 14°C (Table 7.2). A tropical isolate of *Gracilaria coronopifolia* from Hawaii, where temperatures fluctuate between 24 and 27°C, survived for 6 wk in the range of 15–28°C and grew optimally at 20–28°C (McLachlan and Bird 1984). This stenothermal behavior is contrasted by *G. tikvahiae* from Nova Scotia and Florida; both isolates survived in the range of 0–34°C. Tropical algae may lack the capacity to form cold-adapted membranes and enzymes. However, Yarish et al. (1984) report from culture experiments that the amphi-Atlantic tropical species *Solieria filiformis* can be slowly adapted to 0°C. In addition, members of the northeast American tropical to temperate group (including *Grinnellia americana, Lomentaria baileyana, Agardhiella subulata,* and *Gracilaria tikvahiae*) may withstand temperatures lower than 0°C for at least 6 wk (Yarish et al. 1984).

Sublittoral species: heat tolerance. The sublittoral algae of the tropical regions differ little in heat tolerance from their counterparts in temperate regions. Temperatures of 33–35°C represent a general upper survival limit for both these groups (Tables 7.1 and 7.2). The rather narrow survival temperature span of sublittoral **tropical seaweeds** reflects the possibility that a temperature of 33°C was rarely surpassed in the tropical regions since the late Precambrian (700 Ma; Section 4.1). The temperature range of these regions also probably persisted, although restricted in latitudinal width, during the Pleistocene. The present-day lower survival limit of tropical sublittoral algae (5–14°C; Table 7.2) was probably never encountered in earlier times.

Sublittoral species: upper survival limit of low-latitude species. Seaweeds inhabiting the polar regions live all year round at temperatures at or below 0°C, but their upper survival limits are well above this value (Tables 7.2 and 7.3). Heat tolerance in seaweeds is higher in the Arctic than in the Antarctic region because the Antarctic has been a cold-water habitat longer. Seaweeds of the **Antarctic** region do not survive 11–18°C, depending on the species, while the corresponding range for **Arctic** seaweeds is 18–24°C (Tables 7.2 and 7.3). The sporophyte of the Arctic cold temperate kelp *Alaria esculenta* (Fig. 2.12) dies at temperatures higher than 16°C (Sundene 1962; Munda and Lüning 1977), and the sporophyte of the kelplike Antarctic *Himantothallus grandifolius* at 11–13°C (Wiencke and tom Dieck 1989). The sub-Antarctic *Durvillaea antarctica* does not survive 14°C (Delépine and Asensi 1976). The upper survival limit of the gametophytes of the Arctic-endemic kelp *Laminaria solidungula* is at 18°C, while the gametophytes of the Desmarestiales in Antarctica, which here replace the Laminariales, survive 13–16°C, depending on the species (Table 7.3).

Temperature tolerance: no ecotypic differentiation. An extensive search for geographical ecotypes of an algal species, differing in upper temperature tolerance, has yielded a rather uniform behavior of several species over a broad latitudinal range (review in Breeman 1988). The brown alga *Scytosiphon lomentaria,* distributed from northern Norway to the Azores and from British Columbia to Baja California, survived 28–29°C and died at 30°C after two weeks (tom Dieck 1987). The same result was obtained for the red alga *Chondrus crispus,* found from Iceland to Spain, while two North Pacific species, *C. nipponicus* and *C. giganteus,* exhibited a slightly, but distinctly higher temperature tolerance of 1–2°C (Lüning et al. 1987). The red alga *Gigartina teedii,* distributed from Ireland to Greece, uniformly survived 31°C and died at 32°C after two weeks (Guiry et al. 1987b). The red alga *Dumontia contorta,* distributed from Iceland to Brittany, survived 24°C for five weeks and died at 26°C (Rietema and van den Hoek 1984). All these data suggest that the upper temperature tolerance limit in several species is deeply entrenched and not easily changed by the prevailing environmental conditions. This explains the coexistence in a local flora of species that hardly survive the local extreme temperatures with those whose upper temperature limit is 10°C or more above the local seasonal maximum (Table 7.1).

Temperature tolerance: ecotypic differentiation. A species that spans a wide latitudinal range may form temperature ecotypes. For example, in the filamentous brown alga

Table 7.3 Upper survival limit of gametophytes of Laminariales from different biogeographical regions, and of Antarctic Desmarestiales, and temperature limits for formation of gametangia

Type of region	origin	species	upper survival limit (°C)	gametangia (°C)	authors
Antarctic	King George I.	Himantothallus grandifolius	15-16		Wiencke and tom Dieck 1989
Antarctic	King George I.	Phaeurus antarcticus	15-16		Wiencke and tom Dieck 1989
Antarctic	King George I.	Desmarestia anceps	13		Wiencke and tom Dieck 1989
Antarctic-cold temperate	King George I.	Desmarestia menziesii	16-17		Wiencke and tom Dieck 1989
Arctic	Canad.Actic	Laminaria solidungula	18		Bolton and Lüning 1982
Arctic-cold temperate	North Atlantic	Laminaria saccharina, L. digitata	22-23	below 18	Bolton and Lüning 1982
Arctic-cold temperate	North Atlantic	Laminaria longicruris	23	below 20	Egan and Yarish (unpubl.)
cold temperate	Europe	Laminaria hyperborea	21	below 18	Bolton and Lüning 1982
warm temperate	Europe	Saccorhiza polyschides	25	5-23	Norton 1977
warm temperate	South Africa	Laminaria pallida	25	below 18	Branch 1974
warm temperate	South Africa	Ecklonia maxima	25	below 22.5	Bolton and Levitt 1985
warm temperate	South Africa	Ecklonia biruncinata	26	below 22.5	Bolton and Anderson 1987
warm temperate	Eastern Asia	Undaria pinnatifida	27	below 25	Akiyama 1965

Ectocarpus siliculosus the uppermost survival temperature for geographic isolates is 33°C for Port Aransas, Texas; 28°C for Wilmington, North Carolina; 25°C for Woods Hole, Massachusetts; and 23°C for the Canadian Arctic (Bolton 1983). Survival was tested at intervals of 2–3°C, and all isolates survived 0°C. Possibly *E. siliculosus* was an old inhabitant of the warm temperate Atlantic, distributed worldwide in Tethyan times, with sufficient time to form a latitudinal array of thermal ecotypes and colonize the cooling northern regions in the course of the Tertiary. The North Atlantic species *Chondrus crispus* and *Dumontia contorta,* on the other hand, may be Pacific invaders, arriving late in the North Atlantic after the opening of the Bering Strait; for them there may not have been enough geological time for ecotypic differentiation. *Laminaria saccharina,* possibly another Pacific invader, formed temperature ecotypes near its southern boundary where selection pressure would operate on the survival capacity of the sporophytes (Gerard and Du Bois 1988). Sporophytes from Maine, transplanted to the coast of New York state near the southern boundary of the species, died in summer when temperatures exceeded 20°C, while most of the local New York sporophytes survived. This ecotypic differentiation is accompanied by a rather plastic physiological regulation of enzyme levels and photosynthetic capacity in *L. saccharina* (Davison 1987; Davison and Davison 1987) as well as gametophytic growth and reproduction in *L. longicruris* (Egan et al. 1989) regulated by environmental temperature.

Experimental temperature mutants. Examples of a rapid change in the temperature character of *Ulva mutabilis* were obtained after UV irradiation of gametes; while the wild type developed normally at 15°C and 22°C, the mutants grew normally only at 22°C (Løvlie 1978).

7.1.2 Temperature Dependence of Growth and Reproduction

The optimal temperatures for **growth** (Table 7.4) range between 0–10°C for polar species, 10–15°C for various cold temperate species (see also Table 7.1), 10–20°C for warm temperate species, to 15–30°C in warm temperate to tropical species. In several species of the warm-water red algal genus *Gracilaria* it has been noticed that, for short periods, growth continues at or even above the long-term upper survival temperature (McLachlan and Bird 1984). Growth does not occur usually at the lower extreme of the temperature tolerance range in these species.

Stable environments tend to produce stenothermal responses. The tropical pelagic species *Sargassum natans* from the Sargasso Sea, with a broad

Table 7.4 Temperature optima for seaweed growth from different biogeographical regions. G = green alga, B = brown alga, R = red alga. Data for representatives of Laminariales (L) refer to young sporophytes.

region	species	(°C)	authors
Antarctic	B Desmarestia anceps	0	Wiencke and tom Dieck 1989
Antarctic	B Himantothallus grandifolius, Phaeurus antarcticus	0-5	Wiencke and tom Dieck 1989
Arctic	L Laminaria saccharina subsp. longicruris	10	Bolton and Lüning 1982
Arctic-cold temperate	R Lithothamnion glaciale, Clathromorphum circumscriptum	5-10	Adey 1970
Arctic-cold temperate	L Saccorhiza dermatodea	10	Norton 1977
Arctic-cold temperate	R Devaleraea ramentacea	6-10	Rueness and Tananger 1984
cold temperate	B Durvillaea antarctica	12	Delépine and Asensi 1976
cold temperate	L Laminaria digitata, L. saccharina	10-15	Bolton and Lüning 1982
cold temperate	G, B, R diverse spp. from Helgoland (Table 7.1)	5-15	Fortes and Lüning 1980
cold temperate	L Laminaria hyperborea	15	Bolton and Lüning 1982
warm temperate	L Undaria pinnatifida	15-20	Akiyama 1965
warm temperate	R Gelidium pusillum (central Chile)	15-20	Oliger and Santelices 1981
warm temperate	L Saccorhiza polyschides	23	Norton 1977
warm temperate	R Corallina spp., B Sargassum piluliferum	20	Masaki et al. 1981b; Ohno 1979
warm temperate	R Callithamnion byssoides	20-25	Kapraun 1978
warm temperate	R diverse Mediterranean-Atlantic spp.[1]	10-20	Yarish et al. 1984, 1987
warm temperate	R Gracilaria sp. (Adriatic Sea)	15-30	McLachlan and Bird 1984
tropical to temperate	R diverse Northeast American species[2]	15-30	Yarish et al. 1984
tropical to temperate	R Gracilaria tikvahiae	15-30	McLachlan and Bird 1984
tropical to warm temp.	R Solieria tenera	15-30	Yarish et al. 1984
tropical to warm temp.	R diverse spp.[3]	20-28	Yarish and Edwards 1982
tropical	B Sargassum natans	18-30	Hanisak and Samuel 1987
tropical	R Hypnea cervicornis	28	Mshigeni 1976
tropical	R Gracilaria coronopifolia	20-28	McLachlan and Bird 1984

[1]Lomentaria articulata, Cryptopleura ramosa, Callibepharis ciliata, Polyneura hilliae, Halurus equisetifolius, Callophyllis laciniata, Hypoglossum hypoglossoides

[2]Grinnellia americana, Lomentaria baileyana, Agardhiella subulata

[3]Caloglossa leprieurii, Bostrychia radicans, Polysiphonia subtilissima

growth optimum of 18–30°C (Table 7.4), has lost the ability to grow at a temperature of 12°C, which favors the growth of benthic *Sargassum* species from Florida (Hanisak and Samuel 1987).

Growth optima: ecotypic differentiation. Isolates of *Ectocarpus siliculosus,* from polar to warm temperate latitudes, exhibited differences of up to 10°C for optimal growth temperature (Bolton 1983). Similarly, the temperatures for growth optima and cold tolerance in the temperate to tropical genus *Gracilaria* differ by as much as 10°C (McLachlan and Bird 1984). These are examples of Scarlato's (1977) scheme of phylogenetic adaptation during the temperature deterioration in the Tertiary, and correspondingly during the preceding two great Ice Ages of the Phanerozoic (comprising the Paleozoic, Mesozoic, and Cenozoic; Fig. 2.1A). Ecotypes of temperate species growing near the Poles first broadened the temperature range of survival and reproduction to encompass the near-zero temperatures developing in the course of the Tertiary, and then lost the ability to grow and reproduce at higher temperatures. This loss does not always occur, as shown by an example from the marine invertebrates. The snail *Littorina littorea,* which has planktonic larvae, achieved its wide distribution through physiological plasticity, while *L. saxatilis,* a direct developer (without larvae), achieved wide distribution via locally adapted ecotypes for growth with fixed, different temperature optima (Behrens Yamada 1987).

The processes involved in **reproduction** resulting in the formation of algal spores and gametes may be quite sharply tuned to the actual environmental temperature range. High temperature inhibition of gametangia formation in the gametophytes of the Laminariales occurs at values 5–7°C higher values in warm temperate members compared with Arctic–cold temperate representatives (Table 7.3).

Species showing a close dependence on temperature for reproduction in their specific environmental temperatures may be seen from Table 7.5. In some instances temperature instead of daylength may be used as an **environmental signal for a season.** Day-neutral gametophytes of the brown alga *Chorda tomentosa* (Fig. 2.19) arise in late spring from the zoospores of the macroscopic sporophytes. They are prevented from becoming fertile; by the high summer temperatures this secures the production of only one sporophyte generation per same year. A similar response is displayed by the day-neutral gametophytes of *Desmarestia viridis* (Lüning 1980*b*). A temperature-regulated switching response from one generation to the other, due to the restriction of gamete or spore formation within a narrow temperature range, is exhibited by *Ectocarpus siliculosus* from Naples, by the red alga *Callithamnion byssoides* from the North American east coast, and by *Lomentaria articulata* from Brittany (Table 7.5).

A particularly impressive temperature regulation of development is shown by the green alga *Urospora wormskioldii*. Three morphologically different phases (a filamentous phase, a dwarf phase, and a Codiolum phase) are all attainable through reproduction by zoospores without the interference of sexuality. The morphological form obtained is dependent only on the cultivation temperature of the zoospores (Fig. 7.2). In addition, Bachmann et al.

Table 7.5 Temperature limits for reproduction of seaweeds from different biogeographical climates. B = brown alga, R = red alga; DN = day neutral; SDP = short-day plant (details relating to photoperiodic behavior according to Luning 1990b, in the case of *Ectocarpus siliculosus*: Lüning, unpublished)

type of region	species	photoperiodic behavior	formation of gametes (°C)	formation of spores (°C)	authors
Arctic-cold temperate	R *Clathromorphum circumscriptum*			below 3	Adey 1973
Arctic-cold temperate	B *Desmarestia aculeata, D. viridis* (gametophytes)	DN	below 10		Lüning 1980b
Arctic-cold temperate	B *Chorda tomentosa* (gametophytes)	DN	below 8		Maier 1984
Arctic-cold temperate	B *Chorda filum* (gametophytes)	DN	5-15		Novaczek et al. 1986a
Arctic-cold temperate	R *Porphyra miniata* (Conchocelis-phase)	DN		below 5	Chen et al. 1970
Arctic-cold temperate	B *Sphaerotrichia divaricata* (microthalli)	LDP	0-15		Novaczek and McLachlan 1987
cold temperate	B *Stilophora rhizodes* (microthalli)	DN	below 10		Novaczek et al. 1986b
warm temperate	R *Bonnemaisonia hamifera* (Trailliella-phase)	SDP		13-19	Lüning 1981b
warm temperate	B *Ectocarpus siliculosus* (Naples)	DN	above 13	below 19	Müller 1962
warm temperate	R *Callithamnion byssoides*		20-25	15-20	Kapraun 1978
warm temperate	R diverse Mediterranean-Atlantic spp.[1]	DN		10-20	Yarish et al. 1984, 1986
warm temperate	R *Lomentaria articulata*	DN	10-15	10-20	Yarish et al. 1984
warm temperate	R *Polyneura hilliae*	DN	10-20	10-20	Yarish et al. 1984
temperate to tropical	R *Grinnellia americana*	DN	15-30	20-30	Yarish et al. 1984
temperate to tropical	R *Lomentaria baileyana*	DN		20-30	Yarish et al. 1984
temperate to tropical	R *Agardhiella subulata*	DN		15-30	Yarish et al. 1984
warm temp. to trop.	R *Caloglossa leprieurii*	DN	20-28	20-28	Yarish and Edwards 1982
warm temp. to trop.	R *Bostrychia radicans*	DN	20-32.5	20-28	Yarish and Edwards 1982

[1]*Callithamnion tetragonum, Halurus equisetifolius, Callophyllis laciniata, Hypoglossum hypoglossoides*

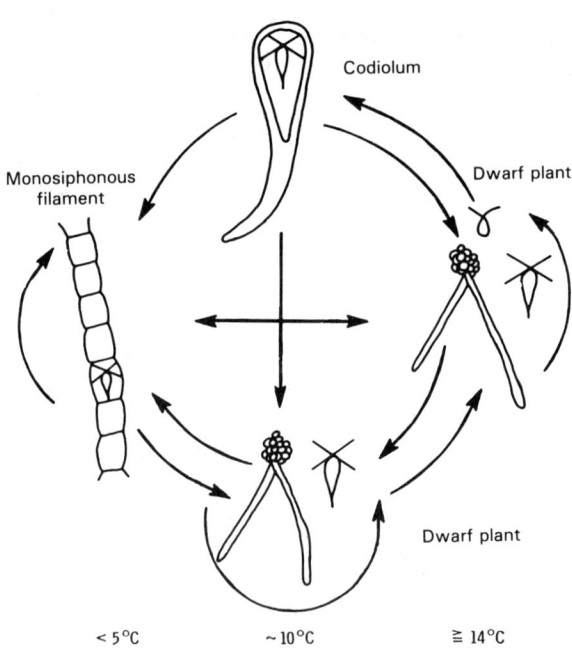

Fig. 7.2 Life history of the green alga *Urospora wormskioldii*, governed by temperature (see text for further explanation). (From Bachmann et al. 1976.)

(1976) reported that the stages differ in their chemical cell-wall composition and even in their enzyme pattern. It is assumed that the temperatures that act on the zoospores determine which genes are activated.

7.1.3 Temperature Dependence of Photosynthesis and Respiration

The relationship between **photosynthesis** and temperature is represented by an optimum curve (Fig. 7.3). The rate of photosynthesis doubles if temperature is increased by 10°C (Fig. 7.3B; $Q_{10} = 2$), provided that photosynthesis is not limited by the light or carbon supply. The photosynthetic rate doubles because the enzymatic reactions that limit photosynthesis at saturating light are temperature-dependent. The temperature optimum of photosynthesis may be seasonally shifted up by 5°C (Fig. 7.3A). The optimum curve has a rapid decline near the species-specific upper lethal limit (Fig. 7.3). At low photon fluence rates the rate of photosynthesis is hardly influenced by temperature (Fig. 7.3B).

For algae growing in colder regions the temperature optima are situated well above environmental temperatures (Table 7.6). This is particularly evident in the Arctic seaweeds and is well known in Arctic terrestrial plants. The Arctic terrestrial plants often exhibit photosynthetic optima at 20°C (Berry and Raison 1981). This apparent discrepancy, between optimum photosynthetic temperature and environmental temperature, may be explained by considering the differential adaptation mechanisms developed to withstand heat or cold. Heat adaptation requires processes like heat stabilization of protein synthesis. Cold adaptation requires high levels of photosynthetic enzymes in order to balance the adverse effect of low temperature on chemical reactions (Berry and Björkman 1980), or the production and activa-

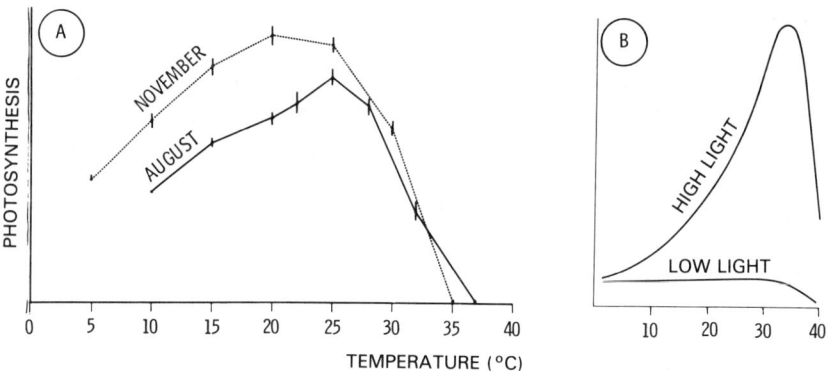

Fig. 7.3 (**A**) Net photosynthesis of the brown alga *Fucus spiralis*, collected in summer and winter, in relation to temperature. (**B**) General dependence of photosynthetic rate on temperature at high and low irradiances. (A from Niemeck and Mathieson 1978; B from Richter 1982.)

Table 7.6 Optimal temperatures for photosynthesis of seaweeds from different biogeographical regions. G = green alga, B = brown alga, R = red alga

type of region	species	summer (°C)	winter (°C)	authors
arctic	G Chaetomorpha sp., B Fucus distichus	20		Healey 1972
antarctic	R Palmaria decipiens, B Himantothallus grandifolius	15		Drew 1977
cold temperate	R Delesseria sanguinea, B Fucus serratus		20	Ehrke 1931
cold temperate	B Fucus spiralis, F. vesiculosus	25	20	Niemeck and Mathieson 1978
cold temperate	R Polysiphonia lanosa	22-24		Fralick and Mathieson 1975
warm temperate	R Gloiopeltis complanata, B Hizikia fusiforme	30	25	Yokohama 1973
warm temp.-trop.	R Bostrychia binderi, Acanthophora spicifera	35		Dawes et al. 1978
warm temp.-trop.	R Polysiphonia subtilissima	27		Fralick and Mathieson 1975
tropical	R Eucheuma spp.	30		Glenn and Doty 1981

tion of new enyzmes or enzymes with modified properties (Descolas-Gros and de Billy 1987). The extent to which the photosynthetic optimum may be shifted, either seasonally or at different latitudes, depends on the balance between these differential adaptation mechanisms. In seaweeds, enzymatic acclimation has been shown, for example, in *Laminaria hyperborea* and *L. saccharina* (Davison 1987; Davison and Davison 1987).

Measurements of **respiration** rates also indicate doubling with 10°C increases in temperature (Q_{10} = 2). A decline in the respiration of *Fucus vesiculosus* occurs at 25°C if measured in winter and at 30°C in summer (Fig. 7.4). Species like *Ulva lactuca, Enteromorpha linza,* and *Ceramium rubrum* exhibit similar relationships between respiration rate and temperature in summer and winter (Fig. 7.4). This means that no seasonal adaptation of respiration occurs in these species and that their respiration increases considerably in summer. A 5°C seasonal shift is seen in the photosynthetic optimum exhibited by *F. spiralis* (Fig. 7.3A).

A seasonal respiration adaptation is seen in *Ascophyllum nodosum* (Fig. 7.4), where respiration increases less steeply with temperature increase in summer (Q_{10} = 1.5) than in winter (Q_{10} = 2). Respiration also depends on growth activity. In spring (5°C water temperature) the actively growing, new frond of *Laminaria hyperborea* respires twice as quickly as the same nongrowing frond in summer (15°C), measured on a dry-weight basis (Lüning 1971). In summer the frond stores carbohydrates. Similar effects may explain the reduced respiration of *Chondrus crispus* in summer (Fig. 7.4).

7.2 SALINITY

The salinity of seawater has not varied much over the last 600 million years, as shown by the composition of seawater salt from ancient salt deposits and by the Paleozoic occurrence of radiolarians, brachiopods, and echinoderms, all with modern relatives exclusively in the marine habitat (Seibold and

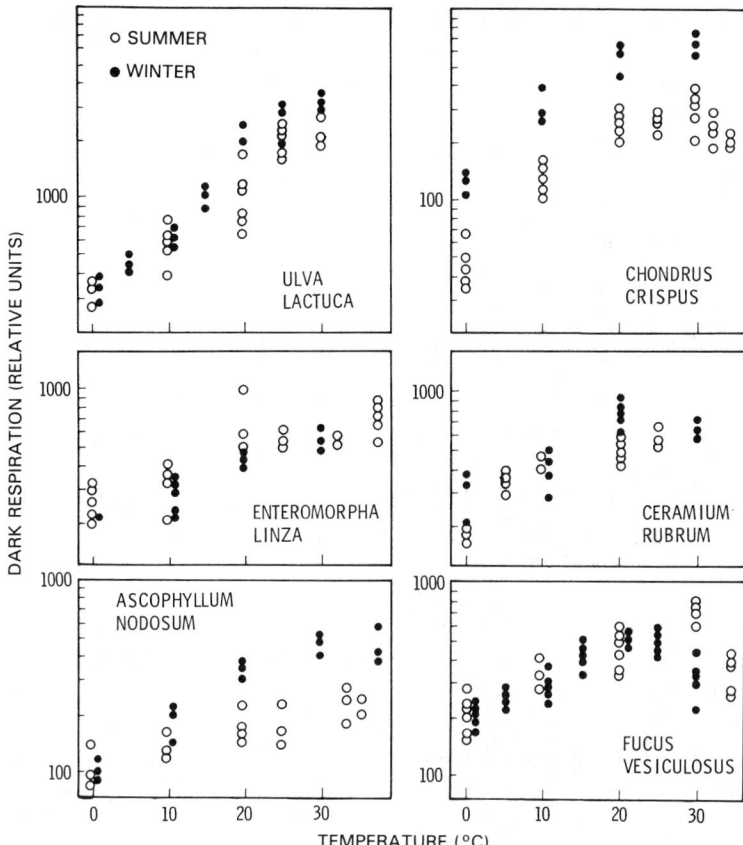

Fig. 7.4 Dark respiration in relation to temperature in several seaweed species, measured in summer and winter. (From Kanwisher 1966.)

Berger 1982). Seawater contains only the most soluble salts that can be delivered by rivers in a relatively short time, for example, the calcium in the oceans within a period of about one million years. Thus marine organisms could adapt to an early established and rather constant salt equilibrium in their medium.

Marine organisms are exposed to an osmotic pressure of 2 MPa (20 bar), corresponding to a salinity of 30‰, while freshwater organisms encounter an osmotic pressure of almost zero in their environment. (For reviews on seaweeds see Gessner 1955–1959; Gessner and Schramm 1971; Wilkinson 1980; Yarish and Edwards 1982; Russell 1986). In order to maintain their turgor, algae had to evolve adaptations to keep their osmotic pressures somewhat higher than the corresponding value in their environment. Osmotic pressures of 2.6–3 MPa are typical for marine algae, whereas freshwater algae have cellular osmotic pressures of 0.5 MPa. Because of these different adaptations a **salinity barrier** prevents most aquatic plants from passing from marine to freshwater habitats and vice versa.

Brackish-water habitats are intermediate between marine and freshwater habitats and represent ephemeral environments in geological terms. They harbor few endemics and have been mainly colonized from the marine side, because of the marked salt sensitivity of freshwater algae.

The **sublittoral** seaweeds of the true marine habitat live in an osmotically constant medium of 30–35‰. Their tolerance to salinity, according to experiments lasting 24 h performed by Biebl (1937, 1938, 1939, 1958), covers a concentration range of 0.5–1.5 times that of normal seawater (16–50‰). The corresponding concentration range is 0.1–3.5 times that of seawater for seaweeds inhabiting the **eulittoral** and **supralittoral** zones. It is obvious that, due to precipitation and evaporation, a wider tolerance range is required in these zones.

The geographical distribution of most sublittoral Northern Atlantic seaweeds ends in the Baltic as salinity has been reduced to half of its normal value. This confirms the validity of Biebl's short-term experiments. Similarly, most of the 17 investigated species of the almost wholly sublittoral red algal genus *Gracilaria* from a wide geographical range grew well at salinities from 15‰ to 60‰ but optimally at salinities around 30‰ (Bird and McLachlan 1986).

At the margins of the brackish-water habitat, for example, in the inner parts of the Baltic and in estuaries generally, algae from the upper eulittoral and supralittoral zones of the marine environment are found. These marine algae are routinely exposed to low salinities, as are those in the marine habitat that receive precipitation while emersed (Wiencke and Davenport 1987).

As an adaptation to **fluctuating salinities** in their habitat, supra- and eulittoral seaweed species shift their internal ion concentrations, mainly K^+, Na^+, Cl^-, and the concentrations of organic osmolytes, such as mannitol in brown algae, floridoside in red algae, and sucrose in green algae. This shift in osmolytes regulates **turgor pressure** arising from the osmotic gradient between the inside and the outside of a cell. (For reviews see Zimmermann 1978; Kauss 1978; Kirst 1990; Kirst and Bisson 1979; Bisson and Gutknecht 1980; Wyn Jones and Gorham 1983; Lobban et al. 1985).

In the **first phase,** lasting 1–15 min in macroalgae, after a **sudden osmotic stress** (hyperosmotic stress from increasing salinity), or hypoosmotic stress from decreasing salinity), rapid water fluxes follow the osmotic gradient and intermittently change the turgor pressure. These changes are dangerous for the alga because its stability, shape, and growth depend on a constant turgor pressure. Therefore, the alga **restores its turgor** pressure in the **second phase** by **osmotic regulation or osmoacclimation.** This process lasts 24–72 h or more in macroalgae and is achieved mainly by stimulating or inhibiting selective ion uptake and, to a lesser extent, by metabolizing low-molecular-weight organic compounds. (See Kirst 1988 for laboratory experiments and measurement of osmotic potentials.)

Organic osmolytes are more important in algae with small vacuoles. They are more useful where higher concentrations of osmolytes are required, because as "compatible solutes" they do not inhibit cellular metabolism and enzyme activities as much as ions do (Reed 1980a; Wiencke and Läuchli 1981; Kirst 1988; Li 1990). The organic osmolytes are preferentially localized within the cytoplasm, while the more toxic inorganic ions are restricted to the vacuole (Wiencke et al. 1983). During osmotic adjustment to hyperosmotic shocks, rapid concentration changes in the inorganic ions often precede the slower synthesis of organic osmolytes, for example, sucrose and proline in *Enteromorpha intestinalis* (Edwards 1987).

One of the few species that has managed the passage from the marine to the freshwater habitat is the red alga *Bangia atropurpurea* (= *fuscopupurea*). It may be gradually transferred from one medium to the other. *B. atropurpurea* only represents a freshwater ecotype (Reed 1980b). On transfer from a freshwater-based medium to saline media, floridoside is rapidly synthesized as an organic osmolyte, while photosynthesis becomes intermittently reduced and then recovers (Reed 1985).

The dependence of **growth rate** on salinity follows an optimum curve. The width of the optimum curve in a particular species or ecotype and the steepness of the growth decline, at higher or lower salinity values, depends on whether the species has a more euryhaline (Figs. 7.5A,B) or stenohaline character (Figs. 7.5C,D).

Seaweed species with maximum growth at lower salinities. There are exceptions to the rule that most seaweeds grow optimally at salinities around 30‰. A *Gracilaria* species from Chilean estuaries showed maximum growth at 15–22‰, and *G. tikvahiae* with its preference for brackish habitats exhibited good growth in the laboratory from 18‰ onwards (Bird and McLachlan 1986). Maximum growth at lower salinities also occurs in estuarine red algae like *Caloglossa leprieurii* and *Bostrychia radicans* (Yarish et al. 1979; Yarish and Edwards 1982); the latter species has a growth optimum salinity as low as 10‰ (Karsten and Kirst 1989). **Sister species** may diverge significantly when the ancestor of one of them moves from marine to brackish water. *Fucus ceranoides* (Fig. 2.37) has a growth optimum at 8‰ and will grow at 17‰, but lethal damage occurs above 25‰ (Khfaji and Norton 1979). The marine sister species *F. vesiculosus* grows well in the range of 8–34‰.

Salinity ecotypes. Ecotypic evolution of seaweeds in the Baltic over the past 3000 years has reponded to the decrease in salinity (reviews in Russell 1985a, 1986). This has been shown, for example, in *Chorda filum* and *Fucus vesiculosus* exhibiting Atlantic and Baltic ecotypes with genetically fixed, differential salinity tolerances and even anatomical traits (Russell 1985b, 1988). Salinity ecotypes have also been found outside the Baltic and in other species. The lower limit for growth in the filamentous brown alga *Pilayella littoralis* is 5.7‰ for marine material but 1.4‰ for material from an estuary. These differences remain after cultivation of many generations in the laboratory (Bolton 1979). The marine ecotype of this species grows in normal or twofold concentration of seawater, but the brackish-water ecotype will not tolerate these concentrations (Reed and Barron 1983). Brackish-water ecotypes with **better growth at low salinities** have also been detected in *Ectocarpus siliculosus* (Russell and Bolton 1975) and in the red algae *Caloglossa leprieurii* and *Bostrychia radicans*

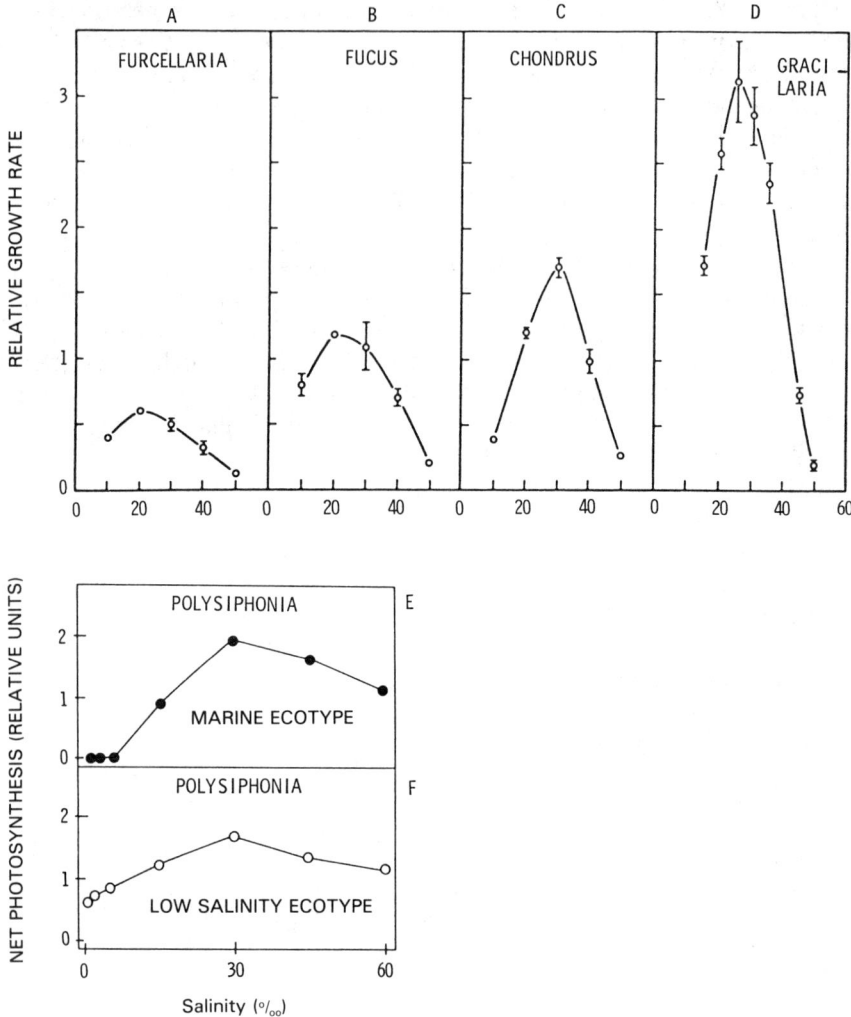

Fig. 7.5 Dependence of growth rate **(A–D)** and net photosynthesis **(E–F)** on salinity. Incubation times at the different salinities: 28 days (A,B), 14 days (C,D), 24 h (E,F). Species names: red algae: *Furcellaria lumbricalis, Gracilaria tikvahiae, Chondrus crispus, Polysiphonia lanosa;* brown alga: *Fucus serratus.* (A–D from Bird et al. 1979; E,F from Reed 1983.)

(Yarish et al. 1979). Localities separated by only a few hundred meters are inhabited by ecotypes of *Enteromorpha linza,* genetically differentiated and adapted to high or low salinities (Innes 1987).

If salinity is suddenly increased for short periods, for example, from 30‰ to 40‰, the rate of **net photosynthesis,** measured as oxygen exchange, may become enhanced. However, after a brief interval, photosynthesis may be finally lower than at normal salinity (Nellen 1966; Gessner and Schramm 1971; Kirst 1981; Wiencke and Davenport 1987). The net gain of photosynthesis also depends on the extent of **respiration,** which is also influenced by salinity (Ogata and Takada 1968; Wilkinson 1980; Kirst 1981). Several species and ecotypes from brackish-water habitats exhibited a fast decline of respiration rate in the range of 10–0‰, but only minor differences in the wide range of 10–50‰ (Dawes et al. 1978).

Most seaweed species inhabiting **brackish-water habitats** probably belong to the euryhaline type. Their **photosynthetic optimum** has not shifted from normal salinity to lower values; rather, the optimal range is very wide; in other words, the rate of photosynthesis declines little with decreasing salinity as, for example, in the brackish-water ecotype of the red alga *Polysiphonia lanosa* (Fig. 7.5F). As a result of this wide range the brackish-water algae have a higher net photosynthetic gain at lower salinities than do stenohaline algae of the normal marine habitat.

When working with lower salinities it must be realized that the damaging effects of seawater diluted with distilled or fresh water may be due to the lower dissolved carbon content of fresh water and/or the reduction in the concentration of calcium and potassium ions. Thus reduced photosynthesis may not be due to lowered salinity (surveys in Dring 1982; Yarish et al. 1979). The low content of dissolved carbon must be compensated for by the addition of bicarbonate or by vigorous aeration (Ogata and Matsui 1965; Gessner and Schramm 1971; Ohno 1976).

7.3 DESICCATION TOLERANCE

One reason for the relatively low diversity of seaweeds (8000 species) may be that they do not undergo desiccation stress. On land, desiccation stress has greatly increased the diversity of plants (review in, e.g., Levitt 1980). However, supra- and eulittoral seaweed species exhibit a remarkable desiccation tolerance. The European fucalean species *Pelvetia canaliculata* will survive emerged for several days after reduction of its water content to a few percent (Schonbeck and Norton 1978). In contrast, sublittoral species like the red algae *Pterothamnion plumula* and *Plocamium cartilagineum* may die if exposed to an atmosphere of 98% humidity (Biebl 1938).

Since desiccation-tolerant seaweeds do not possess any special morphological structures like the stomata of the terrestrial plants, they cannot avoid desiccation but simply tolerate it. Seaweeds of all vertical zones lose their water content like a "gelatin-covered glass plate" (Biebl 1938). The primary factor governing the **velocity** of the **desiccation process** is obviously represented by a species-specific surface–volume ratio called the **specific surface.** A small specific surface is the best way to reduce desiccation, for example, in the thalli of fucalean species and the coarser red algae, and in the voluminous thallus of *Codium*. The thin thalli of *Porphyra columbina,* a New Zealand alga, loses 10% of its water content in 0.1 h in an emersed state, whereas the thick fucalean species *Hormosira banksii* takes 3 h (Table 7.7). Other factors such as cell-wall thickness and composition or water content, which ranges in seaweed species from 80–90% of the fresh weight, may not be as important for differential desiccation tolerance (Dromgoole 1980).

Individual thalli of thin algae with a high specific surface, like *Porphyra, Ulva,* and *Monostroma,* lose their water content very fast. They are better protected against desiccation in dense populations, in which one thallus covers the other in the emersed state. For the vegetation formed by young *Fucus spiralis* in the emersed state, only one-fifth of the total thallus area is exposed to air (Schonbeck and Norton 1979a). Saccate algae, like the red alga *Halosaccion americanum* and the brown alga *Colpomenia peregrina,* reduce drying under desiccating conditions and enhance photosynthesis during emersion by having a central cavity filled with seawater (Oates 1985, 1986, 1988).

Pelvetia canaliculata grows further up the shore than any European fucalean species. It may survive an emersion period of 4–6 days. The emersion period of *Fucus spiralis* is reduced to 1–2 days and for the deeper-growing species, such as *Ascophyllum nodosum, F. vesiculosus,* and *F. serratus,* to a few hours (Figs. 7.6A,B). From their similar specific surfaces it would be expected that these species lose their water content at similar rates (Fig. 7.6C; Kristensen 1968; Schonbeck and Norton 1979a). In fact, a water content of 20–30% of the fresh weight is attained after 4 h on sunny days in these species. The rate of photosynthesis of these fucalean species declines on emersion and is completely restored when resubmerged, provided the thallus has not been desiccated past a species-specific **critical water content.** This is the plant's **differential survival capability.** If emersion causes desiccation past this point, nonreversable damage occurs (Dring and Brown 1982). *Pelvetia canaliculata* and *Fucus spiralis* reach their normal photosynthetic rate 2 h after resubmersion if the water content has been reduced to 10–20% (Fig. 7.7). This critical water content amounts to 30% and 40% in *Fucus vesicu-*

Table 7.7 Drying period (D) required for a water loss of 10%, and surface/volume ratios (S/V) for seaweed species from New Zealand. G = green alga, B = brown alga, R = red alga, W = maximum water content in submersed state. Experimental conditions: 60% air humidity, 22°C, irradiance of 4 W m^{-2}

species	position	thallus part	D (h)	S/V (cm^{-1})	W (%)
R *Porphyra columbina*	middle eulitt. zone	whole thallus	0.1	166	86
B *Sargassum sinclairii*	sublittoral zone	blade	0.2	102	97
R *Gigartina alveata*	lower eulitt. zone	whole thallus	0.4	50	82
B *Ecklonia radiata*	sublittoral zone	blade	0.4	67	83
B *Cystophora retroflexa*	sublittoral zone	thallus branch	0.5	44	86
G *Codium fragile*	lower eulitt. zone	thallus branch	1.9	13	99
B *Durvillaea antarctica*	lower eulitt. zone	blade	2.4	13	93
B *Hormosira banksii*	middle eulitt. zone	blade	3.0	5.5	93
B *Ecklonia radiata*	sublittoral zone	stipe	5.5	3.5	87

After Dromgoole 1980.

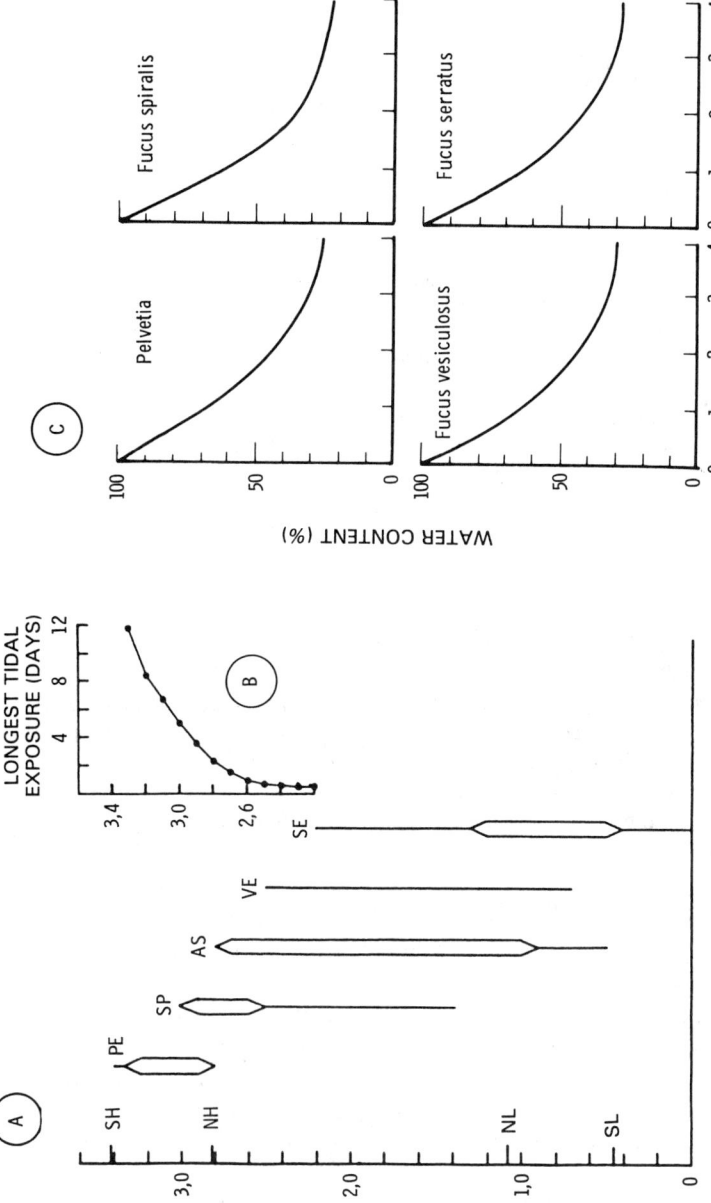

Fig. 7.6 Vertical zonation and water loss in the emersed state. **(A)** Vertical zonation of fucalean species on the Scottish coast. Abbreviations relating to tide levels: S = spring tides; N = neap tides; H = high water; L = low water. Abbreviations relating to species: PE = *Pelvetia canaliculata*; SP = *Fucus spiralis*; AS = *Ascophyllum nodosum*; VE = *Fucus vesiculosus*; SE = *F. serratus*. **(B)** Duration of air exposure in days at neap tides (May to August) for fucalean species growing at different water levels above chart datum. **(C)** Water content in relation to duration of air exposure. (A,B from Schonbeck and Norton 1978; C from Kristensen 1968.)

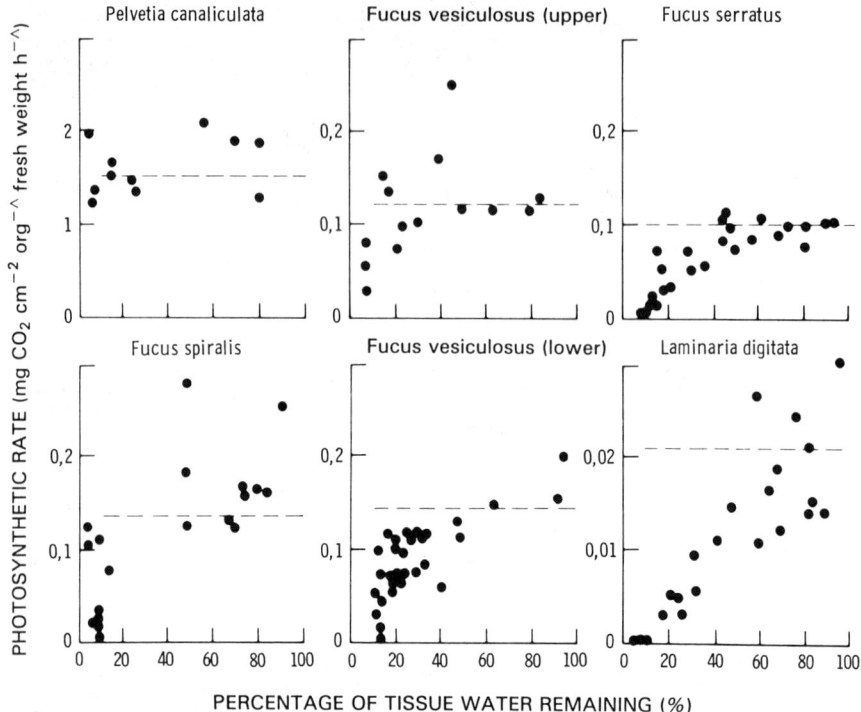

Fig. 7.7 Rates of photosynthesis of water-saturated thalli in fucalean species (2 h of immersion) after drying to different water contents. The broken lines indicate averages of photosynthetic rates, as measured in normal thalli that had not been dried before immersion. Photosynthetic rates refer to unit of fresh weight (*Pelvetia*) or to area (remaining species). In the case of *Fucus vesiculosus,* experimental algae were collected from populations that grew either higher or deeper on the shore. (From Dring 1982.)

losus and *Fucus serratus,* respectively, whereas *Laminaria digitata* from the upper sublittoral zone may not be dried to less than 45% for a full recovery of photosynthesis (Fig. 7.7).

The ability to recover photosynthetic activity following experimental dehydration related well to the vertical distribution of eulittoral seaweed species in California (C. M. Smith and Berry 1986) and New Zealand (M. T. Brown 1987). It appears from these data that photosynthetic recovery is of prime importance in determining the upper limits of seaweed species. Some previous investigators (reviews in Gessner 1955–1959; Gessner and Schramm 1971) had developed similar ideas, but convincing experimental evidence was not available.

If photosynthesis does not attain its full value after reimmersion, it will never recover and the plants will eventually die. Schramm (1968) demonstrated that photosynthesis for Baltic *F. vesiculosus* recovers completely if its water content does not fall below 30%. The experimental plants were from the western Baltic, where the species is exposed to irregular fluctuations of water level but may have to survive emersed for at least 48 h.

The physiological reason why a given fucalean species cannot survive beyond its upper limit in the eulittoral zone can now be partially understood. However, the issue of which cytoplasmic, genetically fixed mechanisms or structures exactly cause these species-specific differences remains largely unsolved. An interesting protective mechanism against **excessive light** induced by desiccation during increasing emersion time was discovered in the intertidal red alga *Porphyra perforata* (Öquist and Fork 1982). In this alga energy transfer from the more vulnerable photosystem II to photosystem I is enhanced by increasing desiccation.

Experimental plants of *Fucus spiralis* transplanted to the next higher zone, where *Pelvetia canaliculata* thrives, will die within 4–8 wk (Schonbeck and Norton 1978). According to these authors, the osmotic stress due to precipitation is not a major factor in causing fucalean zonation, as was formerly thought. Evidence for this view is furnished by the observation that from May to September in sunny weather, after periods of low water and at air temperatures of 10–25°C, the uppermost specimens of *P. canaliculata* and *F. spiralis* died within 3–4 wk. Little or no damage occurred when the weather was rainy, so that precipitation even had an advantageous effect.

Desiccation tolerance may also vary **seasonally** for each of these algae, with higher tolerance in summer than in winter. Individuals growing higher on the shore have a better desiccation tolerance than individuals growing lower on the shore (Fig. 7.7; *Fucus vesiculosus*). Fucalean species may be experimentally "hardened" with regard to desiccation stress within a few days by exposing them to the air in the laboratory for 2–4 h daily (Schonbeck and Norton 1979c).

In addition to desiccation stress due to prolonged emersion, **nutrient stress** may be an important factor, as the entire nutrient supply is obtained during submersion, e.g., *Pelvetia canaliculata* survives on lower levels of nutrients than the deeper-growing *F. spiralis* (Schonbeck and Norton 1979b).

Critical tide levels. At certain tide levels the duration of continuous exposure or submergence increases abruptly, due to the various cycles of the tide, with its daily, monthly, and annual cycles. The critical-tide-level hypothesis expects that certain biological zonal boundaries coincide with such tide levels, emerging, for example, every 24 or 45 days (Doty 1946; Swinbanks 1982). Even a cycle like the 18.6 yr solar-lunar cycle may be of biological importance by shifting the upper limit of the sublittoral zone by 60 cm in the course of 11 yr at British Columbia, or exposing acres of organisms in the upper sublittoral zone of the Bay of Fundy (Swinbanks 1982). These views have been modified, for example, by stressing the additional importance of wave action and topography (Druehl and Green 1982), or refuted by authors suggesting biotic relationships as being more important than tidal emersion in structuring intertidal communities (e.g., Chaloupka and Hall 1984).

7.4 NUTRIENTS

Phosphorus and especially nitrogen normally occur at such low concentrations in seawater that they often become limiting nutrient factors (reviews in DeBoer 1981; Lobban et al. 1985). This is especially evident in the clear "blue" waters of the tropics and in temperate regions in summer if no upwelling or human-caused eutrophication occurs.

Some macroalgae are inhibited in their uptake of **nitrate** by the presence of ammonium (reviews in Hanisak 1983; Lobban et al. 1985; on measurement of uptake see Harlin and Wheeler 1985). Nitrate uptake was approximately 50% slower in the dark than at ambient midday levels in *Macrocystis pyrifera* (Gerard 1982b). The **nitrogen input** to the forest of *Macrocystis integrifolia* in British Columbia was estimated at 60 g N m^{-2} of ocean bottom per year (Wheeler and Druehl 1986). A nitrate concentration of 1-2 μM is required throughout the water column to support growth of *M. pyrifera* at a rate of 4% increase in wet weight per day (Gerard 1982b).

Ammonium, although less abundant in seawater than nitrate and stored at a lower capacity in internal tissue pools, can supply 33–100% of the total in situ nitrogen requirements in several species (Fujita et al. 1988).

In the laboratory the growth rate of *Laminaria saccharina* increases up to a **nitrate concentration** of 10–20 μM (A. R. O. Chapman et al. 1978; Wheeler and Weidner 1983). Above this concentration level an intensive **storage of nitrate** within the thallus may start. The nitrate uptake rate increases linearly with nitrate concentration up to 60 μM, the highest level tested, in *L. bongardiana* (Harrison et al. 1986). A storage of nitrate has also been demonstrated in several other perennial species, such as *Fucus vesiculosus, Ascophyllum nodosum, Codium fragile,* and *Chondrus crispus* (Asare and Harlin 1983). Examples of species with **reduced capacity for nitrate storage** are *Macrocystis pyrifera* (Gerard 1982c), *Pleurophycus gardneri* (Germann et al. 1987), and, among red algae, *Gracilaria secundata* from a eutrophic environment (Lignell and Pedersén 1987).

Macroalgae also take up radioactively labeled amino acids from seawater (Schmitz and Riffarth 1980; Hartmann and Löhr 1983), but due to the low concentrations of organic nitrogen in seawater this may hardly represent a significant source of nitrogen for seaweeds.

During mid-winter, when underwater light is scarce in deeper water and low water temperatures impede the rates of growth and slow the rate of photosynthesis, sublittoral seaweed species start to reproduce and to grow from reserve materials. The ecological factors that favor growth in winter are the **high nutrient content** of the seawater and the **increasing water movement** due to the winter storms. An increase in water movement facilitates ion and gas exchange of the thalli and enhances the mineralization processes (see Section 7.5). After mineralization of senescent plankton and seaweeds in autumn and early winter has occurred, the seawater concentration of **nitrate** rises along the open Atlantic coasts to a concentration of about 10 μM.

In spring, when optimal conditions of light, temperature, and nutrients occur, the phytoplankton exhibits a bloom. This bloom soon reduces the nitrate concentration of the seawater to almost zero. At this time the kelp species are still in an advantageous position, because they may continue to grow by means of **internal nitrate reserves** that were stored in winter; an example is *Laminaria longicruris* growing on the coast of Nova Scotia (Chapman and Craigie 1977).

After depletion of possible internal nutrient reserves, the **growth rates decline** in late spring or early summer, and then the **external nutrient supply** governs the growth activity of seaweeds (e.g., Gagné et al. 1982; Conolly and Drew 1985a,b). Towards late summer and autumn, the **shortening daylength** is used by *L. saccharina* as an environmental signal to induce reproduction, which temporarily reduces vegetative growth, also under nutrient-rich conditions (Lüning 1988). *L. hyperborea* stops growth due to a long-day signal in summer (tom Dieck 1989) and forms, as another photoperiodic response, a new blade in winter (Lüning 1986).

Phosphorous occurs in seawater mainly as orthophosphate (HPO_4^{2-}). Nitrogen and phosporus uptake rates are lower in coarse species with low surface–volume ratios (e.g., *Fucus vesiculosus* and *Phyllophora truncata*) than in filamentous species (e.g., *Cladophora glomerata* and *Ceramium strictum* ssp. *tenuicorne*) or species with numerous hairs (e.g., *Scytosiphon lomentaria* and *Dictyosiphon foeniculaceus;* Wallentinus 1984).

Sulphur is essential for protein metabolism, and in seaweeds it is especially important for the synthesis of their sulphated polysaccharides like agar, carrageenan, and fucose.

Inorganic micronutrients (see, e.g., DeBoer 1981; Fries 1982; McLachlan 1982; Lobban et al. 1985), **hormones** and **growth regulators** (see, e.g., Provasoli and Pinter 1980; Buggeln 1981) have been less intensively investigated. The inorganic and organic **pollution** of coastal areas induced an increasing number of studies on its effects on seaweeds (see, e.g., Munda 1982; Edyvean and Bailey 1984; South and Whittick 1987; Bayne et al. 1988).

Colorless hairs enlarge the surface–volume ratio of algae and possibly serve as nutrient "antennae." They show greater development under nutrient shortage in several species, e.g., the apical hairs in *Fucus* germlings (Schonbeck and Norton 1979d), the hairs in *Codium fragile* (Benson et al. 1983), *Gracilaria verrucosa* (Rueness et al. 1987), and in marine and freshwater chaetophoralean algae (Yarish 1976; Gibson and Whitton 1987).

Carbon is abundantly present in seawater (pH range of 7.8–7.2) in the form of bicarbonate (HCO_3^-), representing 90% of the inorganic carbon supply (reviews in Kremer 1981a,b; Borowitzka 1982; Dring 1982; Bidwell and McLachlan 1985). In seaweed cultures, high CO_2 concentrations are required in aerating gas streams to maintain the bicarbonate concentrations at levels normally found in seawater and to

7.5 WATER MOTION AND HYDROMECHANICAL ADAPTATIONS

The diffusion rate of gases and ions is 10,000 times lower in water than in air. Without water movement most life under water would end immediately. For example, it would take 2000 years for the oxygen concentration at a 10 m water depth to increase exclusively by diffusion from 7 to 9 ml O_2 dm^{-3} (Riedl 1971). The importance of water movement for seaweeds is evident from every culture experiment in which growth rates measured in stagnant medium are compared with those in shaken or aerated medium. (See reviews in Riedl 1971; Schwenke 1971; Neushul 1972; Wheeler and Neushul 1981; Norton et al. 1982; Hiscock 1983; Koehl 1986; Givnish 1986; for methodology see Koehl and Wainwright 1985).

Around an aquatic plant there is a **boundary layer,** a layer of slowly moving fluid through which substances must diffuse to reach the plant from solution (Koehl 1982). The rate of transport depends on the thickness of this layer and can be slower than the potential uptake rate of the plant. Boundary-layer thickness is determined by turbulent mixing and water flow around the plant and may be a few millimeters thick in stagnant water. Flat encrusting algae and microscopic algal germlings live within the boundary layer. For *Macrocystis pyrifera* blades the rate of nitrate uptake is increased by 500% and the rate of photosynthesis by 300%, if current velocity is raised from 0 to about 4 cm s^{-1}, this latter value indicating a saturation level (Wheeler 1982; Gerard 1982a). A further increase in current velocity may reduce photosynthesis again so that the saturation level occurs only over a limited range of current velocities (K. M. Khailov, personal communication; work of Kovardakov and Zavalko). One beneficial effect of **aeration** to seaweed cultures is to reduce diffusion boundary layers and thus increase nutrient uptake (Hanisak 1987).

The various examples of differential floristic composition at wave-exposed and sheltered sites described in the first part of this book (see, e.g., Fig. 2.47) indicate that many seaweed species have adapted to one or the other of these contrasting sets of environmental conditions. **Surface structures** of the thallus, like the spiny, marginal outgrowths of the blades of *Macrocystis pyrifera* (Fig. 2.88F), the wavy margins of *Laminaria saccharina* (Fig. 2.53C) and *L. longicruris* (Fig. 2.79D), the holes in the blade of *Agarum cribrosum* (Figs. 2.10 and 2.79F), and the bullations in the blade of *Hedophyllum sessile* (Armstrong 1989), probably enhance turbulence in laminar-flowing water as it passes the blades and thus secure enhanced exchange of ions and gases. **Long hairs,** found e.g., in the crustose stage of *Scytosiphon lomentaria* (Fig. 6.14A), may serve to penetrate the boundary layer and to supply the crustose thallus, situated at a highly disadvantaged position, with nutrients.

The **mechanical stress** of currents and waves imposing tension, shear, bending, and twisting on the thalli (Koehl 1982, 1986) results in **narrower,** and dissected **blades** in areas with relatively **high wave action,** for example in *L. hyperborea* (Svendsen and Kain 1971) and *Hedophyllum sessile* (Armstrong 1989). In *Nereocystis luetkeana* the narrow, flat blades in fast-flowing water flap with lower amplitude and form a more streamlined bundle than wide, undulate blades, typical of calmer conditions (Koehl and Alberte 1988). A **hydrodynamic streamlining,** with narrower blades and higher elongation rates, was also obtained in individuals of *Laminaria saccharina* subjected to constant longitudinal tension by hanging a weight from a clamp attached to the distal end of the blade (Gerard 1987).

Spores must be adapted to reach the substratum eventually. Most red algal spores seem to be adapted to rapid sinking and localized dispersal (Zechman and Mathieson 1985), although some red algal species like those of *Porphyra* produce buoyant spores (Suto 1950). Small red algal spores have greater chance of fixing quickly to the substratum than bigger spores, which on the other hand, sink faster (Neushul 1972; Okuda and Neushul 1981). At first spores, gametes, and zygotes of green and brown alga attach to the substratum by means of their flagella, but glycoproteins and mucopolysaccharides produced in the Golgi apparatus and excreted as adhesives soon improve the contact and prevent removal by wave action (review in Boney 1981).

The main problem in an experimental approach to "water motion" is its complexity. Of the various physical quantities incorporated by this parameter, current velocity represents only one facet (reviews in Riedl 1964b, 1971). An integrated method to measure the effect of water motion is to follow the erosion of "clod cards" fixed in the sublittoral zone (Doty 1971; Mathieson et al. 1977).

Of the more easily measurable quantities, the force required to remove the attaching disks or haptera of the fucaleans or laminarians, respectively, is on the order of 40 kg cm^{-2} (Schwenke 1971). The stipes of the giant kelps of the family Lessoniaceae (e.g., *Nereocystis luetkeana*) may encounter current velocities of 0.5 m s^{-1} in their habitat. The stipes are extensible, mainly due to the angle of wrap of cellulose fibrils in the stipe cortex (Koehl and Wainwright 1977).

Different strategies are used by kelps to survive in the upper sublittoral zone on wave-beaten shores. *Lessonia nigrescens* (Fig. 5.6B) has a stiff, strong stipe and bends with the flow. In contrast, *Durvillaea antarctica* (Figs. 5.6E,F), a companion seaweed in the same belt along the rocky coast of central and southern Chile, has an elastic stipe and extensible blade. This alga is pulled and deformed by the moving water and thus the stress on the thallus is considerably reduced (Koehl 1982). Flexibility of seaweed thalli enables them to be bent parallel to the flow, bringing them closer to the bottom where the flow is slower.

Gas-filled thallus parts. As pointed out earlier, the evolution of gas-filled thallus parts enhanced the seaweeds' access to light. This is especially evident in the success of the

Lessoniaceae as dominating algae. Other members of the Laminariales have thallus inflations, such as *Egregia menziesii* (Fig. 2.84) and *Chorda* (Figs. 2.9 and 2.19), or gas bladders as do many of the Fucales (e.g., *Fucus vesiculosus*, Fig. 2.23; *Halidrys siliquosa*, Fig. 2.41; and *Sargassum*, Fig. 4.16B). Generally the gas content is similar to air. In the Fucaceae the air bladders are filled with one-third oxygen and two-thirds nitrogen. However, the content of CO_2 may be as high as 1% (review in Dromgoole 1981). The gas in the thallus bladders of two members of the Lessoniaceae, *Pelagophycus porra* and *Nereocystis luetkeana*, consists of up to 10% of the poisonous gas carbon monoxide, its significance being uncertain (Chapman and Tocher 1966; Carefoot 1977). Gas-filled thalli also occur among the smaller algae, such as the brown algae *Scytosiphon lomentaria* and *Colpomenia peregrina* (Fig. 2.62), among the green algae *Enteromorpha* species, and among the red algae *Dumontia contorta* (Fig. 2.15). To the diver the ecological significance of gas-filled thalli becomes obvious when he sees them in an upright position, optimally exposed to light and the open water. Most algal species do not have gas-filled thallus parts but require a minimal turbulence to expose their thalli to the open water. They may be excluded from sheltered sites, where algae with gas-filled thalli may dominate (e.g., *Chorda* species in the inner parts of fjords; Fig. 2.54L). On the other hand, algae with gas-filled thallus parts (e.g., *Fucus vesiculosus*) will be more affected by mechanical wave stress at exposed sites; either they are absent there or if present they reduce or lose their air bladders (review in Norton et al. 1982). Stones and gravel attached to *Fucus vesiculosus* are floated when the alga is three times heavier than the substrate. This is a locally important sedimentary process in the Arctic, potentially confused with ice-rafted sediments (Gilbert 1984).

8 Biotic Factors in the Euphotic Zone. Strategies, Productivity of Seaweeds, and Commercial Uses

As well as abiotic factors, the dynamics in the euphotic zone are governed by biotic factors such as competition between species and the activity of herbivores. (See reviews in Carefoot 1977; Vermeij 1978; Chapman 1979, 1986; Barnes and Mann 1980; Paine 1980; Barnes and Hughes 1982; Gaines and Lubchenco 1982; Mann 1982; Nybakken 1982; Hawkins and Hartnoll 1983; Moore 1983; Denley and Dayton 1985; Russell 1986; Schiel and Foster 1986; Underwood and Denley 1986; Vadas 1985).

8.1 COMPETITION AMONG SEAWEED SPECIES AND THE RELATIONSHIPS BETWEEN SEAWEEDS, HERBIVORES, AND PREDATORS

8.1.1 Eulittoral Zone

Pelvetia canaliculata, the fucalean species growing farthest up in the eulittoral zone of European coasts, is outcompeted if transplanted to the subzone of *F. spiralis,* the next fucalean species to follow in the zonation pattern (Fig. 7.6A; Schonbeck and Norton 1980; Rugg and Norton 1987). Both species can thrive lower down in the subzones of *F. vesiculosus* and of *F. serratus* if these two dominants are experimentally removed. This illustrates the general point that lower growth limits may be determined by **competition** (apart from grazing, see below), whereas the upper limits are often set by abiotic factors, in these cases mainly by desiccation tolerance. In other words, the upper, more stress-tolerant species are confined to their adverse and stressful habitat by more competitive, less stress-tolerant species. An example of an eulittoral species growing lower than usual, when **competitors** are **absent,** is provided by *Fucus serratus* in the western Baltic, where the species inhabits the sublittoral zone, because a dense laminarian vegetation does not exist here. In the mid- and lower eulittoral zones the upper limits of fucalean species may also be set by **grazing** on exposed shores and by competition on sheltered shores (Hawkins and Hartnoll 1985; Underwood and Denley 1986).

Pelvetia canaliculata, "a seaweed that shuns the sea," exhibits an intolerance to prolonged submersion. This may mean that its ability to survive after having lost 96% of its initial water content (see Section 7.3) requires a specialized metabolism requiring periodic emersion. Or it may mean, as inferred from tank experiments, that excesses of potential pathogens or parasites must be curbed by periodic emersion (Rugg and Norton 1987).

The factors influencing the population densities and vertical limits of barnacles, mussels, snails, and seaweeds such as fucalean species and *Chondrus crispus,* which are the main components in the eulittoral zone of Northern Atlantic coasts, have been investigated by cage and exclusion experiments (e.g., Menge 1976; Lubchenco and Menge 1978). The upper limits of the barnacle *Balanus balanoides* and some fucalean species are determined by their desiccation and freezing tolerance and by their tolerance to wave action. At wave-exposed sites the lower limit of the barnacle is governed by competition for space with the mussel *Mytilus edulis.* This has a lower desiccation tolerance than the barnacle, but it outcompetes the barnacle on horizontal and sloping surfaces within five months and on vertical surfaces within two years. At more sheltered sites representatives of the Fucaceae appear with the predaceous limpet *Nucella* (= *Thais*) *lapillus.* The **predator** feeds on barnacles and mussels while avoiding wave-exposed areas. As soon as wave action decreases enough for fucalean species to attach, the thalli exert a "whiplash" effect on barnacles, whose settlement is made more difficult. *N. lapillus* creates new space for settlement of various species by removing barnacles and mussels, and in this way enhances diversity within the complex community (Menge and Sutherland 1976).

Grazers. Important herbivores in the eulittoral zone are the limpets and snails (Lubchenco and Menge 1978, Lubchenco 1982; Choat and Black 1979; Underwood 1979; Andrew 1989). For example, along the Atlantic coast of Europe one mainly deals with the genera *Littorina* and *Patella* (Fig. 2.46), along the North American Pacific coasts with genera including *Katharina, Acmaea, Tegula,* and *Lacuna.* Jernakoff (1985) found that the limpet *Patelloida latistrigata* on the mideulittoral shores of southern New South Wales, Australia, could survive on macroalgae but tended to migrate to bare rock and graze on microalgae including propagules of macroalgae. In a eulittoral belt on the Atlantic shores of Nova Scotia, Canada, the recruitment of *Fucus spiralis* was hardly dependent on the activities of the consumer animals, *Littorina obtusata* and *L. rudis,* but mainly on the presence of a canopy of conspecific adults as the most important regulator of recruitment density (Chapman 1989).

Seaweeds surviving digestion. Algal spores of opportunistic green and brown algae often **survive digestion** and may thus be propagated by **limpets** (Santelices and Correa 1985). The same has been found for fragments of the red alga *Audouinella* (= *Rhodochorton*) *purpureum,* whose small and rather constant turf, 1–1.5 mm high, is continually shorn by the gastropod *Littorina littorea* and the amphipod *Gammarus salinus* (Breeman and Hoeksema 1987).

Grazing pressure and predators. As long as **predators** such as the starfish *Asterias rubens* on the European coasts and *Pisaster ochraceus* on the North American Pacific coast keep the herbivores at a low population density, the dynamic equilibrium between algal vegetation and herbivores is not in danger of being shifted to the side of the limpets (Southward 1964; Dayton 1975). An example with the starfish *Pisaster*, which eats the limpets grazing on the red alga *Endocladia muricata* and ensures its own future supply of mussels, is illustrated in Fig. 8.1.

8.1.2 Sublittoral Zone

Competition between sublittoral seaweed species is mainly a struggle for light and for substratum among crustaceous algae. Having a larger growth form that shades other algae represents the main competitive advantage, repeatedly emphasized in the first part of this book. Examples of dominating **sublittoral canopy algae** include the following cases, the first three being of a more general character.

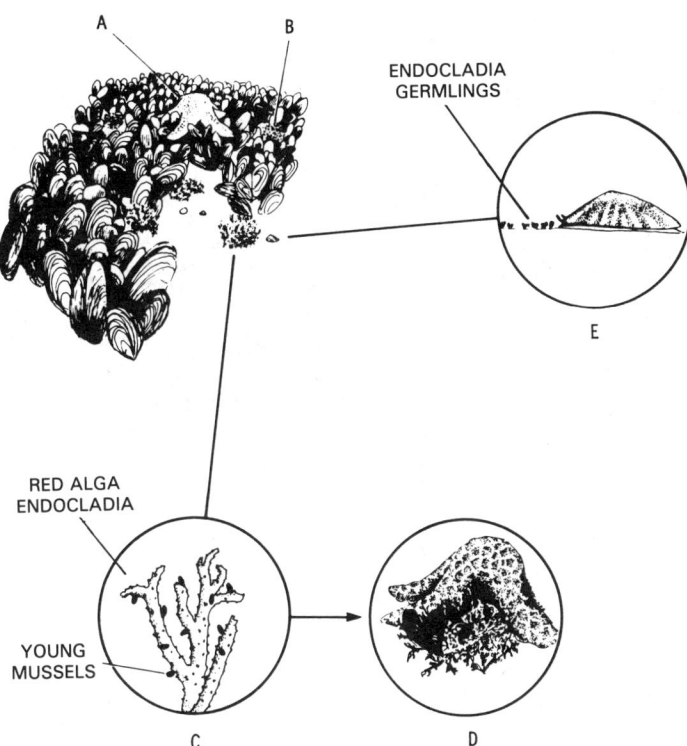

Fig. 8.1 Biotic relations between organisms in the eulittoral zone on the North American Pacific coast. **(A)** Starfish *Pisaster ochraceus* eats mussels. **(B)** Space colonized by the red alga *Endocladia muricata*. **(C)** Larvae of the mussel *Mytilus californianus* settle on *E. muricata*. **(D)** Mussels on and around *E. muricata* are eaten by *Pisaster*. **(E)** Spores and sporelings of *E. muricata* are eaten by the limpet *Tegula funebralis*, which in turn serves as food for the starfish. (From Carefoot 1977; based on Paine 1969, and Dayton 1971.)

(1) The **Fucales** may play this role only in the warm temperate and tropical regions, where their main cooler-water competitors, the **Laminariales,** are at their southern geographical limit or are absent.

(2) The other members of the Laminariaceae may be outcompeted as canopy algae by the giant thalli of the **Lessoniaceae.**

(3) The smaller **understory algae** of the midsublittoral zone (e.g., *Delesseria sanguinea* on the European coasts) begin to flourish as soon as insufficient light prevents kelp species from forming a dense canopy. This happens in the upper part of the lower sublittoral zone (see Section 2.3.3). In other words, the understory algae are more stress-tolerant, in this case to chronic lack of light, but are less competitive at higher levels of light. These algae live in an adverse habitat, like *Pelvetia* are outcompeted in the mideulittoral zone, and have been selected for the stressed environment in the upper eulittoral zone.

(4) *Laminaria digitata* is confined, as a canopy alga with a flexible stipe, to the **upper sublittoral zone** along European coasts because it cannot compete successfully with *L. hyperborea,* which has a rigid stipe and overshadows *L. digitata*. In areas where *L. hyperborea* is absent, as on the North American east coast, in the White Sea, or in the Arctic region in general (Chapter 3), *L. digitata* may take over the role of a canopy alga in the midsublittoral zone.

The dominating herbivores in the sublittoral zone on many coasts are the **sea urchins** (reviews in Lawrence 1975; Vadas 1977). **Limpets** also play an important role in this respect, for example, *Helcion* (= *Patina*) *pellucidus,* which infests *Laminaria hyperborea,* and the crustacean herbivores are represented by amphipods and isopods. Herbivorous shore-**fish** species (reviews in Wheeler 1980; Littler et al. 1983*b*) mainly occur in the tropics (families Acanthuridae, Siganidae, Scaridae, and Pomacentridae), with less importance in warm temperate regions, like western Australia (Kyphosidae), central Chile, or the southern Californian coast. For many fish species the euphotic zone provides shelter for the juveniles as well as food (review in Mann 1982).

Seaweeds, sea urchins, and sea otters. The *first* underwater experiment that clearly demonstrated the close ecological relationships between these two groups was performed by Jones and Kain (1967). The sea urchin *Echinus esculentus* was repeatedly removed at the lower limit of the forest of *Laminaria hyperborea,* where only a sparse vegetation of young sporophytes of this kelp and smaller algae existed. Once the sea urchins had been removed, the algal vegetation began to flourish. Sea urchins use chemoreception to locate preferred algae such as *Alaria* and *Laminaria,* while among the deep-water algae, *Agarum* species and the red alga *Ptilota serrata* are of little interest to them (Himmelman 1986). Sea urchins may almost completely destroy the kelp vegetation of the midsublittoral zone (reviews in Dayton 1985*a;* Schiel and Foster 1986; Harrold and Pearse 1987; Andrew 1989). This was observed along a wide expanse of the coastline of Nova Scotia, where the vegetation was dominated by *Laminaria longicruris*. After the vegetation had been removed, the sea urchins continued to feed on benthic diatoms on the apparently barren rocky surface (A. R. O. Chapman 1981). Along the southern Californian coast, favored for its absence of predators, sea urchins decimated the *Macrocystis pyrifera* beds (Leighton 1971). The **sea otter Enhydra lutris** was a major predator of sea urchins during the 19th century on the North American Pacific coast until its extinction along most of the coast between Alaska and California. It has since been introduced, and its numbers are increasing along the Californian and Oregon coasts (reviews in Foster and Schiel 1985; Schiel and Foster 1986; van Blaricom and Estes 1988) and for British Columbia (Breen et al. 1982). Estes and Steinberg (1988) proposed that kelps evolved in the North Pacific in an environment with a low intensity of herbivory because of the limiting predatory influences of sea otters, their ancestors, and possibly dusignathine odobenids (a group of walrus). Because sea otters must hold their breath to dive, they are less efficient foragers in deeper water where sea urchins density increases. Deepwater kelps such as *Agarum* species may have been forced to evolve more effective chemical defenses against urchins than shallow-water kelps such as *Laminaria* species (Estes and Steinberg 1988).

Sea urchin disease. The large populations of sea urchins along 200 km of coastline in Nova Scotia and locally in southern California, were eradicated, probably by an infectious disease. A denser seaweed vegetation has returned (Pearse and Hines 1979; Miller 1985*a,b;* Miller and Colodey 1983; Chapman 1986). Novaczek and McLachlan (1986) surveyed the area in 1984 and found that algal communities previously recorded returned in shallow water to a 10–15 m depth. In water deeper than this, long-lived red algae such as *Phyllophora truncata* did not return. Their place had been taken by algae that are either resistant to herbivory or, because of their ephemeral nature, are able to occupy a wider variety of habitats. Scheibling (1986) working along the Nova Scotia coastline found that changes in macroalgal cover, species composition, and increases in biomass, density, and size of kelp (*Laminaria*) species characterize the succession from barren rocky surface to three- and four-year-old kelp beds. In the geological past, at the Cretaceous/Tertiary boundary, the echinoids were greatly entrenched (Raup 1975), and the "lights out" hypothesis (see discussion in Section 2.1 on the asteroid-impact hypothesis) would plausibly predict a scarcity of their food, that is, seaweeds.

Sewage outfalls and sea urchins. Near Marseille, in France, a sewage outfall was blamed for the decimation of the seagrass *Posidonia oceanica*. After investigations, using exclosure cages, it was found that the seagrass was growing more rapidly near the sewage outfall but that a large urchin population was responsible for removing the *Posidonia* (Kirkman and Young 1981).

8.1.3 Protection Against Grazing

Seaweeds and herbivores have coevolved, and thus one finds **antigrazer substances** in the algae and feeding strategies in herbivores that avoid toxic algae. Examples of such substances are **sulphuric acid** in the vacuoles of a few *Desmarestia* species, such as *D. viridis* and *D. firma* (R. J. Anderson and

Velimirov 1982); **phenollike** substances in brown algae (reviews in Ragan 1976; Steinberg 1985; Ragan and Glombitza 1986); and **halogenated** substances like the brominated phenols in red algae (Ragan 1981). In the kelp *Alaria marginata* the sporophylls have the highest concentrations of phenolic compounds and are thus better protected from grazers than the vegetative parts (Steinberg 1984).

Germlings of the green alga *Enteromorpha* stop their growth if the swarmers settle on *Dictyota dichotoma* or *Dictyopteris membranacea*. Inhibiting substances might explain the clean appearance of these brown algae in the field, for they are remarkably free of epiphytes (Beth and Merola 1960), although they do have specialist epiphytes such as the brown alga *Myriactula stellulata* (G. Russell, personal communication).

The periwinkle *Littorina littorea* stops feeding as soon as it encounters food to which a 2–5% solution of the polyphenolic extract of *Fucus vesiculosus* or *Ascophyllum nodosum* has been added (Geiselman and McConnell 1981). Sea urchins feed selectively on algae exhibiting a low value of "astringency" (measured as a precipitation of hemoglobin by phenolic substances in an algal extract; R. J. Anderson and Velimirov 1982), as long as a variety of algal species is present (Lawrence 1975; Vadas 1977). Among the Laminariales the species *Agarum cribrosum* may possess a well-functioning defense system, as it is particularly avoided by sea urchins.

Several warm temperate–tropical green algal genera such as *Anadyomene* and *Penicillus,* the red alga *Laurencia,* and brown algae such as *Dictyota* and *Stypopodium* are largely avoided by herbivorous fish and sea urchins. The extracts of these algae have been shown to be toxic to fish (Targett and Mitsui 1979; Littler et al. 1983), but not to smaller grazers such as amphipods and polychaetes (Hay et al. 1988*b*). If the highly toxic brown alga *Stypopodium zonale* is maintained in a well-aerated aquarium, reef-dwelling herbivorous fishes immediately sense the toxins, attempt to jump out, and eventually die within 10 h (Littler et al. 1986). In the field, within a 2 cm radius of *S. zonale*, smaller algae (e.g., *Laurencia* and *Dictyota* species) grow and are protected from grazing fish. Chemically the toxins represent mostly halogenated and nonhalogenated terpenoids, for example, the sesquiterpene caulerpenyen in *Caulerpa prolifera,* and the diterpenoid halimedatrial in *Halimeda* (Norris and Fenical 1982; Paul and Fenical 1983, 1987; Paul and Hay 1986).

The calcified thalli of erect or crustose algae (Section 4.2) act as a **physical protection** against grazers, although specialists, such as the limpet *Acmaea testudinalis* on the North American east coast, do feed on crustose, calcified algae (Steneck 1982). It has often been observed that sea urchins and herbivorous fish feed first on the softer, filamentous, sheetlike algae and then on the coarse, thick, leathery algae, so that finally only the crustose coralline algae remain on the rocky substratum (Lawrence 1975; Littler et al. 1983*a,b*). This last growth form has not only adapted to survive adverse seasons and to optimally absorb light as a flat-light receiver, but it is also successful in surviving in the presence of grazers.

8.2 EPIPHYTES, ENDOPHYTES, ENDOZOANS, AND PARASITES

To obtain a better position in the "struggle for light," smaller, nonparasitic algal species may grow as **epiphytic algae** on larger algae or seagrasses, or as **epizoic algae** on animals, and in most cases they are attached only superficially to the surface of their hosts. Epiphytic algae decrease the growth rate of their host, increase the probability of breakage, and may decrease reproductive output (d'Antonio 1985). Animals (e.g., amphipods) browsing on epiphytic algae thus have a beneficial effect on the host alga or seagrass.

An epizoic, prokaryotic organism, *Prochloron,* which lives on ascidians, became famous for its unique pigment composition. Morphologically it looks like a member of the blue-green algal order Chlorococcales, but it has no phycobilins and has chlorophyll b as well as chlorophyll a (Lewin 1976).

Most epiphytic algae are not specific to their host, but there are notable exceptions. For example, the cushionlike brown alga *Elachista scutulata* occurs exclusively on *Himanthalia elongata.* Among the red algae, *Polysiphonia lanosa* grows on *Ascophyllum nodosum* (occasionally on *Fucus* and only in Europe), *Porphyra nereocystis* on the kelp *Nereocystis luetkeana,* and *Smithora naiadum* on the seagrasses *Zostera marina* and *Phyllospadix scouleri.*

Polysiphonia lanosa anchors its thallus with long, unicellular rhizoids deep in the cortex of *Ascophyllum* and, for this purpose, dissolves the host cells enzymatically (a rare case among epiphytic algae), but no exchange of ^{14}C could be detected between the epiphytic alga and its host in a tracer study (Rawlence 1972). Harlin (1973) has investigated these "obligate" cases of epiphytic algae. She found that the obligate epiphyte *Smithora naiadum* of the seagrass *Phyllospadix scouleri* would also colonize plastic imitations of seagrass. Thus it may be the specific set of environmental conditions, such as the particular light field among the leaves of the seagrass or the special conditions of water motion, to which the epiphytic algae are adapted, rather than dependence on substances derived from the host, as in parasitic algae (see below).

As a **protection against epiphytes,** young actively growing plants of the fucalean species *Ascophyllum nodosum, Himanthalia elongata,* and *Halidrys siliquosa* regularly **cast off their "skin,"** the outer layers of their outermost cell walls (Filion-Myklebust and Norton 1981; Moss 1982). As a result they are often remarkably free of epiphytes. Successful epiphytes include the mainly pennate, epiphytic **diatoms** (review in McIntire and Moore 1977), which may threaten their macrophytic hosts in eutrophic waters by shading and in anoxic conditions at the surface of the macrophyte (Sand-Jensen et al. 1985).

The epiphytic brown algae *Elachista scutulata* and *Herponema* species bypass the skin-shedding mechanism of their host *Himanthalia elongata* by growing in cryptostomata (invaginations with hairs) and other natural breaks in the host thallus surface, where no skin is removed (Russell and Veltkamp

1984). The phenomenon of skin shedding has also been detected in **crustose coralline** algae that grow underneath the kelp canopy. In this case they have to "clean" their surfaces mainly of laminarian gametophytes (Masaki et al. 1981*a*).

Another way of protection against epiphytes may be seen in **continuous erosion** at the distal end of the blades, in, for example, in *Laminaria* species and the seagrass *Posidonia oceanica* (Ott 1979, 1980). In both genera new tissue is produced by a basal blade meristem.

Substratum-specific alga-herbivore relationship. Crustose red algae (e.g., *Lithothamnium* species) give off proteinaceous, **morphogenetic substances** that specifically regulate the metamorphosis of the larvae of the gastropod *Haliotis* (Morse and Morse 1984). This relationship has the advantage for the algal crust of keeping its surface clean due to the abrading and sweeping movements of the *Haliotis* shells in addition to the sloughing mechanism. The biochemical signal substance of the algal crust ensures that *Haliotis* larvae do not settle on other (e.g., foliose) algae but on the more permanent and solid substratum of the algal crust. The mucilaginous signal substance of the alga also controls the metamorphosis of the animal, which has to rely completely on the alga because it has no endogenous mechanism for inducting settlement and metamorphosis. Immediately after settling, the larva incorporates algal pigment into its new shell and thus obtains a camouflage from predators (Morse and Morse 1984).

Actively growing thallus parts have to be protected against excessive covering by **bacteria** (reviews in Sieburth and Tootle 1981; Mann 1982). The **antibiotic** effects of thallus extracts have been discovered for many algal species (Glombitza 1979; Reichelt and Borowitzka 1984; Hornsey and Hide 1985; Baik and Kang 1986; Ballantine et al. 1987; Rosell and Srivastava 1987). **Morphogenetic substances** formed by bacteria, living on the brown alga *Scytosiphon lomentaria*, induce macrothalli in another brown alga, *Dictyosiphon foeniculaceus* (Saga 1986).

Epiphytic animals. These representatives of many invertebrate groups are often incorrectly called "epizoans." They make use of the seaweed thalli or seagrass leaves and stems as a settling space, as a refuge from predators, and as a site where food accumulates, the plant parts functioning as traps for detritus and plankton (review in Hayward 1980). There is evidence that the larvae of several species of epiphytic animals may be chemically attracted by certain algal species (Crisp and Williams 1960), and numerous species-specific relationships are known. For example, the larvae of the polychaete *Spirorbis spirorbis* (= *S. borealis*) settle preferentially on *Fucus serratus* or *F. vesiculosus* (Knight-Jones et al. 1971), and the larvae of the bryozoan *Membranipora membranacea* settle the blades of kelps such as *Laminaria hyperborea* or *Nereocystis luetkeana*. This bryozoan, which may cover a blade within a few months, possibly lives not only on plankton, but partially on exuded kelp photosynthate, as shown by De Burgh and Fankboner (1978) using ^{14}C tracer studies.

Endophytes. Minute, filamentous algae known as endophytes, which inhabit intercellular spaces in various seaweed species, occur among the **brown algae** in the genus

Mikrosyphar (Ectocarpaceae; endophytic in *Polysiphonia* and *Porphyra*). **Green** endophytes such as *Epicladia flustrae* were classified in the green algal family Chaetophoraceae (Yarish 1976; Nielsen 1980; O'Kelley 1982*b*) and are now found in the order Ctenocladales (South and Tittley 1986). The unicellular Codiolum phase of several green algal genera lives endophytically in other algae. For example, the sporophytic Codiolum phase of the filamentous green alga *Spongomorpha aeruginosa* from summer to winter lives intercellularly in the crustose, noncalcified red alga *Haemascharia hennedyi* found in the eulittoral zone. The Codiolum phases of other algal species and species of the green alga *Tellamia* bore into chalk. Several species of the **red alga** *Audouinella* (family Acrochaetiaceae) are endophytic (Garbary 1979) and **endozoic** (e.g., living in hydroids). Since the endophytic and endozoic algae are normally pigmented and may easily be kept in culture without the host, they most likely use the host as a shelter and not as a source of organic material. **Tube-dwelling** marine diatoms may mimic tubular or bladelike seaweed thalli (Lobban 1984, 1985).

Parasitic red algae. Parasitic algae are only known among the red algae. They have lost their pigments and live on the organic material of their algal host, which in consequence exhibits some damage. There may be an exception in the almost colorless brown alga *Herpodiscus durvilleae,* which is probably parasitic on *Durvillaea antarctica* (South 1974). There are about 40 species of parasitic red algae, and the only hosts of these are other red algae (reviews in L. V. Evans et al. 1978; Goff 1982; Bold and Wynne 1985).

Parasitic red algae: adelphoparasites. Ninety percent of the parasitic red algae show a close taxonomic relationship to their red algal host and are called **adelphoparasites** (Feldmann and Feldmann 1958). The parasites still exhibit normal rhodophycean life cycles; for example, *Ceratocolax hartzii* forms pale reddish tufts on *Phyllophora truncata,* and both belong to the family Phyllophoraceae. *Janczewskia* species grow as yellowish cushions a few millimeters in diameter on *Laurencia* and *Chondria* species, all three genera belong to the family Rhodomelaceae. According to one of the hypotheses for explaining these striking phenomena, during the course of phylogenetic history, the adelphoparasites may have started as spores belonging to the host plant and germinated on the latter (i.e., in situ germination).

Parasitic red algae: alloparasites. Alloparasites are not closely related to their hosts; for example, *Harveyella mirabilis* (order Cryptonemiales) forms 1–2 mm hemispherical thalli on species of *Rhodomela* or *Odonthalia* (order Ceramiales). Instead of normal rhodoplasts, *Harveyella* has "proplastids" without thylakoids (Goff 1979*b*), and tracer studies using ^{14}C clearly indicate that the parasite lives on the photosynthate of its host (Kremer 1983). If younger host thalli are heavily infected with a parasite, they may stop growing, as shown in the thalli up to 3 cm long thalli of *Laurencia nipponica* infected by *Janczewskia morimotoi* (Nonomura 1979). In the host–parasite **secondary pit connection** between the parasitic red alga *Choreocolax polysiphoniae* and its host *Polysiphonia confusa,* parasite nuclei are delivered into the cytoplasm of a host cell and control and redirect the physiology of the host for the benefit of the parasite (Goff and Coleman 1985). Red algae form **conjunctor cells** capable of fusing with and transferring nuclei into their own cells. During evolution, red algal epiphytes on red algal hosts probably evolved into partial endophytes, the cytoplasm of closely related hosts providing less incompatible barriers to secondary pit connection and transfer of nuclei into the host's cytoplasm. This hypothesis, by

Goff and Coleman (1985), explains why red algal parasites are often found on closely related taxa and not on brown or green algae.

Marine fungi. These are represented in the sea by about 500 species (1% of total species number), the largest reaching only a few millimeters in size (reviews in Johnson and Sparrow 1961; Kohlmeyer and Kohlmeyer 1979; Goff 1983*b;* Andrews and Goff 1985; Moss 1986). If the Ascomycetes and the Rhodophyta arose from a common marine ancestor, as postulated by Kohlmeyer and Kohlmeyer (1979), the apparent propensity for parasitism in both groups might be explained (Goff 1983*b*). As parasites, some fungi cause damage to seaweeds, for example, the ascomycete *Phycomelaina laminariae,* which forms tarlike dots on the stipes of *Laminaria* species. The ascomycete *Mycosphaerella ascophylli* is an obligate endophyte in the intercellular space of the thalli of *Ascophyllum nodosum* and *Pelvetia canaliculata,* This is known as "mycophycobiosis" (Kohlmeyer and Kohlmeyer 1972). One of the most widely known and unusual fungi is *Labyrinthula,* which was implicated as the causal agent in the "wasting disease" of the seagrass *Zostera marina.*

8.3 STRATEGIES FOR THE SURVIVAL AND GROWTH OF SEAWEEDS

Stable or regularly fluctuating environmental conditions are a prerequisite for the existence of **perennial** algae. They represent the final stage of succession, often observed in recolonization and successional studies, and they may, once established, reach a considerable age.

Maximal age of seaweeds. In the eulittoral zone *Ascophyllum nodosum* may attain an age of 20 years in Norway (Printz 1959) or 30.35 years in the Bay of Fundy (McLachlan, personal communication) calculated from its annually formed air bladders in the main axis. Most of the individuals of fucalean species live for only three to four years once they have escaped the initial high mortality that occurs in a newly forming population (Gunnill 1980). *Fucus vesiculosus* in the Barents Sea may reach seven years of age, and *Cystoseira barbata* in the Black Sea 18–20 years (Khailov, personal communication). Within the dense *Laminaria hyperborea* forest, many individuals exhibit an age of five to seven years, determined by counting the slow-growth lines formed within the stipe during the second half of the year (Kain 1963, 1979). A maximum age of 18 years was recorded for this species. Only the stipe and haptera are perennial in Laminaria species; the blade is shed continuously, as in *L. digitata,* or discontinuously, as in *L. hyperborea* (Fig. 8.2). The northeastern Pacific kelp *Pleurophycus gardneri* (Laminariaceae) sheds its blade in late fall, leaving a stipe and haptera for two weeks before a new blade is formed (Germann 1986). *Pterygophora californica* (Alariaceae), which sheds its sporophylls (lateral blades) and its apical blade each year, may have a stipe and holdfast at least 11 years old (De Wreede 1984, 1986). Individuals of *Ecklonia radiata* (Alariaceae) in New Zealand may attain an age of about 10 years (Novaczek 1981). In *Macrocystis pyrifera* the holdfast is perennial and, after initial high losses, most individuals of a population may live for up to three to four years, some for up to seven years (Rosenthal et al. 1974). *M. integrifolia* often attains an age of four to eight years (Lobban 1978), but only in deeper water, whereas

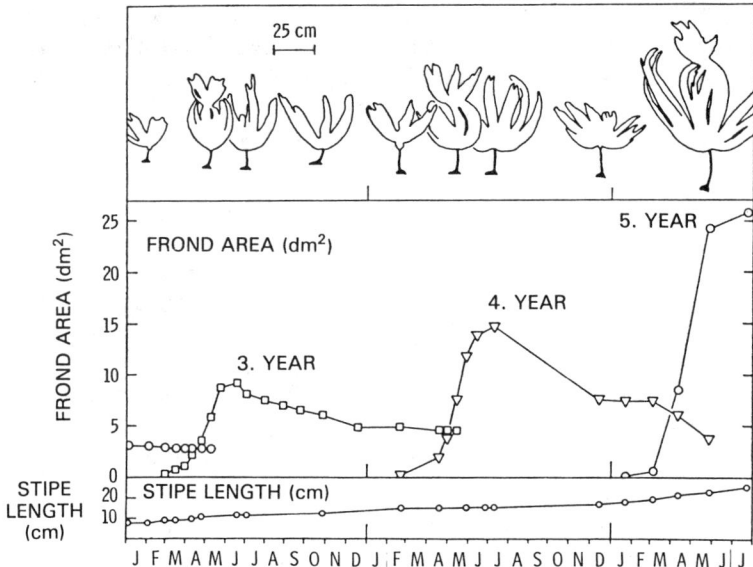

Fig. 8.2 *Laminaria hyperborea*. Growth rhythm of an individual in successive years at 2.5 m deep in the sublittoral zone of Helgoland. (From Lüning 1971.)

plants in shallow beds along open coasts exposed to storms may live for only several months (Druehl and Wheeler 1986). The giant kelps *Pelagophycus porra* and *Nereocystis luetkeana* only live 1–1.5 years (Coyer and Zaugg-Haglund 1982). For red algae, maximum ages of 13 and 18 years have been recorded for *Constantinea subulifera* or *C. rosa-marina,* respectively (Powell 1964; Lindstrom 1980). In these species, which occur on the North American Pacific coast, a determination of age is possible by counting the scars left on the perennial stipe after the annually formed, peltate blade has been cast off. **Demographic studies** determining the effects mainly of recruitment and mortality on the density of a population, have been performed on several seaweed populations (review in A. R. O. Chapman 1986; methodology in A. R. O. Chapman 1985).

In contrast to perennials, **annual algae** dominate as **opportunists** at sites with irregularly fluctuating conditions. Opportunists are fast-growing algae, for example, *Enteromorpha, Ulva, Ceramium* or *Pilayella*. They are, after bacteria and benthic diatoms, the first multicellular algae to appear on substrates that have been experimentally cleared or on newly constructed jetties and new, volcanically formed islands. If conditions are stable, the opportunistic algae cannot compete with the more slowly appearing perennial algae, and so the opportunistic, annual algae may also be regarded as **fugitive species** (Dayton 1975; Russell 1986).

The strategies of annual and perennial algae may be compared to the *r*- and *K*-strategies known from population biology (MacArthur and Wilson; Pianka 1970). These latter terms designate differential stages reached in evolution.

The **r-strategists** have a high growth rate and a high reproduction rate r, but the capacity K of the environment is not fully used. The **K-strategists** use up the capacity of the environment. The ratio of annual primary production P (carbon fixed per unit of substratum area per year) to the actually occurring biomass B (algal weight per unit of substratum area) is smaller in K- than in r-strategists.

One may regard **perennial algae as K-strategists,** since they take optimal advantage of the environmental resources (see Section 8.4) and use a greater part of their annually fixed organic carbon to build up their thallus (low $P:B$ ratio), rather than allocate it to reproduction, as the annual algae do (high $P:B$ ratio). A sporophyte of the kelp *Ecklonia maxima* may produce 3×10^{10} spores per year, but this is only 5.3 cm^3 of spores per plant per year, or 0.2% of the total annual production of this kelp (Joska and Bolton 1987).

The thalli of *Ulva* and *Enteromorpha*, which arise from spores, zygotes, or unfused gametes, grow within a few weeks into thalli that reproduce again, and most of the cell contents are converted into reproductive cells. One may therefore regard such **annual algae as r-strategists,** which exhibit a variable biomass. This may be very high in one month and low in the next month, for example, after a catastrophic attack by herbivores.

Only the perennial algae are able to take maximum advantage of components of the **capacity K** of their environment, that is, the scarce light supply and the nutrients, which they store. As K- strategists the perennial algae build up large thalli and dominate as canopy species (Fig. 1.3).

8.4 SPECIAL GROWTH STRATEGIES OF SUBLITTORAL, PERENNIAL SEAWEEDS: STORAGE, TRANSLOCATION, AND GROWTH IN DARKNESS

For seaweed species adapted to life at higher latitudes or in deeper water, it is important for them to take full advantage of the main light supply occurring in summer and to store their photosynthates, formed during this season, as reserve materials. The mannitol and laminaran content in the blade of *Laminaria hyperborea* increases in summer; both substances are consumed during the light-poor season due to respiration and support of the new blade's growth (Fig. 8.3). The primitive, annual kelp *Saccorhiza polyschides* contains mannitol but no laminaran (Jensen et al. 1985).

From year to year *L. hyperborea* forms a bigger blade (Fig. 8.2), since reserve materials stored in the old frond are mobilized and, together with newly formed photosynthate in spring, are translocated to the growing zone at the base of the new blade (Fig. 8.4). About half of the carbon and nitrogen required by the new blade is supplied by the old blade (Lüning et al. 1973; Davison and Stewart 1983). Thus **storage** of reserve materials and a well-functioning **translocation system** are important prerequisites for survival at conditions where light supply is fluctuating in a predictable way. This has

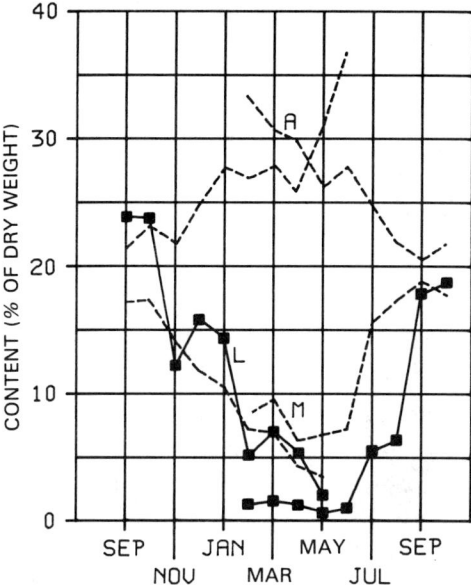

Fig. 8.3 *Laminaria hyperborea* on the coast of North Norway. Seasonal content of mannitol (M, lower pair of dotted lines), laminaran (L, solid lines), and alginic acid (A, upper pair of dashed lines) in the old blade (September to May/June), and in the new blade (from March onward). For each measuring point at least 50 individuals were analyzed. (After Haug and Jensen 1954.)

been intensively investigated in the Laminariales (reviews in Schmitz and Lobban 1976; Schmitz 1981; Floc'h 1982; Buggeln 1983; methodology in Buggeln 1985).

In order to have a blade ready as soon as the main annual supply of light arrives, some algae have the ability for **dark growth.** Dark growth has been mentioned with respect to life in the Arctic and Antarctic regions. *Laminaria hyperborea* forms a blade of considerable size in darkness (Fig. 6.10D). Another kelp that shows growth in darkness is *L. solidungula* (Dunton 1985; Dunton and Schell 1986).

Growth in darkness also occurs in the **red algae** *Constantinea subulifera* (Powell 1964), *Maripelta rotata* (Bowen 1971), and *Delesseria sanguinea* (Lüning 1984). *M. rotata* and *Delesseria sanguinea* form rather small new blades in darkness, but as in most higher plants no chlorophyll a is formed in darkness, in contrast to most unicellular algae (Bowen 1971; Lüning and Schmitz 1988). Long-distance translocation in *D. sanguinea* occurs in June when ^{14}C-labeled photoassimilates move from one frond to another via the stipe (Turner and Evans 1986). In September, labeled assimilates accumulate in the lower, thickened regions of the frond ribs, which persist over the winter, while the remaining blade parts are torn off.

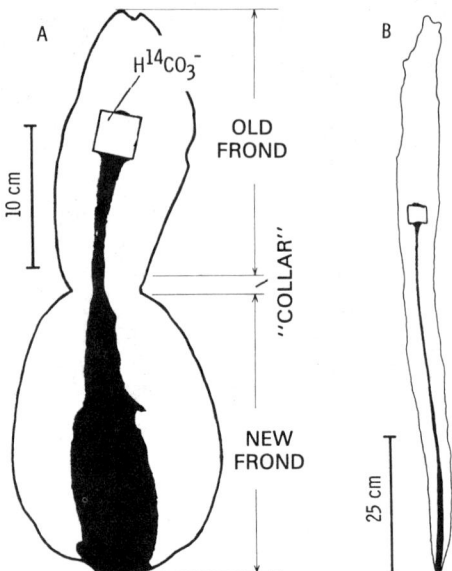

Fig. 8.4 Radioautographs of *Laminaria hyperborea* (**A**) and *L. saccharina* (**B**). Demonstration of long-distance translocation. ^{14}C-labeled bicarbonate was fed at the distal end (quadratic Plexiglas incubation chamber, and stipes removed for purpose of radioautography). Labeled translocate accumulated at the growing zone of the blades. (A from Schmitz et al. 1972; B from Lüning et al. 1973.)

8.5 PRODUCTIVITY AND GROWTH RATES OF SEAWEEDS

The maximum algal **biomass** on rocky substratum in the sublittoral zone may be 16 kg of fresh weight per square meter, and perhaps half this in the mideulittoral zone (Table 8.1). The **annual primary productivity** (carbon fixed per square meter per year; reviews in Barnes and Mann 1980; Mann 1982) may be several times higher than the biomass (see the discussion of *P:B* ratio in Section 8.3); for example, part of the annually fixed carbon is continuously lost in the form of eroding tissue. Kirkman (1984) has calculated that approximately 20 wet kg m^{-2} of fresh *Ecklonia radiata* plant material are lost from the standing crop each year (Table 8.1). In *Laminaria* species erosion occurs at the distal end of the thallus, possibly as a protection against epiphytes (Section 8.2). In *Macrocystis* the erosion takes place in fronds whose apical meristem has ceased growing. Other parts of an alga may be eaten by herbivores, or the whole alga or parts may be removed by wave action.

In the vegetation dominated by annual algae, several generations per year follow each other. For example, the **turnover rate** of the vegetation formed by the filamentous green alga *Cladophora glomerata* in the Baltic is 17 days, while the filamentous brown alga *Pilayella littoralis* takes 36 days. In contrast, the vegetation of the perennial red algae *Furcellaria fastigiata* and *Phyllophora truncata* has a turnover rate of 110 days and 150 days, respec-

Table 8.1 Examples of biomass (fresh weight per m² of rocky area) for seaweeds in temperate regions

dominating species	locality	biomass (kg m^{-2})	authors
(a) Eulittoral zone			
Enteromorpha spp., Ulva lactuca	Helgoland	2-4	Munda and Markham 1982
Ascophyllum nodosum, Fucus vesiculosus	Maine, USA	8	Topinka et al. 1981
(b) Sublittoral zone			
Laminaria spp.	Diverse coasts	4-16	Mann 1972
Ecklonia radiata	SW Australia	6-18	Kirkman 1984
Macrocystis pyrifera	California	4-22	Coon 1982
Cystoseira spp.	Adriatic Sea	2-5	Munda 1979
Iridaea spp., Gelidium spp.	NW Pacific	1-4	Coon 1982

tively (Wallentinus 1978). The tropical calcified green algae *Halimeda incrassata* and *Penicillus capitatus* and the calcified brown alga *Padina sanctae-crucis* renew their biomass every 1–1.5 months (Wefer 1980). In forests of *Macrocystis pyrifera* and beds of *Laminaria* species a turnover of up to six times per year is possible (Mann 1982; Schiel and Foster 1986).

From the foregoing it is clear that the determination of annual primary productivity must take many factors into account. Regular measurements of growth increment (e.g., Mann 1972, for *Laminaria*), or regular in situ measurements of photosynthesis and respiration in algae enclosed in transparent containers (e.g., Schramm and Martens 1976; Hatcher et al. 1977; Johnston et al. 1977) have at least shown the magnitudes with which one has to deal.

According to the data compiled in Table 8.2, **annual primary productivity** in the sublittoral zone, as well as among terrestrial vegetation, ranges through maximum values in **tropical regions** (optimal light and temperature conditions all year round) to high values at temperate latitudes, and finally to minimum values in the **Arctic region** (reduction in photosynthesis due to low temperatures and periodic lack of light). Lower values, compared with sublittoral habitats, are typical for the **eulittoral zone,** due to periodic reduction of photosynthesis during the emersion phases.

The well-known low productivity in the tropical pelagic realm, characterized by its "blue" and nutrient-poor waters, is contrasted with the situation in the sublittoral zone of the benthic realm. High productivity occurs on coral reefs and in benthic *Sargassum* vegetation (Table 8.2). However, in both these cases the nutrients remain in a relatively closed environment, within which they are quickly recirculated.

Primary productivity for **dense terrestrial** vegetation is similar to that of **marine-benthic vegetation** (Table 8.2) because, in both cases, up to about 10 m² of leaf or thallus area may grow on 1 m² of substratum ("leaf area index"; reviews in Whittaker and Likens 1975; Larcher 1980; for seaweeds see Lüning 1969b; Jupp and Drew 1974). The canopy density in a *Macrocystis*

Table 8.2 Examples of annual net primary productivity for marine and terrestrial vegetation. Data for terrestrial vegetation were multiplied by 0.5 for conversion from dry weight to carbon content

Type of region	vegetation type	annual net primary productivity (g C m^{-2} year^{-1})	authors
tropical	rain forest	500-1800	Lieth 1975
	coral reef	2300	Lewis 1981; Wanders 1976-77
	Sargassum (benthic)	2500	Wanders 1976-77
	Corallinaceae	2100	Littler 1973
temperate	terrestrial forests	200-1000	Lieth 1975
	terrestrial grassland	100- 800	Lieth 1975
	Laminaria	1200-1900	Mann 1982
	Macrocystis pyrifera	800-1000	Mann 1982
	Macrocystis integrifolia	1300	Wheeler and Druehl 1986
	seagrasses	500-1000	McRoy and Helfferich 1977
	Fucus, Ascophyllum	300-600	Brinkhuis 1977; Cousens 1981
	benthic diatoms(on muddy substrate)	100-400	McIntire and Moore 1977
	coastal phytoplankton	100-200	Bunt 1975
Arctic	Laminaria solidungula	20	Chapman and Lindley 1981
	coastal phytoplankton	10-25	Nemoto and Harrison 1981

pyrifera forest ranges up to about 12 m^2 blade m^{-2} bottom (Gerard 1984a). Irrespective of morphological differences, the chlorophyll is maximally "packed" in these dense vegetations, with concentrations of up to 1.5–3 g of chlorophyll per square meter of substrate. Under such circumstances practically all incident light is absorbed. An example for seaweed vegetation is found in the upper sublittoral vegetation of the European Atlantic coast, where the brown alga *Himanthalia elongata* has up to 1.4 g of chlorophyll per square meter of rocky substratum (Niell 1981). The **phytoplankton** exhibits values of primary productivity that are an order of magnitude lower than those of the marine-benthic or terrestrial vegetation (Table 8.2). The reason is obvious: most of the quanta in the open ocean are absorbed by water molecules and not by plant pigments.

Global primary productivity. In the sea, phytoplankton produce up to 30 × 10^9 metric tons of carbon per year, and plants on the continents up to 50 × 10^9 metric tons of carbon per year (Bunt 1975; Whittaker and Likens 1975; Whittle 1977; Longhurst 1981). Marine-benthic plants (macroalgae, seagrasses, and microalgae) produce about 3% of the organic carbon produced by phytoplankton. This approximation is arrived at by assuming that the mean rate of production is 1000 g C m^{-2} yr^{-1} along a global coastline of 400,000 km, 2 km wide (Gierloff-Emden 1980). From this global "benthic production area" of 0.8 × 10^6 km^2 would arise a primary productivity of 0.8 × 10^9 metric tons of carbon per year. Note that the global shelf area in the depth range of 0–200 m amounts to 27.5 × 10^6 km^2 (Schopf et al. 1978), but much of this area is not covered by the productive part of the euphotic zone (Section 6.3). Either it is covered with muddy sediment and thus may exhibit a rather low productivity (Table 8.2), or it is characterized by adverse conditions, (e.g., found in the polar regions), or there may only be a shallow euphotic zone due to turbid water. The upper half of the shelf area, which is inhabited by benthic algae and seagrasses, amounts to 2–3% of the total area of the seafloor (Seibold and Berger 1982).

At optimal environmental conditions the highest rates of **net photosynthesis** (per unit of dry weight) are achieved by the sheetlike and filamentous annuals (Table 8.3). These exhibit a high **surface–volume ratio,** and many of their photosynthesizing cells are optimally exposed to light and nutrient supply in the open water. Photosynthetic rates decrease in the more bulky growth forms of the perennials. The thalli for the most part consist of nonphotosynthesizing cells set aside by the plant for purposes of storage, translocation, and long-term stability of the thallus. The surface-volume and **surface–wet weight ratios** are important parameters, since functions such as the intensity of growth and photosynthesis are correlated with these ratios (Khailov 1976, 1978, 1988). The high cost of **calcification** for protection from grazing is evident from the generally low photosynthetic rates of calcified algae. Although crustose algae are "optimal light receivers," they exhibit extremely low photosynthetic rates, probably because their fixed position causes disadvantages in gas exchange and nutrient uptake via the boundary layer.

Examples of growth and elongation rates have been compiled in Table 8.4 for various seaweed species. A better comparison is **relative growth rate** (or specific growth rate μ), for example, as a measure of the change in weight over time. The relative growth rate R may be calculated, assuming no loss of plant material, from the following (G. C. Evans 1972):

$$R = \frac{\ln W_2 - \ln W_1}{T_2 - T_1} = \frac{\ln(W_2/W_1)}{T_2 - T_1},$$

where W_1 = weight at time T_1,
and W_2 = weight at time T_2.

Typical values for R fall in the range of 0.1–0.3 for *Gracilaria* species and 0.02–0.07 for *Chondrus crispus* and *Plocamium cartilagineum* (Kain 1987a). Another possibility is to calculate the percentage increase per day, as the relative growth rate $\times 100$:

$$\% \text{ increase per day} = \frac{100 \ln (W_2/W_1)}{T_2 - T_1},$$

where T_1 and T_2 are time in days.

Maximum values achieved in the field by *Gracilaria* species or juvenile plants of *Macrocystis pyrifera* are of the order of 10% increase per day (Coon 1982; McLachlan and Bird 1986).

Table 8.3 Rates of net photosynthesis, PS (mg C per g dry weight per h), for six life form types of seaweeds. G = green alga, B = brown alga, R = red alga

life form type	examples	PS
(1) Sheet-Tubular-Group	G Ulva, Enteromorpha, B Dictyota	11 - 2
(2) Filamentous-Group	G Cladophora, R Ceramium	7 - 1
(3) Coarsely Branched-Group	G Codium, R Gigartina, Laurencia	3 - 0.5
(4) Thick Leathery-Group	B Fucus, Sargassum, Laminaria, Macrocystis	1 - 0.3
(5) Jointed Calcareous-Group	G Halimeda, R Corallina, Jania, Amphiroa	0.6 - 0.2
(6) Crustose-Group	R Peyssonnelia, Petrocelis, Porolithon	0.1

Compiled after Littler and Arnold 1982; Littler et al. 1983.

Table 8.4 Examples of maximal growth rates of seaweeds under optimal environmental conditions. Parameters for growth: increase in length, or marginal meristem in crustose algae. B = brown alga, G = green alga, R = red alga

species	increase per day (cm)	increase per year (cm)	authors
B Macrocystis pyrifera (stipe)	30		Wilson et al. 1977
B Macrocystis integrifolia (stipe)	6-4		Lobban 1978
B Nereocystis luetkeana (stipe)	12-8		Duncan 1973
B Nereocystis luetkeana (blade)	14		Kain 1987c
B Pelagophycus porra (blade)	7-3		Coyer and Zaugg-Haglund 1982
B Laminaria angustata var. longissima (blade)	13-7		Kain 1979
B Laminaria saccharina, L. japonica (blade)	5-2		Kain 1979
B Laminaria hyperborea, L. digitata (blade)	1		Kain 1979
B Sargassum muticum	4-3		Nicholson et al. 1981
B Fucus spiralis	0.1		Niemeck and Mathieson 1976
B Fucus vesiculosus, Fucus serratus	0.2-0.1	12-4	Knight and Parke 1950, Printz 1926
G Penicillus capitatus	0.1		Wefer 1980
R Neorhodomela larix	0.12		d'Antonio 1986
R Corallina spp.		2-1	Masaki et al. 1981b
R tropical crustose coralline algae		2-1	Adey and Vassar 1975
R Lithophyllum incrustans		0.3	Edyvean and Ford 1987
R cold temperate crustose coralline algae	0.001	0.3	Adey 1970

Annual algae (groups 1 and 2 in Table 8.3) exhibit high net photosynthetic rates, so their relative growth rates are higher than for the perennials (groups 3–6 in Table 8.3). This difference represents just another facet of what has been said about shorter turnover rates or higher P : B ratios in annuals.

8.6 CONSTITUENTS, COMMERCIAL USE, AND DECOMPOSITION OF SEAWEED BIOMASS

Literature: Constituents: Hellebust and Craigie (1978); Stewart (1974). **Ash:** Whyte and Englar (1980). **Low-molecular-weight carbohydrates:** Kremer (1980, 1981a,b).

Polysaccharides: Percival and McDowell (1967); Craigie (1974); McCandless and Craigie (1979); Percival (1979); McCandless (1981). **Lipophilic substances:** Ragan (1981); Sargent and Whittle (1981). **Phenolic substances:** Ragan (1976, 1981); Glombitza (1979); Ragan and Glombitza (1986). **Commercial uses of seaweeds and their constituents:** Glicksman (1969); Levring et al. (1969); Jensen (1972, 1979); Booth (1975); Ryther et al. (1978); Hoppe et al. (1979); V. J. Chapman and Chapman (1980); Waaland (1981); Hoppe and Levring (1982); McHugh (1989). **Algal resources and seaweed farming:** Michanek (1975, 1978, 1983); Bonotto (1976); Naylor (1976); Moss (1977); Tseng (1981a); Caddy (1984); Bird and Benson (1987); Druehl (1988); Druehl et al. (1988a); Yarish et al. (1989). **Japan:** Hasegawa (1976); Miura (1975); Okazaki (1971); Saito (1975). **China:** Tseng (1981a,b). **Thailand:** Edwards et al. (1982). **Indonesia:** Soegiarto (1979). **India:** Chauhan and Mairh (1978). **South Africa:** Simons (1976). **California:** North (1971). **Europe:** Levring (1977). **Ireland:** Guiry and Blunden (1981). **North America:** Yarish et al (1990). **South America:** Oliveira Filho (1981).

The main chemical constituents of the seaweeds will be briefly treated in this section, particularly with regard to their role in food chains and decomposition and to their commercial uses. At present the annual global seaweed harvest amounts to about 3 million metric tons of algal fresh weight, of which around 40% is represented by the *Laminaria* harvest in China (Fig. 8.5). About 10% of the annual global seaweed harvest (fresh weight) consists of *Porphyra,* which is cultivated in Japan, China, and Korea for food purposes (Table 8.5; Fig. 8.6).

The fresh weight of seaweeds consists of 75–90% water. Of the remaining **dry weight,** about 75% is organic matter (see below) and 25% **mineral ash,** consisting mainly of potassium, sodium, magnesium, and calcium ions.

Around the year 1800, at the height of the historical **potash industry,** 400,000 metric tons fresh weight of kelp were burnt annually along the shores of Brittany, Normandy, Scotland, and Norway to obtain potash for the production of glass and soap (Jensen 1979). This industry later came to an end because soda was produced according to a synthetic process. Subsequently the annual kelp harvest boomed to a value of 3 million metric tons of seaweeds (fresh weight) used for the extraction of **iodine.** This amount is similar to today's annual harvest (Table 8.5). Subsequent to the Chernobyl incident, iodine-131 was found in increased quantities in *Fucus* species on the western coast of Canada (Druehl et al. 1988b).

About half of the **organic dry weight** of algae consists of carbon (Westlake 1963); 10% is protein (e.g., in *Laminaria* or *Palmaria;* Haug and Jensen 1954; Morgan et al. 1980), but this value may increase to 20–40% in genera such as *Ulva, Hypnea* (Durako and Dawes 1980), *Porphyra,* and *Palmaria.* The remaining of the organic dry weight is mainly low-molecular-weight carbohydrates and polysaccharides (see below). In *Laminaria* species the content of lipophilic substances may amount to 1%, and the iodine content to 4% of the dry weight (Haug and Jensen 1954).

Low-molecular-weight carbohydrates and algal polysaccharides. The chemical formulae of some of these major substances are given in Fig. 8.7. The **primary photosynthetic products,** besides the readily marked amino and organic acids in ^{14}C tracer

CONSTITUENTS, COMMERCIAL USE, AND DECOMPOSITION

Fig. 8.5 Commercial uses of kelps (A–E: *Laminaria* spp. in Japan; F–H: *L. japonica* in China; I: *Macrocystis pyrifera* in California). **(A)** Historical view of algal harvesting in Japan. **(B)** Harvesting in the sea. **(C)** Drying on the beach. **(D)** Drying on mats. **(E)** Commercial products. **(F)** Scheme of Chinese rope culture. **(G)** Spraying nutrients. **(H)** Harvest. **(I)** Harvesting ship for *Macrocystis* in California. (Photographs A–E by the author; F–H courtesy of C. K. Tseng; I courtesy of J. Woessner.)

Table 8.5 Annually manufactured dry products and global harvest of seaweeds (fresh weight). R = red alga, B = brown alga

	Examples of seaweeds used		Dry product (tons per year)	harvested fresh weight (tons per year)	authors
(a) Phycocolloids					
agar	R	Gelidium, Pterocladia,	7 000	180 000	Jensen 1979, Briand 1988
	R	Gracilaria, Ahnfeltia			
carrageenan	R	Chondrus, Mastocarpus, Eucheuma	15 000	200 000	Jensen 1979, Briand 1988
alginate	B	Macrocystis, Laminaria	18 000	500 000	Jensen 1979, Briand 1988
	B	Ascophyllum nodosum			
(b) Seaweeds as food					
Kombu (Japan)	B	Laminaria japonica, L. angustata	30 000#	150 000	Hasegawa 1976; # 15 % for food
Haidai (China)	B	Laminaria japonica	275 000#	1 300 000	Tseng 1981b; # 50% for food
Wakame (Japan)	B	Undaria pinnatifida and o. spec.	14 000	120 000	Jensen 1979, Tseng 1981a
Nori (Japan)	R	Porphyra tenera and o. species	18 000	220 000	Jensen 1979, Tseng 1981a
Zicai (China)	R	Porphyra haitanensis and o. spec.	7 000	80 000	Tseng 1981a
(c) Seaweeds as supplement to animal fodder					
seaweed meal	B	Ascophyllum nodosum	30 000	100 000	Jensen 1979
(d) Seaweed manure					
Maerl	R	Lithothamnion corallioides		300 000	Blunden et al. 1975
	R	Phymatolithon calcareum			
Sum (a–d)				≈ 3 000 000	

Fig. 8.6 Red alga *Porphyra* (Nori) for food production in Japan. **(A)** Culture field. **(B)** Culture net. **(C)** Drying procedure. **(D,E)** Rice rolled in *Porphyra*. **(F)** Culture of Conchocelis phase (sporophyte) in and on mussels. (Photographs A–C,F courtesy of J. Woessner; D–E by the author.)

studies, are sucrose in green algae, mannitol in brown algae, and the glycosides (substances with both sugar and non sugar parts) floridoside, isofloridoside (in the Bangiophycidae), and digeneaside (in the Ceramiales). **Reserve polysaccharides** including starch are synthesized in green algae, the starchlike laminaran in brown algae, and floridean starch in red algae. The **structural polysaccharides** represent important substances for industry ("phycocolloids," "algal mucilages;" Table 8.5): the sulfated galactans **agar** (whose structural unit is agarose) and **carrageenan** in the red algae (Figs. 8.7F,G), and **alginic acid** and its salts, the **alginates,** in the brown algae (Fig. 8.7B; Table 8.5). The alginate composition is environmentally important because

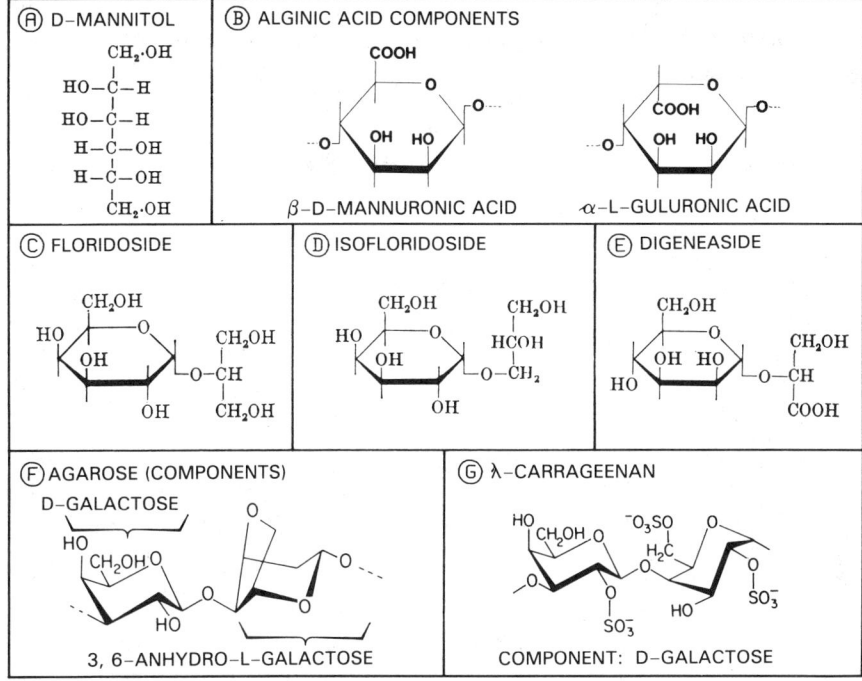

Fig. 8.7 Structural formulae of important seaweed carbohydrates. (Compiled after Percival 1979; Percival and McDowell 1967; McCandless 1981.)

young tissue is rich in mannuronic acid, while the firmer holdfast tissue is rich in guluronic acid (Haug et al. 1974; Cheshire and Hallam 1985).

The three types of phycocolloids, extracted annually from about 900,000 metric tons (fresh weight) of harvested seaweeds, respresent a quarter of the global seaweed harvest (Table 8.5). They are widely used as gelling agents in various industries such as the food industry (see literature cited at beginning of chapter). Also, green algae contain mucilaginous polysaccharides, which are sulfated like the galactans of red algae.

In contrast to the tropical regions, where much of the seaweed biomass is consumed by grazers, at temperate latitudes about 90% of the seaweed biomass is thought to be decomposed and finally mineralized in **detritus food chains** (reviews in Barnes and Mann 1980; Mann 1982). This decomposition is carried out by bacteria and fungi and by the activity of small invertebrate detrital feeders (see below).

The abundance and nutritional value of macroalgae ensure that both the benthic community and the pelagic food chain attain important input. In an Arctic kelp community dominated by *Laminaria solidungula,* up to 50% of the body carbon of mysid crustaceans, a major prey for birds, fish, and mammals, was composed of carbon derived from kelp detritus (Dunton and Schell 1987).

While still alive, macroalgae may exude a small percentage, probably 1–5%, of their current photosynthate. This is quickly mineralized in the water by bacteria (reviews in Brylinski 1977; Pregnall 1983). Another part of the biomass given off by live seaweeds is the continuously sloughed off tissue, for example, as surface layers of the thallus in some fucalean species such as *Ascophyllum nodosum,* and as surface layers of the stipe of *Laminaria setchellii* (Klinger and de Wreede 1989). The sloughing process also occurs in *Laminaria* species at the distal, eroding end of the blade, where a rich bacterial epiflora rapidly hydrolyzes the low-molecular-weight carbohydrates, protein, laminaran, and alginate (Laycock 1974). Lipoid substances like palmitic acid and linoleic acid may amount to 0.2% in the sloughed-off mucilaginous material. These substances are found in the sea foam at the water's surface. Linoleic acid is particularly important as it is an essential substance required by animals unable to synthesize it.

That part of the seaweed biomass entering the water as **dissolved organic carbon** (DOC) is estimated in kelp vegetation to represent 15–30% of the annual primary productivity (Hatcher et al. 1977; Johnston et al. 1977; Newell et al. 1980; Mann 1982). The remaining biomass is decomposed, starting, for example, with individuals detached from the substratum by wave action, broken to pieces, and piled up in submarine troughs. After the erosive mechanical action of sand and water movement and the activity of fungi and bacteria, these large pieces are reduced to fine particles in the millimeter size range known as **particulate organic carbon** (POC). The POC serves as food for **detritus feeders** such as protozoans, turbellarians, nematodes, and amphipods (Robertson and Lucas 1983). The results of further breakdown of organic matter are fed on by the filter feeders such as sponges and ascidians. **Macrodetrivores** (e.g., amphipods) may partially inhibit the decomposition by cropping the rotting weed from the margins of the kelp fronds and thus preventing the saprophagous microbes from completing the decomposition (Bedford and Moore 1984).

Half of the fresh organic material of *Laminaria pallida* becomes decomposed within 10 days at 10°C (Stuart et al. 1981). For the vegetation dominated by *Ecklonia radiata* on the southern Australian coast, it is estimated that the decomposition of detached thalli, which became covered by sediment in troughs of the sublittoral zone, takes two weeks in summer and one to two months in winter (Robertson and Hansen 1982).

Thallus parts of *Macrocystis integrifolia* and *Nereocystis luetkeana* from the North American Pacific coast, enclosed in litter bags exposed in the sublittoral zone, became fragmented to detritus particle sizes of less than 1.5 mm in diameter within five to eight weeks, and 60% of the algal biomass reappeared in the form of bacterial biomass (Albright et al. 1980). Of the South African sublittoral *Ecklonia maxima* vegetation, probably 5% becomes intertidally stranded (Koop et al. 1982); a quarter of the organic carbon is transformed into bacterial biomass within the accumulated kelp masses, while the remaining organic carbon seeps into the sand as an organic solution, where it is mineralized by bacteria within eight days.

Bibliography

Abbott, I. A., 1979: Taxonomy and nomenclature of the type species of *Dumontia* Lamouroux (Rhodophyta). Taxon 28, 563–566.

Abbott, I. A., 1989: Marine algae of the Northwest Hawaiian Islands. Pacific Sci. 43, 223–233.

Abbott, I. A., and Hollenberg, G. J., 1976: Marine algae of California. Stanford University Press, Stanford, Calif.; 827 pp.

Abbott, I. A., and North, W. J., 1972: Temperature influences on floral composition in California waters. Proc. Int. Seaweed Symp. 7 (Sapporo), 72–79.

Abdel-Rahman, M. H., 1982: Photopériodisme chez *Acrochaetium asparagopsis* (Rhodophycées). I. Reponse à une photopériode de jours courts au cours de la formation de tétrasporocystes. II. Influence de l'interruption de la nyctipériode, par un eclairement blanc ou monochromatique, sur la formation des tétrasporocystes. Physiol. Veg. 20, 155–164; C. R. Acad. Sci. Paris 294, 389–392.

Acleto, C., 1973: Las algas marinas del Peru. Bol. Soc. Peru. Bot. 6, 1–164.

Adams, C. G., 1981: An outline of Tertiary paleogeography. In: The evolving earth. Ed. by L. R. M. Cocks. British Museum (Natural History) and Cambridge University Press, pp. 221–235.

Adams, N. M., 1983: Checklist of marine algae possibly naturalized in New Zealand. N. Z. J. Bot. 21, 1–2.

Adams, N. M., and Nelson, W. A., 1985: Marine algae of the Three Kings Islands. Nat. Mus. N. Z. Misc. Ser. 13, 1–29.

Adey, W. H., 1970: The effects of light and temperature on growth rates in boreal-subarctic crustose corallines. J. Phycol. 6, 269–276.

Adey, W. H., 1973: Temperature control of reproduction and productivity in a subarctic coralline alga. Phycologia 12, 111–118.

Adey, W. H., 1976: Crustose coralline algae as microenvironmental indicators for the Tertiary. In: Historical biogeography, plate tectonics, and the changing environment. Ed. by J. Gray and A. J. Boucot. Oregon State University Press, Corvallis, pp. 459–464.

Adey, W. H., and Adey, P. J., 1973: Studies on the biosystematics and ecology of the epilithic crustose Corallinaceae of the British Isles. Br. Phycol. J. 8, 343–407.

Adey, W. H., and Goertemiller, T., 1987: Coral reef algal turfs: master producers in nutrient poor seas. Phycologia 26, 374–386.

Adey, W. H., and MacIntyre, I. G., 1973: Crustose coralline algae: a re-evaluation in the geological sciences. Geol. Soc. Am. Bull. 84, 883–904.

Adey, W. H., and Vassar, J. M., 1975: Colonization, succession, and growth rates of tropical crustose coralline algae (Rhodophyta, Cryptonemiales). Phycologia 14, 55–69.

Adey, W. H., Townsend, R. A., and Boykins, W. T., 1982: The crustose coralline algae (Rhodophyta: Corallinaceae) of the Hawaiian Islands. Smithson. Contrib. Mar. Sci. 15, 1–74.

Afonso-Carrillo, J., and Gil-Rodríguez, M. C., 1982a: Sobre la presencia de un fondo de "maerl" en las Islas Canarias. Collect. Bot. 13, 703–708.

Afonso-Carrillo, J., and Gil-Rodríguez, M. C., 1982b: Aspectos biogéograficos de la flora ficológica marina de las islas Canarias. Actas II Simp. Ibér. Bentos Mar. 3, 41–48.

Afonso-Carrillo, J., Gil-Rodríguez, M. C., Haroun Tabraue, R., Villena Balsa, M., and Wildpret de la Torre, W., 1984a: Adiciones y correcciones al catálogo de algas marinas bentónicas para el Archipiélago Canario. Vieraea 13, 27–49.

Afonso-Carrillo, J., Gil-Rodríguez, M. C., and Wildpret de la Torre, W., 1984b: Algunas consideraciones florístícas, corológicas y ecologicas sobre las algas Corallinaceae (Rhodophyta) de las Islas Canarias. Anal. Biol. 2 (sección especial, 2), 23–37.

Akiyama, K., 1965: Studies of ecology and culture of Undaria pinnatifida (Harv.) Sur. II. Environmental factors affecting the growth and maturation of gametophyte. Bull. Tohoku Reg. Fish. Res. Lab. 25, 143–170.

Albright, L. J., Chocair, J., Masuda, K., and Valdés, M., 1980: In situ degradation of the kelps Macrocystis integrifolia and Nereocystis luetkeana in British Columbia coastal waters. Nat. Can. (Que.) 107, 3–10.

Aleem, A. A., 1951: Algues marines de profundeur des environs d'Alexandrie (Egypte). Bull. Soc. Bot. Fr. 98, 249–252.

Aleem, A. A., 1973: Ecology of a kelp bed in southern California. Botanica Mar. 16, 83–95.

Aleem, A. A., 1978a: A preliminary list of marine algae from Sierra Leone. Botanica Mar. 21, 397–399.

Aleem, A. A., 1978b: Contributions to the study of the marine algae of the Red Sea. I. The algae in the neighbourhood of al-Ghardaqua, Egypt (Cyanophyceae, Chlorophyta and Phaeophyta). II. (Rhodophyta). III. Marine algae from Obhar, in the vicinity of Jeddah, Saudi Arabia. Bull. Fac. Sci. King Abdul Aziz Univ. 2, 73–118.

Allender, B. M., and Kraft, G. T., 1983: The marine algae of Lord Howe Island (New South Wales): The Dictyotales and Cutleriales (Phaeophyta). Brunonia 6, 73–130.

Almaca, C., 1985: Evolutionary and zoogeographical remarks on the Mediterranean fauna of Brachyuran crabs. In: Mediterranean marine ecosystems. Ed. by M. Moraitou-Apostolopoulou and V. Kiortsis. Plenum Press, New York, pp. 347–366.

Almodovar, L. R., and Ballantine, D. L., 1983: Checklist of benthic marine macroalgae plus additional species records from Puerto Rico. Caribb. J. Sci. 19, 7–20.

Alvarez, L. W., Alvarez, W., Asaro, F., and Michel, H. V., 1980: Extraterrestrial cause for the Cretaceous-Tertiary extinction. Experimental results and theoretical interpretation. Science 208, 1095–1108.

Alvarez, M., Gallardo, T., Ribera, M. A., and Garreta, A. G., 1988: A reassessment of Northern Atlantic seaweed biogeography. Phycologia 27, 221–233.

Alvarez, W., Kauffmann, E. G., Surlyk, F., Alvarez, L. W., Asaro, F., and Michel, H. V., 1984: Impact theory of mass extinctions and the invertebrate fossil record. Science 223, 1135–1141.

Amsler, C. D., and Searles, R. B., 1980: Vertical distribution of seaweed spores in a water column offshore of North Carolina. J. Phycol. 16, 617–619.

Anand, P., 1940–1943: Marine algae from Karachi. Part I. Chlorophyceae. Part II. Rhodophyceae. Panjab Univ. Bot. Publ. 8, 1–52 (1940); 9, 1–76 (1943).

Anderson, E. K., and North, W. J., 1966: In situ studies of spore production and dispersal in the giant kelp *Macrocystis*. Proc. Int. Seaweed Symp. 5, 73–86.

Anderson, J. M., 1983: Chlorophyll-protein complexes of a *Codium* species, including a light-harvesting siphonaxanthin-chlorophyll a/b-protein complex, an evolutionary relic of some Chlorophyta. Biochem. Biophys. Acta 724, 370–380.

Anderson, J. M., 1986: Photoregulation of the composition, function, and structure of thylakoid membranes. Annu. Rev. Plant Physiol. 37, 93–186.

Anderson, R. J., 1985: Morphological and taxonomic relationships among branched, ligulate members of the genus *Desmarestia* (Phaeophyceae, Desmarestiales), with particular reference to South African *D. firma*. Can. J. Bot. 63, 437–447.

Anderson, R. J., and Bolton, J. J., 1985: Suitability of the agarophyte *Suhria vittata* (L.) J. Ag. (Rhodophyta: Gelidiaceae) for mariculture: geographical distribution, reproductive phenology, and growth of sporelings in culture in relation to light and temperature. S. Afr. J. Mar. Sci. 3, 169–178.

Anderson, R. J., and Stegenga, H., 1989: Subtidal algal communities at Bird Island, eastern Cape, southern Africa. Botanica Mar. 32, 299–311.

Anderson, R. J., and Velimirov, B., 1982: An experimental investigation of the palatability of kelp bed algae to the sea urchin *Parechinus angulosus* Leske. P. S. Z. N. I. Mar. Ecol. 3, 357–373.

Andrew, N. L., 1989: Contrasting ecological implications of food limitation in sea urchins and herbivorous gastropods. Mar. Ecol. Prog. Ser. 51, 189–193.

Andrews, J. H., and Goff, 1985: Pathology. In: Handbook of phycological methods. Ecological field methods: macroalgae. Ed. by M. M. Littler and D. S. Littler. University Press, Cambridge, pp. 573–598.

Ardré, F., 1970–1971: Contribution à l'etude des algues marines du Portugal. I. La flore (1970). Portugaliae Acta Biol. (B) 10, 137–555; II. Ecologie et chorologie (1971). Bull. Cent. Etud. Rech. Sci. Biarritz 8, 359–574.

Armstrong, S. L., 1989: The behaviour in flow of the morphologically variable seaweed *Hedophyllum sessile* (C. Ag.) Setchell. Hydrobiologia 183, 115–122.

Asare, S. O., and Harlin, M. M., 1983: Seasonal fluctuations in tissue nitrogen for five species of perennial macroalgae in Rhode Island Sound. J. Phycol. 19, 254–257.

Athasaniadis, A., 1987: A survey of the seaweeds of the Aegean Sea with taxonomic studies on species of the tribe Antithmanieae (Rhodophyta). University of Gothenburg, Department of Marine Botany; 174 pp.

Audiffred, P. A. J., and Prud'homme van Reine, W. F., 1985: Marine algae of Ilha do Porto Santo and Deserta Grande (Madeira archipelago). CANCAP Project Contribution No. 40. Bol. Mus. Munic. Funchal 37 (166), 20–51.

Audley-Charles, M. G., and Hallam, A. (Eds.), 1988: Gondwana and Tethys. University Press, Oxford; 317 pp.

Augier, H., 1985: Première contribution à la cartographie des biocénoses marines benthiques de l'Ile de Madère. Bol. Mus. Munic. Funchal 37 (168), 86–129.

Axelrod, D. I., 1984: An interpretation of Cretaceous and Tertiary biota in polar regions. Palaeogeogr., Palaeoclimatol. Palaeoecol. 45 105–147.

Azuara, M. P., and Aparicio, P. J., 1983: In vivo blue-light activition of *Chlamydomonas reinhardii* nitrate reductase. Plant Physiol. 71, 286–290.

Baardseth, E., 1941: The marine algae of Tristan da Cunha. In: Results of the Norwegian Scientific Expedition to Tristan da Cunha 1937–1938, Vol. 9. Ed. by Jacob Dybwad, Oslo, pp. 1–173.

Baca, B. J., Sorensen, L. O., and Cox, E. R., 1979: Systematic list of seaweeds of South Texas. Contrib. Mar. Sci. 22, 179–192.

Bachmann, P., Kornmann, P., and Zetsche, K., 1976: Regulation der Entwicklung und des Stoffwechsels der Grünalge *Urospora* durch die Temperatur. Planta 128, 241–245.

Baik, S. E., and Kang, J. W., 1986: Antimicrobial activity of the volatile and lipid fractions of marine algae. Korean J. Phycol. 1, 293–310.

Baissac, J. de B., Lubet, P. E., and Michel, C. M., 1962: Les biocoenoses benthiques littorales de l'ile Maurice. Recl. Trav. Stn. Mar. Endoume Bull. 25, 253–291.

Baker, K. S., and Smith, R. C., 1982: Bio-optical classification and model of natural waters. 2. Limnol. Oceanogr. 27, 500–509.

Bakus, G. J., 1969: Energetics and feeding in shallow marine waters. Int. Rev. Gen. Exp. Zool. 4, 275–369.

Ballantine, D. L., Gerwick, W. H., Velez, S. M., Alexander, E., and Guevara, P., 1987: Antibiotic activity of lipid-soluble extracts from Caribbean marine algae. Hydrobiologia 151/152, 463–469.

Ballesteros, E., and Martinengo, J., 1982: Catálogo de las algas bentónicas (con exclusion de las Diatomeas) de la costa Catalana. Collect. Bot. 13, 723–765.

Baptista, L. R. M., 1977: Flora marinha de Torres (Chloroyphyta, Xantophyta, Phaeophyta, Rhodophyta). Bol. Fac. Fil. Cienc. Univ. Rio Grande do Sul, sér. Bot. 7 (3), 1–244.

Barber, R. T., and Smith, R. L., 1981: Coastal upwelling ecosystems. In: Analysis of marine ecosystems. Ed. by A. R. Longhurst. Academic Press, London, pp. 31–68.

Barlow, R. G., and Alberte, R. S., 1987: Photosynthetic characteristics of phycoerythrin-containing marine *Synechococcus* spp. II. Time course responses of photosynthesis to photoinhibition. Mar. Ecol. Prog. Ser. 39, 191–196.

Barnes, D. J., 1983: Perspectives on coral reefs. Brian Clouston, Manuka, Australia; 277 pp.

Barnes, R. S. K., and Hughes, R. N., 1982: An introduction to marine ecology. Blackwell, Oxford; 339 pp.

Barnes, R. S. K., and Mann, K. H., 1980: Fundamentals of aquatic ecosystems. Blackwell, Oxford; 229 pp.

Barrales, H. L., and Lobban, C. S., 1974: The comparative ecology of *Macrocystis pyrifera*, with emphasis on the forests of Chubut, Argentina. J. Ecol. 63, 657–677.

Barrett, J., Anderson, J. M., 1980: The P 700–chlorophyll a-protein complex and two major light-harvesting complexes of *Acrocarpia paniculata* and other brown seaweeds. Biochem. Biophys. Acta 590, 309–323.

Bas, C., 1949: Contribucion al conocimiento algologico de la costa catalana. Publ. Inst. Biol. Apl., Barc. 6, 103–127.

Baslow, M. H., 1977: Marine pharmacology. A study of toxins and other biologically active substances of marine origin. Krieger, New York; 327 pp.

Basson, P. W., 1979: Marine algae of the Arabian Gulf coast of Saudi Arabia. Botanica Mar. 22, 47–82.

Basson, P. W., Burchard, J. E., Hardy, J. T., and Price, A. R. G., 1977: Biotopes of the western Arabian Gulf. Aramco Department of Loss Prevention and Environmental Affairs, Dhahran, Saudi Arabia; 284 pp.

Bayne, B. L., Clarke, K. R., and Gray, J. S. (Eds), 1988: Biological effects of pollutants. Mar. Ecol. Prog. Ser. 46, 1–278.

Beale, S. I., 1984: Biosynthesis of photosynthetic pigments. In: Chloroplast biogenesis. Ed. by N. R. Baker and J. Barber. Elsevier, Amsterdam, pp. 133–205.

Bedford, A. P., and Moore, P. G., 1984: Macrofaunal involvement in the sublittoral decay of kelp debris: the detrivore community and species interaction. Estuar. Coast. Shelf Sci. 18, 97–111.

Behrens Yamada, S., 1987: Geographic variation in the growth rates of *Littorina littorea* and *L. saxatilis*. Mar. Biol. 96, 529–534.

Bell, C. J., and Rose, D. A., 1981: Light measurement and the terminology of flow. Plant Cell Environ. 4, 86–96.

Bell, H., P., and MacFarlane, C., 1933: The marine algae of the maritime provinces of Canada. 1. List of species with their distribution and prevalance. Can. J. Res. 9, 265–279.

Belsher, T., Augier, H., Boudouresque, C.-F., and Coppejans, E., 1976: Inventaire des algues marines benthiques de la rade et des iles d'Hyères (Méditerranee France). Trav. Sci. Parc nation. Port-Cros 2, 39–89.

Bennett, I., and Pope, E. C., 1953: Intertidal zonation of the exposed rocky shores of Victoria, together with a rearrangement of the biogeographical provinces of temperate Australian shores. Aust. J. Mar. Freshwater Res. 4, 105–159.

Bennett, I., and Pope, E. C., 1960: Intertidal zonation of the exposed rocky shores of Tasmania and its relationship with the rest of Australia. Aust. J. Mar. Freshwater Res. 11, 182–221.

Benson, E. E., Rutter, J. C., and Cobb, A. H., 1983: Seasonal variation in frond morphology and chloroplast physiology of the intertidal alga *Codium fragile* (Suringar) Hariot. New Phytol. 95, 569–580.

Benson, R. H., Chapman, R. E., and Deck, L. T., 1984: Paleoceanographic events and deep-sea ostracods. Science 224, 1334–1336.

Berggren, W. A., and Hollister, C. D., 1974: Paleogeography, paleobiogeography, and the history of circulation in the Atlantic Ocean. In: Studies in paleo-oceanography. Ed. by W. W. Hay. Society of Economic Paleontologists and Mineralogists, Tulsa, Okla., pp. 126–186.

Berggren, W. A., and Schnitker, D., 1983. Cenozoic marine environments in the North Atlantic and Norwegian-Greenland Sea. In: Structure and development of the Greenland-Scotland ridge. Ed. by H. P. Bott, S. Saxov, M. Talwani, and J. Thiede. Plenum Press, N. Y., pp. 495–548.

Bergh, W. H., 1987: Underlying fracture zone of Astrid Ridge off Antarctica's Queen Maud land. J. Geophys. Res. 92, 475–484.

Bergmann, C., 1847: Über die Verhältnisse der Wärmeökonomie der Thiere zu ihrer Grösse. Abt. Göttinger Studien 1, 595–708.

Berry, J. A., and Raison, J. K., 1980: Responses of macrophytes to temperature. In: Encyclopedia of plant physiology, n. s., vol. 12A. Ed. by O. L. Lange, P. S. Nobel, C. B. Osmond and H. Ziegler. Springer-Verlag, Berlin, pp. 277–338.

Berry, J. R., and Björkman, O., 1980: Photosynthetic response and adaptation to temperature in higher plants. Annu. Rev. Plant Physiol. 31, 491–543.

Bert, T. M., 1986: Speciation in western Atlantic stone crabs (genus *Menippe*): the role of geological processes and climatic events in the formation and distribution of species. Mar. Biol. 93, 157–170.

Berthold, G., 1882: Über die Verbreitung der Algen im Golf von Neapel nebst einem Verzeichnis der bisher beobachteten Arten. Mitt. Zool. Stn. Neapel 3, 393–536.

Beth, K., and Merola, A., 1960: Einige Experimente zum Epiphytismus in Zönosen mariner Algen. Delpinoa 2, 1–14.

Bianchi, C. N., and Morri, C., 1983: Note sul benthos marino costiero dell'isola di Kos (Egeo sud-orientale). Natura-Soc. Ital. Sci. Nat., Mus. Civ. Stor. Nat. Acquario Civ. Milano 74, 96–114.

Bidwell, R. G. S., and McLachlan, J., 1985: Carbon nutrition of seaweeds: photosynthesis, photorespiration, and respiration. J. Exp. Mar. Biol. Ecol. 86, 15–46.

Biebl, R., 1937: Ökologische und zellphysiologische Untersuchungen an Rotalgen der englischen Südküste. Beih. Bot. Z. Abt. A, 57, 381–424.

Biebl, R., 1938: Trockenresistenz und osmotische Empfindlichkeit der Meeresalgen verschieden tiefer Standorte. J. Wiss. Bot. 86, 350–386.

Biebl, R., 1939: Über die Temperaturresistenz von Meeresalgen verschiedener Klimazonen und verschieden tiefer Standorte. J. Wiss. Bot. 88, 389–420.

Biebl, R., 1958: Temperatur- und osmotische Resistenz von Meeresalgen der bretonischen Küste. Protoplasma 50, 217–242.

Biebl, R., 1962: Temperaturresistenz tropischer Meeresalgen. (Verglichen mit jener von Algen in temperierten Meeresgebieten). Botanica Mar. 4, 241–254.

Biebl, R., 1968: Über Wärmehaushalt und Temperaturresistenz arktischer Pflanzen in Westgrönland. Flora, Abt. B, 157, 327–354.

Bird, C. J., and McLachlan, J., 1974: Cold-hardiness of zygotes and embryos of *Fucus* (Phaeophyceae, Fucales). Phycologia 13, 215–225.

Bird, C. J., and McLachlan, J., 1986: The effect of salinity on distribution of *Gracilaria* Grev. (Rhodophyta, Gigartinales): an experimental assessment. Botanica Mar. 29, 231–238.

Bird, C. J., Edelstein, T., and McLachlan, J., 1976: Investigations of the marine algae of Nova Scotia. XII. The flora of Pomquet Harbour. Can. J. Bot. 54, 2726–2737.

Bird, C. J., Greenwell, M., and McLachlan, J., 1983: Benthic marine algal flora of the north shore of Prince Edward Island (Gulf of St. Lawrence), Canada. Aquat. Bot. 16, 315–335.

Bird, C. J., Meer, J. P. van der, and McLachlan, J., 1982: A comment on *Gracilaria verrucosa* (Huds.) Papenf. (Rhodophyta: Gigartinales). J. Mar. Biol. Ass. U. K. 62, 453–459.

Bird, N. L., Chen, L. C.-M., and McLachlan, J., 1979: Effects of temperature, light, and salinity on growth in culture of *Chondrus crispus*, *Furcellaria lumbricalis*, *Gracilaria tikvahiae* (Gigartinales, Rhodophyta), and *Fucus serratus* (Fucales, Phaeophyta). Botanica Mar. 22, 521–527.

Bird, K. T., and Benson, P. H., 1987: Seaweed cultivation for renewable resources. Elsevier, New York; 382 pp.

Bird, K. T., and McIntosh, 1979: Notes on the marine algae of Guatemala. Rev. Biol. Trop. 27, 163-169.

Bisson, M. A., and Gutknecht, J., 1980: Osmotic regulation in algae. In: Plant membrane transport: current conceptual issues. Ed. by W. J. Lucas, R. M. Spanswick, and J. Dainty. North-Holland, Amsterdam, pp. 131-142.

Björkman, O., 1981: Responses to different quantum flux densities. In: Encyclopedia of plant physiology, n. s. vol. 12A. Ed. by O. L. Lange, P. S. Nobel, C. B. Osmond, and H. Ziegler. Springer-Verlag, Berlin, pp. 57-107.

Björn, L. O., 1979: Photoreversibly photochromic pigments in organisms: properties and role in biological light perception. Quart. Rev. Biophys. 12, 1-23..

Blair, S. M., 1983: Taxonomic treatment of the *Chaetomorpha* and *Rhizoclonium* species (Cladophorales; Chlorophyta) in New England. Rhodora 85, 175-211.

Blanc, J.-J., and Molinier, R., 1955: Les formations organogènes construites superficielles en Méditerranee occidentale. Bull. Inst. Oceanogr. Monaco 52 (1067), 1-26.

Blaricam, G. R. van, and Estes, J. A., 1988: The community ecology of sea otters. Springer-Verlag, Berlin; 247 pp.

Bliding, C., 1963: A critical survey of European taxa in Ulvales. I. *Capsosiphon, Percursaria, Blidingia, Enteromorpha*. Op. Bot. Soc. Bot. Lund 8 (3), 1-160.

Bliding, C., 1968: A critical survey of European taxa in Ulvales. II. *Ulva, Ulvaria, Monostroma, Kornmannia*. Bot. Not. 121, 535-629.

Blunden, G., Binns, W. W., and Perks, F., 1975: Commercial collection and utilisation of maerl. Econ. Bot. 29, 140-145.

Blunden, G., Farnham, W. F., Jephson, N., Barwell, C. J., Fenn, R. H., and Plunkett, B. A., 1981: The composition of maerl beds of economic interest in northern Brittany, Cornwall, and Ireland. Proc. Int. Seaweed Symp. 10, 651-656.

Boardman, N. K., 1977: Comparative photosynthesis of sun and shade plants. A. Rev. Pl. Physiol. 28, 355-377.

Börgesen, F., 1903-1908: Marine algae of the Faeröes (1903). The algae-vegetation of the Faeröese coasts (1908). In: Botany of the Faeröes. Part II (1903), pp. 339-532; Part III (1908), pp. 683-834. Det Nordiske Forlag, Kopenhagen.

Börgesen, F., 1913-1920: The marine algae of the Danish West Indies. I. Chlorophyceae. Dan. Bot. Ark. 1 (4), 1-158 (1913). II. Phaeophyceae. Dan. Bot. Ark. 2 (2), 1-66 (1914). III. Rhodophyceae. Dan. Bot. Ark. 3, 1-504 (1915-1920).

Börgesen, F., 1924: Marine algae from Easter Island. In: The natural history of Juan Fernandez and Easter Island. Vol. 2. Ed. by C. Skottsberg. Almquist & Wiksell, Uppsala, pp. 247-309.

Börgesen, F., 1925-1936: The marine algae from the Canary Islands. K. danske Vidensk. Selsk. Biol. Medd. 5 (3) (1925); 6 (2) (1926); 6 (6) (1927); 8 (1) (1929); 9 (1) (1930); 12 (5) (1936).

Börgesen, F., 1934: Some marine algae from the northern part of the Arabian Sea with remarks on their geographical distribution. K. Danske Vidensk. Selsk. Biol. Medd. 9 (6), 1-72.

Börgesen, F., 1937–1938: Contributions to a south Indian marine algal flora. J. Indian Bot. Soc. 16, 1–56 (1937); 17, 205–242 (1938).

Börgesen, F., 1940–1957: Some marine algae from Mauritius. K. Danske Vidensk. Selsk. Biol. Medd. 15 (4), 3–81 (1940); 16 (3), 3–81 (1941); 17 (5), 3–64 (1942); 19 (1), 3–85 (1943); 19 (6), 3–32 (1944); 19 (10), 3–68 (1945); 20 (6), 3–64 (1946); 20 (12), 3–55 (1948); 21 (5), 3–48 (1949); 18 (11), 3–46 (1950); 18 (16), 3–44 (1951); 18 (19), 3–72 (1952); 21 (9), 3–62 (1953); 22 (4), 3–51 (1954); 23 (4), 3–35 (1957).

Börgesen, F., and Jonsson, H., 1905: The distribution of the marine algae of the Arctic and of the northernmost of the Atlantic. Bot. Faeroes 3, Appendix, 1–28.

Bogorad, L., 1975: Phycobiliproteins and complementary chromatic adaptation. A. Rev. Pl. Physiol. 26, 369–401.

Bold, H. C., and Wynne, M. J., 1985: Introduction to the algae. Structure and reproduction. 2d ed.. Prentice Hall Inc., Englewood Cliffs, N. J.; 720 pp.

Bolton, J. J., 1979: Estuarine adaptation in populations of *Pilayella littoralis* (L.) Kjellm. (Phaeophyta, Ectocarpales). Estuar. Coast. Mar. Sci. 9, 273–280.

Bolton, J. J., 1981: Community analysis of vertical zonation patterns on a Newfoundland rocky shore. Aquat. Bot. 10, 299–316.

Bolton, J. J., 1983: Ecoclinal variation in *Ectocarpus siliculosus* (Phaeophyceae) with respect to temperature growth optima and survival limits. Mar. Biol. 73, 131–138.

Bolton, J. J., 1986: Marine phytogeography of the Benguela upwelling region on the west coast of southern Africa: a temperature dependent approach. Botanica Mar. 29, 251–256.

Bolton, J. J., and Anderson, R. J., 1987: Temperature tolerances of two southern African *Ecklonia* species (Alariaceae: Laminariales) and of hybrids between them. Mar. Biol. 96, 293–297.

Bolton, J. J., and Levitt, G. J., 1985: Light and temperature requirements for growth and reproduction in gametophytes of *Ecklonia maxima*. Mar. Biol. 87, 131–135.

Bolton, J. J., and Levitt, G. J., 1987: The influence of upwelling on South African west coast seaweeds. S. Afr. J. Mar. Sci. 5, 319–325.

Bolton, J. J., and Lüning, K., 1983: Optimal growth and maximal survival temperatures of Atlantic *Laminaria* species (Phaeophyta) in culture. Mar. Biol. 66, 89–94.

Bolton, J. J., and Stegenga, H., 1987: The marine algae of Hluleka (Transkei) and the warm temperate/sub-tropical transition on the east coast of southern Africa. Helgoländer Meeresunters. 41, 165–183.

Bolton, J. J., Germann, I., and Lüning, K., 1983: Hybridisation between Atlantic and Pacific representatives of the Simplices section of *Laminaria* (Phaeophyceae). Phycologia 22, 133–140.

Boney, A. D., 1981: Mucilage: the ubiquitous algal attribute. Br. Phycol. J. 16, 115–132.

Bonotto, S., 1976: Cultivation of plants. Multicellular plants. In: Marine ecology. Vol. 3, P. 1. Ed. by O. Kinne. John Wiley, London, pp. 467–529.

Boo, S.-M., 1987: Distribution of marine algae from shore area of Kangwon Province. Korean J. Phycol. 2, 223–235.

Boo, S.-M., and Lee, I. K., 1986: Studies on benthic algal community in the east coast of Korea. 1. Floristic composition and periodicity of a Sokcho rocky shore. Korean J. Phycol. 1, 107–116.

Booth, E., 1975: Seaweeds in industry. In: Chemical oceanography. Ed. by J. P. Riley and G. Skirrow. Academic Press, London, pp. 219–268.

Booth, E., 1981: Some factors affecting seaweed fertilizers. Proc. Int. Seaweed Symp. 8 (Bangor), 661–666.

Borowitzka, M. A., 1977: Algal calcification. Oceanogr. Mar. Biol. Annu. Rev. 15, 189–223.

Borowitzka, M. A., 1982: Mechanisms in algal calcification. Prog. Phycol. Res. 1, 137–177.

Bosence, D. W. J., 1985: The "Coralligène" of the Mediterranean: a Recent analog for Tertiary coralline algal limestones. In: Paleoalgology. Contemporary research and applications. Ed. by D. F. Toomey and M. H. Nitecki. Springer-Verlag, Berlin, pp. 216–225.

Bott, H. P., Saxov, S. Talwani, M., and Thiede, J. (Eds.), 1983: Structure and development of the Greenland-Scotland ridge. Plenum Press, New York; pp. 1–685.

Boudouresque, C. F., 1969: Note préliminaire sur le peuplement algal des biotopes sciaphiles superficiels le long des côtes de l'Algérois et de la Kabylie. Bull. Mus. Hist. Nat. Marseille 29, 165–187.

Boudouresque, C. F., 1971: Contribution à l'étude phytosociologique des peuplements algaux des côtes varoises. Vegetatio 22, 83–184.

Boudouresque, C. F., and Cinelli, F., 1971: Le peuplement algal des biotopes sciaphiles superficiels de mode battu de l'Île d'Ischia (Golfe de Naples, Italie). Pubbl. Stn. Zool. Napoli 39, 1–43.

Boudouresque, C. F., and Cinelli, F., 1976: Le peuplement algal des biotopes sciaphiles superficiels de mode battu en Méditerranée occidentale. Pubbl. Stn. Zool. Napoli 40, 433–459.

Boudouresque, C. F., and Denizot, M., 1975: Révision du genre *Peyssonnelia* (Rhodophyta) en Méditerranée. Bull. Mus. Hist. Nat. Marseille 35, 7–92.

Boudouresque, C. F., Gerbal, M., and Knoepffler-Peguy, M., 1985: L'algue japonaise *Undaria pinnatifida* (Phaeophyceae, Laminariales) en Méditerranée. Phycologia 24, 364–366.

Boudouresque, C.-F., and Perret, N., 1977: Inventaire de la flore marine de Corse (Méditerranée): Rhodophyceae, Phaeophyceae, Chlorophyceae et Bryopsidophyceae. Bibl. Phycol. 25, 1–170.

Bousfield, E. L., and Thomas, M. L. H., 1975: Postglacial changes in distribution of littoral marine invertebrates in the Canadian Atlantic region. Proc. N. S. Inst. Sci. 27 (Suppl. 3), 47–60.

Bouxin, H., and Dizerbo, A. H., 1971: Les algues de l'Archipel des Glenan. Bot. Rhedonica A, 10, 201–226.

Bowen, K. Y., 1971: The growth and development of the deep growing marine alga, *Maripelta rotata* (Daws.) Daws. Univ. Microfilms Inc., Ann Arbor, Mich.; pp. 1–220.

Brady, J., 1982: Biological timekeeping. University Press, Cambridge; 197 pp.

Brain, C. K., 1984: The terminal Miocene event: a critical environmental and evolutionary episode? In: Late Cainozoic palaeoclimates of the Southern Hemisphere. Ed. by J. C. Vogel. Balkema, Rotterdam, pp. 5–20.

Bramwell, D., 1979: Plants and islands. Academic Press, London; 459 pp.
Branch, G., and Branch, M., 1981: The living shores of southern Africa. Struik Publishers, Cape Town; 272 pp.
Branch, M. L., 1974: Limiting factors for the gametophytes of three South African Laminariales. Investl. Rep. Sea Fish. Branch S. Afr. 104, 1–38.
Braud, J.-P., 1974: Étude de quelques paramètres écologiques, biologiques et biochimiques chez un phéophycee des côtes bretonnes, *Laminaria ochroleuca*. Rev. Trav. Inst. Pêches Marit. 38, 115–204.
Breeman, A. M., 1988: Relative importance of temperature and other factors in determining geographic boundaries: experimental and phenological evidence. Helgoländer Meeresunters. 42, 199–241.
Breeman, A. M., 1989: Expected effects of changing seawater temperatures on the geographic distribution of seaweed species. In: Expected effects of climatic change on marine coastal ecosystems. Ed. by J. J. Beukema, W. J. Wolff, and J. J. W. M. Brouns. Kluwer Academic Publishers, Dordrecht, pp. 69–76.
Breeman, A. M., and Hoeksema, B. W., 1987: Vegetative propagation of the red alga *Rhodochorton purpureum* by means of fragments that escape digestion by herbivores. Mar. Ecol. Prog. Ser. 35, 197–201.
Breeman, A. M., and ten Hoopen, A., 1987: The mechanism of daylength perception in the red alga *Acrosymphyton purpuriferum*. J. Phycol. 23, 36–42.
Breeman, A. M., Meulenhof, E. J. S., and Guiry, M. D., 1988: Life history regulation and phenology of the red alga *Bonnemaisonia hamifera*. Helgoländer Meeresunters. 42, 535–551.
Breen, P. A., Carson, T. A., Foster, J. B., and Stewart, E. A., 1982: Change in subtidal community structure associated with British Columbia sea otter transplants. Mar. Ecol. Prog. Ser. 7, 13–20.
Brenchley, P. J. (Ed.), 1984: Fossils and climate. Wiley, Chichester; 352 pp.
Breton-Provencher, M., and Cardinal, A., 1978: Les algues benthiques de James et d'Hudson: état actuel des connaissances et nouvelles données sur les parties méridionales de ces regions. Nat. Can. (Que.) 105, 277–284.
Brett, D. W., and Norton, T. A., 1969: Late glacial marine algae from Greenock and Renfrew. J. Geol. 5, 42–68.
Breuer, G., and Schramm, W., 1989: Changes in macroalgal vegetation of Kiel Bight (Western Baltic Sea) during the past 20 years, Kiel. Meeresforsch. 6 (Suppl.), (Proc. Baltic Marine Biologists Symposium 10), 241–255.
Briand, X., 1988: Exploitation of seaweeds in Europe. In: Aquatic primary biomass (marine macroalgae): biomass conversion, removal and use of nutrients. I. Ed. by P. Morand and E. H. Schulte. COST 48, European Community, Brussels; pp. 53–66.
Bricaud, A., Morel, A., and Prieur, L., 1981: Absorption by dissolved organic matter of the sea (yellow substance) in the UV and visible domains. Limnol. Oceanogr. 26, 43–53.
Briggs, J. C., 1974: Marine zoogeography. McGraw-Hill, New York; 475 pp.
Briggs, J. C., 1984: Centres of origin in biogeography. Biogeogr. Monogr. 1, 1–95.
Brinkhuis, B. H., 1977: Comparisons of salt-marsh fucoid production estimated from three different indices. J. Phycol. 13, 328–335.

Britz, S. J., 1979: Chloroplast and nuclear migration. In: Encyclopedia of plant physiology, n. s. Vol. 7. Ed. by W. Haupt and M. E. Feinleib. Springer-Verlag, Berlin, pp. 163–177.

Britz, S. I., and Briggs, W. R., 1976: Circadian rhythms of chloroplast orientation and photosynthetic capacity in *Ulva*. Plant Physiol. 58, 22–27.

Brodie, J., Guiry, M. D., 1987: Life history and photoperiodic responses in *Cordylecladia erecta* (Rhodophyta) from Ireland. Br. Phycol. J. 22, 300–301.

Brodie, J., Guiry, M. D., 1988: Life history and reproduction of *Botryocladia ardreana* sp. nov. (Rhodymeniales, Rhodophyta) from Portugal. Phycologia 27, 109–130.

Broecker, W. S., 1986: Oxygen isotope constraints on surface ocean temperatures. Quat. Res. 26, 121–134.

Brown, J. H., and Gibson, A. C., 1983: Biogeography. Mosby Company, St. Louis; 643 pp.

Brown, M. T., 1987: Effects of desiccation on photosynthesis of intertidal algae from a southern New Zealand shore. Botanica Mar. 30, 121–127.

Brundin, L., 1966: Transantarctic relationships and their significance, as evidenced by chironomid midges. K. svenska Vetensk. Akad Handl., Ser. 4, 11, 1–472.

Brundin, L., 1972: Phylogenetics and biogeography. Syst. Zool. 21, 69–79.

Brundin, L., 1981: Croizat's panbiogeography versus phylogenetic biogeography. In: Vicariance biogeography: a critique. Symposium of the Systematics Discussion Group of the American Museum of Natural History, May 2–4, 1979. Ed. by G. Nelson and D. E. Rosen. Columbia University Press, New York, pp. 94–158.

Brundin, L., 1988: Phylogenetic biogeography. In: Analytical biogeography. An integrated approach to the study of animal and plant distributions. Ed. by A. A. Myers and P. S. Giller. Chapman of Hall, London, pp. 343–369.

Brylinski, M., 1977: Release of dissolved organic matter by some marine macrophytes. Mar. Biol. 39, 213–220.

Budyko, M. I., and Vinnikov, K. Y., 1977: Global warming. In: Global chemical cycles and their alteration by man. Ed. by W. Stumm. Dahlem Konferenzen, Berlin, pp. 189–205.

Bünning, E., 1973: The physiological clock. Springer-Verlag, Berlin; 285 pp.

Buggeln, R. G., 1981: Morphogenesis and growth regulators. In: The biology of seaweeds. Ed. by C. S. Lobban and M. J. Wynne. Blackwell, Oxford, pp. 627–660.

Buggeln, R. G., 1983: Photoassimilate translocation in brown algae. Prog. Phycol. Res. 2, 283–332.

Buggeln, R. G., 1985: Carbon allocation. In: Handbook of phycological methods. Ecological field methods: macroalgae. Ed. by M. M. Littler and D. S. Littler. University Press, Cambridge, pp. 415–425.

Bunt, J. S., 1975: Primary productivity of marine ecosystems. In: Primary productivity of the biosphere. Ed. by H. Lieth and R. H. Whittaker. Springer-Verlag, New York, pp. 169–183.

Burke, K., 1979: The edges of the ocean: an introduction. Oceanus 22 (3), 2–9.

Butler, J. N., Morris, B. F., Cadwallader, J., and Stoner, A. W., 1983: Studies of

Sargassum and the *Sargassum* community. Bermuda Biol. Stn. Res. Spec. Publ. 22, 1–307.

Buzas, M. A., and Culver, S. J., 1984: Species duration and evolution: benthic Foraminifera on the Atlantic continental margin of North America. Science 225, 829–830.

Buzas, M. A., and Culver, S. J., 1986: Geographic origin of benthic foraminiferal species. Science 232, 775–776.

Cabioch, J., 1969: Les fonds de maerl de la baie de Morlaix et leur peuplement végétal. Cah. Biol. Mar. 10, 139–161.

Cabioch, J., 1981: Premières observations de l'algue Japonaise *Sargassum muticum* (Yendo) Fensholt dans la région de Roscoff. Trav. Stn. Biol. Roscoff, N. S. 27, 1–2.

Cabioch, J., and Magne, F., 1986: Apparition du *Lomentaria hakodatensis* Yendo sur les côtes de France. Cryptogam. Algol. 7, 170.

Caddy, J. F. and Fischer, W. A., 1984: FAO interests in promoting understanding of world seaweed resources, their optimal harvesting, and fishery and ecological interactions. Hydrobiologia 116/117, 355–362.

Caldwell, M. M., 1981: Plant responses to solar ultraviolet radiation. In: Encyclopedia of plant physiology, n. s. Vol. 12A. Ed. by O. L. Lange, P. S. Nobel, C. B. Osmond, and H. Ziegler. Springer-Verlag, Berlin, pp. 169–197.

Caldwell, M. M., 1984: Effects of UV-radiation on plants in the transition region to blue light. In: Blue light effects in biological systems. Ed. by H. Senger. Springer, Berlin, pp. 20–28.

Calkins, J., 1982: The role of solar ultraviolet radiation in marine ecosystems. Plenum Press, New York; 724 pp.

Calkins, J., Thordardottir, T., 1980: The ecological significance of solar UV radiation on aquatic organisms. Nature 283, 563–566.

Calvert, H. E., Dawes, C. J., and Borowitzka, M. A., 1976: Phylogenetic relationships of *Caulerpa* (Chlorophyta) based on comparative chloroplast ultrastructure. J. Phycol. 12, 149–162.

Calvin, N. I., and Ellis, R. J., 1978: Quantitative and qualitative observations on *Laminaria dentigera* and other subtidal kelps of southern Kodiak Island, Alaska. Mar. Biol. 47, 331–336.

Calvin, N. I., and Lindstrom, S. C., 1980: Intertidal algae of Port Valdez, Alaska: species and distribution with annotations. Botanica Mar. 23, 791–797.

Cambridge, M. L., Breeman, A. M., Kraak, S., and van den Hoek, C., 1987: Temperature responses of tropical to warm temperate *Cladophora* species in relation to their distribution in the North Atlantic Ocean. Helgoländer Meeresunters. 41, 329–354.

Campbell, G. S., 1981: Fundamentals of radiation and temperature relations. In: Encyclopedia of plant physiology, n. s. Vol. 12 A. Ed. by O. L. Lange, P. S. Nobel, C. B. Osmond and H. Ziegler. Springer-Verlag, Berlin, pp. 11–40.

Cancino, J. M., Munoz, J., Munoz, M., and Orellana, M. C., 1987: Effects of the bryozoan *Membranipora tuberculata* (Bosc.) on the photosynthesis and growth of *Gelidium rex* Santelices et Abbott. J. Exp. Mar. Biol. Ecol. 113, 105–112.

Cande, S. C., and Mutter, J. C., 1982: A revised identification of the oldest sea-floor spreading anomalies between Australia and Antarctica. Earth Planet. Sci. Lett. 58, 151–160.

Cande, S. C., LaBrecque, J. L., and Haxby, W. F., 1988: Plate kinematics of the South Atlantic: chron 34 to present. J. Geophys. Res. 93 (B11), 13,479–13,492.

Caram, B., and Jónsson, S., 1972: Nouvel inventaire des algues marines de l'Islande. Acta Bot. Isl. 1, 5–31.

Cardinal, A., 1967: Inventaire des algues marines benthiques de la Baie des Chaleurs et de la Baie de Gaspé (Quebec). I. Phéophycees. II. Chlorophycées. III. Rhodophycées. Nat. Can. (Que.) 94, 233–271, 447–469, 735–760.

Cardinal, A., and Villalard, M., 1971: Inventaire des algues marines benthiques de l'estuaire du Saint-Laurent (Québec). Nat. Can. (Que). 98, 887–904.

Carefoot, T., 1977: Pacific seashores. A guide to intertidal ecology. University of Washington Press, Seattle; 208 pp.

Carlquist, S., 1974: Island Biology. Columbia University Press, New York; 660 pp.

Carlton, J. T., 1985: Transoceanic and interoceanic dispersal of coastal marine organisms: the biology of ballast water. Oceanogr. Mar. Biol. Annu. Rev. 23, 313–371.

Carlton, J. T., and Scanlon, J. A., 1985: Progression and dispersal of an introduced alga: *Codium fragile* ssp. *tomentosoides* (Chlorophyta) on the Atlantic coast of North America. Botanica Mar. 28, 155–165.

Carpenter, E. J., and Cox, J. L., 1974: Production of pelagic *Sargassum* and a blue-green epiphyte in the western Sargasso Sea. Limnol. Oceanogr. 19, 429–436.

Carter, A. R., and Anderson, R. J., 1985: Regrowth after experimental harvesting of the agarophyte *Gelidium pristoides* (Gelidiales: Rhodophyta) in the eastern Cape Province. S. Afr. J. Mar. Sci. 3, 111–118.

Casey, R. E., 1977: The ecology and distribution of Recent Radiolaria. In: Oceanic micropalaeontology. Vol. 2. Ed. by A. T. S. Ramsay. Academic Press, New York, pp. 809–846.

Casey, R. E., and Rawson, P. F., 1974: The boreal lower Cretaceous. Seel House Press, Liverpool; 448 pp.

Caspers, H., 1957: Black Sea and Sea of Azov. Mem. Geol. Soc. Am. 67 (1), 801–889.

Castric-Fey, A., Girard-Descatoire, A. et al., 1973: Etagement des algues et des invertébrés sessiles dans l'Archipel de Glenan. Définitions biologiques des horizons bathymétriques. Helgoländer wiss. Meeresunters. 24, 490–509.

Chalon, J., 1905: Liste des algues marines observées jusqu'à ce jour entre l'Embouchure de l'Escaut et La Corogne incl. Iles Anglo-Normandes. Buschmann, Anvers; 258 pp.

Chaloupka, M. Y., and Hall, D. N., 1984: An examination of species dispersion patterns along the intertidal gradient on Macquarie Island (Sub-Antarctica) using a restricted occupancy model. J. Exp. Mar. Biol. Ecol. 84, 133–143.

Chamberlain, Y., M., 1965: Marine algae of Gough Island. Bull. Br. Mus. Bot. 3, 175–232.

Chamberlain, Y., M., Holdgate, M. W., and Wace, N., 1985: The littoral ecology of Gough Island, South Atlantic Ocean. Téthys 11, 302–319.

Chapman, A. R. O., 1973a: Phenetic variability of stipe morphology in relation to season, exposure, and depth in the non-digitate complex of *Laminaria* Lamour. (Phaeophyta, Laminariales) in Nova Scotia. Phycologia 12, 53–57.

Chapman, A. R. O., 1973b: A critique of prevailing attitudes towards the control of seaweed zonation on the seashore. Botanica Mar. 16, 80–82.

Chapman, A. R. O., 1979: Biology of seaweeds. Levels of organization. Edward Arnold, London; 134 pp.

Chapman, A. R. O., 1981: Stability of sea urchin dominated barren grounds following grazing of kelp in St. Margaret's Bay, eastern Canada. Mar. Biol. 62, 307–311.

Chapman, A. R. O., 1985: Demography. In: Handbook of phycological methods. Ecological field methods: macroalgae. Ed. by M. M. Littler and D. S. Littler. University Press, Cambridge, pp. 251–268.

Chapman, A. R. O., 1986: Population and community ecology of seaweeds. Adv. Mar. Biol. 23, 1–161.

Chapman, A. R. O., 1989: Abundance of *Fucus spiralis* and ephemeral seaweeds in a high eulittoral zone: effects of grazers, canopy and substratum type. Mar. Biol. 102, 565–572.

Chapman, A. R. O., and Craigie, J. S., 1977: Seasonal growth in *Laminaria longicruris*: relations with dissolved inorganic nutrients and internal reserves of nitrogen. Mar. Biol. 40, 197–205.

Chapman, A. R. O., and Lindley, J. E., 1980: Seasonal growth of *Laminaria solidungula* in the Canadian high Arctic in relation to irradiance and dissolved nutrient concentrations: a year-round study. Mar. Biol. 57, 1–5.

Chapman, A. R. O., and Lindley, J. E., 1981: Productivity of *Laminaria solidungula* in the Canadian high Arctic. Proc. Int. Seaweed Symp. 10, 247–252.

Chapman, A. R. O., Markham, J. W., and Lüning, K., 1978: Effects of nitrate concentration on the growth and physiology of *Laminaria saccharina* (Phaeophyta) in culture. J. Phycol. 14, 195–198.

Chapman, D. J., and Tocher, R. D., 1966: Occurrence and production of carbon monoxide in some brown algae. Can. J. Bot. 44, 1438–1442.

Chapman, V. J., 1956–1969: The marine algae of New Zealand. I. Myxophyceae and Chlorophyceae. J. Linn. Soc. Lond. Bot. 55, 333–501 (1956); II. Rhodophyceae, issue 1. Bangiophycidae and Florideophycidae (Nemalionales, Bonnemaisoniales, Gelidiales). Cramer, Lehre, 113 pp. (1969); III, issue 4: Gigartinales. Cramer, Vaduz, pp. 279–509.

Chapman, V. J., 1961–1963: The marine algae of Jamaica. 1. Myxophyceae and Chlorophyceae. 2. Phaeophyceae and Rhodophyceae. Bull. Inst. Jamaica Sci. Ser. no. 12, pt. 1, 1–159 (1961); no. 12, pt. 2, 1–201 (1963).

Chapman, V. J., 1971: The marine algae of Fiji. Rev. Algol., sér. 2, 10, 164–171.

Chapman, V. J. (Ed.), 1977: Wet coastal ecosystems. Elsevier, Amsterdam; 428 pp.

Chapman, V. J., and Chapman, D. J., 1976: Life forms in the algae. Botanica Mar. 19, 65–74.

Chapman, V. J., and Chapman, D. J., 1980: Seaweeds and their uses. 3rd ed. Chapman and Hall, London; 334 pp.

Chapman, V. J., and Dromgoole, F. I., 1970: The marine algae of New Zealand, Part III, Issue 2: Florideophycidae: Rhodymeniales. Cramer, Lehre, pp. 115–154.

Chapman, V. J., and Parkinson, P. G., 1974: The marine algae of New Zealand, Part III, Issue 3: Florideophycidae: Cryptonemiales. Cramer, Lehre, 447 pp.

Chauhan, V. D., and Mairh, O. P., 1978: Report on survey of economic seaweeds resources of Saurashtra coast, India. Salt Res. Ind. 14, 21–41.

Chen, L. C.-M., Edelstein, T., Ogata, E., and McLachlan, J., 1970: The life history of *Porphyra miniata*. Can. J. Bot. 48, 385–389.

Cheney, D. P., and Dyer, J. P., 1974: Deep-water benthic algae of the Florida Middle Ground. Mar. Biol. 27, 185–190.

Cheshire, A. C., and Hallam, N. D., 1985: The environmental role of alginates in *Durvillaea potatorum* (Fucales, Phaeophyta). Phycologia 24, 147–153.

Chiang, Y., 1960–1962: Marine algae of northern Taiwan. Taiwania 7, 51–55; 8, 143–180.

Chiang, Y., 1973: Notes on marine algae of Taiwan. Taiwania 18, 13–17.

Chihara, M. 1975: Geographic distribution of marine algae in Japan. In: Advance of phycology in Japan. Ed. by J. Tokida and H. Hirose. VEB Gustav Fischer, Jena, pp. 241–253.

Choat, J. H., and Black, R., 1979: Life histories of limpets and the limpet-laminarian relationship. J. Exp. Mar. Biol. Ecol. 41, 25–50.

Choat, J. H., and Schiel, D. R., 1982: Patterns of distribution and abundance of large brown algae and invertebrate herbivores in subtidal regions of northern New Zealand. J. Exp. Mar. Biol. Ecol. 60, 129–162.

Chopin, T., Floc'h, J.-Y., 1987: Seasonal variations of growth in the red alga *Chondrus crispus* on the Atlantic French coasts. I. A new approach by fluorescence labelling. Can. J. Bot. 65, 1014–1018.

Christensen, T., 1984: Sargassotang, en ny algeslaegt i Danmark. Urt 1984 (4), 99–104.

Christensen, T., 1987: Seaweeds of the British Isles. Vol. 4. Tribophyceae (Xanthophyceae). British Museum (Natural History), London; 36 pp.

Christensen, T., Koch, C., and Thomsen, H. A., 1985: Distribution of algae in Danish salt and brackish waters. Institut for Sporeplanter, Copenhagen. Universitetsbogladen, Copenhagen; 64 pp.

Christie, A. O., and Evans, L. V., 1962: Periodicity in the liberation of gametes and zoospores of *Enteromorpha intestinalis* Link. Nature 4811, 193–194.

Churkin, M., and Trexler, J. H., 1981: Continental plates and accreted oceanic terranes in the Arctic. In: The ocean basins and margins. Vol. 5. Ed. by A. E. M. Nairn, M. Churkin, and F. G. Stehli. Plenum, New York, pp. 1–20.

Cifelli, R., 1976: Evolution of ocean climate and the record of planktonic Foraminifera. Nature 264, 431–432.

Cinelli, F., 1969: Primo contributo alla conoscenca della vegetazione algale bentonica del litorale di Livorno. Pubbl. Stn. Zool. Napoli 37, 545–566.

Cinelli, F., 1971: Alghe bentoniche di profondità raccolte alla Punta pp. Pancracio nell'isola di Ischia (Golfo di Napoli). Giorn. Bot. Ital. 105, 207–236.

Cinelli, F., 1981: Biogeography and ecology of the Sicily Channel. Proc. Int. Seaweed Symp. 10, 235–240.

Cinelli, F., 1985: On the biogeography of the benthic algae of the Mediterranean. In: Mediterranean marine ecosystems. Ed. by M. Moraitou-Apostolopoulou and V. Kiortsis. Plenum Press, New York, pp. 49–56.

Cinelli, F., and Colantoni, P., 1982: Alcune osservazioni sulla zonazione del benthos marino sulle coste rocciose delle isole Galàpagos (Oceano Pacifico). In: Galàpagos, Studi e Ricerche-Spedizione "L. Mares- G. R. S. T. S." Ed. by Gruppo Ricerche Scientifiche e Tecniche Subaquee. Museo Zoologico dell' Universita di Firenze, Florece, pp. 277–300.

Cinelli, F., Fresi, E., Mazella, L., and Ponticelli, M. P., 1979: Deep algal vegetation of the western Mediterranean. Giorn. Bot. Ital. 113, 173–188.

Clark, D. L., 1977: Paleontological response to post-Jurassic crustal plate movements in the Arctic Ocean. In: Paleontological and plate tectonics with special reference to the history of the Atlantic Ocean. Ed. by R. M. West. Milwaukee Public Museum, Special Publications in Biology and Geology, pp. 55–76.

Clark, D. L., 1982: Origin, nature, and world climate effect of Arctic Ocean ice-cover. Nature 300, 321–325.

Clark, D. L., and Kitchell, J. A., 1979: Injection events in ocean history. Nature 278, 669.

Clarke, A., 1983: Life in cold water: the physiological ecology of polar marine ectotherms. Oceanogr. Mar. Biol. Annu. Rev. 21, 341–453.

Clauss, H., 1970: Effect of red and blue light on morphogenesis and metabolism of *Acetabularia mediterranea*. In: Biology of *Acetabularia*. Ed. by J. Brachet and S. Bonotto. Academic Press, London, pp. 177–191.

Clayton, M. N., 1984: Evolution of the Phaeophyta with particular reference to the Fucales. Progr. Phycol. Res. 3, 11–46.

Clayton, M. N., 1985: A critical investigation of the vegetative anatomy, growth, and taxonomic affinities of *Adenocystis*, *Scytothamnus*, and *Splachnidium* (Phaeophyta). Br. Phycol. J. 20, 285–296.

Clayton, M. N., 1986: Culture studies on the life history of *Scytothamnus australis* and *Scytothamnus fasciculatus* (Phaeophyta) with electron microscope observations on sporogenesis and gametogenesis. Br. Phycol. J. 21, 371–386.

Clayton, M. N., 1987: Isogamy and a fucalean type of life history in the Antarctic brown alga *Ascoseira mirabilis* (Ascoseirales, Phaeophyta). Botanica Mar. 30, 447–454.

Clayton, M. N., 1988: Evolution and life histories of brown algae. Botanica Mar. 31, 379–387.

Clayton, M. N., Hallam, N. D., Luff, S. E., and Diggins, T., 1985: Cytology of the apex, thallus development, and reproductive structures of *Hormosira banksii* (Fucales, Phaeophyta). Phycologia 24, 181–190.

Clayton, M. N., Hallam, N. D., and Shankly, C. M., 1987: The seasonal pattern of conceptacle development and gamete maturation in *Durvillaea potatorum* (Durvillaeales, Phaeophyta). Phycologia 26, 35–45.

Clayton, M. N., and King, R. J., 1981: Marine botany: an Australasian perspective. Longman Cheshire, Melbourne; 478 pp.

Clayton, M. N., Shankly, C. M., 1987: The apical meristem of *Splachnidium rugosum* (Phaeophyta). J. Phycol. 23, 296–307.

Clayton, M. N., and Wiencke, C., 1986: Techniques and equipment for culturing Antarctic benthic marine algae, and for preparing specimens for electron microscopy. Ser. Cient. INACH 34, 93–97.

Clayton, M. N., and Wiencke, C., 1990: The anatomy, life history, and development of the Antarctic brown alga *Phaeurus antarcticus* (Desmarestiales, Phaeophyceae). (in press).

Clokie, J. J. P., and Boney, A. D., 1980: *Conchocelis* distribution in the Firth of Clyde: estimates of the lower limits of the photoc zone. J. Exp. Mar. Biol. Ecol. 46, 111–125.

Clokie, J. J. P., Scoffin, T. P., and Boney, A. D., 1981: Depth maxima of Conchocelis and *Phymatolithon rugulosum* on the N. W. Shelf and Rockall Plateau. Mar. Ecol. Prog. Ser. 4, 131–133.

Coachman, L. K., Aagard, K., and Tripp, R. B., 1975: Bering Strait. The Regional physical oceanography. University of Washington Press, Seattle; 172 pp.

Coleman, D. C., and Mathieson, A. C., 1975: Investigations of New England marine algae. VII.Seasonal occurrence and reproduction of marine algae near Cape Cod, Massachusetts. Rhodora 77, 76–104.

Colin, P. I., 1978: Caribbean reef invertebrates and plants. THF Publications, Neptune City, N. J.; 512 pp.

Colinvaux, P. A., 1986: Introduction to ecology. Wiley, New York; 621 pp.

Collins, F. S., 1927: Marine algae from the Bering Strait and Arctic Ocean collected by the Canadian Arctic Expedition, 1913–1918. Rep. Can. Arct. Exped. 1913–1918. Vol. 4 (Bot. B. Marine Algae): 1B-16B.

Collins, F. S., and Hervey, A. B., 1917: The algae of Bermuda. Proc. Acad. Arts Sci. 53, 1–195.

Colman, J., 1946: Marine biology in St. Helena. Proc. Zool. Soc. London 116, 266–281.

Conolly, N. J., and Drew, E. A., 1985a: Physiology of *Laminaria*. III. Effect of coastal eutrophication gradient on seasonal patterns of growth and tissue composition in *L. digitata* Lamour. and *L. saccharina* (L.) Lamour. P. S. Z. N. I. Mar. Ecol. 6, 181–195.

Conolly, N. J., and Drew, E. A., 1985b. Physiology of *Laminaria*. IV. Nutrient supply and daylength, major factors affecting growth of *L. digitata* and *L. saccharina*. P.S.Z.N.I. Mar. Ecol. 6, 299–320.

Conover, J. T., 1964: The ecology, seasonal periodicity, and distribution of benthic plants in some Texas lagoons. Botanica Mar. 7, 4–41.

Coon, D., 1982: Primary productivity of macroalgae in North Pacific America. In: CRC handbook of biosolar resources. Vol. 1, P. 2. Basic principles. Ed. by A. Mitsui and A. C. Black. CRC Press, Boca Raton; Fla., pp. 447–454.

Cooper, M. R., 1977: Eustacy during the Cretaceous: its implications and importance. Palaeogeogr. Palaeoclimatol. Palaeoecol. 22, 1–60.

Cope, E. D., 1896: Primary factors of organic evolution. Chicago, 547 pp.

Coppejans, E., 1975: A preliminary study of the marine algal communities on the islands of Milos and Sikinos (Cyclades-Greece). Bull. Soc. R. Bot. Belg. 107, 387–406.

Coppejans, E., 1982a: Zeewierengids voor de Belgische en Noordfranse kust. Deel II. Beschrijvingen Groen- en Bruinwieren. Stentor 17, 157–254.

Coppejans, E., 1982b: Zeewierengids voor de Belgische en Noordfranse kust. Deel III. Beschrijvingen Roodwieren. Stentor 18, 255–392.

Coppejans, E., 1983: Iconographie d'algues Méditerranéennes. Chlorophyta, Phaeophyta, Rhodophyta. Bibl. Phycol. 63. Cramer, Vaduz; var. pag.

Coppejans, E., and Beeckman, T., 1986: Zeewierengids voor de Belgische en Noordfranse kust. Deel IV. Addendum. Met herwerkte sleutel. Incluant des clefs d'identication et un glossaire en francais. Nationale Plantentuin van Belgie, pp. 393–568.

Coppejans, E., and van der Ben, D., 1980: Zeewierengids voor de Belgische en Noordfranse kust. Belgische Jeugdbond voor Natuurstudie, Gent; 156 pp.

Coppejans, E., and Boudouresque, C. F., 1983: Végétation marine de la Corse (Méditerranée). VI. Documents pour la flore des algues. Botanica Mar. 26, 457–470.

Cordero, P. A., 1976–1979: The marine algae of Batan Island, northern Philippines. Fish. Res. J. Philipp. 1, 3–29 (1976); 2, 19–55 (1977); 3, 13–64 (1979).

Cormaci, M., and Furnari, G., 1979: Flora algale marina della Sicilia orientale: Rhodophyceae, Phaeophyceae e Chlorophyceae. Inf. Bot. Ital. 11, 221–250.

Correa, J., Avila, M., and Santelices, B., 1985: Effects of some environmental factors on growth of sporelings in two species of *Gelidium* (Rhodophyta). Aquaculture 44, 221–227.

Cosson, J., and Olivari, R., 1982: Premiers résultats concernant les possibilités d'hybrididation interspécifiques et intergénériques chez les Laminariales des côtes de la Manche. C. R. Acad. Sci. sér. III, Sci.-Vie. 295, 381–384.

Cosson, J., 1987: Croissance des sporophytes resultant d'hybrididations interspécifiques et intergénériques chez les Laminariales. Cryptogam., Algol. 8, 61–72.

Cott, H. B., 1957: Adaptive coloration in animals. Methuen, London; 508 pp.

Cotton, A. D., 1915: Cryptogams from the Falkland Islands collected by Mrs. Vallentin. J. Linn. Soc. Lond. Bot. 43, 137–231.

Cousens, R., 1981: Variation in annual production by *Ascophyllum nodosum* (L.) Le Jolis with degree of exposure to wave action. Proc. Int. Seaweed Symp. 10, 253–258.

Coutinho, R., and Seeliger, U., 1984: The horizontal distribution of the benthic algal flora in the Patos Lagoon estuary, Brazil, in relation to salinity, substratum, and wave exposure. J. Exp. Mar. Biol. Ecol. 80, 247–257.

Coutinho, R., and Seeliger, U., 1986: Seasonal occurrence and growth of benthic algae in the Patos Lagoon estuary, Brazil. Estuar. Coast. Shelf Sci. 23, 889–900.

Coyer, J. A., and Zaugg-Haglund, A. C., 1982: A demographic study of the elk kelp *Pelagophycus porra* (Laminariales, Lessoniaceae), with notes on *Pelagophycus* x *Macrocystis* hybrids. Phycologia 21, 399–407.

Craigie, J. S., 1974: Storage products. In: Algal physiology and biochemistry. Ed. by W. D. P. Stewart. Blackwell, Oxford, pp. 206–235.

Cribb, A. B., 1973: The algae of the Great Barrier Reefs. In: Biology and geology of coral reefs. Vol. 2. Ed. by O. A. Jones and R. Endean. Academic Press, New York, pp. 47–75.

Cribb, A. B., 1981: Coral reefs. In: Marine Botany: an Australasian perspective. Ed. by M. N. Clayton and R. J. King. Longman Cheshire, Melbourne, pp. 329–345.

Crisp, D. J., and Fischer-Piette, E., 1959: Répartition des principales espèces intercôtidales de la côte atlantique française en 1954–1955. Ann. Inst. Océanogr. 36, 275–387.

Crisp, D. J., and Williams, G. B., 1960: Effect of extracts from fucoids in promoting settlement of epiphytic polyzoa. Nature, Lond. 188, 1206–1207.

Critchley, A. T., Farnham, W. F., and Morrell, S. L., 1983: A chronology of new European sites of attachment for the invasive brown alga, *Sargassum muticum*, 1973–1981. J. Mar. Biol. Ass. U. K. 63, 799–811.

Critchley, A. T., Nienhuis, P. H., and Verschuure, K., 1987: Presence and development of populations of the introduced brown alga *Sargassum muticum* in the southwest Netherlands. Hydrobiologia 151/152, 245–255.

Croizat, L., 1964. Space, time, form: the biological synthesis. Caracas; 881 pp.

Crossland, C. J., and Barnes, D. J., 1976: Acetylene reduction by coral skeletons. Limnol. Oceanogr. 21, 153–155.

Dahl, A. L., 1973*a*: Benthic ecology in a deep reef and sand habitat off Puerto Rico. Botanica Mar. 16, 171–175.

Dahl, A. L., 1973*b*: Surface area in ecological analysis: quantification of benthic coral-reef algae. Mar. Biol. 23, 239–249.

Dahl, A. L., 1979: Marine ecosystems and biotic provinces in the South Pacific area. Proc. Int. Symp. Mar. Biogeogr. Evol. S. Hemisphere. N. Z. DSIR Inform. Ser. 137 (2), 541–546.

Dana, J. D., 1853: On an isothermal oceanic chart, illustrating the geographical distribution of marine animals. Am. J. Sci., ser. 2, 16, 153–167, 314–327.

Dangeard, P., 1949: Les algues marines de la côte occidentale du Maroc. Le Botaniste 34, 89–189.

d'Antonio, C., 1985: Epiphytes on the rocky intertidal red alga *Rhodomela larix* (Turner) C. Agardh: negative effects on the host and food for herbivores? J. Exp. Mar. Biol. Ecol. 86, 197–218.

d'Antonio, C., 1986: Growth and reproduction of the red alga *Rhodomela larix*. Can. J. Bot. 64, 1499–1506.

Darlington, P. J., 1965: Biogeography of the southern end of the World. Harvard University Press, Cambridge, Mass.; 236 pp.

Davis, A. N., and Wilce, R. T., 1987*a*: Floristics, phenology, and ecology of the sublittoral marine algae in an unstable cobble habitat (Plum Cove, Cape Ann, Massachusetts, USA). Phycologia 26, 23–34.

Davis, A. N., and Wilce, R. T., 1987*b*: Algal diversity in relation to physical disturbance: a mosaic of successional stages in a subtidal cobble habitat. Mar. Ecol. Prog. Ser. 37, 229–237.

Davison, I. R., 1987: Adaptation of photosynthesis in *Laminaria saccharina* (Phaeophyta) to changes in growth temperature. J. Phycol. 23, 273–283.

Davison, I. R., and Davison, J. O., 1987: The effect of growth temperature on enzyme activities in the brown alga *Laminaria saccharina*. Br. Phycol. J. 22, 77–87.

Davison, I. R., Dudgeon, S. R., and Ruan, H.-M., 1989: Effect of freezing on seaweed photosynthesis. Mar. Ecol. Prog. Ser. 58, 123–131.

Davison, I. R., Stewart, W. D. P., 1983: Occurrence and significance of nitrogen transport in the brown alga *Laminaria digitata*. Mar. Biol. 77, 107–112.

Dawes, C. J., 1974: Marine algae of the west coast of Florida. University of Miami Press, Coral Gables, Fla.; 201 pp.

Dawes, C. J., 1981: Marine botany. Wiley, New York; 628 pp.

Dawes, C. J., Earle, S. A., and Croley, F. C., 1967: The offshore benthic flora of the southwest coast of Florida. Bull. Mar. Sci. 17, 211–231.

Dawes, C. J., Moon, R. E., Davis, M. A., 1978: The photosynthetic and respiratory rates and tolerances of benthic algae from a mangrove and salt marsh estuary: a comparative study. Estuar. Coast. Mar. Sci. 6, 175–185.

Dawson, E. Y., 1941: The marine algae of the Gulf of California. Allan Hancock Pacif. Exped. 3, 189–462.

Dawson, E. Y., 1953–1963: Marine red algae of Pacific Mexico. I-IV, VI-VIII. Allan Hancock Pacif. Exped. 17, 1–171 (1953); 17, 241–307 (1954); Pacif. Nat. 2, 1–125 (1960); 2, 191–341 (1961); Nova Hedwigia 5, 437–476 (1963); Allan Hancock Pacif. Exped. 26, 1–207 (1962); Nova Hedwigia 6, 401–481 (1963).

Dawson, E. Y., 1954: Marine plants in the vicinity of Nha Trang, Vietnam. Pacif. Sci. 8, 371–481.

Dawson, E. Y., 1957a: Notes on eastern Pacific insular marine algae. Contrib. Sci. L. A. Count. Mus. 8, 1–8.

Dawson, E. Y., 1957b: An annotated list of marine algae from Eniwetok Atoll, Marshall Islands. Pacif. Sci. 11, 92–132.

Dawson, E. Y., 1959: Some algae from Clipperton Island and the Danger Islands. Pacif. Nat. 7, 2–8.

Dawson, E. Y., 1960: A review of the ecology, distribution, and affinities of the benthic flora. Syst. Zool. 9, 93–100.

Dawson, E. Y., 1961a: A guide to the literature and distributions of Pacific benthic algae from Alaska to the Galapagos Islands. Pacif. Sci. 15, 370–461.

Dawson, E. Y., 1961b: Plantas marinas de la zona de las mareas de El Salvador. Pacif. Nat. 2, 388–461.

Dawson, E. Y., 1962a: Una clave illustrada de los géneros de algas benticas del Pacifico de la America central. Pacif. Nat. 3, 168–231.

Dawson, E. Y., 1962b: Additions to the marine flora of Costa Rica and Nicaragua. Pacif. Nat. 3, 375–395.

Dawson, E. Y., 1963: New records of marine algae from the Galápagos Islands. Pacif. Nat. 4, 3–23.

Dawson, E. Y., Acleto, C., and Foldvik, N., 1964: The seaweeds of Peru. Nova Hedwigia (Beih.) 13, 1–111.

Dawson, E. Y., and Foster, M. S., 1982: Seashore plants of California. University of California, Berkeley, 226 pp.

Dawson, E. Y., Neushul, M., and Wildman, R. D., 1960: Seaweeds associated with kelp beds along southern California and northwestern Mexico. Pacif. Nat. 1 (14), 1–90.

Dawson, M. R., West, R. M., Langston, W., and Hutchison, J. H., 1976: Paleogene terrestrial vertebrates: northernmost occurrence, Ellesmere Island, Canada. Science 192, 781–782.

Dayton, P. K., 1971: Competition, disturbance, and community organization: The provision and subsequent utilization of space in a rock intertidal community. Ecol. Monogr. 41, 351–389.

Dayton, P. K., 1973: Dispersion, dispersal, and persistence of the annual intertidal alga, *Postelsia palmaeformis* Ruprecht. Ecology 54, 431–438.

Dayton, P. K., 1975: Experimental evaluation of ecological dominance in a rocky intertidal algal community. Ecol. Monogr. 45, 137–159.

Dayton, P. K., 1985a: Ecology of kelp communities. Annu. Rev. Ecol. Syst. 16, 215–245.

Dayton, P. K., 1985b: The structure and regulation of some South American kelp communities. Ecol. Monogr. 55, 447–468.

Dayton, P. K., Currie, V., Gerrodette, T., Keller, B. D., Rosenthal, R., and Tresca, D. V., 1984: Patch dynamics and stability of some California kelp communities. Ecol. Monogr. 54, 253–289.

Dayton, P. K., and Tegner, M. J., 1984: Catastrophic storms, El Niño, and patch stability in a southern California kelp community. Science 224, 282–285.

Deacon, G. E. R., 1964: antarctic oceanography: The physical environment. In: Biologie Antarctique. Ed. by R. Carrick, M. Holdgate, and J. Prévost. Hermann, Paris, pp. 81–86.

Dean, T. A., Jacobsen, F. R., Thies, K., and Lagos, S. L., 1988: Differential effects of grazing by white sea urchins on recruitment of brown algae. Mar. Ecol. Prog. Ser. 48, 99–102.

DeBoer, J. A., 1981: Nutrients. In: The biology of seaweeds. Ed. by C. S. Lobban and M. J. Wynne. Blackwell, Oxford, pp. 356–392.

De Burgh, M. E., and Fankboner, P. V., 1978: A nutritional association between the bull kelp *Nereocystis luetkeana* and its bryozoan *Membranipora membranacea*. Oikos 31, 69–72.

DeCew, T. C., and West, J. A., 1981: Investigations on the life histories of three *Farlowia* species (Rhodophyta: Crytonemiales, Dumontiaceae) from Pacific North America. Phycologia 20, 342–351.

Deichman, H., and Rosenvinge, L. K., 1908: Note sur la limite supérieure des Fucacées et sur le bord de glace ("isfod") sur les côtes du Groenland. Bot. Tidsskr. 28, 182–184.

DeLaca, T. E., Lipps, J. H., 1976: Shallow-water marine associations, Antarctic Peninsula. Antarct. J. 11, 12–20.

De la Pylaie, A. J. M. B., 1829: Flore de Terre-Neuve et des Iles Saint Pierre et Miclon. Paris; 128 pp.

Delépine, R., 1959: Observations sur quelques *Codium* (Chlorophycees) des côtes françaises. Rev. Gén. Bot. 66, 1–29.

Delépine, R., 1963: Un aspect des études de biologie marine dans les Îles Australes Françaises. Com. Nat. Fr. Rech. Antarct. 3, 1–22.

Delépine, R., 1966: La végétation marine dans l'Antarctique de l'Ouest comparée à celle des Îles Australes Françaises. Conséquences biogéographiques. C. R. Séanc. Biogéogr. Paris, 374, 52–68.

Delépine, R., 1967: Sur un nouveau genre de Chlorophycées antarctiques, *Lambia*. C. R. Acad. Sci. Paris 264, 1410–1413.

Delépine, R., and Asensi, A. 1976: Quelques données experimentales sur l'écophysiologie de *Durvillaea antarctica* (Cham.) Hariot (Phéophycees). Bull. Soc. Phycol. Fr. 21, 65–80.

Delépine, R., Lamb, I. M., and Zimmermann, M. H., 1966: Preliminary report on the marine vegetation of the Antarctic Peninsula. Proc. Int. Seaweed Symp. 5, 107–116.

Dell, R. K., 1972: Antarctic benthos. Adv. Mar. Biol. 10, 1–216.

Dellow, V., 1955: Marine algal ecology of the Hauraki Gulf. Trans. R. Soc. N. Z. 83, 1–91.

De May, D., John, D. M., and Lawson, G. W., 1977: A contribution to the littoral ecology of Liberia. Botanica Mar. 20, 41–46.

Denley, E. J., Dayton, P. K., 1985: Competition among macroalgae. In: Handbook of phycological methods. Ecological field methods: macroalgae. Ed. by M. M. Littler and D. S. Littler. University Press, Cambridge, pp. 511–530.

Denton, G. H., and Hughes, T. J., 1981: The last great ice sheets. Wiley, New York; 484 pp.

Descolas-Gros, C., and de Billy, G., 1987: Temperature adaptation of RuBP carboxylase: kinetic properties in marine Antarctic diatoms. J. Exp. Mar. Biol. Ecol. 108, 147–158.

Dethier, M. N., 1987: The distribution and reproductive phenology of intertidal fleshy crustose algae in Washington. Can. J. Bot. 65, 1838–1850.

Devinny, J. S., 1978: Ordination of seaweed communities: environmental gradients at Punta Bunda, Mexico. Botanica Mar. 21, 357–363.

Devinny, J. S., and Kirkwood, P. D., 1974: Algae associated with kelp beds of the Monterey Peninsula, California. Botanica Mar. 17, 100–106.

De Wreede, R. E., 1984: Growth and age class distribution of *Pterygophora californica* (Phaeophyta). Mar. Ecol. Prog. Ser. 19, 93–100.

De Wreede, R. E., 1986: Demographic characteristics of *Pterygophora californica* (Laminariales, Phaeophyta). Phycologia 25, 11–17.

De Wreede, R. E., and Jones, E. C., 1973: New records of *Sargassum hawaiiensis* Doty and Newhouse (Sargassaceae, Phaeophyta), a deep water species. Phycologia 12, 59–62.

Deysher, L., and Norton, T. A., 1982: Dispersal and colonization in *Sargassum muticum* (Yendo) Fensholt. J. Exp. Mar. Biol. Ecol. 56, 179–195.

Diannelidis, T., Haritonidis, S., and Tsekos, I., 1977: Contribution à l'étude des peuplements des algues benthiques de quelques régions de l'île de Rhodos, Grèce. Botanica Mar. 20, 205–226.

Diapoulis, A., and Haritonidis, S., 1984: Marine algae of the Lesvos Island, Greece. I. Systematic identification and geographical distribution. Thalassographica 7, 95–107.

Diaz-Piferrer, M., 1969: Distribution of the marine benthic flora of the Caribbean Sea. Caribb. J. Sci. 9, 151–178.

Diaz-Piferrer, M., 1981: The genus *Sargassum* in western Atlantic: a biogeographical approach. Proc. Int. Seaweed Symp. 8, 307–312.

tom Dieck, I., 1987: Temperature tolerance and daylength effects in isolates of

Scytosiphon lomentaria (Phaeophyceae) of the North Atlantic and Pacific Ocean. Helgoländer Meeresunters. 41, 307–321.

tom Dieck, I., 1989: Vergleichende Untersuchungen zur Ökophysiologie und Kreuzbarkeit innerhalb der digitaten Sektion der Gattung *Laminaria* (Phaeophyceae). Dissertation; Universität Hamburg; 168 pp.

Dieckmann, G. S., 1980: Aspects of the ecology of *Laminaria pallida* (Grev.) J. Ag. off the Cape Peninsula (South Africa). Botanica Mar. 23, 579–585.

Dietrich, G., Kalle, K., Krauss, W., and Siedler, G., 1980: General oceanography. An introduction. Second edition. Wiley, New York; 626 pp.

Dietrich, G., and Köster, R., 1974a: Geschichte der Ostsee. In: Meereskunde der Ostsee. Ed. by L. Magaard and G. Rheinheimer. Springer-Verlag, Heidelberg, pp. 5–10.

Dietrich, G., and Köster, R., 1974b: Bodengestalt und Bedeckung. In: Meereskunde der Ostsee. Ed. by L. Magaard and G. Rheinheimer. Springer-Verlag, Heidelberg, pp. 11–18.

Dijkema, K. S., and Wolff, W. J. (Eds.), 1983: Flora and vegetation of the Wadden Sea islands and coastal areas. Final report of the section "Flora and vegetation of the islands" of the Wadden Sea Working Group. Balkema, Rotterdam; var. pag.

Dixon, P. S., and Irvine, L. M., 1977: Seaweeds of the British Isles. Vol. 1. Rhodophyta. P. 1. Introduction, Nemaliales, Gigartinales. British Museum (Natural Hististory), London; 252 pp.

Dizerbo, A. H., 1970: Note sur la végétation marine du Cap Frehel en Plévenon (Côtes-du-Nord). Bull. Soc. Sci. Bretagne 45, 171–176.

Donze, M., 1968: The algal vegetation of the Ria de Arosa (N. W. Spain). Blumea 16, 159–183.

Doty, M. S., 1946: Critical tide factors that are correlated with the vertical distribution of marine algae and other organisms along the Pacific coast. Ecology 27, 315–328.

Doty, M. S., 1947: The marine algae of Oregon. Pt. I. Chlorophyta and Phaeophyta. Pt. II. Rhodophyta. Farlowia 3, 1–65, 159–215.

Doty, M. S., 1971: Measurement of water movement in reference to benthic algal growth. Botanica Mar. 14, 32–35.

Doty, M. S., 1979: Status of marine agronomy, with special reference to the tropics. Proc. Int. Seaweed Symp. 9, 35–58.

Doty, M. S., Gilbert, W. J., and Abbott, I. A., 1974: Hawaiian marine algae from seaward of the algal ridge. Phycologia 13, 345–357.

Drach, P., 1949: Premières recherches en scaphandre autonome sur les formations de Laminaires en zone littorale profonde. C. R. Somm. Séances Soc. Biogéogr. 227, 46–49.

Drew, E. A., 1972: Growth of a kelp forest at 60 metres in the Straits of Messina. Mem. Biol. Mar. Oceanogr. 2, 135–157.

Drew, E. A., 1977: The physiology of photosynthesis and respiration in some antarctic marine algae. Br. Antarct. Surv. Bull. 46, 59–76.

Drew, E. A., 1983: Light. In: Sublittoral ecology. The ecology of the shallow sublittoral benthos. Ed. by R. Earll and D. G. Erwin. Clarendon Press, Oxford, pp. 1–57.

Drew, E. A., 1986: *Halimeda*-the sand producing alga. Oceanus 29, 45.
Dring, M. J., 1967a. Effects of daylength on growth and reproduction of the Conchocelis-phase of *Porphyra tenera*. J. Mar. Biol. Ass. U. K. 47, 501–510.
Dring, M. J., 1967b. Phytochrome in red alga, *Porphyra tenera*. Nature 215, 1411–1412.
Dring, M. J., 1981: Chromatic adaptation of photosynthesis in benthic marine algae: an examination of its ecological significance using a theoretical model. Limnol. Oceanogr. 26, 271–284.
Dring, M. J., 1982: The biology of marine plants. Edward Arnold, London; 199 pp.
Dring, M. J., 1984: Photoperiodism and phycology. Prog. Phycol. Res. 3, 159–162.
Dring, M. J., 1986: Pigment composition and photosynthetic action spectra of *Laminaria* (Phaeophyta) grown in different light qualities and irradiances. Br. Phycol. J. 21, 199–207.
Dring, M. J., 1987a: Marine plants and blue light. In: Blue light phenomena and occurrence in plants and microorganisms. Ed. by H. Senger. CRC Press, Boca Raton, Fla., pp. 121–140.
Dring, M. J., 1987b: Stimulation of light-saturated photosynthesis in brown algae by blue light. Br. Phycol. J. 22, 302.
Dring, M. J., 1987c: Light climate in intertidal and subtidal zones in relation to photosynthesis and growth of benthic algae: a theoretical model. In: Plant life in aquatic and amphibious habitats. Ed. by R. M. M. Crawford. Blackwell, Oxford, pp. 23–34.
Dring, M. J., 1988: Photocontrol of development in algae. Annu. Rev. Plant Physiol. Plant Mol. Biol. 39, 157–174.
Dring, M. J., 1989: Stimulation of light-saturated photosynthesis in *Laminaria* (Phaeophyta) by blue light. J. Phycol. 25, 254–258.
Dring, M. J., and Brown, F. A., 1982: Photosynthesis of intertidal brown algae during and after periods of emersion: a renewed search for physiological causes of zonation. Mar. Ecol. Prog. Ser. 8, 301–308.
Dring, M. J., and Lüning, K., 1975a: A photoperiodic response mediated by blue light in the brown alga *Scytosiphon lomentaria*. Planta 125, 25–32.
Dring, M. J., and Lüning, K., 1975b: Induction of two-dimensional growth and hair formation by blue light in the brown alga *Scytosiphon lomentaria*. Z. Pflanzenphysiol. 75, 107–117.
Dring, M. J., and Lüning, K., 1983: Photomorphogenesis of marine macroalgae. In: Encyclopedia of Plant Physiology, n. s., Vol. 16B. Ed. by W. Shropshire and H. Mohr. Springer-Verlag, Berlin, pp. 545–568.
Dring, M. J., and West, J. A., 1983: Photoperiodic control of tetrasporangium formation in the red alga *Rhodochorton purpureum*. Planta 159, 143–150.
Dromgoole, F. I., 1980: Desiccation resistance of intertidal and subtidal algae. Botanica Mar. 23, 149–159.
Dromgoole, F. I., 1981: Form and function of the pneumatocysts of marine algae. I. Variations in the the pressure and composition of internal gases. Botanica Mar. 24, 257–266.
Druehl, L. D., 1968: Taxonomy and distribution of northeast Pacific species of *Laminaria*. Can J. Bot. 46, 539–547.

Druehl, L. D., 1970: The pattern of Laminariales distribution in the northeast Pacific. Phycologia 9, 237–247.
Druehl, L. D., 1973: Marine transplantations. Science 179, 231.
Druehl, L. D., 1979: On the taxonomy of California *Laminaria* (Phaeophyta) J. Phycol. 15, 337–338.
Druehl, L. D., 1981: The distribution of Laminariales in the North Pacific with reference to environmental influences. In: Evolution today. Proceedings of the Second International Congress of Systematic and Evolutionary Biology. Ed. by G. G. E. Scudder and J. L. Reveal. Hunt Institute for Botanical Documentation, Carnegie-Mellon University, Pittsburgh, pp. 55–67.
Druehl, L. D., 1988: Cultivated edible kelp. In: Algae and human affairs. Ed. by C. A. Lembi and J. R. Waaland. University Press, Cambridge; pp. 119–134.
Druehl, L. D., Baird, R., Linwall, A., Lloyd, K. E., and Pakula, S., 1988a: Longline cultivation of some Laminariaceae in British Columbia, Canada. Aquacult. Fish. Managem. 19, 253–263.
Druehl, L. D., Cabot, E. L., and Lloyd, K. E., 1987: Seasonal growth of *Laminaria groenlandica* as a function of plant age. Can. J. Bot. 65, 1599–1604.
Druehl, L. D., Cackette, M., and d'Auria, J. M., 1988b: Geographical and temporal distribution of iodine-131 in the brown seaweed *Fucus* subsequent to the Chernobyl incident. Mar. Biol. 98, 125–129.
Druehl, L., Foottit, R. G., and Masuda, M., 1988c: Morphological affinities of Japanese species of *Laminaria* (Phaeophyta). Phycologia 27, 405–412.
Druehl, L. D., and Green, J. M., 1982: Vertical distribution of intertidal seaweeds as related to patterns of submersion and emersion. Mar. Ecol. Prog. Ser. 9, 163–170.
Druehl, L. D., and Wheeler, W. N., 1986: Population biology of *Macrocystis integrifolia* from British Columbia, Canada. Mar. Biol. 90, 173–179.
Ducker, S. C., 1967: The genus *Chlorodesmis* (Chloroyphyta) in the Indo-Pacific region. Nova Hedwigia 13, 145–182.
Dudgeon, S. R., Davison, I. R., and Vadas, R. L., 1989: Effect of freezing on photosynthesis of intertidal macroalgae: tolerance of *Chondrus crispus* and *Mastocarpus stellatus* (Rhodophyta). Mar. Biol. 101, 107–114.
Duggins, D. O., 1980: Kelp dominated communities: experimental studies on the relationships between sea urchins, their predators, and their algal resources. Dissertation, University of Washington, Seattle; 134 pp.
Duggins, D. O., 1983: Starfish predation and the creation of mosaic patterns in a kelp-dominated community. Ecology 64, 1610–1619.
Duggins, D. O., Simenstad, C. A., Estes, J. A., 1989: Magnification of secondary production by kelp detritus in coastal marine ecosystems. Science 245, 170–173.
Dunbar, M. J., 1951: Eastern arctic waters. A summary of our present knowledge of the physical oceanography of the arctic area, from Hudson Bay to Cape Farewell and from Belle Isle to Smith Sound. Bull. Fish. Res. Bd Can. 88, 131 pp.
Duncan, M. J., 1973: In situ studies of growth and pigmentation of the phaeophycean *Nereocystis luetkeana*. Helgoländer Meeresunters. 24, 510–525.
Dunlap, W. C., Chalker, B. E., and Oliver, J. K., 1986: Bathymetric adaptations of reef-building corals at Davies Reef, Great Barrier Reef, Australia. III. UV-B absorbing compounds. J. Exp. Mar. Biol. Ecol. 104, 239–248.

Dunton, K. H., 1985: Growth of dark-exposed *Laminaria saccharina* (L.) Lamour. and *Laminaria solidungula* J. Ag. (Laminariales: Phaeophyta) in the Alaskan Beaufort Sea. J. Exp. Mar. Biol. Ecol. 94, 181–189.

Dunton, K. H., Reimnitz, E., and Schonberg, S., 1982: An arctic kelp community in the Alaskan Beaufort Sea. Arctic 35, 465–484.

Dunton, K. H., and Schell, D. M., 1986: Seasonal carbon budget and growth of *Laminaria solidungula* in the Alaskan high Arctic. Mar. Ecol. Prog. Ser. 31, 57–66.

Dunton, K. H., and Schell, D. M., 1987: Dependence of consumers on macroalgal (*Laminaria solidungula*) carbon in an arctic kelp community: δ^{13} evidence. Mar. Biol. 93, 615–625.

Durairatnam, M., 1961: Contribution to the study of the marine algae of Ceylon. Fish. Res. Sta. Dept. Fish. Ceylon Bull. 10, 1–181.

Durako, M. J., and Dawes, C. J., 1980: A comparatative study of two populations of *Hypnea musciformis* from the east and west coasts of Florida, USA. I. Growth and chemistry. Mar. Biol. 59, 151–156.

Durham, J. W., and MacNeil, F. S., 1967: Cenozoic migrations of marine invertebrates through the Bering Strait region. In: The Bering land bridge. Ed. by D. M. Hopkins. University Press, Stanford, Cal., pp. 312–325.

Earle, S. A., 1969: Phaeophyta of the eastern Gulf of Mexico. Phycologia 7, 71–254.

Earle, S., 1972: A review of the marine plants of Panama. Bull. Biol. Soc. Wash. 2, 60–87.

Earll, R., and Farnham, W., 1983: Biogeography. In: Sublittoral ecology. The ecology of the shallow sublittoral benthos. Ed. by R. Earll and D. G. Erwin. Clarendon Press, Oxford, pp. 165–208.

Ebeling, A. W., Laur, D. R., and Rowley, R. J., 1985: Severe storm disturbances and reversal of community structure in a southern California kelp forest. Mar. Biol. 84, 287–294.

Edelstein, T., 1964: On the sublittoral algae of the Haifa Bay. Vie Milieu 15, 177–212.

Edelstein, T., Chen, L., and McLachlan, J., 1970: Investigations of the marine algae of Nova Scotia. VIII. The flora of Digby Neck Peninsula, Bay of Fundy. Can. J. Bot. 48, 621–629.

Edgar, G. J., 1984: General features of the ecology and biogeography of Tasmanian subtidal rocky shore communities. Pap. Proc. R. Soc. Tasman. 118, 173–186.

Edgar, G. J., 1986: Biogeographical processes in the Southern Hemisphere marine environment. Actas Segundo Congreso Nacional Sobre Algas Marinas chilénas. Ed. by R. Westermeier. Universidad Austral de Chile, pp. 29–46.

Edgar, G. J., 1987: Dispersal of faunal and floral propagules associated with drifting *Macrocystis pyrifera* plants. Mar. Biol. 95, 599–610.

Edwards, D. M., 1987: Organic solute accumulation in osmotically-stressed *Enteromorpha intestinalis*. Mar. Biol. 95, 583–592.

Edwards, P., 1969: Field and cultural studies on the seasonal periodicity of growth and reproduction of selected Texas benthic marine algae. Contrib. mar. Sci. 14, 59–114.

Edwards, P., 1970: Illustrated guide to the seaweeds and sea grasses in the vicinity of Port Aransas, Texas. Contrib. Mar. Sci. 15 (Suppl.), 1–128.

Edwards, P., Boromthanarat, S., and Tam, D. M., 1982: Seaweeds of economic importance in Thailand. Part 1. Field survey, Thai government statistics and future prospects. Botanica Mar. 25, 237–246.

Edwards, P., and Kapraun, D. F., 1973: Benthic marine algal ecology in the Port Aransas area. Contrib. Mar. Sci. 17, 15–52.

Edwards, R. J., 1979: Tasman and Coral Sea ten year mean temperature and salinity fields, 1967–1976. Commonwealth Scientific and Industrial Research Organization. Division of Fisheries and Oceanography. Repr. no. 46, Marine Laboratory Cronulla, Sydney; pag. var.

Edyvean, R. G. J., and Bailey, I. F., 1984: Heavy metal accumulation by seaweeds in two estuaries on the north-east coast of England. Br. Phycol. J. 19, 193.

Edyvean, R. G. J., and Ford, H., 1987: Growth rates of *Lithophyllum incrustans* (Corallinales, Rhodophyta) from South West Wales. Br. Phycol. J. 22, 139–146.

Edyvean, R. G. J., Terry, L. A., and Picken, G. B., 1985: Marine fouling and its effects on offshore structures in the North Sea-a review. Int. J. Biodeterior. 21, 277–284.

Egan, B., and Yarish, C., 1988: The distribution of the genus *Laminaria* (Phaeophyta) at its southern limit in the western Atlantic Ocean. Botanica Mar. 31, 155–161.

Egan, B., Vlasto, A., Yarish, C., 1989: Seasonal acclimation to temperature and light in *Laminaria longicruris* de la Pyl. (Phaeophyta). J. Exp. Mar. Biol. Ecol. 129, 1–16.

Egerod, L., 1974: Report of the marine algae collected on the fifth Thai- Danish expedition of 1966. Chlorophyceae and Phaeophyceae. Botanica Mar. 17, 130–157.

Ehrke, G., 1931: Über die Wirkung der Temperatur und des Lichtes auf die Atmung und Assimilation einiger Meeres- und Süsswasseralgen. Planta 13, 221–310.

Einarsson,T., Hopkins, D. M., and Doell, R. D., 1967: The stratigraphy of Tjörnes, northern Iceland, and the history of the Bering Land Bridge. In: The Bering Land Bridge. Ed. by D. M. Hopkins. University Press, Stanford, Cal., pp. 312–325.

Eiseman, N. J., 1978: Observations on the marine algae occurring from 30–100 meter depths on the east coast of Florida. J. Phycol. 14 (Suppl.), 25.

Eiseman, N. J., 1979: Marine algae of the east Florida continental shelf. I. Some new records of Rhodophyta, including *Scinaia incrassata* n. sp. (Nemaliales: Chaetangiaceae). Phycologia 18, 355–361.

Eiseman, N. J., and Earle, S. A., 1983: *Johnson-sea-linkia profunda*, a new genus and species of deep-water Chlorophyta from the Bahama Islands. Phycologia 22, 1–6.

Ekman, S., 1953: Zoogeography of the sea. Sidgwick & Jackson, London; 417 pp.

Eldholm, O., and Thiede, J., 1980: Cenozoic continental separation between Europe and Greenland. Palaeogeogr. Palaeoclimatol. Palaeoecol. 30, 243–259.

Ellis, D. V., and Wilce, R. T., 1961: Arctic and subarctic examples of intertidal zonation. Arctic 14, 224–235.

Emiliani, C., 1961: The temperature decrease of surface sea-water in high latitudes and of abyssal-hadal water in open oceanic basins during the past 75 million years. Deep-Sea Res. 8, 144–147.

Engelmann, T. W., 1883: Farbe und Assimilation. Bot. Ztg. 41, 1–13, 17–29.

Engelmann, T. W., 1884: Untersuchungen über die quantitativen Beziehungen zwischen Absorption des Lichtes und Assimilation in Pflanzenzellen. Bot. Ztg. 42, 81–110.

Engler, A., 1879–1882: Versuch einer Entwicklungsgeschichte der Pflanzenwelt seit der Tertiärperiode. Vols. 1 and 2. W. Engelmann, Leipzig; 202 pp. (vol. 1); 386 pp. (vol. 2).

Ercegovic, A., 1948: Sur quelques algues phéophycées peu connues ou nouvelles récoltées dans le bassin de l'Adriatique moyen. Acta Adriat. 3, 3–33.

Ercegovic, A., 1952: Fauna et flora adriatica. Vol. 3. Sur les Cystoseira adriatiques. Split, 212 pp.

Ercegovic, A., 1957*a*: La flore sous-marine de l'îlot de Jabuca. Acta Adriat. 8, 3–130.

Ercegovic, A., 1957*b*: Principes et essai d'un classement des étages benthiques. Recl. Trav. Stn. Mar. Endoume Fac. Sci. Mars. 22, 17–21.

Ercegovic, A., 1959: Les facteurs de sélection et d'isolement dans la genèse de quelques espèces d'algues adriatiques. Int. Rev. ges. Hydrobiol. Hydrogr. 44, 473–483.

Ercegovic, A., 1960: La végétation des algues sur les fonds pêchereux de l'Adriatique. Inst. Oceanogr. Ribar. Split. Izvjesca 6 (4), 1–32.

Ernst, J., 1955: Sur la végétation sous-marine de la Manche d'après des observations en scaphandre autonome. C. R. Acad. Sci. Paris 241, 1066–1068.

Ernst, J., 1959: Studien über die Seichtwasser-Vegetation der Sorrentiner Küste. Pubbl. Stn. Zool. Napoli, Suppl., 30, 470–518.

Estes, J. A., and Steinberg, P. D., 1988: Predation, herbivory, and kelp evolution. Paleobiology 14, 19–36.

Estes, R., and Hutchison, J. H., 1980: Eocene lower vertebrates from Ellesmere Island, Canadian Archipelago. Palaeogeogr. Palaeoclimatol. Palaeoecol. 30, 325–347.

Eston, V. R. de, Migotto, A. E., Oliveira Filho, E. C. de, Rodrigues, A., and Freitas, J. C. de, 1986: Vertical distribution of benthic marine organisms on rocky coasts of the Fernando de Noronha Archipelago (Brazil). Bolm Inst. oceanogr., Sao Paulo, 34, 37–53.

Etcheverry, D. H., 1960: Algas marinas de las islas oceanicas chilenas. (Juan Fernandez, San Felix, San Ambrosio, Pascua). Rev. Biol. Mar. 10, 83–132.

Etcheverry, D. H., 1986: Algas marinas benthonicas de Chile. UNESCO, Montevideo; 397 pp.

Evans, G. C., 1972: The quantitative analysis of plant growth. Blackwell, Oxford; 734 pp.

Evans, L. V., Callow, J. A., and Callow, M. E., 1978: Parasitic red algae: an appraisal. In: Modern approaches to the taxonomy of red and brown algae. Systematics Association Special Volume no. 10. Ed. by D. E. G. Irvine and J. H. Price. Academic Press, London; pp. 87–110.

Evans, L. V., Callow, J. A., and Callow, M. E., 1982: The biology and biochemistry of reproduction and early development in *Fucus*. Prog. Phycol. Res. 1, 67–110.

Evans, R. G., 1957: The intertidal ecology of some localities on the Atlantic coast of France. J. Ecol. 45, 245–271.

Fain, S. R., and Druehl, L. D., 1989: Chloroplast DNA evolution in the chromophytic alga *Macrocystis*. (Manuscript).

Fain, S. R., Druehl, L. D., and Baillie, D. L., 1988: Repeat and single copy sequences are differentially conserved in the evolution of kelp chloroplast DNA. J. Phycol. 24, 292–302.

Fain, S. R., and Murray, S. N., 1982: Effects of light and temperature on net photosynthesis and dark respiration of gametophytes and embryonic sporophytes of *Macrocystis pyrifera*. J. Phycol. 18, 92–98.

Fairbridge, R. W., 1973: Glaciation and plate migration. In: Implications of continental drift to the earth sciences. Vol. 1. Ed. by D. H. Tarling and S. K. Runcorn. Academic Press, London, pp. 503–515.

Falkenberg, P., 1878: Die Meeres-Algen des Golfes von Neapel. Mitt. Zool. Stn. Neapel 1 (3), 218–277.

Faller, A. J., and Woodcock, A. H., 1964: The spacing of windrows of *Sargassum* in the ocean. J. Mar. Res. 22, 22–29.

Farnham, W. F., 1980: Studies on aliens in the marine flora of southern England. In: The shore environment. Vol. 2. Ecosystems. Ed. by J. H. Price, D. E. G. Irvine and W. F. Farnham. Academic Press, London, pp. 875–914.

Farnham, W. F., and Bishop, G. M., 1985: Survey of the Fal Estuary, Cornwall. Progr. Underwater Sci. 10, 53–63.

Farnham, W. F., Fletcher, R. L., and Irvine, L. M., 1973: Attached *Sargassum* found in Britain. Nature, 243, 231–232.

Farnham, W. F., and Lambert, G., 1981: Preliminary observations on the benthic marine algae of Natal, South Africa. Phycologia 20, 103.

Feldmann, J., 1931: Contribution à la flore algologique marine de l'Algérie: Les algues de Cherchell. Bull. Soc. Hist. Nat. Afr. N. 22, 179–254.

Feldmann, J., 1934: Les Laminariacées de la Méditerranée et leur repartition géographique. Bull. Trav. Stn. Aquic. Pêche Castiglione 2, 3–42.

Feldmann, J., 1937a: Les algues marines de la côte des Albères. I-III. Cyanophycées, Chlorophycées, Pheophycées. Rev. Algol. 9, 141–355.

Feldmann, J., 1937b: Recherches sur la végétation marine de la Mediterranée. La côte des Albères. Rev. Algol. 10, 1–339.

Feldmann, J., 1937c: Sur une nouvelle espèce de Laminariacée de Mauritanie, *Ecklonia muratii* nov. sp. Bull. Soc. Hist. Nat. Afr. N. 28, 325–327.

Feldmann, J., 1937–1947: Additions à la flore des algues marines de l'Algérie. Bull. Soc. Hist. Nat. Afr. N. 28, 318–321; 30, 453–464; 33, 230–245; 38, 80–91.

Feldmann, J., 1939–1942: Les algues marines de la côte des Albères. IV. Rhodophycées. Rev. Algol. 11, 247–330; 12, 77–100; Trav. Algol. 1, 29–113.

Feldmann, J., 1943: Contribution à l'étude de la flore marine de profondeur sur les côtes d'Algérie. Bull. Soc. Hist. Nat. Afr. N. 34, 150–167.

Feldmann, J., 1946: La flore marine des îles Atlantides. Mem. Soc. Biogéogr. 8, 395–435.

Feldmann, J., 1951: La flore marine de l'Afrique du nord. C. R. Somm. Séances Soc. Biogéogr. 243, 103–108.

Feldmann, J., 1954: Inventaire de la flore marine de Roscoff. Trav. Stn. Biol. Roscoff 6 (Suppl.), 1–152.

Feldmann, J., 1955: La zonation des algues sur la côte atlantique du Maroc. Soc. Sci. Nat. Phys. Maroc 35, 9–17.

Feldmann, J., 1958: Origine et affinités du peuplement végétal benthique de la Méditerranée. Rapp. P.-v. Reun. Cons. Perm. Int. Explor. Mer N. S. 14, 515–518.

Feldmann, J., and Feldmann, G., 1942: Recherches sur les Bonnemaisoniacées et leur alternance de générations. Ann. Sci. Nat., ser. 11, Botan. 3, 75–175.

Feldmann, J., and Feldmann, G., 1958: Recherches sur quelques Floridées parasites. Rev. Gen. Bot. 65, 49–124.

Feldmann, J., and Magne, F., 1964: Additions à l'inventaire de la flore marine de Roscoff. Algues, champignons, lichens. Editions de la Station Biologique de Roscoff, 28 pp.

Fernandez, C., and Niell, F. X., 1982: Zonación del fitobentos intermareal de la región de Cabo Penas (Asturias). Invest. Pesq. 46, 121–141.

Field, J. G., Griffiths, C. L., Jarman, N., Zoutendyk, P., Velimirov, B., and Bowes, A., 1980: Variation in structure and biomass of kelp communities along the southwest Cape coast. Trans. R. Soc. S. Afr. 44, 145–203.

Filion-Myklebust, C., and Norton, T. A., 1981: Epidermis shedding in the brown seaweed *Ascophyllum nodosum* (L.) Le Jolis, and its ecological significance. Mar. Biol. Lett. 2, 45–51.

Fischer-Piette, E., 1932: Répartition des principales espèces fixées sur les rochers battus des côtes et des îles de la Manche, de Lannion à Fécamp. Ann. Inst. Océanogr. 12, 105–213.

Fischer-Piette, E., 1959: Contribution à l'écologie intercôtidale du Detroit de Gibraltar. Bull. Inst. Océanogr. Monaco 1145, 1–32.

Fischer-Piette, E., 1963: La distribution des principaux organismes intercôtidaux nord-ibériques en 1954–1955. Ann. Inst. Océanogr. Monaco 40, 165–312.

Fischer-Piette, E., and Lahondère, C., 1973: Evolutions récentes de populations de Fucacées de nos côtes sud-ouest. Le Botaniste 56, 5–17.

Fleming, C. A., 1979: The geological history of New Zealand and its life. University Press, Auckland; 141 pp.

Fletcher, A., 1980: Marine and maritime lichens of rocky shores: their ecology, physiology and biological interactions. In: The shore environment. Vol. 2. Ecosystems. Ed. by J. H. Price, D. E. G. Irvine, and W. F. Farnham. Academic Press, London, pp. 789–842.

Fletcher, R. L., 1987: Seaweeds of the British Isles. Vol. 3. Fucophyceae (Phaeophyceae). Part 1. British Museum (Natural History), London; 359 pp.

Floc'h, J.-Y., 1982: Uptake of inorganic ions and their long distance transport in Fucales and Laminariales. In: Synthetic and degradative processes in marine macrophytes. Ed. by L. M. Srivastava. Walter de Gruyter, Berlin, pp. 139–166.

Flohn, H., 1984: Climate evolution in the Southern Hemisphere and the equatorial region during the Late Cenozoic. In: Late Cainozoic palaeoclimates of the Southern Hemisphere. Ed. by J. C. Vogel. Balkema, Rotterdam, pp. 5–20.

Forbes, R. B. (Ed.), 1975: Contributions to the geology of the Bering Sea Basin and adjacent regions. Geol. Soc. Am. Spec. Pap. 151, 1–213.

Foreman, R. E., 1984: Studies on *Nereocystis* growth in British Columbia, Canada. Proc. Int. Seaweed Symp. 11, 325–332.

Fork, D. C., 1963: Observations on the function of chlorophyll a and accessory pigments. In: Photosynthetic mechanisms in green plants, NAS-NRC, Publ. no. 1145, Washington, D.C., S. 352–361.

Fortes, M. D., and Lüning, K., 1980: Growth rates of North sea macroalgae in relation to temperature, irradiance and photoperiod. Helgoländer Meeresunters. 34, 15–29.

Foster, M. S., 1975: Algal succession in a *Macrocystis pyrifera* forest. Mar. Biol. 32, 313–329.

Foster, M. S., De Vogelaere, A. P., Harold, C., Pearse, J. S., and Thum, A. B., 1988: Causes of spatial and temporal patterns in rocky intertidal communities of central and northern California. Mem. Calif. Acad. Sci. 9, 1–45.

Foster, M. S., and Schiel, D. R., 1985: The ecology of giant kelp forests in California: a community profile. U. S. Fish. & Wildlife Service Biol. Rep. 85 (7.2), Washington, D.C.; 152 pp.

Frakes, L. A., 1979: Climates throughout geologic time. Elsevier, Amsterdam; 310 pp.

Fralick, R. A., and Mathieson, A. C., 1975: Physiological ecology of four *Polysiphonia* species (Rhodophyta, Ceramiales). Mar. Biol. 29, 29–36.

Franz, D. R., and Merrill, A. S., 1980: The origins and determinants of distribution of molluscan faunal groups on the shallow continental shelf of the northwest Atlantic. Malacologia 19, 227–248.

Frazer, A. W. J., Brown, M. T., and Bannister, P., 1988: The frost resistance of some littoral and sub-littoral algae from southern New Zealand. Botanica Mar. 31, 461–464.

Fredj, G., 1972: Compte rendu de plongée en SP 300 sur les fonds à *Laminaria rodriguezii* Bornet de la Pointe de Revellata (Corse). Bull. Inst. Oceanogr. Monaco 71 (1421), 1–42.

French, C. S., 1960: The chlorophylls in vivo and in vitro. In: Handbuch der Pflanzenphysiologie, vol. V, pt. 1. Ed. by W. Ruhland. Springer-Verlag, Berlin, pp. 252–297.

Fricke, H., and Meischner, D., 1985: Depth limits of Bermudan scleractinian corals: a submersible survey. Mar. Biol. 88, 175–187.

Fricke, H. W., and Schuhmacher, H., 1983: The depth limits of Red Sea stony corals: an ecophysiological problem. (A deep diving survey by submersible). P. S. Z. N. I.: Mar. Ecol. 4, 163–194.

Fries, L., 1982: Vanadium as an essential element for some marine macroalgae. Planta 154, 393–396.

Fritsch, F. E., 1959–1961: The structure and reproduction of the algae. Vols. 2 (1959) and 2 (1961). University Press, Cambridge; 791 pp. (1959); 939 pp. (1961).

Fuhrer, B., Christianson, I. G., Clayton, M. N., and Allender, B. M., 1981: Seaweeds of Australia. Reed, Sydney; 112 pp.

Fujita, K., 1978: Pre-Cenozoic tectonic evolution of northeast Siberia. J. Geol. 86, 159–172.

Fujita, R. M., Wheeler, P. A., and Edwards, R. L., 1988: Metabolic regulation of ammonium uptake by *Ulva rigida* (Chlorophyta): a compartmental analysis of the rate-limiting steps for uptake. J. Phycol. 24, 560–566.

Funahashi, S., 1973: Distribution of marine algae in the Japan Sea, with reference to the phytogeographical positions of Vladivostok and Noto Peninsula districts. J. Fac. Sci. Hokkaido Univ. Ser. V (Botany) 10, 1–31.

Funk, G., 1927: Die Algenvegetation des Golfs von Neapel. Nach neueren ökologischen Untersuchungen. Pubbl. Stn. zool. Napoli 7 (Suppl.), 1–507.

Funk, G., 1951: Konstanz und Veränderlichkeit der Algenvegetation von Neapel. Pubbl. Stn. Zool. Napoli 23, 17–51.

Funk, G., 1955: Beiträge zur Kenntnis der Meeresalgen von Neapel. Zugleich mikrophotographischer Atlas. Pubbl. Stn. Zool. Napoli 25 (Suppl.), 1–178.

Funk, G. 1957: Fruktifikationszeiten der Meeresalgen bei Neapel. Pubbl. Stn. Zool. Napoli 29, 126–138.

Furmanczyk, K., Zielinsky, K., 1982: Distribution of macroalgae groupings in shallow waters of Admiralty Bay (King George Island, South Shetland Islands, Antarctic), plotted with the help of air photograph analysis. Polish Polar Res. 3, 41–47.

Furnari, G., 1984: The benthic marine algae of Southern Italy. Floristic and geobotanic considerations. Webbia 38, 349–369.

Furuya, M., 1983: Photomorphogenesis in ferns. In: Encyclopedia of plant physiology, n. s. Vol. 16B. Ed. by W. Shropshire and H. Mohr. Springer-Verlag, Berlin, pp. 569–600.

Gabrielson, P. W., and Kraft, G. T., 1984: The marine algae of Lord Howe Island (N.S.W.): the family Solieriaceae (Gigartinales, Rhodophyta). Brunonia 7, 217–251.

Gabrielson, P. W., Scagel, R. F., and Widdowson, T. B., 1987: Keys to the benthic marine algae of British Columbia, northern Washington, and southeast Alaska. Univ. B. C. Dept. Bot. Ser. IV. Phycol. Contrib. no. 2, pp. 1–197.

Gagne, J. A., Mann, K. H., and Chapman, A. R. O., 1982: Seasonal patterns of growth and storage in *Laminaria longicruris* in relation to differing patterns of availability of nitrogen in the water. Mar. Biol. 69, 91–101.

Gaines, S. D., and Lubchenco, J., 1982: A unified approach to marine plant-herbivore interactions. II. Biogeography. Annu. Rev. Ecol. Syst. 13, 111–138.

Galbraith, R. G., and Boehler, T., 1974: Subtidal marine biology of California. Naturegraph Publishers, Healdsburg, Calif.; 128 pp.

Gantt, E., 1981: Phycobilisomes. Annu. Rev. Plant Physiol. 32, 327–347.

Garbary, D., 1976: Life-forms of algae and their distribution. Botanica Mar. 19, 97–106.

Garbary, D., 1979: A revised species concept for endophytic and endozoic members of the Acrochaetiaceae (Rhodophyta). Bot. Not. 132, 451–455.

Garbary, D., 1987: A critique of traditional approaches to seaweed distribution in light of the development of vicariance biogeography. Helgoländer Meeresunters. 41, 235–244.

Garbary, D. J., Hansen, G. I., and Scagel, R. F., 1980–1982: The marine algae of British Columbia and northern Washington: division Rhodophyta (red algae). Class Bangiophyceae. Class Florideophyceae, orders Acrochaetiales and Nemaliales. Syesis 13, 137–195; 15 (Suppl.), 1–102.

Garbary, D. J., Hansen, G. I., and Scagel, R. F., 1984: Additions to the marine algae of Barkley Sound, Vancouver Island, British Columbia. Syesis 17, 43–45.

Gartner, S., and Keany, J., 1978: The terminal Cretaceous event: a geological problem with an oceanographic solution. Geology 6, 708–712.

Garwin, L., 1988: Of impacts and volcanoes. Nature 336, 714–716.

Gates, D. M., 1979: Biophysical ecology. Springer-Verlag, Berlin; 640 pp.

Gauthier, B., Cardinal, A., and Himmelman, J. H., 1980: Limites amont de distribution des algues marines dans l'estuaire du Saint-Laurent (Québec), et addition de quelques espèces à la flore de cette région. Nat. Can. (Que). 107, 195–197.

Gayral, P., 1958: Algues de la côte atlantique marocaine. Société des sciences naturelles et physiques du Maroc, Rabat; 523 pp.

Gayral, P., 1966: Les algues des côtes françaises (Manche et Atlantique). Editions Doin, Paris; 632 pp.

Gealey, W. K., 1988: Plate tectonic evolution of the Mediterranean - Middle East region. Tectonophysics 155, 285–306.

Geider, R. J., Osborne, B. A., Raven, J. A., 1985: Light dependence of growth and photosynthesis in *Phaeodactylum tricornutum* (Bacillariophyceae). J. Phycol. 21, 609–619.

Geider, R. J., Osborne, B. A., and Raven, J. A., 1986: Growth, photosynthesis and maintenance metabolic cost in the diatom *Phaeodactylum tricornutum* at very low light levels. J. Phycol. 21, 609–619.

Geiger, R., 1965: Die Atmosphäre der Erde. Karte 1: Jährliche Sonnenstrahlung. Perthes, Darmstadt.

Geiselman, J. A., and McConnell, O. J., 1981: Polyphenols in brown algae *Fucus vesiculosus* and *Ascophyllum nodosum*: chemical defenses against the marine herbivorous snail, *Littorina littorea*. J. Chem. Ecol. 7, 1115–1133.

Gentry, A. H., 1982: Neotropical floristic diversity: phytogeographical connections between Central and South America, Pleistocene climatic fluctuations, or an accident of the Andean orogeny? Ann. Mo. Bot. Gard. 69, 557–593.

Gerard, V. A., 1982a: In situ water motion and nutrient uptake by the giant kelp *Macrocystis pyrifera*. Mar. Biol. 69, 51–54.

Gerard, V. A., 1982b: In situ rates of nitrate uptake by giant kelp, *Macrocystis pyrifera* (L.) C. Agardh: tissue differences, environmental effects, and predictions of nitrogen-limited growth. J. Exp. Mar. Biol. Ecol. 62, 211–224.

Gerard, V. A., 1982c: Growth and utilization of internal nitrogen reserves by the giant kelp *Macrocystis pyrifera* in a low-nitrogen environment. Mar. Biol. 66, 27–35.

Gerard, V. A., 1984a: The light environment in a giant kelp forest: influence of *Macrocystis pyrifera* on spatial and temporal variability. Mar. Biol. 84, 189–195.

Gerard, V. A., 1984b: Physiological effects of El Niño on giant kelp in southern California. Mar. Biol. Lett. 5, 317–322.

Gerard, V. A., 1986: Photosynthetic characteristics of giant kelp (*Macrocystis pyrifera*) determined in situ. Mar. Biol. 90, 473–482.

Gerard, V. A., 1987: Hydrodynamic streamlining of *Laminaria saccharina* Lamour. in response to mechanical stress. J. Exp. Mar. Biol. Ecol. 107, 237–244.

Gerard, V. A., 1988: Ecotypic differentiation in light-related traits of the kelp *Laminaria saccharina*. Mar. Biol. 97, 25–36.

Gerard, V. A., and Du Bois, K. R., 1988: Temperature ecotypes near the southern boundary of the kelp *Laminaria saccharina*. Mar. Biol. 97, 575–580.

Gerloff, J., and Geissler, U., 1974: Eine revidierte Liste der Meeresalgen Griechenlands. Nova Hedwigia 22, 721–793.

Germann, I., 1986: Growth phenology of *Pleurophycus gardneri* (Phaeophyceae, Laminariales), a deciduous kelp of the northeast Pacific. Can. J. Bot. 64, 2538–2547.

Germann, I., Druehl, L. D., and Hoeger, U., 1987: Seasonal variation in total and soluble tissue nitrogen of *Pleurophycus gardneri* (Phaeophyceae: Laminariales) in relation to environmental nitrate. Mar. Biol. 96, 413–423.

Gerwick, W. H., and Lang, N. J., 1977: Structural, chemical, and ecological studies on iridescence in *Iridaea* (Rhodophyta). J. Phycol. 13, 121–127.

Gessner F., 1955–1959: Hydrobotanik. Die physiologischen Grundlagen der Pflanzenverbreitung im Wasser. I. Energiehaushalt (1955). II. Stoffhaushalt (1959). VEB Deutscher Verlag der Wissenschaften, Berlin; 517 pp. (1955); 701 pp. (1959).

Gessner, F., and Hammer, L., 1967: Die litorale Algenvegetation an den Küsten von Ost-Venezuela. Int. Rev. Ges. Hydrobiol. Hydrogr. 52, 657–692.

Gessner, F., and Schramm, W., 1971: Salinity. Plants. In: Marine ecology. Vol. 1, P. 2. Ed. by O. Kinne. Wiley-Interscience, London, pp. 705–820.

Giaccone, G., 1967: Osservazioni sul genere *Palmophyllum*. Lav. Ist. Bot. Giard. Col. Palermo 22, 318–326.

Giaccone, G., 1968a: Raccolte di fitobenthos nel Mediterraneo orientale. G. Bot. Ital. 102, 217–228.

Giaccone, G., 1968b: Aspetti della biocenosi coralligena in due stazioni dei bacini occidentale ed orientale del Mediterraneo. G. Bot. Ital. 102, 537–541.

Giaccone, G., 1969a: Note sistematiche ed osservazioni fitosociologiche sulle Laminariales del Mediterraneo occidentale. G. Bot. Ital. 193, 457–474.

Giaccone, G., 1969b: Raccolte di fitobenthos sulla banchina continentale Italiana. G. Bot. Ital. 103, 485–514.

Giaccone, G., 1971: Significato biogeografico ed ecologico di specie algali delle coste Italiane. Natura e Montagna 4, 41–47.

Giaccone, G., 1972: Struttura, ecologia e corologia dei popolamenti a Laminarie dello stretto di Messina e del Mare di Alboran. Mem. Biol. Mar. Oceanogr. N. S. 2, 37–59.

Giaccone, G., 1973: Elementi di botanica marina. II. Chiavi di determinazione per le alghe e le angiosperme marine del Mediterraneo. Pubbl. Ist. Bot. Univ. Trieste, ser. didat., pp. 1–358.

Giaccone, G., 1974: Tipologia delle comunita fitobentoniche del Mediterraneo. Mem. Biol. Mar. Oceanogr. 4, 149–168.

Giaccone, G., 1978: Revisione della flora del Mare Adriatico. Annuar. Parco Mar. Miramare Staz. Controllo Trieste 1977 (Suppl.), 1–118.

Giaccone, G., and Bruni, A., 1971: Le Cistoseire delle coste Italiane. I. Contributo. Ann. Univ. Ferrara, N. S., Sezione IV, Botanica, 4, 45–70.

Giaccone, G., and Bruni, A., 1973: Le Cistoseire e la vegetazione sommersa del Mediterraneo. Atti Ist. Veneto Sci. Lett. Arti 131, 59–103.

Giaccone, G., Colonna, P., Graziano, C., Mannino, M., Tornatore, E., Cormaci, M., Furnari, G., and Scammacca, B., 1985: Revisione della flore marina di Sicilia e isole minore. Boll. Accad. Gioenia Sci. Nat. 18,,463–484.

Giaccone, G., and Rizzi Longo, L., 1976: Revisione della flora dello Stretto di Messina. (Note storiche, bionomiche e corologiche). Mem. Biol. Mar., Oceanogr. 6, 69–123.

Giaccone, G., and Sortino, M., 1974: Zonazione della vegetazione marina delle Isole Egadi (Canale di Sicilia). Lav. Ist. Bot. Palermo 25, 166–183.

Gibson, C. E., 1987: Adaptations in *Oscillatoria redekei* at very slow growth rates-Changes in growth efficiency and phycobilin content. Br. Phycol. J. 22, 187–191.

Gibson, G., and Clayton, M. N., 1987: Sexual reproduction, early development and branching in *Notheia anomala* (Phaeophyta) and its classification in the Fucales. Phycologia 26, 363–373.

Gibson, M. T., and Whitton, B. A., 1987: Hairs, phosphatase activity and environmental chemistry in *Stigeoclonium*, *Chaetophora*, and *Draparnaldia* (Chaetophorales). Br. Phycol. J. 22, 11–22.

Gierloff-Emden, H. G., 1980: Geographie des Meeres. Ozeane und Küsten. Teil 1. Teil 2. de Gruyter, Berlin; 1310 pp.

Gilbert, R., 1984: The movement of gravel by the alga *Fucus vesiculosus* (L.) on an Arctic intertidal flat. J. Sediment. Petrol. 54, 463–468.

Gili, J. M., and Ros, J., 1984: Study and cartography of the benthic communities of Medes Islands (NE Spain). P. S. Z. N. I. Mar. Ecol. 6, 219–238.

Gilmartin, M., 1960: The ecological distribution of the deep water algae of Eniwetok Atoll. Ecology 41, 210–221.

Gilmartin, M., 1966: Ecology and morphology of *Tydemania expeditionis*, a tropical deep-water siphonous green alga. J. Phycol. 2, 100–105.

Gil-Rodríguez, M. C., and Afonso-Carrillo, J., 1980: Catálogo de las algas marinas bentónicas (Cyanophyta, Chlorophyta, Phaeophyta y Rhodophyta) para el Archipiélago Canario. Aula de Cultura de Tenerife; 65 pp.

Gil-Rodríguez, M. C., Haroun Tabraue, R., Afonso-Carrillo, J., and Wildpret de la Torre, W., 1985: Adiciones al catalogo de algas marinas bentónicas para el Archipiélago Canario. II. Vieraea 15, 101–112.

Gislén, T., 1929–1930: Epibioses of the Gullmarfjord. K. Svenska Vetensk. Akad. Skr. Naturskydd. 1929 (3), 1–113; 1930 (4), 1–380.

Givnish, T. J. (Ed.), 1986: On the economy of plant form and function. University Press, Cambridge; 717 pp.

Gladenkov, Yu., B., 1979: Comparison of late Cenozoic molluscan assemblages in northern regions of the Atlantic and Pacific Oceans. Int. Geol. Rev. 21, 880–890.

Gladenkov, Yu., B., 1986: Some aspects of the late Cenozoic history of the Bering Strait in light of stratigraphic data on Iceland. In: Beringia in the Cenozoic era. Ed. by V. L. Kontrimavichus. Russian Translation Series 28. Balkema, Rotterdam, pp. 36–45.

Glazer, A. N., 1982: Phycobilisomes: structure and dynamics. Ann. Rev. Microbiol. 36, 173–198.

Glenn, E. P., and Doty, M. S., 1981: Photosynthesis and respiration of the tropical red seaweeds, *Eucheuma striatum* (Tambalang and Elkhorn varieties) and *E. denticulatum*. Aquat. Bot. 10, 353–364.

Glicksman, M., 1969: Food gum technology. Academic Press, London; 273 pp.

Glombitza, K.-W., 1979: Antibiotics from algae. In: Marine algae in pharmaceutical science. Ed. by H. A. Hoppe, T. Levring, and Y. Tanaka. De Gruyter, Berlin, pp. 303-342.

Glover, H. E., Keller, M. D., and Guillard, R. R. L., 1986: Light quality and oceanic ultraphytoplankters. Nature 319, 142-143.

Gobi, C., 1878: Die Algenflora des Weissen Meeres und der demselben zunächst liegenden Theile des Nördlichen Eismeeres. Mém. Acad. imp. Sci. St. Petersbourg, Sér. VII, 24 (1), 1-92.

Goedheer, J. C., 1970: On the pigment system of brown algae. Photosynthetica 4, 97-106.

Goedheer, J. C., 1979: Carotenoids in the photosynthetic apparatus. Ber. Dt. Bot. Ges. 92, 427-436.

Goff, L. J., 1979a: The biology of *Harveyella mirabilis* (Cryptonemiales, Rhodophyceae). VI. Translocation of photoassimilated ^{14}C. J. Phycol. 15, 82-87.

Goff, L. J., 1979b: The biology of *Harveyella mirabilis* (Cryptonemiales, Rhodophyceae). VII. Structure and proposed function of host-penetrating cells. J. Phycol. 15, 87-100.

Goff, L. J., 1982: The biology of parasitic red algae. Prog. phycolog. Res. 1, 289-369.

Goff, L. J. (Ed.), 1983a: Algal symbiosis. A continuum of interaction strategies. Cambridge University Press, New York; 216 pp.

Goff, L. J., 1983b: Marine algal interactions: epibiosis, endobiosis, parasitism and disease. In: Proceedings of the Joint China-U. S. Phycology Symposium. Ed. by C. K. Tseng. Science Press, Beijing, pp. 221-274.

Goff, L. J., and Coleman, A. W., 1985: The role of secondary pit connections in red algal parasitism. J. Phycol. 21, 438-508.

Gomez Garreta, A., Ribera Siguan, A., and Seoane Camba, J. A., 1982: Aportación al estudio fenológico de las algas de la Isla de Mallorca. Boll. Soc. Hist. Nat. Baleares 26, 37-62.

Goodwin, T. W., 1974: Carotenoids and biliproteins. In: Algal physiology and biochemistry. Ed. by W. D. P. Stewart. Blackwell, Oxford, pp. 176-205.

van Goor, A. C. J., 1923: Die holländischen Meeresalgen. Verh. K. Akad. Wet., sect. 2, 23, 1-232.

Gordon, W. A., 1974: Physical controls on marine biotic distribution in the Jurassic period. In: Paleogeographic provinces and provinciality. Ed. by C. A. Ross. Society of Economic Paleontologists and Mineralogists. Tulsa, Oklahaoma; pp. 136-147.

Goreau, T. F., and Goreau, N. I., 1973: The ecology of Jamaican coral reefs. II. Geomorphology, zonation, and sedimentary phases. Bull. Mar. Sci. 23, 398-464.

Gosz, J. R., Holmes, R. T., Likens, G. E., and Bormann, F. H., 1978: The flow of energy in a forest ecosystem. Sci. Am. 238, 92-102.

Götting, K.-J., Kilian, E. F., and Schnetter, R., 1982: Einfrüng in die Meeresbiologie 1. Marine Organismen-Marine Biogeographie. Vieweg, Braunschweig; 179 pp.

Gradstein, F. M., and Srivastava, S. P., 1980: Aspects of Cenozoic stratigraphy and paleoceanography of the Labrador Sea and Baffin Bay. Palaeogeogr. Palaeoclimatol. Palaeoecol. 30, 261-295.

Graham, D., and Patterson, B. D., 1982: Responses of plants to low, nonfreezing temperatures: proteins, metabolism, and acclimation. Annu. Rev. Plant. Physiol. 33, 347-372.

Grant-Mackie, J. A., 1979: Cretaceous-Recent plate tectonic history and paleoceanographic development of the Southern Hemisphere. In: Proceedings of the International Symposium on Marine Biogeography and Evolution in the Southern Hemisphere, Auckland, New Zealand, July 1978. N. Z. DSIR Inform. Ser. 137 (1), 27–42.

Gray, J., Boucot, A. J., and Berry, W. B. N., 1981: Communities of the past. Hutchinson Ross, Stroudsburg, Penn.; 623 pp.

Gressel, J., and Rau, W., 1983: Photocontrol of fungal development. In: Encyclopedia of plant physiology, n. s. Vol. 16B. Ed. by W. Shropshire and H. Mohr. Springer-Verlag, Berlin, pp. 603–639.

Grigg, R. W., 1983: Community structure, succession, and development of coral reefs in Hawaii. Mar. Ecol. Prog. Ser. 11, 1–14.

Güven, K. C., and Ötzig, F., 1971: Über die marinen Algen an den Küsten der Türkei. Botanica Mar. 14, 121–128.

Guiler, E. R., 1959: Intertidal belt-forming species on the rocky coasts of northern Chile. Pap. Proc. R. Soc. Tasman. 93, 33–58.

Guimarães, S. M. P. B., Braga, M. R. A., Cordeiro-Marino, M., and Pedrini, A. G., 1986: Morphology and taxonomy of *Jolyna laminarioides*, a new member of the Scytosiphonales (Phaeophyceae) from Brazil. Phycologia 25, 99–108.

Guimarães, S. M. P. B., Cordeiro-Marino, M., and Yamaguishi-Tomita, N., 1981: Deep water Phaeophyceae and their epiphytes from northeastern and southeastern Brazil. Revta Bras. Bot. 4, 95–113.

Guiry, M. D., 1974: A preliminary consideration of the taxonomic position of *Palmaria palmata* (Linnaeus) Stackhouse = *Rhodymenia palmata* (Linnaeus) Greville. J. Mar. Biol. Ass. U. K. 54, 509–528.

Guiry, M. D., 1975: An assessment of *Palmaria palmata* forma *mollis* (S. et G.) comb. nov. (= *Rhodymenia palmata* forma *mollis* S. et G.) in the eastern North Pacific. Syesis 8, 245–261.

Guiry, M. D., 1978: A consensus and bibliography of Irish seaweeds. Cramer, Vaduz; 287 pp.

Guiry, M. D., 1982: *Devaleraea*, a new genus of the Palmariaceae (Rhodophyta) in the North Atlantic and North Pacific. J. Mar. Biol. Ass. U. K. 62, 1–13.

Guiry, M. D., 1984: Photoperiodic and temperature responses in the growth and tetrasporogenesis of *Gigartina acicularis* (Rhodophyta) from Ireland. Helgoländer Meeresunters. 38, 335–347.

Guiry, M. D., and Blunden, G., 1981: The commercial collection and utilisation of seaweeds in Ireland. Proc. Int. Seaweed Symp. 10, 675–680.

Guiry, M. D., and Cunningham, E. M., 1984: Photoperiodic and temperature responses in the reproduction of north-eastern Atlantic *Gigartina acicularis* (Rhodophyta: Gigartinales). Phycologia 23, 357–367.

Guiry, M. D., and Hollenberg, G. J., 1975: *Schottera* gen nov. and *Schottera nicaeensis* (Lamour. ex Duby) comb. nov. (= *Petroglossum nicaeense* (Lamour. ex Duby) Schotter) in the British Isles. Br. Phycol. J. 10, 149–164.

Guiry, M. D., Kee, W. R., and Garbary, D. J., 1987a: Morphology, temperature, and photoperiodic responses in *Audouinella botryocarpa* (Harvey) Woelkerling (Acrochaetiaceae, Rhodophyta) from Ireland. G. Bot. Ital. 121, 229–246.

Guiry, M. D., and Maggs, C. A., 1985: Reproduction and life history of *Meredithia microphylla* (J. Ag.) J. Ag. (Kallymeniaceae, Rhodophyta) from Ireland. G. Bot. Ital. 118, 105–125.

Guiry, M. D., Tripodi, G., and Lüning, K., 1987b: Biosystematics, genetics, and upper temperature tolerance of *Gigartina teedii* (Rhodophyta) from the Atlantic and Mediterranean. Helgoländer Meeresnunters. 41, 283–295.

Guiry, M. D., and West, J. A., 1983: Life history and hybridization studies on *Gigartina stellata* and *Petrocelis cruenta* (Rhodophyta) in the North Atlantic. J. Phycol. 19, 474–494.

Guiry, M. D., West, J. A., Kim, D.-H., and Masuda, M., 1984: Reinstatement of the genus *Mastocarpus* Kützing (Rhodophyta). Taxon 33, 53–63.

Gunnill, F. C., 1980: Demography of the intertidal brown alga *Pelvetia fastigiata* in southern California, USA. Mar. Biol. 59, 169–179.

Gwinner, E., 1986: Circannual rhythms. Endogenous annual clocks in the organization of seasonal processes. Springer, Berlin; 154 pp.

Hackett, H. E., 1969: Marine algae in the atoll environment: Maldive Islands. Proc. Int. Seaweed Symp. 6, 187–191.

Hackett, H. E., 1977: Marine algae known from the Maldive Islands. Atoll Res. Bull. 210, 1–30.

Hagevang, T, Eldholm, O., and Aalstad, I., 1983: Pre-23 magnetic anomalies between Jan Mayen and Greenland-Senja fracture zones in the Norwegian Sea. Mar. Geophys. Res. 5, 345–363.

Hagmeier, A., 1930: Die Besiedelung des Felsstrandes und der Klippen von Helgoland. Teil I. Der Lebensraum. Wiss. Meeresunters. (Abt. Helgoland) 15 (18a), 1–35.

Halbach, 1980: Ökologische Anpassungsstrategien an variable Umwelten. Biologie in unserer Zeit 10, 1–10.

Hallam, A. (Ed.), 1973: Atlas of palaeobiogeography. Elsevier, Amsterdam; 531 pp.

Hallam, A., 1981: Relative importance of plate movements, eustasy, and climate in controlling major biogeographical changes since the Early Mesozoic. In: Vicariance biogeography. Ed. by G. Nelson and D. E. Rosen. Columbia University Press, New York, pp. 303–340.

Halldal, P., 1968: Photosynthetic capacities and photosynthetic action spectra of endozoic algae of the massive coral *Favia*. Biol. Bull. Mar. Biol. Lab. Woods Hole 134, 411–424.

Hällfors, G., Niemi, A., Ackefors, H., Lassig, J., and Leppäkoski, E., 1981: Biological oceanography. In: The Baltic Sea. Ed. by A. Voipio. Elsevier, Amsterdam, pp. 219–274.

Hambrey, M. J., and Harland, W. B. (Eds.), 1981: Earth's pre-Pleistocene glacial record. University Press, Cambridge; 1024 pp.

Hamel, G., 1924–1930: Floridées de France. I-VI. Rev. Algol. Fr. 1, 278–292 (I), 427–457 (II); 2, 39–67 (III); 280–309 (IV); 3, 99–158 (V); 5, 61–109 (VI).

Hamel, G., 1930–1931: Chlorophycées des côtes françaises. Rev. Algol. Fr. 5, 1–54, 383–430; 6, 9–73.

Hamel, G., 1931-1939: Pheophycées de France. Wolf, Paris, 432 pp.

Hamm, D., and Humm, H. J., 1976: Benthic algae of the Anclote estuary. II. Bottom-dwelling species. Fla. Sci. 39, 209-229.

Hanisak, M. D., 1979: Growth patterns of *Codium fragile* spp. *tomentosoides* in response to temperature, irradiance, salinity, and nitrogen source. Mar. Biol. 50, 319-332.

Hanisak, M. D., 1981: Methane production from the red seaweed *Gracilaria tikvahiae*. Proc. Int. Seaweed Symp. 10, 681-686.

Hanisak, M. D., 1983: The nitrogen relationship of marine macroalgae. In: Nitrogen in the marine environment. Ed. by E. J. Carpenter and D. G. Capone. Academic Press, London, pp. 699-730.

Hanisak, M. D., 1987: Cultivation of *Gracilaria* and other macroalgae in Florida for energy production. In: Seaweed cultivation for renewable resources. Ed. by K. T. Bird and P. H. Benson. Elsevier, Amsterdam, pp. 191-218.

Hanisak, M. D., and Blair, S. M., 1988: The deep-water macroalgal community of the East Florida continental shelf (USA). Helgoländer Meeresunters. 42, 133-163.

Hanisak, M. D., and Samuel, M. A., 1987: Growth rates in culture of several species of *Sargassum* from Florida, USA. Hydrobiologia 151/152, 399-404.

Hannach, G., and Waaland, J. R., 1986: Environment, distribution, and production of *Iridaea*. Aquat. Bot. 26, 51-78.

Hansen, G. I., 1980: A morphological study of *Fimbrifolium*, a new genus in the Cystocloniaceae (Gigartinales, Rhodophyta). J. Phycol. 16, 207-217.

Hansen, G. I., Garbary, D. J., Oliveira, J. C., and Scagel, R. F., 1981: New records and range extensions of marine algae from Alaska. Syesis 14, 115-123.

Hansen, T. A., 1978: Larval dispersal and species longevity in Lower Tertiary gastropods. Science 199, 885-887.

Haq, B. U., Hardenbol, J., and Vail, P. R., 1987: Chronology of fluctuating sea levels since the Triassic. Science 235, 1156-1167.

Haq, B. U., and Boersma, A., 1978: Introduction to marine micropaleontology. Elsevier, New York; 376 pp.

Harder, R., 1923: Über die Bedeutung von Lichtintensität und Wellenlänge für die Assimilation farbiger Algen. Z. Bot. 15, 305-355.

Harder, R., and Bederke, B., 1957: Über Wachstumsversuche mit Rot- und Grunalgen (*Porphyridium cruentum*, *Trailliella intricata*, *Chlorella pyrenoidosa*) in verschiedenfarbigem, energiegleichem Licht. Arch. Mikrobiol. 28, 153-172.

Haritonidis, S., and Tsekos, I., 1975: Marine algae of northern Greece. Botanica Mar. 18, 203-221.

Haritonidis, S., and Tsekos, I., 1976: Marine algae of the Greek west coast. Botanica Mar. 19, 273-286.

Harkin, E., 1981: Fluctuations in epiphyte biomass following *Laminaria hyperborea* canopy removal. Proc. Int. Seaweed Symp. 10, 303-308.

Harlin, M. M., 1973: "Obligate" algal epiphyte: *Smithora naiadum* grows on a synthetic substrate. J. Phycol. 9, 230-232.

Harlin, M. M., and Wheeler, P. A., 1985: Nutrient uptake. In: Handbook of phycological methods. Ecological field methods: macroalgae. Ed. by M. M. Littler and D. S. Littler. University Press, Cambridge, pp. 33–51.

Harm, W., 1981: Biological effects of ultraviolet radiation. University Press, Cambridge; 216 pp.

Harms, J., 1984: Influence of water temperature on larval development of *Elminius modestus* and *Semibalanus balanoides* (Crustacea, Cirripedia). Helgoländer Meeresunters. 38, 123–134.

Harris, G. P., 1978: Photosynthesis, productivity and growth: the physiological ecology of phytoplankton. Arch. Hydrobiol. Beih. Ergebn. Limnol. 10, 1–171.

Harris, G. P., and Piccinin, B., B., 1977: Photosynthesis by natural phytoplankton populations. Arch. Hydrobiol. 80, 405–457.

Harrison, P. J., Druehl, L. D., Lloyd, K. E., and Thompson, P. A., 1986: Nitrogen uptake kinetics in three year-classes of *Laminaria groenlandica* (Laminariales: Phaeophyta). Mar. Biol. 93, 29–35.

Harrold, C., and Pearse, J. S., 1987: The ecological role of echinoderms in kelp forests. Echin. Stud. 2, 137–233.

Harrold, C., Watanabe, J., and Lisin, S., 1988: Spatial variation in the structure of kelp forest communities along a wave exposure gradient. P. S. Z. N. I. Mar. Ecol. 9, 131–156.

Hartmann, T., and Löhr, E., 1983: Decarboxylation of environmental L-leucine by marine red algae. Mar. Biol. 78, 7–12.

den Hartog, C., 1959: The epilithic algal communities occurring along the coast of the Netherlands. Wentia 1, 3–241.

den Hartog, C., 1964: Typologie des Brackwassers. Helgoländer Meeresunters. 10, 377–390.

den Hartog, C., 1967: Brackish water as an environment for algae. Blumea 15, 31–43.

den Hartog, C., 1968: The littoral environment of rocky shores as a border between the sea and the land and between the sea and the fresh water. Blumea 16, 375–393.

den Hartog, C., 1970: The sea-grasses of the world. North-Holland Publishing Company, Amsterdam; 275 pp.

den Hartog, C., 1973: Preliminary survey of the algal vegetation of salt-marshes, a littoral border environment. Hydrobiol. Bull. 7, 3–14.

den Hartog, C., 1987: "Wasting disease" and other dynamic phenomena in *Zostera* beds. Aquat. Bot. 27, 3–14.

Hasegawa, Y., 1976: Progress of *Laminaria* cultivation in Japan. J. Fish. Res. Bd. Can. 33, 1002–1006.

Hatcher, B. G., Chapman, A. R. O., and Mann, K. H., 1977: An annual carbon budget for the kelp *Laminaria longicruris*. Mar. Biol. 44, 85–96.

Hatcher, B. G., Kirkman, H., and Wood, W. F., 1987: Growth of the kelp *Ecklonia radiata* near the northern limit of its range in Western Australia. Mar. Biol. 95, 63–73.

Hatcher, B. G., and Larkum, A. W. D., 1983: An experimental analysis of factors controlling the standing crop of the epilithic algal community on a coral reef. J. Exp. Mar. Biol. Ecol. 69, 61–84.

Haug, A., and Jensen, A., 1954: Seasonal variations in the chemical composition of

Alaria esculenta, Laminaria saccharina, Laminaria hyperborea and *Laminaria digitata* from northern Norway. Norsk institutt for tang- og tareforskning, rep. no. 4, pp. 1–14.

Haug, A., Larsen, B., and Smidsrod, O., 1974: Uronic acid sequence in alginate from different sources. Carbohyd. Res. 32, 217–225.

Haupt, W., 1983: Movements of chloroplasts under the control of light. Progr. Phycol. Res. 2, 227–281.

Hawkes, M. W., and Scagel, R. F., 1986: The marine algae of British Columbia and northern Washington: division Rhodophyta (red algae), class Rhodophyceae, order Palmariales. Can. J. Bot. 64, 1148–1173.

Hawkes, M. W., Tanner, C. E., and Lebednik, P. A., 1978: The benthic marine algae of northern British Columbia. Syesis 11, 81–115.

Hawkins, S. J., and Hartnoll, R. G., 1983: Grazing of intertidal algae by marine invertebrates. Oceanogr. Mar. Biol. Annu. Rev. 21, 195–282.

Hawkins, S. J., and Hartnoll, R. G., 1985: Factors determining the upper limits of intertidal canopy-forming algae. Mar. Ecol. Prog. Ser. 20, 265–271.

Haxo, F. T., and Blinks, L. R., 1950: Photosynthetic action spectra of marine algae. J. Gen. Physiol. 33, 389–422.

Hay, C. H., 1979a: Nomenclature and taxonomy within the genus *Durvillaea* Bory (Phaeophyceae: Durvilleales Petrov). Phycologia 18, 191–202.

Hay, C. H., 1979b: A phytogeographical account of the southern bull kelp *Durvillaea* spp. Bory 1826 (Durvilleales Petrov 1965). In: Proceedings of the International Symposium on Marine Biogeography and Evolution in the Southern Hemisphere, Auckland, New Zealand, July 1978. N. Z. DSIR Inform. Ser. 137 (2), 443–453.

Hay, C. H., 1986: A new species of *Macrocystis* C. Ag. (Phaeophyta) from Marion Island, southern Indian Ocean. Phycologia 25, 241–252.

Hay, C. H., 1987: The Asian kelp *Undaria pinnatifida* (Phaeophyta: Laminariales) found in a New Zealand harbour. N. Z. J. Bot. 25, 329–332.

Hay, C. H., 1988: The occurrence of *Durvillaea antarctica* (Durvillaeales, Phaeophyta) at South Georgia, South Atlantic Ocean. Phycologia 27, 424–427.

Hay, C. H., Adams, N. M., and Parsons, M. J., 1985: The marine algae of the subantarctic islands of New Zealand. A list of species. Nat. Mus. N. Z. Misc. Ser. 11, 1–70.

Hay, C. H., Luckens, P. A., 1987: The Asian kelp *Undaria pinnatifida*, (new record, Phaeophyta, Laminariales) found in a New Zealand harbour. N. Z. J. Bot. 25, 329–332.

Hay, M. E., 1986: Functional geometry of seaweeds: ecological consequences of thallus layering and shape in contrasting light environments. In: On the econonmy of plant form and function. Ed. by T. J. Givnish. University Press, Cambridge, pp. 635–666.

Hay, M. E., and Gaines, S. D., 1984: Geographic differences in herbivore impact: do Pacific herbivores prevent Caribbean seaweeds from colonizing via the Panama Canal? Biotropica 16, 24–30.

Hay, M. E., Paul, V., Lewis, S. M., Gustavson, K., Tucker, J., and Trindell, R. N., 1988a: Can tropical seaweeds reduce herbivory by growing at night? Diel patterns of growth, nitrogen content, herbivory, and chemical versus morphological defenses. Oecologia (Berlin) 75, 233–245.

Hay, M. E., Renaud, P. E., and Fenical, W., 1988b: Large mobile versus small sedentary herbivores and their resistance to seaweed chemical defenses. Oecologia (Berlin) 75, 246–252.

Hayashida, F., 1983: Synecological studies of a brown alga, *Ecklonia cava* Kjellmann, forming aquatic forest. I. Benthic marine algal vegetation at survey area. J. Fac. Mar. Sci. Technol. Tokai Univ. 16, 207–212.

Hayashida, F., 1984: Synecological studies of a brown alga, *Ecklonia cava* Kjellmann, forming aquatic forest. II. On growth of *Ecklonia cava*. J. Fac. Mar. Sci. Technol. Tokai Univ. 18, 275–280.

Hayward, P. J., 1980: Invertebrate epiphytes of coastal marine algae. In: The shore environment. Vol. 2. Ecosystems. Ed. by J. H. Price, D. E. G. Irvine, and W. F. Farnham. Academic Press, London; pp. 761–787.

Hazel, J. E., 1970: Geographic distribution and temperature and depth tolerances of Recent Arctic-North Atlantic benthonic ostracodes with particular emphasis on the sublittoral species off the northeastern United States. Geol. Surv. Prof. Pap. 529-E, 1–21.

Healey, F. P., 1972: Photosynthesis and respiration of some Arctic seaweeds. Phycologia 11, 267–271.

Hedgpeth, J. W., 1957a: Marine biogeography. Mem. Geol. Soc. Am. 67, 359–382.

Hedgpeth, J. W., 1957b: Classification of marine environments. Geol. Soc. Am. Mem. 67 (1), 17–28.

Hedgpeth, J. W., 1969: Distribution of selected groups of marine invertebrates in waters south of 35 S latitude. Introduction to Antarctic Zoogeography. Antarct. Map Folio Ser.-Folio 11, 1–9.

Hedgpeth, J. W., 1970: Marine biogeography of the Antarctic regions. In: Antarctic ecology. Ed. by M. Holdgate. Academic Press, London, pp. 67–96.

Hedgpeth, J. W., 1976: Prologue: At sea with provinces and plates. In: Historical biogeography, plate tectonics, and the changing environment. Ed. by J. Gray and A. J. Boucot. Oregon State University Press, Corvallis; pp. 2–7.

Heijs, F. M. L., 1987: Community structure and seasonality of macroalgae in some mixed seagrass meadows from Papua New Guinea. Aquat. Bot. 27, 139–158.

Hellebust, J. A., and Craigie, J. S. (Eds.), 1978: Handbook of phycological methods. 2. Physiological and biochemical methods. University Press, Cambridge; 512 pp.

Hempel, G., 1985: On the biology of polar seas, particularly the Southern Ocean. In: Marine biology of polar regions and effects of stress on organisms. Ed. by J. S. Gray and M. E. Christiansen. Wiley, Chichester, pp. 3–33.

Henley, W. J., and Ramus, J., 1989: Optimization of pigment content and the limits of photoacclimation for *Ulva*. Mar. Biol. (in press.)

Hennig, W., 1966: Phylogenetic systematics. University of Illinois Press, Urbana; 263 pp.

Henry, E. C., 1984: Syringodermatales ord. nov. and *Syringoderma floridana* sp. nov. (Phaeophyceae). Phycologia 23, 419–426.

Henry, E. C., 1987a: The life history of *Phyllariopsis brevipes* (= *Phyllaria reniforme*) (Phyllariaceae, Laminariales, Phaeophyceae), a kelp with dioecious but sexually monomorphic gametophytes. Phycologia 26, 17–22.

Henry, E. C., 1987b: Primitive characters and a photoperiodic response in *Saccorhiza dermatodea* (Laminariales, Phaeophyceae). Br. Phycol. J. 22, 23–31.

Henry, E. C., and Müller, D. G., 1983: Studies on the life history of *Syringoderma phinneyi* sp. nov. (Phaeophyceae). Phycologia 22, 387–393.

Henry, E. C., and South, G. R., 1987: *Phyllariopsis* Gen. nov. and a reappraisal of the Phyllariaceae Tilden 1935 (Phaeophyceae, Laminariales). Phycologia 26, 9–16.

Herman, Y., 1974: Marine geology and oceanography of the Arctic seas. Springer, New York; 397 pp.

Herman, Y., 1983: Arctic paleoceanography in Late Neogene time and its relationship to global climates. Oceanology 23, 81–87.

Herman, Y., and Hopkins, D. M., 1980: Arctic oceanic climate in late Cenozoic time. Science 209, 557–562.

Herron, E. M., Dewey, J. F., and Pitman, W. C., 1973: Plate tectonics model for the evolution of the Arctic. Geology 2, 377–380.

Heywood, R. B., and Whitaker, T. M., 1984: The Antarctic marine flora. In: Antarctic ecology. Vol. 2. Ed. by R. M. Laws. Academic Press, London, pp. 373–419.

Hickey, L. J., West, R. M., Dawson, M. R., and Choi, D. K., 1983: Arctic terrestrial biota: paleomagnetic evidence of age disparity with mid-northern latitudes during the Late Cretaceous and early Tertiary. Science 221, 1153–1156.

Hickman, C. S., and Lipps, J. H., 1985: Geological youth of Galápagos Islands confirmed by marine stratigraphy and paleontology. Science 227, 1578–1580.

Hillis-Colinvaux, L., 1980: Ecology and taxonomy of *Halimeda*: primary producer of coral reefs. Adv. Mar. Biol. 17, 1–327.

Hillis-Colinvaux, L., 1986a: Deep water populations of *Halimeda* in the economy of an atoll. Bull. Mar. Sci. 38, 155–169.

Hillis-Colinvaux, L., 1986b: Historical perspectives on algae and reefs: have reefs been misnamed? Oceanus 29, 43–48.

Hillson, C. J., 1977: Seaweeds. A color-coded, illustrated guide to common marine plants of the east coast of the United States. Keystone Books, Pennsylvania State University Press, University Park and London; 194 pp.

Himmelman, J. H., 1986: Population biology of green sea urchins on rocky barrens. Mar. Ecol. Prog. Ser. 33, 295–306.

Hiscock, K., 1983: Water movement. In: Sublittoral ecology. The ecology of the shallow sublittoral benthos. Ed. by R. Earll and D. G. Erwin. Clarendon Press, Oxford, pp. 58–96.

Hiscock, K., 1985: Aspects of the ecology of rocky sublittoral areas. In: The ecology of rocky coasts. Essays presented to J. R. Lewis, D.Sc.Ed., by P. G. Moore and R. Seed. Columbia University Press, New York, pp. 290–328.

Hiscock, K., and Mitchell, R., 1980: The description and classification of sublittoral epibenthic ecosystems. In: The shore environment. Vol. 2. Ecosystems. Ed. by J. H. Price, D. E. G. Irvine, and W. F. Farnham. Academic Press, London, pp. 323–370.

Hiscock, S., 1979: A field key to the British brown seaweeds. Fld. Stud. Council Occ. Publ. 5; 44 pp.

Hiscock, S., 1986: A field key to the British red seaweeds. Fld. Stud. Council Occ. Publ. 13; 101 pp.

Hiscock, S., Barratt, L., and Ormond, R., 1984: The marine algae of Dhofar, Oman-an upwelling system in the Arabian Sea. Br. Phycol. J. 19, 194.

Hoch, E., 1983: Fossil evidence of early Tertiary North Atlantic events viewed in European context. In: Structure and development of the Greenland-Scotland ridge. Ed. by H. P. Bott, S. Saxov, M. Talwani and J. Thiede. Plenum Press, New York, pp. 401–415.

van den Hoek, C., 1969: Algal vegetation-types along the open coasts of Curacao, Netherland Antilles. Proc. K. Ned. Akad. Wet. Sect. C, 72, 537–577.

van den Hoek, C., 1975: Phytogeographic provinces along the coasts of the northern Atlantic Ocean. Phycologia 14, 317–330.

van den Hoek, C., 1982a: Phytogeographic distribution groups of benthic marine algae in the North Atlantic Ocean. A review of experimental evidence from life history studies. Helgoländer Meeresunters. 35, 153–214.

van den Hoek, C., 1982b: The distribution of benthic marine algae in relation to the temperature regulation of their life histories. Biol. J. Linn. Soc. 18, 81–144.

van den Hoek, C., 1984: World-wide longitudinal seaweed distribution patterns and their possible causes, as illustrated by the distribution of rhodophytan genera. Helgoländer Meeresunters. 38, 227–257.

van den Hoek, C., 1987: The possible significance of long-range dispersal for the biogeography of seaweeds. Helgoländer Meeresunters. 41, 261–272.

van den Hoek, C., 1988: The emergence of a new chlorophytan system and Dr. Kornmann's contribution thereto. Helgoländer Meeresunters. 42, 339–383.

van den Hoek, C., Admiraal, W., Colijn, F., and Jonge, V. N. de, 1979: The role of algae and seagrasses in the ecosystem of the Wadden Sea: a review. In: Flora and vegetation of the Wadden Sea. Ed. by W. J. Wolff. Balkema, Rotterdam, pp. 3/9–3/206.

van den Hoek, C., Breeman, A. M., Bak, R. P. M., and van Buurt, G., 1978: The distribution of algae and gorgonians in relation to depth, light attenuation, water movement, and grazing pressure in the fringing coral reef of Curaçao, Netherland Antilles. Aquat. Bot. 5, 1–46.

van den Hoek, C., Colijn, F., Cortel-Breeman, A. M., and Wanders, J. B. W., 1972: Algal vegetation-types along the shores of inner bays and lagoons of Curaçao, and of the Lagoon Lac (Bonaire), Netherland Antilles. Verh. Ned. Akad. Wet. Afd. Natuurk. Reeks 2, Deel 61, no. 2, 1–72.

van den Hoek, C., Cortel-Breeman, A. M., and Wanders, J. B. W., 1975: Algal zonation in the fringing coral reef of Curaçao, Netherland Antilles, in relation to zonation of corals and gorgonians. Aquat. Bot. 1, 269–308.

van den Hoek, C., and Donze, M., 1966: The algal vegetation of the rocky Côte Basque (SW France). Bull. Cent. Etud. Rech. Sci. Biarritz 6, 289–319.

van den Hoek, C., and Donze, M., 1967: Algal phytogeography of the European Atlantic coasts. Blumea 15, 63–89.

van den Hoek, C., Wanders, J. B. W., and Cortel-Breeman, A. M., 1981: The role of benthic algae in the coral reef of Curaçao, Netherland Antilles. Proc. Int. Seaweed Symp. 8, 353–359.

Hoffmann, A. J., 1987: The arrival of seaweed propagules at the shore: a review. Botanica Mar. 30, 151–165.

Hoffmann, C., 1940: Die Vegetation der Nord- und Ostsee. In: Die Tierwelt der Nord- und Ostsee. Ed. by G. Grimpe and E. Wagler. Akademische Verlagsgesellschaft, Leipzig, pp. Ic1–1c32.

Højerslev, N. K., 1982: Yellow substance in the sea. In: The role of solar ultraviolet radiation in marine ecosystems. Ed. by J. Calkins. Plenum Press, New York, pp. 263–281.

Holmes, M. G., and Klein, W. H., 1987: The light and temperature environments. In: Plant life in aquatic and amphibious habitats. Ed. by R. M. M. Crawford. Blackwell, Oxford, pp. 3–22.

Hommersand, M. H., 1972: Taxonomic and phytogeographic relationships of warm temperate marine algae occurring in Pacific North America and Japan. Proc. Int. Seaweed Symp. 7, 66–71.

Hommersand, M. H., 1986: The biogeography of the South African marine red algae. Botanica Mar. 29, 257–270.

Hoopen, A. ten, Bos, S., and Breeman, A. M., 1983: Photoperiodic response in the formation of gametangia of the long-day plant *Sphacelaria rigidula*. Mar. Ecol. Prog. Ser. 13, 285–289.

Hooper, R. G., 1987: Benthic algal communities of an extremely isolated offshore shoal-Virgin Rocks, Grand Banks of Newfoundland. Br. Phycol. J. 22, 316.

Hooper, R. G., 1988: Natural environmental stresses and mass mortalities in northeast North American marine communities. Br. Phycol. J. 23, 289.

Hooper, R. G., Henry, E. C., and Kuhlenkamp, R., 1988: *Phaeosiphoniella cryophila* gen. et sp. nov., a third member of the Tilopteridales (Phaeophyceae). Phycologia 27, 395–404.

Hooper, R. G., and South, R., 1977: Distribution and ecology of *Papenfussiella callitricha* (Rosenv.) Kylin (Phaeophyceae, Chordariaceae). Phycologia 16, 153–157.

Hooper, R. G., South, G. R., and Whittick, A., 1980: Ecological and phenological aspects of the marine phytobenthos of the Island of Newfoundland. In: The shore environment. Vol. 2. Ecosystems. Ed. by J. H. Price, D. E. G. Irvine, and W. F. Farnham. Academic Press, London, pp. 395–423.

Hopkins, D. M. (Ed.), 1967: The Bering land bridge. Stanford University Press, Stanford, Calif.; 351 pp.

Hopkins, D. M., and Marincovich, L., 1984: Whale biogeography and the history of the Arctic Basin. Groningen Rijksuniversitet, Netherlands, Work of the Arctic Centre, no. 8, pp. 7–24.

Hoppe, H. A., and Levring, T., 1982: Marine algae in pharmaceutical science. Vol. 2. De Gruyter, New York; 309 pp.

Hoppe, H. A., Levring, T., and Tanaka, Y., 1979: Marine algae in pharmaceutical science. Vol. 1. De Gruyter, New York; 807 pp.

Hörnig, I., and Schnetter, R., 1988: Notes on *Dictyota dichotoma*, *D. indica* and *D. pulchella* spec. nova (Phaeophyta). Phyton (Austria) 28, 277–291.

Hornsey, I. S., and Hide, D., 1985: The production of antimicrobial compounds by British marine algae. IV. Variation of antimicrobial activity with algal generation. Br. Phycol. J. 20, 21–25.

Howard, K. L., and Menzies, R. J., 1969: Distribution and production of *Sargassum* in the waters off the Carolina coast. Botanica Mar. 12, 244–254.

Howe, M. A., 1914: The marine algae of Peru. Mem. Torrey Bot. Club 15, 1–185.

Hruby, T., and Norton, T. A., 1979: Algal recolonization on rocky shores in the Firth of Clyde. J. Ecol. 67, 65–77.

Hsü, K. J., 1972: When the Mediterranean dried up. Sci. Am. 227, 27–36.

Hsü, K. J., 1978: When the Black Sea was drained. Sci. Am. 238, 52–63.

Hsü, K. J., et al., 1977: History of the Mediterranean salinity crisis. Nature 267, 399–403.

Humbeck, K., Schumann, R., and Senger, H., 1984: The influence of blue light on the formation of chlorophyll-protein complexes in *Scenedesmus*. In: Blue light effects in biological systems. Ed. by H. Senger. Springer, Berlin, pp. 359–365.

Humm, H. J., 1969: Distribution of marine algae along the Atlantic coast of North America. Phycologia 7, 43–53.

Humm, H. J., 1979: The marine algae of Virginia. University Press, Charlottesville, Va.; 263 pp.

Humphries, C. J., 1979: Endemism and evolution in Macaronesia. In: Plants and islands. Ed. by D. Bramwell. Academic Press, London, pp. 171–199.

Humphries, C. J., and Parenti, L. R., 1986: Cladistic biogeography. Clarendon Press, Oxford; 98 pp.

Hurdle, B. G. (Ed.), 1986: The Nordic seas. Springer-Verlag, New York; 777 pp.

Hutchins, L. W., 1947: The bases for temperature zonation in geographical distribution. Ecol. Monogr. 17, 325–335.

Huth, K., 1979: Einfluss von Tageslänge und Beleuchtungsstärke auf den Generationswechsel bei *Batrachospermum moniliforme*. Ber. Dt. bot. Ges. 92, 467–472.

Huvé, H., 1955: Présence de *Laminaria rodriguezii* Bornet sur les côtes françaises de la Méditerranée. Recl. Trav. Stn. Mar. Endoume Bull. 15, 74–89.

Huvé, H., 1958: Contribution à l'etude des peuplements des Phyllariacees du détroit de Messine. C. I. E. S. M. Extr. Rapp. Proc. Verb. N. S. 14, 525–533.

Huvé, H., 1972: Aperçu sur la distribution en Mer Egee de quelques espèces du genre *Cystoseira* (Phéophycées, Fucales). Bull. Soc. Phycol. Fr. 17, 22–37.

Ingle, J. C., 1967: Foraminiferal biofacies variation and the Miocene-Pliocene boundary in southern California. Bull. Am. Paleontol. 52, 217–394.

Ingle, J. C., 1977: Summary of Late Neogene planktic foraminiferal biofacies, biostratigraphy, and paleoceanography of the marginal North Pacific Ocean. In: Proceedings of the First International Congress on Pacific Neogene Stratigraphy. Kaiyo Shuppan Company, Tokyo, pp. 177–182.

Innes, D. J., 1987: Genetic structure of asexually reproducing *Enteromorpha linza* (Ulvales: Chlorophyta) in Long Island Sound. Mar. Biol. 94, 459–467.

Irvine, D. E. G., 1982: Seaweeds of the Faroes. 1: The flora. Bull. Br. Mus. Nat. Hist. (Bot.) 10, 109–131.

Irvine, D. E. G., Guiry, M. D., Tittley, I., and Russell, G., 1975: New and interesting marine algae from the Shetland isles. Br. Phycol. J. 10, 57–71.

Irvine, L. M., 1983: Seaweeds of the British Isles. Vol. 1. Rhodophyta. Part 2A. Cryptonemiales (sensu strictu), Palmariales, Rhodymeniales. British Museum (Natural History), London; 120 pp.

Isaac, W. E., 1971: A third list of Kenya marine algae. J. E. Afr. Nat. Hist. Soc. 28, 1–23.

Isaac, W. E., and Chamberlain, Y. M., 1958: Marine algae of Inhaca Island and of the Inhaca Peninsula. II. J. S. Afr. Bot. 24, 123–158.

Jaasund, E., 1965: Aspects of the marine algal vegetation of north Norway. Bot. Gothoburg. 4, 1–174.

Jaasund, E., 1969–1977: Marine algae of Tanzania. I-VIII. Botanica Mar. 12, 255–274; 13, 59–79; 20, 333–338, 405–425, 509–520.

Jaasund, E., 1976: Intertidal seaweeds in Tanzania. A field guide. University of Tromsö, Norway; 160 pp.

Jackson, B. A., and Kaufmann, K. W., 1987: *Diadema antillarum* was not a keystone predator in cryptic reef environments. Science 235, 687–689.

Jackson, G. A., and Winant, C. D., 1983: Effect of a kelp forest on coastal currents. Cont. Shelf Rep. 2, 75–80.

Jacquotte, R., 1962: Etude des fonds de maerl de Méditerranée. Recl. Trav. Stn. mar. Endoume 26, 141–235.

Janke, K., 1986: Die Makrofauna und ihre Verteilung im Nordost-Felswatt von Helgoland. Helgoländer Meeresunters. 40, 1–55.

Jeffrey, S. W., 1968: Two spectrally distinct components in preparations of chlorophyll c. Nature 220, 1032–1033.

Jeffrey, S. W., 1981: Responses to light in aquatic plants. In: Encyclopedia of plant physiology, n. s. Vol. 12A. Ed. by O. L. Lange, P. S. Nobel, C. B. Osmond and H. Ziegler. Springer-Verlag, Berlin, pp. 249–276.

Jeffrey, S. W., and Humphrey, G. F., 1975: New spectrophotometric equations for determining chlorophylls a, b, c_1, and c_2 in higher plants, algae, and natural phytoplankton. Biochem. Physiol. Pflanzen (BPP) 167, 191–194.

Jensen, A., 1972: The nutritive value of seaweed meal for domestic animals. Proc. Int. Seaweed Symp. 7, 7–14.

Jensen, A., 1978: Chlorophylls and carotenoids. In: Handbook of phycological methods. 2. Physiological and biochemical methods. Ed. by J. A. Hellebust and J. S. Craigie. University Press, Cambridge, pp. 59–70.

Jensen, A., 1979: Industrial utilization of seaweeds in the past, present, and future. Proc. Int. Seaweed Symp. 9, 17–34.

Jensen, A., Indergaard, M., and Holt, T. J., 1985: Seasonal variation in the chemical composition of *Saccorhiza polyschides*. Botanica Mar. 28, 375–381.

Jensen, J. B., 1974: Morphological studies in Cystoseiraceae and Sargassaceae (Phaeophyceae). Univ. Calif. Publs. Bot. 68, 1–61.

Jerlov, N. G., 1951: Optical studies of ocean waters. Rep. Swed. Deep- Sea Exped. 3, 1–59.

Jerlov, N. G., 1974: A simple method for measuring quanta irradiance in the ocean. Rep. Kjob. Univ. Inst. Fys. Oceanogr. 24, 1–7.

Jerlov, N. G., 1976: Marine optics. Elsevier, Amsterdam; 231 pp.

Jerlov, N. G., 1977: Classification of sea water in terms of quanta irradiance. J. Cons. Explor. Mer 37, 281–287.

Jerlov, N. G., 1978: The optical classification of sea water in the euphotic zone. Rep. Kjob. Univ. Inst. Fys. Oceanogr. 36, 1–46.

Jerlov, N. G., and Nygard, K., 1969: A quanta and energy meter for photosynthetic studies. Rep. Kjob. Univ. Inst. Fys. Oceanogr. 10, 1–19.

Jernakoff, P., 1985: Interactions between the limpet *Patelloida latistrigata* and algae on an intertidal rock platform. Mar. Ecol. Prog. Ser. 23, 71–78.

Jerzmanska, A., and Kotlarczyk, J., 1976: The beginnings of the Sargasso assemblage in the Tethys? Palaeogeogr. Palaeoclimatol. Palaeoecol. 20, 297–306.

Johannes, R. E., Wiebe, W. J., Crossland, C.J., Rimmer, D. W., and Smith, S. V., 1983: Latitudinal limits of coral reef growth. Mar. Ecol. Prog. Ser. 11, 105–111.

Johansen, H. W., 1971a: Effects of elevation changes on benthic algae in Prince William Sound. In: The great Alaska earthquake of 1964: Biology. National Academy of Sciences, Washington, D. C., pp. 35–68.

Johansen, H. W., 1981a: Coralline algae, a first synthesis. CRC Press, Boca Raton, Fla.; 239 pp.

John, D. M., 1971: The distribution and net productivity of sublittoral populations of attached macrophytic algae in an estuary on the Atlantic coast of Spain. Mar. Biol. 11, 90–97.

John, D. M., 1986: Coastal vegetation. Littoral and sub-littoral marine vegetation. In: Plant ecology in West Africa. Ed. by G. W. Lawson. Wiley, Chichester, pp. 195–246.

John, D. M., and Lawson, G. W., 1974: Observations on the marine algal ecology of Gabon. Botanica Mar. 17, 249–254.

John, D. M., Lieberman, D., and Lieberman, M., 1977: A quantitative study of the structure and dynamics of benthic subtidal algal vegetation in Ghana (West Africa). J. Ecol. 65, 497–521.

John, D. M., and Price, J. H., 1979: The marine benthos of Antigua (Lesser Antilles). I. Environment, distribution, and ecology. Botanica Mar. 22, 313–331.

Johnson, C. R., and Mann, K. H., 1986: The crustose coralline alga, *Phymatolithon* Foslie, inhibits the overgrowth of seaweeds without relying on herbivores. J. Exp. Mar. Biol. Ecol. 96, 127–146.

Johnson, B. D., Powell, C. McA., amd Veevers, J. J., 1980: Early spreading history of the Indian Ocean between India and Australia. Earth Planet. Sci. Lett. 47, 131–143.

Johnson, T. W., and Sparrow, F. K., 1961: Fungi in oceans and estuaries. Cramer, Weinheim; 668 pp.

Johnston, C. S., Jones, R. G., and Bunt, R. D., 1977: A seasonal carbon budget for a laminarian population in a Scottish sea-loch. Helgoländer Meeresunters. 30, 527–545.

Johnston, R. (Ed.), 1976: Heavy metal contamination in the sea. Academic Press, London; 729 pp.

Joly, A. B. (1965): Flora marinha do litoral norte do estado de São Paulo e regiões circunvizinhas. Bol. Filos. Ciênc. Letr. Univ. São Paulo 294 (Bot. 21), 1–393.

Joly, A. B., and Oliveira Filho, E. C. de, 1967: Two Brazilian Laminarias. Publ. Inst. Pesq. Mar. 4, 1–13.

Jones, N. S., and Kain, J. M., 1967: Subtidal algal colonization following the removal of *Echinus*. Helgoländer Meeresunters. 15, 460–466.

Jones, O. A., and Endean, R., 1973–1976: Biology and geology of coral reefs. Vol. II. Biology 1 (1973); Biology 2 (1976). Academic Press, New York and London; 480 pp. (1973); 435 pp. (1976).

Jones, R., and Kraft, G. T., 1984: The genus *Codium* (Codiales, Chlorophyta) at Lord Howe Island (N.S.W.). Brunonia 7, 253–276.

Jones, W. E., 1974: Changes in the seaweed flora of the British Isles. In: The changing flora and fauna of Britain. Ed. by D. L. Hawksworth. Academic Press, London, pp. 97–113.

Jones, W. E., and Demetropoulos, A., 1968: Exposure to wave action: measurements of an important ecological parameter on rocky shores of Anglesey. J. Exp. Mar. Biol. Ecol. 2, 46–63.

Jonker, F. P., 79: *Prototaxites* in the lower Devonian. Paleontographica, Abt. B, 171, 39–56.

Jonsson, S., 1970: Meeresalgen als Erstbesiedler der Vulkaninsel Surtsey. Schr. Naturw. Ver. Schlesw.-Holst. Sonderband 1970, 21–28.

Jonsson, S., and Gunnarsson, K., 1982: Marine algal colonization of Surtsey. Surtsey Res. Progr. Rep. 9, 33–45.

Joosten, A. M. T., and van den Hoek, C., 1986: World-wide relationships between red seaweed floras: a multivariate approach. Botanica Mar. 29, 195–214.

Jorde, I., and Klavestad, N., 1963: The natural history of the Hardangerfjord. 4. The benthonic algal vegetation. Sarsia 9, 1–99.

Joska, M. A. P., and Bolton, J. J., 1987: In situ measurement of zoospore release and seasonality of reproduction in *Ecklonia maxima* (Alariaceae, Laminariales). Br. Phycol. J. 22, 209–214.

Jupp, B. P., and Drew, E. A., 1974: Studies on the growth of *Laminaria hyperborea* (Gunn.) Fosl. I. Biomass and productivity. J. Exp. Mar. Biol. Ecol. 15, 185–196.

Kaestner, A., 1984: Lehrbuch der speziellen Zoologie. Teil I. Wirbellose. VEB Gustav Fischer, Jena; 621 pp.

Kain, J. M., 1962: Aspects of the biology of Laminaria hyperborea. I. Vertical distribution. J. Mar. Biol. Ass. U. K. 42, 377–385.

Kain, J. M., 1963: Aspects of the biology of *Laminaria hyperborea*. II. Age, weight, and length. J. Mar. Biol. Ass. U. K. 43, 129–151.

Kain, J. M. (Mrs. N. S. Jones), 1964: Aspects of the biology of *Laminaria hyperborea*. III. Survival and growth of gametophytes. J. Mar. Biol. Ass. U. K. 44, 415–433.

Kain, J. M. (Mrs. N. S. Jones), 1969: Aspects of the biology of *Laminaria hyperborea*. V. Comparison with early stages of competitors. J. Mar. Biol. Ass. U. K. 49, 455–473.

Kain, J. M. (Mrs. N. S. Jones), 1975a: The biology of *Laminaria hyperborea*. VII. Reproduction of the sporophyte. J. Mar. Biol. Ass. U. K. 55, 567–582.

Kain, J. M. (Mrs. N. S. Jones), 1975b: Algal recolonization of some cleared subtidal areas. J. Ecol. 63, 739–765.

Kain, J. M. (Mrs. N. S. Jones), 1976a: New and interesting marine algae from the Shetland Isles. II. Hollow and solid stiped *Laminaria* (Simplices). Br. Phycol. J. 11, 1–11.

Kain, J. M. (Mrs. N. S. Jones), 1976b: The biology of *Laminaria hyperborea*. VIII. Growth on cleared areas. J. Mar. Biol. Ass. U. K. 56, 267–290.

Kain, J. M. (Mrs. N. S. Jones), 1979: A view of the genus *Laminaria*. Oceanogr. Mar. Biol. Ann. Rev. 17, 101–161.

Kain, J. M. (Jones), 1987a: Seasonal growth and photoinhibition in *Plocamium cartilagineum* (Rhodophyta) off the Isle of Man. Phycologia 26, 88–99.

Kain, J. M. (Jones), 1987b: Photoperiod and temperature as triggers in the seasonality of *Delesseria sanguinea*. Helgolander Meeresunters. 41, 355–370.

Kain, J. M. (Jones), 1987c: Patterns of relative growth in *Nereocystis luetkeana* (Phaeophyta). J. Phycol. 23, 181–187.

Kain, J. M. (Mrs. N. S. Jones), and Svendsen, P., 1969: A note on the behaviour of *Patina pellucida* in Britain and Norway. Sarsia 38, 25–30.

Kajimura, M., 1981: *Streptophyllopsis*, a new genus of Laminariaceae, Phaeophyta, from Japan. Mem. Fac. Sci. Shimane Univ. 15, 75–87.

Kajimura, M., 1987: Typification of *Streptophyllopsis kuroshioensis* (Segawa) Kajimura (Phaeophyta, Laminariaceae). Jap. J. Phycol. (Sorui) 35, 19–21.

Kamura, S., and Choonhabandit, S., 1986: Distribution of benthic marine algae on the coasts of Khang Khao and Thai Ta Mun, Sichang Islands, the Gulf of Thailand. Galaxea 5, 97–114.

Kang, J. W., 1966: On the geographical distribution of marine algae in Korea. Bull. Pusan Fish. Coll. 7 (1/2), 1–125.

Kanwisher, J., 1957: Freezing and drying in intertidal algae. Biol. Bull. Mar. Biol. Lab. Woods Hole 116, 275–285.

Kanwisher, J. W., 1966: Photosynthesis and respiration in some seaweeds. In: Some contemporary studies in marine sciences. Ed. by H. Barnes. George Allen & Unwin, London, pp. 407–420.

Kappen, L., 1981: Ecological significance of resistance to high temperature. In: Encyclopedia of plant physiology, n. s., Vol. 12A. Ed. by O. L. Lange, P. S. Nobel, C. B. Osmond, and H. Ziegler. Springer-Verlag, Berlin, pp. 439–474.

Kapraun, D. F., 1978: Field and culture studies on growth and reproduction of *Callithamnion byssoides* (Rhodophyta, Ceramiales) in North Carolina. J. Phycol. 14, 21–24.

Kapraun, D. F., 1980: Floristic affinities of North Carolina inshore benthic marine algae. Phycologia 19, 245–252.

Kapraun, D. F., 1980–1984: An illustrated guide to the benthic marine algae of coastal North Carolina. I. Rhodophyta (1980). II. Chlorophyta and Phaeophyta (1984). University of North Carolina Press, Chapel Hill; 206 pp. (1980); 174 pp. (1984).

Kapraun, D. F., and Martin, D. J., 1987: Karyological studies of three species of *Codium* (Codiales, Chlorophyta) from coastal North Carolina. Phycologia 26, 228–234.

Karsten, U., Wiencke, C., and Kirst, G. O., 1989: The β-dimethylsulphonio propionate (DMSP) content of macroalgae from Antarctica and southern Chile. Botanica Mar. 32, 143–146.

Karsten, U., and Kirst, G. O., 1989: The effect of salinity on growth, photosynthesis, and respiration in the estuarine red alga *Bostrychia radicans* Mont. Helgoländer Meeresunters. 43, 61–66.

Katada, M., and Satomi, M., 1975: Ecology of marine algae. In: Advance of phycology in Japan. Ed. by J. Tokida and H. Hirose. VEB Gustav Fischer, Jena, pp. 211–239.

Kauss, H., 1978: Osmotic regulation in algae. In: Progress in phytochemistry. Vol. 5. Pergamon Press, Oxford, pp. 1–27.

Kautsky, N., Kautsky, H., Kautsky, U., and Waern, M., 1986: Decreased depth penetration of *Fucus vesiculosus* (L.) since the 1940's indicates eutrophication of the Baltic Sea. Mar. Ecol. Prog. Ser. 28, 1–8.

Kawai, H., and Kurogi, M., 1985: On the life history of *Pseudochorda nagaii* (Pseudochordaceae fam. nov.) and its transfer from the Chordariales to the Laminariales. Phycologia 24, 289–296.

Keast, A., 1973: Contemporary biotas and the separation sequence of the southern continents. In: Implications of continental drift to the earth sciences. Vol. 1. Ed. by D. H. Tarling, D. H. and S. K. Runcorn. Academic Press, London, pp. 309–343.

Keast, J. A., 1983: In the steps of Alfred Russel Wallace: biogeography of the Asian-Australian interchange zone. In: Evolution, time and space: the emergence of the biosphere. Ed. by R. W. Sims, J. H. Price and P. E. S. Whalley. Academic Press, London, pp. 367–407.

Keats, D. W., South, G. R., and Steele, D. H., 1985: Algal biomass and diversity in the upper subtidal at a pack-ice disturbed site in eastern Newfoundland. Mar. Ecol. Prog. Ser. 25, 151–158.

Keats, D. W., Steele, D. H., and South, G. R., 1984: Depth-dependent reproductive output of the green sea urchin, *Strongylocentrotus droebachiensis* (O. F. Muller), in relation to the nature and availability of food. J. Exp. Mar. Biol. Ecol. 80, 77–91.

Keigwin, L. D., 1978: Pliocene closing of the isthmus of Panama, based on biostratigraphic evidence from nearby Pacific Ocean and Caribbean Sea cores. Geology 6, 630–634.

Keller, G., 1988: Extinction, survivorship and evolution of planktic Foraminifera across the Cretaceous/Tertiary boundary at El Kef, Tunisia. Mar. Micropal. 13, 239–263.

Kendrick, R. E., and Kronenberg, G. H. M., 1986: Photomorphogenesis in plants. Nijhoff, Dordrecht; 580 pp.

Kennelly, S. J., 1987a: Physical disturbances in an Australian kelp community. I. Temporal effects. Mar. Ecol. Prog. Ser. 40, 145–153.

Kennelly, S. J., 1987b: Physical disturbances in an Australian kelp community. II. Effects on understorey species due to differences in kelp cover. Mar. Ecol. Prog. Ser. 40, 155–165.

Kennett, J. P., 1977: Cenozoic evolution of Antarctic glaciation, the circum- Antarctic Ocean, and their impact on global paleoceanography. J. Geophys. Res. 82, 3843–3860.

Kennett, J. P., 1982: Marine geology. Prentice-Hall, Englewood Cliffs, N.J.; 813 pp.

Kent, D. V., and Gradstein, F. M., 1986: A Jurassic to recent chronology. In: The western North Atlantic region. Ed. by P. R. Vogt and B. E. Tucholke, Geol. Soc. Am., DNAG-series, Vol. M; pp. 45–50.

Kenny, R., and Haysom, N., 1962: Ecology of rocky shore organisms at Macquarie Island. Pacif. Sci. 16, 245–263.

Kent, D. V., McKenna, M. C., Opdyke, N. D., Flynn, J. J., and MacFadden, B. J., 1984: Technical comment: Arctic biostratigraphic heterochroneity. Science 224, 173–174.

Ketchum, B. H. (Ed.), 1983: Estuaries and enclosed seas. Elsevier, Amsterdam; 500 pp.

Khailov, K. M., 1976: The relationships between weight, length, age, and intensity of photosynthesis and organotrophy of macrophytes in the Barents Sea. Botanica Mar. 19, 335–339.

Khailov, K. M., 1978: Changes of the mass, length, and metabolism of simple and composite thalli of marine macrophytes in their ontogenetic series. Botanica Mar. 21, 313–321.

Khailov, K. M., 1988: Physico-chemical modeling of the relationship between the external structure of the thallus and interactions with the environment in marine macrophytes. Trudy-Bot. Inst. Akad. Nauk SSSR., ser. 2, 35, 1085–1091.

Khfaji, A. K., and Norton, T. A., 1979: The effects of salinity on the distribution of *Fucus ceranoides*. Estuar. Coast. Mar. Sci. 8, 433–439.

Khoja, T. M., 1987: New records of marine algae for the Red Sea coast of Saudi Arabia. Botanica Mar. 30, 167–176.

Kim, E.-A., Lee, H. B., and Lee, I. K., 1986: Marine algal vegetation of Samchonpo, south coast of Korea. Korean J. Bot. 29, 175–183.

Kim, H. S., Lee, I. K., Koh, C. H., Kim, I. H., Suh, Y. B., and Sung, N., 1983: Studies on the marine benthic communities in inter- and subtidal zones. I. Analysis of benthic community structures at Aninjin, eastern coast of Korea. Proc. Coll. Natur. Sci. SNU 8, 71–108.

Kim, Y. H., and Lee, J. H., 1981: Intertidal marine algal community and species composition of Wolseong area, east coast of Korea. Korean J. Bot. 24, 145–158.

Kim, Y. H., and Lee, J. H., 1985: The structure analysis of intertidal algal community in Muchangpo, western coast of Korea. Korean J. Bot. 28, 149–164.

King, R. J., 1981: Mangroves and saltmarsh plants. In: Marine botany: an Australasian perspective. Ed. by M. N. Clayton and R. J. King. Longman Cheshire, Melbourne, pp.308–328.

Kingsbury, J. M., 1969: Seaweeds of Cape Cod and the Islands. Chatham Press, Chatham, Mass.; 212 pp.

Kinsey, D. W., and Davies, P. J., 1979: Effects of elevated nitrogen and phosphorus on coral reef growth. Limnol. Oceanogr. 24, 935–940.

Kirk, J. T. O., 1977: Thermal dissociation of fucoxanthin-protein binding in pigment complexes from chloroplasts of *Hormosira* (Phaeophyta). Plant Sci. Lett. 9, 373–380.

Kirk, J. T. O., 1983: Light and photosynthesis in aquatic ecosystems. University Press, Cambridge; 399 pp.

Kirkman, H., 1981: The first year in the life history and the survival of the juvenile marine macrophyte, *Ecklonia radiata* (Turn.) J. Agardh. J. Exp. Mar. Biol. Ecol. 55, 243–254.

Kirkman, H., 1984: Standing stock and production of *Ecklonia radiata* (C. Ag.) J. Agardh. J. Exp. Mar. Biol. Ecol. 76, 119–130.

Kirkman, H., 1985: Community structure in seagrasses in southern western Australia. Aquat. Bot. 21, 363–375.

Kirkman, H., 1989: Growth, density and biomass of *Ecklonia radiata* at different depths and growth under artificial shading off Perth, Western Australia. Aust. J. Mar. Freshwater Res. 40, 169–177.

Kirkman, H., and Young, P. C., 1981: Measurement of health and echinoderm grazing on *Posidonia oceanica* (L.) Delile. Aquat. Bot. 10, 329–338.

Kirst, G. O., 1981: Photosynthesis and respiration of *Griffithsia monilis* (Rhodophyceae): effect of light, salinity, and oxygen. Planta 151, 281–288.

Kirst, G. O., 1988: Turgor pressure regulation in marine macroalgae. Organic osmotica: their role as "compatible solutes" in response to salinity. In: Experimental phycology. A laboratory manual. Ed. by C. S. Lobban, D. J., Chapman, and B. P. Kremer. University Press, Cambridge, pp. 203–216.

Kirst, G. O., 1990: Salinity tolerance of eukaryotic marine algae. Annu. Rev. Plant Physiol. 41.

Kirst, G. O., and Bisson, M. A., 1979: Regulation of turgor pressure in marine algae: ions and low-molecular-weight organic compounds. Aust. J. Plant Physiol. 6, 539–556.

Kitchell, J. A., and Clark, D. L., 1982: Late Cretaceous-Paleogene paleogeography and paleocirculation: evidence of north polar upwelling. Palaeogeogr. Palaeoclimatol. Palaeoecol. 40, 135–165.

Kitching, J. A. 1941: Studies in sublittoral ecology. III. *Laminaria* forest on the west coast of Scotland; a study of zonation in relation to wave action and illumination. Biol. Bull. Mar. Biol. Lab. Woods Hole 80, 324–337.

Kitching, J. A., 1987: Ecological studies at Lough Hyne. Adv. Ecol. Res. 17, 115–186.

Kjellman, F. R., 1877: Ueber die Algenvegetation des Murmanschen Meeres an der Westküste von Nowaja Semlja und Wajgatsch. Nova Acta R. Soc. Sci. Upsal. Ser. III, 1877, 1–86.

Kjellman, F. R., 1878: Über Algenregionen und Algenformationen im östlichen Skager Rack. Bih. K. svenska Vetensk. Akad. Handl. 5 (6), 1–35.

Kjellman, F. R., 1883: The Algae of the Arctic Sea. Kongl. Boktryckeriet, Stockholm; 350 pp.

Kjellman, F. R., 1889: Om Beringhafvets Algflora. K. Svenska Vetensk. Akad. Handl. 23 (8), 1–58.

Kjellman, F. R., 1903: Über die Meeresalgen-Vegetation von Beeren Eiland. Ark. Bot. 1, 1–6.

Kjellman, F. R., 1906: Zur Kenntnis der marinen Algenflora von Jan Mayen. Ark. Bot. 5 (14), 1–29.

Klavestad, N., 1978: The marine algae of the polluted inner part of the Oslofjord. A survey carried out 1962–1966. Botanica Mar. 21, 71–97.

Klinger, T., and De Wreede, R. E., 1988: Stipe rings, age, and size in populations of *Laminaria setchellii* Silva (Laminariales, Phaeophyta) in British Columbia, Canada. Phycologia 27, 234–240.

Knight, M., and Parke, M., 1950: A biological study of *Fucus vesiculosus* L. and *Fucus serratus* L. J. Mar. Biol. Ass. U. K. 29, 439–514.

Knight-Jones, E. W., Bailey, J. H., and Isaac, M. J., 1971: Choice of algae by larvae of *Spirorbis*, particularly of *Spirorbis spirorbis*. In: Fourth European Marine Biology Symposium. Ed. by D. J. Crisp. University Press, Cambridge, pp. 89–104.

Knox, G. A., 1960: Littoral ecology and biogeography of the southern oceans. Proc. R. Soc. Lond. B 152, 577–624.

Knox, G. A., 1963: The biogeography and intertidal ecology of the Australasian coasts. Oceanogr. Mar. Biol. Annu. Rev. 1, 341–404.

Knox, G. A., 1970: Antarctic marine ecosystems. In: Antarctic ecology. Ed. by M. Holdgate, Academic Press, London, pp. 67–96.

Knox, G. A., 1975: The marine benthic ecology and biogeography. In: Biogeography and ecology in New Zealand. Ed. by G. Kueschel. Junk, the Hague, pp. 353–403.

Knox, G. A., 1980: Plate tectonics and the evolution of intertidal and shallow- water benthic biotic distribution patterns of the Southwest Pacific. Palaeogeogr. Palaeoclimatol. Palaeoecol. 31, 153–196.

Koch, W., 1951: Historisches zum Vorkommen der Rotalge *Trailliella intricata* (Batters) bei Helgoland. Arch. Mikrobiol. 16, 78–79.

Koehl, M. A. R., 1982: The interaction of moving water and sessile organisms. Sci.. Am. 247, 124–132.

Koehl, M. A. R., 1986: Seaweeds in moving water: form and mechanical function. In: On the economy of plant form and function. Ed. by T. J. Givnish. University Press, Cambridge, pp. 603–634.

Koehl, M. A. R., and Alberte, R. S., 1988: Flow, flapping, and photosynthesis of *Nereocystis luetkeana*: a functional comparison of undulate and flat blade morphologies. Mar. Biol. 99, 435–444.

Koehl, M. A. R., and Wainwright, S. A., 1977: Mechanical adaptations of a giant kelp. Limnol. Oceanogr. 22, 1067–1071.

Koehl, M. A. R., and Wainwright, S. A., 1985: Biomechanics. In: Handbook of phycological methods. Ecological field methods: macroalgae. Ed. by M. M. Littler and D. S. Littler. University Press, Cambridge, pp. 291–313.

Koeman, R. T. P., and van den Hoek, C., 1981: The taxonomy of *Ulva* (Chlorophyceae) in the Netherlands. Br. Phycol. J. 16, 9–53.

Koeman, R. T. P., and van den Hoek, C., 1982a: The taxonomy of *Enteromorpha* (Chlorophyceae) in the Netherlands. 1. The section *Enteromorpha*. Arch. Hydrobiol. 63, (Suppl.) 279–330.

Koeman, R. T. P., and van den Hoek, C., 1982b: The taxonomy of *Enteromorpha* (Chlorophyceae) in the Netherlands. 2. The section Proliferae. Cryptogam. Algol. 3, 37–70.

Koeman, R. T. P., and van den Hoek, C., 1984: The taxonomy of *Enteromorpha* (Chlorophyceae) in the Netherlands. 3. The sections Flexuosae and Clathratae and an addition to the section Proliferae. Cryptogam. Algol. 5, 21–61.

Kohlmeyer, J., and Kohlmeyer, E., 1972: Is *Ascophyllum nodosum* lichenized? Botanica Mar. 15, 109–112.

Kohlmeyer, J., and Kohlmeyer, E., 1979: Marine mycology. The higher fungi. Academic Press, New York; 690 pp.

Koop, K., Newell, R. C., and Lucas, M. I., 1982: Biodegradation and carbon flow based on kelp (*Ecklonia maxima*) debris in a sandy beach microcosm. Mar. Ecol. Prog. Ser. 7, 315–326.

Kondratyev, K. Ya, 1954: Radiant solar energy (in Russian). Leningrad.

Kontrimavichus, V. L. (Ed.), 1986: Beringia in the Cenozoic era. Russian Translation Series 28. Balkema, Rotterdam; 724 pp.

Kornas, J., 1972: Corresponding taxa and their ecological background in the forests of temperate Eurasia and North America. In: Taxonomy, phytogeography and evolution. Ed. by D. H. Valentine. Academic Press, London; 431 pp.

Kornas, J. E., Pancer, E., and Brzyski, B., 1960: Studies on sea-bottom vegetation in the Bay of Gdansk off Rewa. Fragm. flor. geobot. Ann. 6, 1–92.

Kornmann, P., 1986: *Porphyra yezoensis* bei Helgoland-eine entwicklungsgeschichtliche Studie. Helgoländer Meeresunters. 40, 327–342.

Kornmann, P., and Sahling, P.-H., 1974: Prasiolales (Chlorophyta) von Helgoländ. Helgoländer Meeresunters. 26, 99–133.

Kornmann, P., and Sahling, P.-H., 1977: Meeresalgen von Helgoland. Benthische Grun-, Braun- und Rotalgen. Helgoländer wiss. Meeresunters. 29, 1–289.

Kornmann, P., and Sahling, P.-H., 1978: Die *Blidingia*-Arten von Helgoland (Ulvales, Chlorophyta). Helgoländer wiss. Meeresunters. 31, 391–413.

Kornmann, P., and Sahling, P.-H., 1980: Kalkbohrende Mikrothalli bei *Helminthocladia* und *Scinaia* (Nemaliales, Rhodophyta). Helgolander Meeresunters. 34, 31–40.

Kornmann, P., and Sahling, P.-H., 1983: Meeresalgen von Helgoland: Ergänzung. Helgoländer Meeresunters. 36, 1–65.

Kosswig, C., 1956: Beitrag zur Faunengeschichte des Mittelmeeres. Pubbl. Stn. Zool. Napoli 28, 78–88.

Kowallik, W., 1982: Blue light effects on respiration. Annu. Rev. Plant Physiol. 33, 51–72.

Kowallik, W., and Schurmann, R., 1984: Chlorophyll a/chlorophyll b ratios of *Chlorella vulgaris* in blue or red light. In: Blue light effects in biological systems. Ed. by H. Senger. Springer, Berlin, pp. 352–358.

Kozloff, E. N., 1983: Seashore life of the northern Pacific coast. An illustrated guide to northern California, Oregon, Washington, and British Columbia. University of Washington Press, Seattle; 370 pp.

Kraft, G. T., and Olsen-Stojkovich, J., 1985: *Avrainvillea calithina* (Udoteaceae, Bryopsidales), a new green alga from Lord Howe Island, NSW, Australia. Phycologia 24, 339–345.

Krassilov, V. A., 1986: Role of the Bering land connections in the formation of the Cenozoic flora of East Asia and North America. In: Beringia in the Cenozoic era. Ed. by V. L. Kontrimavichus. Russian Translation Series 28. Balkema, Rotterdam; pp. 164–172.

Krause, H., 1988: Photoinhibition of photosynthesis. An evaluation of damaging and protective mechanisms. Physiologia Pl. 74, 566–574.

Kremer, B. P., 1980: Taxonomic implications of algal photoassimilate patterns. Br. Phycol. J. 15, 399–409.

Kremer, B. P., 1981a: Carbon metabolism. In: The biology of seaweeds. Ed. by C. S. Lobban and M. J. Wynne. Blackwell, Oxford, pp. 493–533.

Kremer, B. P., 1981b: Aspects of carbon metabolism in marine macroalgae. Oceanogr. Mar. Biol. Annu. Rev. 19, 41–94.

Kremer, B. P., 1983: Carbon economy and nutrition of the alloparasitic red alga *Harveyella mirabilis*. Mar. Biol. 76, 231–239.

Kremer, B. P., Kuhbier, H., and Michaelis, H., 1983: Die Ausbreitung des Brauntanges *Sargassum muticum* in der Nordsee. Natur Mus. 113, 125–130.

Krishnamurthy, V., and Yoshi, H. V., 1970: A check-list of Indian marine algae. Cent. Salt Mar. Chem. Res. Inst. Bhavnagar India, 1970, 1–36.

Kristensen, I., 1968: Surf influence on the thallus of fucoids and the rate of desiccation. Sarsia 34, 69–82.

Kristoffersen, Y., and Huseby, E. S., 1985: Multi-channel seismic reflection measurements in the Eurasian Basin, Arctic Ocean, from U.S. ice station FRAM IV. Tectonophysics 114, 103–115.

Krümmel, O., 1891: Die nordatlantische Sargassosee. Petermanns geogr. Mitt. 37, 129–141.

Kuckuck, P., 1894: Bemerkungen zur marinen Algenvegetation Helgolands. Wiss. Meeresunters. (Helgoland) 1, 225–263.

Kuckuck, P., 1897: Über marine Vegetationsbilder. Ber. Dt. Bot. Ges. 15, 441–447.

Kufer, W., and Björn, G. S., 1989: Photochromism of the cyanobacterial light harvesting biliprotein phycoerythrocyanin. Physiologia Pl. 75, 389–394.

Kühlmann, D. H. H., 1982: Darwin's coral reef research-a review and tribute. P. S. Z. N. I. Mar. Ecol. 3, 193–212.

Kühlmann, D., 1985: Coral reefs of the world. Arco, New York; 185 pp.

Kühnemann, O., 1969: Vegetación marina de la ría de Puerto Deseado. Opera Lilloana 17, 1–123.

Kühnemann, O., 1972: Bosquejo fitogeográfico de la vegetación marina del litoral Argentino. Physis 31, 117–142, 295–325.

Küppers, U., and Weidner, M., 1980: Seasonal variation of enzyme activities in *Laminaria hyperborea*. Planta 148, 222–230.

Kumke, J., 1973: Beiträge zur Periodizität der Oogon-Entleerung bei *Dictyota dichotoma* (Phaeophyta). Z. Pflanzenphysiol. 70, 191–210.

Kusel, H., 1972: Contribution to the knowledge of the seaweeds of Cuba. Botanica Mar. 15, 186–198.

Kussakin, O. G., 1961: (On the characteristic of the fauna and flora in the intertidal of the Kurilen; in Russian). Invest. Far East Seas USSR. 7, 312–343.

Kussakin, O. G., 1977: Intertidal ecosystems of the seas of the USSR. Helgoländer Meeresunters. 30, 243–262.

Kyle, D. J., Osmond, C. B., and Arntzen, C. J. (Eds.), 1987: Photoinhibition. Elsevier, Amsterdam; 316 pp.

Kylin, H., 1925: The marine red algae in the vicinity of the Biological Station at Friday Harbor, Wash. Lunds Univ. Arsskr. N. F., Avd. 2, 21 (9), 1–87.

Kylin, H., 1944–1949: Die Rhodophyceen der schwedischen Westküste. Die Phaeophyceen der schwedischen Westküste. Die Chlorophyceen der schwedischen

Westküste. Lunds Univ. Årsskr. N. F., Avd. 2, 40 (2), 1–104 (1944); 43 (4), 1–99 (1947); 45 (4), 1–79 (1949).

LaBrecque, J. L., Barker, P., 1981: The age of the Weddell basin. Nature 290, 489–492.

Lakowitz, K., 1929: Die Algenflora der gesamten Ostsee (ausschl. Diatomeen). Friedländer, Danzig; 474 pp.

Lamb, M., and Zimmermann, M. H., 1964: Marine vegetation of Cape Ann, Essex County, Massachusetts. Rhodora 66, 217–254.

Lamb, M., and Zimmermann, M. H., 1977: Benthic marine algae of the Antarctic peninsula. Am. Geophys. Union Antarct. Res. Ser. 23 (4), 130–229.

Lancelot, A., 1961: Recherches biologiques et océanographiques sur les vegetaux marins des côtes françaises entre la Loire et la Gironde. Muséum National d'Histoire de Cryptogamie, Paris; 210 pp.

Land, J. van der, 1989: The need for new concepts of reef ecology. Neth. J. Sea Res. 23, 231–238.

Lang, J. C., 1974: Biological zonation at the base of a reef. Am. Scient. 62, 271–281.

Larcher, W., 1980: Ökologie der Pflanzen auf physiologischer Grundlage. Eugen Ulmer, Stuttgart; 399 pp.

Larcher, W., and Bauer, H., 1981: Ecological significance of resistance to low temperature. In: Encyclopedia of plant physiology, n.s., Vol. 12A. Ed. by O. L. Lange, P. S. Nobel, C. B. Osmond, and H. Ziegler. Springer-Verlag, Berlin; pp. 403–437.

Larkum, A. W. D., McComb, A. J., and Shepherd, S. A. (Eds), 1989: Biology of seagrasses. A treatise on the biology of seagrasses with special reference to the Australian region. Elsevier, Amsterdam; 885 pp.

Larkum, A. W. D., and Barrett, 1983: Light-harvesting processes in algae. Adv. bot. Res. 10, 1–219.

Larkum, A. W. D., Drew, E. A., and Crossett, R. N., 1967: The vertical distribution of attached marine algae in Malta. J. Ecol. 55, 361–371.

Larkum, A. W. D., Cox, G. C., Hiller, R. G., Parry, D. L., and Dibbayawan, T. P., 1987: Filamentous cyanophytes containing phycourobilin and in symbiosis with sponges and an ascidian of coral reefs. Mar. Biol. 95, 1–13.

Larkum, A. W. D., McComb, A. J., Shepherd, S. A., 1989: Biology of seagrasses. A treatise on the biology of seagrasses with special reference to the Australian region. Elsevier Science Publishers, New York; 885 pp.

Lattin, G. de, 1967: Grundriss der Zoogeographie. Gustav Fischer, Stuttgart; 602 pp.

Laubier, L., 1966: Le Coralligène des Albères. Monographie biocénotique. Ann. Inst. Océanogr. Paris 43, 137–316.

Law, R., and Lewis, D. H., 1983: Biotic environment and the maintenance of sex-some evidence from mutualistic symbioses. Biol. J. Linn. Soc. 20, 249–276.

Lawrence, J. M., 1975: On the relationships between marine plants and sea urchins. Oceanogr. Mar. Biol. Annu. Rev. 13, 213–286.

Lawrence, J. M., 1986: Proximate composition and standing crop of *Durvillaea antarctica* (Phaeophyta) in the Bay of Morbihan, Kerguelen (South Indian Ocean). Mar. Ecol. Prog. Ser. 33, 1–5.

Laws, R. M., 1984: Antarctic ecology. Vol. 20 Academic Press, London: 851 pp.

Lawson, G. W., 1956: Rocky shore zonation on the Gold Coast. J. Ecol. 44, 153–170.

Lawson, G. W., 1966: The littoral ecology of West Africa. Oceanogr. Mar. Biol. Annu. Rev. 4, 405–448.

Lawson, G. W., 1978: The distribution of seaweed floras in the tropical and subtropical Atlantic Ocean: a quantitative approach. J. Linn. Soc. Lond. Bot. 76, 177–193.

Lawson, G. W., 1980: A check-list of East African seaweeds (Djibouti to Tanzania). Department of Biological Sciences, University of Lagos, Lagos, Nigeria; pag. var.

Lawson, G. W., 1986: Plant ecology in West Africa. Ed. by G. W. Lawson. Wiley, Chichester; 357 pp.

Lawson, G. W., 1988: Seaweed distribution patterns as revealed by ordination with reference to the Atlantic and southern Oceans. Helgoländer Meeresunters. 42, 187–197.

Lawson, G. W., and John, D. M., 1977: The marine flora of the Cap Blanc peninsula: its distribution and affinities. J. Linn. Soc. Lond. Bot. 75, 99–118.

Lawson, G. W., and John, D. M., 1982: The marine algae and coastal environment of tropical West Africa. Nova Hedwigia (Beih.) 70, 1–455.

Lawson, G. W., John, D. M., and Price, J. H., 1975: The marine algal flora of Angola: its distribution and affinities. J. Linn. Soc. Lond. Bot. 70, 307–324.

Lawson, G. W., and Norton, T. A., 1971: Some observations on littoral and sublittoral zonation at Teneriffe (Canary Isles). Botanica Mar. 14, 116–120.

Lawver, L. A., Sclater, J. G., and Meinke, L., 1985: Mesozoic and Cenozoic reconstructions of the South Atlantic. Tectonophysics 114, 233–234.

Laycock, R. A., 1974: The detrital food chain based on seaweeds. I. Bacteria associated with the surface of *Laminaria* fronds. Mar. Biol. 25, 223–231.

Lebednik, P. A., Weinmann, F. C., and Norris, R. E., 1971: Spatial and seasonal distributions of marine algal communities at Amchitka Island, Alaska. BioScience 21, 656–660.

Leclerc, J. C., Couté, A., and Dupuy, P., 1983: Le climat annuel de deux grottes et d'une église du Poitou, ou vivent des colonies pures d'algues sciaphiles. Cryptogam. Algol. 4, 1–19.

Lee, B. D., and Titlyanov, E. A., 1978: Light adaptation of benthic plants. III. Photosynthetic pigment content of macrophytes from variously illuminated habitats. Biol. Morya (Vladivost.) 1978 (2), 47–55.

Lee, H. B., and Lee, I. K., 1981: Flora of benthic marine algae in Gyeonggi Bay, western coast of Korea. Korean J. Bot. 3, 107–138.

Lee, I. K., 1982: *Halosaccion americanum* sp. nov. (Rhodophyta, Palmariaceae) in Pacific North America. Jap. J. Phycol. (Sorui) 30, 265–271.

Lee, I. K., and Kang, J. W., 1986: A check list of marine algae in Korea. Korean J. Phycol. 1, 311–325.

Lee, R. K. S., 1973: General ecology of the Canadian Arctic benthic marine algae. Arctic 26, 32–43.

Lee, R. K. S., 1980: A catalogue of the marine algae of the Canadian Arctic. Natl. Mus. Can. Nat. Sci. Publ. Bot. 9, 1–83.

Lee, T. F., 1977: The seaweed handbook. An illustrated guide to seaweeds from North Carolina to the Arctic. Mariners Press, Boston; 217 pp.

Leighton, D. L., 1971: Grazing activities of benthic invertebrates in southern California kelp beds. Nova Hedwigia (Beih.) 32, 421–453.

Levinton, J. S., 1982: Marine ecology. Prentice-Hall, Englewood Cliffs, N.J.; 526 pp.

Levitt, J., 1980: Responses of plants to environmental stresses. Vol. 1. Chilling, freezing, and high temperature stresses. Vol. 2. Water, radiation, salt, and other stresses. Academic Press, New York; 497 pp. (vol. 1); 607 pp. (vol. 2).

Levring, T., 1937: Zur Kenntnis der Algenflora der norwegischen Westküste. Lunds Univ. Årsskr. N. F. Avd. 2, 33 (8), 1–147.

Levring, T., 1940: Studien über die Algenvegetation von Blekinge, Sudschweden. Hakan Ohlsson, Lund; 178 pp.

Levring, T., 1941: Die Meeresalgen der Juan Fernandez-Inseln. In: The natural history of Juan Fernandez and Easter Island. Vol. 2. Ed. by C. Skottsberg. Almquist & Wiksell, Uppsala, pp. 601–670.

Levring, T., 1960: Contributions to the marine algal flora of Chile. Lunds Univ. Årsskr. N. F. Avd. 2, 56 (10), 1–83.

Levring, T., 1974: The marine algae of the archipelago of Madeira. Bol. Mus. Munic. Funchal 28 (125), 5–111.

Levring, T., 1977: Potential yields of marine algae-with emphasis on European species. In: The marine plant biomass of the Pacific northwest coast. Ed. by R. W. Krauss. Oregon State University Press, pp. 251–270.

Levring, T., Hoppe, H. A., and Schmidt, O. J., 1969: Marine algae. A survey of research and utilization. Cram, de Gruyter, Hamburg; 421 pp.

Lewin, R. A., 1976: Prochlorophyta as a proposed new division of algae. Nature 261, 697–698.

Lewin, R. A., Cheng, L., (Eds), 1989: *Prochloron* - a microbial enigma. Chapman and Hall, New York; 196 pp.

Lewis, J. A., 1983: Floristic composition and periodicity of subtidal algae on an artificial structure in Port Phillip Bay (Victoria, Australia). Aquat. Bot. 15, 257–274.

Lewis, J. A., 1984–1987: Checklist and bibliography of benthic marine macroalgae recorded from northern Australia. I. Rhodophyta. II. Phaeophyta. III. Chlorophyta. Defence Science and Technology Organisation. Reports Material Research Laboratories, Melbourne, Victoria. MRL-R-912, 97 pp. (1984); MRL-R-962, 40 pp. (1985); MRL-R-1063, 58 pp. (1987).

Lewis, J. E., and Norris, J. N., 1987: A history and annotated account of the benthic marine algae of Taiwan. Smithson. Contrib. Mar. Sci. 29, 1–38.

Lewis, J. B., 1981: Coral reef ecosystems. In: Analysis of marine ecosystems. Ed. by A. R. Longhurst. Academic Press, London, pp. 127–158.

Lewis, J. R., 1955: The mode of occurrence of the universal intertidal zones in Great Britain. J. Ecol. 43, 270–290.

Lewis, J. R., 1964: The ecology of rocky shores. English Universities Press, London, 323 pp.

Lewis, S. M., 1985: Herbivory on coral reefs: algal susceptibility to herbivorous fishes. Oecologia (Berlin) 65, 370–375.

L'Hardy-Halos, M.-T., 1972: Recherches en scaphandre autonome sur le peuplement végétal de l'infralittoral rocheux: La Baie de Morlaix (Nord-Finistère). Bull. Soc. Sci. Bretagne 47, 177–192.

Lidholm, J., Gustafsson, P., and Öquist, G., 1987: Photoinhibition of photosynthesis and its recovery in the green alga *Chlamydomonas reinhardii*. Plant Cell Physiol. 28, 1133–1140.

Lieth, H., 1975: Primary production of the major vegetation units of the world. In: Primary productivity of the biosphere. Ed. by H. Lieth and R. H. Whittaker. Springer-Verlag, New York, pp. 203–213.

Lignell, A., and Pedersén, M., 1987: Nitrogen metabolism in *Gracilaria secundata*. Hydrobiologia 151/152, 431–441.

Lim, B.-L., Kawai, H., Hori, H., and Osawa, S., 1986: Molecular evolution of 5S ribosomal RNA from red and brown algae. Jap. J. Genet. 61, 169–176.

Lindauer, V. W., Chapman, V. J., and Aitken, M. (1961): The marine algae of New Zealand. II. Phaeophyceae. Nova Hedwigia 3, 129–350.

Lindstrom, S. C., 1977: An annotated bibliography of the benthic marine algae of Alaska. Alaska Department of Fish and Game Data Report no. 31, pp. 1–172.

Lindstrom, S., 1980: New blade initiation in the perennial red alga *Constantinea rosa-marina* (Gmelin) Postels et Ruprecht (Cryptonemiales, Dumontiaceae). Jap. J. Phycol. (Sorui) 28, 141–150.

Lindstrom, S., 1985: Nomenclatural and taxonomic notes on *Dilsea* and *Neodilsea*. (Dumontiaceae, Rhodophyta). Taxon 34, 260–266.

Lindstrom, S., 1987a: Possible sister groups and phylogenetic relationships among selected North Pacific and North Atlantic Rhodophyta. Helgoländer Meeresunters. 41, 245–260.

Lindstrom, S., 1987b: *Orculifilium denticulatum* (Dumontiaceae, Rhodophyta), a new genus and species from the northeast Pacific. Phycologia 26, 129–137.

Lindstrom, S., 1988: The Dumontieae, a resurrected tribe of red algae (Dumontiaceae, Rhodophyta). Phycologia 27, 89–102.

Lindstrom, S. C., Scagel, R. F., 1979: Some new distribution records of marine algae in southeastern Alaska. Syesis 12, 163–168.

Lindstrom, S. C., South, G. R., 1989: Evidence of species relationships in the Palmariaceae (Palmariales, Rhodophyta) based on starch gel electrophoresis. Crypt. Bot. 1, 32–41.

Lipkin, Y., 1972: Marine algal and sea-grass flora of the Suez Canal. Israel J. Zool. 21, 405–446.

Lipkin, Y., 1975: A history, catalogue and bibliography of Red Sea seagrasses. Israel J. Bot. 24, 89–105.

Lipkin, Y., and Safriel, U., 1971: Intertidal zonation on rocky shores at Mikhmoret (Mediterranean, Israel). J. Ecol. 59, 1–30.

Littler, D. S., Littler, M. M., Bucher, K. E., Norris, J. N., 1989: Marine plants of the Caribbean. A field guide from Florida to Brazil. Smithsonian Institution Press; Blue Ridge Summit, PA; 272 pp.

Littler, M. M., 1973: The productivity of Hawaiian fringing-reef crustose Corallinaceae and an experimental evaluation of production methodology. Limnol. Oceanogr. 18, 946–952.

Littler, M. M., and Arnold, K. E., 1982: Primary productivity of marine macroalgal functional-form groups from southwestern North America. J. Phycol. 18, 307–311.

Littler, M. M., and Littler, D. S., 1981: Intertidal macrophyte communities from Pacific Baja California and the upper Gulf of California: relatively constant vs. environmentally fluctuating systems. Mar. Ecol. Prog. Ser. 4, 145–158.

Littler, M. M., and Littler, D. S., 1984: Models of tropical reef biogenesis: the contribution of algae. Prog. Phycol. Res. 3, 323–364.

Littler, M., M., Littler, D. S., 1988: Structure and role of algae in tropical reef communities. In: Algae and human affairs. Ed. by C. A. Lembi and J. R. Waaland. University Press, Cambridge; pp. 29–56.

Littler, M. M., Littler, D. S., Blair, S. M., and Norris, J. N., 1985: Deepest known plant life discovered on an uncharted seamount. Science 227, 57–69.

Littler, M. M., Littler, D. S., Blair, S. M., and Norris, J. N., 1986: Deep-water plant communities from an uncharted seamount off Salvador Island, Bahamas: distribution, abundance, and primary productivity. Deep-Sea Res. 33, 881–892.

Littler, M. M., Littler, D. S., and Taylor, P. R., 1983a: Evolutionary strategies in a tropical barrier reef system: functional-form groups of marine macroalgae. J. Phycol. 19, 229–237.

Littler, M. M., Littler, D. S., and Taylor, P. R., 1987: Animal-plant defense associations: effects on the distribution and abundance of tropical reef macrophytes. J. Exp. Mar. Biol. Ecol. 105, 107–121.

Littler, M. M., Taylor, P. R., and Littler, D. S., 1983b: Algal resistance to herbivory on a Caribbean barrier reef. Coral Reefs 2, 111–118.

Littler, M. M., Taylor, P. R., Littler, D. S., 1986: Plant defense associations in the marine environment. Coral Reefs 5, 63–71.

Livermore, R. A., and Smith, A. G., 1985: Some boundary conditions for the evolution of the Mediterranean region. In: Geological evolution of the Mediterranean Basin. Ed. by D. J. Stanley and F.-C. Wezel. Springer-Verlag, New York, pp. 83–98.

Lobban, C. S., 1978: The growth and death of the *Macrocystis* sporophyte (Phaeophyceae, Laminariales). Phycologia 17, 196–212.

Lobban, C. S., 1984: Marine tube-dwelling diatoms of eastern Canada: descriptions, checklist, and illustrated key. Can. J. Bot. 62, 778–794.

Lobban, C. S., 1985: Marine tube-dwelling diatoms of the Pacific coast of North America. I. *Berkeleya, Haslea, Nitzschia*, and *Navicula* sect. *Microstigmaticae*. Can. J. Bot. 63, 1779–1784.

Lobban, C. S., Harrison, P. J., and Duncan, M. J., 1985: The physiological ecology of seaweeds. University Press, Cambridge; 242 pp.

Lobban, C. S., and Wynne, M. J. (Eds.), 1981: The biology of seaweeds. Blackwell, Oxford; 786 pp.

Longhurst, A. R. (Ed.), 1981: Analysis of marine ecosystems. Academic Press, London; 741 pp.

Longhurst, A. R., and Pauly, D., 1987: Ecology of tropical oceans. Academic Press, San Diego; 407 pp.

López-Figueroa, F., Lindemann, P., Braslawski, S. E., Schaffner, K., and

Schneider-Poetsch, H. A. W., 1989: Detection pf a phytochrome-like protein in macroalgae. Botanica Acta 102, 178–180.

López-Figueroa, F., and Niell, F. X., 1989: Red and blue-light photoreceptors controlling chlorophyll a synthesis in the red alga *Porphyra umbilicalis* and in the green alga *Ulva rigida*. Physiol. Pl. 76, 391–397.

López-Figueroa, F., Pérez, R., and Niell, F. X., 1990: Effects of red and far-red light pulses on the chlorophyll and biliprotein accumulation in the red alga *Corallina elongata*. J. Photochem. Photobiol. B4, 185–193.

Lorenz, J. R., 1863: Physicalische Verhältnisse und Vertheilung der Organismen im Quarnerischen Golfe. Wien; 379 pp.

Lovlie, A., 1978: On the genetic control of cell cycles during morphogenesis in *Ulva mutabilis*. Devl. Biol. 64, 164–177.

Lubchenco, J., 1982: Effects of grazers and algal competitors on fucoid colonization in tide pools. J. Phycol. 18, 544–550.

Lubchenco, J., and Menge, B. A., 1978: Community development and persistence in a low rocky intertidal zone. Ecol. Monogr. 59, 67–94.

Lüning, K., 1969a: Growth of amputated and dark-exposed individuals of the brown alga *Laminaria hyperborea*. Mar. Biol. 2, 218–223.

Lüning, K., 1969b: Standing crop and leaf area index of the sublittoral *Laminaria* species near Helgoland. Mar. Biol. 3, 282–286.

Lüning, K., 1970: Tauchuntersuchungen zur Vertikalverbreitung der sublitoralen Helgoländer Algenvegetation. Helgoländer wiss. Meeresunters. 21, 271–291.

Lüning, K., 1971: Seasonal growth of *Laminaria hyperborea* under recorded underwater light conditions near Helgoland. In: Fourth European Marine Biology Symposium. Ed. by D. J. Crisp. University Press, Cambridge; pp. 347–361.

Lüning, K., 1979: Growth strategies of three *Laminaria* species (Phaeophyceae) inhabiting different depth zones in the sublittoral region of Helgoland (North Sea). Mar. Ecol. Prog. Ser. 1, 195–207.

Lüning, K., 1980a: Critical levels of light and temperature regulating the gametogenesis of three *Laminaria* species (Phaeophyceae). J. Phycol. 16, 1–15.

Lüning K., 1980b: Control of algal life-history by daylength and temperature. In: The shore environment. Vol. 2. Ecosystems. Ed. by J. H. Price, D. E. G. Irvine, and W. F. Farnham. Academic Press, London, pp. 915–945.

Lüning, K., 1981a: Light. In: The biology of seaweeds. Ed. by C. S. Lobban and M. J. Wynne. Blackwell, Oxford, pp. 326–355.

Lüning, K. 1981b: Photomorphogenesis of reproduction in marine macroalgae. Ber. Dt. Bot. Ges. 94, 401–417.

Lüning, K., 1981c: Egg release in gametophytes of *Laminaria saccharina*: induction by darkness and inhibition by blue light and U.V. Br. Phycol. J. 16, 379–393.

Lüning, K., 1984a: Growth and lack of chlorophyll a in a dark-cultivated *Delesseria sanguinea*. Br. Phycol. J. 19, 269–273.

Lüning, K., 1984b: Temperature tolerance and biogeography of seaweeds: the marine algal flora of Helgoland, North Sea, as an example. Helgoländer Meeresunters. 38, 305–317.

Lüning, K., 1986: New frond formation in *Laminaria hyperborea* (Phaeophyta): a photoperiodic response. Br. Phycol. J. 21, 269–273.

Lüning, K., 1988: Photoperiodic control of sorus formation in the brown alga *Laminaria saccharina*. Mar. Ecol. Prog. Ser. 45, 137–144.

Lüning, K., Chapman, A. R. O., and Mann, K. H., 1978: Crossing experiments in the non-digitate complex of *Laminaria* from both sides of the Atlantic. Phycologia 17, 293–298.

Lüning, K., and tom Dieck, I, 1990: Circannual growth rhythm in a kelp, *Pterygophora californica*. Acta Botanica 2.

Lüning, K., and Dring, M. J., 1975: Reproduction, growth and photosynthesis of gametophytes of *Laminaria saccharina* grown in blue and red light. Mar. Biol. 29, 195–200.

Lüning, K., and Dring, M. J., 1979: Continuous underwater light measurement near Helgoland (North Sea) and its significance for characteristic light limits in the sublittoral region. Helgoländer wiss. Meeresunters. 32, 403–424.

Lüning, K., and Dring, M. J., 1985: Action spectra and spectral quantum yield of photosynthesis in marine macroalgae with thin and thick thalli. Mar. Biol. 87, 119–129.

Lüning, K., and Freshwater, W., 1988: Temperature tolerance of northeast Pacific marine algae. J. Phycol. 24, 310–315.

Lüning, K., Guiry, M. D., and Masuda, M., 1987. Upper temperature tolerance of North Atlantic and North Pacific geographic isolates of the red alga *Chondrus*. Helgoländer Meeresunters. 41, 297–306.

Dring, M. J., 1985: Action spectra and spectral quantum yield of photosynthesis in marine macroalgae with thin and thick thalli. Mar. Biol. 87, 119–129.

Lüning, K., and Müller, D. G., 1978: Chemical interaction in sexual reproduction of several Laminariales (Phaeophyceae): release and attraction of spermatozoids. Z. Pflanzenphysiol. 89, 333–341.

Lüning K., and Neushul M., 1978: Light and temperature demands for growth and reproduction of laminarian gametophytes in Southern and Central California. Mar. Biol. 45, 297–309.

Lüning, K., and Schmitz, K., 1988: Dark growth of the red alga *Delesseria sanguinea* (Ceramiales): lack of chlorophyll, photosynthetic capability, and phycobilisomes. Phycologia 27, 72–77.

Lüning, K., Schmitz, K., and Willenbrink, J., 1973: CO_2 fixation and translocation in benthic marine algae. III. Rates and ecological significance of translocation in *Laminaria hyperborea* and *L. saccharina*. Mar. Biol. 23, 275–281.

Lund, S., 1942: On *Colpomenia peregrina* Sauv. and its occurrence in Danish waters. Rep. Dan. Biol. Stn. 47, 1–16.

Lund, S., 1949: Immigration of algae into Danish waters. Nature 164, 616.

Lund, S., 1951: Marine algae from Jörgen Brönlunds Fjord in eastern North Greenland. Meddr. Grønland 128 (4), 1–26.

Lund, S., 1959: The marine algae of East Greenland. I. Taxonomical part. II. Geographic distribution. Meddr. Grønland 156 (1), 1–247; 156 (2), 1–67.

Luther, H., 1951: Verbreitung und Ökologie der hoheren Wasserpflanzen im Brackwasser der Ekenäs-Gegend in Südfinnland. Acta Bot. Fenn. 49, 231–370.

MacArthur, R. H., and Connell, J. H., 1970: The biology of populations. Wiley, New York; 200 pp.

MacArthur, R. H., and Wilson, E. O., 1967: The theory of island biogeography. Princeton University Press, Princeton, N.J.; 203 pp.

MacNeil, S., 1965: Evolution and distribution of the genus *Mya*, and tertiary migrations of Mollusca. Prof. Pap. Geol. Surv. 483–G, 1–51.

Maegawa, M., Yokohama, Y., and Aruga, Y., 1987: Critical light conditions for young *Ecklonia cava* and *Eisenia bicyclis* with reference to photosynthesis. Hydrobiologia 151/152, 447–455.

Maggs, C. A., 1987: A long-day photoperiodic response in *Atractophora* (Rhodophyta). Br. Phycol. J. 22, 307.

Maggs, C. A., Freamhainn, M. T., and Guiry, M. D., 1983: A study of the marine algae of subtital cliffs in Lough Hyne (Ine), Co. Cork. Proc. R. Ir. Acad. 83B, 251–266.

Maggs, C. A., and Guiry, M. D., 1982: Morphology, phenology and photoperiodism in *Halymenia latifolia* Kütz. (Rhodophyta) from Ireland. Botanica Mar. 15, 589–599.

Maggs, C. A., and Guiry, M. D., 1985: Life history and reproduction of *Schmitzia hiscockiana* sp. nov. (Rhodophyta, Gigartinales) from the British Isles. Phycologia 24, 297–310.

Maggs, C. A., and Guiry, M. D., 1987a: An Atlantic population of *Pikea californica* (Dumontiaceae, Rhodophyta). J. Phycol. 23, 170–176.

Maggs, C. A., and Guiry, M. D., 1987b: Environmental control of macroalgal phenology. In: Plant life in aquatic and amphibious habitats. Ed. by R. M. M. Crawford. Blackwell, Oxford, pp. 359–373.

Maggs, C. A., McLachlan, J. L., and Saunders, G. W., 1989: Infrageneric taxonomy of *Ahnfeltia* (Ahnfeltiales, Rhodophyta). J. Phycol. 25, 351–368.

Maggs, C. A., and Pueschel, C. M., 1989: Morphology and development of *Ahnfeltia plicata* (Rhodophyta): Proposal of Ahnfeltiales ord. nov. J. Phycol. 25, 333–351.

Maier, I., 1984: Culture studies on *Chorda tomentosa* Lyngbye (Phaeophyta, Laminariales). Br. Phycol. J. 19, 95–106.

Maier, I., and Clayton, M. N., 1989: Oogenesis in *Durvillaea potatorum* (Durvillaeales, Phaeophyta). Phycologia 28, 271–274.

Maier, I., and Müller, D. G., 1981: Observations on antheridium fine structure and spermatozoid release in *Laminaria digitata* (Phaeophyceae). Phycologia 21, 1–8.

Maier, I., and Müller, D. G., 1986: Sexual pheromones in algae. Biol. Bull. 170, 145–175.

Makienko, V. F., 1975: Macroalgae of the Bay of Vostok Bay, Sea of Japan, (in Russian). Biol. Morya (Vladivost.) 2, 45–57.

Mann, K. H., 1972: Ecological energetics of the seaweed zone in a marine bay on the Atlantic coast of Canada. II. Productivity of the seaweeds. Mar. Biol. 14, 199–209.

Mann, K. H., 1982: Ecology of coastal waters. A systems approach. Blackwell, Oxford; 322 pp.

Mann, K. H., and Breen, P. A., 1972: The relation between lobster abundance, sea urchins, and kelp beds. J. Fish. Res. Bd. Can. 29, 603–605.

Marcus, N. H., 1980: Photoperiodic control of diapause in the marine calanoid copepod *Labidocera aestiva*. Biol. Bull. 159, 311–318.

Margulies, M. M., 1970: Changes in absorbance spectrum of the diatom *Phaeodactylum tricornutum* upon modification of protein structure. J. Phycol. 6, 160–164.

Marincovich, L., Brouwers, E. M., and Carter, L. D., 1985: Early Tertiary marine fossils from northern Alaska: implications for Arctic Ocean paleogeography and faunal evolution. Geology 13, 770–773.

Markham, J. W., 1973: Observations on the ecology of *Laminaria sinclairii* on three northern Oregon beaches. J. Phycol. 9, 336–341.

Markham, J. W., and Celestino, J. L., 1976: Intertidal marine plants of Clatsop County, Oregon. Syesis 9, 253–266.

Markham, J. W., and Munda, I. M., 1980: Algal recolonization in the rocky eulittoral at Helgoland, Germany. Aquat. Bot. 9, 33–71.

Marr, J. W. S., 1927: Plants collected during the British Arctic Expedition, 1925. J. Bot. (Brit. and For.) 65, 272–277.

Marshall, L. G., 1988: Extinction. In: Analytical biogeography. An integrated approach to the study of animal and plant distributions. Ed. by A. A. Myers and P. S. Giller. Chapman & Hall, London, pp. 219–254.

Marshall, L. G., Butler, R. F., Drake, R. E., Curtis, G. H., and Tedford, R. H., 1979: Calibration of the great American interchange. Science 204, 272–279.

Masaki, T., Fujita, D., and Akioka, H., 1981a: Observation on the spore germination of *Laminaria japonica* on *Lithophyllum yessoense* (Rhodophyta, Corallinaceae) in culture. Bull. Fac. Fish. Hokkaido Univ. 32, 349–356.

Masaki, T., Miyata, M., Akioka, H., and Johansen, H. W., 1981b: Growth rates of *Corallina* (Rhodophyta, Corallinaceae) in Japan. Proc. Int. Seaweed Symp. 10, 607–612.

Masuda, M., 1982: A systematic study of the tribe Rhodomeleae (Rhodomelaceae, Rhodophyta). J. Fac. Sci. Hokkaido Univ. Ser. V (Botany) 12, 209–400.

Mathieson, A. C., and Burns, R. L., 1971: Ecological studies of economic red algae. I. Photosynthesis and respiration of *Chondrus crispus* Stackhouse and *Gigartina stellata* (Stackhouse) Batters. J. Exp. Mar. Biol. Ecol. 7, 197–206.

Mathieson, A. C., and Dawes, C. J., 1975: Seasonal studies of Florida sublittoral marine algae. Bull. Mar. Sci. 25, 46–65.

Mathieson, A. C., and Hehre, E. J., 1982: The composition, seasonal occurrence and reproductive periodicity of the Phaeophyceae (brown algae) in New Hampshire. Rhodora 84, 411–437.

Mathieson, A. C., and Norall, T. L., 1975a: Photosynthetic studies of *Chondrus crispus*. Mar. Biol. 33, 207–213.

Mathieson, A. C., and Norall, T. L., 1975b: Physiological studies of subtidal red algae. J. Exp. Mar. Biol. Ecol. 20, 237–247.

Mathieson, A. C., Tveter, E., Daly, M., and Howard, J., 1977: Marine algal ecology in a New Hamshire tidal rapid. Botanica Mar. 20, 277–290.

Matoo, A., Hoffmann-Falk, H., Marder, J. B., and Edelmann, M., 1984: Regulation of protein metabolism: coupling of photosynthetic electron transport and in vivo degradation of the rapidly metabolized 32 kDa protein of the chloroplast. Proc. Natl. Acad. Sci. USA 81, 1380–1384.

May, V., and Larkum, A. W. D., 1981: A subtidal transect in Jervis Bay, New South Wales. Aust. J. Ecol. 6, 439–457.

Mayhoub, H., 1976a: Cycle de développement du *Calosiphonia vermicularis* (J. Ag.)

Sch. (Rhodophycées, Gigartinales). Mise en evidence d'une réponse photopériodique. Bull. Soc. Phycol. Fr. 21, 48.

Mayhoub, H., 1976*b*: Recherches sur la végétation marine de la côte syrienne. Doctoral dissertation, Université de Caen; 286 pp.

Maykut, G. A., and Grenfell, T. C., 1975: The spectral distribution of light beneath first-year sea ice in the Arctic Ocean. Limnol. Oceanogr. 20, 554–563.

McCandless, E. L., 1981: Polysaccharides of seaweeds. In: The biology of seaweeds. Ed. by C. S. Lobban and M. J. Wynne. Blackwell, Oxford, pp. 559–588.

McCandless, E. L., and Craigie, J. S., 1979: Sulfated polysaccharides in red and brown algae. Annu. Rev. Plant Physiol. 30, 41–78.

McCoy, E. D., and Heck, K. L., 1976: Biogeography of corals, seagrasses, and mangroves: an alternative to the center of origin concept. Syst. Zool. 25, 201–210.

McCree, K. J., 1981: Photosynthetically active radiation. In: Encyclopedia of plant physiology, n.s., Vol. 12A. Ed. by O. L. Lange, P. S. Nobel, C. B. Osmond, and H. Ziegler. Springer-Verlag, Berlin, pp. 41–55.

McHugh, D. J., 1989: Production and utilization of products from commercial seaweeds. FAO Fish. Tech. Pap. 288; 189 pp.

McIntyre, C. D., and Moore, W. W., 1977: Marine littoral diatoms: ecological considerations. In: Biology of diatoms. Botanical Monographs 13. Ed. by D. Werner. University of California Press, Berkeley, pp. 333–371.

McIntyre, C. D. et al. (CLIMAP project members), 1976: The surface of the ice-age earth. Science 191, 1131–1137.

McIntyre, C. D. et al., 1981: Seasonal reconstructions of the earth's surface at the last glacial maximum. Geol. Soc. Am. Map Chart Ser. MC-36, pag. var.

McKenna, M. C., 1973: Sweepstakes, filters, corridors, Noah's arks, and beached Viking funeral ships in palaeogeography. In: Implications of continental drift to the earth sciences. Vol. 1. Ed. by D. H. Tarling and S. K. Runcorn. Academic Press, London, pp. 295–308.

McKenna, M. C., 1975: Fossil mammals and Early Eocene North Atlantic land continuity. Ann. Mo. Bot. Gard. 62, 335–353.

McKenna, M. C., 1980: Eocene paleolatitude, climate, and mammals of Ellesmere Island. Palaeogeogr. Palaeoclimatol. Palaeoecol. 24, 169–208.

McKenna, M. C., 1983: Cenozoic paleogeography of North Atlantic land bridges. In: Structure and development of the Greenland-Scotland ridge. Ed. by H. P. Bott, S. Saxov, M. Talwani and J. Thiede. Plenum Press, New York, pp. 351–399.

McLachlan, J., 1973: Growth media-marine. In: Handbook of phycological methods. Culture methods and growth measurements. Ed. by J. R. Stein. University Press, Cambridge, pp. 25–51.

McLachlan, J., 1979: *Gracilaria tikvahiae* sp. nov. (Rhodophyta, Gigartinales, Gracilariaceae), from the northwestern Atlantic. Phycologia 18, 19–23.

McLachlan, J., 1982: Inorganic nutrition of marine macro-algae in culture. In: Synthetic and degradative processes in marine macrophytes. Ed. by L. M. Srivastava. Walter de Gruyter, Berlin, pp. 71–97.

McLachlan, J., and Bidwell, R. G. S., 1983: Effects of colored light on the growth and metabolism of *Fucus* embryos and apices in culture. Can. J. Bot. 61, 1993–2003.

McLachlan, J., and Bird, C. J., 1984: Geographical and experimental assessment of the distribution of species of *Gracilaria* in relation to temperature. Helgoländer Meeresunters. 38, 319–334.

McLachlan, J., and Bird, C. J., 1986: *Gracilaria* (Gigartinales, Rhodophyta) and productivity. Aquat. Bot. 26, 27–49.

McLachlan, J., Chen, L., C.-M., and Edelstein, T., 1969: Distribution and life history of *Bonnemaisonia hamifera* Hariot. Proc. Int. Seaweed Symp. 6, 245–249.

McLean, J. H., 1962: Sublittoral ecology of kelp beds of the open coast area near Carmel, California. Biol. Bull. 122, 95–114.

McRoy, C. P., 1968: The distribution and biogeography of *Zostera marina* (eelgrass) in Alaska. Pacif. Sci. 22, 507–513.

McRoy, C. P., and Helfferich, C. (Eds.), 1977: Seagrass ecosystems. Marcel Dekker, New York; 314 pp.

Meeks, J. C., 1974: Chlorophylls. In: Algal physiology and biochemistry. Ed. by W. D. P. Stewart. Blackwell, Oxford; pp. 161–175.

van der Meer, J., 1980: The life history of *Halosaccion ramentaceum*. Can. J. Bot. 59, 433–436.

van der Meer, J., 1986: Genetic contributions to research on seaweeds. Prog. Phycol. Res. 4, 1–38.

van der Meer, J., and Bird, C. J., 1985: *Palmaria mollis* stat. nov.: a newly recognized species of *Palmaria* (Rhodophyceae) from the northeast Pacific Ocean. Can. J. Bot. 63, 398–403.

Meinesz, A, 1979: Contribution à l'étude de *Caulerpa prolifera* (Forsskal) Lamouroux (Chlorophycée, Caulerpale). I. Morphogenèse et croissance dans une station des côtes continentales françaises de la Méditerranée. Botanica Mar. 22, 27–39.

Meinesz, A., 1980: Connaissance actuelles et contribution à l'étude de la reproduction et du cycle des Udotéacées (Caulerpales, Chlorophytes). Phycologia 19, 110–138.

Meinesz, A., et al., 1983: Normalisation des symboles pour la cartographie des biocénoses benthiques littorales de Méditerranée. Ann. Inst. oceanogr. Paris 59, 155–172.

Meñez, E. G., and Calumpong, H. P., 1981: Phycological results of the Smithsonian Institution-Philippines expeditions of 1978 and 1979 in Central Visayas, Philippines. Proc. Fourth Int. Coral Reef Symp. Manila 2, 379–384.

Meñez, E. G., and Mathieson, A. C., 1981: The marine algae of Tunisia. Smithson. Contrib. Mar. Sci. 10, 1–59.

Menge, B. A., 1976: Organization of the New England rocky intertidal community: role of predation, competition, and environmental heterogeneity. Ecol. Monogr. 46, 355–393.

Menge, B. A., and Sutherland, J. P., 1976: Species diversity gradients: synthesis of the roles of predation, competition, and temporal heterogeneity. Am. Nat. 110, 351–369.

Menzies, R. J., George, R. Y., and Rowe, G. T., 1973: Abyssal environment and ecology of the world oceans. John Wiley, New York; 488 pp.

Menzies, R. J., Zaneveld, J. S., and Pratt, R. M., 1967: Transported turtle grass as a source of organic enrichment of abyssal sediments off North Carolina. Deep-Sea Res. 14, 111–112.

Mercer, J. H., 1983: Cenozoic glaciation in the Southern Hemisphere. Annu. Rev. Earth Planet. Sci. 11, 99–132.

Mergner, H., 1979: Quantitative ökologische Analyse eines Rifflagunenareals bei Aqaba (Golf von Aqaba, Rotes Meer). Helgoländer wiss. Meeresunters. 32, 476–507.

Mergner, H., and Svoboda, A., 1977: Productivity and seasonal changes in selected reef areas in the Gulf of Aqaba (Red Sea). Helgoländer wiss. Meeresunters. 30, 383–399.

Meslin, R., 1964: Sur la naturalisation du *Codium fragile* (Suring.) Hariot et son extension aux côtes de Normandie. Bull. Lab. Marit. Dinard 49/50, 110–117.

Meyen, S. V., 1987: Fundamentals of palaeobotany. Chapman & Hall, London; 432 pp.

Michanek, G., 1975: Seaweed resources of the ocean. Food and Agriculture Organization of the United Nations. FAO Fish. Tech. Pap. no. 138, 1–127.

Michanek, G., 1978: Trends in applied phycology. With a literature review: seaweed farming on an industrial scale. Botanica Mar. 21, 469–475.

Michanek, G., 1979: Phytogeographic provinces and seaweed distribution. Botanica Mar. 22, 375–391.

Michanek, G., 1983: World resources of marine plants. In: Marine ecology. Vol. V, P. 2. Ed. by O. Kinne. John Wiley, New York, pp. 795–837.

Migita, S., 1984: Intergeneric and interspecific hybridization between four species of *Eisenia* and *Ecklonia*. Bull. Dac. Fish. Nagasaki Univ. 56, 15–20.

Miller, R. J., 1985a: Succession in sea urchin and seaweed abundance in Nova Scotia, Canada. Mar. Biol. 84, 275–286.

Miller, R. J., 1985b: Seaweeds, sea urchins, and lobsters: a reappraisal. Can. J. Fish. Aquat. Sci. 42, 2061–2072.

Miller, R. J., and Colodey, A. G., 1983: Widespread mass mortalities of the green sea urchin in Nova Scotia, Canada. Mar. Biol. 73, 263–267.

Mishkind, M., Mauzerall, D., and Beale, S. I., 1979: Diurnal variation in situ of photosynthetic capacity in *Ulva* is caused by a dark reaction. Plant Physiol. 64, 896–899.

Mitchell-Thomé, R. C., 1976: Geology of the Middle Atlantic islands. Bornträger, Berlin; 382 pp.

Miura, A., 1975: *Porphyra* cultivation. In: Advance of phycology in Japan. Ed. by J. Tokida and H. Hirose. VEB Gustav Fischer, Jena, pp. 273–304.

Möbius, K., 1877: Die Auster und die Austernwirthschaft. Wiegand, Hempel und Parey, Berlin; 126 pp.

Moe, R. L., 1985: *Gainia* and Gainiaceae, a new genus and family of crustose marine Rhodophyceae from Antarctica. Phycologia 24, 419–428.

Moe, R. L., and Henry, E. C., 1982: Reproduction and early development of *Ascoseira mirabilis* Skottsberg (Phaeophyta), with notes on Ascoseirales Petrov. Phycologia 21, 55–66.

Moe, R. L., and Silva, P. C., 1977: Antarctic marine flora: uniquely devoid of kelps. Science 196, 1296–1208.

Moe, R. L., and Silva, P. C., 1981: Morphology and taxonomy of *Himanthothallus* (including *Phaeoglossum* and *Phyllogigas*), an Antarctic member of the Desmarestiales (Phaeophyceae). J. Phycol. 17, 15–29.

Mohr, H., and Schäfer, E., 1979: Uniform terminology for radiation: a critical comment. Photochem. Photobiol. 29, 1061–1062.

Mohr, H., Schäfer, E., and Shropshire, W., 1983: Description of light fields used in research on photomorphogenesis. In: Encyclopedia of plant physiology, n.s. Vol. 16B. Ed. by W. Shropshire and H. Mohr. Springer-Verlag, Berlin, pp. 761–763.

Mohr, J. L., Wilimovsky, N. J., and Dawson, E. Y., 1957: An arctic Alaskan kelp bed. Arctic 10, 45–52.

Molinier, R., 1960: Êtude des biocoenoses marines du Cap Corse. I, II. Vegetatio 9, 121–192 (I); 217–312 (II).

Mercer, 1983: Cenozoic glaciations in the southern hemisphere. Ann. Rev. Earth Plaet. Sci. 11, 99–132.

Montfort, C., 1933: Über Lichtempfindlichkeit und Leistungen roter Tiefseealgen und Grottenflorideen an freier Meeresoberfläche. Protoplasma 19, 385–413.

Montfort, C., Felgner, I., and Müller, L., 1952: Differenzfilterversuche über die wirksamen Strahlenbereiche bei der Chlorophyllzerstörung durch Sonnenlicht im Gewebe von Tiefen-Laminarien. Z. Bot. 40, 173–186.

Moore, H. B., 1972: Aspects of stress in the tropical marine environment. Adv. Mar. Biol. 10, 217–269.

Moore, L. B., 1961: Distribution patterns in New Zealand seaweeds. Tuatara 9, 18–23.

Moore, P. G., 1983: Biological interactions. In: Sublittoral ecology. The ecology of the shallow sublittoral benthos. Ed. by R. Earll and D. G. Erwin. Clarendon Press, Oxford, pp. 125–143.

Moore, P. G., and Seed, R. (Eds.), 1985: The ecology of rocky coasts. Essays presented to J. R. Lewis, D.Sc. Hodder & Stoughton, London; 467 pp.

Moorjani, S. A., 1977: The ecology of marine algae of the Kenya coast. Ph.D. thesis, University of Nairobi; 285 pp.

Moraitou-Apostolopoulou, M., and Kiortsis, V. (Eds.), 1985: Mediterranean marine ecosystems. Plenum Press, New York; 407 pp.

Morel, A., and Smith, R. C., 1974: Relation between total quanta and total energy for aquatic photosynthesis. Limnol. Oceanogr. 19, 591–600.

Moreno, C. A., Sutherland, J. P., and Jara, H. F., 1984: Man as a predator in the intertidal zone of southern Chile. Oikos 42, 155–160.

Morgan, D. C., and Smith, H., 1981: Non-photosynthetic responses to light quality. In: Encyclopedia of plant physiology, n.s. Vol. 12A. Ed. by O. L. Lange, P. S. Nobel, C. B. Osmond and H. Ziegler. Springer-Verlag, Berlin, pp. 109–134.

Morgan, K. C., Wright, J. L. C., and Simpson, F. J., 1980: Review of chemical constituents of the red alga *Palmaria palmata* (dulse). Econ. Bot. 34, 27–50.

Morrissey, J., 1980: Community structure and zonation of macroalgae and hermatypic corals on a fringing reef flat of Magnetic Island (Queensland, Australia). Aquat. Bot. 8, 91–139.

Morse, A. N. C., and Morse, D. E., 1984: Recruitment and metamorphosis of *Haliotis* larvae induced by molecules uniquely available at the surfaces. of crustose red algae. J. Exp. Mar. Biol. Ecol. 75, 191–215.

Morton, B., and Morton, J., 1983: The seashore ecology of Hong Kong. University Press, Hong Kong; 350 pp.

Morton, J. E., 1973: The intertidal ecology of the British Solomon Islands. I. The zonation patterns of the weather coast. Phil. Trans. R. Soc. Lond. B, 265, 491–537.

Morton, J. E., and Miller, M., 1968: The New Zealand sea shore. Collins, London, 638 pp.

Moss, B. L., 1982: The control of epiphytes by *Halidrys siliquosa* (L.) Lynbg. (Phaeophyta, Cystoseiraceae). Phycologia 21, 185–191.

Moss, B., Tovey, D., and Court, P., 1981: Kelps as fouling organisms on North Sea platforms. Botanica Mar. 24, 207–209.

Moss, J. R., 1977: Essential considerations for establishing seaweed extraction factories. In: The marine plant biomass of the Pacific northwest coast. Ed. by R. W. Krauss. Oregon State University Press; pp. 301–314.

Moss, S. T. (Ed.), 1986: The biology of marine fungi. University Press, Cambridge; 382 pp.

Moustakas, M., 1981: Contribution to the study of the marine flora of Crete Island, Greece. Sci. Ann. Fac. Phys. Math. Univ. Thessaloniki 21, 263–274.

Mshigeni, K. E., 1976: Effects of the environment on developmental rates of sporelings of two *Hypnea* species (Rhodophyta: Gigartinales). Mar. Biol. 36, 99–103.

Mshigeni, K. E., 1983: Algal resources, exploitation and use in East Africa. Prog. phycolog. Res. 2, 387–419.

Müller, D. G., 1962: Über jahres - und lunarperiodische Erscheinungen bei einigen Braunalgen. Botanica Mar. 4, 140–155.

Müller, D. G., 1981: Sexuality and sex attraction. In: The biology of seaweeds. Ed. by C. S. Lobban and M. J. Wynne. Blackwell, Oxford, pp. 661–674.

Müller, D. G., 1984: Culture studies on the life history of *Adenocystis utricularis* (Phaeophyceae, Dictyosiphonales). Phycologia 23, 87–94.

Müller, D. G., 1988: Demonstration of sexual pheromones in Laminariales and Desmarestiales. In: Experimental phycology. A laboratory manual. Ed. by C. S. Lobban, D. J. Chapman, and B. P. Kremer. University Press, Cambridge, pp. 243–250.

Müller, D. G., Maier, I., and Gassmann, G., 1985: Survey on sexual pheromone specificity in Laminariales (Phaeophyta). Phycologia 24, 475–477.

Müller, D. G., and Schmid, C. E., 1988: Qualitative and quantitative determination of pheromone secretion in female gametes of *Ectocarpus siliculosus*. Biol. Chem. Hoppe-Seyler 369, 647–653.

Müller, S., and Clauss, H., 1976: Aspects of photomorphogenesis in the brown alga *Dictyota dichotoma*. Z. Pflanzenphysiol. 78, 461–465.

Munda, I. M., 1972a: General features of the benthic algal zonation around the Icelandic coast. Acta Naturalia Islandica 21. Museum of Natural History, Reykjavik; 36 pp.

Munda, I. M., 1972b: On the chemical composition, distribution and ecology of some common benthic marine algae from Iceland. Botanica Mar. 15, 1–45.

Munda, I. M., 1973: The production of biomass in the settlements of benthic marina algae in the northern Adriatic. Botanica Mar. 15, 218–244.

Munda, I. M., 1975: Hydrographically conditioned floristic and vegetation limits in Icelandic coastal waters. Botanica Mar. 18, 223–235.
Munda, I. M., 1979: Some fucacean associations from the vicinity of Rovinj, Istrian coast, northern Adriatic. Nova Hedwigia 31, 607–666.
Munda, I. M., 1980: Survey of the benthic algal vegetation of the Borgarfjördur, Southwest Iceland. Nova Hedwigia 32, 855–918.
Munda, I. M., 1982: The effects of organic pollution on the distribution of fucoid algae from the Istrian coast (vicinity of Rovinj). Acta Adriat. 23, 329–337.
Munda, I. M., 1987: Characteristic features of the benthic algal vegetation along the Snaefellsnes peninsula (southwest Iceland). Nova Hedwigia 44, 399–448.
Munda, I. M., and Lüning, K., 1977: Growth performance of *Alaria esculenta* off Helgoland. Helgoländer wiss. Meeresunters. 29, 311–314.
Munda, I. M., and Markham, J. W., 1982: Seasonal variations of vegetation patterns and biomass constituents in the rocky eulittoral of Helgoland. Helgoländer Meeresunters. 35, 131–151.
Murakami, A., and Fujita, 1983: Occurrence of photoreversible absorption changes in dissociated allophycocyanin and phycocyanin. Photochem. Photobiol. 38, 605–608.
Murase, N., Maegawa, M., and Kida, W., 1989: Photosynthetic characteristics of several species of Rhodophyceae from different depths in the coastal area of Shima Peninsula, central Japan. Jap. J. Phycol. (Sorui) 37, 213–220.
Murray, S. N., and Littler, M. M., 1981: Biogeographical analysis of intertidal floras of southern California. J. Biogeogr. 8, 339–351.
Murthy, M. S., Bhattacharya, M., and Radia, P., 1978: Ecological studies on the intertidal algae at Okha (India). Botanica Mar. 21, 381–386.
Muscatine, L., 1980: Productivity of zooxanthellae. In: Productivity of the sea. Ed. by P. G. Falkowski. Plenum Press, New York, pp. 381–400.
Mutter, J. C., Hegarthy, K. A., Cande, S. C., and Weissel, J. K., 1985: Breakup between Australia and Antarctica: A brief review in the light of new data. Tectonophysics 114, 255–279.
Myers, A. A., and Giller, P. S., 1988: Analytical biogeography. An integrated approach to the study of animal and plant distributions. Chapman & Hall, London; 578 pp.

Nairn, A. E. M., Kanes, W. H., and Stehli, F. G. (Eds.), 1977: The ocean basins and margins. Vol. 4A. The eastern Mediterranean. Plenum Press, New York; 503 pp.
Nairn, A. E. M., Kanes, W. H., and Stehli, F. G. (Eds.), 1978: The ocean basins and margins. Vol. 4B. The western Mediterranean. Plenum Press, New York; 447 pp.
Nairn, A. E. M., Kanes, W. H., and Stehli, F. G. (Eds.), 1981: The ocean basins and margins. Vol. 4A. The Arctic Ocean. Plenum Press, New York; 672 pp.
Nagai, M., 1940–1941: Marine algae of the Kurile Islands. J. Fac. Agric. Hokkaido (Imp.) Univ. 46, 1–310.
Nam, K. W., 1986: On the marine benthic algal community of Chukdo in eastern coast of Korea. Korean J. Phycol. 1, 185–202.
Nasr, A. H., 1940: The marine algae of Alexandria. II. A study of the occurrence of

some marine algae on the Egyptian Mediterranean coast. Fouad Inst. Hydrobiol. Fish. Not. Mem. 37, 1–10.

Natour, R. M., Gerloff, J., and Nizamuddin, M., 1979: Algae from the Gulf of Aqaba, Jordan. I. Chlorophyceae and Phaeophyceae. II. Rhodophyceae. Nova Hedwigia 31, 39–54; 31, 69–90.

Naylor, J., 1976: Production, trade and utilization of seaweeds and seaweed products. FAO Fish. Tech. Pap. no. 159, 1–73.

Navarro, F. de P., and Bellon Uriarte, L., 1945: Catálogo de la Flora del Mare de Baleares (con exclusion de las Diatomeas). Boln. Inst. Ssp. Oceanogr. Not. Res., ser. II, 124, 161–298.

Nellen, U. R., 1966: Über den Einfluss des Salzgehaltes auf die photosynthetische Leistung verschiedener Standortformen von *Delesseria sanguinea* und *Fucus serratus*. Helgoländer wiss. Meeresforsch. 13, 288–313.

Nelson, C. M., 1978: *Neptunea* (Gastropoda: Buccinacea) in the Neogene of the North Pacific and adjacent Bering Sea. Veliger 21, 203–215.

Nelson, G., and Rosen, D. E. (Eds.), 1981: Vicariance biogeography. A critique. Symposium of the Systematics discussion group of the American Museum of Natural History, May 2–4, 1979. Columbia University Press, New York; 593 pp.

Nelson, G., and Platnick, N., 1981: Systematics and biogeography. Cladistics and vicariance. Columbia University Press, New York; 567 pp.

Nelson, W. A., and Ryan, K. G., 1986: *Palmophyllum umbracola* sp. nov. (Chlorophyta) from offshore islands of northern New Zealand. Phycologia 25, 168–177.

Nelson, W. A., and Adams, N. M., 1984: Marine algae of the Kermadec Islands. Nat. Mus. N. Z. Misc. Ser. 10, 1–29.

Nelson, W. A., and Adams, N. M., 1987: Marine algae of the Bay of Islands area. Nat. Mus. N. Z. Misc. Ser. 16, 1–47.

Nemoto, T., and Harrison, G., 1981: High latitude ecosystems. In: Analysis of marine ecosystems. Ed. by A. R. Longhurst. Academic Press, London, pp. 95–126.

Neumann, D., 1981: Tidal and lunar rhythms. In: Handbook of behavioral neurobiology. Vol. 4. Ed. by J. Aschoff. Plenum Press, pp. 351–380.

Neushul, M., 1965a: Scuba diving studies of the vertical distribution of benthic marine algae. Acta Univ. Gothoburg. 3, 161–176.

Neushul, M., 1965b: Diving observations of sub-tidal Antarctic marine vegetation. Botanica Mar. 8, 234–243.

Neushul, M., 1967: Studies of subtidal marine vegetation in western Washington. Ecology 48, 83–94.

Neushul, M., 1968: Benthic marine algae. Am. Geograph. Soc. Antarct. Map Folio Ser. 10, 9–10.

Neushul, M., 1971: The species of *Macrocystis* with particular reference to those of North and South America. Nova Hedwigia (Beih.) 32, 211–222.

Neushul, M., 1972: Functional interpretation of benthic marine algal morphology. In: Contributions to the systematics of benthic marine algae of the North Pacific. Ed. by I. A. Abbott and M. Kurogi. Japanese Society of Phycology, Kobe, pp. 47–74.

Neushul, M., 1977: The domestication of the giant kelp, *Macrocystis*, as a marine plant biomass producer. In: The marine plant biomass of the Pacific Northwest coast. Ed. by R. W. Krauss. Oregon State University Press, Corvallis, pp. 163–181.

Newell, R. C., Lucas, M. I., Velimirov, B., and Seiderer, L. J., 1980: Quantitative significance of dissolved organic losses following fragmentation of kelp (*Ecklonia maxima* and *Laminaria pallida*). Mar. Ecol. Prog. Ser. 2, 45–59.

Newroth, P. R., 1971: The distribution of *Phyllophora* in the North Atlantic and Arctic regions. Can. J. Bot. 49, 1017–1024.

Newroth, P. R., and Taylor, A. R. A., 1971: The nomenclature of the North Atlantic species of *Phyllophora* Greville. Phycologia 10, 93–97.

Newton, L., 1931: A handbook of the British seaweeds. British Museum, London; 487 pp.

Nicholson, N., Hosmer, H., Bird, K., Hart, L., Sandlin, W., Shoemaker, C. and Sloan, C., 1981: The biology of *Sargassum muticum* (Yendo) Fensholt at Santa Catalina Island. Proc. Int. Seaweed Symp. 8, 416–424.

Nicholson, N. L., 1979: Evolution within *Macrocystis:* northern and southern hemisphere taxa. In: Proceedings of the International Symposium on Marine Biogeography and Evolution in the Southern Hemisphere, Auckland, New Zealand, July 1978. N. Z. DSIR Information Series 137, 433–441.

Niell, F. X., 1981: Photosynthetic liposoluble pigments in seaweeds, physiological and ecological meaning. Proc. Int. Seaweed Symp. 10, 333–338.

Niell, F. X., 1978: Catálogo florístico y fenológico de las algas superiores y cianofíceas bentónicas de las Rías Bajas Gallegas. Invest. Pesq. 42, 365–400.

Nielsen, R., 1980: A comparative study of five marine Chaetophoraceae. Br. Phycol. J. 15, 131–138.

Niemeck, R. A., and Mathieson, A. C., 1976: An ecological study of *Fucus spiralis* L. J. Exp. Mar. Biol. Ecol. 24, 33–48.

Niemeck, R. A., and Mathieson, A. C., 1978: Physiological studies of intertidal fucoid algae. Botanica Mar. 21, 221–227.

Nienburg, W., 1925: Die Besiedelung des Felsstrandes und der Klippen von Helgoland. 2. Die Algen. Wiss. Meeresunters. (Abt. Helgoland) 15 (19), 1–15.

Nienhuis, P. H., 1970: The benthic algal communities of flats and salt marshes in the Grevelingen, a sea-arm in the south-western Netherlands. Neth. J. Sea Res. 5, 20–49.

Niklas, K. J., 1976: Chemotaxonomy of *Prototaxites* and evidence for possible terrestrial adaptation. Rev. Palaeobot. Palynol. 22, 1–17.

Nilsen, T. H., 1983: Influence of the Greenland-Scotland Ridge on the geological history of the North Atlantic and Norwegian-Greenland Sea areas. In: Structure and development of the Greenland-Scotland Ridge. Ed. by M. P. H. Bott, S. Saxov, M. Talwani and J. Thiede. Plenum Press, New York; pp. 457–478.

Nizamuddin, M., 1962: Classification and the distribution of the Fucales. Botanica Mar. 4, 191–203.

Nizamuddin, M., 1968: Observations on the order Durvilleales J. Petrov 1965. Botanica Mar. 11, 115–117.

Nizamuddin, M., 1970: Phytogeography of the Fucales and their seasonal growth. Botanica Mar. 13, 131–139.

Nizamuddin, M., and Gessner, F., 1970: The marine algae of the northern part of the Arabian Sea and of the Persian Gulf. Meteor Forschungsergeb. D, 6, 1–42.

Nizamuddin, M., and Lehnberg, W., 1970: Studies on the marine algae of Paros and Sikinos islands, Greece. Botanica Mar. 8, 116–130.

Nizamuddin, M., West, J. A., and Menez, E. G., 1978: A list of marine algae from Libya. Botanica Mar. 22, 465–476.

Nizamuddin, M., and Womersley, H. B. S., 1960: Structure and systematic position of the Australian brown alga, *Notheia anomala*. Nature 187, 673–674.

Nonomura, A. M., 1979: Development of *Janczewskia morimotoi* (Ceramiales) on its host *Laurencia nipponica* (Ceramiales, Rhodophyta). J. Phycol. 15, 154–162.

Norris, R. E., and Aken, M. E., 1985: Marine benthic algae new to South Africa. S. Afr. J. Bot. 51, 55–65.

Norris, R. E., and Bucher, K. E., 1982: Marine algae and seagrasses from Carrie Bow Cay, Belize. In: The Atlantic barrief reef ecosystem at Carrie Bow Cay, Belize, Vol. 1. Structure and communities. Ed. by K. Rützler and I. G. Macintyre. Smithsonian Institution Press, Washington, pp. 167–223.

Norris, R. E., and Conway, E., 1974: *Fucus spiralis* L. in the northeast Pacific. Syesis 7, 79–81.

Norris, R. E., and Fenical, W., 1982: Chemical defenses in tropical marine algae. In: The Atlantic barrief reef ecosystem at Carrie Bow Cay, Belize, I. Structure and communities. Ed. by K. Rützler and I. G. Macintyre. Smithsonian Institution Press, Washington, pp. 417–431.

North, W. J. (Ed.), 1971: The biology of giant kelp beds (*Macrocystis*) in California. Nova Hedwigia (Beih.) 32, 1–600.

Norton, I. O., Sclater, J. G., 1979: A model for the evolution of the Indian Ocean and the breakup of Gondwanaland. J. geophys. Res. 84, 6803–6830.

Norton, T. A., 1968: Underwater observations on the vertical distribution of algae at St. Mary's, Isles of Scilly. Br. Phycol. J. 3, 585–588.

Norton, T. A., 1969: Growth form and environment in *Saccorhiza polyschides*. J. Mar. Biol. Ass. U. K. 49, 1025–1045.

Norton, T. A., 1977: Experiments on the factors influencing the geographical distributions of *Saccorhiza polyschides* and *Saccorhiza dermatodea*. New Phytol. 78, 625–635.

Norton, T. A., 1985a: The zonation of seaweeds on rocky shores. In: The ecology of rocky coasts. Essays presented to J. R. Lewis, D.Sc. Ed. by P. G. Moore and R. Seed. Columbia University Press, New York, pp. 7–21.

Norton, T. A., 1985b: Provisional atlas of the marine algae of Britain & Ireland. Biological Records Centre. Institute of Terrestrial Ecology, Huntingdon; pag. var.

Norton, T. A., and Mathieson, A. C., 1983: The biology of unattached seaweeds. In: Progress in phycological research. Vol. 2. Ed. by F. E. Round and V. J. Chapman. Elsevier, London, pp. 333–386.

Norton, T. A., Mathieson, A. C., and Neushul, M., 1982: A review of some aspects of form and function in seaweeds. Botanica Mar. 25, 501–510.

Norton, T. A., and Burrows, E. M., 1969: Studies on marine algae of the British Isles. 7. *Saccorhiza polyschides* (Lightf.) Batt. Br. Phycol. J. 4, 19–53.

Norton, T. A., Hiscock, K., and Kitching, J. A., 1977: The ecology of Lough Ine. XX. The *Laminaria* forest at Carriga Thorna. J. Ecol. 65, 919–941.

Norton, T. A., and Milburn, J. A., 1972: Direct observations on the sublittoral marine algae of Argyll, Scotland. Hydrobiologia 40, 55–68.

Norton, T. A., and Powell, H. T., 1979: Seaweeds and rocky shores of the Outer Hebrides. Proc. R. Soc. Edinb. 77B, 141–153.

Novaczek, I., 1981: Stipe growth rings in *Ecklonia radiata* (C. Ag.) J. Ag. (Laminariales). Br. Phycol. J. 16, 363–371.

Novaczek, I., 1984: Response of gametophytes of *Ecklonia radiata* (Laminariales) to temperature in saturating light. Mar. Biol. 82, 241–246.

Novaczek, I., Bird, C. J., and McLachlan, J., 1986a: The effect of temperature on development and reproduction in *Chorda filum* and *C. tomentosa* (Phaeophyta, Laminariales) from Nova Scotia. Can. J. Bot. 64, 2414–2420.

Novaczek, I., Bird, C. J., and McLachlan, J., 1986b: Culture and field study of *Stilophora rhizodes* (Ehr.) J. Ag. (Phaeophyceae, Chordariales) from Nova Scotia. Br. Phycol. J. 21, 407–416.

Novaczek, I., Bird, C. J., and McLachlan, J., 1987: Phenology and temperature tolerance of the red algae *Chondria baileyana, Lomentaria baileyana, Griffithsia globifera,* and *Dasya baillouviana* in Nova Scotia. Can. J. Bot. 65, 57–62.

Novaczek, I., and McLachlan, J., 1986c. Recolonization of the sublittoral habitat of Halifax County, following the demise of sea urchins. Botanica Mar. 29, 69–73.

Novaczek, I., and McLachlan, J., 1987. Correlation of temperature and daylength response of *Sphaerotrichia divaricata* (Phaeophyta, Chordariales) with field phenology in Nova Scotia and distribution in eastern North America. Br. Phycol. J. 22, 215–219.

Nultsch, W., 1986: Allgemeine Botanik. Thieme, Stuttgart; 530 pp.

Nultsch, W., and Pfau, J., 1979: Occurrence and biological role of light-induced chromatophore displacements in seaweeds. Mar. Biol. 51, 77–82.

Nultsch, W., Pfau, J., and Rüffer, U., 1981: Do correlations exist between chromatophore arrangement and photosynthetic activity in seaweeds? Mar. Biol. 62, 111–117.

Nultsch, W., Pfau, J., and Materna-Weide, M., 1987: Fluence and wavelength dependence of photoinhibition in the brown alga *Dictyota dichotoma*. Mar. Ecol. Prog. Ser. 41, 93–97.

Nultsch, W., Rüffer, U., and Pfau, J., 1979: Chromatophorenanordnungen in emersen Thalli von *Fucus vesiculosus* unter verschiedenen Lichtbedingungen. Helgoländer Meeresunters. 32, 228–238.

Nultsch, W., Rüffer, U., and Pfau, J., 1984: Circadian rhythms in the chromatophore movements of *Dictyota dichotoma*. Mar. Biol. 81, 217–222.

Nunns, A. G., 1983: Plate tectonic evolution of the Greenland-Scotland Ridge and surrounding regions. In: Structure and development of the Greenland-Scotland Ridge.Ed. by M. P. H. Bott, S. Saxov, M. Talwani and J. Thiede. Plenum Press, New York; pp. 11–30.

Nurul Islam, A. K. M., 1976: Contribution to the study of the marine algae of Bangladesh. Cramer, Vaduz; 253 pp.

Nybakken, J. W., 1982: Marine ecology. An ecological approach. Harper & Row, New York; 446 pp.

Oates, B. R., 1985: Photosynthesis and amelioration of desiccation in the intertidal saccate alga *Colpomenia peregrina*. Mar. Biol. 89, 109–119.

Oates, B. R., 1986: Components of photosynthesis in the intertidal saccate alga *Halosaccion americanum* (Rhodophyta, Palmariales). J. Phycol. 22, 217–223.

Oates, B. R., 1988: Water relations of the intertidal saccate alga *Colpomenia peregrina* (Phaeophyta, Scytosiphonales). Botanica Mar. 31, 57–63.

Öquist, G., and Fork, D. C., 1982: Effects of desiccation on the excitation energy distribution from phycoerythrin to the two photosystems in the red alga *Porphyra perforata*. Physiol. Plant. 56, 56–62.

Ogata, E., and Matsui, T., 1965: Photosynthesis in several marine plants of Japan as affected by salinity, drying and pH, with attention to their growth habitat. Botanica Mar. 8, 199–217.

Ogata, E., and Takada, H., 1968: Studies on the relationship between the respiration in some marine plants in Japan. J. Shimonoseki Collect. Fish. 16, 67–88.

Ogawa, H., and Machida, M., 1976–1977: Marine algae of the Oshika Peninsula. I. Chlorophyceae and Phaeophyceae. II. Rhodophyceae. Tohoku J. Agric. Res. 27, 145–154 (1976); 28, 151–165 (1977).

Ohad, J., Kyle, D. J., and Arntzen, C. J., 1984: Membrane protein damage and repair: removal and replacement of inactivated 32 kilodalton polypeptides in chloroplast membranes. J. Cell Biol. 99, 481–485.

Ohno, M., 1976: Some observations on the influence of salinity on photosynthetic activity and chloride ion loss in several seaweeds. Int. Rev. Ges. Hydrobiol. Hydrogr. 61, 665–672.

Ohno, M., 1979: Culture and field survey of *Sargassum piluliferum*. Rep. Usa Mar. Biol. Inst. 1, 25–32.

Ohno, M., and Arasaki, S., 1969: Examination of the dark treatment at spore stage of seaweeds. Bull. Jap. Soc. Phycol. 17, 37–42.

Ohno, M., Mairh, O. P., 1982: Ecology of green alga Ulvaceae occurring on the coast of Okha, India. Rep. Usa Mar. Biol. Inst. 4, 1–8.

Okamura, K., 1932: The distribution of marine algae in Pacific waters. Rec. Oceanogr. Works Japan 4, 30–150.

Okazaki, A., 1971: Seaweeds and their uses in Japan. Tokai University Press, Tokyo; 165 pp.

Okazaki, M., Pentecost, A., Tanaka, Y., and Miyata, M., 1986: A study of calcium carbonate deposition in the genus *Padina* (Phaeophyceae, Dictyotales). Br. Phycol. J. 21, 217–224.

O'Kelly, C. J., 1982a: Chloroplast pigments in selected marine Chaetophoraceae and Chaetosiphonaceae (Chlorophyta): the occurrence and significance of siphonoxanthin. Botanica Mar. 25, 133–137.

O'Kelly, C. J., 1982b: Observations on marine Chaetophoraceae. III. The structure, reproduction, and life history of *Endophyton ramosum*. Phycologia, 21, 247–257.

Okuda, T., and Neushul, M., 1981: Sedimentation studies of red algal spores. J. Phycol. 17, 113–118.

Oliger, P., and Santelices, B., 1981: Physiological ecology studies on Chilean Gelidiales. J. Exp. Mar. Biol. Ecol. 53, 65–75.

Oliveira Filho, E. C. de, 1976: Deep water marine algae from Espiritu Santo State (Brazil). Bol. Bot. Univ. Sao Paulo 4, 73–80.

Oliveira Filho, E. C. de, 1981: Marine phycology and exploitation of seaweeds in South America. Proc. Int. Seaweed Symp. 10, 97–112.

Oliveira Filho, E. C. de, Pirani, J. R., and Giulietti, A. M., 1983: The Brazilian seagrasses. Aquat. Bot. 16, 251–267.

Oliveira Filho, E. C. de, and Ugadim, Y., 1976: A survey of the marine algae of Atol das Rocas (Brazil). Phycologia 15, 41–44.

Ollivier, G., 1929: Étude de la flore de la côte d'Azur. Ann. Inst. Océanogr. N. S. 7, 53–173.

Olsen, J. L., Stam, W. T., Bot, P. V. M., and van den Hoek, C., 1987: scDNA-DNA hybridization studies in Pacific and Caribbean isolates of *Dictyosphaeria cavernosa* (Chlorophyta) indicate a long divergence. Helgoländer Meeresunters. 41, 377–383.

Olsen-Stojkovich, J., West, J. A., and Lowenstein, J. M., 1986: Phylogenetics and biogeography in the Cladophorales complex (Chlorophyta): some insights from immunological distance data. Botanica Mar. 29, 239–249.

Oltmanns, F., 1892: Ueber die Cultur- und Lebensbedingungen der Meeresalgen. Wiss. Bot. 23, 349–440.

Oltmanns, F., 1922–1923: Morphologie und Biologie der Algen. 2d ed. (1st ed. 1905). 3 vols. Fischer, Jena; 459 pp.; 439 pp.; 558 pp.

Orris, P. K., 1980: A revised species list and commentary on the macroalgae of the Chesapeake Bay in Maryland. Estuaries 3, 200–206.

Osborne, B. A., and Raven, J. A., 1986: Growth light level and photon absorption by cells of *Chlamydomonas rheinhardii*, *Dunaliella tertiolecta* (Chlorophyceae, Volvocales), *Scenedesmus obliquus* (Chlorophyceae, Chlorococcales) and *Euglena viridis* (Euglenophyceae, Euglenales). Br. Phycol. J. 21, 303–313.

Ott, J. A., 1979: Persistence of a seasonal growth rhythm in *Posidonia oceanica* (L.) Delile under constant conditions of temperature and illumination. Mar. Biol. Lett. 1, 99–104.

Ott, J. A., 1980: Growth and production in *Posidonia oceanica* (L.) Delile. P. S. Z. N. I. Mar. Ecol. 1, 47–64.

Ott, F. D., 1973: The marine algae of Virginia and Maryland including the Chesapeake Bay area. Rhodora 75, 258–296.

Owen, H. G., 1983: Atlas of continental displacement, 200 Mio y to the present. University Press, Cambridge; 159 pp.

Oza, R. M., 1976: Culture studies on induction of tetraspores and their subsequent development in the red alga *Falkenbergia rufolanosa* Schmitz. Botanica Mar. 20, 29–32.

Paine, R. T., 1969: The *Pisaster-Tegula* interaction: prey patches, predator food preference, and intertidal community structure. Ecology 50, 950–961.

Paine, R. T., 1979: Disaster, catastrophe, and local persistance of the sea palm *Postelsia palmaeformis*. Science 205, 685–687.

Paine, R. T., 1980: Food webs: linkage, interaction strength, and community infrastructure. J. Anim. Ecol. 49, 667–685.

Paine, R. T., Slocum, C. J., and Duggins, D. O., 1979: Growth and longevity in the crustose red alga *Petrocelis middendorfii*. Mar. Biol. 51, 185–192.

Palmisano, A. C., SooHoo, J. B., White, D. C., Smith, G. A., Stanton, G. R., and Burckle, L. H., 1985: Shade adapted diatoms beneath Antarctic sea ice. J. Phycol. 21, 664–667.

Palmisano, A. C., and Sullivan, C. W., 1983: Physiology of sea ice diatoms. II. Dark survival of three polar diatoms. Can. J. Microbiol. 29, 157–160.

Pankow, H., 1971: Algenflora der Ostsee. I. Benthos (Blau-, Grün-, Braun- und Rotalgen). Gustav Fischer, Stuttgart; 419 pp.

Pankow, H., and Festerling, E., Festerling, H., 1971: Beitrag zur Kenntnis der Algenflora der mecklenburgischen Küste (sudliche Ostsee: Lübecker Bucht-Darss). Int. Rev. Ges. Hydrobiol. Hydrogr. 56, 241–263.

Papenfuss, G. F., 1940: A revision of the South African marine algae in Herbarium Thunberg. Symb. Bot. Upsal. 4 (3), 1–17.

Papenfuss, G. F., 1942: Studies on South African Phaeophyceae. I. *Ecklonia maxima, Laminaria pallida, Macrocystis pyrifera*. Am. J. Bot. 29, 15–24.

Papenfuss, G. F., 1961: Nils Eberhard Svedelius: a chapter in the history of phycology. Phycologia 1, 172–182.

Papenfuss, G. F., 1964a: Catalogue and bibliography of Antarctic and Sub-antarctic benthic marine algae. Am. Geophys. Union, Antarct. Res. Ser. 1, 1–76.

Papenfuss, G. F., 1964b: Problems in the taxonomy and geographical distribution of Antarctic marine algae. In: Biologie antarctique. Ed. by R. Carrick, M. Holdgate, and J. Prévost. Hermann, Paris, pp. 155–160.

Papenfuss, G. F., 1968: A history, catalogue, and bibliography of Red Sea benthic algae. Israel J. Bot. 17, 1–118.

Papenfuss, G. F., 1972: On the geographical distribution of some tropical marine algae. Proc. Int. Seaweed Symp. 7, 45–51.

Papenfuss, G. F., 1976: Landmarks in Pacific North American marine phycology. In: Marine algae of California. Ed. by I. A. Abbott and G. J. Hollenberg. Stanford University Press, Stanford, Calif., pp. 21–46.

Papenfuss, G. F., 1977: Review of the genera of Dictyotales (Phaeophycophyta). Bull. Jap. Soc. Phycol. 25, 271–287.

Parke, M. W., 1931: Manx algae. University Press, Liverpool; 155 pp.

Parke, M., and Dixon, P. S., 1976: Check-list of British marine algae-third revision. J. Mar. Biol. Ass. U. K. 56, 527–594.

Parker, B. C., 1971: Studies of translocation in *Macrocystis*. Nova Hedwigia (Beih.) 32, 191–195.

Parker, B. C., and Dawson, E. Y., 1965: Non-calcareous marine algae from California Miocene deposits. Nova Hedwigia 10, 273–295.

Parker, J., 1960: Seasonal changes in cold-hardiness of *Fucus vesiculosus*. Biol. Bull. Mar. Biol. Lab. Woods Hole, 119, 474–478.

Parr, A. E., 1939: Quantitative observations on the pelagic *Sargassum* vegetation of the western North Atlantic. Bull. Bingham Oceanogr. Collect. 7, 1–94.

Parsons, M. J., 1985a: New Zealand seaweed flora and its relationships. N. Z. J. Mar. Freshw. Res. 19, 131–138.

Parsons, M. J., 1985b: Biosystematics of the cryptogamic flora of New Zealand: algae. N Z. J. Bot. 23, 663–675.

Parsons, M. J., and Fenwick, G. D., 1984: Marine algae and a marine fungus from Open Bay Islands, Westland, New Zealand. N. Z. J. Bot. 22, 425–432.

Paul, V. J., and Fenical, W., 1983: Isolation of halimedatrial: chemical defense adaptation in the calcareous reef-building alga *Halimeda*. Science 221, 747–749.

Paul, V. J., and Fenical, W., 1987: Natural products chemistry and chemical defense in tropical marine algae of the phylum Chlorophyta. In: Bioorganic marine chemistry. Vol 1. Ed. by P. J. Scheuer. Springer, Berlin, pp. 1–29.

Paul, V. J., and Hay, M. E., 1986: Seaweed susceptibility to herbivory: chemical and morphological correlates. Mar. Ecol. Prog. Ser. 33, 255–264.

Payri, C. E., 1987: Zonation and seasonal variation of the commonest algae on Tiahura reef (Moorea Island, French Polynesia). Botanica Mar. 30, 141–149.

Pearse, J. S., and Hines, A. H., 1979: Expansion of a central California kelp forest following the mass mortality of sea urchins. Mar. Biol. 51, 83–91.

Peckol, P., and Searles, R. B., 1983: Effects of seasonality and disturbance on population development in a Carolina shelf community. Bull. Mar. Sci. 33, 67–86.

Pedersen, P. M., 1976: Marine, benthic algae from southernmost Greenland. Meddr. Grønland 199 (3), 1–79.

Pedersen, P. M., 1984: Studies on primitive brown algae (Fucophyceae). Op. Bot. 74, 1–76.

Percival, E., 1979: The polysaccharides of green, red, and brown seaweeds: their basic structure, biosynthesis, and function. Br. Phycol. J. 14, 103–117.

Percival, E., and McDowell, R. H., 1967: Chemistry and enzymology of marine algal polysaccharides. Academic Press, London; 219 pp.

Pérès J. M., 1967a: The Mediterranean benthos. Oceanogr. Mar. Biol. Annu. Rev. 5, 449–533.

Pérès J. M., 1967b: Les biocoenoses benthiques dans le système phytal. Recl. Trav. Stn. Mar. Endoume 58 (42), 1–113.

Pérès, J. M., 1982a: Specific pelagic assemblages. In: Marine ecology. Vol. 5, Pt. 1. Ed. by O. Kinne. John Wiley, New York, pp. 313–372.

Pérès, J. M., 1982b: Major benthic assemblages. In: Marine ecology. Volume 5, Pt. 1. Ed. by O. Kinne. John Wiley, New York, pp. 373–522.

Pérès, J. M., and Molinier, R., 1957: Compte-rendu du colloque tenu à Gènes par le comité du Benthos de la Commission internationale pour l'Exploration scientifique de la mer Méditerranée. Recl. Trav. Stn. Mar. Endoume 13 (22), 5–15.

Pérès, J. M., and Picard, J., 1958: Recherches sur les peuplements benthiques de la Méditerranée nord-orientale. Ann. Inst. Océanogr. Monaco 34, 213–291.

Pérès, J. M., and Picard, J., 1964: Nouveau manuel de bionomie benthique de la mer Méditerranée. Recl. Trav. Stn. Mar. Endoume 31 (47), 5–137.

Perestenko, L. P., 1980: Wodorosli Zaliva Petra Welikogo (Algae of the Peter the Great Bay; in Russian). Izd. Akad. Nauk SSSR, Leningrad; 232 pp.

Pérez, R., Kaas, R., and Barbaroux, O., 1984: Culture expérimentale de l'algue *Undaria pinnatifida* sur les côtes de France. Sci. et Pêche Bull. Inst. Pêches Marit. 343, 1–16.

Pérez, R., Lee, J. Y., and Juge, C., 1981: Observations sur la biologie de l'algue *Undaria pinnatifida* (Harvey) Suringar introduite accidentellement dans l'Étang de Thau. Sci. et Peche Bull. Inst. Pêches Marit. 315, 1–12.

Pérez-Cirera, J. L., 1975: Catálogo florístico de las algas bentónicas de la Rîa de Corme y Lage, NO. de España. An. Inst. Bot. Cavanilles 32, 5–87.

Peters, A, F., 1988; Culture studies of a sexual life history in *Myriotrichia clavaeformis* (Phaeophyceae, Dictyosiphonales). Br. Phycol. J. 23, 299–306.

Peters, A., and Müller, D. G., 1986: Life-history studies—a new approach to the taxonomy of ligulate species of *Desmarestia* (Phaeophyceae) from the Pacific coast of Canada. Can. J. Bot. 64, 2192–2196.

Petrov, Ju. E., 1963: Development of conceptacles in *Ascoseira mirabilis* Skottsberg and the origin of Cyclosporeae (in Russian). Bot. Zhurn. 48, 1298–1309.

Petrov, Ju. E., 1965: De positione familiae Durvilleacearum et systematica classis cyclosporophycearum (Phaeophyta). Nov. Sist. Vysshikh. Rast.; novitates systematicae plantarum non vascularium. Akademiva Nauk SSSR. Botanicheskii Institut, Moscow, 70–72.

Petrov, Ju. E., 1967: The development of conceptacles (scaphidia) in *Ascophyllum nodosum* (L.) Le Jolis and *Durvillea antarctica* (Chamisso) Hariot. Bot. Zh. SSSR 52, 348–350.

Petrov, Ju. E., 1974: Synoptic key of the Laminariales and Fucales in the seas of the USSR (in Russian). Nov. Sist. Nizhnikh Rast. 11, 153–169.

Petrov, K. M., 1967: Vertical distribution of the phytobenthos in the Black Sea and Caspian Sea (in Russian). Oceanologia 7, 314–320.

Pfau, J., Hahnelt, D., and Nultsch, W., 1988: A new dual-beam microphotometer for determination of action spectra of light-induced phaeoplast movements in *Dictyota dichotoma*. J. Plant Physiol. 133, 572–579.

Pham-Hoang, H., 1962: Contribution à l'étude du peuplement du littoral du Vietnam (sud). Ann. Fac. Sci. Saigon 1962, 249–350.

Phillips, R. C., and McRoy, C. P., 1980: Handbook of seagrass biology: an ecosystem perspective. Garland STPM Press, New York; 353 pp.

Phillips, R. C., Vadas, R. L., and Ogden, N., 1982: The marine algae and seagrasses of the Miskito Bank, Nicaragua. Aquat. Bot. 13, 187–195.

Phinney, H. K., 1977: The macrophytic marine algae of Oregon. In: The marine plant biomass of the Pacific Northwest coast. Ed. by R. W. Krauss. Oregon State University Press, Corvallis, pp. 93–115.

Pianka, E. R., 1970: On r- and K-selection. Am. Nat. 104, 592–597.

Pickard, G. L., Emery, K. O., 1982: Descriptive physical oceanography. Enlarged Edition. Pergamon Press, New York; 449 pp.

Pickard, J., and Seppelt, R. D., 1984: Phytogeography of Antarctica. J. Biogeogr. 11, 83–102.

Pielou, E. C., 1977: The latitudinal spans of seaweed species and their patterns of overlap. J. Biogeogr. 4, 299–311.
Pielou, E. C., 1978: Latitudinal overlap of seaweed species: evidence for quasi-sympatric speciation. J. Biogeogr. 5, 227–238.
Pielou, E. C., 1979: Biogeography. Wiley, New York; 351 pp.
Pignatti, S., 1962: Associazioni di alghe marine sulla costa Veneziana. Atti Accad. Naz. Lincei Mem. Classe Sci. Mat. Nat. 32 (3), 1–134.
Pirt, S. J., 1982: Maintenance energy: a general model for energy-linked and energy-sufficient growth. Arch. Mikrobiol. 133, 300–302.
Pitman, W. C., and Talwani, M., 1972: Sea-floor spreading in the North Atlantic. Bull. Geol. Soc. Am. 83, 619–646.
Platnick, N. I., and Nelson, G., 1978: A method of analysis for historical biogeography. Syst. Zool. 27, 1–16.
Polanshek, A. R., and West, J. A., 1975: Culture and hybridization studies on *Petrocelis* (Rhodophyta) from Alaska and California. J. Phycol. 11, 434–439.
Polderman, P. J. G., 1979: The saltmarsh algal communities in the Wadden area with reference to their distribution and ecology in N. W. Europe. I. The distribution and ecology of the algal communities. J. Biogeogr. 6, 225–266.
Polderman, P. J. G., and Polderman-Hall, R. A., 1980: Algal communities in Scottish saltmarshes. Br. Phycol. J. 15, 59–71.
Pomerol, C., 1982: The Cenozoic era. Tertiary and Quaternary. Ellis Horwood Ltd., Chichester. Distributed by John Wiley & Sons, New York; 272 pp.
Por, F. D., 1971: One hundred years of Suez Canal - a century of Lessepsian migration: a retrospect and viewpoints. Syst. Zool. 20, 138–159.
Por, F. D., 1978: Lessepsian migration. The influx of Red Sea biota into the Mediterranean by way of the Suez Canal. Springer-Verlag, Heidelberg, 228 pp.
Por, F. D., and Dimentman, C., 1985: Continuity of Messinian biota in the Mediterranean basin. In: Geological evolution of the Mediterranean basin. Ed. by D. J. Stanley and F.-C. Wezel. Springer-Verlag, New York, pp. 545–557.
Por, F. D., and Dor, I. (Eds.), 1984: Hydrobiology of the mangal. The ecosystem of the mangrove forests. Junk, The Hague; 260 pp.
Post, E., 1963: Zur Verbreitung und Ökologie der *Bostrychia-Caloglossa* Assoziation. Int. Rev. Ges. Hydrobiol. Hydrogr. 48, 47–152.
Postels, A., and Ruprecht, F., 1840: Illustrationes algarum. St. Petersburg.
Powell, H. T., 1957: Studies in the genus *Fucus* L. II. Distribution and ecology of *Fucus distichus* L. emend. Powell in Britain and Ireland. J. Mar. Biol. Ass. U. K. 36, 663–693.
Powell, H. T., 1981: The occurrence of *Fucus distichus* subsp. *edentatus* in Macduff harbour, Scotland-the first record for mainland England. Br. Phycol. J. 16, 139.
Powell, J. H., 1986: A short day photoperiodic response in *Constantinea subulifera*. Am. Zool. 26, 479–487.
Powell, C. McA., Roots, S. R., and Veevers, J. J., 1988: Pre-breakup continental extension in East Gondwanaland and the early opening of the eastern Indian Ocean. Tectonophysics 155, 261–283.

Powell, N. A., 1968: Bryozoa (polyzoa) of Arctic Canada. J. Fish. Res. Bd. Can. 25, 2269–2320.

Powles, S. B., 1984: Photoinhibition of photosynthesis induced by visible light. Ann. Rev. Plant Physiol. 35, 15–44.

Pregnall, A. M., 1983: Release of dissolved organic carbon from the estuarine intertidal macroalga *Enteromorpha prolifera*. Mar. Biol. 73, 37–42.

Prescott, G. W., 1979: A contribution to a bibliography of Antarctic and subantarctic algae. Cramer, Vaduz; 312 pp.

Prézelin, B., 1981: Light reactions in photosynthesis. Can. Bull. Fish. Aquat. Sci. 210, 1–43.

Prézelin, B. B., and Boczar, B. A., 1986: Molecular basis of cell absorption and fluorescence in phytoplankton: potential applications to studies in optical oceanography. Prog. Phycol. Res. 4, 351–464.

Price, I. R., Larkum, A. W. D., and Bailey, A., 1976: Check list of marine benthic plants collected in the Lizard Island area. Aust. J. Plant Physiol. 3, 3–8.

Price, J. H., 1971: The shallow sublittoral marine ecology of Aldabra. Phil. Trans. R. Soc. Lond. B 260, 123–171.

Price, J. H., and Farnham, W. F., 1982: Seaweeds of the Faroes. 3: Open shores. Bull. Br. Mus. Nat. Hist. (Bot.) 10, 153–225.

Price, J. H., and John, D. M., 1979: The marine benthos of Antigua (Lesser Antilles). II. An annotated list of algal species. Botanica Mar. 22, 327–321.

Price, J. H., and John, D. M., 1980: Ascension Island, South Atlantic: a survey of inshore macroorganisms, communities, and interactions. Aquat. Bot. 9, 251–278.

Price, J. H., Tittley, I., and Honey, S. I., 1977: The benthic marine flora of Lincolnshire and Cambridgeshire: a preliminary review. Naturalist 102, 3–20, 91–104.

Price, J. H., Tittley, I., and Richardson, W. D., 1979: The distribution of *Padina pavonica* (L.) Lamour. (Phaeophyta: Dictyotales) on British and adjacent European shores. Bull. Br. Mus. Nat. Hist. (Bot.) 7 (1), 1–67.

Pringle, J. D., 1986: Swarmer release and distribution of life-cycle phases of *Enteromorpha intestinalis* (L.) (Chlorophyta) in relation to environmental factors. J. Exp. Mar. Biol. Ecol. 100, 97–111.

Printz, H., 1926: Die Algenvegetation des Trondhjemsfjordes. Skr. Norske Vidensk. Akad. I. Mat.-Naturv. Kl. 5, 1–274.

Printz, H., 1953: On some rare or recently immigrated marine algae on the Norwegian coast. Nytt Mag. Bot. 1, 135–151.

Printz, H., 1959: Phenological studies of marine algae along the Norwegian coast. I. *Ascophyllum nodosum* (L.) Le Jol. II. *Fucus vesiculosus* L. Skr. Norske Vidensk. Akad. I. Mat.-Naturv. Kl. 4, 1–28.

Provasoli, L., and Pinter, I. J., 1980: Bacteria induced polymorphism in an axenic laboratory strain of *Ulva lactuca* (Chlorophyceae). J. Phycol. 16, 196–201.

Prud'homme van Reine, W. F., 1982: A taxonomic revision of the European Sphacelariaceae (Sphacelariales, Phaeophyceae). E. J. Brill, Leiden University Press, Leiden; 293 pp.

Prud'homme van Reine, W. F., 1984: *Neomeris* in the Cape Verde Islands, a new record for the eastern coast of the Atlantic Ocean (Dasycladales: Chlorophyceae). Cour. Forsch.-Inst. Senckenberg 68, 139–142.

Prud'homme van Reine, W. F., and van den Hoek, C., 1988: Biogeography of Capeverdean seaweeds. Cour. Forsch.-Inst. Senckenberg. 105, 35–49.
Prud'homme van Reine, W. F., and Lobin, W., 1986: Katalog der von den Kapverdischen Inseln beschriebenen Taxa von Algen (Algae: Chlorophyceae, Phaeophyceae & Rhodophyceae). Cour. Forsch.-Inst. Senckenberg 81, 85–88.
Prud'homme van Reine, W. F., 1988: Phytogeography of seaweeds of the Azores. Helgoländer Meeresunters. 42, 165–185.

Quége, N., 1988: *Laminaria* (Phaeophyta) no Brasil, una perspectiva econômica. Ph.D. dissertation, Universidade de São Paulo; 230 pp.
Quast, J. C., 1971: Some physical aspects of the inshore environment, particularly as it affects kelp bed fishes. Nova Hedwigia (Beih.) 32, 229–240.

Rabinowitch, E. I., 1945–1956. Photosynthesis and related processes. Vol. 1 (1945); Vol. 2, Pt. 1 (1951), Pt. 2 (1956). Interscience Publishers, New York; 599 pp; 1208 pp; 2088 pp.
Ragan, M. A., 1976: Physodes and the phenolic compounds of brown algae. Composition and significance of physodes in vivo. Botanica Mar. 19, 145–154.
Ragan, M. A., 1981: Chemical constituents of seaweeds. In: The biology of seaweeds. Ed. by C. S. Lobban and M. J. Wynne. Blackwell, Oxford, pp. 589–626.
Ragan, M. A., and Glombitza, K.-W., 1986: Phlorotannins, brown algal polyphenols. Progr. Phycol. Res. 4, 129–241.
Ramirez, M. E., Müller, D. G., and Peters, A. F., 1986: Life history and taxonomy of two populations of ligulate *Desmarestia* (Phaeophyta) from Chile. Can. J. Bot. 64, 2948–2954.
Ramsay, A. T. S. (Ed.), 1977: Oceanic micropalaeontology. Vol. 1 and 2. Academic Press, New York; 1454 pp.
Ramus, J., 1978: Seaweed anatomy and photosynthetic performance: the ecological significance of light guides, heterogenous absorption and multiple scatter. J. Phycol. 14, 352–362.
Ramus, J., 1981: The capture and transduction of light energy. In: The biology of seaweeds. Ed. by C. S. Lobban and M. J. Wynne. Blackwell, Oxford; pp. 458–492.
Ramus, J., 1983: A physiological test of the theory of complementary chromatic adaptation. II. Brown, green, and red seaweeds. J. Phycol. 19, 173–178.
Ramus, J., 1985: Light. In: Handbook of phycological methods. Ecological field methods: macroalgae. Ed. by M. M. Littler and D. S. Littler. University Press, Cambridge, pp. 33–51.
Ramus, J., Beale, S. I., Mauzerall, D., and Howard, K. L., 1976: Changes in photosynthetic pigment concentration in seaweeds as a function of water depth. Mar. Biol. 37, 223–229.
Ramus, J., Lemons, F., and Zimmerman, C., 1977: Adaptation of light-harvesting pigments to downwelling light and the consequent photosynthetic performance of the eulittoral rockweeds *Ascophyllum nodosum* and *Fucus vesiculosus*. Mar. Biol. 42, 293–303.
Ramus, J., and van der Meer, J. P., 1983: A physiological test of the theory of

complementary chromatic adaptation. I. Color mutants of a red seaweed. J. Phycol. 19, 86–91.
Raup, D. M., 1975: Taxonomic diversity estimation using rarefaction. Paleobiology 1, 333–342.
Ravanko, O., 1968: Macroscopic green, brown, and red algae in the southwestern archipelago of Finland. Acta Bot. Fenn. 79, 1–50.
Ravanko, O., 1972: The physiognomy and structure of the benthic macrophyte communities on the rocky shores in the southwestern archipelago of Finland (Seili Islands). Nova Hedwigia 23, 363–403.
Raven, J. A., 1984a: Energetics and transport in aquatic plants. A. R. Liss, New York; 587 pp.
Raven, J. A., 1984b: A cost-benefit analysis of photon absorption by photosynthetic unicells. New Phytol. 98, 593–625.
Raven, J. A., 1986: Evolution of plant life forms. In: On the economy of plant form and function. Ed. by T. J. Givnish. University Press, Cambridge, pp. 421–492.
Raven, J. A., and Sprent, J. I., 1989: Phototrophy, diazotrophy, and palaeoatmospheres: biological catalysis and the H, C, N, and O cycles. J. Geol. Soc. Lond. 146, 161–170.
Rawlence, D. J., 1972: An ultrastructural study of the relationship between rhizoids of *Polysiphonia lanosa* (L.) Tandy (Rhodophyceae) and tissue of *Ascophyllum nodosum* (L.) Le Jolis (Phaeophyceae). Phycologia 11, 279–290.
Rawson, P. F., 1974: Lower Cretaceous (Ryazanian-Barremian) marine connections and cephalopod migrations between the Tethyan and boreal realms. In: The boreal lower Cretaceous Ed. by R. Casey, P. F. Rawson, Seel House Press, Liverpool, pp. 131–144.
Rayss, T., and Dor, I., 1963: Nouvelle contribution à la connaissance de la flore marine de la Mer Rouge. Bull. Sea Fish. Res. St. Haifa 34, 11–42.
Reed, D. C., Laur, D. R., and Ebeling, A. W., 1988: Variation in algal dispersal and recruitment: the importance of episodic events. Ecol. Monogr. 58, 321–335.
Reed, R. H., 1980a: The influence of salinity upon cellular mannitol concentration of the euryhaline marine alga *Pilayella littoralis* (L.) Kjellm. (Phaeophyta, Ectocarpales): preliminary observations. Botanica Mar. 23, 603–605.
Reed, R. H., 1980b: On the conspecifity of marine and freshwater *Bangia* in Britain. Br. Phycol. J. 15, 411–416.
Reed, R. H., 1983: The osmotic responses of *Polysiphonia lanosa* (L.) Tandy from marine and estuarine sites: evidence for incomplete recovery of turgor. J. Exp. Mar. Biol. Ecol. 68, 169–193.
Reed, R. H., 1985: Osmoacclimation in *Bangia atropurpurea* (Rhodophyta, Bangiales): the osmotic role of floridoside. Br. Phycol. J. 20, 211–218.
Reed, R. H., and Barron, A., 1983: Physiological adaptation to salinity change in *Pilayella littoralis* from marine and estuarine sites. Botanica Mar. 26, 409–416.
Rehault, J.-P., Boillot, G., and Mauffret, A., 1985: The western Mediterranean basin. In: Geological evolution of the Mediterranean basin. Ed. by D. J. Stanley and F.-C. Wezel. Springer-Verlag, New York, pp. 101–129.
Reichelt, J. L., and Borowitzka, M. A., 1984: Antimicrobial activity from marine algae: results of a large-scale screening programme. Proc. Int. Seaweed Symp. 11, 158–168.

Reinke, J., 1889a: Algenflora der westlichen Ostsee deutschen Antheils. VI. Bericht d. Komm. zur Unters. d. deutsch. Meere in Kiel. Parey, Berlin; 101 pp.

Reinke, J., 1889b: Atlas deutscher Meeresalgen. Parey, Berlin; 70 pp.

Reiskind, J. B., Beer, S., Bowes, G., 1989: Photosynthesis, photorespiration and ecophysiological interactions in marine macroalgae. Aquat. Bot. 34, 131–152.

Remane, A., 1933: Verteilung und Organisation der benthonischen Mikrofauna der Kieler Bucht. Wiss. Meeresunters. Kiel. 21/22, 161–122.

Remane, A., 1955: Die Brackwasser-Submergenz und die Umkomposition der Coenosen in Belt- und Ostsee. Kieler Meeresforsch. 11, 59–73.

Remane, A., and Schlieper, C., 1971: Biology of brackish water. Schweizerbart'sche Verlagshandlung, Stuttgart; 372 pp.

Remmert, H., 1980: Arctic animal ecology. Springer-Verlag, New York; 288 pp.

Renoux-Meunier, A., 1965: Étude de la végétation algale du Cap Saint-Martin (Biarritz). Bull. Cent. Étud. Rech. Sci. Biarritz 5, 379–557.

Rentschler, H. G., 1967: Photoperiodische Induktion der Monosporenbildung bei *Porphyra tenera* Kjellm. (Rhodophyta-Bangiophyceae). Planta (Berlin) 76, 65–74.

Rey, J. R., 1984: Experimental tests on island biogeographic theory. In: Ecological communities: conceptual issues and the evidence. Ed. by D. R. Strong, D. Simberloff, L. G. Abele, and A. B. Thistle. University Press, Princeton, pp. 101–112.

Rheinheimer, G., 1981: Mikrobiologie der Gewässer. Gustav Fischer, Stuttgart; 251 pp.

Rice, E. L., Chapman, A. R. O., 1985. A numerical study of *Fucus distichus* (Phaeophyta). J. Mar. Biol. Ass. U. K. 65, 433–459.

Richardson, N., 1970: Studies on the photobiology of *Bangia fuscopurpurea* J. Phycol. 6, 215–219.

Richardson, W. D., 1975: The marine algae of Trinidad, West Indies. Bull. Br. Mus. Nat. Hist. (Bot.) 5, 71–143.

Richter, G., 1982: Stoffwechselphysiologie der Pflanzen. Physiologie und Biochemie des Primär- und Sekundärstoffwechsels. Georg Thieme Verlag, Stuttgart; 592 pp.

Ricker, R. W., 1987: Taxonomy and biogeography of Macquarie Island seaweeds. British Museum of Natural History, London; 344 pp.

Ricketts, E. G., Calvin, J., Hedgpeth, J., and Phillips, D. W., 1985: Between Pacific tides. 5th ed., revised by J. Hegpeth. Stanford University Press, Palo Alto; 652 pp.

Riechert, R., and Dawes, C. J., 1986: Acclimation of the green alga *Caulerpa racemosa* var. *uvifera* to light. Botanica Mar. 29, 533–537.

Riedl, R., 1964a: 100 Jahre Litoralgliederung seit Josef Lorenz, neue und vergessene Gesichtspunkte. Int. Rev. Ges. Hydrobiol. Hydrogr. 49, 281–305.

Riedl, R. 1964b: Die Erscheinungen der Wasserbewegung und ihre Wirkung auf Sedentarier im mediterranen Felslitoral. Helgoländer Meeresunters. 10, 155–186.

Riedl, R., 1966: Biologie der Meereshöhlen. Parey, Hamburg; 636 pp.

Riedl, R., 1967: Die Tauchmethode, ihre Aufgaben und Leistungen bei der Erforschung des Litorals; eine kritische Untersuchung. Helgoländer Meeresunters. 15, 294–351.

Riedl, R., 1971: Water movement. Animals. In: Marine ecology. Vol. 1, Pt. 2. Ed. by O. Kinne. Wiley-Interscience, London, pp. 1123–1156.

Riedl, R., 1983: Fauna und Flora der Adria. Vol. 3. Aufl. Parey, Hamburg; 836 pp.

Rietema, H., and Klein, A. W. O., 1981: Environmental control of the life cycle of *Dumontia contorta* (Rhodophyta) kept in culture. Mar. Ecol. Prog. Ser. 4, 23–29.

Rietema, H., and van den Hoek, C., 1984: Search for possible latitudinal ecotypes in *Dumontia contorta*. Helgoländer Meeresunters. 38, 389–399.

Rigg, G. B., and Miller, R. C., 1949: Intertidal plant and animal zonation in the vicinity of Neah Bay, Washington. Proc. Calif. Acad. Sci. Fourth Ser. 26, 323–352.

Rivkin, R. B., and Putt, M., 1987: Photosynthesis and cell division by Antarctic microalgae: comparison of benthic, planktonic, and ice algae. J. Phycol. 23, 223–229.

Robberecht, R., and Caldwell, M. M., 1978: Leaf epidermal transmittance of ultraviolet radiation and its implications for plant sensitivity to ultraviolet-radiation induced injury. Oecologia 32, 277–287.

Roberts, M., 1978: Active speciation in the taxonomy of the genus *Cystoseira* C. Ag. In: Modern approaches to the taxonomy of red and brown algae. Ed. by D. E. G. Irvine and J. H. Price. Academic Press, London; pp. 399–422.

Robertson, A. I., and Hansen, J. A., 1982: Decomposing seaweed: a nuisance or a vital link in coastal food chains? CSIRO Div. Fish. Res. Rep. 1980–1981, pp. 75–83.

Robertson, A. I., and Lucas, J. S., 1983: Food choice, feeding rates, and the turnover of macrophyte biomass by a surf-zone inhabiting amphipod. J. Exp. Mar. Biol. Ecol. 72, 99–124.

Robinson, N., 1966: Solar radiation. Elsevier, Amsterdam; 347 pp.

Rodriguez, G., 1959: The marine communities of Margarita Island, Venezuela. Bull. Mar. Sci. Gulf Caribb. 9, 237–280.

Rodriguez, J. J., 1889: Algas de las Baleares. An. Soc. Esp. Hist. Nat. 18, 199–274.

Rögl, F., and Steininger, F. F., 1984: Neogene Paratethys, Mediterranean, and Indo-Pacific seaways. Implications for the paleobiogeography of marine and terrestrial biotas. In: Fossils and climate. Ed. by P. J. Brenchley. Wiley, Chichester, pp. 171–200.

Rosell, K.-G., and Srivastava, L. M., 1987: Fatty acids as antimicrobial substances in brown algae. Hydrobiologia 151/152, 471–475.

Rosen, B. R., 1982: Darwin, coral reefs, and global geology. BioScience 32, 519–525.

Rosen, B. R., 1984: Reef coral biogeography and climate throught the late Cenozoic: just islands in the sun or a critical pattern of islands? In: Fossil and climate. Ed. by P. J. Brenchley. Wiley, Chichester; pp. 201–262.

Rosen, B. R., 1988: Progress, problems, and patterns in the biogeography of reef corals and other tropical marine organisms. Helgoländer Meeresunters. 42, 269–301.

Rosen, D. E., 1976: A vicariance model of Caribbean biogeography. Syst. Zool. 24, 431–464.

Rosen, D. E., 1985: Geological hierarchies and biogeographic congruence in the Caribbean. Ann. Mo. Bot. Gard. 72, 636–659.

Rosenthal, R. J., Clarke, W. D., and Dayton, P. K., 1974: Ecology and natural history of a stand of giant kelp, *Macrocystis pyrifera*, off Del Mar, California. Fish. Bull. 72, 670–684.

Rosenvinge, L. K., 1898: I. Deuxième mémoire sur les algues marines du Groenland. Meddr. Grønland 20, 1–125.

Rosenvinge, L. K., 1910: On the marine algae from north-east Greenland collected by the "Danmark-Expedition". Meddr. Grønland 43, 91–133.

Rosenvinge, L. K., 1924: A botanical trip to Jan Mayen by Johannes Gandrup. 3. Marine algae. Dansk Bot. Ark. 4 (5), 1–35.

Rosenvinge, L. K., 1909–1931: The marine algae of Denmark. 1. Rhodophyceae. K. Danske Vidensk. Selsk. Skr. 7 Raekke. I, 1–152 (1909); II 153–284 (1917); III, 285–488 (1924); IV, 489–630 (1931).

Rosenvinge, L. K., and Lund, S., 1941–1950: The marine algae of Denmark. 2. Phaeophyceae. K. Danske Vidensk. Selsk. Biol. Skr. 1, 1–79 (4, 1941); 2, 1–59 (6, 1943); 4, 1–99 (5, 1947); 6, 1–80 (2, 1950).

Ross, M. I., and Scotese, C. R., 1988: A hierarchical tectonic model of the Gulf of Mexico and Caribbean Region. Tectonophysics 155, 139–168.

Roth, A. A., Clausen, C. D., Yahiku, P. Y., Clausen, V. E., and Cox, W. W., 1982: Some effects of light on coral growth. Pacif. Sci. 36, 65–81.

Round, F. E., 1981: The ecology of algae. University Press, Cambridge; 653 pp.

Rouland, D., Xu, S. H., and Schindele, F., 1985: Upper mantle structure in the southeast Indian Ocean: A surface wave investigation. Tectonophysics 114, 281–292.

Rowley, D. B., and Lottes, A. L., 1988: Plate-kinematic reconstructions of the North Atlantic and Arctic: Late Jurassic to Present. Tectonophysics 155, 73–120.

Rueness, J., 1977: Norsk algeflora. Universitetsforlaget, Oslo; 266 pp.

Rueness, J., 1978: Hybridization in red algae. In: Modern approaches to the taxonomy of red and brown algae. Ed. by D. E. G. Irvine and J. H. Price. Academic Press, London, pp. 247–262.

Rueness, J., 1985: Japansk drivtang, *Sargassum muticum*. Biologisk forurensning av europeiske farvann. Blyttia 43, 71–74.

Rueness, J., 1989: *Sargassum muticum* and other introduced Japanese macroalgae: Biological pollution of European coasts. Mar. Pollut. Bull. 20, 173–176.

Rueness, J., and Asen, P. A., 1982: Field and culture observations on the life history of *Bonnemaisonia asparagoides* (Woodw.) C. Ag. (Rhodophyta) from Norway. Botanica Mar. 25, 577–587.

Rueness, J., Mathisen, H. A., and Tananger, T., 1987: Culture and field observations on *Gracilaria verrucosa* (Huds.) Papenf. (Rhodophyta) from Norway. Botanica Mar. 30, 267–276.

Rueness, J., and Tananger, T., 1984: Growth in culture of four red algae from Norway with potential for mariculture. Hydrobiologia 116/117, 303–307.

Rüffer, U., Nultsch, W., and Pfau, J., 1978: Untersuchungen zur lichtinduzierten Chromatophorenverlagerung bei *Fucus vesiculosus*. Helgoländer Meeresunters. 31, 333–346.

Rugg, D. A., and Norton, T. A., 1987: *Pelvetia canaliculata*, a high-shore seaweed that shuns the sea. In: Plant life in aquatic and amphibious habitats. Ed. by R. M. M. Crawford. Blackwell, Oxford; pp. 347–358.

Ruprecht, F. J., 1852: Neue oder unvollständig bekannte Pflanzen aus dem nördlichen Theile des Stillen Ozeans. Mém. Acad. Imp. Sci. St. Petersbourg 7, 55–82.

Russell, G., 1973: Phytosociological studies on a two-zone shore. II. Community structure. J. Ecol. 61, 525–536.

Russell, G., 1985a: Recent evolutionary changes in the algae of the Baltic Sea. Br. Phycol. J. 20, 87–104.
Russell, G., 1985b: Some anatomical and physiological differences in *Chorda filum* from coastal waters of Finland and Great Britain. J. Mar. Biol. Ass. U. K. 65, 343–349.
Russell, G., 1986: Variation and natural selection in marine macroalgae. Oceanogr. Mar. Biol. Annu. Rev. 24, 309–377.
Russell, G., 1987: Salinity and seaweed vegetation. In: Plant life in aquatic and amphibious habitats. Ed. by R. M. M. Crawford. Blackwell, Oxford, pp. 35–52.
Russell, G., 1988: The seaweed flora of a young semi-enclosed sea: the Baltic. Salinity as a possible agent of a flora divergence. Helgoländer Meeresunters. 42, 243–250.
Russell, G., and Bolton, J. J., 1975: Euryhaline ecotypes of *Ectocarpus siliculosus* (Dillw.) Lyngb. Estuar. Coast. Mar. Sci. 3, 91–94.
Russell, G., and Veltkamp, C. J., 1984: Epiphyte survival on skin-shedding macrophytes. Mar. Ecol. Prog. Ser. 18, 149–153.
Rützler, K., and Macintyre, I. G. (Eds.), 1982: The Atlantic barrief reef ecosystem at Carrie Bow Cay, Belize, I. Structure and communities. Smithsonian Institution Press, Washington; 539 pp.
Ruyter van Steveninck, E. D., and Breeman, A. M., 1987: Deep-water vegetations of *Lobophora variegata* (Phaeophyceae) in the coral reef of Curacao: population dynamics in relation to mass mortality of the sea urchin *Diadema antillarum*. Mar. Ecol. Prog. Ser. 36, 81–99.
Ruyter van Steveninck, E. D., Kamermans, P., and Breeman, A. M., 1988: Transplant experiments with two morphological forms of *Lobophora variegata* (Phaeophyceae). Mar. Ecol. Prog. Ser. 49, 191–194.
Ryther, J. H., de Boer, J. A., and Lapointe, B. E., 1978: Cultivation of seaweeds for hydrocolloids, waste treatment, and biomass for energy conversion. Proc. Int. Seaweed Symp. 9, 1–16.

Saga, N., 1986: Regulation of life cycle in epiphytic brown alga *Dictyosiphon foeniculaceus*. Sci. Pap. Inst. Algol. Res. Fac. Sci. Hokkaido Imp. Univ. 8, 31–61.
Saifullah, S. M., 1973: A preliminary survey of the standing crop of seaweeds from Karachi coast. Botanica Mar. 16, 139–144.
Saito, Y., 1975: *Undaria*. In: Advance of phycology in Japan. Ed. by J. Tokida and H. Hirose. VEB Gustav Fischer, Jena, pp. 304–320.
Salisbury, F. B., 1981: Responses to photoperiod. In: Encyclopedia of plant physiology, New Series, Vol. 12A. Ed. by O. L. Lange, P. S. Nobel, C. B. Osmond, and H. Ziegler. Springer-Verlag, Berlin; pp. 135–167.
Sanbonsuga, Y., and Neushul, M., 1978: Hybridization of *Macrocystis* (Phaeophyta) with other float-bearing kelps. J. Phycol. 14, 214–224.
Sand-Jensen, K., 1988a: Minimum light requirements for growth in *Ulva lactuca*. Mar. Ecol. Prog. Ser. 50, 187–193.
Sand-Jensen, K., 1988b: Photosynthetic responses of *Ulva lactuca* at very low light. Mar. Ecol. Prog. Ser. 50, 195–201.
Sand-Jensen, K., Revsbech, N. P., and Barker Jorgensen, B., 1985: Microprofiles of

oxygen in epiphyte communities on submerged macrophytes. Mar. Biol. 89, 55–62.

Santelices, B., 1980: Phytogeographic characterization of the temperate coast of Pacific South America. Phycologia 19, 1–12.

Santelices, B., 1989: Algas marinas de Chile. Distribución, ecología, utilización, diversidad. Ediciones Universidad Católica de Chile. Santiago, Chile; 399 pp.

Santelices, B., and Abbott, I. A., 1978: New records of marine algae from Chile and their effect on phytogeography. Phycologia 17, 213–222.

Santelices, B., Abbott, I. A., 1987: Geographic and marine isolation: an assessment of the marine algae of Easter Island. Pacific Sci. 41, 1–20.

Santelices, B., Castilla, J. C., Cancino, J., and Schmiede, P., 1980: Comparative ecology of *Lessonia nigrescens* and *Durvillaea antarctica* (Phaeophyta) in Central Chile. Mar. Biol. 59, 119–132.

Santelices, B., and Correa, J., 1985: Differential survival of macroalgae to digestion by intertidal herbivore molluscs. J. Exp. Mar. Biol. Ecol. 88, 183–191.

Sargent, J. R., and Whittle, K. J., 1981: Lipids and hydrocarbons in the marine food web. In: Analysis of marine ecosystems. Ed. by A. R. Longhurst. Academic Press, London, pp. 491–533.

Sarnthein, M., et al., 1982: Atmospheric and oceanic circulation patterns of N. W. Africa during the past 25 million years. In: The geology of the northwest African continental margin. Ed. by U. von Rad et al. Springer-Verlag, Berlin, pp. 545–604.

Sartoni, G., 1986: Algal flora and its vertical distribution on the Gesira cliff (central-southern Somalia). Webbia 39, 355–377.

Sauberer, F., and Härtel, O., 1959: Pflanze und Strahlung. Akademische Verlagsgesellschaft Geest und Portig, Leipzig; 268 pp.

Sauvageau, C., 1918: Recherches sur les laminaires des côtes de France. Mém. Acad. Sci. Inst. Fr. 56, 1–240.

Savin, S. M., 1977: The history of the earth's surface temperature during the past 100 million years. Annu. Rev. Earth Planet. Sci. 5, 319–355.

Savin, S. M., Douglas R. G., and Stehli F. G., 1975: Tertiary marine paleotemperatures. Bull. Geol. Soc. Am. 86, 1499–1510.

Scagel, R. F., 1957: An annotated list of the marine algae of British Columbia and northern Washington (including keys for genera). Natl. Mus. Can. Bull. 150, Biol. Ser. 52, Ottawa, 1–289.

Scagel, R. F., 1967: Guide to common seaweeds of British Columbia. British Columbia Provincial Museum, Handbook no. 27. Victoria, B.C.; 330 pp.

Scagel, R. F., 1973: Marine benthic plants in the vicinity of Bamfield, Barkley Sound, British Columbia. Syesis 6, 127–145.

Scagel, R. F., Gabrielson, P. W., Garbary, D. J., Golden, L., Hawkes, M. W., Lindstrom, S. C., Oliveira, J. C., and Widdowson, T. B., 1989: A synopsis of the benthic marine algae of British Columbia, southeast Alaska, Washington and Oregon. University of British Columbia, Dept. of Botany, Vancouver; 532 pp.

Scarlato, O. A., 1977: Bivalve molluscs and temperature as an agent determining their geographical distribution. Malacologia 16, 247–250.

Scheibling, R., 1986: Increased macroalgal abundance following mass mortalities of sea urchins (*Strongylocentrotus droebachiensis*) along the Atlantic coast of Nova Scotia. Oecologia 68, 186–198.

Scheltema, R. S., 1971: The dispersal of the larvae of shoal-water benthic invertebrate species over long distances by ocean currents. In: Fourth European marine biology symposium. Ed. by D. J. Crisp. University Press, Cambridge, pp. 7–28.

Schiel, D. R., and Foster, M. S., 1986: The structure of subtidal algal stands in temperate waters. Oceanogr. Mar. Biol. Annu. Rev. 24, 265–307.

Schlichter, D., Weber, W., and Fricke, H. W., 1985: A chromophore system in the hermatypic, deep-water coral *Leptoseris fragilis* (Antozoa: Hexocorallia). Mar. Biol. 89, 143–147.

Schlichter, D., Fricke, H. W., and Weber, W., 1986: Light harvesting by wavelength transformation in a symbiotic coral of the Red Sea twilight zone. Mar. Biol. 91, 403–407.

Schmid, R., 1976: Septal pores in *Prototaxites*, an enigmatic Devonian plant. Science 191, 287–288.

Schmid, R., 1984: Blue light effects on morphogenesis and metabolism in *Acetabularia*. In: Blue light effects in biological systems. Ed. by H. Senger. Springer, Berlin, pp. 419–432.

Schmid, R., 1988: Twofold effect of blue light on a circadian rhythm in *Acetabularia*. J. Interdiscipl. Cycle Res. 17, 99–107.

Schmid, R., Idziak, E.-M., and Tünnermann, M., 1987: Action spectrum for the blue-light-dependent morphogenesis of hair whorls in *Acetabularia mediterranea*. Planta 171, 96–103.

Schmid, R., and Koop, H.-U., 1983: Properties of the chloroplast movement during the circadian chloroplast migration in *Acetabularia mediterranea*. Z. Pflanzenphysiol. 112, 351–357.

Schmidt, O. C., 1931: Die marine Vegetation der Azoren, in ihren Grundzügen dargestellt. Bibl. Bot. 102. Ed. by L. Diels. Schweizerbartsche Verlagsbuchhandlung, Stuttgart; 116 pp.

Schmidt, O. C., 1957: Die marine Vegetation Afrikas, in ihren Grundzügen dargestellt. Willdenowia 1, 709–756.

Schmitz, K., 1981: Translocation. In: The biology of seaweeds. Ed. by C. S. Lobban and M. J. Wynne. Blackwell, Oxford, pp. 534–558.

Schmitz, K., and Lobban, C. S., 1976: A survey of translocation in Laminariales (Phaeophyta). Mar. Biol. 36, 207–216.

Schmitz, K., and W. Riffarth, 1980: Carrier-mediated uptake of L-leucine by the brown alga *Giffordia mitchellii*. Z. Pflanzenphysiol. 96, 311–324.

Schmitz, K., Lüning, K., and Willenbrink, J., 1972: CO_2-Fixierung und Stofftransport in benthischen marinen Algen. II. Zum Ferntransport ^{14}C-markierter Assimilate bei *Laminaria hyperborea* und *Laminaria saccharina*. Z. Pflanzenphysiol. 67, 418–429.

Schneider, C. W., Suyemoto, M. M., and Yarish, C., 1979: An annotated checklist of Connecticut seaweeds. State Geolog. Nat. Hist. Surv. Dep. Environ. Prot. Bull. 108, 1–20.

Schnetter, R., 1976–1978: Marine Algen der karibischen Küsten von Kolumbien. I.

Phaeophyceae. II. Chlorophyceae. Bibl. Phycol. 24 (1976); 42 (1978); Cramer, Vaduz; 125 pp. (1976); 198 pp. (1978)

Schnetter, R., and Bula Meyer, G., 1982: Marine Algen der Pazifikküste von Kolumbien. Chlorophyceae, Phaeophyceae, Rhodophyceae. Bibl. Phycol. 60. Cramer, Vaduz; 287 pp.

Schnetter, R., Hörnig, I., and Weber-Peukert, G., 1987: Taxonomy of some North Atlantic *Dictyota* species. Hydrobiologia 151/152, 193–197.

Schnitker, D., Jorgensen, J. B., 1990: Late glacial and Holocene diatom successions in the Gulf of Maine: response to climatologic and oceanographic change. In: Evolutionary biology of the marine algae of the North Atlantic. Ed. by G. R. South and D. Garbary. Springer-Verlag, Berlin.

Schoener, A., 1988: Experimental island biogeography. In: Analytical biogeography. An integrated approach to the study of animal and plant distributions. Ed. by A. A. Myers and P. S. Giller. Chapman & Hall, London, pp. 483–512.

Scholl, D. W., Buffington, E. C., and Marlow, M. S., 1975: Plate tectonics and the structural evolution of the Aleutian-Bering Sea region. Geol. Soc. Am. Spec. Pap. 151, 1–31.

Schonbeck, M. W., and Norton, T. A., 1978: Factors controlling the upper limits of fucoid algae on the shore. J. Exp. Mar. Biol. Ecol. 31, 303–313.

Schonbeck, M. W., and Norton, T. A., 1979a: An investigation of drought avoidance in intertidal fucoid algae. Botanica Mar. 22, 133–144.

Schonbeck, M. W., and Norton, T. A., 1979b: The effects of brief periodic submergence on intertidal fucoid algae. Estuar. Coast. Mar. Sci. 8, 205–211.

Schonbeck, M. W., and Norton, T. A., 1979c: Drought-hardening in the upper-shore seaweeds *Fucus spiralis* and *Pelvetia canaliculata*. J. Ecol. 67, 687–696.

Schonbeck, M. W., and Norton, T. A., 1979d: The effects of diatoms on the growth of *Fucus spiralis* germlings in culture. Botanica Mar. 22, 233–236.

Schonbeck, M. W., and Norton, T. A., 1980: Factors controlling the lower limits of fucoid algae on the shore. J. Exp. Mar. Biol. Ecol. 43, 131–150.

Schopf, T. J. M., 1970: Relation of floras of the Southern Hemisphere to continental drift. Taxon 19, 657–674.

Schopf, T. J. M., 1977: Pattern of evolution: a summary and discussion. In: Patterns of evolution as illustrated by the fossil record. Ed. by A. Hallam. Elsevier, Amsterdam, pp. 547–561.

Schopf, T. J. M., 1980: Paleoceanography. Harvard University Press, Cambridge, Mass.; 341 pp.

Schopf, T. J. M, 1981: Cretaceous endings. Science 211, 571–572.

Schopf, T. J. M., Fisher, J. B., and Smith, C. A. F., 1978: Is the marine latitudinal diversity gradient merely another example of the species area curve? In: Marine organisms. Ed. by B. Battaglia and J. A. Beardmore. Plenum Press, New York; pp. 365–386.

Schotter, G., 1968: Recherches sur les Phyllophoracées. Notes posthumes publiées par Jean Feldmann et Marie-France Magne. Bull. Inst. Oceanogr. Monaco 67, 1–99.

Schrader, H.-J., and Fenner, J., 1976: Norwegian Sea Cenozoic diatom biostratigraphy and taxonomy. Part I. Norwegian Sea Cenozoic diatom biostratigraphy. In:

Initial Reports of the Deep Sea Drilling Project, Vol. 38. Ed. by M. Talwani et al. US. Government Printing Office, Washington, D.C., pp. 921–1099.

Schramm, W., 1968: Ökologisch-physiologische Untersuchungen zur Austrocknungs- und Temperaturresistenz an *Fucus vesiculosus* L. der westlichen Ostsee. Int. Rev. Ges. Hydrobiol. Hydrogr. 53, 469–510.

Schramm, W., and Martens, V., 1976: Ein Messystem für *in situ* Untersuchungen zum Stoff- und Energieumsatz in Benthosgemeinschaften. Kieler Meeresforsch., Sonderheft 3, 1–6.

Schuhmacher, H., and Zibrowius, H., 1985: What is hermatypic? A redefinition of ecological groups in corals and other organisms. Coral Reefs 4, 1–9.

Schweitzer, H.-J., 1983: Die Unterdevonflora des Rheinlandes. 1. Teil. Paleontographica, Abt. B, 189, 1–138.

Schwenke, H., 1969: Meeresbotanische Untersuchungen in der westlichen Ostsee als Beitrag zu einer marinen Vegetationskunde. Int. Rev. Ges. Hydrobiol. Hydrogr. 54, 35–94.

Schwenke, H., 1971: Water movement. Plants. In: Marine ecology. Vol. 1, Pt. 2. Ed. by O. Kinne. Wiley-Interscience, London, pp. 1091–1121.

Schwenke, H., 1974: Die Benthosvegetation. In: Meereskunde der Ostsee. Ed. by L. Magaard and G. Rheinheimer. Springer-Verlag, Berlin, pp. 131–146.

Sclater, J. G., Hellinger, S., and Tapscott, C., 1977: The paleobathymetry of the Atlantic Ocean from the Jurassic to the Present. J. Geol. 85, 509–552.

Scoffin, T. P., Stoddart, D. R., Tudhope, A. W., and Woodroffe, C., 1985: Rhodoliths and coralliths of Muri Lagoon, Rarotonga, Cook Islands. Coral Reefs 4, 71–80.

Scotese, C. R., Gahagan, L. M., and Larson, R. L., 1988: Plate tectonics reconstruction of the Cretaceous and Cenozoic ocean basins. Tectonophysics 155, 27–48.

Sculthorpe, C. D., 1967: The biology of aquatic vascular plants. Edward Arnold, London; 610 pp.

Seagrief, S. C., 1984: A catalogue of South African green, brown and red marine algae. Mem. Bot. Surv. S. Afr. 47, 1–72.

Seagrief, S. C., 1988: Marine algae. In: A field guide to the eastern Cape coast. Ed. by R. A. F. Lubke, F. W. Gess, and M. N. Bruton. Wildlife Society of Southern Africa, Grahamstown, pp. 35–72.

Searles, R. B., 1978: The genus *Lessonia* Bory (Phaeophyta, Laminariales) in southern Chile and Argentinia. Br. Phycol. J. 13, 361–381.

Searles, R. B., 1984: Seaweed biogeography of the mid-Atlantic coast of the United States. Helgoländer Meeresunters. 38, 259–271.

Searles, R. B., and Schneider, C. W., 1987: Observations on the deep-water flora of Bermuda. Hydrobiologia 151/152: 261–266.

Searles, R. B., Hommersand, M. H., and Amsler, C. D., 1984: The occurrence of *Codium fragile* subsp. *tomentosoides* and *C. taylori* (Chlorophyta) in North Carolina. Botanica Mar. 27, 185–187.

Searles, R. B., and Schneider, C. W., 1978: A checklist and bibliography of North Carolina seaweeds. Botanica Mar. 21, 99–108.

Searles, R. B., and Schneider, C. W., 1980: Biogeographic affinities of the shallow and deep water benthic marine algae of North Carolina. Bull. Mar. Sci. 30, 732–736.

Sears, J. R., and Cooper, R. A., 1978: Descriptive ecology of offshore, deep-water, benthic algae in the temperate western North Atlantic Ocean. Mar. Biol. 44, 309–314.

Sears, J. R., and Wilce, R. T., 1975: Sublittoral, benthic marine algae of southern Cape Cod and adjacent islands: seasonal periodicity, associations, diversity, and floristic composition. Ecol. Monogr. 45, 337–365.

Segawa, S., 1971: Colored illustrations of marine algae of Japan (in Japanese). Hoikuska Publ. Co., Osaka; 175 pp.

Segerstrale, S. G., 1957: Baltic Sea. Mem. Geol. Soc. Am. 67, (1), 751–800.

Seibold, E., and Berger, W. H., 1982: The sea floor. An introduction to marine geology. Springer-Verlag, Berlin; 288 pp.

Senger, H. (Ed.) 1980: The blue light syndrome. Springer-Verlag, Berlin; 665 pp.

Senger, H. (Ed.), 1984: Blue light effects in biological systems. Springer, Berlin; 538 pp.

Senger, H. (Ed.), 1987: Blue light phenomena and occurrence in plants and microorganisms. CRC Press, Boca Raton, Fla.; 405 pp.

Seoane-Camba, J. A., 1965: Estúdios sobre las algas bentónicas en la costa sur de la Península Ibérica (litoral de Cádiz). Invest. Pesq. 29, 3–216.

Seoane-Camba, J. A., 1969: Algas bentónicas de Menorca en los herbarios Thuret-Bornet y Sauvageau del Muséum National d'Histoire Naturelle de Paris. Invest. Pesq. 33, 213–260.

Seoane-Camba, J. A., 1975: Algas bentónicas Españolas en los herbarios Thuret-Bornet y Sauvageau del Muséum National d'Histoire Naturelle de Paris. II. Algas de Cataluña y Baleares (excepto Menorca). An. Inst. Bot. Cavanîlles 32, 33–51.

Serman, D., Span, A., Pavletic, Z., and Antolic, B., 1981: Phytobenthos of the island of Lokrum. Acta Bot. Croat. 40, 167–182.

Setchell, W. A., 1893: On the classification and geographic distribution of the Laminariaceae. Trans. Conn. Acad. Arts Sci. 9, 333–375.

Setchell, W. A., 1899: Algae of the Pribilof Islands. In: Fur seals and fur seal islands of the North Pacific Ocean. Pt. 3. Ed. by D. S. Jordan. Washington, D.C.; pp. 589–596.

Setchell, W. A., 1917: Geographical distribution of the marine algae. Science 45, 197–204.

Setchell, W. A., 1920: The temperature interval in the geographical distribution of marine algae. Science 53, 187–190.

Setchell, W. A., 1922: Cape Cod in its relation to the marine flora of New England. Rhodora 24, 1–11.

Setchell, W. A., and Gardner, N. L., 1920–1925: The marine algae of the Pacific coast of North America. II. Chlorophyceae. III. Melanophyceae. Univ. Calif. Publs. Bot. 8, 139–374 (1920); 8, 383–739 (1925).

Sève, M. de, Cardinal, A., and Goldstein, M., E., 1979: Les algues marines benthiques des Îles-de-la-Madelaine. Proc. N. S. Inst. Sci. 29, 223–233.

Shackleton, N., and Boersma, A., 1981: The climate of the Eocene ocean. J. Geol. Soc. Lond. 138, 153–157.

Shameel, M., Afaq-Husain, S., and Shahid-Husain, S., 1989: Addition to the knowledge of seaweeds from the coast of Lasbela, Pakistan. Botanica Mar. 32, 177–180.

Shanab, S., Jacques, R., and Magne, F., 1988: Croissance et ramification du thalle de *Bachelotia antillarum* cultivé en éclairements monochromatiques. Plant Physiol. Biochem. 26, 303–311.

Shanab, S.,and Abdel Rahman, M. H., 1988: Action de la photopériode sur la croissance de la Phéophycée *Bachelotia antillarum*. Cryptogam., Algol. 9, 87–100.

Shannon, L. V., 1985: The Benguela ecosystem. Part I. Evolution of the Benguela, physical features, and processes. Oceanogr. Mar. Biol. Annu. Rev. 23, 105–182.

Sheath, R. G., and Harlin, M. M., 1988: Freshwater and marine plants of Rhode Island. Kendall/Hunt Publishing Company; Dubuque, IA; 160 pp.

Shepherd, S. A., and Womersley, H. B. S., 1970: The sublittoral ecology of West Island, South Australia: 1. Environmental features and algal ecology. Trans. R. Soc. S. Aust. 94, 105–137.

Shepherd, S. A., and Womersley, H. B. S., 1971: Pearson Island Expedition 1969: 7. The subtidal ecology of benthic algae. Trans. R. Soc. S. Aust. 95, 155–167.

Shepherd, S. A., and Womersley, H. B. S., 1976: The subtidal algal and seagrass ecology of St. Francis Island, South Australia. Trans. R. Soc. S. Aust. 100, 177–191.

Shepherd, S. A., and Womersley, H. B. S., 1981: The algal and seagrass ecology of Waterloo Bay, South Australia. Aquat. Bot. 11, 305–371.

Sheppard, C. R., 1987: Coral species of the Indian Ocean and adjacent seas: a synonymized compilation and some regional distribution patterns. Atoll Res. Bull. 307, 1–32.

Sheppard, C. R. C., Jupp, B. P., Sheppard, A. L. S., and Bellamy, D. J., 1978: Studies on the growth of *Laminaria hyperborea* (Gunn.) Fosl. and *Laminaria ochroleuca* De la Pylaie on the French Channel coast. Botanica Mar. 21, 109–116.

Shibata, K., and Haxo, F. T., 1969: Light transmission and spectral distribution through epi- and endozoic layers in the brain coral, *Favia*. Biol. Bull. 136, 461–468.

Shropshire, W., and Mohr, H. (Eds.), 1983: Photomorphogenesis. Encyclopedia of plant physiology, n. s. Vols. 16A, 16B. Springer-Verlag, Berlin; 456 pp., 832 pp.

Sieburth, J., and Tootle, J. L., 1981: Seasonality of microbial fouling on *Ascophyllum nodosum* (L.) Lejol., *Fucus vesiculosus* L., *Polysiphonia lanosa* (L.) Tandy, and *Chondrus crispus* Stackh. J. Phycol. 17, 57–64.

Siefermann-Harms, D., 1985: Carotenoids in photosynthesis. I. Location in photosynthetic membranes and light-harvesting function. Biochim. Biophys. Acta 811, 325–355.

Silva, P. C., 1955: The dichotomous species of *Codium* in Britain. J. Mar. Biol. Ass. U. K. 34, 565–577.

Silva, P. C., 1957: *Codium* in Scandinavian waters. Svensk Bot. Tidskr. 51, 117–134.

Silva, P. C., 1962: Comparison of algal floristic patterns in the Pacific with those in the Atlantic and Indian oceans, with special reference to *Codium*. Proc. Ninth Pac. Sci. Cong. 1957, 4, 201–216.

Silva, P., 1966: Status of our knowledge of the Galápagos benthic marine algal flora prior to the Galápagos International Scientific Project. In: The Galápagos.

Proceedings of the Symposia of the Galápagos International Scientific Project. Ed. by R. I. Bowman. University of California Press, Berkeley; pp. 149–156.

Silva, P., 1979: The benthic algal flora of central San Francisco Bay. In: San Francisco Bay, the urbanized estuary. Ed. by T. J. Conomos. Pacific Division, AAAS, San Francisco, pp. 287–345.

Silva, P. C., and Johansen, H. W., 1986. A reappraisal of the order Corallinales (Rhodophyceae). Br. Phycol. J. 21, 245–254.

Silva, P. C., Menez, E. H., and Moe, R. L., 1987: Catalog of the benthic algae of the Philippines. Smiths. Contrib. Mar. Sci. 27, 1–179.

Simmons, H. G., 1906: Remarks about the relations of the floras of the Northern Atlantic, the Polar Sea, and the Northern Pacific. Bot. Zbl. 19, 149–194.

Simon, M., 1984: Présence d'une Pheophycée arctique *Omphalophyllum ulvaceum* Rosenvinge, dans la région de Roscoff (Finistère, France). Cah. Biol. Mar. 25, 355–359.

Simons, R. H., 1976: Seaweeds of southern Africa: guide-lines for their study and identification. Fish. Bull. S. Afr. 7, 1–113.

Simonsen, R., 1968: Zur Küstenvegetation der Sarso-Inseln im Roten Meer. Meteor Forschungsergeb. D, 3, 57–66.

Sims, R. W., Price, J. H., and Whalley, P. E. S. (Eds.), 1983: Evolution, time, and space: the emergence of the biosphere. Academic Press, London; 492 pp.

Sisson, R. F., 1976: Adrift on a raft of *Sargassum*. A picture story. Natl. Geogr. 14, 188–199.

Sivalingam, P. M., 1977: Marine algal distribution in Penang Island. Bull. Jap. Soc. Phycol. 25, 202–209.

Sivalingam, P. M., 1982: Biofuel-gas production from marine algae. Jap. J. Phycol. (Sorui) 30, 207–212.

Skinner, S., and Womersley, H. B. S., 1983: New records (possibly introductions) of *Striaria*, *Stictyosiphon* and *Arthrocladia* (Phaeophyta) for Southern Australia. Trans. R. Soc. S. Aust. 107, 59–68.

Skottsberg, C., 1907: Zur Kenntnis der subantarktischen und antarktischen Meeresalgen. I. Phaeophyceen. Wissenschaftliche Ergebnisse der Schwedischen Südpolarexpedition 1901–1903, 4: 1–172.

Skottsberg, C., 1921–1923: Botanische Ergebnisse der Schwedischen Expedition nach Patagonien und dem Feuerlande 1907–1909. Marine algae. 1. Phaeophyceae. 2. Rhodophyceae. K. svenska Vetensk. Akad. Handl. 61 (11), 1–56 (1921); 63 (8), 1–70 (1923).

Skottsberg, C., 1941*a*: Communities of marine algae in sub-Antarctic and Antarctic waters. Kungl. Svensk. Vetenskap. Handl. (3), 19 (4), 1–92.

Skottsberg, C., 1941*b*: Marine algal communities of the Juan Fernandez Islands, with remarks on the composition of the flora. In: The natural history of Juan Fernandez and Easter Island. Vol. 2. Ed. by C. Skottsberg. Almquist & Wiksell, Uppsala, pp. 671–696.

Skottsberg, C., 1964: Antarctic phycology. In: Biologie antarctique. Ed. by R. Carrick, M. Holdgate, and J. Prévost. Hermann, Paris, pp. 147–154.

Smith, A. G., Hurley, A. M., and Briden, J. C., 1981: Phanerozoic paleocontinental world maps. University Press, Cambridge; pag. var.

Smith, C. M., and Berry, J. A., 1986: Recovery of photosynthesis after exposure of intertidal algae to osmotic and temperature stresses: comparatative studies of species with different distribution limits. Oecologia (Berlin) 70, 6–12.

Smith, G. M., 1947: On the reproduction of some Pacific coast species of *Ulva*. Am. J. Bot. 31, 80–87.

Smith, H., and Morgan, D. C., 1981: The spectral characteristics of the visible radiation incident upon the surface of the earth. In: Plants and the daylight spectrum. Ed. by H. Smith. Academic Press, London, pp. 4–20.

Smith, R. C., and Baker, K. S., 1979: Penetration of UV-B and biologically effective dose-rates in natural waters. Photochem. Photobiol. 29, 311–323.

Smith, R. C., Baker, K. S., Holm-Hansen, O., and Olson, R., 1980: Photoinhibition of photosynthesis in natural waters. Photochem. Photobiol. 31, 585–592.

Smith, R. C., Baker, K. S., 1982: Assessment of the influence of enhanced UV-B on marine primary productivity. In: The role of solar ultraviolet radiation in marine ecosystems. Ed. by J. Calkins. Plenum Press, New York; pp. 509–538.

Smithsonian Meteorological Tables, 1951: Duration of daylight. Smithson. Misc. Collect. 114, 507–512.

Soegiarto, A., 1979: Indonesian seaweed resources: their utilization and management. Proc. Int. Seaweed Symp. 9, 463–471.

Song, C. B., 1986: An ecological study of the intertidal macroalgae in Kwangyang Bay, southern coast of Korea. Korean J. Phycol. 1, 203–223.

Sonnenfeld, P., 1985: Models of Upper Miocene evaporite genesis in the Mediterranean region. In: Geological evolution of the Mediterranean basin. Ed. by D. J. Stanley and F.-C. Wezel. Springer-Verlag, New York; pp. 323–346.

Sourie, R., 1954: Contribution à l'étude écologique des côtes rocheuses du Sénégal. Mém. Inst. Fr. Afr. Noire 38, 1–342.

Sousa, W. P., Schroeter, S. C., and Gaines, S. D., 1981: Latitudinal variation in intertidal algal community structure: the influence of grazing and vegetative propagation. Oecologia (Berlin) 48, 297–307.

South, G. R., 1974: *Herpodiscus* gen. nov. and *Herpodiscus durvilleae* (Lindauer) comb. nov., a parasite of *Durvillea antarctica* (Chamisso) Hariot endemic to New Zealand. J. R. Soc. N. Z. 4, 455–461.

South, G. R., 1976: A check-list of marine algae of eastern Canada: First revision. J. Mar. Biol. Ass. U.K. 56, 817–843.

South, G. R., 1979: Biogeography of benthic marine algae of the southern oceans. In: Proceedings of the International Symposium on Biogeography and Evolution in the Southern Hemisphere, Auckland, New Zealand, July 1978, pp. 85–108.

South, G. R., (Ed.), 1983*a*: Biogeography and ecology of the island of Newfoundland. W. Junk, The Hague; 723 pp.

South, G. R., 1983*b*: Benthic marine algae. In: Biogeography and ecology of the island of Newfoundland. Ed. by G. R. South. W. Junk, The Hague, pp. 385–420.

South, G. R., 1984: A checklist of marine algae of eastern Canada, second revision. Can. J. Bot. 62, 680–704.

South, G. R., and Cardinal, A., 1973: Contributions to the flora of marine algae of eastern Canada. 1. Introduction, historical review, and key to the genera. Nat. Can. (Que.) 100, 605–630.

South, G. R., 1987: Biogeography of the benthic marine algae of the North Atlantic Ocean—an overview. Helgoländer Meeresunters. 41, 273–282.

South, G. R, and Adams, N. M., 1976: Marine algae of the Kaikoura coast. Nat. Mus. N. Z. Misc. Ser. 1, 1–71.

South, G. R., and Hooper, R. G., 1980: A catalogue and atlas of the benthic marine algae of the island of Newfoundland. Memorial Univ. Nfld. Occas. Pap. Biol. 3, 1–136.

South, G. R., and Hooper, R. G., Irvine, L. M., 1972: The life history of *Turnerella pennyi* (Harv.) Schmitz. Br. Phycol. J. 7, 221–233.

South, G. R., and Tittley, I., 1986: A checklist and distributional index of the benthic marine algae of the North Atlantic Ocean. Huntsman Marine Laboratory and British Museum (Natural History), St. Andrews and London; 76 pp.

South, G. R., Tittley, I., Farnham, W. F., and Keats, D. W., 1988: A survey of the benthic marine algae of southwestern New Brunswick, Canada. Rhodora 90, 419–451.

South, G. R., and Whittick, A., 1987. Introduction to phycology. Blackwell, Oxford; 341 pp.

Southward, A. J., 1964: Limpet grazing and the control of vegetation on rocky shores. In: Grazing in terrestrial and marine environments. Ed. by D. J. Crisp. Blackwell, Oxford; pp. 265–273.

Southward, A. J., and Butler, E. I., 1972: A note on further changes of sea temperature in the Plymouth area. J. Mar. Biol. Ass. U. K. 52, 931–937.

Span, A., 1980: Composition et zonation de la flore et végétation benthique de l'île de Hvar (Adriatique moyenne). Acta Adriat. 21, 169–194.

Spicer, R. A., Wolfe, J. A., and Nichols, D. J., 1987: Alaskan Cretaceous-Tertiary floras and arctic origins. Paleobiology 13, 73–83.

van der Spoel, S., and Heyman, R. P., 1983: A comparative atlas of zooplankton. Biological patterns in the oceans. Springer-Verlag, Berlin; 186 pp.

van der Spoel, S., and Pierrot-Bults, A. C., 1979: Zoogeography and diversity of plankton. Edward Arnold, London; 401 pp.

Srivastava, S., 1985: Evolution of the Eurasian basin and its implication to the motion of Greenland along Nares Strait. Tectonophysic 114, 29–53.

Srivastava, S., and Tapscott, C. R., 1986: Plate kinematics of the North Atlantic. In: The geology of North America. Vol. M. The Western North Atlantic Region. (A Decade of North American Geology). Ed. by P. R. Vogt and B. E. Tucholke. Geol. Soc. Am., Boulder, Colo., pp. 379–404.

Stapleton, J. C., 1988: Occurrence of *Undaria pinnatifida* (Harvey) Suringar in New Zealand. Jpn J. Phycol. 36, 178–179.

Stam, W. T., Bot, P. V. M., Boele-Bos, S. A., van Rooij, J. M., and van den Hoek, C., 1988: Single-copy DNA-DNA hybridizations among five species of *Laminaria* (Phaeophyceae): phylogenetic and biogeographic implications. Helgoländer Meeresunters. 42, 251–267.

Stanley, D. J., and Wezel, F.-C. (Eds.), 1985: Geological evolution of the Mediterranean basin. Springer-Verlag, New York; 598 pp.

Stanley, S. M., 1986: Earth and life through time. W. H. Freeman, New York; 690 pp.

Steele, D. H., 1983: Marine ecology and zoogeography. In: Biogeography and ecology

of the island of Newfoundland. Ed. by G. R. South. W. Junk, The Hague, pp. 421–465.
Steemann-Nielsen, E., 1975: Marine photosynthesis. Elsevier, Amsterdam; 114 pp.
Stegenga, H., 1986: The Ceramiaceae (excl. *Ceramium*) (Rhodophyta) of the South West Cape Province, South Africa. Cramer, Berlin; 149 pp.
Stegenga, H., and Mol, I., 1983: Flora van de Nederlandse Zeewieren. Koninklijke Nederlandse Natuurhistorische Vereniging, Amsterdam; 263 pp.
Steinberg, P. D., 1984: Algal chemical defense against herbivores: allocation of phenolic compounds in the kelp *Alaria marginata*. Science 223, 405–407.
Steinberg, P. D., 1985: Feeding preferences of *Tegula funebralis* and chemical defenses of marine brown algae. Ecol. Monogr. 55, 333–349.
Steneck, R. S., 1982: A limpet-coralline alga association: adaptations and defenses between a selective herbivore and its prey. Ecology 63, 507–522.
Steneck, R. S., 1983: Escalating herbivory and resulting adaptive trends in calcareous algal crusts. Paleobiology 9, 44–61.
Steneck, R. S., 1986: The ecology of coralline algal crusts: convergent patterns and adaptive strategies. Annu. Rev. Ecol. Syst. 17, 273–303.
Steneck, R. S., 1990: Herbivory and the evolution of nongeniculate coralline algae (Rhodophyta, Corallinales) in the North Atlantic and North Pacific. In: Evolutionary biogeography of the marine algae of the North Atlantic. Ed. by G. R. South and D. Garbary. Springer-Verlag, Berlin. (in press).
Steneck, R. S., and Paine, R. T., 1986: Ecological and taxonomic studies of shallow-water encrusting Corallinaceae (Rhodophyta) of the boreal northeastern Pacific. Phycologia 25, 221–240.
Stephenson, T. A., 1948: The constitution of the intertidal fauna and flora of South Africa. Part III. Ann. Natal Mus. 11, 207–324.
Stephenson T. A., and Stephenson A., 1949: The universal features of zonation between tide-marks on rocky coasts. J. Ecol. 37, 289–305.
Stephenson, T. A., and Stephenson, A., 1972: Life between tidemarks on rocky shores. Freeman, San Francisco; 425 pp.
Sterrer, W. (Ed.), 1986: Marine fauna and flora of Bermuda. A systematic guide to the identification of marine organisms. Wiley, New York; 746 pp.
Stevens, G. R., 1980: Southwest Pacific faunal palaeobiogeography in Mesozoic and Cenozoic times: a review. Palaeogeogr. Palaeoclimatol. Palaeoecol. 31, 153–196
Stewart, W. D. P. (Ed.), 1974: Algal physiology and biochemistry. Blackwell, Oxford; 989 pp.
Stock, J., Molnar, P., 1987: Revised history of the early Tertiary plate motion in the southwest Pacific. Nature 325, 495–499.
Stoddart, D. R., 1969: Ecology and morphology of recent coral reefs. Biol. Rev. 44, 433–498.
Stoner, A. W., and Greening, H. S., 1984: Geographic variation in the macrofaunal associates of pelagic *Sargassum* and some biogeographic implications. Mar Ecol. Prog. Ser. 20, 185–192.
Strauch, F., 1972: Phylogenese, Adaptation und Migration einiger nordischer Molluskengenera (*Neptunea*, *Panomya*, *Cyrtodaria* und *Mya*). Abh. Senckenb. Naturforsch. Ges. 531, 1–211.

Strömgren, T., 1975: Linear measurements of growth of shells using laser diffraction. Limnol. Oceanogr. 20, 845–848.

Strömgren, T., and Nielsen, M. V., 1986: Effect of diurnal variations in natural irradiance on the apical length growth and light saturation of growth in five species of benthic macroalgae. Mar. Biol. 90, 467–474.

Stuart, V., Lucas, M. I., and Newell, R. C., 1981: Heterotrophic utilisation of particulate matter from the kelp *Laminaria pallida*. Mar. Ecol. Prog. Ser. 4, 337–348.

Stuessy, T. F., Foland, K. A., Sutter, J. F., Sanders, R. W., and Silva, M., 1984: Botanical and geological significance of potassium-argon dates from the Juan Fernández Islands. Science 225, 49–51.

Suda, M., 1987: Marine algae from the coast of Iwaki City, Fukushima Prefecture. Jap. J. Phycol. (Sorui) 35, 22–22.

Summerhayes, C. P., and Shackleton, N. J. (Eds), 1986: North Atlantic Palaeoceanography. Blackwell, Oxford; 473 pp.

Sundene, O., 1953: The algal vegetation of Oslofjord. Skr. Norske Vidensk. Akad. I. Mat.-Naturv. Kl. 2, 1–244.

Sundene, O., 1962: The implications of transplant and culture experiments on the growth and distribution of *Alaria esculenta*. Nytt Mag. Bot. 9, 155–174.

Sundene, O., 1964: The ecology of *Laminaria digitata* in Norway in view of transplant experiments. Nytt Mag. Bot. 11, 83–107.

Suto, S., 1950: Studies on shedding, swimming, and fixing of spores of seaweeds. Bull. Jap. Soc. Sci. Fish. 16, 1–9.

Svane, I., and Gröndahl, F., 1987: Epibioses of the Gullmar Fjord: an underwater stereophotographical analysis in comparison with the investigations of Torsten Gislén 1926–29. Abstr. 22nd EMBS Barcelona, August 1987.

Svedelius, N., 1906: Über die Algenvegetation eines ceylonischen Korallenriffes mit besonderer Rücksicht auf ihre Periodizität. In: Botaniska studier tillägnade F. R. Kjellman. Almquist & Wiksells, Uppsala, pp. 184–221.

Svedelius, N., 1924: On the discontinuous geographical distribution of some tropical and subtropical marine algae. Ark. Bot. 19 (3), 1–70.

Svendsen, P., 1959: The algal vegetation of Spitsbergen. Skr. norsk Polarinst. no. 116, pp. 1–49.

Svendsen, P., and Kain, J. M., 1971: The taxonomic status, distribution, and morphology of *Laminaria cucullata* sensu Jorde and Klavestad. Sarsia 46, 1–22.

Sweeney, B. M., 1969: Rhythmic phenomena in plants. Academic Press, London; 147 pp.

Sweeney, B. M., 1977: Chronobiology (circadian rhythms). In: The science of photobiology. Ed. by K. C. Smith. Plenum Press, New York, pp. 209–226.

Sweeney, J. F., 1985: Comments about the age of the Canada Basin. Tectonophysics 114, 1–10.

Swinbanks, D. D., 1982: Intertidal exposure zones: a way to subdivide the shore. J. Exp. Mar. Biol. Ecol. 62, 69–86.

Tait, R. V., 1972: Elements of marine ecology. An introductory course. Butterworths, London; 314 pp.

Talbot, F., and Steene, R., 1987: Reader's Digest book of the Great Barrier Reef. Reader's Digest, Sydney; 384 pp.

Talwani, M., and Eldholm, O., 1977: Evolution of the Norwegian-Greenland Sea. Bull. Geol. Soc. Am. 88, 969–999.

Tardent, P., 1979: Meeresbiologie. Eine Einführung. Thieme, Stuttgart; 381 pp.

Targett, N. M., and Mitsui, A., 1979: Toxicity of subtropical marine algae using fish mortality and red blood cell hemolysis for bioassays. J. Phycol. 15, 181–185.

Taylor, P. R., Littler, M. M., and Littler, D. S., 1986: Escapes from herbivory in relation to the structure of mangrove island macroalgal communities. Oecologia 69, 481–490.

Taylor, T. N., 1981: Paleobotany. An introduction to fossil plant biology. McGraw-Hill, New York; 569 pp.

Taylor, W. R. A., 1930: A synopsis of the marine algae of Brazil. Rev. Algol. 5, 1–35.

Taylor, W. R. A., 1935: Marine algae from the Yucatan peninsula. Carnegie Inst. Washington Publ. no. 461, pp. 115–124.

Taylor, W. R. A., 1938: Algae collected by the "Hassler," "Albatros," and Schmitt expeditions. II. Marine algae from Uruguay, Argentinia, the Falkland Islands, and the Strait of Magellan. Pap. Mich. Acad. Sci. Arts Lett. 24, 127–164.

Taylor, W. R., 1945: Pacific marine algae of the Allan Hancock Expeditions to the Galapagos Islands. Allan Hancock Pacif. Exped. 12, 1–528.

Taylor, W. R., 1950: Plants of Bikini and other northern Marshall Islands. University of Michigan Press, Ann Arbor, Mich.; 277 pp.

Taylor, W. R., 1954: The cryptogamic flora of the Arctic. II. Algae: non-planktonic. Bot. Rev. 20, 363–399.

Taylor, W. R., 1957: Marine algae of the northeastern coast of North America. 2. Edition. University of Michigan Press, Ann Arbor; 509 pp.

Taylor, W. R., 1960: Marine algae of the eastern tropical and subtropical coasts of the Americas. University of Michigan Press, Ann Arbor; 870 pp.

Taylor, W. R., 1966: Records of Asian and Western Pacific algae, particularly from Indonesia and the Philippines. Pacif. Sci. 20, 342–359.

Taylor, W. R., 1969: Notes on the distribution of West Indian marine algae particularly in the Lesser Antilles. Contrib. Univ. Mich. Herb. 9, 125–203.

Taylor, W. R., 1970: Marine algae of Dominica. Smithson. Contrib. Bot. 3, 1–16.

Taylor, W. R., and Bernatowicz, A. J., 1969: Distribution of marine algae about Bermuda. Bermuda Biol. Stn. Res. Spec. Publ. 1, 1–42.

Teal, J., and Teal, M., 1975: The Sargasso Sea. Little, Brown, Boston; 206 pp.

Termier, H., and Termier, G., 1960: Atlas de paléogéographie. Masson, Paris; 99 pp.

Terry, L. A., and Moss, B. L., 1980: The effect of photoperiod on receptacle initiation in *Ascophyllum nodosum*. Br. Phycol. J. 15, 291–301.

Terry, L. A., and Picken, G. B., 1986: Algal fouling in the sea. In: Algal biofouling. Ed. by L. V. Evans and K. D. Hoagland. Elsevier, Amsterdam, pp. 179–199.

Thiede, J., 1979: History of the North Atlantic Ocean: evolution of an asymmetrical zonal paleo-environment in a latitudinal ocean basin. Deep Drilling Results in the Atlantic Ocean. Continental Margins and Paleoenvironment. Maurice-Ewing-series (AGU) 3, 275–296.

Thiede, J., 1980: Palaeo-oceanography, margin stratigraphy, and palaeophysiography of the Tertiary North Atlantic and Norwegian-Greenland Seas. Phil. Trans. R. Soc. Lond. A, 294, 177–185.

Thiede, J., and Eldholm, O., 1983: Speculations about the paleodepth of the Greenland-Scotland Ridge during late Mesozoic and Cenozoic times. In: Structure and development of the Greenland-Scotland Ridge. Ed. by M. H. Bott, S. Saxov, M. Talwani and J. Thiede. Plenum Press, New York; pp. 445–456.

Thierstein, H. R., and Berger, W. H., 1978: Injection events in ocean history. Nature 276, 461–466.

Thom, R. M., 1980: A gradient in benthic intertidal algal assemblages along the southern California coast. J. Phycol. 16, 102–108.

Thomas, B. A., and Spicer, R. A., 1987: The evolution and palaeobiology of land plants. Crom Helm, London; 309 pp.

Thomas, M. L. H., 1985: Littoral community structure and zonation on the rocky shores of Bermuda. Bull. Mar. Sci. 37, 857–870.

Thorpe, J. P., 1982: The molecular clock hypothesis: biochemical evolution, genetic differentiation and systematics. Annu. Rev. Ecol. Syst. 13, 139–168.

Thunell, R. C., 1979: Pliocene-Pleistocene paleotemperature and paleosalinity history of the Mediterranean Sea: results from DSPD sites 125 and 132. Mar. Micropaleontol. 4, 173–187.

Thunell, R., and Belyea, P., 1982: Neogene planktonic foraminiferal biogeography of the Atlantic Ocean. Micropaleontology 28, 381–398.

Tilman, D., 1982: Resource competition and community structure. University Press, Princeton, N.J.; 296 pp.

Tischler, W., 1984: Einführung in die Ökologie. Gustav Fischer, Stuttgart; 420 pp.

Titlyanov, E. A., Kolmakov, P. V., Lee, B. D., and Horvath, I., 1978: Functional states of the photosynthetic apparatus of the marine green alga *Ulva fenestrata* during the day. Acta Bot. Acad. Sci. (Hung.) 24, 167–177.

Titlyanov, E. A., Kolmakov, P. V., Leletkin, V. A., and Voskoboinikov, G. M., 1987: A new type of light adaptation in aquatic plants (in Russian). Biol. Morya (Vladivost.) 1987 (2), 48–57.

Titlyanov, E. A., Le, N. H., Nechai, E. G., Butorin, P. V., Hoang, T. L. 1983: Daily changes in physiological parameters of photosynthesis and dark respiration in the seaweeds of the genus *Sargassum* from southern Vietnam (in Russian). Biol. Morya (Vladivost.) 1983 (3), 39–48.

Titlyanov, E. A., and Lee, B. D., 1978: Adaptation of benthic plants to light. IV. Variations of photosynthetic pigment content and ratio in green algae under changing light conditions (in Russian). Biol. Morya (Vladivost.) 1978 (4), 36–41.

Titlyanov, E. A., Zvalinsky, V. I., Yadikin, A. A., Li, B. D., and Chernova, S. I., 1977: Adaptation of benthic plants to light. II. Resistance of the green alga *Ulva fenestrata* to high light intensity (in Russian). Biol. Morya (Vladivost.) 1977 (2), 3–10.

Tittley, I., 1986: Seaweed communities on the artificial coastline of South Eastern England. 2. Open sea-shores. Trans. Kent Fld. Club 10, 55–67.

Tittley, I., Farnham, W. F., and Gray, P. W. G., 1982: Seaweeds of the Faroes. 2: Sheltered fjords and sounds. Bull. Br. Mus. Nat. Hist. (Bot.) 10, 133–151.

Tittley, I., Farnham, W. F., South, G. R., and Keats, D., 1987: Seaweed communities of the Passamaquoddy region, southern Bay of Fundy, Canada. Br. Phycol. J. 22, 313.

Tittley, I., Farnham, W. F., Fletcher, R. L., Morrell, S., and Bishop, G., 1985: Sublittoral seaweed assemblages of some northern Atlantic islands. Progr. Underwater Sci. 10, 39–52.

Tittley, I., Farnham, W. F., Hooper, R. G., and South, G. R., 1989: Sublittoral seaweed assemblages (2): a transatlantic comparison. Progr. Underwater Sci. 13, 185–205.

Tokida, J., 1954: The marine algae of southern Saghalien. Mem. Fac. Fish. Hokkaido Univ. 2, 1–264.

Tokida, J., Nakamura, Y., and Druehl, L. D., 1980: Typification of species of *Laminaria* (Phaeophyta, Laminariales) described by Miyabe, and taxonomic notes on the genus on Japan. Phycologia 19, 317–328.

Tomlinson, P. B., 1974: Vegetative morphology and meristem dependence—the foundation of productivity in seagrasses. Aquaculture 4, 107–130.

Tomlinson, P. B. (Ed.), 1986: The botany of mangroves. University Press, Cambridge; 413 pp.

Toomey, D. F., 1981: Organic-buildup constructional capability in lower Ordovician and Late Paleozoic mounds. In: Communities of the past. Ed. by J. Gray, A. J. Boucot and W. B. N. Berry. Hutchinson Ross, Stroudsburg, Penn. pp. 35–68.

Toomey, D. F., and Nitecki, M. H., 1985: Paleoalgology. Contemporary research and applications. Springer-Verlag, Berlin; 376 pp.

Topinka, J., Tucker, L., and Korjeff, W., 1981: The distribution of fucoid macroalgal biomass along central coastal Maine. Botanica Mar. 24, 311–319.

Tranter, D. J., 1982: Interlinking of physical and biological processes in the Antarctic Ocean. Oceanogr. Mar. Biol. Annu. Rev. 20, 11–35.

Tremblay, C., and Chapman, A. R. O., 1980: The local occurrence of *Agarum cribrosum* in relation to the presence or absence of its competitors and predators. Proc. N. S. Inst. Sci. 30, 165–170.

Trono, G. C., 1968–1969: The marine benthic algae of the Caroline Islands. Micronesica 4, 137–206 (1968); 5, 25–121 (1969).

Trono, G. C., and Ganzon-Fortes, E. T., 1980: An illustrated seaweed flora of Calatagan, Batangas, Philippines. Bautista, Manila; 114 pp.

Tsekos, I., and Haritonidis, S., 1977: A survey of the marine algae of the Ionian islands, Greece. Botanica Mar. 20, 47–65.

Tseng, C. K., 1981a: Commercial cultivation. In: The biology of seaweeds. Ed. by C. S. Lobban and M. J. Wynne. Blackwell, Oxford, pp. 680–725.

Tseng, C. K., 1981b: Marine phycoculture in China. Proc. Int. Seaweed Symp. 10, 123–152.

Tseng, C. K. (Ed.), 1983: Common seaweeds of China. Science Press, Beijing; 316 pp.

Tseng, C. K., 1983–1984: Oceanographic factors and seaweed distribution. Oceanus 26, 48–56.

Tseng, C. K., and Chang, C. F., 1963: A preliminary analytical study of the Chinese marine algal flora. Oceanol. Limnol. Sin. 5, 245–253.

Tseng, C. K., and Chang, C. F., 1964: An analytical study of the marine algal flora of the western Yellow Sea coast. II. Phytogeographical nature of the flora. Oceanol. Limnol. Sin. 6, 152–168.

Tseng, C. K., Chang, C. F., and Xia, B., 1980: Studies on some marine red algae from Hong Kong. In: Proceedings of the First International Marine Biological Workshop: the marine flora and fauna of Hong Kong and southern China. Ed. by B. S. Morton and C. K. Tseng. University Press, Hong Kong, pp. 57–84.

Tsuda, R. T., and Tobias, W. J., 1977: Marine benthic algae from the northern Mariana Islands. Bull. Jap. Soc. Phycol. 25, 46–51, 155–158.

Tsuda, R. T., and Wray, F. O., 1977: Bibliography of marine benthic algae in Micronesia. Micronesica 13, 85–120.

Turner, N. C., and Kramer, P. J. (Eds.), 1980: Adaptation of plants to water and high temperature stress. Wiley, New York; 457 pp.

Turner, R. D., and Evans, L. V., 1986: Structural and physiological aspects of translocation in the red alga *Delesseria*. Br. Phycol. J. 21, 338.

Tyler, J. E. (Ed.), 1977: Light in the sea. Dowden, Hutchinson & Ross, Stroudsburg, Pennsylvania; 384 pp.

Ugadim, Y., 1973–1976: Algas marinhas bentonicas do litoral Sul do Estado de São Paulo e do litoral do Estado do Parana. I. Divisao Chlorophyta. Bol. Bot. Univ. São Paulo 1, 11–77 (1973*a*); II. Divisao Phaeophyta. Port. Acta Biol. Ser. B. 12, 69–131 (1973*b*); III. Rhodophyta (3). *Ceramium* (Ceramiaceae-Ceramiales). Bol. Zool. Biol. Mar. (Nova Ser.) 30, 691–712 (1973*c*); III. Divisao Rhodophyta (1): Goniotrichales, Bangiales, Nemalionales e Gelidiales. (2): Cryptonemiales, Gigartinales e Rhodymeniales. Ceramiales (Rhodophyta) do litoral Sul do Estado de São Paulo e do litoral do Estado do Parana (Brasil). Bol. Bot. Univ. São Paulo 2, 93–137 (1974); 3, 115–163 (1975); 4, 133–172 (1976).

Ugadim, Y., and Pereira, S. M. B., 1978: Deep-water marine algae from Brazil, collected by the Recife Commission. I. Chlorophyta. Cienc. Cult. (São Paulo) 30, 839–842.

Umaheswara Rao, M., and Kaliaperumal, N., 1987: Diurnal periodicity of spore-shedding in some red algae of Visakhapatnam coast. J. Exp. Mar. Biol. Ecol. 106, 193–199.

Umaheswara Rao, M., and Sreeramulu, T., 1964: An ecological study of some intertidal algae at Visakhapatnam coast. J. Ecol. 52, 595–616.

Umaheswara Rao, M., and Sreeramulu, T., 1970: An annotated list of marine algae of Visakhapatnam (India). J. Linn. Soc. Lond. Bot. 63, 23–45.

Underwood, A. J., 1979: The ecology of intertidal gastropods. Adv. Mar. Biol. 16, 111–210.

Underwood, A. J., and Denley, E. J., 1984: Paradigms, explanations, and generalizations on models for the strudure of intertidal communities on rocky shores. In: Ecological communities: conceptual issues and the evidence. Ed. by D. R. Strong, D. Simberloff, L. G. Abele, and A. B. Thistle. University Press, Princeton, pp. 151–180.

Untawale, A. G., and Jagtap, T. G., 1989: Observations on marine macrophytes of the Republic of Sechelles. Botanica Mar. 32, 115–119.

Vadas, R. L., 1977: Preferential feeding: an optimization strategy in sea urchins. Ecol. Monogr. 47, 337–371.

Vadas, R. L., 1979: Seaweeds: an overview; ecological and economical importance. Experientia 35, 429–432.

Vadas, R. L., 1985: Herbivory. In: Handbook of phycological methods. Ecological field methods: macroalgae. Ed. by M. M. Littler and D. S. Littler. University Press, Cambridge, pp. 531–572.

Vadas, R. L., and Steneck, R. S., 1988: Zonation of deep water benthic algae in the Gulf of Maine. J. Phycol. 24, 338–346.

Vail, P. R., and Hardenbol, J., 1979: Sea-level changes during the Tertiary. Oceanus 22, 71–79.

Valentine, D. H. (Ed.)., 1972: Taxonomy, phytogeography, and evolution. Academic Press, London; 431 pp.

Valentine, J. W., 1973: Evolutionary paleoecology of the marine biosphere. Prentice Hall, Englewood Cliffs, N.J.; 511 pp.

Valentine, J. W., and Jablonski, D. J., 1983: Speciation in the shallow sea: general patterns and biogeographic controls. In: Evolution, time, and space: the emergence of the biosphere. Ed. by R. W. Sims, J. H. Price, and P. E. S. Whalley. Academic Press, London, pp. 201–226.

Valet, G., 1969: Contribution à l'étude des Dasycladales. Cytologie et reproduction. Révison systematique. Nova Hedwigia 17, 551–644.

Vanney, J.-R., and Gennesseaux, M., 1985: Mediterranean seafloor features: overview and assessment. In: Geological evolution of the Mediterranean basin. Ed. by D. J. Stanley and F.-C. Wezel. Springer-Verlag, New York, pp. 3–32.

Vareschi, E., and Fricke, H., 1986: Light responses of a scleractinian coral (*Pleroga sinuosa*). Mar. Biol. 90, 395–402.

Vasiliu, F., and Bodeanu, N., 1972: Repartition et quantité d'algues rouges du genre *Phyllophora* sur la plate forme continentale Roumaine de la Mer Noire. Cercetari Mar. I. R. C. M. 3, 47–52.

Velasquez, G. T., Trono, G. C., and Doty, M. S., 1975: Algal species reported from the Philippines. Philipp. J. Sci. 101, 115–169.

Velimirov, B., 1982: Sugar and lipid components in sea foam near kelp beds. P. S. Z. N. I. Mar. Ecol. 3, 97–107.

Velimirov, B., Field, J. G., Griffiths, C. L., and Zoutendijk, P., 1977: The ecology of kelp bed communities in the Benguela upwelling system. Analysis of biomass and spatial distribution. Helgoländer Meeresunters. 30, 495–518.

Verlaque, M., 1981: Contribution à la flore des algues marines de Méditerranée: especes nouvelles pour la Mediterranée occidentale. Botanica Mar. 24, 559–568.

Verlaque, M., 1984: Biologie des juvéniles de l'oursin herbivore *Paracentrotus lividus* (Lamarck): sélectivité du broutage et impact de l'espèce sur les communautès algales de substrat rocheux en Corse (Méditerranée, France). Botanica Mar. 27, 401–424.

Vermeij, G. J., 1978: Biogeography and adaptation. Patterns of marine life. Harvard University Press, Cambridge, Mass.; 332 pp.

Vermeij, G. J., 1983: Intimate associations and coevolution in the sea. In: Coevolution. Ed. by D. J. Futuyama and M. Slatkin. Sinauer, Sunderland, Mass., pp. 311–327.

Viera-Rodríguez, M. A., Audiffred, P. A. J., Gil-Rodríguez, M. C., Prud'homme van Reine, W. F., and Afonso-Carrillo, J., 1987: Adiciones al catálogo de algas marinas bentonicas para el Archipiélago Canário. III. Vieraea 17, 227–235.

Villouta, E., and Santelices, B., 1984: Estructura de la comunidad submareal de *Lessonia* (Phaeophyta, Laminariales) en Chile norte y central. Rev. Chil. Hist. Nat. 57, 111–122.

Villouta, E., and Santelices, B., 1986: *Lessonia trabeculata* sp. nov. (Laminariales, Phaeophyta), a new kelp from Chile. Phycologia 25, 81–86.

Vince-Prue, D., 1975: Photoperiodism in plants. McGraw-Hill, London; 444 pp.

Vince-Prue, D., 1983: Photomorphogenesis and flowering. In: Encyclopedia of plant physiology, n. s. Vol. 16B. Ed. by W. Shropshire and H. Mohr. Springer-Verlag, Berlin, pp. 457–490.

Vince-Prue, D., 1985: Photoperiodism and hormones. In: Encyclopedia of plant physiology, n. s. Vol. 11B. Ed. by R. P. Pharis, and D. M. Reid. Springer-Verlag, Berlin; pp. 308–364.

Vinogradova, K. L., 1973: Species composition in the littoral zone of the northwestern part of the Bering Sea (in Russian). Nov. Sist. Nizhnikh Rast. 10, 32–44.

Virville, A. D. de, 1966: La flore marine de la presqu'île de Quiberon. Rev. Gén. Bot. 69, 89–152.

Vogel, J. C. (Ed.), 1984: Late Cenozoic palaeoclimates of the Southern Hemisphere. Balkema, Rotterdam; 536 pp.

Vogt, P. R., 1986: Seafloor topography, sediments, and paleoenvironments. In: The Nordic seas. Ed. by B. G. Hurdle. Springer Verlag, New York, pp. 237–410.

Voipio, A. (Ed.), 1981: The Baltic Sea. Elsevier, Amsterdam; 418 pp.

Voskresenskaya, N. P., 1952. Blue light and carbon metabolism. Annu. Rev. Plant Physiol. 23, 219–234.

Voskresenskaya, N. P., 1972. Effect of spectral composition of light on the relations of substances formed in photosynthesis (in Russian). Dokl. Akad. Nauk SSSR 86, 429–434.

Vozzhinskaja, V. B., 1964: Marine macrophytes of the coasts of Sachalin (in Russian). Tr. Inst. Okean. Akad. Nauk SSSR 69, 330–440.

Vozzhinskaja, V. B., 1965: Distribution of algae along the shores of western Kamchatka. Akad. Nauk Oceanol. 5, 123–127.

Vroman, M., 1968: Studies on the flora of Curaçao and other Caribbean islands. Vol. 2. The marine algal vegetation of St. Martin, St. Eustatius, and Saba (Netherland Antilles). Natuurwetenschappelijke Studiekring voor Suriname en de Nederlands Antillen, Utrecht, no. 52, pp. 1–120.

Vroman, M., Stegenga, H., 1988: An annotated checklist of the marine algae of the Caribbean islands Aruba and Bonaire. Nova Hedwigia 46, 433–480.

Waaland, J. R., 1977: Common seaweeds of the Pacific coast. Douglas, Vancouver; 120 pp.

Waaland, J. R., 1981: Commercial utilization. In: The biology of seaweeds. Ed. by C. S. Lobban and M. J. Wynne. Blackwell, Oxford, pp. 726–741.

Waaland, J. R., Dickson, L. G., and Carrier, J. E., 1987: Conchocelis growth and photoperiodic control of conchospore release in *Porphyra torta* (Rhodophyta). J. Phycol. 23, 399–406.

Waern, M., 1952: Rocky-shore algae in the Öregrund archipelago. Acta Phytogeogr. Suec. 30, 1–298.

Wallentinus, I., 1978: Productivity studies on Baltic macroalgae. Bot. Mar. 21, 365–380.

Wallentinus, I., 1979: Environmental influences on benthic macro-vegetation in the Trosa-Askö area, northern Baltic proper. II. The ecology of macroalgae and submersal phanerogams. Contrib. Askö Lab. 25, 1–210.

Wallentinus, I., 1981: Phytobenthos. In: Assessment of the effects of pollution on the natural resources of the Baltic Sea, 1980. Ed. by T. Melvasolo, J. Pawlak, K. Grasshoff, L. Thorell, and A. Tsiban. Baltic Sea Environ. Proc. 5B, 322–342.

Wallentinus, I., 1984: Comparisons of nutrient uptake rates for Baltic macroalgae with different thallus morphologies. Mar. Biol. 80, 215–225.

Wanders, J. B. W., 1976–1977: The role of benthic algae in the shallow reef of Curaçao (Netherland Antilles). I. Primary productivity in the coral reef. II. Primary productivity of the *Sargassum* beds on the north-east coast submarine plateau. III. The significance of grazing. Aquat. Bot. 2, 235– 270 (1976); 2, 327–335 (1976); 3, 357–390 (1977).

Wang, P. (Ed.), 1985: Marine micropaleontology of China. China Ocean Press, Beijing and Springer-Verlag, Berlin; 332 pp.

Webb, K. L., DuPaul, W. D., Wiebe, W., Sottile, W., and Johannes, R. E., 1975: Enewetak (Eniwetok) Atoll: aspects of the nitrogen cycle on a coral reef. Limnol. Oceanogr. 20, 198–210.

Weber-Peukert, G., and Schnetter, R., 1982: Floristische und ökologische Untersuchungen über benthische Meeresalgengesellschaften der Küste Asturiens (Spanien). 1. Ein Beitrag zur Pflanzengeographie der iberischen Nordküste. Nova Hedwigia 36, 65–80.

Weber-van Bosse, A., 1913–1928: Liste des algues du Siboga. Siboga-Exped. Monogr. 59A (1913), 59B (1921), 59C (1923), 59D (1928). Brill, Leiden; pag. var.

Wefer, G., 1980: Carbonate production by algae *Halimeda*, *Penicillus* and *Padina*. Nature 285, 323–324.

Weissel, J. K., Hayes, D. E., Herron, E. M., 1977: Plate tectonic synthesis, the displacements between Australia, New Zealand, and Antarctica since Late Cretaceous. Mar. Geol. 25, 231–277.

Weisscher, F. C. M., 1983: Marine algae from Selvagem Pequena (Salvage Islands). Bol. Mus. Munic. Funchal 35, 41–80.

Wells, J. W., 1957: Coral reefs. Mem. Geol. Soc. Am. 67, 609–631.

Wennicke, H., and Schmid, R., 1987: Control of the photosynthetic apparatus of *Acetabularia mediterranea* by blue light. Plant Physiol. 84, 1252–1256.

West, J. A., 1968: Morphology and reproduction of the red alga *Acrochaetium pectinatum* in culture. J. Phycol. 4, 89–99.

Westermeier, R., Rivera, P. J., Chacana, M., and Gomez, I., 1987: Biological bases for management of *Iridaea laminarioides* Bory in southern Chile. Hydrobiologia 151/152, 313–328.

Westermeier, R., and Rivera, P., 1989: Rocky shore community along an estuary of the south of Chile: zonation pattern and temporal variability. P. S. Z. N. I. Mar. Ecol. (in press).

Westlake, D. F., 1963: Comparisons of plant productivity. Biol. Rev. 38, 385–425.

Wheeler, A., 1980: Fish-algal relations in temperate waters. In: The shore environment, Vol. 2. Ecosystems. Ed. by J. H. Price, D. E. G. Irvine, and W. F. Farnham. Academic Press, London, pp. 677–698.

Wheeler, W. N., 1982: Nitrogen nutrition of *Macrocystis*. In: Synthetic and degradative processes in marine macrophytes. Ed. by L. M. Srivastava. Walter de Gruyter, Berlin, pp. 121–135.

Wheeler, W. N., and Druehl, L. D., 1986: Seasonal growth and productivity of *Macrocystis integrifolia* in British Columbia, Canada. Mar. Biol. 90, 181–186.

Wheeler, W. N., and Neushul, M., 1981: The aquatic environment. In: Encyclopedia of Plant Physiology, n. s. Vol. 12A. Ed. by O. L. Lange, P. S. Nobel, C. B. Osmond, and H. Ziegler. Springer-Verlag, Berlin, pp. 229–247.

Wheeler, W. N., and Weidner, M., 1983: Effects of external inorganic nitrogen on metabolism, growth, and activities of key carbon and nitrogen assimilatory enzymes of *Laminaria saccharina* (Phaeophyceae) in culture. J. Phycol. 19, 91–96.

Whittaker, R. H., and Likens, G. E., 1975: The biosphere and man. In: Primary productivity of the biosphere. Ed. by H. Lieth and R. H. Whittaker. Springer-Verlag, New York, pp. 305–328.

Whittle, K. J., 1977: Marine organisms and their contribution to organic matter in the ocean. Mar. Chem. 5, 381–411.

Whyte, J. N. C., and Englar, J. R., 1980: Seasonal variation in the inorganic constituents of the marine alga *Nereocystis luetkeana*. Part. I. Metallic elements. Part II. Non-metallic elements. Botanica Mar. 23, 13–24.

Wicksten, M. K., 1983: Camouflage in marine invertebrates. Oceanogr. Mar. Biol. Annu. Rev. 21, 177–193.

Widdowson, T. B., 1971: A taxonomic revision of the genus *Alaria* Greville. Syesis 4, 11–49.

Widdowson, T. B., 1973–1974: The marine algae of British Columbia and northern Washington: revised list and keys. Part I. Phaeophyceae (brown algae). Part II. Rhodophyceae (red algae). Syesis 6, 81–96 (1973); 7, 143–186 (1974).

Wiebe, W. J., Johannes, R. R., and Webb, K. L., 1975: Nitrogen fixation in a coral reef community. Science 188, 257–259.

Wiedmann, J., 1979: Aspects of the Cretaceous in Europe. International Union of Geological Sciences, Ser. A, no. 6. Schweizerbart'sche Verlagsbuchhandlung, Stuttgart; 545 pp.

Wiencke, C., 1988: Notes on the development of some benthic marine macroalgae of King George Island, Antarctica. Ser. Cient. INACH 37, 23–47.

Wiencke, C., and Clayton, M. N., 1990: Sexual reproduction, life history, and early development of the Antarctic brown alga *Himantothallus grandifolius* (Desmarestiales, Phaeophyceae). (in press).

Wiencke, C., and Davenport, J., 1987: Respiration and photosynthesis in the intertidal alga *Cladophora rupestris* (L.) Kütz. under fluctuating salinity regimes. J. Exp. Mar. Biol. Ecol. 114, 183–197.

Wiencke, C., and tom Dieck, I., 1989: Temperature requirements for growth and temperature tolerance of macroalgae endemic to the Antarctic region. Mar. Ecol. Prog. Ser. 54, 189–197.

Wiencke, C., and tom Dieck, I., 1989: Temperature requirements for growth and survival of macroalgae from Antarctica to southern Chile. Mar. Ecol. Prog. Ser. 59, 157–170.

Wiencke, C., and Läuchli, A., 1981: Inorganic ions and floridosid as osmotic solutes in *Porphyra umbilicalis*. Z. Pflanzenphysiol. 103, 247–258.

Wiencke, C., Stelzer, R., and Läuchli, A., 1983: Ion compartmentation in *Porphyra umbilicalis* determined by electron-probe X-ray microanalysis. Planta 159, 336–341.

Wilce, R. T., 1959: The marine algae of the Labrador Peninsula and northwest Newfoundland (ecology and distribution). Nat. Mus. Can. Bull. 158, 1–103.

Wilce, R. T., 1959–1965: Studies in the genus *Laminaria*. I. *Laminaria cuneifolia* J. G. Agardh: a review. II. *Laminaria groenlandica* Rosenvinge. III. A revision of the North Atlantic species of the Simplices section of *Laminaria*. Bot. Not. 112, 158–174 (1959); 113, 203–209 (1960); Bot. Gothoburg. 3, 247–256 (1965).

Wilce, R. T., 1963: Studies on benthic marine algae in north-west Greenland. Proc. Int. Seaweed Symp. 4, 280–287.

Wilce, R. T., 1967: Heterotrophy in arctic sublittoral seaweeds: an hypothesis. Botanica Mar. 10, 185–197.

Wilce, R. T., Maggs, C. A., 1989: Reinstatement of the genus *Haemescharia hennedyi* (Rhodophyta, Haemeschariaceae fam. nov.) for *H. polygyna* and *H. hennedyi* comb. nov. (= *Petrocelis hennedyi*). Can. J. Bot. 67, 1465–1479.

Wilcox, H. A., 1977: The ocean food and energy farm project. Proc. Second Ship Technol. Res. Symp. 1977, 269–278.

Wildgoose, P. B., and Blunden, G., 1981: Effects of maerl in agriculture. Proc. Int. Seaweed Symp. 8, 754–759.

Wiley, E. O., 1981: Phylogenetics. The theory and practice of phylogenetic systematics. John Wiley, New York; 439 pp.

Wilkinson, M., 1980: Estuarine benthic algae and their environment: a review. In: The shore environment. Vol. 2. Ecosystems. Ed. by J. H. Price, D. E. G. Irvine, and W. F. Farnham. Academic Press, London, pp. 425–486.

Wilkinson, M., 1982: Marine algae from Glamorgan. Br. Phycol. J. 17, 101–106.

Wilkinson, M., and Burrows, E. M., 1972: The distribution of marine shell-boring green algae. J. Mar. Biol. Ass. U. K. 52, 59–65.

Williamson, M., 1988: Relationship of species number to area, distance, and other variables. In: Analytical biogeography. An integrated approach to the study of animal and plant distributions. Ed. by A. A. Myers and P. S. Giller. Chapman & Hall, London, pp. 91–115.

Wilson, B. R., and L. M. Marsh, 1980: Coral reef communities at the Houtman

Abrolhos, western Australia, in a zone of biogeographic overlap. Proc. Int. Symp. Ma. Biogeogr. S. Hemisphere N. Z. DSIR Inform. Ser. 137, 259–278.

Wilson, J. S., Bird, C. J., McLachlan, J., and Taylor, A. R. A., 1979: An annotated checklist and distribution of benthic marine algae in the Bay of Fundy. Memorial Univ. Nfld. Occas. Pap. Biol. 2, 1–65.

Wilson, K. C., Haaker, P. L., and Hanan, D. A., 1977: Kelp restoration in southern California. In: The marine plant biomass of the Pacific northwest coast. Ed. by R. W. Krauss. Oregon State University Press, Corvallis; pp. 183–202.

Winge, Ö., 1923: The Sargasso Sea, its boundaries and vegetation. Rep. Dan. Oceanogr. Exped. 1908–10 Medit. and adj. Seas, 3 (2), 1–34.

Woelkerling, W. J., 1972: Some algal invaders of the northwestern fringes of the Sargasso Sea. Rhodora 74, 295–298.

Woelkerling, W. J., 1983: A taxonomic reassessment of *Lithophyllum* (Corallinaceae, Rhodophyta) based on studies of R. A. Philippi's original collections. Br. Phycol. J. 18, 299–328.

Woelkerling, W. J., 1986: The genus *Litholepis* (Corallinaceae, Rhodophyta): taxonomic status and disposition. Phycologia 25, 253–261.

Woelkerling, W. J., 1988: The coralline red algae: an analysis of the genera and subfamilies of nongeniculate Corallinaceae. Oxford University Press, London and Oxford; 268 pp.

Woelkerling, W. J., and Irvine, L. M., 1986: The typification and status of *Phymatolithon* (Corallinaceae, Rhodophyta). Br. Phycol. J. 21, 55–80.

Woelkerling, W. J., Chamberlain, Y. M., Silva, P. C., 1985: A taxonomic and nomenclatural reassessment of *Tenarea*, *Titanoderma*, and *Dermatolithon* (Corallinaceae, Rhodophyta) based on studies of type and other critical specimens. Phycologia 24, 317–337.

Wolfe, J. A., 1980: Tertiary climates and floristic relationships at high latitudes in the Northern Hemisphere. Palaeogeogr. Palaeoclimatol. Palaeoecol. 30, 313–323.

Wolfe, J. A., and Upchurch, G. R., 1986: Vegetation, climatic, and floral changes at the Cretaceous-Tertiary boundary. Nature, Lond. 324, 148–152.

Womersley, H. B. S., 1954: The species of *Macrocystis* with special reference to those on southern Australian coasts. Univ. Calif. Publs. Bot. 27, 109–132.

Womersley, H. B. S., 1958: Marine algae from Arnhem Land, northern Australia. In: Records of the American-Australian Scientific Expedition to Arnhem Land. Vol. 3. Ed. by R. L. Specht and C. P. Mountford. Melbourne University Press, Melbourne, pp. 139–161.

Womersley, H. B. S., 1959: The marine algae of Australia. Bot. Rev. 25, 545–614.

Womersley, H. B. S., 1967: A critical survey of the marine algae of southern Australia. II. Phaeophyta. Aust. J. Bot. 15, 189–270.

Womersley, H. B. S., 1971: *Palmoclathrus*, a new deep water genus of Chlorophyta. Phycologia 10, 229–233.

Womersley, H. B. S., 1981a: Marine ecology and zonation of temperate coasts. Biogeography of Australasian marine macroalgae. In: Marine botany: an Australasian perspective. Ed. by M. N. Clayton and R. J. King. Longman Cheshire, Melbourne, pp. 211–240, 292–307.

Womersley, H. B. S., 1981b: Aspects of the distribution and biology of Australian marine macro-algae. In: The biology of Australian plants. Ed. by J. S. Pate and A. J. McComb. University of Western Australia Press, Nedlands, pp. 294–305.

Womersley, H. B. S., 1984–1987: The marine benthic flora of southern Australia. Pt. 1 (1984). Pt. 2 (1987). Govt. Printer, Adelaide; 329 pp. (1984); 484 pp. (1987).

Womersley, H. B. S., and Bailey, A., 1969: The marine algae of the Solomon Islands and their place in biotic reefs. Phil. Trans. R. Soc. Lond. B 255, 433–442.

Womersley, H. B. S., and Bailey, A., 1970: Marine algae of the Solomon Islands. Phil. Trans. R. Soc. Lond. B 259, 257–352.

Womersley, H. B. S., Edmonds, S. J., 1952: Marine coastal zonation in southern Australia in relation to a general scheme of classification. J. Ecol. 40, 84–90.

Womersley, H. B. S., and Edmonds, S. J., 1958: General account of the intertidal ecology of South Australian coasts. Aust. J. Mar. Freshwater Res. 9, 217–260.

Wood, W. F., 1987: Effect of solar ultra-violet radiation on the kelp *Ecklonia radiata*. Mar. Biol. 96, 143–150.

Woodruff, F., Savin, S. M., and Douglas, R. G., 1981: Miocene stable isotope record: a detailed deep Pacific Ocean study and its paleoclimatic implications. Science 212, 665–668.

Worrest, R. C., 1983: Impact of solar ultraviolet-B radiation (290–320 nm) upon marine microalgae. Physiol. Plant. 58, 428–434.

Wray, J. L., 1971: Algae in reefs through time. Proc. N. Am. Soc. Convention, Chicago 1969 (I), 1358–1373.

Wray, J. L., 1977: Calcareous algae. Elsevier, Amsterdam; 185 pp.

Wright, P. J., Clayton, M. N., Chudek, J. A., Foster, R., and Reed, R. H., 1987: The carbohydrate altritol in *Bifurcariopsis capensis*, *Hormosira banksii*, *Notheia anomala* and *Xiphophora chondrophylla* (Fucales, Phaeophyta) from the southern hemisphere. Phycologia 26, 429–434.

Wyn Jones, R. G., and Gorham, J., 1983: Osmoregulation. In: Encyclopedia of plant physiology, n. s. Vol. 12C. Ed. by O. L. Lange, P. S. Nobel, C. B. Osmond, and H. Ziegler. Springer-Verlag, Berlin, pp. 35–58.

Wynne, M. J., 1970: Marine algae of Amchitka Island (Aleutian Islands). I. Delesseriaceae. Syesis 3, 95–144.

Wynne, M. J., 1982: Observations on four species of Delesseriaceae (Rhodophyta) from the South Sandwich Islands, the Antarctic. Contrib. Univ. Mich. Herb. 15, 325–337.

Wynne, M. J., 1983: The current status of genera in the Delesseriaceae. Botanica Mar. 26, 437–450.

Wynne, M. J., 1985–1987: Records and notes on Alaskan marine algae. I (1985). II (1987). British Columbia Provincial Museum, no. 2, 1–3 (1985); Contrib. Univ. Mich. Herb. 16, 223–232 (1987).

Wynne, M. J., 1986a: A checklist of benthic marine algae of the tropical and subtropical western Atlantic. Can. J. Bot. 64, 2239–2281.

Wynne, M. J., 1986b: Report on a collection of benthic marine algae from the Namibian coast (southwestern Africa). Nova Hedwigia 43, 311–355.

Wynne, M. J., Lindstrom, S. C., and Calvin, N. I., 1982: Occurrence of *Omphalo-*

phyllum ulvaceum Rosenv. (Phaeophyta, Pogotrichaceae) in the North Pacific. Syesis 15, 65–66.

Yabu, H., and Sanbonsuga, Y., 1985: Mitosis in the gametophytes and young sporophytes of *Macrocystis angustifolia* (Bory). Jap. J. Phycol. (Sorui) 33, 1–4.
Yabu, H., and Sanbonsuga, Y., 1987: Chromosome count in *Macrocystis integrifolia* Bory. Bull. Fac. Fish. Hokkaido Univ. 38, 339–342.
Yamada, Y., and Tanaka, T., 1944: Marine algae in the vicinity of Akkeshi Marine Biological Station. Sci. Pap. Inst. Algol. Res. Fac. Sci. Hokkaido Imp. Univ. 3, 47–77.
Yarish, C., 1976: Polymorphism of selected marine Chaetophoraceae (Chlorophyta). Br. Phycol. J. 11, 29–38.
Yarish, C., Breeman, A. M., and C. van den Hoek, 1984. Temperature, light, and photoperiod responses of some Northeast American and West European endemic rhodophytes in relation to their geographic distribution. Helgoländer Meeresunters. 38, 273–304.
Yarish, C., Breeman, A. M., and C. van den Hoek, 1986. Survival strategies and temperature responses belonging to different biogeographic distribution groups. Botanica Mar. 24, 215–230.
Yarish, C., and Edwards, P., 1982: A field and cultural investigation of the horizontal and seasonal distribution of estuarine red algae of New Jersey. Phycologia 21, 112–124.
Yarish, C., Edwards, P., and Casey, S., 1979: A culture study of salinity responses in ecotypes of two estuarine red algae. J. Phycol. 15, 341–346.
Yarish, C., and Egan, B., 1987: Biological studies of *Laminaria longicruris* and its aquaculture potential in Long Island Sound. In: Pollution and water resources. Ed. by G. J. Halasi-Kun. University Seminar series; Columbia; pp. 194–218.
Yarish, C., Kirkman, H., and Lüning, K., 1987. Lethal exposure times and preconditioning to upper temperature limits of some temperate North Atlantic red algae. Helgoländer Meeresunters. 41, 323–327.
Yarish, C., Penniman, C. A., and van Patten, P. (Eds), 1990: Economically important marine plants of the Atlantic: their biology and cultivation. Connecticut Sea Grant Publications, Groton, CT 158 pp.
Yarish, C., Brunkhuis, B., Egan, B., and Garcia-Ezquivel, Z., 1990. Morphological and physiological bases for *Laminaria* selection protocols in Long Island Sound aquaculture. In: Economically important marine plants of the Atlantic: their biology and cultivation. Ed. by C. Yarish, C. A. Penniman, and P. van Patten, Connecticut Sea Grant Publications, Groton, CT. pp. 53–94.
Yokohama, Y., 1973: A comparative study on photosynthesis-temperature relationships and their seasonal changes in marine benthic algae. Int. Rev. Ges. Hydrobiol. Hydrogr. 58, 463–472.
Yokohama, Y., 1981: Distribution of the green-light-absorbing pigments siphonaxanthin and siphonein in marine green algae. Botanica Mar. 24, 637–640.
Yokohama, Y., 1983: A xanthophyll characteristic of deep-water green algae lacking siphonaxanthin. Botanica Mar. 26, 45–48.

Yokohama, Y., and Kageyama, A., 1977: A carotenoid characteristic of chlorophycean seaweeds living in deep coastal waters. Botanica Mar. 20, 433–436.

Yokohama, Y., and Misonou, T., 1980: Chlorophyll a:b ratios in marine benthic algae. Jap. J. Phycol. 28, 219–223.

Yoneshigue, Y., 1985: Taxonomie et écologie des algues marines dans la region de Cabo Frio (Rio de Janeiro, Brésil). Thèse de Docteur d'État-Sciences Université d'Aix-Marseille II; 466 pp.

Yoshida, T., 1963: Studies on the distribution and drift of the floating seaweeds. Bull. Tohoku Reg. Fish. Res. Lab. 23, 141–185.

Yoshida, T., Nakajima, Y., and Nakata, Y., 1985: Preliminary check-list of marine benthic algae of Japan. I. Chlorophyceae and Phaeophyceae. II. Rhodophyceae. Jap. J. Phycol. 33, 57–74, 249–275.

Young, P. C., and Kirkman, H., 1975: The seagrass communities of Moreton Bay, Queensland. Aquat. Bot. 1, 191–202.

Yu, M.-H., Glazer, A. N., Spencer, K. G., and West, J. A., 1981: Phycoerythrins of the red alga *Callithamnion*. Variation in phycoerythrobilin and phycourobilin content. Plant Physiol. 68, 482–488.

Zaneveld, J. S., 1966a: The occurrence of benthic marine algae under shore fast-ice in the western Ross Sea, Antarctica. Proc. Int. Seaweed Symp. 5, 217–234.

Zaneveld, J. S., 1966b: Vertical zonation of antarctic and subantarctic marine algae. Antarct. J. U. S. 1, 211–213.

Zaneveld, J. S., 1968: Sub-ice observations of Ross Sea benthic marine algae. Antarct. J. U. S. 3, 127–128.

Zaneveld, J. S., 1969: Iconography of Antarctic and sub-Antarctic benthic marine algae. Part I. Chlorophycophyta and Chrysophycophyta. Cramer, Lehre, W. Germany; pag. var.

Zaneveld, J. S., 1972: The benthic marine algae of Delaware, U.S.A. Ches. Sci. 12, 120–138.

Zaneveld, J. S., and Willis, W. M., 1974–1976: The marine algae of the American coast between Cape May, New Jersey, and Cape Hatteras, North Carolina. II. The Chlorophycophyta. III. The Phaeophycophyta. Botanica Mar. 17, 65–81 (1974); 19, 33–46 (1976).

Zechman, F. W., and Mathieson, A. C., 1985: The distribution of seaweed propagules in estuarine, coastal, and offshore waters of New Hampshire, U.S.A. Botanica Mar. 18, 283–294.

Zenkevitch, L., 1963: Biology of the seas of the U.S.S.R. George Allen & Unwin, London; 955 pp.

Ziegler, P. A., 1977: Geology and hydrocarbon provinces of the North Sea. GeoJournal 1 (1), 7–32.

Ziegler Page, J., and Kingsbury, J. M., 1968: Culture studies on the marine green alga *Halicystis parvula-Derbesia marina*. II. Synchrony and periodicity in gamete formation and release. Am. J. Bot. 55, 1–11.

Zimmermann, U., 1978: Physics of turgor- and osmoregulation. Annu. Rev. Plant Physiol. 29, 121–148.

Zinova, A. D., 1950: About some pecularities of the flora of the White sea (in Russian). Tr. V. G. O. 2, 231–252.

Zinova, A. D., 1953: Brown algae of the northern seas of the U.S.S.R. (in Russian). Izdatel'stvo Akademii Nauk SSSR, Moscow, Leningrad; 224 pp.

Zinova, A. D. 1955: Red algae of the northern seas of the U.S.S.R. (in Russian). Izdatel'stvo Akademii Nauk SSSR, Moscow, Leningrad; 249 pp.

Zinova, A. D., 1957: Marine algae of the eastern part of the Soviet sector of the Arctic (in Russian). Trudy Inst. Okeanol. 23, 146–167.

Zinova, A. D., 1958: Composition and character of the algal flora near the shores of the Antarctic continent and in the vicinity of the Kerguelen and Macquarie Islands (in Russian). Inform. Byull. Soviet Antarktic Ekspeds. 3, 47–49.

Zinova, A. D., 1959: Marine algae of southern Sachalin and the southern Kurile islands (in Russian). Issledov. Morei SSSR. 6, 141–161.

Zinova, A. D., 1967: Green, brown and red algae of the southern seas of the U.S.S.R. (in Russian). Izdatel'stvo Akademii Nauka SSSR, Moscow, Leningrad; 398 pp.

Zinova, E. S., 1925: The algae of the Karasee (in Russian). Trudy Leningr. Obshch. Estest. 55, 53–116.

Zinova, E. S., 1929: The algae of Novaya Zemlja (in Russian). Issledov. Morei SSSR.; Edinaia Gidro-Meterol. Sluzhba, Gosudarstven. Gidrobiol. Inst. 10, 41–128.

Zinova, E. S., 1933: The algae of Murmansk in the vicinity of the island of Kildine and their commercial use (in Russian). 18, 49–74.

Zinova, E. S., 1934: New investigations on the algae of the White Sea along the coast of Lietnaia und their commercial utilization (in Russian). idem 20, 65–85.

Zinova, E. S., 1940: The algae of the Commander islands (in Russian). Trans. Pacif. Comm. Acad. Sci. USSR. 5, 1–64.

Zinova, E. S., 1954a: Algae of the Sea of Ochotsk (in Russian). Trudy Bot. Inst. Akad. Nauk SSSR., ser. 2, 9, 259–310.

Zinova, E. S., 1954b: Marine algae of the Tatar Strait (in Russian). 311–364. Trudy Bot. Inst. Akad. Nauk SSSR, ser. 2, 9, 311–364.

Zinova, E. S., 1954c: Marine algae of southeastern Kamchatka (in Russian). Trudy Bot. Inst. Akad. Nauk SSSR, ser. 2, 9, 365–400.

Zinova, E. S., 1956: Marine algae of the Arctic Sea (in Russian). Trudy Bot. Inst. W. L. Komarowa Acad. Sci. URSS C. II, Plantae Cryptogamae, 11, 39–51.

Zinsmeister, W. J., 1982: Late Cretaceous-early Tertiary molluscan biogeography of the southern circum-Pacific. J. Paleontol. 56, 84–102.

Zinsmeister, W. J., and Feldman, R. M., 1984: Cenozoic high latitude heterochroneity of southern hemisphere marine faunas. Science 224, 281–283.

TAXONOMIC OVERVIEW OF GENERA

The following list contains the names of those Recent genera of the Chlorophyta, Phaeophyta and Rhodophyta that have been mentioned in this book. The systematic arrangement follows basically Bold and Wynne (1985), with modifications according to, for example, Clayton (1984), South and Tittley (1986), Woelkerling (1988), and Womersley (1984–1987).

Division Chlorophyta (Green algae)
  Order Chlorococcales: *Chlorochytrium, Palmoclathrus, Palmophyllum*
  Order Prasiolales: *Prasiola, Rosenvingiella*
  Order Ulotrichales: *Ulothrix*
  Order Ctenocladales: *Entocladia, Epicladia*
  Order Trentepohliales: *Tellamia*
  Order Ulvales
    Family Monostromaceae: *Monostroma*
    Family Ulvaceae: *Blidingia, Enteromorpha, Ulva*
  Order Cladophorales
    Family Cladophoraceae: *Chaetomorpha, Cladophora, Rhizoclonium*
    Family Anadyomenaceae: *Anadyomene, Microdictyon*
  Order Acrosiphoniales: *Acrosiphonia, Spongomorpha, Urospora*
  Order Caulerpales
    Family Codiaceae: *Codium, Johnson-sea-linkia*
    Family Udoteaceae: *Chlorodesmis, Halimeda, Penicillus, Tydemania, Udotea*
    Family Caulerpaceae: *Caulerpa*
    Family Derbesiaceae: *Bryobesia, Bryopsis, Derbesia, Lambia*
    Family Phyllosiphoniaceae: *Ostreobium*
  Order Siphonocladales: *Boodlea, Dictyosphaeria, Siphonocladus, Struvea, Valonia*
  Order Dasycladales: *Acetabularia, Batophora, Bornetella, Cymopolia, Dasycladus, Neomeris*

Division Phaeophyta (Brown algae). Class Phaeophyceae
  1. Order Ectocarpales
    Family Ectocarpaceae: *Bachelotia, Ectocarpus, Giffordia, Mikrosyphar Pilayella, Spongonema*
    Family Ralfsiaceae: *Analipus, Basispora, Pseudolithoderma, Ralfsia*
  2. Order Chordariales
    Family Myrionemataceae: *Leptonematella*
    Family Elachistaceae: *Elachista, Herpodiscus*
    Family Spermatochnaceae: *Stilophora*
    Family Chordariaceae: *Chordaria, Eudesme, Monosiphon, Papenfussiella, Sphaerotrichia*
    Family Chordariopsidaceae: *Chordariopsis*
    Family Splachnidiaceae: *Splachnidium*
  3. Order Cutleriales: *Cutleria, Zanardinia*
  4. Order Sphacelariales: *Cladostephus, Halopteris, Sphacelaria, Stypocaulon*
  5. Order Syringodermatales: *Syringoderma*
  6. Order Tilopteridales: *Phaeosiphoniella*

7. Order Dictyotales: *Dictyopteris, Dictyota, Lobophora, Pachydictyon, Padina, Stypopodium, Taonia, Zonaria*
8. Order Dictyosiphonales
    Family Myriotrichiaceae: *Myriotrichia*
    Family Striariaceae: *Stictyosiphon*
    Family Punctariaceae: *Adenocystis, Litosiphon, Punctaria*
    Family Pogotrichaceae: *Omphalophyllum*
    Family Dictyosiphonaceae: *Dictyosiphon*
    Family Chnoosporaceae: *Chnoospora*
9. Order Scytosiphonales: *Colpomenia, Hydroclathrus, Petalonia, Scytosiphon*
10. Order Sporochnales: *Arthrocladia, Carpomitra, Sporochnus*
11. Order Desmarestiales: *Desmarestia, Himantothallus, Phaeurus*
12. Order Laminariales
    Family: Pseudochordaceae: *Pseudochorda*
    Family Chordaceae: *Chorda*
    Family Phyllariaceae: *Phyllariopsis, Saccorhiza*
    Family Laminariaceae: *Agarum, Arthrothamnus, Costaria, Cymathere, Hedophyllum, Kjellmaniella, Laminaria, Pleurophycus, Thalassiophyllum*
    Family Lessoniaceae: *Dictyoneurum, Dictyoneuropsis, Lessonia, Lessoniopsis, Macrocystis, Nereocystis, Pelagophycus, Postelsia*
    Family Alariaceae: *Alaria, Ecklonia, Egregia, Eisenia, Pterygophora, Undaria*
13. Order Notheiales: *Notheia*
14. Order Durvillaeales: *Durvillaea*
15. Order Ascoseirales: *Ascoseira*
16. Order Fucales
    Family Hormosiraceae: *Hormosira*
    Family Seirococcaceae: *Axillariella, Cystosphaera, Marginariella, Phyllospora, Scytothalia, Seirococcus*
    Family Fucaceae: *Ascophyllum Fucus, Hesperophycus, Hizikia, Pelvetia, Pelvetiopsis, Xiphophora*
    Family Himanthaliaceae: *Himanthalia*
    Family Sargassaceae: *Antophycus, Carpophyllum, Hizikia, Oerstedtia, Sargassum, Turbinaria*
    Family Cystoseiraceae: *Acystis, Acrocarpia, Bifurcaria, Bifurcariopsis, Carpoglossum, Caulocystis, Coccophora, Cystophora, Cystoseira, Halidrys, Hormophysa, Landsburgia, Myagropsis, Myriodesma, Platythalia, Scaberia, Stolonophora*

Division Rhodophyta (Red algae). Class Rhodophyceae
1. Subclass Bangiophycidae
    Order Bangiales: *Bangia, Porphyra, Porphyropsis, Smithora*
2. Subclass Florideophycidae
    1. Order Batrachospermales: *Batrachospermum*
    2. Order Palmariales: *Devaleraea, Halosaccion, Palmaria*
    3. Order Nemaliales
        Family Acrocaetiaceae: *Audouinella*
        Family Nemaliaceae: *Nemalion*
        Family Helminthocladiaceae: *Dermonema, Helminthocladia, Helminthora, Liagora*
        Family Chaetangiaceae: *Galaxaura, Scinaia*
        Family Naccariaceae: *Atractophora*
    4. Order Gelidiales: *Gelidiella, Gelidium, Pterocladia*
    5. Order Bonnemaisoniales: *Asparagopsis, Bonnemaisonia, Delisea*
    6. Order Cryptonemiales
        Family Dumontiaceae: *Constantinea, Dilsea, Dumontia, Neodilsea*
        Family Acrosymphytaceae: *Acrosymphyton*
        Family Cryptonemiaceae: *Aeodes, Grateloupia, Halymenia, Polyopes*

Family Endocladiaceae: *Endocladia, Gloiopeltis*
Family Kallymeniaceae: *Callophyllis, Kallymenia, Meredithia*
Family Choreocolaceae: *Choreocolax, Harveyella*
Family Peyssonneliaceae: *Peyssonnelia*
7. Order Corallinales
   Family Corallinaceae: *Amphiroa, Archaeolithothamnion, Arthrocardia, Calliarthron, Clathromorphu*m, *Corallina, Dermatolithon, Haliptilon, Hydrolithon, Jania, Leptophytum, Lithophyllum, Lithoporella, Lithothamnion, Melobesia, Mesophyllum, Millepora, Neogoniolithon, Phymatolithon, Porolithon, Pseudolithophyllum, Tenarea, Titanoderma*
8. Order Hildenbrandiales: *Hildenbrandia*
9. Order Gigartinales
   Family Calosiphoniaceae: *Calosiphonia, Schmitzia*
   Family Sarcodiaceae: *Nizymenia*
   Family Furcellariaceae: *Furcellaria, Halarachnion, Neurocaulon*
   Family Cruoriaceae: *Cruoria*
   Family Haemaschariaceae: *Haemascharia*
   Family Solieriaceae: *Eucheuma, Opuntiella, Sarcodiotheca, Turnerella*
   Family Polyideaceae: *Polyides*
   Family Caulacanthaceae: *Caulacanthus, Catenella*
   Family Cystocloniaceae: *Cystoclonium, Fimbrifolium, Rhodophyllis*
   Family Hypneaceae: *Hypnea*
   Family Plocamiaceae: *Plocamium*
   Family Phacelocarpaceae: *Phacelocarpus*
   Family Sphaerococcaceae: *Sphaerococcus*
   Family Rissoellaceae: *Rissoella*
   Family Gracilariaceae: *Curdiea, Gracilaria*
   Family Phyllophoraceae: *Ahnfeltia, Ceratocolax, Gymnogongrus, Phyllophora, Schottera*
   Family Petrocelidaceae: *Mastocarpus*
   Family Gigartinaceae: *Chondrus, Gigartina, Iridaea, Petrocelis*
10. Order Rhodymeniales
    Family Champiaceae: *Champia, Gastroclonium, Lomentaria*
    Family Rhodymeniaceae: *Botryocladia, Chrysymenia, Fauchea, Fryella, Leptosomia, Maripelta, Rhodymenia*
11. Order Ceramiales
    1. Family Ceramiaceae: *Ballia, Callithamnion, Centroceras, Ceramium, Georgiella, Plumaria, Pterothamnion, Ptilota, Scagelia, Spondylothamnion, Spyridia, Wrangelia*
    2. Family Delesseriaceae: *Acrosorium, Botryocarpa, Caloglossa, Cryptopleura, Delesseria, Dotyella, Drachiella, Hypoglossum, Martensia, Membranoptera, Myriogramme, Neuroglossum, Nitophyllum, Pantoneura, Phycodrys, Polyneura*
    3. Family Dasyaceae: *Dasya, Dictyurus, Heterosiphonia, Thuretia*
    4. Family Rhodomelaceae: *Amansia, Acanthophora, Bostrychia, Brongniartella, Chondria, Digenea, Janczewskia, Laurencia, Neorhodomela, Odonthalia, Polysiphonia, Rhodomela, Pterosiphonia, Vidalia*

INDEX

Italicized numbers refer to illustrations and/or legends. # refers to figure showing geographic distribution

Aasfontein, 259
Abrolhos Islands, 216, 265, 267, 272
Abudefduf saxatilis, 207
Abyssal, 4, *4*
Acanthophora, 222
  spicifera (VAHL) BOERG., 209
    photosynthesis, temperature optimum, 332
Acanthuridae, 216, 350
Acanthurus monroviae, 207
Acetabularia, 193, 201
  acetabulum (L.) SILVA (= *mediterranea*), *104*
    blue light effects on hair and cap formation, 316
    chloroplast movements, circadian rhythm, 317
    growth, light quality, 316
  crenulata LAMOUR., 195
Acinetospora, 198
Acmaea, 143, 348
  testudinalis, 352
Acrocarpia, 243
  paniculata (TURN.) ARESCH., 270
  robusta (J. AG.) WOMERSLEY, 271
Acrochaetiaceae, 355
Acropora, *218*-219, 233
  palmata, *219*, 222
Acrosiphonia, *68*, 177
  arcta (DILLW.) J. AG., 71, 127, *170*
    temperature optimum, growth, 323
    temperature tolerance, 323
  pacifica (MONT.) J. AG., 251
Acrosorium uncinatum (TURN.) KYL., 87
Acrosymphyton purpuriferum (J. AG.) SJÖST., *116*
  Hymenoclonium phase, 312
  photoperiodism, 312–313
Acystis, 243
Adak Island, *138*
Adélie coast, 249
Adelphoparasites, 355

Adenocystis utricularis (BORY) SKOTTS., 244, 248, 250–253, 272
Adriatic Sea, 98–99, *100*, 102, *105*, 105–106, 108, 115, 117
Aegean Islands, 98, 117
Aeodes orbitosa (SUHR) SCHM., *260*
Aeration, 344
Africa:
  North Africa, *see* Mediterranean
  Northwestern Africa, 85, 89–90, 252
  paleoceanography, paleobiogeography, 237, *238–239*, 239
  Southern Africa, 86, 235, 243, 258–259, *260–261*, 261–265, *263*
    Agulhas province, 259, *263*, 263–265
    commercial uses of seaweeds, 365
    paleoceanography, paleobiogeography, 258–259
    southwestern Africa province, 258–259, *260*, 261–263
  tropical:
    East Africa, 226–228
    West Africa, 203, 205–208, 257
Agar, 193, 368, *369*
  world production, 367
Agardhiella subulata (AG.) KRAFT et WYNNE, 125, 326
  temperature:
    growth, optimum, 328
    reproduction, 330
    tolerance, 326
Agaricia agaricites, *219*
  lamarckii, *220*
Agarophyte, 256, 263
Agarose, 368, *369*
Agarum, 136, *144*
  cribrosum (MERT.) BORY, 41, 43, #*46*, 82, 125, *127*, 128–129, *130–131*, 131, 136, *138*, 142, 161, 174, 180, 188
    antigrazer substances, 352

489

Agarum, cribrosum (MERT.) BORY (*Continued*)
   grazing, 351–352
   holes in blade, enhanced turbulence, 344
   fimbriatum HARV., *138*, 145, *144–145*
Age of individual algae, 76, 143, 148, 152, 167, 172, 286, 356–357
Aglaozonia phase, *see Cutleria multifida*
Agulhas Current, 226, 259
Ahnfeltia:
   commercial use, 367
   durvillaei (BORY) J. AG., 256
   fastigiata (POST. et RUPR.) MAKIENKO, 42, 48
   plicata (HUDS.) FRIES, 41–42, #48, 67, 183–184
      life form type, 314
      temperature tolerance, 323
   tobuchiensis (KANNO et MATSUI) MAK., 302
      light saturation, photosynthesis, 302
Ahnfeltiales, 42
Alabama, 123, 133
Alaria, 40, 48–49, 136, *138–139*, *144*, 186
   crassifolia KJELLM., 155, *156*, 157, *158*, *160*
   crispa KJELLM., *138*
   esculenta (L.) GREV., 40–42, #47, 49, 66, 68, *68*, 73, *74–75*, 78, 80, 85, 94, 126–129, 130–131, 143, 155, 175, 181–184, 186
      forma *dolichorachis*, 155
      forma *grandifolia*, 41–42, 49, 181, 188
      forma *pylaii*, 42, 49, 178
      photosynthesis, blue-light effect, 316
      temperature tolerance, 68, 326
   fistulosa POST. et RUPR., 136, *138–139*, *140*, 142, 151
      grazing, 350
   marginata POST. et RUPR., *138*
      phenolic compounds, 352
   nana SCHRADER, *138–139*
   praelonga KJELL., 136, *138–139*
   taeniata KJELLM., 136, *138–139*
   tenuifolia SETCH., 136, *138*
Alariaceae, 151, #*138*
Alaska, 29, 134–135, 139, 141, 166
Alaska Current, 139, 156
Alaska Peninsula, 141
Albatross Deep Sea Expedition, 278
Albedo, 28, 32
Alboran, Sea of, 90
Alcyonium digitatum, *74–75*
   glomeratum, 87

Aldabra, 225
Aleutian Current, 156
Aleutian Islands, 136, *138*, 139, 156, 160
Aleutian Ridge, 135
Alexander Archipelago, 141
Alexandria, 116
Algal depth limits, lower, *see* Deep-water algae
Algal mucilages, 368
Algal resources, 365
Algal ridge, *see* Coral reefs, algal ridge
Algeria, 90, 98, 111, 117
Alginate, 152, 368, *369*, 370
   world production, 367
Alginic acid, 359, 367–369, *369*
   seasonal course, 359
Alloparasites, 355
Amami, *15*
Amansia glomerata C. AG., 233
Amazon, 221–222
Amchitka Island, *138*, 139
Amerasian Basin, 25
America:
   Central and South America:
      commercial uses of seaweeds, 365
      paleoceanography, paleobiogeography, 237, *238–239*, 239
      temperate coasts:
         Argentina / Brazil, 256–258
         Chile / Peru, 252–253, *254–255*, 255–256
      tropical coasts:
         Atlantic, 196–197, 211, 216–217, *218–219*, 219–222, *220–222*, 223
         Pacific, 233–234, 274
   North America:
      Arctic, 164, *172*, 185–188
      North Atlantic, 123–134, 350
         Cape Hatteras / Cape Kennedy, 132–134
         history of algal investigation, 124–125
         Newfoundland / Cape Hatteras, 125–132
         paleobiogeography, paleoceanography, 125–126
      North Pacific, 102 ff, *104–107*, *109*, *110*, *112–116*, 351
         Alaska / Central California, 139–145, 348–349, *349*
         Cenozoic cooling phases, 135
         Central California / Baja California, 145–154
         history of algal investigation, 134–135
         paleoceanography, paleobiogeography, 135
         persistence of cool-water aspect, 135
Amino acids, 342, 365

INDEX   491

Ammonium, 342
Amphi-Atlantic species, 41, *56–58*, 57–58, *194–195*, 196, 205
Amphibolis, 203, *204*
 antarctica, 270
 griffithii, 270
Amphiequatorial distribution, 241, *242*, 253
Amphioceanic species, 41, *42–50*, 146, 155, 185–186
Amphi-Pacific species, 157, 159
Amphipods, 348, 350, 353, 370
Amphiroa, 193
 foliacea LAMOUR., 231
 photosynthetic rate, 364
 rigida LAMOUR., *104*
Anadyomene, 221
 antigrazer substances, 352
 menziesii HARV., *195*
 stellata (WULFEN) C. AG., *104*, 195
Analipus japonicus (HARV.) WYNNE (= *Heterochordaria abietina*), 155, *158*
Angola, 203, 258
Annuals, 314, 357–358
Antarctic, *15*, 235–237, *236*, *238–239*, 239–240, 245–250, 253, 256, 265
 circum-Antarctic distribution, 245, *249*
 Circumpolar Antarctic Current (West Wind Drift), 63, 193, 236–237, *236*, 239–241, 250–252, 253, 259, 266, 272, 274
 convergence, *15*, 235, *236*, 245, 250
 dark growth, 359
 divergence, 236–237, *236*
 eastern Antarctic, 245–246, 248–250
 East Wind Drift, 236, *236*
 endemism, 246
 glaciation, beginning of, 240
 history of algal investigation, 245–246
 ice, *see* Ice
 innovations in Tertiary, 240
 light damage, unicellular algae, 304
 paleoceanography, paleobiogeography, 34, 237–240, *238–239*
 Peninsula, zonation, 248. *See also* Antarctic, western Antarctic
 Polar Front, 235, *236*
 southern limit of seaweed distribution, 249–250
 temperature:
  growth optima of seaweeds, 246, 328
  photosynthesis, optimum, 332
  supercooled state, 325
  tolerance of seaweeds, 173, 246, 324, 326–327
 Tertiary flora, 239

 western Antarctic, 240, 245–248, *247*, *249*
 West Wind Drift, *see* Antarctic, Circumpolar Antarctic Current (West Wind Drift)
Antennarius marmoratus, 223
Anthozoans, 87
Antibiotics, 354
Antigrazer substances, *see* Grazing, antigrazer substances
Antilles, *see* Greater Antilles; Lesser Antilles
Antipatharian corals, *219*, 220, *222*
Antipode Islands, 250–251
Antithamnion, 270
 boreale, see Scagelia pylaisei (= *Scagelia corallina*)
 plumula, see Pterothamnion plumula
Antophycus longifolius (TURN.) KÜTZ (= *Sargassum linifolium*), 243, 264
Aphanius, 105
Aphotic zone, 4
Aquaculture, 95–97, *366–367*
Arabia, 228, 252
Arabic Sea, 228, 243, 264
Aragonite, 200
Aral Sea, 98, 121–123
 paleoceanography, *100–101*, 122–123
Arbacia lixula, 112
Archaeolithothamnion, 233
Archaeolithophyllum, 200
Arctic:
 age of individuals, 167
 Alaskan Arctic, 166, 185–186
 -boreal fauna, 125
 brackish water, 173
 Canadian Arctic, 164, 166, 169, *170*, 171, *172*, 185–187
 -cold temperate species, 40–41, *42–50*, 43–50, 125, 139, 142, 166, 180, 185–186, 188
 circumpolar distribution, 40–41, *42–50*, 43–50, 173, 175–177
 currents, 166
 daylength, *168*
 dark growth, 359
 dark tolerance, 171
 -endemic species, 173, *174–177*, 180, 184, 188, 246
 European Arctic, 180–184
 expeditions, 164–165
 freshwater hypothesis, 31
 glaciation, start of, 164
 growth:
  in darkness, 171
  rates, 167
 history of algal research, 164–165
 ice, *see* Ice

Arctic (*Continued*)
  light, 167, *168*, 169, *170*, 171, 181
  limits of, *12*, *14–15*, *64–65*, 165–166
  northernmost seaweed species, 174–175, 177, 181
  -origin hypothesis, 29–30
  paleoceanography, paleobiogeography, 22–23, 25, *25–27*, 28–35, *36–37*, 38–39, 164
  permanent ice, 164, *165*, 181, 187, *190*
  primary productivity, 167, 361–362
  R/P-Quotient, 147
  Russian Arctic, 164, 173, *182*, 182–185
  silt, *170*, 172
  size of seaweeds, 167
  temperature:
    adaptation to, 167
    in Arctic waters, 167
    growth optima of seaweeds, 328
    photosynthesis, optimum, 332
    reproduction, 330
    tolerance of seaweeds, 324, 326–327
  Tertiary flora, 29–30, 239
  transport of stones and gravel by floating algae, 346
  water motion, 172
Argentina, 252, 256
Armeria maritima, 69
Arthrocardia, 199
Arthrocladia villosa (HUDS.) DUBY, *115*, 121, 129, 274
Arthropoda, 240
Arthrothamnus bifidus (GMEL.) RUPR., 156
Ascension, 203, 208
Ascidians, 353, *263*
Ascomycetes, 356
Ascophyllum nodosum (L.) LE JOL., 41, #*52*, 53–54, *66*, 68, 70, *72–73*, 86–87, 129, *130–131*, 179–180, 183, 187, 224, 243
  age, maximal, of individuals, 356
  antigrazer substances, 352
  biomass, 361
  casting of outer layers, 353, 370
  commercial use, 367
  desiccation, 338, *339*
  endophytic fungus, *Mycosphaerella*, 356
  epiphytic alga, *Polysiphonia lanosa*, 353
  growth rate, diurnal changes, 304
  nitrate storage, 342
  pheromone, Finavarrene, 319
  photoperiodism, 312
  pigment content, acclimation of, 307
  primary productivity, 362
  respiration, seasonal rates, 332, *333*
  temperature:
    growth, optimum, 323
    tolerance, 323
Ascoseirales, *243*, 244, 246
Ascoseira mirabilis SKOTTSB., 243–245, 246, *247*, 248, 250
Ash, mineral, 365
Asia:
  Arctic, 164, 184–185
  Far East, Kamchatka/South China, South Japan, 154–157, *158*, 159–161, *160*, *162–163*, 163
  paleobiogeography, paleoceanography, 22–25, *26*, 34–35, *36–37*, 38–39, 154, 225–226, 228
  South Asia, Southeast Asia, tropical, 225, 230
Asparagopsis armata HARV., *96*
  Falkenbergia phase, *96*
  introduction to Europe, 95
  photoperiodism, 312
Asperococcus turneri (SM.) HOOK. (= *bullosus*) LAMOUR., 274
Associations, 106–108
Asterias rubens, 348
Asteroid-impact hypothesis, 29–30
Astringency, 352
Astroides calycularis, 210
Atlantic, *see* Africa, Southern Africa, America, Europe/Northwest Africa, paleoceanography; Tropics
Atlantic Spitsbergen Current, 181
Atolls, 234
Atractophora hypnoides P. CROUAN et H. CROUAN:
  compensation light level, growth, 300
  light damage, growth, 304
  light saturation, growth, 302
  photoperiodism, 312
Attenuation, *see* Light, attenuation
Attu Island, *138*
Auckland Islands, 250–251, 253
Audouinella (incl. *Rhodochorton*), 68
  asparagopsis (CHEMIN) DIXON: photoperiodism, 312
  botryocarpa (HARV.) WOELKERLING, 312
    photoperiodism, 312
  efflorescens (J. AG.) PAPENF., 177
    endophytic, endozoic, 355
  membranacea (MAGN.) PAPENF., *67*, 79
  pectinata (KYL.) PAPENF., 312
  purpurea (LIGHTF.) WOELKERLING, 139, 178
    photoperiodic ecotypes, 312
    photoperiodism, 312

INDEX **493**

Australia, 230, 264
 eastern Australia, 231, 265, 272
 Fucales, 242–245, 267, *269*, 270
 land bridge to Indonesia, in Pleistocene, 242
 northern Australia, 230
 paleoceanography, paleobiogeography, 237, *238–239*, 239, 265–266
 regions and water temperatures, 266
 southern Australia, 192, 235, 242, 244, 261, 263, 265–267, *268–269*, 270–271, *271*
 tropical, 216, 225–226
 western Australia, 230, 265, 267, 270
Avicennia, 208
 germinans, 206
 marina, 208
 nitida, 222
Axillariella, 243, 265
 constricta (J. AG.) SILVA, 264
Azores, 90–92

Bachelotia antillarum (GRUN.) GERLOFF, 206, *207*
 branch formation in blue light, 316
 photoperiodism, long-day plant, 312
Bacteria, 354, 369–370
Baffin Bay, 180, 187
Baffin Island, 166, 185–186
Bahamas, 221
 light, algal depth limit, 284
Bahía Tethys, 256
Baja California, *15*, 134, *146*, 152–154, 234, 242
Balanus, 65
 balanoides, see Semibalanus balanoides
 crenatus, 142
 glandula, 142
 nigrescens, *271*
 tintinnabulum, 206
Balearic Basin, 102
Balearic Islands, 98, 111, 115
Balistidae, 216
Ballia, 270
 scoparia (J. D. HOOK. et HARV.) HARV., 256
Bamfield Marine Station, 135
Bangia, 66
 atropurpurea (= *fuscopupurea*) (ROTH) C. AG., 70, 79, 84, 108, 133, *158–159*, 206, 230

 osmotic adjustment, 335
 photoperiodism, 309, 312
 temperature tolerance, 325
Banks Island, 33, 187
Banyuls, marine biological station, 99, 117
Barents Sea, 181–184
Barnacles, 5, 65, 70, *71*, 88, 262, *263*, 270, 348
Basispora africana JOHN et LAWSON, 206
Basque coast, *64*, 87–89, 126
Bathyal, 4, *4*
Bathypelagic zone, 3, *4*
Batophora oerstedi J. AG., *202*
Batrachospermum moniliforme:
 photoperiodism, long-day plant, 309, 312
Bay of Bengal, 228
Bay of Chaleurs, 128
Bay of Fundy, 123, 129, 131
Bay of Gaspé, 128
Bear Island, 180, 182
Beaufort Sea, 185–186
Belgium, 62
Belomorsk, 184
Belt Sea, 83
Benguela Current, 63, 203, 252, 258–259
Benthic realm, 3, *4*
Benthos, 3, 4, *4*
 phytobenthos, 3
 zoobenthos, 3
Bergmann's size rule, 32
Beringa Island, 186
Bering land bridge, 11, 26–27, 29, 34–40, 186, 203
Bering Sea, 15, 134, 154, 166, 185, 182–183
Bering Strait, 34–39, 125, 139, 153, 237, 258
Bermagui, 15
Bermudas, 208, 214, 216
Biarritz, 87, *88*
Bifurcaria, 56, 243
 bifurcata ROSS, 41, #*61*, 73, 86, 89–91
 brassicaeformis (KUETZ.) BARTON, 264
Bifurcariopsis, 243
 capensis (ARESCH.) PAPENFUSS, 246
Biocenosis, 107–108
Biologische Anstalt Helgoland, 69
Biomass, 360–361
Bird Island, 264
Birma, 230
Bivalvia, *see* Mussels
Black Sea, 98, 121–123
 paleoceanography, *100–101*, 122–123
Blidingia, 66, 88, 93
 dispersal, 93

Blidingia (Continued)
  minima (NÄG. ex KÜTZ.) KYL., 68, 70, 82, 84
    temperature tolerance, 323
Bodega Bay, 135
Bodega Marine Station, 135
Bohai Gulf, 161
Bonnemaisonia:
  asparagoides (WOODW.) C. AG., *87*, *113*
    Hymenoclonium phase, 233
  hamifera HARIOT, 20, *#20*, 21, *96*
    introduction to Europe, 95
    photoperiodism, 308, *310*, 310–312, 313
    temperature tolerance, 322–323
    Trailliella phase, 20, 21, 310–311, *310*, *313*
Boodlea composita (HARV.) BRAND, *194*, 230–231
Boreal Gulf, 25, 32
Boring algae, *6*, 8, 71, *210*, 212–213, 220, 355
Bornetella oligospora SOLMS-LAUBACH, 202, 230
Bosporus, 122
Bostrychia, 209, 230
  binderi HARV., 332
    photosynthesis, temperature optimum, 332
  radicans MONT., 222
    salinity, growth rate, ecotypes, 335
    temperature:
      growth, optimum, 328
      reproduction, 330
      tolerance, 326
Botryocarpa prolifera GREV., *260*
Botryocladia:
  botryoides (WULF.) J. FELDM., 112, *113*
  obovata (SONDER) KYL., 268
  occidentalis (BOERG.) KYL., 133
Boundary layer, 344
Bounty Islands, 250–251
Bouvet Island, *236*, 245
Brackish water:
  areas, 82–85, 121–123, 173, 182–183, 187, 222, 334–335, 337
  submergence, 83–84
Brazil, 216–217, 222, 252, 256–257
Brazil Current, *63*, 217
Bristol Channel, light percentage depth, 286
British Columbia, 134, 139, 142
British Honduras, 221
British Isles, 60–61, 64, *65–66*, *88*, *339*
Brittany, 62, *64*, *72–73*, 85, *87*
  light, algal depth limit, 284
  potash industry, historical, 365

temperature tolerance of seaweeds, 324
Brock Island, 181, 187
Brongniartella byssoides (GOOD. et WOODW.) SCHMITZ, *67*, 78
Bryobesia cylindrocarpa HOWE, 220
Bryopsis, 206
  hypnoides LAMOUR., 323
    temperature tolerance, 323
  mucosa, 108
  plumosa (HUDS.) C. AG., *67*
Bryozoans, *71*, 76, 77, 78, *87*, 117, 166, 211, *271*
Burma, *see* Birma

Calcite, 200
Calcification, 193, 197–201, 213–214, *215*, 363
Calcium carbonate, 198, 200–201, 212
California, 134, 142, 145–146, 147–148, *144–151*, 151–153, 157, 159, 252
  commercial use of seaweeds, 365, *367*
California Current, *63*, 135, 139, 145, 234, 252
Callao, 252
Calliarthron:
  yessoense (YENDO) MANZA, 158
Callibepharis ciliata (HUDS.) KUETZ., 328
  temperature optimum, growth, 328
Callithamnion:
  byssoides ARNOTT ex HARV. in HOOK.:
    temperature:
      growth, optimum, 328
      reproduction, 329–330
  kirillianum A. ZIN. et ZABERZH., 123
  roseum (ROTH) LYNGB., 293
    phycourobilin, 293
Callophyllis, *144*, 145
  cristata (C. AG.) KÜTZ. (= *Euthora cristata*), 41–42, *#49*, 175, 178, 181, 183
  flabellulata HARV., *144–145*
  laciniata (HUDS.) KÜTZ., *87*
    temperature:
      growth, optimum, 328
      reproduction, 330
  lambertii (TURN.) GREV., 268, 270
Caloglossa, 209
  leprieurii (MONT.) J. AG.:
    salinity, growth rate, ecotypes, 335
    temperature:
      growth, optimum, 328
      reproduction, 330
      tolerance, 326
Calosiphonia vermicularis (J. AG.) SCHMITZ:
  photoperiodism, 312

INDEX **495**

Calothrix, 212
  crustacea THUR. ex BORN. et FLAH. (= scopulorum), 69
Caloundra, 267
Campbell Island, 250–251
Canada, see America
Canadian Arctic, 164, 166, 169, *170*, 171, 181, 185–187
Canaries Current, 62, *63*, 203, 252
Canary Islands, 29–30, 90–92
Canary province, *65*, 90–92
Cannonball Sea, 27
Canopy algae, *6–7*
Capacity K, 358
Cape Ann, 131
Cape Agulhas, 255–259, 261
Cape Bismarck, 175
Cape Blanc, 90, 206
Cape Breton Island, 128
Cape Canaveral, see Cape Kennedy
Cape Cod, *15*, 123, 131–132
Cape Farvel, 179–180
Cape Frio *15*, 217, 257
Cape of Good Hope, 197, 258, 261
Cape Hatteras, *15*, 123, 125, 131–133
Cape Inubo, 154–155
Cape Jakan, 164
Cape Kennedy, *15*, 123, 133
Cape May, 123
Cape Mendocino, *138*
Cape Olyutorsky, 15, 154, 166, 182, 185
Cape Romano, 133
Cape Verde, *15*, 85, 90, 203
Cape Verde Islands, 90–92
Carbohydrates, 365, 367–370, *369*
Carbon, 343
  translocation in kelp, 358–359, *360*
Caribbean, 91, 189, 216–217, *218–219*, 219–222, *220–222*
Carolina region, *15*, *65*, 123, 132–134
Caroline Islands, 225
Carpathian algal and fish deposits, 201, 223
Carpoglossum, 243
Carpomitra costata (STACKH.) BATT., 78, 86
Carpophyllum, 243, 273
  scalare, see *Oerstedtia scalaris*
Carrageenan, 368, *369*
  world production, 367
Caspian Sea, 85, 98, 121–123
  paleoceanography, *100–101*, 122–123
Casting of outer layers, 353, 370
Catenella, 209, 222
Caulacanthus, 273
  ustulatus (TURN.) KÜTZ., *88*, 91, 263

Caulerpa, 91, 121, 193, 201, 206, 209, 214, 220–222, 227, 228, 267, 270, 272–273
  cactoides (TURN.) C. AG., 272
  hedleyi WEBER VAN BOSSE, 270
  lentillifera J. AG., *263*
  ollivieri DOSTAL, 121
  prolifera (FORSSK.) LAMOUR., 115, 120–121, *121*, 133
    antigrazer substance, Caulerpenyen, 352
  racemosa (FORSSK.) J. AG., 121, 231, 263, 268
    pigment content, acclimation of, 306
    var. *clavifera* (TURN.) WEBER VAN BOSSE, 268
  serrulata (FORSSK.) J. AG. emend. BOERG., 227
  sertularioides (GMEL.) HOWE, *194*
  vesiculifera HARV., 268
Caulerpales, 98, 192–193, *194*, 199, 202, 221, *221*, 233
  deep-water adaptation, siphonaxanthin, siphonein, *288*, 289–290
Caulerpenyen, 352
Caulocystis, 243, 267
  cephalornithos (ARESCH.) LABILL., 272
  uvifera (C. AG.) ARESCH., 271
Central American (Panama) land bridge, *196*, 234
Central Pacific Islands, 226, 231–232
Centroceras clavulatum (C. AG. in KUNTH) MONTAGNE, 91, 206, *207*, 230, 255–256, 263, 270
Ceramiales, 174, *195*, 200, *207*, 262–263, 267, 355, 368
Ceramium, 241
  deslongchampsii CHAUV. in DUBY, *71*, 73
  leutzelburgii SCHMIDT, 219
  opportunists, 357
  photosynthetic rate, 364
  rubrum (HUDS.) C. AG., *71*, 73, 105, 256
    respiration, seasonal rates, 332, *333*
    temperature optimum, growth, 323
    temperature tolerance, 323
  tenuicorne (KÜTZ.) WAERN, 84–85
    nutrient uptake, 343
Ceratocolax hartzii ROSENV., 40, 177, 355
Ceylon, see Sri Lanka
Chaetomorpha:
  antennina (BORY) KÜTZ., 206, 230
  capillaris (KÜTZ.) BOERG. (= *tortuosa*), *71*, 177
  melagonium (WEB. et MOHR) KÜTZ., 41–42, 46, #*50*, 67, 177–178, 187

Chaetomorpha (Continued)
   temperature tolerance, 322–323
   photosynthesis, temperature optimum, 332
Chaetophorales:
   deep-water, siphonaxanthin, 290
   hairs, nutrient uptake, 343
Chaetopteris plumosa, see Sphacelaria, plumosa
Chagos Islands, 229
Chalk boring algae, 355
Chamaephyceae, 314
Chamaesipho columna, 270
Champia:
   lumbricalis (ROTH) DESV., 260, 262
   viridis C. AG., 268
Channel, *see* English Channel
Chantransia stage, photoperiodism, 309
Chatham Islands, 274
Chernobyl incident, iodine-131 in *Fucus*, 365
Chesapeake Bay, 132
Chichagof Island, *138*
Chile, 244, 251–252, 255–256, 345
Chiloe Island, *15*, 252
China, 154, 161, *162*, 163, 225, 264
   commercial uses of seaweed, 365, *366–367*
   paleoceanography, 161
Chlorella vulgaris, pigments, light quality effects, 307
Chlorochytrium:
   "*inclusum*", 177
   schmitzii ROSENV., 176
Chlorococcales, 116, 198, 207, 353
Chlorodesmis, 231, 233, 272
   caespitosa J. AG., 233
   fastigiata (C. AG.) DUCKER, 231
Chloromytilus meridionalis, 260
Chlorophyll, *see* Photosynthesis, pigments
Chnoospora implexa (HERING) J. AG., 230
Chondria:
   algal parasite, *Janczewskia*, 355
   armata OKAM., *229*
   polyrhiza COLL. et HERV., 219
Chondrides, 200
Chondrus:
   armatus (HARV.) OKAM., *156*
   commercial use, 367
   crispus STACKH., 41, 57, #*57*, 67, 71, 76, 87, 89, 128–129, *130–131*, 183, 188
     absorption, action spectrum, photosynthesis, *291*, 293
     biotic relations, 348
     freezing tolerance, 323
     greenish color, 296
     growth rate, diurnal changes, 304
     nitrate storage, 342
     pigment content, acclimation of, 307
     relative growth rate, 363
     respiration, seasonal rates, 332, *333*
     salinity, growth rate, *336*
     temperature optimum, growth, 323
     temperature tolerance, 323, 326–327
   giganteus YENDO, 57
     temperature tolerance, 326
   nipponicus YENDO, 57
     temperature tolerance, 326
   pinnulatus (HARV.) OKAMURA, 57
   yendoi, see Iridaea, cornucopiae
Chorda:
   filum (L.) STACKH., 41, #*45*, 72–75, 81, *81*, 128, 136, *163*, 180, 186
   gas-filled thallus parts, 345–346
   salinity ecotypes, 335
   temperature:
     reproduction, 329–330
     tolerance, 323
   tomentosa LYNGB., 41, #*50*, 53–54, 177, 180
Chordaria, 198
   flagelliformis (O. F. MÜLL.) C. AG., 41, #*43*, 45, 127–128, 181, 184, 186–187
Chordariales, 244, 262
Chordariopsis capensis (KUETZ.) KYL., 262
Choreocolax polysiphoniae REINSCH, 355
Chromatic adaptation, *see* Photosynthesis, chromatic adaptation
Chronobiology, *see* Circadian, Circatidal, Circalunar, Circannual rhythms
Chrysymenia ventricosa (LAMOUR.) J. AG., *114*
Chthamalus, 70, 230
   antennatus, 270
   dalli, 142
   dentatus, 206
   stellatus, 88, 108, *109*, 206
Chukchi Sea, 166, 182–185
Circadian rhythms:
   chloroplast movement, 317
   gamete release, 317–318
Circalittoral zone, 9, 10, 113
Circalunar rhythms, *see* Lunar periodicity
Circannual rhythms, 320
Circatidal rhythms, 320
Cladistic biogeography, 10
Cladophora, 193
   glomerata (L.) KÜTZ., 84
     nutrient uptake, 343
     turnover rate, 360
   loroxanthin, 290

photosynthetic rate, 364
prolifera (ROTH) KÜTZ., 121
rupestris (L.) KÜTZ., 71, 129, 180
  temperature tolerance, 323
Cladophorales, 192–193, 198
  deep-water, siphonaxanthin, 290
Clathromorphum, 200
  circumscriptum (STRÖMF.) FOSL.:
  reproduction, temperature, 330
Climax community, 70
Clipperton Island, 233–234
Clod carts, 345
Coccophora, 243
  langsdorfii (TURN.) GREV., 157
Codiolum phase, 71
  as endophytic alga, 355
  photoperiodism, 309, *310*, 312
  reproduction, temperature, 329, *330*
Codium, 192, 267, 272–273
  bursa (L.) C. AG., 86, *112*
  carolinianum SEARLES, 97, 133
  corallioides (KÜTZ.) SILVA, 88, 90, 115, *115*, 116
  decorticatum (WOODW.) HOWE, 97
    pigment content, acclimation of, 306
  dimorphum SVED., 255
  duthiae P. C. SILVA, *268*
  dwarkense BOERG., 228
  effusum (RAFINESQUE) DELLE CHIAJE (= *difforme*), 116
  fragile (SUR.) HARIOT, 96, 116, 139
    absorption spectrum, *292*
    desiccation, 337–338
    hairs, nutrient uptake, 343
    introduction to other coasts, 96–97, 116
    light saturation, growth, photosynthesis, 302
    nitrate storage, 342
    pigment content, acclimation of, 307
    temperature tolerance, 323
  mamillosum HARV., 233
    photosynthetic rate, 364
  taylori SILVA, *195*
  tomentosum STACKH., 97
  vermilara (OLIVI) CHIAJE, 97, 116
Cold temperate species, 41, 55–59, *56–62*
Colombia, 216, 233
Colpomenia, 206
  peregrina SAUV., *96*
    desiccation, 338
    gas-filled thallus parts, 346
    introduction to Europe, 95–96
  sinuosa (MERT.) DERB. et SOL., *112*, 273
Commander Islands, 154

Commercial uses, 364–365, *366–367*, 368–370
Communities, 106–108
Compensation depth, *see* Light, compensation depth (point)
Competition, 78, 129, 347–350
Conchocelis phase, 79, 213, 310, *312*, 314
  aquaculture, *368*
  photoperiodism, *312*, 314
Connecticut, 123
Constantinea:
  rosa-marina (GMEL.) POST. et RUPR.:
    age, maximal, of individuals, 356
  simplex SETCH., 148
  subulifera SETCH., 145
    age, maximal, of individuals, 356
    dark growth, 359
    photoperiodism, 312
Constituents, chemical, 364–365, 368–370
Continental drift, *see* Paleoceanography, paleobiogeography
Continental islands, 232
Convergence, 135
Cook Inlet, 141
Coos Bay, *138*
Copenhagen, 93
Cope's rule, 32
Coralligéne, 116–117
Corallina, 148, 206
  cuvieri, see *Haliptilon roseum*
  elongata ELLIS et SOLAND. (= *mediterranea*), *109*, *112*
    growth rate, 364
  officinalis L., *67*, 71, 76, 88, *88*, 129, *149*, 255
    freezing tolerance, 324
    temperature optimum, growth, 328
    temperature tolerance, 323
    var. *chilensis*, *149*, 255
    photosynthetic rate, 364
  pilulifera POST. et RUPR., 155, *158*, 159, 161
  pinnatifolia (MANZA) DAWS., 159
Corallinaceae, 199–200, 211, 233
Corallinales, *see* Crustose coralline algae
Coralline algal ridge, 214
Corallith, 200
Corallium rubrum, 117
Coral reefs, *190*, 209–216, *210*, *215*, 205, 208–216, 227, 228–231, 234, 267
  algal ridge, 214
  atolls, 210
  barrier reefs, 210
  depth limit, 211, 213
  distribution, *190*, 227

Coral reefs (*Continued*)
  fringing reefs, 210
  grazing, 214–216, *215*
  Great Barrier Reef, 216, 225, 231, 272
  hermatypic corals, *190*, 200, 205, 208, 209–213, *210*, *215*, *218–219*, 222, 226, 233, 267, 287
  light requirement, 211, 287
  macroalgae, 212–216, *170*
  nitrogen, atmospheric, fixation of, 212
  northernmost coral reefs, 228
  paleoceanography, paleobiogeography, 205, 210–212
  primary productivity, 212, 361–362
  southernmost coral reefs, 272
  ultraviolet, tolerance, 304–305
  zonation, 217, *218–219*, 219–221
Cordylecladia erecta (GREV.) J. AG., 312
  photoperiodism, 312
Coriolis effect, 16
Corsica, 98, 111, 117
  light, algal depth limit, 284
Cortez province, 153–154
Cosmopolitan species, 105, 241
Costaria costata (C. AG.) SAUNDERS, *136*, *138*, *140*, 142, *144–146*, 157, *158*, 161
Costa Rica, 216, 233
Côte d'Azur, 99
Côte des Albères, 99, 115
Crepidula fornicata, 95
Cretaceous / Tertiary boundary, 31
Crete, 115
Critical tide levels, 341
Critical water content, *see* Desiccation, critical water content
Crossing experiments, 47–48, 51–52
Crozet Islands, *236*, 250–251
Cruoria:
  arctica, *174*, 176
  pellita (LYNGB.) FRIES, 74–75, 79
  rosea, *see Turnerella pennyi*
Crustaceans:
  herbivores, 350
  long-distance dispersal, 93
Crustose algae, 6, 7, 67, 79, 143, 187, 220. See *also* Crustose coralline algae
  brown algae, 181, 187, *188*
  growth rates, 143
  life form type, 314
  red algae, 74–75, 79, 143, 188
Crustose coralline algae, 62, 67, 70–71, 79, 88, 115, 117, 122, 134, 193, 196, 199–200, 206, 211–213, *215*, 216, *218–219*, 220, 222, *263*, 316, 352

adaptation
  low-light habitats, 286
  antigrazer protection, 352
  casting of outer layers, 354
  epiphytism, attraction of gastropod larvae, 354
  evolution, 199–200
  growth rates, 213, 364
  light at depth limit, 285, 287, 300
  primary productivity, 362
Cryptic stages, 93, 309
Cryptochrome, 314
Cryptonemiales, *174*, 193, 199, *207*, *263*, 355
Cryptopleura ramosa (HUDS.) KYL. ex NEWTON, 76, 78, *87*
  temperature optimum, growth, 328
Ctenocladales, 355
Cucumaria, 74–75
Curaçao, 217, *218–219*, 219–222
Currents, *see* Ocean currents
Current velocities, 345
Cutleria multifida (SM.) GREV., 105
Cyanophytes, 198, 211
  chromatic adaptation, 296
  photoinhibition, recovery from, 303
Cymathere triplicata POST. et RUPR., *136*, *138*
Cymodocea, 203, *204–205*, 227
  nodosa, *120*, 121, 206
  serrulata, 227
Cymopolia barbata (L.) LAMOUR., 202
Cyprideis, 105
Cyrtodaria, 35, 37
Cystoclonium purpureum (HUDS.) BATT.:
  temperature optimum, growth, 323
  temperature tolerance, 323
Cystophora, 243, 264–265, 267, *269*, 270–271
  fibrosa SIMONS, 264, 267
  intermedia J. AG., 270–271, *271*
  platylobium (MERT.) J. AG., 270
  retroflexa (LABILL.) J. AG.:
    desiccation, 338
  torulosa (R. BROWN ex TURN.) J. AG., 271–272
    freezing tolerance, 324
Cystophyllum, 157, 267
Cystoseira, 98, 106, 111, 113, *146*, 201, 206, 227, 228, 230, 243, 264
  abies-marina C. AG., 91
  adriatica SAUV., 111
  age, maximal, of individuals, 356
  amentacea VALIANTE, 106

baccata (GMEL.) SILVA (= *fibrosa*), 88
barbata J. AG., 111, 122
biomass, 361
corniculata HAUCK, 106
crinita (DESFONT.) BORY ex MONT., 111, *112*
mediterranea SAUV., 106, 111, *112*
myrica (GMEL.) J. AG., 227, *227*
osmundacea (TURN.) C. AG., 143, *146*, 148, 153
setchellii GARDN., *147*
spicata ERCEGOVIC, 106
spinosa SAUV., 113
stricta (MONT.) SAUV., 106, *109*, 111
tamariscifolia (HUDS.) PAPENF.(= *ericoides*), 88, *88*, 106
trinodis (FORSSK.) C. AG., 231
zosteroides (TURN.) C. AG. (= *opuntioides*), 113, *116*
Cystoseiraceae, 242–243
Cystoseirites, 201
Cystosphaera, 243, *247*, 265
jacquinotii SKOTTSB., 248

Danish West Indies, *196*
Dark growth, 293, 359
Darkness:
 lack of chlorophyll synthesis, red algae, 293
 tolerance, 171
Dasya baillouviana (GMEL.) MONT., 128
Dasycladales, 193, 197, *197*, 198–199, 202
Dasycladus vermicularis (SCOPOLI) KRASSER (= *clavaeformis*), 115, *202*
Davis Strait, 166, 180
Decomposition, algal biomass, 215, 369–370
Deep-water algae, 6–7, 78–79, 107, 113, *114–116*, 117, *118–119*, 129, 145, 148, 155, 178, 181, 188, 221, *221*, 222, 233, 248, 257, 270, 286
De Geer Route, 32
Delaware, 123
Delesseria, 57
 decipiens J. AG., 57
 lancifolia (J. D. HOOK.) J. AG., 249
 sanguinea (HUDS.) LAMOUR., 41, 57–59, #62, 67, 77, 78, *83*, 87, 183
  absorption, action spectrum, photosynthesis, *291*, 293
  competition, 350
  dark, lack of chlorophyll synthesis, 293, 359
  dark growth, 293, 359
  light at depth limit, 285

light saturation, photosynthesis, 302
photoinhibition, 302
photoperiodism, 312
photosynthesis *vs.* light curve, *297*
temperature:
 growth, optimum, 323
 photosynthesis, optimum, 332
 tolerance, 323
translocation, 359
Delesseriaceae, 262, 265, 267
Delesserites, 200
Delisea pulchra (GREV.) MONT., 248, *268*
Demographic studies, 357
Dendrophylliidae, 211
Denmark, 61, 64, 68
Depth limits, lower, *see* Deep-water algae
Derbesia:
 marina (LYNGB.) SOLIER, 131
 tenuissima (DE NOT. in MOR. et DE NOT.) CROUAN, 318
 Halicystis phase, gamete release, light induction, 318
Dermonema:
 dichotomum HARV., 228, *229*
 frappieri (MONT. et MILL.) BOERG., 230
Desiccation, 337–338, *339–340*, 340–341
 nutrient stress, 341
 photosynthesis, 338, 340–341, *340*
 water content, 338, *339–340*
Desmarestene, 319
Desmarestia, 117, 246, 256
 aculeata (L.) LAMOUR., 41–42, #44, 46, 78, *127*, *170*, 171, 177, *179*, 181–182, 184, 186–187, 201
 temperature:
  growth optimum, 323
  reproduction, 330
  tolerance, 322–323
 anceps MONT., *147*
  temperature optimum, growth, 328
  temperature tolerance, gametophytes, 327
 firma SKOTTSB. in NORDENSKJÖLD, 262
  sulphuric acid, 351
 ligulata (LIGHTF.) LAMOUR., 78, 92, 143, *156*, 262
 menziesii J. AG.:
  temperature tolerance, gametophytes, 327
 rossii HOOK. et HARV., 272
 viridis (O. F. MÜLL.) LAMOUR., 41–42, #45, 46, 67, 78, 105, 117, 139, *163*, 177, 181, 184, 187
  sulphuric acid, 351
  temperature

Desmarestia, *viridis* (O.F. Müll.) LAMOUR (*Continued*)
   reproduction, 329–330
   willii REINSCH, 272
   tolerance, 323
Desmarestiales, 244, 246, 248
Detrital feeders, 369–370
Detritus food chains, 369–370
Devaleraea ramentacea (L.) GUIRY (= *Halosaccion ramentaceum*), 64, *68*, 129, 173–176, *#176*, 178, 181, 184–187
   light saturation, growth, 302
   temperature optimum, growth, 328
Diadema antillarum, 207, 219, *219*, 220
Diatoms, 353
   as epiphytes, 353
   fossil, 126
   primary productivity, 362
   tube-dwelling, 355
Dichocoenia stokesii, 219
Dictyoneuropsis reticulata (SAUND.) G. M. SMITH, *138*
Dictyoneurum californicum RUPR., *138*, *146*, 153
Dictyopteren, *319*
Dictyopteris, 193, 207
   delicatula LAMOUR., *194*, 206, 220
   justii LAMOUR., 222, *222*
   membranacea (STACKH.) BATT., 86, *87*, *88*, 92
      growth-inhibiting substances, 352
   plagiogramma (MONT.) VICK., 233
Dictyosiphon:
   foeniculaceus (HUDS.) GREV., 41, *#44*, 45, 84, 127, 128, 175, 179, 184, 186
      morphogenetic substances, 354
      nutrient uptake, hairs, 343
   hirsutus, see *Scytothamnus, hirsutus*
Dictyosiphonales, 244
Dictyosphaeria cavernosa (FORSSK.) BOERG., 192, *195*
Dictyota, 91, 193, *196*, 214, 270
   acutiloba J. AG., 233
   antigrazer substances, 352
   dichotoma (HUDS.) LAMOUR., 17, *#18*, 19, 78, 86, *87*, 92–93, *112*, 133, 206, 220, 256
      chloroplast movements, circadian rhythm, 317
      growth, light quality, 316
      growth-inhibiting substances, 352
      egg release, lunar periodism and light phase, 318–320, *320*
      pheromone, dictyopterene, *319*

      photoinhibition, recovery from, 303
      photosynthesis, blue-light effect, 316
      pigment content, acclimation of, 306
   flabellata (COLL.) SETCH. et GARDN., *147*
   menstrualis (HOYT) SCHNETTER, HÖRNIG et WEBER-PEUKERT, *18*, 19
      photosynthetic rate, 364
   pulchella HÖRNIG et SCHNETTER (= *divaricata*), *194*
Dictyotales, 147, 153, 193, 201, 221, 267, 272
Dictyurus:
   fenestratus DICKINSON, 205–207, *207*
   occidentalis J. AG., *195*
Diffusion, 343
Digenea simplex (WULFEN) C. AG., *104*, 154
Dilsea:
   californica (J.AG.) KUNTZE, 45
   carnosa (SCHMIDEL et ESPER) O. KUNTZE (= *edulis*), 76
   integra (KJELLM.) ROSENV., 45, 173, *174*, *#176*, 178, 184
Dinoflagellates, 210, 212
Diploria, 218
   clivosa, 219, *219*, 222
   strigosa, 219
Disko Bay, 180
Dispersal:
   long-distance, 92–98, 229, 232, 241–242, 264, 266, 272–274
   short-distance, 93, 97, 241
   stepping stones, *see* Stepping stones
Displacement of marine biota, 38–39
Dissolved organic carbon, 370
Diversity, 189, *190*, 191–192, 197, 209, 226, 258, 267
   number of seaweed species:
      Antarctic, 246
      Arctic, 164, 192
      Black Sea, 121
      British Isles, 192
      California, 192
      Carolina region, 132
      China, 161
      Florida, 192
      Galápagos Islands, 234
      Georgia, 133
      Greenland, 175, 178
      India, 228
      Japan, 156
      Korea, 161
      Maldive Islands, 229
      Malaysia / Indonesia, 192
      Maryland / Virginia, 192
      Mediterranean, 102

INDEX **501**

Newfoundland, 126, 192
Novaya Zemlya, 184
Sea of Japan, 156
South America, Pacific, temperate, 252
southern Africa, 192, 258, 267
southern Australia, 102, 192
sub-Antarctic Islands, 250
tropical regions, 192, 196, 205, 217, 234
West Africa, 205
World, 3
Western North Atlantic, 133
Diving, 73
  submersibles, 221
Dixon Entrance, *138*
Dotyella hawaiiensis (DOTY et WAINWRIGHT) WOMERSLEY et SHEPLEY, 233
Drachiella spectabilis ERNST et J. FELDM., 87
Drake Strait, 237, 239
Drifting algae, *see* Floating algae
Dry weight, 365
Dumontia:
  contorta (GMEL.) RUPR. (= *incrassata*), 41–42, 44, *#48, 67*
    gas-filled thallus parts, 346
    photoperiodism, 312
    temperature tolerance, 326–327
  filiformis, *see Dumontia, contorta*
  incrassata, *see Dumontia, contorta*
Durvillaea, 243–244, 246, 251, 274
  antarctica (CHAMISSO in CHORIS) HARIOT, 241, 244, 251–252, 254–256, 258, 262, 272–274
    algal parasite, *Herpodiscus*, 355
  chathamensis HAY, 251, 274
  desiccation, 338
  mechanical adaptation, 345
  potatorum (LABILL.) ARESCH., *269*, 271–272, 274
  temperature optimum, growth, 328
  temperature tolerance, 326
  willana LINDAUER, 273–274
Durvillaeales, 243–244, 246

Earthquake, 142
East Australian Current, *63*
East Cape, *15*
East China Sea, *163*
Easter Island, 226
Eastern Pacific barrier, 233
East Gondwana, *see* Gondwanaland
East Greenland Current, *63*, 166
East Siberian Sea, 164, 182–183

Eburneopecten, 29
Echinoderms, 85, 240
Echinometra lacunter, *219*
*Echinus esculentu*s, 350
Ecklonia, 159, *160*, 161, 228, 242, 259, 261, 267, 270–271
  biruncinata (BORY) PAPENF., 228, 259, 264
    temperature demands, gametophytes fertile, 327
    temperature tolerance, gametophytes, 327
  cava KJELLM., 157, *158*, 159, *160*, *162*
  kurome OKAM., 159, 161
  maxima (OSBECK) PAPENF., 86, 258–259, *260*, 262
    decomposition rate, 370
    spore production per year, 358
    temperature demands, gametophytes fertile, 327
    temperature tolerance, gametophytes, 327
  muratii FELDM., 86, 90, 261
    crossing the equator, 86
  radiata (C. AG.) J. AG., 216, 228, 261, 264, 267, *269*, *271*, 272–273
    age, maximal, of individuals, 356
    biomass, 361
    decomposition rate, 370
    desiccation, 338
    distal erosion, 360
    ultraviolet, tolerance, 305
  stolonifera OKAM., 159, 161
Ecotypes:
  light, 392
  salinity, 335, *336*, 337
  temperature:
    growth, 329
    tolerance, 326–327
Ectocarpene, 318, *319*
Ectocarpus:
  breviarticulatus J. AG., 206
  siliculosus (DILLW.) LYNGB., 133
    pheromone, ectocarpene, *319*
    salinity ecotypes, 335
    temperature:
      optimum growth, ecotypes, 329
      reproduction, 329–330
      tolerance, ecotypes, 327
Ecuador, 233
Edge effect, 152
Egregia menziesii (TURN.) ARESCH., *138*, 143, *146*, 147–148, *148*, 153
  gas-filled thallus parts, 345
Egypt, 98
Eisenia, 159, 198, 242, 253

Eisenia (Continued)
  arborea ARESCH., *138*, 159, 147–148, *148-149*, 153, 253
  bicyclis (KJELLM.) SETCH., 121, 123, *123*, 157, *158*, 159, *160*
  cokeri HOWE, 159, 253
Elachista fucicola (VELL.) ARESCH., 177, 353
  epiphytism, 353
Ellesmere Island, 30, 185–187
Elminius modestus, 70, 95
El Niño, 152, 253
El Salvador, 179
Emerson enhancement effect, *see* Photosynthesis, Emerson enhancement effect
Endemism, 40, 85, 91–92, 102, 105–106, 136, 139, 146, 154, 159, 163, 173, 205, 232–233, 246, 251–252, 258–259, 264, 267, 272, 274
Endocladia muricata (POST. et RUPR.) J. AG., 142, *144–145*, 148, *349*
  biotic relations, 349, *349*
Endogenous clocks, 320
Endophytes, 6, 8, 177, 187, 343–356
Endosymbiontic algae, 210, *210*, 211–213
Endozoans, 6, 8, 79
Enewetak Atoll, 210
Engelmann's hypothesis, *see* Photosynthesis, Engelmann's hypothesis
England, *see* British Isles
English Channel, *15*, 92
Enhalus, 204
Enhydra lutris, 142, 351
Enteromorpha, 68, 70, 82–83, 88, 93–94, 132, 184, 206, *207*, 219, 230, 241, 248
  biomass, 361
  clathrata (ROTH) GREV., 230
  compressa (L.) GREV., 198, 255
  dispersal, 93
  gas-filled thallus parts, 346
  inhibition by other algae, 352
  intestinalis (L.) LINK, 68, 256
    gamete release, lunar periodicity, 320
    osmotic adjustment, 335
  linza (L.) J. AG.:
    respiration, seasonal rates, 332, *333*
    salinity ecotypes, 336
  opportunists, 357
  photosynthetic rate, 364
  prolifera (O. F. MÜLL.) J. AG., 68
  r-strategy, 358
  succession, 93
Entocladia, 71

Enzymes, 298, 327, 332
Ephemerophyceae, 314
Epicladia flustrae REINKE (= *Endoderma flustrae*), 355
Epicontinental seaways, *see* Cannonball Sea; Obik Sea; Turgai Strait
Epifauna, *see* Epiphytes, animals
Epiphytes, 6, 8, 353–354
  algae, 74–75, 76, 225, 264, 353–354
  animals, 354
  inhibiting substances, 352
  protection against epiphytes, 353–354
Epizoans, 6, 8, 353
  algae, 353
Equatorial Countercurrent, 197, 203
Equatorial crossing in Cenozoic, 39, 86, 153, 159–160, 203, 205
Equatorial Islands, 231
Equatorial submergence, 117
Equilibrium theory of insular biogeography, 232
Eriocheir sinensis, 95
Erosion of thallus, 354, 360, 370
Eschara foliacea, 87
Estuaries, 82, 132, 335
Étage circalittoral (French), *see* Circalittoral zone
Étage infralittoral (French), *see* Infralittoral zone
Étang de Thau, 117
Eucheuma, 230
  commercial use, 367
  denticulatum (N. BURMAN) COLL. et HERV. (= *muricatum*), 202
  isiforme (C. AG.) J. AG., *195*
  temperature optimum, photosynthesis, 332
Eudesme virescens (CARM. ex HARV. in HOOK.) J. AG., 41, *#43*, 185
Eugonophyllum, 199
Euhaline environment, 82
Eulittoral, 3, 4–5, *6*
  primary productivity, 167, 361–362
Eunicella verrucosa, 87
Eupomacentrus planifrons, 219
Euphotic zone, 3
Eureca Sound formation, 30, 186
Europe/Northwest Africa:
  Arctic, 180–184
  commercial uses of seaweeds, 365
  Mediterranean, *see* Mediterranean
  Northern France/Northwest Africa, 85–123
  Northern Norway/Northern France, 60–85
    deep-water algae, 78–79
    eulittoral, 70–73
    history of algal investigation, 69

marine biological stations, 68–69
sublittoral, 73–85
supralitoral, 69–70
paleobiogeography, paleoceanography, 22–23, *24–25*, 25, *26–27*, 28–35, *36–37*, 38–39
Euryhaline, 54, 82, 335, *336*
Eurythermal, 17, 54, 321, *322*
Eusmilia fastigiata, 219
Euthora cristata, see *Callophyllis, cristata*
Eutrophication, 84
Evolution of algae, 198–203
Extinction depth, see Light, extinction depth
Exudation, 370

Faeroe Islands, 33, 60, 64, *69*, *74–75*, 80–82, 94
Falkenbergia phase, see *Asparagopsis armata*
Falkland Islands, *249*, 251–252, 258
Farasan Archipelago, 227
Farlowia, photoperiodism, 312
Fauchea repens (C. AG.) MONT., *114*
Favia, *210*, 213
Festuca rubra, 69
Fiji, 226
Fimbrifolium dichotomum (LEPECH.) G. HANSEN (= *Rhodophyllis dichotoma*), 41–42, *#53*, 131
Finavarrene, *319*
Finland, see Baltic
Fish:
 coastal, 85, 105, 226
 evolution, 215–216
 herbivorous, 196, 207, 214–216, 219–220, 272, 350, 352
 Sargasso Sea, 223–224, *223*
Fjord effect, 82
Fjords, *74–75*, 80–82, 173–174, 176–181
Flamborough Head, 64
Flavoprotein, 314
Floating algae, 94, *96*, 157, 217, 222–225, 237, 241–242, 264, 266, 272–274, 346
 transport of stones and gravel, 346
Floral elements, 103, *104*, 105
Florida, 123, 133–134, 216
Florida Current, 217
Floridean starch, 368
Floridoside, 368, *369*
Food, seaweeds as, world production, 367
Foraminifera, 34, 103, 126, 135, *196*
Fortaleza, 221
Fort Ross, *138*
Fossils:
 algae, 136, *137*, 197–202

uncalcified, 136, *137*, 200–201, 223–224
invertebrates, 29, 34–38
land plants, 34
vertebrates, 30, 32
Fouling communities, 94
France:
 Atlantic coast and Channel, 62, 64, 85–89
 Mediterranean coast, 98, *109–110*, 120
Franklin Expedition, 164
Franz Josef Fjord, 175–176
Franz Joseph Land, 180–181
Fraser Island, *15*, 272
Freezing tolerance, 322–325
French Polynesia, 226, 230
Freshwater:
 algae, photoperiodism, 309
 realm, 82
Friday Harbor Laboratories, 135
Frost hardening, 324
Fryella gardneri (SETCH.) KYL., 145
Fucales, 70, 106, 148, 193, *195*, 200, 227, 242–243, 248, 264–265, 267, 270
 age, maximum, individuals, 356
 chloroplast movements, 317
 competition, 347–348, 350
 distribution of genera, world-wide, 243
 evolution, 242, 244–245
 fossil remains, 223–224
 origin in Gondwanaland, see Fucales, evolution
Fucoides, 201
Fucose, 284
Fucoserratene, *319*
Fucoxanthin, see Photosynthesis, pigments
Fucus, 55, 93, 106, 108, *130–131*, 243
 ceranoides L., 41, 56, *#59*
  salinity, growth rate, 335
 Chernobyl incident, iodine-131 in *Fucus*, 365
 distichus L., 41–42, *#47*, 55–56, *68*, 70, *127*, 141, 157, 184
  freezing tolerance, 323
  photosynthesis, temperature optimum, 332
  subspecies *anceps*, 41, *#47*, 70
  epiphytism, 353
 evanescens AG., 41–42, *#47*, 55–56, *68*, 93, 129, 157, *170*, 175, 178–179, 181–182, 185–187
 frost hardening, 324
 gardneri SILVA, 56, 143
 growth rate:
  diurnal changes, 304
 hairs, nutrient uptake, 343
 photosynthesis:

Fucus, photosynthesis (*Continued*)
   blue-light effect, 316
   rate, 364
  polarity induction, 317
  primary productivity, 362
  serratus L., 41, 56, *#57*, 58, *65–67*, 70, *71*, 83–84, 86, 89, 128, 183
   competition, 347
   desiccation, 338, *339–340*
   epiphytic animals, 354
   growth rate, 364
   light saturation, growth, photosynthesis, 302
   salinity, growth rate, *336*
   temperature:
    growth, optimum, 323
    photosynthesis, optimum, 332
    tolerance, 323
  spiralis L., 41–42, 55, *#56*, 57, *65–66*, 68, *68*, 70, *72–73*, 88–91, 105, 129, 188
   competition, 348
   desiccation, 338, *339–340*, 341
   growth rate, 364
   photosynthesis, seasonal rates, *331*, 332
   temperature:
    growth, optimum, 323
    photosynthesis, optimum, 332
    tolerance, 323
  vesiculosus L., 41, *#52*, 53–54, *65–66*, *68*, 70, *72–73*, 82–84, 89–90, 128–129, *130–131*, 132, 179–180, 183, 187, 224
   age, maximal, of individuals, 356
   antigrazer substances, 352
   biomass, 361
   competition, 347
   desiccation, 338, *339–340*, 340–341
   epiphytic animals, 354
   freezing tolerance, 324
   gas-filled thallus parts, 345–346
   growth rate, 364
   light saturation, growth, photosynthesis, 302
   pigment content, acclimation of, 306–307
   nitrate storage, 342
   nutrient uptake, 343
   respiration, seasonal rates, 332, *333*
   salinity:
    ecotypes, 335
    growth rate, 335
   temperature:
    growth, optimum, 323
    photosynthesis, optimum, 332
    tolerance, 323, 325
  virsoides J. AG., 86, 105, *105*, 106, 108

Fugitive species, 357
Fujian, 161
Fungi, marine, 356, 369
Fungia, 211
Furcellaria lumbricalis (HUDS.) LAMOUR. (= *fastigiata*), 40, 128
  salinity, growth rate, *336*
  turnover rate, 360–361

Gabon, 203
Galactans, 368
Galactose, *369*
Galápagos Islands, 233–234
Galaxaura, 91, 193, *194*, 200
  falcata KJELLM., *158*
  obtusata (ELL. et SOL.) LAMOUR., *194*
Gametes, *see* Reproduction, gametes
Gas-filled thalli, 94, 143, 151, 345–346
Gastroclonium:
  clavatum (ROTHPL.) ARDISS., 108
  coulteri (HARV.) KYL., 148
Gastropods, 78, 93, 111, 125, 143, 206, 256, 262–263, 270, 348, 350, 352, 354
  attraction of larvae by crustose coralline algae, 354
  long-distance dispersal, 93
Gelidiella acerosa (FORSSK.) FELDM. et HAMEL. 227, 230–231
Gelidium, 193, 206, 273
  amansii LAMOUR., *158*, 171
  arbuscula (MONT.) BOERG., 91
  biomass, 361
  canariense (GRUNOW) SEOANE-CAMBA (= *cartilagineum* var. *canariensis*), 91
  commercial use, 367
  chilense (MONT.) SANTELICES et MONTALVA, 255
  crinale (TURN.) DESMAZIERES, *194*, 257
  light saturation, growth, 302
  divaricatum MARTENS, 161
  nudifrons GARDN., *149*
  pristoides (TURN.) KÜTZ., 263, *263*
  pteridifolium NORRIS, HOMMERSAND et FREDERICQ, 264
  pusillum (STACKH.) LE JOL. (= *reptans*), 208, 230
   temperature optimum, growth, 323
  sesquipedale (CLEM.) BORN. et THUR., 88, *88*
Gelling agents, phycocolloids, 369
Geolittoral zone, 84
Geological time scale, 22, *24*
Georgia, 123, 132–133
Georgiella confluens (REINSCH) KYL., *249*

INDEX 505

Germany, *see* Baltic; Helgoland; North Sea
Ghana, *207*
Giant kelps, *see* Laminariales, giant kelps
Gibraltar, *see* Strait of Gibraltar
Giffordia:
  duchassaingiana (GRUNOW) TAYLOR, 215, 219
  ovata (KJELLM.) KYL., 177, 187
Gigartina, 45, 273
  acicularis (WULF.) LAMOUR.:
    photoperiodism, 312
  alveata J. AG., desiccation, 338
  canaliculata HARV., 148
  corymbifera (KÜTZ.) J. AG., *149*
  papillata, *see Mastocarpus, papillatus*
  papillosa (BORY) SETCH. et GARDN., *249*
  photosynthetic rate, 364
  radula (ESP.) J. AG., *260*
  stiriata (TURN.) ARESCH., *260*
  teedii (ROTH) LAMOUR.:
    temperature tolerance, 326
  tenella HARV., *158*
Gigartinales, 193, 200
Gironde, *64*
Global, *see* World
Gloiopeltis:
  complanata (HARV.) YAMADA:
    photosynthesis, temperature optimum, 332
  furcata (POST. et RUPR.) J. AG., 155, *156*, 157, 158, 161
Glossopteris, 237
Gold Coast, 206, *207*
Gondwana flora, 237
Gondwanaland, 25, 237, *238–239*, 239, 242, 264, 266
Gorbovy Islands, 184
Gorgonia, *218, 222*
Gough Island, 252, 258
Gracilaria, 48, 193, 206, 256
  commercial use, 367
  coronopifolia J. AG., 325
    temperature optimum, growth, 328
    temperature tolerance, 325
  cultivation, 256
  ecotypes, optimum growth temperature, 329
  foliifera (FORSSK.) BOERG., 228
  greenish color, 296
  mammillaris (MONT.) HOWE, 133
  percentage increase per day, 364
  relative growth rate, 363
    salinity, growth rate, *336*
  salinity tolerance, 334
  secundata HARV., 342

nitrate storage, 342
tikvahiae MCLACHLAN, 48
  temperature optimum, growth, 328
  temperature tolerance, 325–326
verrucosa (HUDS.) PAPENF. (= *confervoides*), 47–48, 92, *194*, 263
  hairs, nutrient uptake, 343
Grateloupia, 148, 193
  elliptica HOLMES, *158*
  filicina (LAMOUR.) C. AG., *147*, *194*
  turuturu YAMADA, 302
    light saturation, photosynthesis, 302
Grazing, 78, 129, 142, 148, 153, 196, 200, 207, 214–216, *215*, 256, 347–353
  algal growth at night to escape, 214
  antigrazer substances, 351–352
  paleobiology, 200
  seaweeds surviving digestion, 348
  toxic metabolites of algae, 214
Great Australian Bight, 270
Great Barrier Reef, *see*, Coral reefs, Great Barrier Reef
Greater Antilles, 216
Great Ice Ages, *see* Ice, Great Ice Ages
Greece, 98–99
Greenland, 33, 39, 94, 166, 174–180, *178*, 187, 246
  temperature tolerance of seaweeds, 324
Greenland–Scotland Ridge, 33, 126
Grinnellia americana (C. AG.) HARV., 326
  temperature:
    growth, optimum, 328
    reproduction, 330
    tolerance, 326
Growth:
  in darkness, 171, 250
  light:
    quality, 316–317
    quantity, 298–301
  maintenance light requirement, 299–300
  rate, 152, 298–301
    diurnal changes, 304
    photon fluence rate, 298–300
  regulators, 343
  relative growth rate, 363–364
  specific growth rate, 300
  salinity, 335, *336*
  temperature optimum, 323, 328–329
Guadeloupe Island, 153, 233, 243
Guatemala, 79
Guayana Current, 203, 217
Gulf of Akaba, 227
Gulf of Alaska, 142, 186
Gulf of Califorma, 153

Gulf of Gabes, 99
Gulf of Guayaquil, *15*, 234, 252
Gulf of Guinea, 203, 207
Gulf of Lion, 106
Gulf of Maine, 126
Gulf of Mexico, 123, 133, 189, *195*
Gulf of Naples, 98, 112, 115
Gulf of St. Lawrence, 123, 128
Gulf of Trieste, 99
Gulf Stream, 63, 217, 224
Guluronic acid, 369, *369*
Gunnarea capensis, *260*, 262
Gymnocodiaceae, 200
Gymnodinium microadriaticum, 212
Gymnogongrus:
    antarcticus SKOTTSB., 249
    crenulatus (TURN.) J. AG. (= *norvegicus*), *105*, 112

Haemascharia hennedyi (HARV.) MAGGS et WILCE (= *Petrocelis hennedyi*), 71, 188
    endophytism, Codiolum phase of green alga, 355
Haidai (Chin.), 367
Hairs:
    nutrient uptake, 343–344
    two-dimensional growth, blue-UV effect, *315*, 316
Hadal, 4, *4*
Halarachnion ligulatum (WOODW.) KÜTZ., 67, 115
Halidrys, 243
    dioica GARDN., 56, 147, *147*, 148
    siliquosa (L.) LYNGB., 41, 56, 59, #*61*, 73
        casting of outer layers, 353
        gas-filled thallus parts, 345
        temperature tolerance, 323
Halifax, *130–131*
Halimeda, 198–199, 209, 213–214, 227
    discoidea DECAISNE, 154, 233
    incrassata (ELLIS) LAMOUR., 230
        turnover rate, 361
    light at depth limit, 285, 287
    opuntia (L.) LAMOUR., 209, 214, 219, *227*, 231
    photosynthetic rate, 364
    tuna (ELLIS et SOL.) LAMOUR., *104*, 231
Halimedatrial, 214, 352
Haliotis, 354
    morphogenetic substances of crustose coralline algae, 354
Haliptilon roseum (= *Corallina cuvieri*), 270
Halodule, 203, *204*, 205, *205*, 222
    wrightii, 206

Halogenated substances, 352
Halophila, *204–205*, 222
    decipiens, 196
    ovalis, 231, 270
    stipulacea, 104, 120, *120*
Halophytes, 82, 132
Halopteris:
    filicina (GRAT.) KÜTZ., 78, 86, 87
    scoparia, see Stypocaulon scoparium
Halosacciocolax, 40
Halosaccion:
    americanum I. K. LEE, 141, *144–145*, 148, 155
    desiccation, 338
    glandiforme (GMEL.) RUPR., 155, *156*
    ramentaceum, see Devaleraea ramentacea
Halurus equisetifolius (LIGHTF.) KUETZ., 328
    temperature:
        growth, optimum, 328
        reproduction, 330
Halymenia, 193
    actinophysa HOWE, 206, *207*
    agardhii DE TONI, *195*
    latifolia KÜTZ.:
        photoperiodism, 312
Hamada, 155
Haplospora globosa KJELLM., 29
Hardanger Fjord, 82
Harveyella mirabilis (REINSCH) REINKE, 355
Hawaii, 226–227, 231, 233
Heard Islands, *236*, 245, *249*, 250
Hedophyllum sessile (C. AG.) SETCH., 138, *140*, 141
    mechanical adaptation, 344
Helcion (= *Patina*) *pellucidus*, 78, 350
Helgoland, 61, *66–67*, 68–69, 71, 77, 78
    Biologische Anstalt Helgoland, 69
    daylength, seasonal course underwater, 285
    light:
        at algal depth limit, 284
        seasonal course underwater, 285
    temperature tolerance, 322
Helminthocladia calvadosii (LAMOUR. ex DUBY) SETCH., 79, 93
Hemiphanerophyceae, 314, 316
Herbivores, *see* Grazing
Hermatypic corals, *see* Coral reefs, hermatypic corals
Herpodiscus durvilleae (LINDAUER) SOUTH, 355
Herponema, 353
    epiphytism, 353
Hesperophycus harveyanus (DECAISNE) SETCH. et GARDN., 147, *147*, 243

Heterochordaria abietina, see *Analipus japonicus*
Heterosiphonia plumosa (ELLIS) BATT., 87
Heterozostera, *204*
  tasmanica, 270
High-latitude origin of taxa, 29–30, 240
Higher plants:
  dispersal to Central America, 196
  light saturation, photosynthesis, 302
Hildenbrandia:
  lecannellieri HARIOT, 249, 251
  rubra (SOMMERF.) MENEGH.
  (= *prototypus*), 70, 82–83, *127*, 129
Himanthalia elongata (L.) S. F. GRAY, 41, 56, 59–60, *72–74*, 87, 89, 143, 183, 242–243
  casting of outer layers, 353
  chlorophyll per ground unit, 362
  epiphytic alga, *Elachista*, 353
Himantothallus grandifolius A. D. ZINOVA, 246, *247*, 249–250
  light saturation, growth, 302
  temperature:
    growth, optimum, 328
    photosynthesis, optimum, 332
    tolerance, 326–327
Histrio histrio, 223, *223*
Hizikia fusiforme (HARV.) OKAM., 157, *158*, 243
  photosynthesis, temperature optimum, 332
Hluleka, 265
Hokkaido, 157, 159, 160–161, 186
Holothurians, *74–75*
Homeostasis, 193
Hongkong, 154–155, 157, *163*, 230
Honshu, 155, 157, *159*, 160
Hopkins Marine Station, 135
Hormones, 343
Hormophysa, 243, 267
  triquetra (C. AG.) KÜTZ., 227, 230
Hormosira banksii (TURN.) DEC., 242–243, 244, *269*, 270–271, 273
  desiccation, 338
  freezing tolerance, 324
Hormosirene, *319*
Hudson Bay, 166
Hudson Strait, 166, 187
Humboldt Current, see Peru (Humboldt) Current
Hungerfordia, 201
Hydrocharitaceae, *205*
Hydroclathrus clathratus (C. AG.) HOWE, 230
Hydroids, endozoic algae, *67*, 79, *271*, 355
Hydrolithon boergesenii (FOSL.) FOSL., *219*, 220
Hydrolittoral zone, 84

Hymenoclonium phase, *see Bonnemaisonia, asparagoides*
Hypnea, 206
  cervicornis J. AG.:
    temperature optimum, growth, 328
  musciformis (WULFEN) LAMOUR., *104*, *194*, 196, 219
    protein content, 365
  spicifera (SUHR) HARV., *263*
Hypnophyceae, 314
Hypoglossum:
  hypoglossoides (STACKH.) F. COLLINS et HERV. (= *woodwardii*), 78
  temperature:
    growth, optimum, 328
    reproduction, 330
  tenuifolium (HARV.) J. AG., 222

Ibiza, see Balearic Islands
Ice:
  Ages, see Paleoceanography, paleobiogeography
  Great Ice Ages, 23, *24*, 239
  drift ice, 126, 179, 181, 248
  limits of sea ice:
    on Northern Hemisphere, *24–25*, *165*, 181
    on Southern Hemisphere, 235–236, *236*, 240
  macroalgae under ice, 169, *169–170*, 171–172, *172*, 181, 187, 249–250
  pack ice, 180
  permanent ice, 164, *165*, 181, 187, 235–236, *236*, 240
  scouring, 126, 171–172, *172*
  shelf ice, 240
Icebergs, *165*, 179, 240
Ice-foot, 171, *172*
Iceland, 34, 39, 60, *64*, *68*, 166, 181
Iceland–Faeroe Ridge, 33
Iceland–Greenland Ridge, 33
Igloolik, *170*
Independence-Fjord, 177
India, 225, 228, 237
  commercial uses of seaweeds, 365
Indonesia, 225–226, 230
  commercial uses of seaweeds, 365
Indo-Pacific, see Africa; America; Asia; Australia; Tropics
Infralittoral zone, 9–10
Introduced species, 94–97, 117, 274
Iodine, commercial use, 365
Ireland, *15*, 60–62, *64*, 85, *88*
  commercial uses of seaweeds, 365

Iridaea, 143, 251
  biomass, 361
  capensis J. AG.. 260
  cordata (TURN.) BORY, 143
  cornucopiae POST. et RUPR. (= *Chondrus yendoi*), 155, *158–159*
  laminarioides BORY (= *boryana*), *254*, 255–256
  obovata KÜTZ., 248–249, *249*, 250
Irish Moss, *see Chondrus, crispus*
Isfjorden, 181
Island:
  hopping, 231
  theory, 232, 269
Isofloridoside, 368, *369*
Isopods, 273, 350
Isothermic submergence, 117
Isotherms, *see* Temperature, isotherms
Israel, 98, 115
Italy, 98
Izu Peninsula, *159*, 160

Janczewskia, 355
  morimotoi TOKIDA, 355
Jania, 193, 206, 219
  fastigiata HARV., 270
  photosynthetic rate, 364
  rubens (L.) LAMOUR., 112, 206, 230, 256
Jan Mayen, 180–181
Japan, 154, 156–157, *158–160*, 159–161, *162*, 186, 243, 264
  commercial uses of seaweeds, 159, *162*, 365, *366–368*, 367–368
Japan Sea, *see* Sea of Japan
Jerlov water type, *see* Light, Jerlov water type
Jetties, 132
Johnson-sea-linkia profunda EISEMAN et EARLE, 221, *221*
Jolyna laminarioides GUIMARÃES, 257
Jörgen Brönlunds Fjord, 174–175, 177
Juan de Fuca Strait, *138*
Juan Fernandez Islands, 252, 255
Juglans, 33
Julescraneia grandicornis, *137*

Kai River, 265
Kallymenia:
  antarctica HARIOT, *249*
  microphylla, see Meredithia microphylla
  reniformis (TURN.) J. AG., 78, 87
  rosacea, see Turnerella, pennyi
  schmitzii DE TONI, 188
Kamchatka, 135, 154–155, *156*, 157
Kanin Peninsula, 184

Kara Sea, 182–184
Katharina, 348
Kattegat, 83
Kenia, 225
Kerguélen Islands, *236*, *249*, 250–251, 253
Kermadec Islands, 265, 274
Kildine Island, 183
King Island, 271
King George Sound, 270
King George Island, 324
  temperature tolerance of seaweeds, 324
Kjellmaniella gyrata (KJELLM.) MIYABE, 156
Kodiak Island, *138*, 141–142
Kola Fjord, *15*, 166, 182
Kola Peninsula, 183
Kombu (Jap.), 367
Konyam Bay, 185
Korea, 154–157, 161, 186, 243
K-strategy, 357–358
Kurile Islands, 136, 154–155
Kuroshio Current, *63*, 135, 139, 154–155, 157, 161
Kyphosidae, 350

Labyrinthula, 356
  macrocystis, 76
Labrador, *124*, 126, 166, *172*, 185, 187
Labrador Current, *63*, 126, 128, 166, 188
Labrador Sea, 32, 126
Laccadive Islands, 229
Lacuna, 348
Laguncularia racemosa, 206
La Jolla, 135, 148
Lambia antarctica (SKOTTSB.) DELEPINE, 189
Laminaran, 358, 359, 370
Laminaria, 48–53, 129, 131, 136, *138*, 139, 140, *144–145*, 155, 157, 159–160, 171, 181, 222, 242, 257, 259
  abyssalis JOLY et C. DE OLIVEIRA FILHO, 222, 257
  alginate, 367, 368, *369*
  angustata KJELLM., 157, 159, *160*
  commercial use, *366–367*, 367
  var. *longissima* MIYABE, *see Laminaria longissima*
  biomass, 361
  bongardiana POST. et RUPR. (= *L. groenlandica*), 40, 49, *138*, *140*, 142–143, 155
  nitrate uptake, 342
  brasiliensis JOLY et C. DE OLIVEIRA FILHO, 49, 222, 257
  cichorioides MIYABE, 159

commercial use, 365, *366–367*
coriacea MIYABE, 159
dentigera KJELLM., 49–50, 136, *138–140*, 142, 155, *156*, 186
diabolica MIYABE, 159
digitata (HUDS.) LAMOUR., 7, 40–41, 49–50, #51, 52–53, *65–68*, 68, 73, *74–75*, 77, 78, 81, 84–85, 94, 127–129, *130–131*, 131, 136, 143, 178, 181–182, 184, 188, 356
  blade shedding, 356
  competition, 78, 350
  forma *nigripes*, 188
  fungal parasites, 356
  grazing, 350–351
  growth rate, 364
  light requirement, 285, 301
  temperature demands, gametophytes fertile, 327
  temperature optimum, growth, 328
  temperature tolerance, 323, 327
ephemera SETCH., 50, 136, *138*
erosion, distal blade end, 354, 360
farlowii SETCH., 136, 148, *149*, 153
groenlandica ROSENV., see *Laminaria, bongardiana*
hyperborea (GUNN.) FOSL., 7, 40–41, 44, 49, 52, #58, 59, *65–67*, 73, *74–75*, 76, 77, 78, 81, *81*, 83–86, *87*, 89, 94, 129, 136, 179, 181, 183–184, 186
  age, maximal, of individuals, 356
  blade, rhythm of growth and shedding, 356, *357*
  competition, 78, 350
  dark growth, *309*, 359
  epiphytic bryozoans, 354
  growth rate, 364
  K-strategy, 358
  life form type, 314
  light:
    at depth limit, 281, 284–286
    tolerance, 303
  mechanical adaptation, 344
  photoinhibition, 302
  photoperiodism, 308, *309*, 312–313, 343
  photosynthesis, enzymatic acclimation, 332
  pigment content, acclimation of, 296, 306
  respiration, seasonal rates, 332, *333*
  reserve materials, 358, *359*
  seasonal course, reserve materials, 358, *359*
  temperature demands, gametophytes fertile, 327
  temperature optimum, growth, 323

  temperature tolerance, 323, 327
  translocation, 358–359, *360*
japonica ARESCH., 117, 157, *158*, 159, *160*, 161, *162*
  commercial use, 365, *366–367*
  growth rate, 364
  light saturation, growth, 302
longicruris PYL., 41, 51–52, 82, 128–129, *130–131*, 131, 167, *170*, 180, 188, 327
  grazing, 351
  temperature demands, 327–328
longipedalis OKAMURA, 159
longipes BORY, 50, *138*, 139, *140*, 141, 143
longissima MIYABE (= *angustata* var. *longissima*), 157, 159, *162*
  growth rate, 364
maintenace light requirement, 300
ochotensis MIYABE, 51, 159
ochroleuca PYL., 49, 52, 86, 89–92, 117, *118–119*, 205, 241, *247*, 256, 261
  crossing the equator, 86
pallida GREV., 49, 86, 257–258, *260*, 261, *261*, 262
  decomposition rate, 370
  temperature demands, gametophytes fertile, 327
  temperature tolerance, gametophytes, 327
photosynthesis:
  blue-light effect, 316
  rate, 364
  versus-light curve, *297*
primary productivity, 362
protein content, 365
religiosa MIYABE, 159, 161
rodriguezii BORNET, 50, 53, 115, *116*, 117, *118–119*, 139, 143
  light at depth limit, 284
saccharina (L.) LAMOUR., 41–42, 44, 46, #46, 50–53, *66–68*, *72–73*, *74–75*, 77, 78, 81, *81*, 84–86, 89, 94, 128–129, 131, 136, *138*, *140*, 142, 155–156, *156*, 159, 167, 177, *178*, 182–184
  absorption, action spectrum, photosynthesis, *291*
  ecotypes:
    light, 392
    temperature tolerance, 327
  forma *faeroensis*, *74–75*, 81
  gametophytes, fertilization, blue-UV effect, *315*, 316–317
  growth rate, 364
  mechanical adapation, 344–345
  nitrate demands, 342
  photoperiodism, 312, 343

Laminaria (Continued)
  photosynthesis:
    effectiveness *vs.* water depth, 294
    enzymatic acclimation, 332
    temperature demands, gametophytes fertile, 327
    temperature optimum, growth, 328
    temperature tolerance, 322–323, 327
    wavy margin, enhanced turbulence, 344
  sachalinensis MIYABE, 159
  section:
    Digitatae, 49–51, 159
    Fasciatae, 50–51, 159
    Simplices, 50–51, 159
  schinzii FOSLIE, 257, 261
  setchellii SILVA, 40, 44, 49, 136, *138–140*, 142, *146*, 153, 155
  sinclairii (HARV.) FARLOW, ANDERSON et EATON, 50, *138*, 139, 143
  solidungula J. AG., 30, 39, 50, 127, 136, *170*, 171, 173, 174–175, #*175*, 178–181, 184, 187–188
    dark growth, 359
    importance for food chain, 369
    light exposure, annual, 281
    primary productivity, 362
    temperature tolerance, 326–327
  subgenus:
    Rhizomaria, 50, 136, *138*
    Solearia, 50, 136, *138*, 159
  synchronization of fertilization, 318
  turnover rate, 361
  yendoana MIYABE, 159
  yezoensis MIYABE, 49, 136, *138*, *140*, 142, 155, 159, *160*
Laminariaceae, *138*, 151, 156, 159–160, 242, 257
Laminariales, 62, 90, 98–99, *118–119*, 134, 136, *137–139*, 139, *140–141*, 143, *144*, *146*, 148, *149–151*, 152–155, 159–161, *159–160*, 162–163, 242, *242*, 246, *247*, 250–253, *254*, 257–259, *260–261*, 261–263, 267, *269*, 270, 272–274, *273*, 359
  age of individuals, 148, 152, 172, 356–357
  ancestral taxa, 160–161
  chloroplast movements, 317
  competition, 350
  dark growth, 359, *360*
  evolution, 23, 29, 33–34, 135–136
  fossil remains, 136, *137*
  gas-filled thallus parts, 345

giant kelps, *6*, 7, 32, 136, *137–138*, *141*, *144*, 145, *146*, 148, *148*, *149–151*, 151–153
holdfast:
  giant kelp, 152
  Phototropismus, 237
reserve materials, 358–359, *359*
rhizomatous species, 50, 136, 160–161
Southern Hemisphere, 242, *242*, 250–251, *254*, 257–258, *260–261*, 261–262, 267, *269*, 272–274, *273*
sporangial sori, 143–144
translocation, 358–359, *360*
temperature tolerance, gametophytes, 327
Lamoxirene, 257, 318, *319*
Land bridge:
  Australia/Indonesia, in Pleistocene, 242
  of Beringia, *see* Bering land bridge
  of Central America (Panama), 11, 34, 38, *191*, 193, 196, 208, 211, 234
  of Gibraltar, *100–101*, 102–103
  of Suez, 103–104, *191*, 193, 208
  Thulean, 33
Landsburgia, 243, 272
La Paz, *15*
Laptev Sea, 182
Laurasia, 23, *238*
Laurencia, *65*, 193, 206, 219, 273
  antigrazer substances, 352
  algal parasite, *Janczewskia*, 355
  caspica A. ZIN. et ZABERZH., 123
  ceylanica J. AG., 229
  nipponica YAMADA:
    algal parasite *Janczewskia morimotoi*, 355
  obtusa (HUDS.) LAMOUR., *112*, 231
  papillosa (FORSSK.) GREV., *196*, *219*
  photosynthetic rate, 364
  pinnatifida (HUDS.) LAMOUR., 71, *88*, 93, 111, 148
  undulata YAMADA, 108
Leaf area index, 362
Lebanon, 111
Lecanora, *69*
Lena River, 183
Leptonematella fasciculata (REINKE) SILVA, 177
Leptophytum:
  foecundum (KJELLM.) ADEY, 178
  laeve (STRÖMF.) ADEY, 131, 178
Leptoseris fragilis, 213
Leptosomia simplex, *see Palmaria, decipiens*
Lessepsian immigrants, *see* Mediterranean, Lessepsian immigrants
Lesser Antilles, 216–217, 224

Lessonia, 136, 242, 251–253, 256, 267
  corrugata LUCAS, 253, 267, 271
  flavicans BORY (= *fuscescens*), 250–251, 253, *254*, 256
  nigrescens BORY, 253, *254*, 255–256, 274
    mechanical adaptation, 345
  trabeculata VILLOUTA et SANTELICES, 253, *254*, 256
  vadosa SEARLES, 253, 256
  variegata J. AG., 251, 253, 273, *273*, 274
Lessoniaceae, *138*, 143, 145, 242, 257
  competition, 7, 350
Lessoniopsis littoralis (FARL. et SETCH.) REINKE, *138*, 143, *144–145*, 147, 253
Lethal limits, 13
Lewis system of zonation, 8, *9*
Liagora, 91, 193, 200
  farinosa LAMOUR., 194
  viscida (FORSSK.) C. AG., *104*
Liberia, 203
Libya, 98
Lichens, 5, 62, 69, *69*, 88, 90, 206
Lichina pygmaea, 90
Lietnaia, 184
Life form types, 6–7, 7–8, 171, 314, 316, 364
  photosynthetic rates, 364
  physiognomic, 364
  seasonal, 314, 316
Light:
  adaptation, low-light habitats, 286, 298
  absorption, 278, 280, *289*, *291–292*, *294*
  attenuation, 278
  chromatic adaptation, *see* Photosynthesis, chromatic adaption
  coastal water, 278, *279–280*, *282*, 294
  compensation depth (point), 3, 286, 297, *297*
  as environmental signal, *see* Photomorphogenesis; Photoperiodism
  euphotic zone, 3, *283*
  exposure, 276, 285
  extinction depth, 3, *6–7*, *282–283*, 284–287
  gelbstoff (gilvin), 280
  geographical variation, *282–284*, 286
  green window, *see* Photsynthesis, green window
  Jerlov water type, 99, 110, 115, 278, *279–280*, 281–282, *282*, 284, 294
  kelp forest, 151
  maintenance levels, 298
  measurement, 276
  Mediterranean, 99, 110
  oceanic water, 278, *279–280*, *282*, 295
  optical water type, *see* Light, Jerlov water type
  percentage depth, *282*, 284–287
  photic zone, *see* Light, euphotic zone
  photon fluence rate, 110, 113, 171, 221, 276, 281
    corals, inside, 213
    at lower algal depth limit, 221, 281, 286–287
    tropics, 281
  photosynthetic active radiation (PAR, PhAR), 281, *282–284*, 286
  protection, 303–305
  quality and vertical algal distribution, 233
  quantity, *see* Light, photon fluence rate
  reflected portion 278
  scattering, 278
  seasonal course, 285
  Secchi depth, 99, 184
  solar constant, 281
  spectral distribution, 278, *280*, 280–281
  transmittance, 278, *279*
  ultraviolet, *279–280*, 304–305
    ozone, 305
    protection, 304–305
    UV-B, 305
    UV-C, 305
  West Africa, 205, 207–208
  yellow substance, *see* Light, gelbstoff (*gilvin*)
Ligia italica, 108
Limnetic realm, *see* Freshwater, realm
Limpets, *see* Gastropods
Linan Current, 161
Linoleic acid, 370
Lipophilic substances, 365
Lithophyllum, 125, 200, 213, 248
  byssoides (LAMK.) FOSL., 108, 111
  dentatum (KUETZ.) FOSLIE, 79
  growth rates, 364
  imitans FOSL., *149*
  incrustans PHIL., 79, 111
  intermedium FOSL., *219*
  lichenoides PHILIPPI (= *tortuosum*), 88–89, 108, *109–110*, 111
  solutum, *see* Lithothamnium corallioides
  tortuosum, *see* Lithophyllum, lichenoides
Lithoporella, 49
Lithothamnion, 145, 200, 233, 248
  calcareum, *see* Phymatolithon calcareum
  commercial use, 367
  corallioides CROUAN frat., 79–80, 121
  glaciale KJELLM., 79, *127*, 131, *179*, 184, 200
    temperature optimum, growth, 328
  philippii FOSL., 115
  sonderi HAUCK, 79

Lithophyllum, sonderi HAUCK (*Continued*)
    annual light exposure, 281
Litosiphon groenlandicus S. LUND, 177
Littoral, 8, 108
Littoral zone, *see* Eulittoral
Littoral fringe, 8, *9*
Littorina, 65, 143, *263*
    knysnaensis, *263*
    littorea, 70, *71*, 129, 329, 348, 352
    mariae, 70
    neritioides (= *Melaraphe neritioides*), 91, 108–109
    obtusata, 70, 129
    peruviana, 255
    punctata, 206
    rudis, 348
    saxatilis, 69, 329
Living fossils, 193, 201–202, 257
Lobophora variegata (LAMOUR.)
    WOM RSLEY (= *Pocockiella variegata*), 220–222, 230, 272
Logy Bay Marine Laboratory, 124
Loire, *64*, 126
Lomentaria:
    articulata (HUDS.) LYNGB., 71, 112, *113*
        temperature:
            growth, optimum, 328
            reproduction, 329–330
    baileyana (HARV.) FARL., 125, 128
        temperature:
            growth, optimum, 328
            reproduction, 330
            tolerance, 326
    clavellosa (TURN.) GAILL., *67*
    hakodatensis YENDO, 153
    orcadensis (HARV.) COLL. ex TAYLOR, *67*
Long Island Sound, 131
Lord Howe Island, 265, 272
Loose-lying algae, 187
Los Angeles Basin, 34, 135
Lower algal depth limits, *see* Deep-water algae
Louisiana, 123, 133
Lunar periodicity, 318–320, *320*
Lusitania province, 65, 85–90
Lusitanic algae, 86, *87–88*, 89–90
Lyngbia, 219
Lytechinus anamesus, 153

Maas, 82
Macaronesian Islands, 91–92
    endemism, 92
    paleobiogeography, 91–92
Mackenzie King Island, 187
Macquarie Islands, *236*, *249*, 250–251

Macroalgae, 3
Macrocystis, 7, 136, *149*, 153, 242, 251–253, 256, 267
    angustifolia BORY, 151, #*242*, 259, 261, 267, *269*, 271
    erosion of thallus, 360
    growth rate, 364
    integrifolia BORY, #*138*, 145, *146*, 151, #*242*, 253, 261
    age, maximal, of individuals, 356
    decomposition rate, 370
    growth rate, 364
    nitrate storage, 342
    nitrogen input, 342
    primary productivity, 362
    laevis C. H. HAY, 251
    photosynthetic rate, 364
    pyrifera (L.) C. AG., 136, *137*, #*138*, 145, *146*, 148, *148*, *149–150*, 151–153, 159, 237, 241, #*242*, *247*, 250–251, 253, 256, 258, 261, 267, 271–274
    age, maximal, of individuals, 356
    biomass, 361
    freezing tolerance, 324
    grazing, 351
    leaf area index, 362
    light saturation, photosynthesis, 302
    mechanical adaptation, 344
    percentage increase per day, 364
    photoinhibition, acclimation, 302
    nitrate uptake and demands, 342, 344
    primary productivity, 362
    short-distance dispersal, 93
    spiny outgrowths, enhanced turbulence, 344
    turnover rate, 361
    water motion, 344
Macrodetrivores, 370
Macrophytobenthos, 3
Macroporella, 198
Madagascar, 226
Madeira, 90–92
Madracis mirabilis, *218–219*, 219
Maerl, 79–80, *80*, 89, 121
    commercial use, 80, 367
Magdalena Bay, *15*, 153
Maine, 123, 131
Majorca, 117
Malaya, 226
Malay Archipelago, 230, 232
Malaysia, 225
Maledive Islands, 225, 229
Malmö, 93
Mammals:
    marine:

Cenozoic biogeography, 135
terrestrial:
　Cenozoic biogeography:
　　Africa/Eurasia, *100–101*
　　Eurasia, 38
Manchuria, 186
Mangrove, 206, 208–209, 222, 226, 228
　seaweeds, 209
Mannitol, 358, 359, 368, *369*
Mannuronic acid, 369, *369*
Manukau, *15*
Marginariella, 243, 272
Marine biological stations, 69, 124–125, 134–135
Mariane Islands, 225
Marion Islands, 250–251
Maripelta rotata (DAWS.) DAWS., *147*, 148
　dark growth, 359
Maritime zone, *9*, 69
Marshall Islands, 210, 225, 229, 232–233
Martensia pavonia (J. AG.) J. AG., *195*
Maryland, 123, 132–133
Massachusetts, 123, 131
Mastocarpus, 45
　commercial use, 367
　jardinii (J. AG.) WEST in GUIRY, WEST, KIM et MASUDA, 45
　pacificus (KJELLM.) PERESTENKO, 45
　papillatus (C. AG.) KÜTZ., 45, 142–143, *144–145*
　　Petrocelis phase, 143, *144–145*
　stellatus (STACKH. in WITH.) GUIRY in GUIRY, WEST, KIM et MASUDA, 45, *66*, 70, 71, 76, 89, *127*
　　freezing tolerance, 323
　　light saturation, photosynthesis, 302
　　photoperiodism, 312
Matotschkin Strait, 184
Mauritania, 85, *88*, 90, 206, 208
Mauritius, 225–227
Meandrina meandrites, *219*
Mechanical stress, 344
Mediterranean:
　Atlantic floral element, 105, 112
　biocenoses, 107–108, 109, 110–113
　Coralligène, *see* Coralligène
　cosmopolitan floral element, 105
　distribution groups, 102–106
　diversity of species, 102, 192, 226
　endemics, 106
　glacial relics, 105, 112
　history of seaweed investigations, 99
　Indo-Pacific floral element, 103
　introduced species, 117

Laminariales, 99, 117, 118–119
Lessepsian immigrants, 103–104
light, 99, 110
marine biological stations, 99
neoendemics, 106
Messinian event, 100–101, 102–103, 122
Messinian species, 105
paleobiogeography, paleooceanography, 98, 100–101, 102–106, 122–123
paleoendemics, 106
Paratethys, *see* Paratethys Sea
photophilous algae, 111
province, 65
salinity, 99
sciaphilous algae, 111
sea caves, 119–120
sea grasses, 120–121, *120*
submarine banks, 114–115, 116–117
temperature, 99–100
Tethys Sea, *see* Tethys Sea
tidal range, 99
Melanesia, 225–226, 230
Melaraphe neritioides, *see* *Littorina, neritiodes*
Melville Island, 187
Membranipora membranacea, 77, 78
　preferential settling, 354
Membranoptera, 54
　alata (HUDS.) STACKH., 41, 53, 54, *#54*, *67*, 76, 180
　　temperature optimum, growth, 323
　　temperature tolerance, 323
Meredithia microphylla (J. AG.) J. AG. (= *Kallymenia microphylla*), 87
　photoperiodism, 312
Mesohaline environment, 82
Mesopelagic zone, 3, *4*
Mesophyllum, 200, 233, 256
　lichenoides (ELLIS) LEMOINE, 117
Mesotaenium, chloroplast movements, 317
Messina, *see* Strait of Messina
Messinian event, *100–101*, 102–103
Messinian species, 105
Metasequoia, 33
Mexico, 153, 233
Microdictyon, 221
　loroxanthin, 290
　setchellianum HOWE, 233
Micronesia, 225–226, 230, 232
Micronutrients, 343
Microphytobenthos, 3
Microspathodon frontatus, 207
Microthalli, 93
Mid-Atlantic Ridge, 208
Midcontinental Sea, *see* Cannonball Sea

Migration, *see* Dispersal
Mikrosyphar, 355
Millepora, 218–219
  alcicornis, 214
Mineral ash, 365
Mississippi, 133
Modiolus modiolus, 74–75
Molluscs, 117, 125, 161, 226, 240. See also Gastropods; Mussels
Monacanthidae, 216
Monosiphon caspicus (HENCK.) VOLK., 123
Monostroma, 181, 248
  desiccation, 338
  grevillei (THUR.) WITTR., 84
    photoperiodism, 308, *310*, 312
  groenlandicum J. AG., 179
  hariotii GAIN, 249
  life form type, 314
  maintenace light requirement, 299
  undulatum WITTR.:
    photoperiodism, 312
    temperature tolerance, 323
Monotypic, 110, 264
Monsoon, *see* Tropics, monsoon
Montastrea, *218*, 219
  annularis, 219
  cavernosa, 219
Monterey Peninsula, 135, 142, *146*, 152
Moorea Island, 230
Morocco, *64*, 85–86, *88*, 89–91, 205
Morphogenetic substances, 354
Mossâmedes, *15*, 203, 258
Mougeotia, chloroplast movements, 317
Mounds, 210
Mozambique, 225
Mucilage, 368, 370
Mud, *see* Soft substrata
Multifidene, *319*
Murmansk, 166, 182–184
Mussels, 29, 70, *71*, 74–75, 125, 263, 348–349, *349*
Mya, 35, *36–37*
Myagropsis, 243
Mycetophyllia lamarckana, *219*
Mycophycobiosis, 356
Mycosphaerella ascophylli, 356
Myriodesma, 243, *271*
  harveyanum NIZAMUDDIN et WOMERSLEY, 271
  quercifolium (BORY) J. AG., 270
Myriogramme mangini, *see Neuroglossum, ligulatum*
Myriotrichia clavaeformis HARV., 312
  photoperiodism, long-day plant, 309, 312

Mytilus:
  californianus, 142–143, *349*
  crenatus, 260
  edulis, 70, 348
  galloprovincialis, 111
  perna, see Perna perna

Namibia, 259, 261
Naples, marine biological station, 99
  temperature tolerance of seaweeds, 324
Natal, *15*, 225, 259, 264–265
Navigation, seaweeds as means of, 152
Nekton, 3
Nemaliales, 193, 200
Nemalion, 273
  helminthoides (VELL. in WITH.) BATT., 108, 159
  vermiculare SUR., *158*, 159
Nematodes, 370
Neoagardhiella baileyi, see Sarcodiotheca gaudichaudii
Neoaustral taxa, 239
Neodilsea integra (KJELLM.) ZINOVA, *see Dilsea, integra*
Neodilsea yendoana TOKIDA, 45
Neogoniolithon, 200, 203, 213
  notarisii (DUFOUR) SETCH. et MASON, 108
Neomeris, 193, 197, *#197*
  annulata DICKIE, *195*, *#197*, 230
  bilimbata KOSTER, *#197*
  cokeri HOWE, *#197*
  dumetosa LAMOUR., *#197*
  mucosa HOWE, *#197*
  stipitata HOWE, *#197*
  van bosseae HOWE, *#197*
Neorhodomela larix (URN.) MASUDA (= *Rhodomela larix*), 185, 364
  growth rate, 364
Neptunea, 35, *36*
Nereocystis luetkeana (MERT.) POST. et RUPR., 81, 136, *137–138*, *140–141*, 142–143, *144–145*, 145, *146*, 147, 151–152
  age, maximal, of individuals, 356
  current velocities, 345
  decomposition rate, 370
  epiphytic algae, 353
  epiphytic bryozoans, 354
  gas-filled thallus parts, carbon monoxide, 345–346
  growth rate, 364
  mechanical adaptation, 344–345
Netherlands, 62, 68, 71
Netherlands Siboga Expedition, 230

INDEX 515

Neurocaulon foliosum (MENEGH.) ZAN.
  (= *reniforme*), 116
Neuroglossum:
  binderianum KÜTZ., 260
  ligulatum (REINSCH) SKOTTSB.(=
    Myriogramme mangini), 248, *249*
New Amsterdam Island, 261
New Brunswick, 123
Newfoundland, *15*, 123–124, *124*, 126, *127*,
  128, *179*, 188
  Grand Banks, 94, 127–128
  temperature conditions, 127
New Guinea, 226, 231
New Hampshire, 121, 123
New Jersey, 123, 132
New South Wales, 265, 270–272
  grazing, 348
New York, 123
New Zealand, 235, 237, 239–240, 244, *247*,
  250–251, 261, 264–267, 272–274, *273*
  desiccation, 338
  freezing tolerance of seaweeds, 324
  paleoceanography, paleobiogeography, 237,
    238–239, 239
Niger River, 206
Nitophyllum bonnemaisoni GREV., *87*
Nitrate, 342–343
Nitrogen:
  atmospheric, fixation of, 212
  translocation in kelp, 358
Nizymenia australis SONDER, 268
Nodules, 79
Nori (Jap.), 367, *368*
Normandy, potash industry, historical, 365
North America, *see* America
North Atlantic, *see* Atlantic
North Atlantic Current (Drift), 62, *63*, 166, 181,
  183, 224
North Cape Current, 183–184
North Carolina, 123, 133, 217
Northeast Passage, 164
North Equatorial Current, *63*, 197, 203, 224,
  264
North Pacific, *see* America, Asia
North Pacific Current, 139, 157
North Sea, 39, 60–61, 64, 68–69, 94, *100–101*,
  126
Northwestern Africa, *see* Africa
Northwest Passage, 164
Norway, 61, 64, *72–73*, *81*, 94, 166, 182
  potash industry, historical, 365
Norwegian–Greenland Sea, 32–33, 40
Notheia, 243–245
  anomala HARV. et BAILEY, 244

Notheiales, 243–244
Nothofagus, 237
Nova Scotia, 123, 129, *130–131*
  grazing, 348
Novaya Zemlya, 165, 171, 182–184, 246
Nucella (= *Thais*) *lapillus*, 348
Number of seaweed species, *see* Diversity
Nunivak Island, *15*, 134, 166, 185
Nutrients, 212–213, 216, 225, 341–343, 361,
  366–367

Obik Sea, 28, 31
Ocean currents, *63*
Oceanic Islands, 232
Ocean Point, 29
Odobenids, 351
Odonthalia, 54
  algal parasite, *Harveyella mirabilis*, 355
  dentata (L.) LYNGB., 41, 54, #*55*, 181,
    183, 185
Oerstedtia scalaris (SUHR) JENSEN (=
  Carpophyllum scalare), 243, 264
Okha, 228
Okhotsk Sea, *see* Sea of Okhotsk
Oligohaline environment, 82
Oman, 228, 264
Omphalophyllum ulvaceum ROSENV., 125,
  176, 180, 187
Opportunists, 357–358
Opuntiella californica (FARL.) KYL., 145
Oregon, 134, 139, 142–143, 252
Organic acids, 365
Organic carbon, 370
Organic dry weight, 365
Orinoco River, 221
Oscillatoria redekei, respiration rate, low light,
  298
Oslofjord, 82
Osmotic regulation, *see* Salinity, osmotic
  regulation
Ostracods, 32, 105
Ostrea, 230
Ostreobium, 210, 213, 221
Otago Peninsula, *273*
Oyashio Current, 154, 156, 161
Oyster thief, 96
Ozone and ultraviolet radiation, 305
Ozophora, 57

Pachydictyon coriaceum (HOLMES) OKAM.,
  53
Pachytheca, 201
Pacific, *see* America; Asia; Tropics,
  paleoceanography

516 INDEX

Pack ice, *see* Ice, pack ice
Padina, 91, 154, 193, 195, 201, 206, 214, *227*
  australis HAUCK (= gymnospora), *194*, 272
  commersonii BORY, 230
  gymnospora (KÜTZ.) SONDER, 231, 256
  pavonica (L.) LAMOUR., 86, 91, 112, *227*
  sanctae-crucis BOERG., 222
    turnover rate, 361
  tenuis, 230
Pakistan, 225
Paleoaustral species, 239
Paleoceanography, paleobiogeography, *see*
    Africa; America; Antarctic; Arctic; Asia;
    Australia; Europe; Mediterranean; Tropics
  land bridges, *see* Land bridges
  sea levels, *see* Sea levels, paleoceanographical
  temperatures, *see* Temperature,
    paleotemperatures
Paleocystophora, 201
Paleohalidrys, 201
Paleosiphonia, 200
Palmaria:
  *decipien*s (REINSCH) RICKER (=
    Leptosomia simplex), 248, *249*, 250
    photosynthesis, temperature optimum, 332
    greenish color, 296
  mollis (SETCH. et GARDN.) VAN DER
    MEER et BIRD, 41, 47, 144–145
  palmata (L.) O. KUNTZE (= Rhodymenia
    palmata), 41–42, 46–47, #49, 68, 71,
    74–75, 76, 144, *174*, 181–182, 184
    protein content, 365
Palmariales, 174
Palmer Arch, 240
Palmoclathrus stipitatus WOMERSLEY, 116,
  268, 270, 274
Palmophyllum, 270
  crassum (NACCARI) RABENHORST, *114*,
    115–116, *116*, 270, 274
  umbracola NELSON et RYAN, 274
Panama, *196*, 233–234
Panama Canal, 196
Pangaea, 23, 25, *238*
Panomya, 35, *37*
Panthalassa, 23
Pantoneura:
  baerii (POST. et RUPR.) KYL. #*133*, 135,
    141, 173, *174*, 175, #*177*, 184
  juergensii (J. AG.) KYL., 134, 173
Pantropical species, *see* Tropics, pantropical
  species
Papenfussiella callitricha (ROSENV.) KYL.,
  125, 180
Papua New Guinea, 225

Paracentrotus lividus, 111
Parasitic plants:
  algae, 355–356
  fungi, 356
Paratethys Sea, 100–101, 122–123
Parry Expedition, 164
Particulate organic carbon, 370
Patagonia, 244, 256
Patella, *65*, 108, *263*, 348
  aspera, 108
  cochlear, *260*, *263*
  granularis, *260*, 263
  safiana, 206
Patelloidea latistrigata, 348
Patina, *see* Helcion
Patos Lagoon, 257
P/B ratio, 358
Pedobesia, 199
Pelagic Realm, 3, *4*
Pelagophycus porra (LEMAN) SETCH., 136,
    137, *148*, *150*, 151–153
  age, maximal, of individuals, 356
  gas-filled thallus parts, carbon monoxide,
    345–346
  growth rate, 364
Pelagos, 3, *4*
Peleponnesos, 120
Pelvetia, 56, 243
  canaliculata (L.) DECAISNE et THURET,
    41, 59, #*60*, *65–66*, 68, *72–73*, 89, 93,
    183
    competition, 347–348, 350
    desiccation, 337–338, *339–340*, 341
    endophytic fungus, *Mycosphaerella*, 356
  fastigiata (J. AG.) DE TONI, 56, 143, 148
  siliquosa TSENG et CHANG, 161
  wrightii OKAM., 155, 157
Pelvetiopsis, 243
Penicillus, 199, 213
  antigrazer substances, 352
  capitatus LAMARCK, 121, *121*
    growth rate, 364
    turnover rate, 361
Percentage depth, *see* Light, percentage depth
Percentage increase per day, 363–364
Perciformes, 215
Perennials, 314, 356
Periwinkles, 70, 91, 348, 352
Permian Great Ice Age, 237
Perna perna (= *Mytilus perna*), 206, *263*
Persian (= Arabian) Gulf, 208, 225, 227
Peru, 252
Peru (Humboldt) Current, *63*, 232, 234, 237,
  252–253

Petalonia, 68
  fascia (O. F. MÜLL.) O. KUNTZE, 71, 94, 133, 187, 241, 257
    photoperiodism, 312
    temperature optimum, growth, 323
    temperature tolerance, 323
  zosterifolia (REINKE) O. KUNTZE:
    photoperiodism, 312
Petrocelis:
  hennedyi, see *Haemascharia hennedyi*
  middendorfii, see *Mastocarpus, papillatus*
  photosynthetic rate, 364
Petroglossum nicaeensis, see *Schottera nicaeense*
Peyssonnelia, 98, 221
  capensis MONTAGNE, 264
  photosynthetic rate, 364
  squamaria (GMEL.) DECAISNE, 113, *113*
Peyssonneliaceae, 200
Phacelocarpus labillardieri (MERT.) J. AG., 268
Phaeoglossum, see *Himanthothallus grandifolius*
Phaeosiphoniella cryophila HOOPER, HENRY et KUHLENKAMP, 29
Phaeurus antarcticus SKOTTSB., 246, *247*, 250
  temperature optimum, growth, 328
  temperature tolerance, gametophytes, 327
Phanerophyceae, 314
Phenolic substances, 352, 365
Pheromones, 318–319, *319*
Philippines, 225–226
Phospherus, 343
Photoinhibition, 152, 302–304
Photomorphogenesis, 308, *315*, 316–317
  blue/UV effects, 307, 314, *315*, 316–317
  sensor pigments, 317
Photon fluence rate, *see* Light, photon fluence rate
Photoperiodism, 308–314
  critical daylength, *311*, 232, 234, *234*
  ecotypes, 311–313
  long-day plants, 308–309
  night-break regime, 312–314
  seasonal window, 311, *313*
  sensor pigments, 314
  short-day plants, 308, *309–311*, 313
Photophilous algae, 10, 111
Photorespiration, 297
Photosynthesis:
  action spectrum, 287, 291, *291*, 293–295
  afternoon depression, 303–304
  black algae, 289
  chromatic adaptation, 295–296
  critical water content, 338, *340*
  Emerson enhancement effect, 293–294, *294*
  Engelmann's hypothesis, 295–296
  enzymes, 298
  green window, 289
  irradiance adaptation, 295–296
  light saturation, 301–302, 331
  net rates, 363–364
  photodamage, 304
  photoinhibition, 152, 302–304
    acclimation, 302
    action spectrum, 303
    recovery, 303
  pigments, 287, *288–289*, 289–296, *291–292*, 294
    absorption spectrum, *289, 291-292, 294*
    acclimation, 296, 305–307
    carotenes, carotenoids, *288*, 289–291, 304, 314
    fucoxanthin, 287, *288–289*
    loroxanthin, 290
    siphonaxanthin, siphonein, *288*, 289–290
    chlorophylls, 287, *288*, 289, 291–292
    no chlorophyll formation in darkness, red algae, 359
    per ground unit, 362
    light quality effects, 307, 314
    phycobilins, phycobiliproteins, 287, *289*, 292–293, 296
    phycocyanins, 293
    phycoerythrins, 293
    regulation, *see* Photosynthesis, pigments, acclimation
  phycobilisomes, 292
  rates, 363–364
  temperature, 331–332, *331*
  ultraviolet:
    impact, 290, 292
    protection, 304–305
    tolerance, 304–305
  vs. light curves, 296–298, *297*
  water motion, 344
Phototropism, 317
pH range, seawater, 343
Phycobilins, *see* Photosynthesis, pigments
Phycobilisomes, 292
Phycochromes, 296
Phycocolloids, 368–369, *369*
  world production, 367
Phycocyanins, 293
Phycodrys, 54
  riggii GARDN. (= *serratiloba* = *fimbriata*), 155, *156*
  rubens (L.) BATT. (= *sinuosa*), 41–42, 53–54, #*55*, *74–75*, 76, 78, *87*, 94, 131, 167, 175, 178–179, 181–185

Phycodrys (Continued)
  temperature tolerance, 323
Phycoerythrins, 293
Phycomelaina laminariae, 356
Phyllaria, see *Phyllariopsis*
Phyllariaceae, 29
Phyllariopsis, 257
  brevipes (C. AG.) HENRY et SOUTH
    (= *Phyllaria reniformis*), 29, 90, 117
    118–119
    photoperiodism, 312
  purpurascens (C. AG.) HENRY et SOUTH
    (= *Phyllaria purpurascens*), 29, 90,
    117, *118*
Phyllogigas, see *Himanthothallus grandifolius*
Phylloid green algae, 210
Phyllophora, 56–57, 243
  antarctica A. et E. S. GEPP, 249
  crispa (HUDS.) DIXON (= *rubens*), 78, 106
  nervosa (DECAISNE) GREV., 106, 112,
    113, 122
  pseudoceranoides (S. G. GMEL.) NEWR. et
    A. R. A. TAYLOR (= *membranifolia*),
    41–42, 57, *#58*, *67*, 128, 188
    temperature optimum, growth, 323
    temperature tolerance, 322–323
  traillii HOLM. ex BATT., 67
  truncata (PALL.) ZINOVA (= *brodiaei*),
    41–42, 47, *#53*, 56, 122, 128, 131,
    142, 171, 177–179, 184–185, 187
    algal parasite, *Ceratocolax*, 355
    forma *interrupta*, 188
    grazing, 351
    nutrient uptake, 343
    temperature tolerance, 323
    turnover rate, 361
Phyllospadix, 204
  epiphytism, 353
  scouleri, 145
  torreyi, 145
Phyllospora:
  comosa (LABILL.) C. AG., *269*, 271–272
Phylogenetic biogeography, 239
Phymatolithon, 184, 200
  calcareum (PALL.) ADEY et MCKIBBIN,
    79–80, *80*, 121
    commercial use, 367
  laevigatum (FOSL.) FOSL., 71
  lenormandii (ARESCH. in J. AG.) ADEY,
    70–71
  purpureum (P. et H. CROUAN)
    WOELKERLING et L.IRVINE (=
    polymorphum), 71
  rugulosum ADEY, 79

Phytoplankton:
  light saturation, photosynthesis, 302
  photoinhibition, 304
  primary productivity, 362
Phytosociology, 106
Pigments, see Photosynthesis, pigments
Pilayella:
  littoralis (L.) KJELLM., 84, 177–178, 181,
    184, 225
    salinity ecotypes, 335
    turnover rate, 360
    opportunists, 357
Pisaster ochraceus, 348–349, *349*
Plankton, 3
Plantago maritima, 69
Platforms, 94
Plate tectonics, see Paleoceanography,
  paleobiogeography
Platythalia, 243
  angustifolia SONDER, 271
Plectonema, 213
Pleurophycus gardneri SETCH. et SAUND.,
  #138, 142
  blade shedding, 356
  nitrate storage, 342
Plexaura, 219, *219*
Plocamium, 270
  cartilagineum (L.) DIXON (= *coccineum*),
    67, 77, 105, *105*, 112, 139, *147*
    desiccation, 337
    relative growth rate, 363
  corallorhiza (TURN.) HARV. in HOOK. et
    HARV., *263*, 264
  rigidum BORY, 264
Plumaria elegans (BONNEM.) SCHMITZ, 71
Plymouth, seawater temperatures, 92
Plymouth Laboratory of the Marine Biological
  Association of the United Kingdom, 69
Pneumatocyst, 143
Pneumatophore, 209
Pocillophora, 211
Pocillophoridae, 211
Pocockiella variegata, see *Lobophora variegata*
Point Conception, *15*, 134, *138*, 139, 145, *146*,
  147, 153
Poland, see Baltic
Polar Circle, 166
Polarity induction, 317
Pollution, 343
Polychaetes, 79, 117, 166, *260*, *263*
  long-distance dispersal, 93
Polydora ciliata, 67
Polyhaline environment, 82
Polyides rotundus (HUDS.) GREV.:

temperature optimum, growth, 323
temperature tolerance, 323
Polyna, 187
Polynesia, 226, 230, 233
Polyneura:
  gmelinii (LAMOUR.) KYL., 78, *87*
  laciniata (LIGHTF.) P. DIXON (= *hilliae*), 328
    temperature:
      growth, optimum, 328
      reproduction, 330
  latissima (HARV.) KYL., *144–145*, 146
Polyopes constrictus (TURN.) J. AG., *260*, 263
Polyphenols, *see* Phenolic substances
Polysaccharides, 365–370, *369*
Polysiphonia, 273
  apiculata HOLLENBERG, 233
  arctica J. AG., 177–178, 181, 184
  caspica KÜTZ., 123
  confusa HOLLENBERG, 355
  elongata (HUDS.) SPRENG., 78
  endophytic algae, 355
  lanosa (L.) TANDY, 225
    epiphytism, 353
    photosynthesis, temperature optimum, 332
    salinity, photosynthesis, ecotypes, *336*, 337
  scopulorum HARV., 220
  subtilissima MONT., 328
    temperature optimum:
      growth, 328
      photosynthesis, 332
  urceolata (LIGHTF. ex DILLW.) GREV., 47, *67*, *71*, 73, 76, 78, 94, 128, 180
    temperature optimum, growth, 323
    temperature tolerance, 323
  violacea (ROTH) SPRENG.,123
Pomacentridae, 350
Pomatoceros triqueter, 79
Pomatoleios crosslandi, *263*
Population biology, 357
Porites astreoides, 219
Poritidae, 211
Po River, 102
Porolithon, 200, 213, 218, 233
  pachydermum (WEB. VAN BOSSE et FOSL.) FOSL., *215*, 219, *219*, *222*
  photosynthetic rate, 364
Porphyra, *66*, *68*, 139, 143, *158*, 248, 251
  commercial use, 365, 367, *368*
  capensis KÜTZ., *260*, 262
  columbina MONT., 256
    desiccation, 337–338
  crispata KJELLM., 230
    desiccation, 338

endophytic algae, 355
haitanensis T. J. CHANG et ZHENG BAOFU:
  commercial use, 367
leucosticta THUR. in LE JOL., 108, 133
light saturation, photosynthesis, 302
maintenace light requirement, 299
miniata (C. AG.) C. AG., 179
  reproduction, temperature demands, 330
nereocystis ANDERS:
  epiphytism, 353
ochotensis NAGAI, 156
protein content, 365
perforata J. AG., 341
  desiccation tolerance, 341
tenera, 171
  commercial use, 367
torta KRISHNAMURTHY, 312
  photoperiodism, 312
umbilicalis (L.) J. AG., 70, 94
  absorption, action spectrum, photosynthesis, *291*, 293
  pigment content, acclimation of, 307
  temperature optimum, growth, 323
  temperature tolerance, 323
vietnamensis TANAKA et P.-H. HO, 230
Porphyropsis coccinea (J. AG. ex ARESCH.) ROSENV., *67*
Port Burwell, *172*
Port Clarence, 185
Port Elizabeth, 264
Port Macquarie, 271
Portugal, 85, *88*, 89–90
Portugal Current, 62
Posidonia, 204, 270
  australis, 270
  oceanica, 120, *120*
    circannual rhythm, growth activity, 320
    erosion, distal blade end, 354
    grazing, 351
Posidonia mats, 120
Postelsia palmaeformis RUPR., *#138*, 143, *144–145*, 148
  short-distance dispersal, 93
Potamogetonaceae, *205*
Potash industry, historical, 365
Prasiola, *66*, *68*, 69, 84, 248, 251
  borealis REED, 139
Prasiolales, 62
Predators, 212, 348–351, *349*
Primary productivity, 360–363
  global, macroalgae, phytoplankton, 362–363
Prince Edward Island, 128, *236*, 251
Prince of Wales Island, 142

Prince William Sound, 141–142
Prochloron, 353
Production / biomass ratio, 358
Productivity, see Primary productivity
Proteins, 365, 379
Protosalvinia, 201
Prototaxites, 201
Protozoans, 370
Pseudochorda nagaii (TOKIDA) INAGAKI, 160–161
Pseudolithoderma, 181
Pseudolithophyllum, 200
  expansum (PHILIPPI) LEMOINE, *114*, 115, 117
Pseudoperennials, 314
Pseudopterogorgia, 219, *219*
  acerosa, 222
Psychrosphere, 32
Pterocladia, 148, 193
  americana TAYLOR, 219
  capillacea (S. G. GMEL.) BORN. et THUR., *147*, *194*
  commercial use, 367
Pterosiphonia:
  complanata (CLEM.) FALKENB., *88*
  parasitica (HUDS.) FALKENB., *87*
Pterothamnion plumula (ELLIS) NAEG. (= Antithamnion plumula), 337
Pterygophora californica RUPR., #*138*, *146*, 148, *148*
  age, maximal, of individuals, 356
  blade shedding, 356
  circannual rhythm, growth activity, 320
Ptilota:
  plumosa (HUDS.) C. AG., 183
  serrata KÜTZ., 41, 47, #*54*, 129, 131, 175, 181, 183–184, 188
  light saturation, photosynthesis, 302
Puerto Rico, temperature tolerance of seaweeds, 324
Pumice, as floating substratum, 94
Punctaria:
  glacialis ROSENV., 177
  plantaginea (ROTH) GREV., 178
Punta Eugenia, 153
Pycnopodia helianthioides, 142
Pyura stolonifera, 263

Qingdao, *162*
Québec, 123
Queensland, 265, 267, 272

Radiolaria, 34
Ralfsia:

  africana (= expansa) (J. AG.) J. AG., 206
  fungiformis (GUNN.) SETCH. et GARDN., 187, *188*
  verrucosa (ARESCH.) J. AG., 70, 88, *144–145*, 255
Recruitment, 148
Red coral, 117
Red Crag, 34
Red drop, 293
Red Sea, 208, 225–227, *227*
Regions, biogeographical, 11, *12*, *14–16*, 65
Regressions of the sea, 28, 32
Relative growth rate, see Growth, relative growth rate
Relict taxa, 29–30
Repopulation, see Succession
Reproduction:
  boundaries (limits), geographical, 13
  fertilization, synchronization of, 318–320
  gametes:
    attraction, pheromones, 318–319, *319*
    release, circadian rhythm, 317–318
  lunar periodism, 319–320, *320*
  daylength, see Photoperiodism
  temperature, 329–331
Reserve materials, 358–359
  seasonal course, 359
Resistance, see Tolerance
Resources, 365
Respiration, *297*, 298
  acclimatisation, 298
  maintenance rate, 298
  salinity, 337
  temperature, 332, *333*
Rhabdoporella, 198
Rhine River, 82, *100–101*
Rhipocephalus, 199
Rhizoclonium, *68*, 123
- riparium (ROTH) HARV., 82
Rhizomaria, see Laminaria, subgenus
Rhizophora, 208–209
  mangle, 209, 222
  racemosa, 206
Rhode Island, 123
Rhodochorton, see Audouinella
Rhodolith, 79–80, *179*, 200, 208
Rhodomela:
  algal parasite, Harveyella mirabilis, 355
  confervoides (HUDS.) SILVA (= subfusca, incl. forma lycopodioides), 41–42, 47, #*56*, 73, 85, 177–178, 181, 184
  temperature tolerance, 323
  crassicaulis, see Chondria armata

INDEX 521

larix, see *Neorhodomela larix*
lycopodioides, see *Rhodomela confervoides*
Rhodophyllis:
  dichotoma, see *Fimbriofolium dichotomum*
  divaricata (STACKH.) PAPENF., *87*
Rhodymenia:
  ardissonei FELDM. (= *corallicola*), *116*
  californica var. *attenuata* (DAWS.) DAWS., *149*
  palmata, see *Palmaria palmata*
Rias, 89
Rikuchu National Park, *159*
Rio de Janeiro, *15*, 217, 257, 222
Rio de la Plata, *15*
Rio Grande Do Sul, 256
Rissoella verruculosa (BERTOLINI) J. AG., 106, 108, *109–110*
Robe, *15*
Rockall, 73
Rockall Bank, 79
Roscoff, Station Biologique, 69
Rosenvingiella polyrhiza (ROSENV.) SILVA, 69
Ross Sea, 245–246, 249
Rovinj, marine biological station, 99
R:P index, 191–192, 259
r-strategy, 71, 357–358
Rugose corals, 211
Russia, see USSR
Russian Arctic, 164, 173, *182*, 182–185
Ryukyu Archipelago, *15*, 155, 227

Saccorhiza, 257
  dermatodea (PYL.) J. AG., 29, 41, *#51*, 127, 178, 184, 188
    photoperiodism, 312
    temperature optimum, growth, 328
  polyschides (LIGHTF.) BATT. (= *bulbosa*), 17, *#18*, 29–30, 41, *#59*, 72–73, 78, 81, 86, 89–91, 117, *118–119*, 257
    temperature demands, gametophytes fertile, 327
    temperature optimum, growth, 328
    temperature tolerance, gametophytes, 327
St. Croix, 217
St. Francis Island, *271*
St. Helena, 203, 208
St. John, 217
St. Lawrence Bay, 185
St. Lawrence Island, 185
St. Lawrence Gulf, 123, 128
St. Paul Island, 261
St. Thomas, 217
Sakhalin, 154–156

Salicornia herbacea, 82
Salinity, 332–337, *336*
  Arctic, 173
  brackish water, typology, 82
  ecotypes, 335–337, *336*
  growth, 335, *336*
  organic osmolytes, 334
  osmotic regulation, 334
  photosynthesis, 336–337, *336*
  tolerance, 82, 85, 333–337, *336*
  tropical regions, 219, 227
  turgor pressure, 334
Salt marshes, 82, 132–133
Salvage Islands, 90–91
San Benedicto Island, 233
San Diego Province, 153
Sand-inhabiting seaweeds, 121, 143, 200
Sandy Cape, *15*, 272
San Salvador, 221
Santa Margarita Island, *138*, 153
Sarcodiotheca gaudichaudii (MONT.) GABRIELSON (= *Neoagardhiella baileyi*), *144–145*
Sargassaceae, 242–243
Sargasso Sea, 133, 222–225
  algal biomass, productivity, 225
  paleobiology, 223–224
Sargassum, 106, 113, 133, 154–155, 157, *158*, 193, *196*, 198, 201, 214, 217, 221, 223, *223*, 224–225, 227, *227*, 228–230, 267, 270–273, 243
  agardhianum J. AG., 147, *147*
  bracteolosum J. AG., 270, *271*
  compensation light level, growth, 300
  crassifolium J. AG., 231
  cymosum C. AG., 256
    light saturation, growth, 302
  filipendula C. AG., 133, *195*, 206, *207*
  flavifolium KÜTZ., 97
  fluitans BOERG., 133, 223–224
  gas-filled thallus parts, 345
  hawaiiensis DOTY et NEWHOUSE, 202, 233
  heterophyllum (TURN.) C. AG., 264
  hornschuchii C. AG., 106, 113, *114*, 115, *116*
  latifolium (TURN.) C. AG., *227*
  linifolium (TURN.) J. AG., 106
  longifolium, see *Antophycus longifolius*
  muticum (YENDO) FENSHOLT, *96*, 97, 117, 143
    introduction to eastern Pacific and eastern Atlantic, 97
    gamete release, lunar periodicity, 320
    growth rate, 364

Sargassum, muticum (YENDO) FENSHOLT (*Continued*)
   natans (L.) J. MEYEN, 133, 223–224, *223*
      light saturation, growth, 302
      temperature optimum, growth, 328
   nigrifolium YENDO, *158*
      photosynthetic rate, 364
      primary productivity, 361
   piluliferum C. AG., temperature optimum, growth, 247
   platycarpum MONT., *222*
   polyceratium MONT., 219
   polycystum C. AG., 231
   rigidulum KÜTZ. *170*
   ringgoldianum HARV., *158*
   serratifolium C. AG., *158*
   sinclairii HOOK et HARV., desiccation, 338
      temperature optimum, growth, 329
   thunbergii (MERT.) O. KUNTZE, 157, 161, *162*
   vulgare C. AG., 97, *104*, 106, 115, *196*, 206, *207*, 208
Sarmatic Inland Sea, 101, 123
Sarso Island, *227*
Saudi Araba, *227*
Scaberia, 243
Scagelia pylaisei (MONT.) WYNNE (= *Scagelia corallina* = *Antithamnion boreale*), 175
Scaridae, 216, 350
Scenedesmus obliquus, pigments, light quality effects, 307
Schmitzia hiscockiana MAGGS et GUIRY, photoperiodism, 312
Schottera nicaeensis (LAMOUR. ex DUBY) GUIRY et HOLLENBERG (= *Petroglossum nicaeense*), 86, *109*, 111, 112–113, *113*
Sciaphilous algae, 111
Scinaia forcellata BIVONA, 79, 93
Scleractinian corals, *see* Coral reefs, hermatypic corals
Scoresby Sound, 179, 181
Scripps Institution of Oceanography, 135
Scytosiphonales, chloroplast movements, 317
Scytosiphon lomentaria (LYNGB.) LINK, 68, 71, 94, 106, 108, *109*, 127–128, 139, *163*, 178, 187, 241
   day-neutral strains, 312
   gas-filled thallus parts, 346
   hairs:
      formation, blue-UV effect, *315*, 316
      nutrient uptake, 344
   life form type, 314

morphogenetic substances, 354
nutrient uptake, hairs, 343
photoperiodic ecotypes, 311–312
photoperiodism, 308–309, *310–311*, 311–312, 314
temperature tolerance, 323, 326
two-dimensional growth, blue-UV effect, *315*, 316
Scytothalia, 243
   dorycarpa (TURN.) GREV., 270, *271*
Scytothamnus, 244
   australis (J. AG.) HOOK. et HARV., 244, 312
      freezing tolerance,, 324
      photoperiodism, 312
   fasciculatus (HOOK. et HARV.) COTTON, 244, 272
   hirsutus SKOTTSBERG, 244
Sea caves, 119–120
Sea foam, 370
Seagrass disease, 76, 356
Seagrasses, 3, 62, 68, 76, 104, 120–121, *196*, 197, 203, 222, 270, 272
   distribution, World, *204–205*
   floral history, 203
   primary productivity, 362
Sea of Alboran, 90
Sea of Japan, 154, 156–157, 160–161
Sea levels, paleoceanographical, 22, *24*, 32, 34, 102, 211–212
Sea of Okhotsk, 154–155
Sea otter, *see Enhydra lutris*
Sea palm, *see Postelsia palmaeformis*
Seasonality, 228, 230–231
Sea urchins, 111–112, 129, *141*, 142, 153, 207, 214, *219*, 350–352
   disease of, 129, 351
Seaweed farming, 365, *366–368*
Seaweed meal, world production, 367
Secchi depth, *see* Light, secchi depth
Seirococcaceae, 243, 265
Seirococcus, 243
Semibalanus balanoides, 52, 66, 70, 84, 100, 129
Senegal, 64, 85, *88*, 90, 203, 206
Sensor pigments, *see* Photomorphogenesis; Photoperiodism, sensor pigments
Septiger, *158*
Serraticardia, 255
Sewage outfall, 351
Seymour Island, 240
Shanghai, 154
Shark Bay, *15*, 270
Shelf, 3, 8, 25

Shetland Islands, 82, 93
Shumagin Island, *138*
Siberia, 34, 38, 164
Siboga Expedition, 230
Sicily, 98, 111, *114–115*, 117, *119*, 224
Siderastrea siderea, 219
Siganidae, 350
Silt, *170*, 172
Simplices, *see Laminaria*, section, Simplices
Sinai Peninsula, 208
Siphonaxanthin, *see* Photosynthesis, pigments
Siphonocladales, 193, *194*, 198, 221
  deep-water, siphonaxanthin, 290
Siphonocladus tropicus J. AG., *194*
Sitka, *174*
Skagerrak, 83
Sloughing off tissue, *see* Casting of outer layers
Smithora naiadum (ANDERS.) HOLLENB., epiphytism, 353
Snails, *see* Gastropods
Snares, 250–251
Soft substrata, 68, 120–121, 182, 187, 222
Solearia, *see Laminaria*
Solenoporaceae, 199, 211
Solieriaceae, 272
Solieria filiformis (KUETZ.) GABRIELSON (= *tenera*), 325
  temperature tolerance, 325
Solomon Islands, 225, 232
Somalia, 225, 228, 252
South Carolina, 123, 133
South China Sea, *163*
South Equatorial Current, *63*, 203, 217, 226
Southern Africa, *see* Africa, Southern Africa
Southern Australia, *see* Australia, southern Australia
South Georgia, 244–245, *247*, *249*, 250–251
South Orkney Islands, *236*, 245
South Shetland Islands, *236*, 245, *247*, *249*
South Sandwich Islands, *236*, 245, *247*
Southwestern Africa Province, 258
Spain:
  Atlantic coast, *64*, 85–89
  Mediterranean coast, 98
Spanish Sahara, *see* Western Sahara
Spartina:
  alterniflora, 132
  townsendii, 82
Speciation, 46, 192
Species:
  diversity, *see* Diversity
  duration, geological, 126, 198–199
Specific growth rate, *see* Growth, specific growth rate

Sphacelaria:
  arctica HARV., 176–178, 187
  cirrosa (ROTH) AG., 201
  plumosa LYNGB. (= *Chaetopteris plumosa*), 177, 187
  rigidula KÜTZ., photoperiodism, long-day plant, 309, 312
  tribuloides MENEGH., 219
Sphacelariaceae, 62, 98
Sphaerococcus coronopifolius STACKH., 87, *105*
Sphaerotrichia divaricata (C. AG.) KYL., 312
  photoperiodism, long-day plant, 309, 312
  reproduction, temperature demands, 330
Spirorbis spirorbis (= *borealis*), 354
  preferential settling, 354
Spitsbergen, 32, 39, 62, *64*, 80–81, 166, 182
Spitsbergen Expedition, Swedish, 165
Spitsbergen Polar Current, 181
Splachnidium rugosum (L.) GREV., 244, 258, 262
Split, marine biological station, 99
Spondylothamnion multifidum (HUDS.) NÄG., 87
Sponges, *271*
Spongomorpha, 178, 187, 248
  aeruginosa (L.) HOEK, 127
  Codiolum phase, endophytism, 355
  arcta see *Acrosiphonia*, *arcta* (DILLW.) J. AG.
Spongonema tomentosum (HUDS.) KÜTZ., 123
Spores:
  Rotalgen, Sinkgeschwindigkeit, 261
  shedding, diurnal rhythm, 318
  sinking velocity, 345
Sporochnus pedunculatus (HUDS.) C. AG., *116*, 121
Spyridia filamentosa (WULF.) HARV. in HOOK., 209, 233
Squamariaceae, 199–200
Sri Lanka, 225, 228, *229*
Standing crop, 251
Starfish, 142, 348–349, *349*
Station Biologique de Roscoff, 69
Stenohaline, 335, *336*
Stenothermal, 17, 321, *322*
Stephenson System of zonation, 8, *9*
Stepping stones, 241–242, 271–272
Stewart Island, 273
Stichopathes, 220
Stictyosiphon tortilis (RUPR.) REINKE, 177–179
Stilophora rhizodes (TURN.) J. AG., 330
  reproduction, temperature demands, 330

Stolonophora, 243
Strait of Belle Isle, 15, *124*, 128, 166, 188
Strait of Gibraltar, 90, 98–99, *100–101*, 102–103
Strait of Messina, 98, 102, 105, 117, *119*
Strait of Sicily, 98, 102, 105, *114–115*
Streptophyllopsis, 161
   kuroshioensis (SEGAWA) KAJIMURA, 160
Stress escape (evasion), 322
Stress-tolerant species, 347
Stromatolites, 198, 210
Stromatopores, 211
Strongylocentrotus, 142
   droebachiensis, 129
   franciscanus, 153
   purpuratus, 153
Struvea, 221
   pulcherrima (J. E. GRAY) MURRAY et BOODLE, 195
Stypocaulon scoparium (L.) KUETZ. (= *Halopteris scoparia*), 91, *112*
Stypopodium zonale (LAMOUR.) PAPENF., 222, 352
   antigrazer substances, 352
Sub-Antarctic Islands, *15*, 235, 244–245, *249*, 250–253
Sub-Arctic region, 166
Sublittoral, 4–7, 10
   primary productivity, 167, 361–362
Submergence, *see* Brackish water, submergence; Equatorial submergence; Isothermic submergence
Submersibles, *see* Diving, submersibles
Subtropical Convergence, 235–237
Succession, 70, 71, *71*, 76, 78, 94, 97, *140–141*, 142, *214*, 215, 357
Sucrose, 368
Suez Canal, 98, 104, 197
Suhria vittata (L.) J. AG., 259
Sulphur, 343
Sumatra, 230
Supralittoral, 4, 108
Surface structures of thalli, 344
Surtsey, 70, 94
Suwannee Straits, 134
Svalbard, *see* Spitsbergen
Swakopmund, 259
Sweden, 61, 83
Swedish Antarctic Expedition, 245
Swedish Spitsbergen Expedition, 165
Syngnathidae, 224
Syngnathus incompletus, 224
Syngnathus pelagicus, 224

Syringoderma, *319*
   phinnei E. C. HENRY et D. G. MÜLLER, *319*
Syringodermatales, *319*
Syringodium, 204, 222

Tabulate corals, 211
Taiwan, *15*, 225
Tampa Bay, 133
Tampico, 133
Tanzania, 225–226
Taonia atomaria (WOODW.) J. AG., 92, *105*
Tasmania, 235, 250, 265–267, 271–272
   paleoceanography, paleobiogeography, 237, *238–239*, 239
Tasman Sea, 266
Tatar Strait, 154
Taxon cycle, 232
Tegula, 348
   funebralis, *349*
Teleplanic larvae, 93
Tellamia, 71, 355
Temperature:
   Gold Coast, 206
   isotherms, 11, *12*, 13, *14*, 16–17, *18–20*, 19–21
   mutants, 327
   paleotemperatures, 22, *24–25*, 32–34, 38–39, 189, 211, 242, 264
   photosynthesis, *see* Photosynthesis, temperature
   reproduction, *see* Reproduction, temperature
   respiration, *see* Respiration, temperature
   secular changes, *92*, 92–93
   southern Africa, 259
   southern Australia, 266
   sub-Antarctic islands, 250
   tolerance, 13, 16–21, 62, 85, 321–328, *322*
      cold, 322–325
      ecotypic differentiation, 326–327
      eulittoral, sublittoral algae, 325–326
      exposure time, 325
      freezing, 322–325
      heat, 322–327
      geographical gradient, 324–328
      Laminariales, gametophytes, 327
      seasonal adaptation, 321, *322*, 323
      supercooled state, 325
      tropical regions, 206, 209, 219, 221, 227, 234
   tropical stability since Precambrian, 326
Tenarea, 200
   tortuosa see Lithophyllum, lichenoides undulosa BORY, 108

INDEX 525

Terrestrial vegetation:
  annual light exposure, 285
  primary productivity, 361–362
Tethys Sea, 25, 27, 28, *100–101*, 102–103, 117, *191*, 193, 199, 202, 211, 223–224, 235
Texas, 123, 133
Thailand, 225, 230
  commercial uses of seaweeds, 365
Thais:
  lapillus, see *Nucella*
  nodosa, 206
Thalassia, *204–205*
  hemprichii, *196*, 231
  testudinum, *196*, 217, 222
Thalassiophyllum clathrus (GMEL.) POST. et RUPR. 136, *138*, *140*, 156, 186
Thalassodendron, *203–204*
Thermocline, 228
Thermosphere, 32
Thule, 180
Thulean land bridge, see Land bridges, Thulean
Thuretia quercifolia DECAISNE, *268*
Tides, 4–5, *6*, 99, 181, 196, 208, 234
  critical tide levels, 341
Tierra del Fuego, 240, 249–250, 256
Tilia, 33
Tilopteridales, 29–30
Tilopteris mertensii (TURNER in SM.) KUETZ., 201
Titanoderma caspicum (FOSL.) WOELKERLING (= *Dermatolithon caspicum*), 123
Titanophora, 200
Tjörnes layers, 34
Tokyo, 154
Tolerance, see Temperature, tolerance; Light, photoinhibition; Salinity, tolerance
Topolobampo, *15*, 153
Torres Strait, 231
Toxic substances, see Grazing, antigrazer substances
Trailliella phase, see *Bonnemaisonia hamifera*
Transequatorial migrations, see Equatorial crossing in Cenozoic
Transgression of the sea, 28, 32
Transhatteran species group, 125
Translocation, 358–359, *360*
Transmission, see Light, transmission
Tristan da Cunha, 94, *249*, 250, 252, 258, 261
Trondheimfjord, 82
Tropics:
  Atlantic Africa, 203, 205–208, *207*
  America (Caribbean), 216–217, *218–219*, 219–222, *220–222*
  America (Pacific), 233–234
  coral reefs, 209–216
  discontinuous distribution, *195*, 196–197, *202*
  Fucales, genera, 243
  Indo-Pacific-Atlantic distributions, 196
  Indo-West Pacific, 211, 225–233, *227*, *229*
  mangroves, 208–209
  monsoon, 228–229, 252
  paleoceanography, paleobiogeography, 22, 189, 191, *191*, 193, 196, 210–211
  pantropical species, 103, *104*, 192–193, *194–195*, 202, 232, 263
  primary productivity, 361–362
  R:P index, 191–192
  Sargasso Sea, see Sargasso Sea
  temperature:
    growth optima, seaweeds, 328
    photosynthesis, optimum, 332
    reproduction, 329
    stability since Precambrian, 326
    tolerance, seaweeds, 324–326
Trottoirs, 89, *110*, 111
Tsushima Current, 161
Tsuga, 33
Tsuyazaki, 313
Tuamotu Archipelago, 226, 231
Tube-dwelling diatoms, 355
Tunisia, 98, 99, *114–115*, 117
Turbellarians, 370
Turbinaria, 193, 214, *227*, 230, 243, 267
  decurrens BORY, 227
  ornata (TURN.) J. AG., 230–231
  turbinata (L.) KUNTZE, *195*
Turbulence, 344
Turf, 212, 219, 263
Turgai Strait, *27*, 31
Turgor pressure, see Salinity, turgor pressur
Turkey, 98
Turnerella:
  mertensii (POST. et RUPR.) SCHMITZ, 173
  pennyi (HARV.) SCHM., 173–175, *#175*, 178, 181, 187–188
Turnover rate, 360–361
Two-dimensional growth, blue-UV effect, *315*, 316
Tydemania expeditionis WEBER VAN BOSSE, 199, 202, 233
Tyrrhenian Sea, 102, 117

Udotea, 199, 214, 220
  petiolata (TURRA) BOERG., *104*, 113, 115

Ulothrix, 66, 68, 70, 82, 93–94, 123, 139, 178–179, 248, 251
  dispersal, 93
  pseudoflacca WILLE, *156*
Ulotrichales, 198
Ultraviolet, *see* Light, ultraviolet
Ulva, 68, 71, *71*, 82, 208, 230, 241
  absorption, action spectrum, photosynthesis, *291–292*
  biomass, 361
  chloroplast movements, circadian rhythm, 317
  curvata (KÜTZ.) DE TONI:
    pigment content, acclimation of, 306
  desiccation, 338
  fasciata DELILE., 206, 207
  gamete release, lunar periodicity, 320
  fenestrata POST. et RUPR.:
    light saturation, photosynthesis, 302
  lactuca L., 67, 68, 71, 77, 105, 228
    life form type, 314
    light requirement, growth, 301
    pigment content, acclimation of, 307
    respiration rate, 297–298
    seasonal, 332, *333*
    temperature:
      growth, optimum, 323
      tolerance, 323
  mutabilis FÖYN, 327
    temperature mutants, 327
  olivascens P. DANG., 115, *115*
  opportunists, 357
  photosynthesis:
    effectiveness versus water depth, 294
    rates, 364
  pigment content, acclimation of, 306–307
  protein content, 365
  rigida C. AG., 91, 255
  r-strategy, 358
Ulvales, 62
  deep-water, siphonaxanthin, 290
Umnak Island, *138*
Undaria:
  commercial use, 367
  pinnatifida (HARV.) SURINGAR, 117, 157, *158*, *160*, 161, *162*
  temperature:
    fertility, gametophytes, 327
    growth, optimum, 323
    tolerance, gametophytes, 327
Understory algae, 6, 7
  competition, 350
Ungava Bay, *172*, 187
Upwelling areas, 89–90, 145, 153, 205, 216, 228, 242, 252, 257–259, 262
Urospora, 66, 70, 94, 178, 248, 251

penicilliformis (ROTH) ARESCH., 84, *156*, 179, 184
wormskioldii (MERT. in HORNEM.) ROSENV.:
  temperature, reproduction, 329, *330*, 331
Uruguay, 252
USA, *see* America
USSR, *see* Asia; Baltic; Black Sea; Europe; Russian Arctic; Siberia
UV, *see* Light, ultraviolet

Valonia, 91
  macrophysa KÜTZ., *116*, 202
  utricularis (ROTH) C. AG., 89, *104*, 111–112
Vancouver Island, 135–136, *138*, 186
Vaucheria, 123
Vega Expedition, 164, 185
Venezuela, 216–217, 221
Verkhoyansk Mountains, 34
Vermetus cristatus, 111
Verrucaria, 65, 251
  maura, 69, *69*, 88, 90
  symbalana, 108
Vertebrates, Arctic-origin hypothesis, 30
Vicariance biogeography, 10
Victoria, Australia, 208, 235, 266–267, 271
Victoria Island, 187
Vidalia volubilis (L.) J. AG., *116*
Vietnam, 225, 230
Virginia, 123, 132–133
Virgin Islands, 217
Virgin Rocks, Newfoundland Grand Banks, 94
Viridiene, *319*
Vivipary, 203, 209
Vladivostok, 154, 161, 186
Volcanic rock, as floating substratum, 94

Wales, *see* British Isles
Walrus, 351
Washington (St.), 134, 142, 155
Wasting disease, *see* Seagrass disease
Water content, fucalean species, *339*
Water motion, *see* Wave action
Wave action, 66, 68, 80–82, 111, 152–153, 343–345
  as influenced by kelp beds, 152
Weddellian species, 240
Wenchow, *15*, 154
West Australian Current, *63*
Western Port Bay, 208
Western Sahara, 85, 90
West Gondwana, *see* Gondwanaland
West Greenland Current, 166, 180

West Wind Drift, *see* Antarctic, Circumpolar Antarctic Current
Whiplash effect, 348
White Sea, 182–184, 350
Windrows, 224
Wood, as floating substratum, 94
World:
  coastline, 362
  harvest of seaweeds, 367, 369
  number of seaweed species, 3
  primary productivity, 362–363
Wrangelia penicillata (C. Ag.) C. AG., *104*

Xanthoria parietina, 69
Xiphophora, 243
  chondropylla (R. BROWN ex TURN.) MONT. ex HARV., 251, *273*
Xylocarpus, 208

Yakutat Bay, *138*
Yellow Sea, 161, 163
Yenisei River, 165

Yugoslavia, *see* Adriatic Sea

Zanardinia prototypus NARDO, *115*
Zhejiang, 161
Zonaria, 91, 270, *271*
  farlowii SETCH. et GARDNER, *147*
  tournefortii (LAMOUR.) MONT., 133
Zonation, 3–10
  critical tide levels, 341
  systems:
    Classical system, 8, *9*
    Genoa system, 8, *9*, 10
    Lewis system, 8, *9*
    Stephenson system, 8, *9*
Zoobenthos, *see* Benthos
Zooxanthellae, 210, 226
Zostera, *204*
  marina, 68, 76, 120, 122, 132, *144–145*, 145, 356
  epiphytism, 353
  seagrass disease, 76, 356
  noltii (= *nana*), 68, 76, 120, *120*, 122–123